PRECALCULUS

PRECALCULUS

DENNIS CARRIE
Golden West College

HOUGHTON MIFFLIN COMPANY   BOSTON
Dallas   Geneva, Illinois
Palo Alto   Princeton, New Jersey

Cover Art:

Circle and Ten Triangles, Yellow-Black by Pol Bury (1976).
Woodcut, 27½ x 22⁹⁄₁₆″.
Collection, The Museum of Modern Art, New York. Abby Aldrich
Rockefeller Fund.
Photograph © 1990 The Museum of Modern Art, New York.

ISBN: 0-395-44464-0

Library of Congress Catalog Card Number: 89-80925

Printed in the U.S.A.

BCDEFGHI -D-99876543210

To my parents, Arthur and Evaline Carrie

Preface

PURPOSE AND PREREQUISITES

The goal of PRECALCULUS is to provide the student with the mathematical skills and maturity necessary for success in the study of calculus. The basic prerequisite is either college intermediate algebra or two years of high school algebra. It is not assumed that the student has maintained complete retention of all the prerequisite material; thus, a review of basic algebra and geometry is provided in Chapter 1, as well as in the appendix.

EMPHASIS

The book emphasizes three areas of study: algebra skills, the concept of function, and the construction and interpretation of graphs.

1. *Algebra skills.* Chapter 1 contains a review of the basic techniques for solving equations, inequalities, and absolute value inequalities. Also included in this chapter are topics from plane analytic geometry: the distance formula, slope, lines, circles, and basic graphs. Frequent opportunities occur throughout the book for manipulating and simplifying a variety of fractional, exponential, logarithmic, and trigonometric expressions similar to those encountered in calculus. The appendix, which is divided into five sections, presents review material on factoring, fractions, completing the square, exponents, and geometry.

2. *Functions.* Chapter 2 introduces the general properties of real functions and sets the stage for the study of elementary functions that will follow. Functional notation in both its analytic and geometric interpretation is firmly stressed. Also discussed are general techniques for graphing functions, symmetry, composition, and inverse functions. In Chapters 3 to 7 the student

learns specific properties of polynomial, rational, exponential, logarithmic, and trigonometric functions.

3. *Graphing.* The geometric interpretation of concepts is a major focus of the text and is stressed throughout by a substantial number of illustrations and numerous graphing exercises. The student progresses from graphing basic curves (including lines, circles, parabolas, and cubics) to the elementary functions (including polynomial, rational, exponential, logarithmic, and trigonometric functions). A special section is devoted to graphing reciprocal functions, and two sections to constructing the graphs of rational functions. Limit notation is introduced and used to describe the geometric behavior of functions near vertical asymptotes and far from the origin. Symmetry and the techniques of shifting, reflecting, and adding graphs are discussed and used in examples and exercises. Conic sections, polar curves, and parametric equations are also covered.

ORGANIZATION

The following chart indicates the suggested order in which chapters are to be covered:

$$1 \longrightarrow 2 \longrightarrow 3 \longrightarrow 5 \longrightarrow 6 \longrightarrow 7 \begin{cases} \longrightarrow 8 \\ \longrightarrow 9 \\ \longrightarrow 10 \\ \longrightarrow 11 \end{cases}$$

$$3 \downarrow 4$$

Care has been taken to provide a logical development of the material while allowing the instructor a considerable amount of flexibility in planning course content. Thus, Chapter 4 may be covered at any time after Chapter 3, and Chapters 8 to 11 may be covered in any order after Chapter 7.

The following list identifies the core and optional sections for each chapter.

Chapter	Core Sections	Optional Sections
1	1.1–1.2, 1.4–1.7	1.3
2	2.1–2.6	
3	3.1, 3.3–3.6	3.2
4	4.1–4.4	4.5–4.6
5	5.1–5.6	5.7
6	6.1–6.6	6.7
7	7.1–7.5, 7.7	7.6
8	8.1–8.2	8.3–8.5
9	9.1–9.3	9.4–9.5
10	10.1–10.4	10.5–10.8
11	11.1–11.3	11.4–11.5

EXAMPLES

Each section contains worked-out examples that further explain new concepts and illustrate applications. Wherever appropriate, figures are provided to help students gain a clear understanding of the solution. By design, examples correlate well with the exercises given at the end of each section.

EXERCISES

Exercise sets are structured to develop both manipulative and thinking skills, leading the student to a deeper understanding of the material. Problems typically range from routine to difficult, with those of greater difficulty or of an optional nature appearing at the end of the exercise set. Problems of similar type are paired off so that most odd-numbered problems match up with the next even-numbered problem. Each chapter ends with a carefully selected collection of review exercises designed to provide good preparation for chapter tests. Answers to the odd-numbered exercises and all of the review exercises are given in the back of the book.

CALCULATORS

It is assumed that each student has access to a scientific calculator. General instructions for its use are included throughout the text, without over-emphasis. Although this text does not require graphics calculators (or personal computers with graphics software), they are recommended for checking work, experimentation, and demonstration.

SUPPLEMENTS

1. *Solutions Manual.* I would like to thank George Wilson for his many contributions to the preparation of the Solutions Manual. The Solutions Manual contains solutions to *all* the exercises in the text. Essential steps for solving more difficult problems are shown, and all required proofs are written out in complete form.

2. *Instructor's Manual with Testing Program.* The Instructor's Manual was written by Patricia Confort of Roger Williams College, Rhode Island. It contains four alternate printed tests for each chapter of the text, as well as alternate cumulative tests and final exams. Tests are set up in both free-response and multiple choice format.

3. *Instructor's Computerized Test Generator.* The Computerized Test Generator was also written by Patricia Confort. The Test Generator disk can be used with the Apple II® family of computers, IBM® PC and compatible computers, and the Macintosh family of computers.

4. *Printed Test Bank.* The Test Bank contains all items in the Computerized Test Generator. Instructors can use the printed Test Bank to select specific items from the Test Generator database. Instructors without access to a computer can select items from the test bank to be included on a test being prepared by hand.

ACKNOWLEDGMENTS

I wish to thank the reviewers who provided me with many helpful comments and suggestions for improving the text.

Sharon Abramson, SUNY-Nassau Community College
Donald J. Albers, Menlo College
David F. Anderson, University of Tennessee-Knoxville
B.R. Asrabadi, Nicholls State University, Los Angeles
Orville Bierman, University of Wisconsin-Eau Claire
Leonard Chastkofsky, University of Georgia
B.M. Friel, Humboldt State University
Richard A. Gibbs, Fort Lewis College
Patricia Gilbert, Diablo Valley College
Gerald K. Goff, Oklahoma State University
Raymond P. Guzman, Pasadena City College
Pamela A. Hager
Samuel Isaak, University of South Florida
Robert W. King, University of Southern Mississippi
John Kroll, Old Dominion University
Andrew Matchett, University of Wisconsin-LaCrosse
Frank P. Mathur, California State Polytechnic University
Michael J. Mears, Manatee Community College
Ann Megaw, University of Texas
Ross M. Rueger, College of the Sequoias
George W. Schultz, St. Petersburg Junior College
Rebecca W. Stamper, Western Kentucky University

C.H. Tjoelker, California State University-Sacramento
Charles R. Wall, Trident Technical College
Glorya Welch, Cerritos College
Timothy R. Wilson, Honolulu Community College

Special recognition goes to my colleague Sr. Frances Teresa for her positive feedback while using the book in its preliminary form.

I would like to express my appreciation to the editors and staff at Houghton Mifflin Company for their support and fine work in producing a very attractive book.

The inspiration for writing this book has come from the many students I have taught over the years. It has been a privilege to explore their questions and ideas with them. The many conversations we had helped immensely in the development of this text.

Finally, I want to give my sincere thanks to Rebecca Combs, who typed the manuscript with incredible care and accuracy and supplied me with much encouragement and good humor along the way.

Dennis Carrie

Contents

CHAPTER 1 Introduction

1.1 REAL NUMBERS

Our work begins with the fundamental objects of algebra and calculus: the real numbers. We start with a geometric interpretation of the real numbers by identifying them with points on a line. Next, we describe some special types of real numbers and examine their properties. Finally, we introduce absolute value, square roots, and cube roots.

Geometric Interpretation of the Real Numbers. An ordinary straight line serves as a picture of the real numbers in the following way. First, orient the line horizontally and select one point to represent 0. We call this point the **origin**. To the right of the origin, select any convenient point to represent 1. The distance between 0 and 1 establishes 1 unit of length. See Figure 1.1(a).

Next, recall that the nonzero real numbers divide into two separate types: the **positive** numbers and the **negative** numbers. Identify a positive real number a with the point on the line whose distance is a units to the right of the origin. If a is a negative number, then identify it with the point located $-a$ units to the left of the origin. (Be careful when interpreting the symbol $-a$. If a is a negative number, then $-a$ will be positive and hence can represent a distance.) Figure 1.1(b) shows the result, which we call the **real line** or the **number line**.

The identification between the real numbers and points on the real line is called a *one-to-one correspondence*. This means that each real number corresponds to exactly one point on the real line, and, conversely, each point on the real line corresponds to exactly one real number. The point identified with the number a is called the **graph** of a.

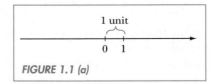

FIGURE 1.1 (a)

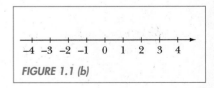

FIGURE 1.1 (b)

Example 1

Graph 3/2 and −2.25 on the real line.

SOLUTION See Figure 1.2. ∎

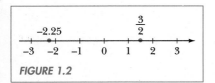

FIGURE 1.2

Sets. The word **set** in mathematics refers to a collection or group of objects. Each object in a particular set is called an **element** of the set. As an example, the collection of all known planets in our solar system forms a set with nine elements: Mercury, Venus, Earth, Mars, Jupiter, Saturn, Uranus, Neptune, and Pluto. Although a set may consist of almost anything, our primary concern is with sets whose elements are real numbers.

The following list defines the language and symbols used when dealing with sets.

- **R** designates the set of all real numbers.
- **Infinite set** is a set with an unlimited number of elements. For example, **R** is an infinite set.
- **Finite set** is a set with a limited (finite) number of elements. The set of known planets in our solar system is a finite set (nine elements).
- ∈ means *is an element of*. We write $y \in A$ to say y is an element of set A. For example, $0 \in \mathbf{R}$ and $1 \in \mathbf{R}$.
- $\{a, b, c\}$ designates the set consisting of elements a, b, and c. In general, a finite set may be represented by listing its elements separated by commas inside braces, { }.
- $\{x: ---\}$ means *the set of all x's such that* ---, where conditions on x appear in the dashed section. We refer to this as **set builder notation**. *Unless stated otherwise, we assume that variables in this book represent real numbers.* Thus the elements of $\{x: x^2 = 9\}$ consist of two real numbers, 3 and −3. Any letter may be used for the variable, so $\{t: t^2 = 9\}$ and $\{x: x^2 = 9\}$ represent the same set.
- $A = B$ The set A equals set B means that A and B each contain exactly the same elements. Thus $\{x: x^2 = 9\} = \{3, -3\}$ and $\{x: x = x\} = \mathbf{R}$.
- ∅ designates the set with no elements, called the **empty set**. For instance, $\{x: x = x + 1\} = \varnothing$. Note that it is incorrect to write $\{\varnothing\}$ for the empty set; the set $\{\varnothing\}$ is not empty because it contains (the set) ∅ as an element, $\varnothing \in \{\varnothing\}$.
- **Z** designates the set of **integers**. We may list this set as follows:

$$\mathbf{Z} = \{\ldots, -4, -3, -2, -1, 0, 1, 2, 3, 4, \ldots\}$$

It is impossible to list every element in **Z**, so we use three dots, . . . , to indicate *and so forth*.

- **N** designates the set of positive integers, or *natural numbers*.

$$\mathbf{N} = \{1, 2, 3, 4, 5, \ldots\}$$

Example 2

Use set builder notation to represent the set of even integers.

SOLUTION   Since an even integer is a multiple of 2, we have

$$\{x : x = 2n, n \in \mathbf{Z}\}$$

represents the set of even integers. ∎

We may call any set of real numbers a **subset** of **R**. Thus the set of integers **Z** and the set of positive integers **N** are both subsets of **R**. Two other subsets of **R**, called the **rational** and **irrational** numbers, play a major role in mathematics.

Rational and Irrational Numbers

A real number x is called *rational* if and only if there are integers p and q such that $x = p/q$ and $q \neq 0$.

A real number is said to be *irrational* if and only if it is not rational.

It follows from these definitions that every real number is either rational or irrational (but not both). Let us examine some typical rational numbers.

Rational Number	Expressed as $\frac{p}{q}$	Decimal Representation
$5\frac{4}{33}$	$\frac{169}{33}$	$5.121212\cdots = 5.\overline{12}$
-2.37	$\frac{-237}{100}$	$-2.3700\cdots = -2.37\overline{0}$
$\frac{1}{7}$	$\frac{1}{7}$	$0.142857142857\cdots = 0.\overline{142857}$
8	$\frac{8}{1}$	$8.000\cdots = 8.\overline{0}$

Notice that the decimal representation for each of these rational numbers eventually repeats itself. (The bar over certain digits indicates that the digits are repeated forever in the decimal expansion.) This suggests the following general result, which we state without proof.

A real number is rational if and only if it has a repeating decimal representation.

Example 3

Find a rational number r between 107/333 and 12/37.

SOLUTION   Consider the decimal expansions

$$\frac{107}{333} = 0.\overline{321}$$

$$\frac{12}{37} = 0.\overline{324}$$

We see that one possible choice for r is $0.322000000\cdots$. This number qualifies as a rational number because it has a repeating decimal representation, $r = 0.322\overline{0}$. In fact, we may write r as a quotient of integers, $r = 322/1000$. ■

Example 3 demonstrates one instance of an important fact about **R**: *between any two distinct real numbers one can always find a rational number.*

Examples of rational numbers are relatively easy to find, but what about irrational numbers? One of the most famous irrational numbers is pi, denoted by the Greek symbol π. Pi is the real number we get when we divide the circumference of any circle by the length of its diameter. For part of its decimal representation, we have

$$\pi = 3.14159265358979323846\cdots$$

Notice that this decimal is infinite and nonrepeating. Any irrational number is characterized by the fact that its decimal representation is infinite and nonrepeating.

Similar to the situation for rational numbers, it is possible to find an irrational number between any two distinct real numbers. In fact, it turns out that the irrational numbers are even more abundant in a certain mathematical sense than the rational numbers. Later in this section we list other examples of irrational numbers.

Inequality Symbols.   The position of points on the real line establishes a natural ordering of the real numbers from left to right. We use this ordering to give a geometric interpretation of the *less than* symbol, denoted by $<$.

Definition of the Less Than Symbol, $<$
If a, $b \in \mathbf{R}$, then $a < b$ means $b - a$ is positive, or equivalently, a is to the left of b on the real line.

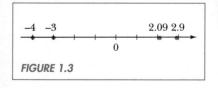

FIGURE 1.3

For example, $-4 < -3$ and $2.09 < 2.9$. See Figure 1.3.
Other inequality symbols can be described in terms of $<$ and $=$.

$$
\begin{array}{lll}
a > b & \text{is equivalent to} & b < a \\
a \leq b & \text{means} & a < b \quad \text{or} \quad a = b \\
a \geq b & \text{means} & a > b \quad \text{or} \quad a = b
\end{array}
$$

In words, we say *greater than* for the symbol $>$, *less than or equal to* for $\leq$, and *greater than or equal to* for $\geq$. Examples using these symbols include $-3 > -4$, $2.09 \leq 2.9$, and $8 \geq 8$.

Example 4

Use mathematical symbols to write (a) x is a nonnegative real number; (b) the set of all nonnegative real numbers.

SOLUTION (a) We write $x \geq 0$. Notice that we do not need to say "and $x \in \mathbf{R}$" since we always assume that variables represent real numbers in this book (unless stated otherwise).
(b) $\{x : x \geq 0\}$ ∎

Example 5

Describe the set $\{x : x^2 \geq 0\}$.

SOLUTION We assert that if x is *any* real number, then its square must be nonnegative. This is certainly true if x is a positive number. Furthermore, if x is negative, then $x^2 > 0$ because a negative number multiplied by a negative number is positive. Finally, if x is zero, then $x^2 = 0^2 = 0 \geq 0$. Thus we conclude that $\{x : x^2 \geq 0\} = \mathbf{R}$. ∎

We can use inequality symbols to indicate that one number is between two others on the real line. For example, to say π is between 3.14 and 3.15, we write

$$3.14 < \pi \quad \text{and} \quad \pi < 3.15$$

This can be shortened to the statement $3.14 < \pi < 3.15$. In general, we have the following:

$$
a < b < c \quad \text{is equivalent to} \quad a < b \text{ and } b < c
$$

Similar statements hold for $a \leq b \leq c$, $a \leq b < c$, and $a < b \leq c$.

Absolute Value. The absolute value of a real number a, denoted by $|a|$, has two equivalent definitions.

Definition of Absolute Value

Geometric: $|a|$ represents the distance between a and the origin on the real line.

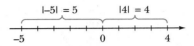

Algebraic: $|a| = \begin{cases} a & \text{if } a \text{ is positive or zero} \\ -a & \text{if } a \text{ is negative} \end{cases}$

Remember that $-a$ is positive when a is a negative number. Thus it follows from either the geometric or algebraic definitions that $|a|$ is always nonnegative for any real number a, $|a| \geq 0$.

Example 6

Simplify (a) $-|4 - 7|$, (b) $|2 - \pi|$.

SOLUTION (a) We have $-|4 - 7| = -|-3| = -3$.

(b) Recall that $\pi \approx 3.14$. (The symbol $\approx$ means *is approximately equal to*.) Therefore $2 - \pi \approx -1.14$ is a negative number. Hence, by the algebraic definition of absolute value, we have

$$|2 - \pi| = -(2 - \pi) = \pi - 2 \ \blacksquare$$

Example 7

Simplify if possible (a) $|x^2 + 1|$, (b) $|x^2 - 1|$.

SOLUTION (a) Recall from Example 1.5 that $x^2 \geq 0$ for all $x \in \mathbf{R}$. It follows that $x^2 + 1$ is always positive. Therefore

$$|x^2 + 1| = x^2 + 1$$

(b) Since x is unknown, x^2 could be less than 1 or greater than or equal to 1. Hence $x^2 - 1$ could be negative, zero, or positive. Therefore the expression $|x^2 - 1|$ cannot be simplified further without making assumptions about x. $\blacksquare$

Example 8

Find all x satisfying (a) $|x| = 5$, (b) $|3 - x| = 5$, (c) $|x| = -5$.

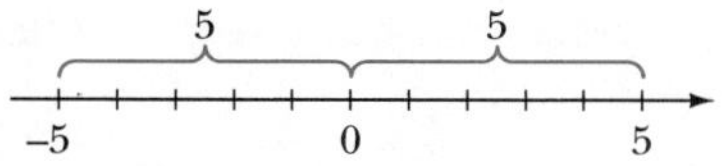

SOLUTION (a) By the geometric definition of absolute value, $|x| = 5$ means that the distance between x and the origin must equal 5. (See Figure 1.4.) It follows that $x = -5$ or $x = 5$. Formally, we write

$$|x| = 5 \qquad \text{implies} \qquad x = -5 \text{ or } x = 5$$

FIGURE 1.4

(b) Again, by the geometric definition of absolute value, $|3 - x| = 5$ means that the distance between the quantity $3 - x$ and the origin must equal 5. It follows that $3 - x$ must be -5 or 5. We write

$$|3 - x| = 5 \qquad \text{implies} \qquad 3 - x = -5 \text{ or } 3 - x = 5$$
$$-x = -8 \qquad\qquad -x = 2$$
$$x = 8 \qquad\qquad x = -2$$

Therefore $x = 8$ or $x = -2$.

(c) The absolute value of any real number cannot be negative. Hence there is no real number x that satisfies $|x| = -5$. ∎

Fundamental Properties of Absolute Value
For real numbers a and b,

1. $|ab| = |a||b|$ 2. $\left| \dfrac{a}{b} \right| = \dfrac{|a|}{|b|}$, provided $b \neq 0$

The proofs of these properties are outlined in the exercises.

Example 9

Simplify $|-4x|$.

SOLUTION We have $|-4x| = |-4||x| = 4|x|$. ∎

Example 10

Simplify $\dfrac{|2x - 6|}{|3 - x|}$ (assume $x \neq 3$).

SOLUTION We have

$$\frac{|2x - 6|}{|3 - x|} = \frac{|2(x - 3)|}{|-1(x - 3)|} = \frac{|2||x - 3|}{|-1||x - 3|}$$

$$= \frac{|2|}{|-1|} = 2 \quad \blacksquare$$

Square Roots. For every positive real number a, there are two real numbers whose square is equal to a. We refer to these numbers as square roots of a. For example, both 5 and -5 are square roots of 25 because $5^2 = 25$ and $(-5)^2 = 25$. Zero has exactly one square root, which is zero. Negative numbers do not possess real square roots since the square of any real number is never negative. (See Example 5.)

We write $\sqrt{a}$ to represent the unique *nonnegative* square root of a. Thus $\sqrt{25} = 5$, not ± 5.

Definition of the Square Root Symbol, $\sqrt{}$
If a is a nonnegative real number, then $\sqrt{a} = b$ means $b^2 = a$ and b is *nonnegative*.

As an important consequence of this definition, we have

If x is any real number, then $\sqrt{x^2} = |x|$.

Example 11

Simplify (a) $\sqrt{(-3)^2}$, (b) $\sqrt{(x+1)^2}$.

SOLUTION (a) $\sqrt{(-3)^2} = \sqrt{9} = 3$

(b) $\sqrt{(x+1)^2} = |x+1|$ ∎

Fundamental Properties of Square Roots
If a and b are nonnegative real numbers, then

1. $\sqrt{ab} = \sqrt{a}\,\sqrt{b}$

2. $\sqrt{\dfrac{a}{b}} = \dfrac{\sqrt{a}}{\sqrt{b}}$, provided $b \neq 0$

Example 12

Simplify (a) $\sqrt{56}$, (b) $\dfrac{\sqrt{18x^2}}{\sqrt{2}}$.

SOLUTION (a) $\sqrt{56} = \sqrt{4(14)} = 2\sqrt{14}$

(b) $\dfrac{\sqrt{18x^2}}{\sqrt{2}} = \sqrt{\dfrac{18x^2}{2}} = \sqrt{9x^2} = \sqrt{9}\sqrt{x^2} = 3|x|$ ∎

A common mistake when working with square roots is to think that $\sqrt{a^2 + b^2}$ reduces to $a + b$. This is incorrect because

$$(a + b)^2 = a^2 + 2ab + b^2$$

Thus when a and b are nonzero, $a + b$ is *not* a square root of $a^2 + b^2$.

Example 13

Simplify $\sqrt{4^2 + 6^2}$.

SOLUTION $\quad \sqrt{4^2 + 6^2} = \sqrt{16 + 36} = \sqrt{52} = \sqrt{4 \cdot 13} = 2\sqrt{13}$

Note that the answer is *not* $4 + 6$. ■

A **prime number** is a positive integer greater than 1 that is not divisible by any positive integer other than 1 and itself. The first six prime numbers are 2, 3, 5, 7, 11, and 13. If p is any prime number, then one can show that $\sqrt{p}$ is irrational. Therefore $\sqrt{2}, \sqrt{3}, \sqrt{5}$, and so on are all irrational numbers. (For a proof that $\sqrt{2}$ is irrational see the appendix in *Cosmos*, by Carl Sagan (Random House, New York, 1980).)

Note that $\sqrt{2} \approx 1.41$ and $\sqrt{3} \approx 1.73$. Memorize the approximate values for $\sqrt{2}$ and $\sqrt{3}$ because they occur frequently in our work.

Cube Roots. For any real number a there is a unique real number whose cube is equal to a. This number is called the **cube root** of a, denoted by $\sqrt[3]{a}$.

Definition of Cube Root
If a is a real number, then $\sqrt[3]{a} = b$ means $b^3 = a$.

Note that $\sqrt[3]{a}$ exists even if a is negative.

Example 14

Simplify (a) $\sqrt[3]{64}$, (b) $\sqrt[3]{-64}$, (c) $\sqrt[3]{x^3}$.

SOLUTION (a) $\sqrt[3]{64} = 4$ since $4^3 = 64$
(b) $\sqrt[3]{-64} = -4$ since $(-4)^3 = -64$
(c) $\sqrt[3]{x^3} = x$ ■

The fundamental rules for simplifying cube roots are similar to those for square roots.

Example 15

Simplify (a) $\sqrt[3]{12} \; \sqrt[3]{-18}$, (b) $\sqrt[3]{54}$

SOLUTION (a) $\sqrt[3]{12} \; \sqrt[3]{-18} = \sqrt[3]{12(-18)} = \sqrt[3]{-216} = -6$
(b) $\sqrt[3]{54} = \sqrt[3]{27 \cdot 2} = 3\sqrt[3]{2}$ ■

EXERCISES 1.1

In Exercises 1 and 2, graph each number on the real line.

1. (a) $\dfrac{3}{4}$

   (b) -1.75

   (c) $-\pi$

   (d) $\sqrt{2}$

2. (a) $-2\dfrac{2}{3}$

   (b) $\dfrac{\pi}{2}$

   (c) -1.4

   (d) $\sqrt{3}$

In Exercises 3 to 8, rewrite the given set by listing its elements inside braces. For example,

$$\{x : x^2 = 9\} = \{3, -3\}$$

If the set consists of all real numbers, write **R**; if it is empty, write $\varnothing$.

3. (a) $\{x : x + 6 = 0\}$
   (b) $\{x : x \cdot 6 = 6 \cdot x\}$

4. (a) $\{x : 4x = 1\}$
   (b) $\{x : x + 4 = 4 + x\}$

5. (a) $\{x : x^2 < 0\}$
   (b) $\{y : 5y = 0\}$

6. (a) $\left\{x : \dfrac{2}{x} = 0\right\}$
   (b) $\{y : y^2 = 36\}$

7. (a) $\{n : n \in \mathbf{Z} \text{ and } -3 < n < 2\}$
   (b) $\{n : n \in \mathbf{N} \text{ and } -3 < n < 2\}$
   (c) $\{p : p \text{ is prime and } p \le 30\}$

8. (a) $\{m : m \in \mathbf{Z} \text{ and } -2 \le m \le 2\}$
   (b) $\{m : m \in \mathbf{N} \text{ and } -2 \le m \le 2\}$
   (c) $\{q : q \text{ is prime and } 20 < q < 40\}$

In Exercises 9 to 12, use set builder notation to represent the collection of real numbers described.

9. (a) Odd integers
   (b) Positive even integers

10. (a) Nonnegative even integers
    (b) Positive odd integers

11. Even integer multiples of π

12. Odd integer multiples of $\pi/2$

In Exercises 13 and 14, answer true or false.

13. (a) $6 \le 7$
    (b) $-9 \ge -10$
    (c) $4 < 4$
    (d) $\sqrt{2} > 2$

14. (a) $-1.9 < -2$
    (b) $\sqrt{3} < 2$
    (c) $\pi > 3.1$
    (d) $-5 \ge -5$

In Exercises 15 to 18, translate into mathematical symbols using inequalities or set builder notation.

15. (a) x is positive.
    (b) The set of all positive real numbers.

16. (a) x is negative.
    (b) The set of all negative real numbers.

17. (a) y is between 0 and 2.
    (b) The set of all real numbers between 0 and 2.

18. (a) n is between 1 and -3.
    (b) The set of all real numbers between 1 and -3.

In Exercises 19 to 21, determine if the given number is rational or irrational.

19. (a) $.\overline{123}$
    (b) $\sqrt{7}$
    (c) -13
    (d) $1 + \dfrac{\pi}{10}$

20. (a) $(\sqrt{5})^2$
    (b) $\sqrt{16}$
    (c) $\sqrt{23}$
    (d) $-\pi/10$

21. (a) $.202002000200002 \cdots$ (Assume this follows the obvious pattern.)
    (b) 3.14159
    (c) $.1234567891011121314 \cdots$ (the natural numbers taken in order)

In Exercise 22, answer true or false.

22. (a) If $x \in \mathbf{R}$, then x is either rational or irrational.
    (b) There are no real numbers that are both rational and irrational.

In Exercises 23 to 26, find a real number x satisfying the given inequality statement such that (a) x is rational; (b) x is irrational.

23. $1 < x < 2$ [*Hint* for part (b): see Exercise 19(d).]

24. $-1 < x < 0$ [*Hint* for part (b): see Exercise 20(d).]

25. $1.\overline{14} < x < 1.\overline{15}$ 26. $3.\overline{3} < x < 3.34\overline{3}$

In Exercises 27 to 36, simplify the given expression if possible. Assume variables represent real numbers.

27. (a) $|-4|$
    (b) $-|4 - 9|$
    (c) $|x^2|$
    (d) $|-a|$
    (e) $|x + 1| - 1$

28. (a) $|3 - 5|$
    (b) $-|-6|$
    (c) $|-x^2 - 1|$
    (d) $|a^2 + b^2|$
    (e) $|x + y| - y$

29. (a) $|a^2 + b^2| + a^2 - b^2$
    (b) $|a + b||b + a|$

30. (a) $|x + y| - |-x - y|$
    (b) $|a - b||b - a|$

31. (a) $|\pi - 3|$
    (b) $|1 - \sqrt{2}|$

32. (a) $|\sqrt{3} - \pi|$
    (b) $|\sqrt{3} - \sqrt{2}|$

33. $\dfrac{|x - 5|}{|5 - x|}$ (assume $x \neq 5$)

34. $\dfrac{|x - 5|}{|15 - 3x|}$ (assume $x \neq 5$)

35. $\dfrac{|2h|}{h}$, assume (a) $h > 0$; (b) $h < 0$

36. $\dfrac{|3x - 3y|}{x - y}$, assume (a) $x > y$; (b) $x < y$

In Exercises 37 to 40, suppose $a < 0$ and $b > 0$. Simplify the given expression if possible.

37. (a) $|a| + |b|$
    (b) $|a| + a$
    (c) $\dfrac{|a|}{a}$

38. (a) $|b| + b$
    (b) $a - |a|$
    (c) $\dfrac{|b|}{b}$

39. (a) $|ab|$
    (b) $|1 - a|$
    (c) $|a - b|$

40. (a) $\left|\dfrac{a}{b}\right|$
    (b) $|b - a|$
    (c) $|a^2|$

In Exercises 41 and 42, answer true or false.

41. $|x_2 - x_1|^2 = (x_2 - x_1)^2$
42. $y \leq |y|$

In Exercises 43 to 48, solve for x.

43. (a) $|x| = 4$
    (b) $|-2x| = 5$
    (c) $|x| = -2$

44. (a) $|-x| = 6$
    (b) $|3x| = 8$
    (c) $|-x| = -3$

45. (a) $|x + 1| = 5$
    (b) $|3 - 2x| = 8$
    (c) $|2 - 5x| = 0$

46. (a) $|6 - x| = 7$
    (b) $|3x + 4| = 5$
    (c) $|7 - 4x| = 0$

47. (a) $|x| = |2x + 1|$
    (b) $|x + 1| = |x - 1|$

48. (a) $|1 - 2x| = |1 - 3x|$
    (b) $|x| = |x + 1|$

In Exercises 49 to 56, simplify if possible.

49. (a) $\sqrt{10^2 - 4(3)(3)}$
    (b) $\sqrt{(-6)^2}$
    (c) $\sqrt{(x - 3)^2}$
    (d) $\sqrt{-4}$

50. (a) $\sqrt{3^2 - 4(2)(1)}$
    (b) $\sqrt{-9}$
    (c) $\sqrt{(x + 2)^2}$
    (d) $\sqrt{(-2)^2}$

51. (a) $\sqrt{3^2 + 9^2}$
    (b) $\sqrt{\dfrac{-x}{4}}$
    (c) $\sqrt{0}$

52. (a) $\sqrt{2^2 + 4^2}$
    (b) $\sqrt{-16a}$
    (c) $\sqrt{8^2 - 4(4)(4)}$

53. (a) $\sqrt{6}\sqrt{24}$
    (b) $\dfrac{\sqrt{3xy^2}}{\sqrt{12x}}$
    (c) $\sqrt{x^2 + y^2}$

54. (a) $\dfrac{\sqrt{15}}{\sqrt{45}}$
    (b) $\dfrac{\sqrt{27x^2}}{\sqrt{3}}$
    (c) $\sqrt{4n^2 - 16m^2}$

55. (a) $\sqrt[3]{(x + 1)^3}$
    (b) $\sqrt[3]{16}\sqrt[3]{-4}$
    (c) $\sqrt[3]{40}$

56. (a) $\sqrt[3]{8y^3}$
    (b) $\sqrt[3]{81}$
    (c) $\sqrt[3]{-100}\sqrt[3]{-80}$

In Exercises 57 and 58, for the number a, find consecutive integers n and $n + 1$ such that $n < a < n + 1$.

57. (a) $a = \sqrt{5}$
    (b) $a = \sqrt[3]{100}$

58. (a) $a = \sqrt{30}$
    (b) $a = \sqrt[3]{55}$

In Exercises 59 and 60, using $\sqrt{2} \approx 1.414$ and $\sqrt{3} \approx 1.732$, approximate the given number to two decimal places.

59. (a) $\dfrac{\sqrt{2}}{2}$
    (b) $\dfrac{1 + \sqrt{3}}{2}$

60. (a) $\dfrac{\sqrt{3}}{2}$
    (b) $\dfrac{1 + \sqrt{2}}{2}$

61. Prove that the square of any odd integer is odd. (*Hint:* Any odd integer may be represented by the expression $2n + 1$, where $n \in \mathbf{Z}$.)

62. Evaluate (a) $\sqrt[3]{\sqrt{64}}$, (b) $\sqrt{\sqrt[3]{64}}$. Is it true in general that $\sqrt[3]{\sqrt{a}} = \sqrt{\sqrt[3]{a}}$?

63. Prove $|ab| = |a||b|$ in the following cases.
    (a) $a > 0$ and $b > 0$ (*Hint:* begin by explaining why $|ab| = ab$, and then explain why $|a||b| = ab$.)
    (b) $a < 0$ and $b < 0$ [see the hint for part (a)].
    (c) $a > 0$ and $b < 0$ (*Hint:* first show that $|ab| = -ab$, and then show that $|a||b| = -ab$.)
    (d) $a = 0$ or $b = 0$

64. Prove $\left|\dfrac{a}{b}\right| = \dfrac{|a|}{|b|}$ for cases (a), (b), and (c) in Exercise 63.

1.2 EQUATIONS

In this section we review several basic techniques for solving equations. Let us begin with the following definition.

Definition of Real Solution
A **real solution** of an equation is a real number such that when it is substituted into the equation for the variable, we obtain the same number on both sides of the equation.

Informally, we say a real solution is a real number that *works* in the equation (or *satisfies* the equation).

Example 1

Determine if the given number is a solution of the equation

$$x + \frac{1}{x-2} = \frac{x-1}{x-2}. \text{ (a) } x = 1, \text{ (b) } x = 2.$$

SOLUTION (a) To check $x = 1$, we substitute 1 into the equation for x.

$$x + \frac{1}{x-2} \overset{?}{=} \frac{x-1}{x-2}$$

$$1 + \frac{1}{1-2} \qquad \frac{1-1}{1-2}$$

$$1 + \frac{1}{-1} \qquad \frac{0}{-1}$$

$$1 - 1 \qquad 0$$

$$0$$

Yes

Since we get the same number on both sides of the equation, we conclude that $x = 1$ is a solution.

(b) To check $x = 2$, we substitute 2 into the equation for x.

$$x + \frac{1}{x-2} \overset{?}{=} \frac{x-1}{x-2}$$

$$
\begin{array}{c|c}
2 + \dfrac{1}{2-2} & \dfrac{2-1}{2-2} \\[2ex]
2 + \dfrac{1}{0} & \dfrac{1}{0} \\[2ex]
\text{Undefined} & \text{Undefined} \\
\end{array}
$$

$$\text{No}$$

Since division by zero is undefined, we conclude that $x = 2$ is not a solution. Note that we do not say that an undefined expression equals another undefined expression. ■

We now consider methods for solving four types of equations: linear, quadratic, fractional, and ones containing square roots. The solutions for these equations involve a sequence of steps, each step consisting of a simplification or an application of an operation. Following is a summary of the allowable operations.

Operations for Solving Equations

Operation	Original Equation	New Equation
Addition	$a = b$	$a + c = b + c$
Multiplication by $c \neq 0$	$a = b$	$ac = bc$
Squaring	$a = b$	$a^2 = b^2$
Square root	$a = b$	$\sqrt{a} = \sqrt{b}$ (provided $a, b \geq 0$)
Zero Product Rule	$ab = 0$	$a = 0 \quad or \quad b = 0$

Remarks

1. The squaring operation does not necessarily produce an *equivalent* equation. (Equivalent equations are ones that have the same solutions.) Therefore, when using this operation, always check answers (see Example 10).
2. The square root of an expression is real only if the expression is nonnegative. Hence the square root operation may be applied to an equation only when both sides are nonnegative. See Examples 5, 7, and 9.

Linear Equations. An equation whose variable appears to the first power only is called a **linear equation**. Thus

$$2x - 3 = 0 \qquad 6(2x - 3) = 5x + 7$$

are examples of linear equations. An equation involving terms such as x^2, $\sqrt{x}$, or $1/x$ does not qualify as linear.

Solving Linear Equations
Use the addition and multiplication operations to isolate the variable on one side of the equation.

Example 2

Solve for x: $6(2x - 3) = 5x + 7$.

SOLUTION We present our solutions to equations using a *line-by-line* format. This means that each succeeding line is an equation obtained from the previous one by the application of an acceptable operation or simplification. Thus

$6(2x - 3) = 5x + 7$	Original equation
$12x - 18 = 5x + 7$	Simplification
$12x - 5x = 18 + 7$	Addition of $-5x + 18$
$7x = 25$	Simplification
$x = \dfrac{25}{7}$	Multiplication by 1/7

Therefore the solution is $x = 25/7$. ■

Example 3

Solve for b: $a(2b - 3) = 5b + c$.

SOLUTION We refer to this type of equation as a **literal equation**. Literal equations contain a variable and one or more constants represented by letters. In this case we think of b as the variable and a and c as constants. This equation is linear as well and can be solved using the same procedure executed in Example 2.

$$a(2b - 3) = 5b + c$$
$$2ab - 3a = 5b + c$$
$$2ab - 5b = 3a + c$$

$$(2a - 5)b = 3a + c$$

$$b = \frac{3a + c}{2a - 5}$$

Therefore the solution is $b = (3a + c)/(2a - 5)$. We assume that $2a - 5 \neq 0$; otherwise this expression would be undefined. ■

Quadratic Equations. An equation with the variable x written in the form

$$ax^2 + bx + c = 0$$

where a, b, and c are constants and $a \neq 0$, is called a **quadratic equation**. There are three methods for solving quadratic equations.

Solving Quadratic Equations

$$ax^2 + bx + c = 0$$

Method 1:   Factor: use the Zero Product Rule.
Method 2:   If the equation is written in the form $x^2 = k$ or $(x - h)^2 = k$, use the square root operation.
Method 3:   Use the quadratic formula,

$$x = \frac{-b \pm \sqrt{b^2 - 4ac}}{2a}$$

Method 3, the quadratic formula, works on any quadratic equation. The derivation of this formula is in Section A.3 of the appendix.

Example 4

Solve $(x - 1)(x + 2) = 4$.

SOLUTION   First, rewrite the equation in the form $ax^2 + bx + c = 0$.

$$(x - 1)(x + 2) = 4$$
$$x^2 + 2x - x - 2 = 4$$
$$x^2 + x - 2 = 4$$
$$x^2 + x - 6 = 0$$

The expression on the left factors,

$$(x + 3)(x - 2) = 0$$

By the Zero Product Rule, we have

$$x + 3 = 0 \qquad or \qquad x - 2 = 0$$
$$x = -3 \qquad \qquad \qquad x = 2$$

Therefore we have two solutions, $x = -3$ or $x = 2$. You can check that both solutions work in the original equation. ■

Example 5

Solve $(x - 2)^2 = 9$.

SOLUTION   Notice that both sides of this equation are nonnegative. [Why is $(x - 2)^2 \geq 0$?] Therefore we may apply the square root operation.

$$\sqrt{(x - 2)^2} = \sqrt{9}$$
$$|x - 2| = 3$$

By the definition of absolute value, we have

$$x - 2 = 3 \qquad or \qquad x - 2 = -3$$
$$x = 5 \qquad \qquad \qquad x = -1$$

Thus we find that $x = 5$ or $x = -1$. ■

Example 6

Solve $-\dfrac{1}{2}x^2 - 2x + 1 = 0$.

SOLUTION   Multiply both sides by -2 to eliminate the fraction.

$$-2\left(-\frac{1}{2}x^2 - 2x + 1\right) = -2(0)$$
$$x^2 + 4x - 2 = 0$$

Once again we have a quadratic equation. Unfortunately, the left side does not factor, so we use the quadratic formula. In this case, $x^2 + 4x - 2 = 0$, so we have $a = 1$, $b = 4$, and $c = -2$. Thus

$$x = \frac{-4 \pm \sqrt{4^2 - 4(1)(-2)}}{2(1)}$$
$$= \frac{-4 \pm \sqrt{24}}{2} = \frac{-4 \pm 2\sqrt{6}}{2}$$
$$= -2 \pm \sqrt{6}$$

The symbol $\pm$ means *plus or minus*. Therefore we have two solutions: $x = -2 + \sqrt{6}$ or $x = -2 - \sqrt{6}$. ■

Example 7

Solve $x^2 + 4 = 0$.

SOLUTION   We may write the equivalent equation

$$x^2 = -4$$

We know that this equation has no real solution because $x^2 \geq 0$ for all $x \in \mathbf{R}$. Therefore we say the original equation has *no real solution*. Notice that we did not apply the square root operation because one side of the equation is negative. Had we done so, we would have obtained $\sqrt{-4}$ on the right side, which is not a real number. ■

As we mentioned, the quadratic formula may be used to solve any quadratic equation. However, the formula may produce a square root of a negative number. In this case, we conclude that the original equation has no real solution. For example, if we apply the quadratic formula to

$$x^2 + x + 1 = 0$$

we obtain

$$x = \frac{-1 \pm \sqrt{-3}}{2}$$

Because of the appearance of $\sqrt{-3}$, we conclude that $x^2 + x + 1 = 0$ has no real solutions.

Fractional Equations.   An equation that contains one or more fractions is called a **fractional equation**. To solve such an equation, we first eliminate the fractions.

Solving Fractional Equations
Step 1   Eliminate fractions by applying the multiplication operation.
Step 2   Solve.
Step 3   Check answers. Remember that division by zero is invalid.

Recall that multiplying a fraction by its denominator cancels the denominator.

$$b\left(\frac{a}{b}\right) = a$$

Therefore, when one or more fractions appear in an equation, they can be eliminated if each term is multiplied by the least common denominator of all the fractions.

Example 8

Solve $x + \dfrac{1}{x - 2} = \dfrac{x - 1}{x - 2}$.

SOLUTION We first eliminate the fractions by using the multiplication operation. In this case, the least common denominator of all the fractions is just $(x - 2)$. Hence we shall multiply both sides of the equation by $(x - 2)$.

$$(x - 2)\left(x + \frac{1}{x - 2} \right) = (x - 2)\left(\frac{x - 1}{x - 2} \right)$$

$$x(x - 2) + 1 = x - 1$$

$$x^2 - 2x + 1 = x - 1$$

$$x^2 - 3x + 2 = 0$$

Once again we have a quadratic equation. We can solve it by factoring and using the Zero Product Rule.

$$(x - 1)(x - 2) = 0$$

$$x - 1 = 0 \quad or \quad x - 2 = 0$$

$$x = 1 \qquad\qquad x = 2$$

Finally, we must check the answers. This was done in Example 1, where we found that $x = 1$ works but $x = 2$ does not. Therefore, the only solution to the original equation is $x = 1$. ∎

Example 9

Solve $\dfrac{x^2}{a^2} + \dfrac{y^2}{b^2} = 1$ for y. Assume that a and b are positive.

SOLUTION First eliminate fractions.

$$\frac{x^2}{a^2} + \frac{y^2}{b^2} = 1$$

$$a^2 b^2 \left(\frac{x^2}{a^2} + \frac{y^2}{b^2} \right) = a^2 b^2 (1)$$

$$b^2 x^2 + a^2 y^2 = a^2 b^2$$

Considering all letters other than y as constants, we have a quadratic equation in y. Rewrite the equation so that y^2 is isolated on one side.

$$a^2 y^2 = a^2 b^2 - b^2 x^2$$

$$y^2 = \frac{a^2 b^2 - b^2 x^2}{a^2}$$

Now apply the square root operation. (We assume $x^2 \leq a^2$; otherwise y^2 would equal a negative number, implying that no real solution exists.)

$$\sqrt{y^2} = \sqrt{\frac{a^2 b^2 - b^2 x^2}{a^2}}$$

$$|y| = \sqrt{\frac{b^2(a^2 - x^2)}{a^2}}$$

Since a and b are positive numbers, we may write

$$|y| = \frac{b}{a}\sqrt{a^2 - x^2}$$

Finally, this gives us

$$y = \frac{b}{a}\sqrt{a^2 - x^2} \quad \text{or} \quad y = -\frac{b}{a}\sqrt{a^2 - x^2} \quad \blacksquare$$

Equations Containing Square Roots. We usually solve an equation containing a square root by applying the squaring operation. In such a situation, we must be careful to check our answers.

Solving Equations Containing Square Roots
Step 1 Isolate a square root term on one side of the equation.
Step 2 Square both sides.
Step 3 Solve.
Step 4 Check answers.

In case the equation contains more than one square root term, steps 1 and 2 must be repeated until all the square roots are eliminated.

Example 10

Solve $\dfrac{x}{2\sqrt{x+1}} + \sqrt{x+1} = 1$.

SOLUTION This is a fractional equation containing square roots. First, we eliminate the fraction by multiplying both sides of the equation by $2\sqrt{x+1}$.

$$2\sqrt{x+1}\left(\frac{x}{2\sqrt{x+1}} + \sqrt{x+1}\right) = 2\sqrt{x+1}\,(1)$$

$$x + 2(x+1) = 2\sqrt{x+1}$$

Notice that $(\sqrt{x+1})(\sqrt{x+1})$ simplifies to $(x+1)$. Further simplification on the left side yields

$$3x + 2 = 2\sqrt{x+1}$$

Now, we square both sides.

$$(3x+2)^2 = (2\sqrt{x+1})^2$$

$$9x^2 + 12x + 4 = 4(x+1)$$

$$9x^2 + 12x + 4 = 4x + 4$$

$$9x^2 + 8x = 0$$

We can factor the left side and then use the Zero Product Rule.

$$x(9x+8) = 0$$

$$x = 0 \quad\quad or \quad\quad 9x + 8 = 0$$

$$x = -\frac{8}{9}$$

Finally, we must check our answers.

Check that $x = 0$:

$$\frac{x}{2\sqrt{x+1}} + \sqrt{x+1} \overset{?}{=} 1$$

$$\frac{0}{2\sqrt{0+1}} + \sqrt{0+1}$$

$$\frac{0}{2} + 1$$

$$1$$

Yes

Check that $x = -8/9$:

$$\frac{x}{2\sqrt{x+1}} + \sqrt{x+1} \overset{?}{=} 1$$

$$\frac{-\dfrac{8}{9}}{2\sqrt{\dfrac{-8}{9}+1}} + \sqrt{\dfrac{-8}{9}+1}$$

$$\frac{-\dfrac{8}{9}}{2\sqrt{\dfrac{1}{9}}} + \sqrt{\dfrac{1}{9}}$$

$$\frac{-8/9}{2/3} + \frac{1}{3}$$

$$-\frac{4}{3} + \frac{1}{3}$$

$$-1$$

No

Therefore the correct solution is $x = 0$. ■

We end this section with a brief remark concerning division. We wish to avoid dividing both sides of an equation by the variable or an expression containing the variable; in doing so, we might be dividing by zero. As a simple example, consider the equation

$$x^2 = 4x$$

Dividing by x yields $x = 4$. But this is not the entire solution. The correct way to solve the equation is as follows.

$$x^2 = 4x \quad \text{is equivalent to} \quad x^2 - 4x = 0$$
$$x(x - 4) = 0$$
$$x = 0 \quad or \quad x - 4 = 0$$
$$x = 4$$

Dividing by the variable can result in a lost solution or even greater difficulties. See Exercises 63 and 66.

EXERCISES 1.2

In Exercises 1 to 46, find all real solutions for the given equation.

1. (a) $(x - 2)(x + 3) = (x - 4)(x - 5)$
   (b) $2 - (x - 1)^2 = 3 - (x + 1)(x - 1)$
   (c) $x^2 + 7x + 12 = 0$
   (d) $x + 1 = x - 1$

2. (a) $19 - (y + 5)(3y - 1) = 0$
   (b) $7y^2 = 4y$
   (c) $x(x + 1) = 2(x + 1)$
   (d) $2y = 3y$

3. (a) $(2y + 7)^2 = 36$
   (b) $y^2 = -9$

4. (a) $(4 - x)^2 = 18$
   (b) $x^2 + 4 = 4$

5. (a) $(x - 2)^2 = 3 - x^2$
   (b) $m^2 - m = 1$
   (c) $n^2 + n + 1 = 0$

6. (a) $(3x - 2)^2 = 3(2x - 1)$
   (b) $m^2 + m = 1$
   (c) $n(n + 2) = n - 2$

7. $\dfrac{x}{3} + \dfrac{1}{x} = \dfrac{4}{3}$

8. $\dfrac{1}{x} = 2x$

9. $\dfrac{x}{x + 1} = \dfrac{x + 1}{x}$

10. $\dfrac{x}{x - 1} - \dfrac{2}{x + 1} = \dfrac{4}{3}$

11. $\sqrt{x} - \dfrac{1}{2\sqrt{x}} = 0$

12. $\dfrac{x}{2\sqrt{x + 1}} + \sqrt{x + 1} = 0$

13. $x\sqrt{x + 3} + \dfrac{x^2}{\sqrt{x + 3}} = 0$

14. $\dfrac{\sqrt{x + 1}}{\sqrt{x + 2}} + \dfrac{\sqrt{x + 2}}{\sqrt{x + 1}} = 0$

15. $x + \dfrac{1}{x} = 0$

16. $\dfrac{1}{x} - \dfrac{2}{x} = 0$

17. $\dfrac{x + 1}{x} = \dfrac{1}{x}$

18. $x + \dfrac{1}{x} = \dfrac{1 - x}{x}$

19. $1 + \dfrac{1}{x + 3} + \dfrac{3}{x(x + 3)} = \dfrac{2}{x}$

20. $1 - \dfrac{1}{x + 2} = \dfrac{2}{x(x + 2)}$

21. $x^3 - x^2 - 3x = 0$

22. $x^3 + x^2 - 2x = 0$

23. $y^2 + 12 = y^4$ (*Hint:* substitute $u = y^2$ and solve for u first.)

24. $y^4 - 11y^2 + 18 = 0$ (See hint for Exercise 23.)

25. $\dfrac{3\sqrt{x}\sqrt{x + 1}}{2} + \dfrac{x\sqrt{x}}{2\sqrt{x + 1}} = 0$

26. $|\sqrt{x} - 2| = 5$

27. $|x^2 + x - 1| = 1$

28. $|x^2 - 4x| = 2$

29. $\left|\dfrac{x}{x+1}\right| = \dfrac{1}{2}$

30. $\left|x - \dfrac{4}{x}\right| = 3$

31. (a) $\sqrt{2x+5} = 1$
    (b) $\sqrt{y+9} = 5$

32. (a) $2\sqrt{7-x} = 3$
    (b) $\sqrt{6y} + 12 = 8$

33. $\sqrt{x^2+16} + 3x = 0$

34. $\sqrt{3x^2+4} - 2x = 0$

35. $\dfrac{x+1}{x} = \dfrac{x}{1-x}$

36. $\dfrac{x}{x+2} = \dfrac{x+2}{x}$

37. (a) $\sqrt{-y} = y$
    (b) $\sqrt{x^2} = x$

38. (a) $\sqrt{y} = -y$
    (b) $\sqrt{x^2} = -x$

39. $\dfrac{1}{2\sqrt{x+4}} = 1$

40. $1 - \sqrt{-x} = \dfrac{1}{2}$

41. $\dfrac{x}{2\sqrt{x-1}} + \sqrt{x-1} = 2$

42. $\dfrac{x}{2\sqrt{x+1}} + \sqrt{x+1} = -1$

43. $\dfrac{\sqrt{x+1}}{\sqrt{x-1}} - \dfrac{\sqrt{x-1}}{\sqrt{x+1}} = 1$

44. $\dfrac{1}{\sqrt{x+1}} - \dfrac{x}{2(x+1)\sqrt{x+1}} = 1$

45. $\sqrt{x} - 1 = \sqrt{x-1}$

46. $\dfrac{1+\sqrt{x}}{1-\sqrt{x}} = \dfrac{1}{1+\sqrt{x}}$

In Exercises 47 to 62, solve for the indicated variable.

47. (a) $ax + b = cx + d$; x
    (b) $\pi r^2 + 2rl = A$; l

48. (a) $m(V + v) = M(V + v)$; v
    (b) $RT = rt + 10$; t

49. (a) $K = \dfrac{1}{2}mv^2$; v
    (b) $\pi r^2 + 2rh = 10$; r

50. (a) $F = \dfrac{GmM}{x^2}$; x
    (b) $\dfrac{1}{2}mv^2 + kv = 5$; v

51. $x^2 + y^2 = 2xy$; y

52. $x^2 - y = y^2 - x$; y

53. (a) $\dfrac{1}{A} = \dfrac{1}{B} + \dfrac{1}{C}$; C
    (b) $x = \dfrac{y+2}{y-3}$; y

54. (a) $\dfrac{x+a}{b-a} = \dfrac{x}{b}$; x
    (b) $x = \dfrac{y-1}{2y+1}$; y

55. $\dfrac{x^2}{a^2} - \dfrac{y^2}{b^2} = 1$; y

56. $\dfrac{x^2}{a^2} + \dfrac{y^2}{a^2} = 1$; x

57. $\dfrac{b}{c + \dfrac{b}{a}} = b$; b

58. $\dfrac{t+x}{y - \dfrac{t}{a}} = b$; t

59. $|x + a| = b$; x

60. $\left|\dfrac{a+b}{x}\right| = c$; x

61. $x = a(y - k)^2$; y

62. $a(y - k)^2 = (x - h)^2$; x

In Exercises 63 to 66, explain what is incorrect in the given "solution" and then write the correct solution.

63.
$$x^2 - x = x$$
$$x(x - 1) = x$$
$$x - 1 = 1$$
$$x = 2$$

64.
$$x^2 + x = 2$$
$$x(x + 1) = 2$$
$$x = 2 \quad \text{or} \quad x + 1 = 2$$
$$x = 2 \qquad\qquad\quad x = 1$$

65.
$$x^2 + 4 = 9$$
$$\sqrt{x^2 + 4} = \sqrt{9}$$
$$|x| + 2 = 3$$
$$|x| = 1$$
$$x = 1 \quad \text{or} \quad x = -1$$

66.
$$x(x - 5) = (x + 1)(x - 5)$$
$$x = x + 1$$
$$0 = 1$$
$$\text{No solution?}$$

67. What is wrong with the following "proof" that $1 = 0$? (There is only one mistake.)

$$\text{Let} \quad x = -\dfrac{1}{2}$$

Multiply by 2	$2x = -1$
Add x^2	$x^2 + 2x = x^2 - 1$
Add 1	$x^2 + 2x + 1 = x^2$
Factor	$(x + 1)^2 = x^2$
Square root	$\sqrt{(x+1)^2} = \sqrt{x^2}$
Simplify	$x + 1 = x$
Add $-x$	$1 = 0$

In this section we solve a variety of problems presented verbally. Each problem involves information that is related to a particular quantity that we wish to find. Of course, no single mechanical process will answer every verbal problem, but we can suggest a general procedure to follow.

General Procedure for Problem Solving

Step 1 Read the problem carefully. Answer the question, "What quantity am I asked to find?" Define a variable to represent this quantity: Let x = the unknown.

Step 2 Translate the given information into an equation for the unknown defined in step 1. To help with this, answer the question, "How is the given information related to the unknown?" Organizing the data in a chart or diagram is recommended. Several preliminary results may be necessary before you obtain the desired equation.

Step 3 Solve the equation found in step 2.

Step 4 Check the answer with the original statement of the problem. Is the answer reasonable?

The first step in this procedure cannot be overemphasized. Beginning every problem by restating in your own words what you are asked to find will prevent two of the most common errors: not knowing where to begin and following a process that does not answer the question. The second step obviously requires the most thought and hence is the most difficult. Experience, practice, and desire are the essential ingredients for success. To help provide these elements, we offer the following examples.

Our beginning example concerns pure numerical manipulations. Three machines, A, B, and C, perform certain operations on a given real number as follows:

Machine	Operation
A	Doubles the number.
B	Adds 1 to the number.
C	Takes the reciprocal of the number.

Thus if we input a real number a into one of these machines, it will output another real number according to the following table:

Machine	Input	Output
A	a	$2a$
B	a	$a + 1$
C	$a, a \neq 0$	$\dfrac{1}{a}$

Note that machines A and B will accept any real number as an input, but machine C will accept only nonzero real numbers. We shall refer to these machines as **function machines**.

Let us connect the function machines A, B, and C in series to form a simple computer. Figure 1.5 illustrates the situation. A real number entering this computer will first be operated on by machine A, the result is then fed to B, and the result from B is fed to C.

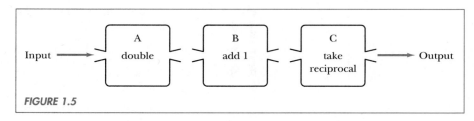

FIGURE 1.5

Thus if 3 enters this computer, the final output is 1/7.

$$3 \xrightarrow[\text{doubles}]{A} 6 \xrightarrow[\text{adds 1}]{B} 7 \xrightarrow[\text{takes reciprocal}]{C} 1/7$$

Example 1

Suppose a certain number enters the computer described in Figure 1.5. If the final output is 14, what was the original input number?

SOLUTION   Step 1.   Let x = the original input number.
Step 2.   In Figure 1.6 we illustrate what happens to x.

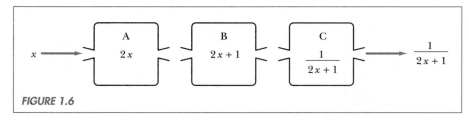

FIGURE 1.6

Since the final output is 14, we must have

$$\frac{1}{2x + 1} = 14$$

Step 3. Now solve the equation.

$$(2x + 1)\left(\frac{1}{2x + 1}\right) = (2x + 1)14$$

$$1 = 28x + 14$$

$$-13 = 28x$$

$$28x = -13$$

$$x = \frac{-13}{28}$$

Step 4. We check this answer as follows (see Figure 1.7).

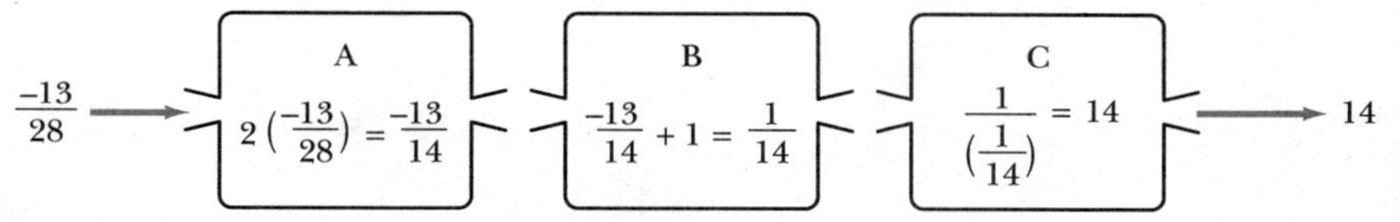

$$\frac{-13}{28} \longrightarrow \boxed{\begin{array}{c} A \\ 2\left(\frac{-13}{28}\right) = \frac{-13}{14} \end{array}} \longrightarrow \boxed{\begin{array}{c} B \\ \frac{-13}{14} + 1 = \frac{1}{14} \end{array}} \longrightarrow \boxed{\begin{array}{c} C \\ \frac{1}{\left(\frac{1}{14}\right)} = 14 \end{array}} \longrightarrow 14$$

FIGURE 1.7

Therefore $x = \dfrac{-13}{28}$ was the original input number producing a final output of 14. ∎

Example 2

A wire of fixed length is bent to form an isosceles right triangle. If the area of the triangle is 24 square units, find the length of the wire.

SOLUTION   Step 1.   The problem asks us to find the length of the original piece of wire. We shall let L represent this length.

Step 2.   We draw a picture of the situation in Figure 1.8(a). How is the triangle related to L? Clearly, the perimeter of the triangle must be the same as L. But the given information concerns the area of the triangle, not its perimeter. So, what do we do?

First, label the length of one of the legs of the triangle with x. Since the triangle is isosceles, the other leg is also x. Next, the Pythagorean Theorem tells us that

$$\text{Length of hypotenuse} = \sqrt{x^2 + x^2} = \sqrt{2x^2} = \sqrt{2}|x| = \sqrt{2}x$$

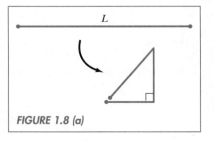

FIGURE 1.8 (a)

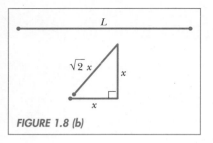

FIGURE 1.8 (b)

See Figure 1.8(b). (Note that $|x| = x$ since x represents a positive quantity.) Now, the length of the wire may be expressed in terms of x as follows:

$$L = x + x + \sqrt{2}x = 2x + \sqrt{2}x$$

Next, find x. We use the given data that the area of the triangle equals 24 square units to set up an equation for x.

$$\text{Area of triangle} = \frac{1}{2}(\text{base})(\text{height})$$

$$24 = \frac{1}{2}(x)(x)$$

Step 3. We solve our equation for x.

$$24 = \frac{1}{2}x^2 \qquad \text{is equivalent to} \qquad x^2 = 48$$

$$\sqrt{x^2} = \sqrt{48}$$

$$|x| = 4\sqrt{3}$$

Since x is a positive distance, we conclude that $x = 4\sqrt{3}$. Therefore

$$L = 2x + \sqrt{2}x = 2(4\sqrt{3}) + \sqrt{2}(4\sqrt{3})$$

$$= 8\sqrt{3} + 4\sqrt{6}$$

Step 4. Using a calculator, we find that the answer for L is approximately 23.65. This is a reasonable number (not too small or too big) for the perimeter of an isosceles triangle with area 24. ■

Example 3

Consider the two right triangles in Figure 1.9. Sides a and b are parallel, and the base of the larger triangle is 10. Find the base of the smaller triangle if its area is half the area of the larger triangle.

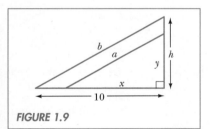

FIGURE 1.9

SOLUTION Step 1. We let $x =$ base of smaller triangle.

Step 2. We label the heights of the smaller and larger triangles with y and h, respectively. It follows that

$$\text{Area of smaller triangle} = \frac{1}{2}xy$$

$$\text{Area of larger triangle} = \frac{1}{2}(10)h = 5h$$

The given information says the area of the smaller triangle is half the area of the larger triangle. Thus

$$\frac{1}{2}xy = \frac{1}{2}(5h) \qquad (1)$$

How can we use this equation to find x? Notice the other piece of data given in the problem: sides a and b are parallel. It follows that both triangles are similar. Therefore the ratios of corresponding sides in these triangles are proportional.

$$\frac{y}{b} = \frac{x}{10}$$

Thus $y = xb/10$. Substituting this into Equation (1), we have

$$\frac{1}{2}x\left(\frac{xb}{10}\right) = \frac{1}{2}(5b)$$

Step 3. Solving this equation for x, we find that

$$\frac{x^2 b}{20} = \frac{5b}{2}$$

$$x^2 = 50$$

$$\sqrt{x^2} = \sqrt{50}$$

$$x = 5\sqrt{2} \qquad \text{since } x \text{ is positive.}$$

Step 4. We expect the value of x to be between 0 and 10. We have $x = 5\sqrt{2} \approx 7.07$, so this is a reasonable answer. You can compute the corresponding value of y in terms of b and confirm that the area of the smaller triangle is half the larger. ∎

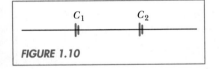

FIGURE 1.10

A **capacitor** is an object with the ability to store electrical charge; the value of a capacitor measures its ability to store charges. If two capacitors with values C_1 and C_2 are connected in **series** (one after another) (see Figure 1.10), they are equivalent to one capacitor with value C given by

$$\frac{1}{C} = \frac{1}{C_1} + \frac{1}{C_2}$$

Example 4

An unknown capacitor is connected in series with another of value 12. If this arrangement is equivalent to one capacitor with value 3, find the unknown capacitor's value.

SOLUTION Let C = value of the unknown capacitor. We have

$$\frac{1}{3} = \frac{1}{C} + \frac{1}{12}$$

Now solve this equation.

$$12C\left(\frac{1}{3}\right) = 12C\left(\frac{1}{C} + \frac{1}{12}\right)$$

$$4C = 12 + C$$

$$3C = 12$$

$$C = 4$$

Therefore the unknown capacitor has value 4. ■

If three capacitors with values C_1, C_2, and C_3 are connected in series, then they are equivalent to one capacitor with value C given by

$$\frac{1}{C} = \frac{1}{C_1} + \frac{1}{C_2} + \frac{1}{C_3}$$

EXERCISES 1.3

1. Suppose a certain number is input into the computer described in Figure 1.5. Find this input number if the corresponding output is (a) 10; (b) $-1/10$.

In Exercises 2 to 6, consider the function machines C, D, E, and F described as follows:

Machine	Operation
C	Takes the reciprocal of the number.
D	Decreases the number by 2.
E	Triples the number.
F	Squares the number.

2. Construct a computer by connecting D and E in series.
   (a) Find the output number if the input is 1.
   (b) Find the input number if the output is 10.

3. Construct a computer by connecting D, E, and F in series.
   (a) Find the output number if the input is 7.
   (b) Find the input number if the output is 4.
   (c) This computer will never output a negative number. Why?

4. Construct a computer by connecting C, D, and E in series.
   (a) Find the output number if the input is 5.
   (b) Find the input number if the output is 6.

5. Construct a computer by connecting D, C, F, and D in series (Figure 1.11).
   (a) Find the output number if the input is 10.
   (b) Find the input number if the output is 10.

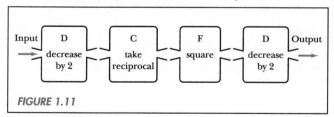

FIGURE 1.11

6. Construct a computer by connecting C, D, E, and F in series.
   (a) Find the output number if the input is 1.
   (b) Find the input number if the output is 0.
   (c) This computer will never output a negative number. Why?

7. Let G be the function machine that adds the input number to the reciprocal of the number.

(a) If the input number is *a*, find the output in terms of *a*.

(b) If the output number is *a*, find the input in terms of *a*.

8. A wire of fixed length is bent to form a square. If the area of the square is 225 square units, how long is the wire?

9. Repeat Exercise 8 if the area of the square is 625 square units.

10. A wire of length 50 is bent to form a rectangle. If the length of the rectangle is twice its width, find the width of the rectangle.

11. A wire of fixed length is bent to form an isosceles right triangle. If the area of the triangle is 36 square units, find the length of the wire.

12. A rancher is to use 800 feet of chain link fencing to make two pens, one square and one rectangular. If the rectangle is twice as long as it is wide and both the square and the rectangle have the same width, find the dimensions of the two pens.

In Exercises 13 and 14, a wire is cut into two pieces. One piece is bent into a square, the other is bent to form the legs (only) of an isosceles right triangle (see Figure 1.12).

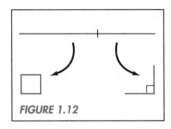

FIGURE 1.12

13. If the original wire was 4 feet in length and the area of the square and triangle are equal, find the length of the piece used for the square.

14. If the square and triangle both have area equal to 100 square inches, how long was the original wire?

15. A wire of length 28 inches is bent to form the legs (only) of a right triangle. If the area of the right triangle is 96 square inches, where was the bend in the wire made?

16. Suppose 60 feet of fencing is used to enclose a rectangular plot of land with an area of 200 square feet. Find the dimensions of the rectangle.

17. A farmer wishes to use 108 feet of chain link fencing to enclose a rectangular area of 400 square feet. The fencing will be used for only three sides of the rectangle, the fourth side being established by a wall (Figure 1.13). Find the possible dimensions for the rectangle.

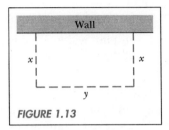

FIGURE 1.13

In Exercises 18 and 19, find *x* and *y* in the two right triangles shown in Figure 1.14 if their areas satisfy the given conditions.

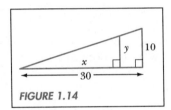

FIGURE 1.14

18. The area of the smaller triangle is half the area of the larger triangle.

19. The area of the smaller triangle is three-fourths the area of the larger triangle.

20. Figure 1.15 shows two triangles with parallel bases. Find *y*, the height of the smaller triangle, if its area is one-fourth the area of the larger triangle.

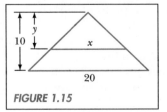

FIGURE 1.15

21. Repeat Exercise 20, assuming that the area of the smaller triangle is half the area of the larger triangle.

22. A square and a circle have the same perimeters. If the area of the square is 9 square units, find the area of the circle. Which contains the larger area?

23. The length of a rectangular photograph is 3 inches more than its width. If a 1-inch border is added to the edge of the photograph, the area of this larger rectangle (photograph plus border) is 70 square inches. Find the dimensions of the photograph.

24. A border of constant width is added to a 15- by 20-inch poster. If the area of the border is 200 square inches, how wide is the border?

25. The height h in feet above the ground of a certain object thrown straight up in the air is given by $h = 80t - 16t^2$, where t is time measured in seconds.
    (a) When will the object be 64 feet above the ground?
    (b) When will the object hit the ground?

26. A man launches a projectile from the top of a 120-foot-high building. The height of the projectile above the ground is given by $h = 120 + 8t - 16t^2$, where t is the time in seconds after the launch.
    (a) When will the projectile be 60 feet above the ground?
    (b) When will the projectile hit the ground?

27. An unknown capacitor is connected in series with another of value 10. This arrangement is equivalent to one capacitor with value 2. Find the value of the unknown capacitor.

28. An unknown capacitor is connected in series with two others having values of 12 and 15. This arrangement is equivalent to one capacitor with value 5. Find the value of the unknown capacitor.

29. An unknown capacitor connected in series with two others having values of 4 and 12 produces an arrangement equivalent to one capacitor with value 3/2. Suppose the unknown capacitor is now connected in series to one of value 10. This arrangement is equivalent to one capacitor of what value?

30. Suppose two capacitors with values C_1 and C_2 are connected in series, producing an arrangement equivalent to one capacitor with value C. Find C in terms of C_1 and C_2.

31. Suppose three capacitors with values C_1, C_2, and C_3 are connected in series, producing an arrangement equivalent to one capacitor with value C. Find C in terms of C_1, C_2 and C_3.

32. (a) A square is inscribed in a circle of radius 10. Find the area of the region outside the square and inside the circle.
    (b) Repeat part (a) if the circle has radius a.

33. (a) A circle and a square have equal areas. Find the radius of the circle if the length of the side of the square is 10.
    (b) Repeat part (a) if the length of the side of the square is a.

34. A water tank in the shape of an inverted cone has radius $r = 1$ foot and height $h = 6$ feet. At what water level will the tank be half full? (*Hint:* the volume of a cone is $\frac{1}{3}\pi r^2 h$. How is x related to y in Figure 1.16?)

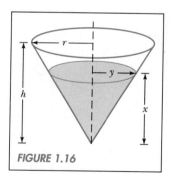

FIGURE 1.16

35. Repeat Exercise 34 if the tank has radius $r = 2$ feet and the height h remains 6 feet.

1.4 INEQUALITIES

Intervals. A connected or unbroken segment of real numbers on the real line is called an **interval**. We represent an interval using any one of three equivalent methods: interval notation, set builder notation, or a graph. Thus, if $a < b$, we have the following:

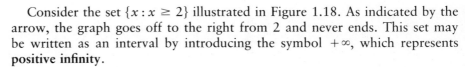

Interval Notation	Set Builder Notation	Graph
(a, b)	$\{x : a < x < b\}$	
$[a, b]$	$\{x : a \leq x \leq b\}$	
$(a, b]$	$\{x : a < x \leq b\}$	
$[a, b)$	$\{x : a \leq x < b\}$	

The interval (a, b) is called an **open** interval; $[a, b]$ is a **closed** interval; $(a, b]$ and $[a, b)$ are **half-open** (also **half-closed**) intervals.

When using interval notation, we always write the numbers in the order they appear on the real line, the smaller number on the left and the larger number on the right.

Example 1

Write the set $\{x : -4 \leq x < -2\}$ in interval notation and graph.

SOLUTION $\{x : -4 \leq x < -2\} = [-4, -2)$

The graph is shown in Figure 1.17. We draw a closed circle at -4 to indicate that this number is included in the set. An open circle at -2 indicates that this number is *not* included. ∎

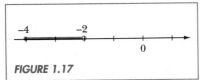

FIGURE 1.17

Consider the set $\{x : x \geq 2\}$ illustrated in Figure 1.18. As indicated by the arrow, the graph goes off to the right from 2 and never ends. This set may be written as an interval by introducing the symbol $+\infty$, which represents **positive infinity**.

$$\{x : x \geq 2\} = [2, +\infty)$$

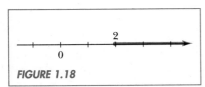

FIGURE 1.18

Notice that we put a parenthesis around $+\infty$, not a bracket. Positive infinity is not a real number and hence cannot be included in the set $\{x : x \geq 2\}$.

Similarly, the symbol $-\infty$ represents **negative infinity**. We can use it to write the set $\{x : x \leq 2\}$ as an interval, $(-\infty, 2]$.

Example 2

Write the set in interval notation and graph: (a) $\{x : x > -1\}$, (b) $\{x : x < 0\}$, (c) **R**.

SOLUTION (a) $\{x : x > -1\} = (-1, +\infty)$ See Figure 1.19.
(b) $\{x : x < 0\} = (-\infty, 0)$ See Figure 1.20.
(c) $\mathbf{R} = (-\infty, +\infty)$ See Figure 1.21.

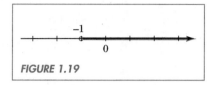

FIGURE 1.19

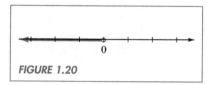

FIGURE 1.20

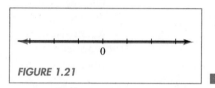

FIGURE 1.21

Intervals can be combined by using the operations of set **union**, designated by $\cup$, and set **intersection**, designated by $\cap$.

Definition of Set Union and Set Intersection
If A and B are sets, then

$$A \cup B = \{x : x \in A \quad or \quad x \in B\}$$

$$A \cap B = \{x : x \in A \quad and \quad x \in B\}$$

The words *or* and *and* are important logical connectives that have distinct meanings. The statement $x \in A$ *or* $x \in B$ means that x must be an element of either A or B, but not necessarily both. However, $x \in A$ *and* $x \in B$ means that x must be an element of both A and B simultaneously. As a simple example, we have

$$\{1, 2\} \cup \{2, 3\} = \{1, 2, 3\} \quad and \quad \{1, 2\} \cap \{2, 3\} = \{2\}$$

Example 3

Graph the given set of real numbers and represent the set using interval notation: (a) $\{x : x < 2 \text{ and } x > -1\}$, (b) $\{x : x \leq 0 \text{ or } x \geq 1\}$.

SOLUTION   (a) By definition of *and*, we have

$$\{x : x < 2 \text{ and } x > -1\} = \{x : x < 2\} \cap \{x : x > -1\}$$

$$= (-\infty, 2) \cap (-1, +\infty)$$

$$= (-1, 2) \qquad \text{See Figure 1.22.}$$

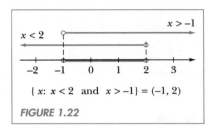

$$\{x : x < 2 \text{ and } x > -1\} = (-1, 2)$$

FIGURE 1.22

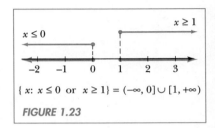

$$\{x : x \leq 0 \text{ or } x \geq 1\} = (-\infty, 0] \cup [1, +\infty)$$

FIGURE 1.23

(b) By definition of *or*, we have

$$\{x : x \leq 0 \text{ or } x \geq 1\} = \{x : x \leq 0\} \cup \{x : x \geq 1\}$$

$$= (-\infty, 0] \cup [1, +\infty) \qquad \text{See Figure 1.23.} \blacksquare$$

We now turn our attention to the topic of solving inequalities. The solution for an inequality consists of the set of all real numbers that yield a true statement when substituted for the variable in the original inequality. To help find the solution set, we may use the following operations:

Operations for Solving Inequalities

Operation	Original Inequality	New Inequality
Addition	$a < b$	$a + c < b + c$
Multiplication by $c \neq 0$	$a < b$	$\begin{cases} ac < bc & \text{if } c > 0 \\ ac > bc & \text{if } c < 0 \end{cases}$
Squaring	$0 \leq a < b$	$0 \leq a^2 < b^2$
Square root	$0 \leq a < b$	$0 \leq \sqrt{a} < \sqrt{b}$

Remarks

1. If we multiply an inequality by a negative number, then the direction of the inequality symbol must be reversed. For example, multiplying the following inequality by -1, we have

$$-x < 6 \text{ is equivalent to } x > -6$$

The reason we change the direction of the inequality symbol can be seen by considering any two real numbers a and b. If a is to the left of b on the real line, $a < b$, then taking opposites reverses the order: we have $-a$ on the right of $-b$, $-a > -b$. See Figure 1.24.

2. The squaring and square root operations are applied only when both sides of the inequality are nonnegative.

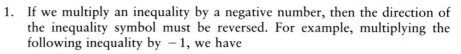

FIGURE 1.24

Linear Inequalities. We classify an inequality as **linear** if the variable appears to the first power only (not in any denominators or inside square roots, etc.)

Example 4

Solve $\dfrac{3 - 2x}{4} < \dfrac{4}{3}$.

SOLUTION We begin by multiplying both sides of the inequality by 12 to clear the fractions.

$$12\left(\frac{3 - 2x}{4}\right) < 12\left(\frac{4}{3}\right)$$

$$9 - 6x < 16$$

$$9 - 6x - 9 < 16 - 9$$

$$-6x < 7$$

Our next step is to multiply both sides by $-1/6$. Remember that this requires that we change the direction of the inequality symbol.

$$-\frac{1}{6}(-6x) > -\frac{1}{6}(7)$$

$$x > -\frac{7}{6}$$

Each step in the argument is *reversible*, which means we could start at the bottom line, $x > -7/6$, and argue backward through the same steps to the top line, $\dfrac{3 - 2x}{4} < \dfrac{4}{3}$. Therefore

$$\frac{3 - 2x}{4} < \frac{4}{3} \qquad \text{is equivalent to} \qquad x > -\frac{7}{6}$$

For the solution set, we write $\left\{ x : x > -\dfrac{7}{6} \right\}$. This set can also be written as the interval $(-7/6, +\infty)$. ∎

Example 5

Solve $-2 < 1 - 3x < 7$.

SOLUTION   By definition, the inequality splits.

$$-2 < 1 - 3x < 7$$

$$-2 < 1 - 3x \quad and \quad 1 - 3x < 7$$
$$-3 < -3x \qquad\qquad\quad -3x < 6$$
$$1 > x \qquad\qquad\qquad x > -2$$

The solution set is $\{ x : x < 1 \ and \ x > -2 \} = \{ x : -2 < x < 1 \} = (-2, 1)$. ∎

Method of Analysis of Signs.   The method of analysis of signs applies to inequalities of the form

$$\boxed{\text{Expression in } x} > 0 \qquad \boxed{\text{Expression in } x} < 0$$

where the sign $(+ \ \text{or} \ -)$ of the expression in x can be easily determined for all x. The following examples illustrate how the method works.

Example 6

Solve $\dfrac{x + 1}{x - 2} > 0$.

SOLUTION   Since zero is on one side, we interpret the inequality as asking where the sign of $(x + 1)/(x - 2)$ is positive. Let us begin by examining the sign of each term involved. First, consider $x + 1$.

$$x + 1 \text{ is positive:} \quad x + 1 > 0 \text{ if and only if } x > -1$$
$$x + 1 \text{ is negative:} \quad x + 1 < 0 \text{ if and only if } x < -1$$

We may illustrate these results on a number line.

Sign of $(x + 1)$:

$$\begin{array}{ccccc|ccccccccc} - & - & - & - & + & + & + & + & + & + & + & + & + \\ \hline & & & -1 & & 0 & & & & & & & \end{array}$$

Next, consider the term $x - 2$. We find

Sign of $(x - 2)$:

$$\begin{array}{cccccccccc|cccc} - & - & - & - & - & - & - & - & - & - & + & + & + & + \\ \hline & & & & & 0 & & 2 & & & & & & \end{array}$$

Now, the sign of the quotient can be found by drawing a third number line beneath the two number lines for $x + 1$ and $x - 2$.

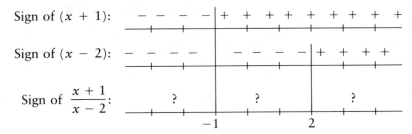

To find the sign at any point on the third line, look at the quotient of the signs in the two lines above it. For example, if x is in the interval $(-\infty, -1)$, the sign of $(x + 1)/(x - 2)$ is the quotient of two negatives, which is positive; $(-)/(-)$ is $(+)$. The complete analysis of signs yields

$$\text{Sign of } \frac{x + 1}{x - 2}: \quad + + + + \;|\; - - - - \;|\; + + + +$$
$$\qquad\qquad\qquad\qquad -1 \qquad\qquad 2$$

We find that $(x + 1)/(x - 2) > 0$ for $x < -1$ or $x > 2$. Note that when $x = -1$, $(x + 1)/(x - 2) = 0$, and when $x = 2$, $(x + 1)/(x - 2)$ is undefined. Hence we do not include $x = -1$ or $x = 2$ in the solution set. We write the answer in set builder notation or interval notation.

$$\{x : x < -1 \text{ or } x > 2\} = (-\infty, -1) \cup (2, +\infty) \;\blacksquare$$

Warning

You may have been tempted to solve the inequality $(x + 1)/(x - 2) > 0$ by first multiplying by $(x - 2)$. Trying this, you might have written

$$(x - 2)\frac{(x + 1)}{(x - 2)} > (x - 2)0$$

$$x + 1 > 0$$

$$x > -1$$

This is not the correct answer! What went wrong? The difficulty is that the quantity $x - 2$ is unknown: it could be positive or negative, depending on the value of x. Therefore we do not know whether or not to change the direction of the inequality symbol when multiplying by $x - 2$. This uncertainty necessitates splitting the problem into two cases, one for $x - 2 > 0$ and one for $x - 2 < 0$. Rather than going through these arguments, we suggest you avoid multiplying an inequality by an unknown quantity.

An expression of the form $ax + b$, where $a \neq 0$, is called a **linear term** in x. If we can rewrite an inequality so that it has a product or quotient of linear terms on one side and zero on the other side, then the solution will follow from the analysis of signs.

Example 7

Solve $\dfrac{2}{x + 1} \leq x$.

SOLUTION Rewrite the inequality so that zero is on one side.

$$\frac{2}{x + 1} - x \leq 0$$

To do the analysis of signs on the left side, we rewrite it as a product and quotient of linear terms.

$$\frac{2}{x + 1} - \frac{x(x + 1)}{x + 1} \leq 0$$

$$\frac{2 - x - x^2}{x + 1} \leq 0$$

$$\frac{(2 + x)(1 - x)}{x + 1} \leq 0$$

The analysis of signs follows.

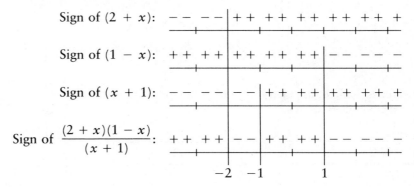

We find that $(2 + x)(1 - x)/(x + 1) < 0$ for x in $(-2, -1) \cup (1, +\infty)$. We must also check the value of $(2 + x)(1 - x)/(x + 1)$ when $x = -2$, -1, or 1. The value is zero when $x = -2$ or 1, and undefined when $x = -1$. Thus we include $x = -2$ and $x = 1$ in the final solution set.

$$\{x : -2 \leq x < -1 \text{ or } x \geq 1\} = [-2, -1) \cup [1, +\infty) \blacksquare$$

Inequalities with Square Roots.   When appropriate, we shall use the squaring operation to remove square roots from inequalities. However, just as with equations, we must be careful to check our answers.

Example   8

Solve (a) $\sqrt{x} > 4$, (b) $\sqrt{x} < 4$, (c) $\sqrt{x} > -1$.

SOLUTION   (a) Squaring both sides of $\sqrt{x} > 4$, we obtain $x > 16$. This answer makes sense because the square root of a number greater than 16 is greater than 4. Therefore the solution set is $\{x : x > 16\} = (16, +\infty)$.

(b) Squaring both sides of $\sqrt{x} < 4$, we obtain $x < 16$. Note that the square root of a number less than 16 is less than 4 *provided* the number is not negative. We need $x \geq 0$ for $\sqrt{x}$ to make sense. Therefore the solution set is $\{x : 0 \leq x < 16\} = [0, 16)$.

(c) We do not apply the squaring operation to this inequality since one side is negative. However, we note that $\sqrt{x} \geq 0$ for all $x \geq 0$ and undefined when $x < 0$. Hence $\sqrt{x} > -1$ for all $x \geq 0$. Therefore the solution set is $\{x : x \geq 0\} = [0, +\infty)$. ■

EXERCISES 1.4

In Exercises 1 and 2, write the set in interval notation and graph.

1. (a) $\{x : -2 < x < 2\}$
   (b) $\{y : 0 < y \leq 1\}$
   (c) $\{x : x \geq -\sqrt{3}\}$
   (d) $\left\{t : t < \dfrac{\pi}{2}\right\}$

2. (a) $\{x : 2 < x < 3\}$
   (b) $\{y : -1 \leq y < 0\}$
   (c) $\{x : x \leq \sqrt{2}\}$
   (d) $\{t : t > -\pi\}$

In Exercises 3 and 4, use interval notation to represent the set of real numbers shown.

3. (a)

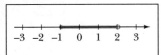

(b)

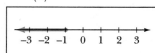

4. (a)

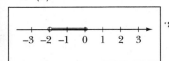

(b)

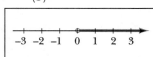

In Exercises 5 to 10, graph the given set of real numbers and then represent the set using interval notation. If the set is empty, write $\varnothing$.

5. (a) $\{x : x < -2 \text{ and } x > -4\}$
   (b) $\{y : y > \sqrt{3} \text{ or } y < \sqrt{2}\}$
   (c) $\left\{t : t > 0 \text{ and } t \leq \dfrac{1}{2}\right\}$

6. (a) $\left\{y : y < \dfrac{3}{2} \text{ and } y > -1\right\}$
   (b) $\{x : x > \pi \text{ or } x < 0\}$
   (c) $\left\{t : t \geq -\dfrac{1}{2} \text{ and } t < 1\right\}$

7. (a) $\{x : x < -1 \text{ or } x > -2\}$
   (b) $\{t : t < 2 \text{ and } t > 1\}$

8. (a) $\{y : y > \sqrt{2} \text{ or } y < 2\}$
   (b) $\{t : t > 2 \text{ and } t < \sqrt{3}\}$

9. (a) $\{x : \sqrt{x^2} = x\}$
   (b) $\{x : \sqrt{x^2} = |x|\}$

10. (a) $\{x : \sqrt{x^2} = -x\}$
    (b) $\{x : |x^2| = x^2\}$

11. Fill in the blank with the appropriate inequality symbol:
  (a) If $-12x + 1 < 0$, then $-12x$ _____ -1.
  (b) If $\frac{2}{3}x < -8$, then x _____ -12.
  (c) If $-3x < 6$, then x _____ -2.
  (d) If $x - 2 > -3$, then x _____ -1.

In Exercises 12 to 18, answer true or false.

12. If $x > y$, then $x - 3 < y - 3$.

13. If $x > 0$ and $y < 0$, then $xy < 0$.

14. If $x < 0$ and $y < 0$, then $xy > 0$.

15. If $x < 0$, then $-x > 0$.

16. If $x < 1$ and $1 < y$, then $x < y$.

17. If $x \in \mathbf{R}$, then $x^2 > 0$.

18. If $x < -1$, then $1 < -x$.

In Exercises 19 to 32, solve the inequalities. Write the solution set in interval notation.

19. $3 - \frac{1}{2}x < 8$

20. $6x + 1 \geq \frac{2}{3}$

21. $\frac{4}{3}x - x < \frac{5 + x}{2}$

22. $x < 22 - 6x$

23. $-10 \leq 4 + 2x \leq 26$

24. $4 < \frac{3}{2} - x < 9$

25. $\frac{1 - 3x}{2} < 1 < 2x + 1$

26. $6 < \frac{4(1 - 2x)}{3} < 6 - 2x$

27. $5x + 2 < 0$ or $3x > 0$

28. $2x - 9 > 0$ or $-3x > 0$

29. $\frac{x}{2} < 2$ and $1 - x \leq 3$

30. $6x > 2x + 1$ and $x + 1 \leq 2$

31. $x(x + 1) < (x + 1)(x + 2)$

32. $x < x + 2$

In Exercises 33 to 44, use the method of analysis of signs to solve the inequality. Write the solution set in interval notation.

33. $\frac{2}{x + 1} < 0$

34. $\frac{1}{3 - x} < 0$

35. $\frac{x}{x + 1} > 0$

36. $\frac{2x + 3}{x - 1} < 0$

37. $x(2x^2 - 3x - 2) \leq 0$

38. $(x + 1)(x^2 + 5x + 6) \geq 0$

39. $\frac{1}{(x - 4)^2} > 0$

40. $\frac{1}{x^2 - 4} < 0$

41. $x^2(2x + 1) > 0$

42. $(x^2 + 1)(x^2 - 1) > 0$

43. $\frac{x - 1}{(x + 3)^2} \leq 0$

44. $\frac{x}{(x - 2)^2} \geq 0$

In Exercises 45 to 54, rewrite the inequality so that zero is on one side and the other side is a product or quotient of linear terms. Solve using the method of analysis of signs. Write the solution set in interval notation.

45. $4x^3 > 2x^2$

46. $\frac{1}{x} > \frac{1}{2}$

47. $(3x + 4)(x + 2) \leq 16$

48. $x^2(x + 1) \geq 2x$

49. $\frac{1}{x} < 1$

50. $\frac{1}{x} < \frac{1}{x^2}$

51. $\frac{1}{x - 1} > \frac{1}{x + 1}$

52. $\frac{3}{2 - x} \geq 1$

53. $\frac{1}{x} \leq x$

54. $\frac{2}{3x} < \frac{1}{x + 2}$

In Exercises 55 and 56, solve each inequality separately and then combine answers for the final solution.

55. $\frac{1}{x} < -5$ or $\frac{1}{x} > 5$

56. $-5 < 4 - \frac{3}{x}$ and $4 - \frac{3}{x} < 5$

In Exercises 57 to 62, suppose k is a positive constant; solve for x in terms of k.

57. $3x - 4 < k$

58. $-k < 3 - 2x$

59. $-k < \frac{1}{2}x + 1 < k$

60. $-k < \frac{1}{4} - x < k$

61. $\frac{1}{x} < k$

62. $\frac{1}{x} > k$

In Exercises 63 to 68, solve for x. Be sure to check answers.

63. (a) $\sqrt{x} > 5$
  (b) $\sqrt{2x} < 5$
  (c) $\sqrt{3x} > -5$

64. (a) $\sqrt{2x} \geq 1$
  (b) $\sqrt{3x} \leq 9$
  (c) $-\sqrt{x} < 2$

65. $\sqrt{x + 1} \geq 0$ 66. $\sqrt{x - 1} > 0$

67. $\sqrt{x^2 + 1} < x + 1$ 68. $\sqrt{x^2 + 1} > x + 1$

69. Find all real numbers that are greater than their reciprocals.

70. Half of a number is greater than itself. What real numbers have this property?

71. The cube of a number is less than its square. What real numbers have this property?

72. Find all real numbers satisfying the following condition: the reciprocal of the number is less than twice the reciprocal of one more than the number.

73. Explain what is *wrong* with each of the following notations:

(a) $(1, -2)$ (b) $[1, +\infty]$

(c) $-3 < x < -4$ (d) $(0, -\infty)$

1.5   ABSOLUTE VALUE INEQUALITIES

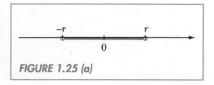

FIGURE 1.25 (a)

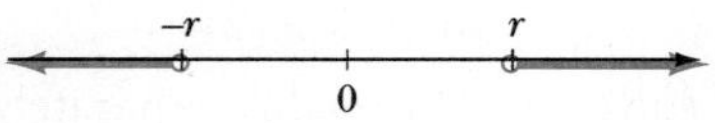

FIGURE 1.25 (b)

Recall the geometric interpretation of absolute value:

$|x|$ represents the distance between x and the origin on the real line.

If r is a positive number, then the inequality $|x| < r$ means that x must be at a distance less than r units from the origin. We conclude that x is somewhere between $-r$ and r on the real line. The graph for all x satisfying $|x| < r$ is in Figure 1.25(a).

If $|x| > r$, then the distance between x and the origin must be greater than r units. Assuming r is positive, this means that x must be somewhere to the right of r or to the left of $-r$ on the real line. The graph for all x satisfying $|x| > r$ is in Figure 1.25(b).

Example 1

Use absolute value to write one inequality statement that represents the set of real numbers shown in (a) Figure 1.26 and (b) Figure 1.27.

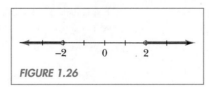

FIGURE 1.26

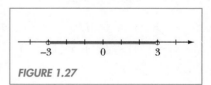

FIGURE 1.27

SOLUTION   (a) This is the set of all real numbers whose distance to the origin is greater than 2 units, $\{x : |x| > 2\}$.

(b) This is the set of all real numbers whose distance to the origin is less than 3 units, $\{x : |x| < 3\}$. ■

With this geometric interpretation in mind, we state the rules for removing absolute values in an inequality.

> **Rules for Removing Absolute Value**
> Let r be a positive real number and suppose $\square$ represents an unknown expression.
>
> $$|\square| < r \text{ is equivalent to } -r < \square < r$$
>
> $$|\square| > r \text{ is equivalent to } \square < -r \text{ or } r < \square$$

The rules remain unchanged if we use the inequality symbols $\leq$ or $\geq$.

Note in the following examples how the rules for removing absolute value split the problem into two inequalities connected by *and* or *or*. Each inequality is solved separately and then recombined with *and* or *or* for the final solution set.

Example 2

Solve $|3x + 1| < 4$.

SOLUTION

$$|3x + 1| < 4 \qquad \text{Given}$$

$$-4 < 3x + 1 < 4 \qquad \text{By rule}$$

$$
\begin{array}{ll}
-4 < 3x + 1 \qquad and \qquad 3x + 1 < 4 & \quad \text{By definition} \\
-5 < 3x \qquad\qquad\qquad\quad 3x < 3 & \\
-\dfrac{5}{3} < x \qquad\qquad\qquad\quad x < 1 & \quad \text{See Figure 1.28.}
\end{array}
$$

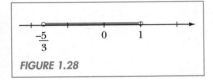

FIGURE 1.28

Therefore the solution set is $\left\{ x : -\dfrac{5}{3} < x \text{ and } x < 1 \right\} = \left\{ x : -\dfrac{5}{3} < x < 1 \right\} = \left(-\dfrac{5}{3}, 1 \right).$ ■

Example 3

Solve $|1 - 2x| \geq \dfrac{3}{2}$.

SOLUTION

$$|1 - 2x| \geq \dfrac{3}{2} \qquad \text{Given}$$

$$
\begin{array}{ll}
1 - 2x \leq -\dfrac{3}{2} \qquad or \qquad \dfrac{3}{2} \leq 1 - 2x & \quad \text{By rule} \\[2mm]
-2x \leq -\dfrac{5}{2} \qquad\qquad \dfrac{1}{2} \leq -2x & \\[2mm]
x \geq \dfrac{5}{4} \qquad\qquad\quad -\dfrac{1}{4} \geq x & \quad \text{See Figure 1.29.}
\end{array}
$$

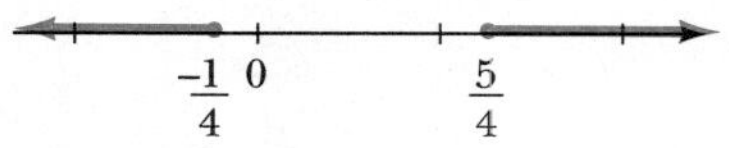

FIGURE 1.29

Therefore the solution set is

$$\left\{ x : x \ge \frac{5}{4} \text{ or } x \le -\frac{1}{4} \right\} = \left(-\infty, -\frac{1}{4} \right] \cup \left[\frac{5}{4}, +\infty \right). \blacksquare$$

Warning

It is incorrect to write the solution set to this problem as $[5/4, -1/4]$. This notation is meaningless because $5/4$ is not less than $-1/4$. Furthermore, from the graph of the solution set, we see that it cannot be combined into one interval. Similarly, the description $5/4 \le x \le -1/4$ is wrong.

When r is a positive constant, the inequality $|x - a| < r$ has an important geometric interpretation. Solving for x, we obtain $\{ x : a - r < x < a + r \} = (a - r, a + r)$.

$|x - a| < r$ means that the distance between x and a on the real line is less than r units.

Example 4

Describe the set of points in Figure 1.30 using one inequality statement.

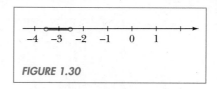

FIGURE 1.30

SOLUTION   The graph represents the set of all real numbers at a distance less than $1/2$ unit from -3. Thus we have $\{ x : |x - (-3)| < 1/2 \}$ or $\{ x : |x + 3| < 1/2 \}$. $\blacksquare$

Recall that if x is unknown, then $\sqrt{x^2} = |x|$. Consequently, the square root operation may change an inequality to one involving absolute value.

Example 5

Solve $(x - 1)^2 > 2$.

SOLUTION   Note that both $(x - 1)^2$ and 2 are nonnegative. Therefore we may apply the square root operation.

$$(x - 1)^2 > 2$$
$$\sqrt{(x - 1)^2} > \sqrt{2}$$
$$|x - 1| > \sqrt{2}$$
$$x - 1 < -\sqrt{2} \qquad or \qquad \sqrt{2} < x - 1$$
$$x < 1 - \sqrt{2} \qquad | \qquad 1 + \sqrt{2} < x$$

Therefore the solution set is $\{x : x < 1 - \sqrt{2} \ or \ x > 1 + \sqrt{2}\} = (-\infty, 1 - \sqrt{2}) \cup (1 + \sqrt{2}, +\infty).$ ■

EXERCISES 1.5

In Exercises 1 and 2, use absolute value to write one inequality statement that represents the set of real numbers shown.

1. (a)

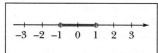

(b)

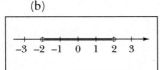

2. (a)

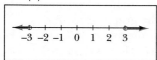

(b)

In Equations 3 to 24, solve the inequalities. If possible, write the solution set in interval notation.

3. $|x| \geq 4$

4. $|x| < 2$

5. $|x| > 3 \ and \ |x| < 4$

6. $|x| < 2 \ or \ |x| \geq 3$

7. $|4 - x| < 3$

8. $|2 - x| > 3$

9. $|2x - 3| > 0$

10. $|2x - 3| < 0$

11. $\left| x - \dfrac{3}{2} \right| \leq \dfrac{1}{10}$

12. $\left| \dfrac{4 - 5x}{3} \right| \geq 15$

13. $|1 - 2x| \leq x$

14. $|2x + 1| > x$

15. $|x| < -9$

16. $|x| > -9$

17. $\left| \dfrac{1}{x} \right| \geq 2$

18. $\left| \dfrac{1}{x} - \dfrac{3}{4} \right| \leq \dfrac{1}{4}$

19. $|(2x + 1) - 3| < .1$

20. $|5x - 10| < .1$

21. $\left| \dfrac{x}{x + 1} \right| > 1$

22. $\left| \dfrac{1}{x - 2} \right| < 1$

23. $\left| |x| - 2 \right| < 1$

24. $\left| |x| + 3 \right| > 4$

In Exercises 25 to 28, describe the set of points shown using absolute value inequalities.

25.

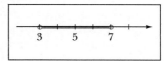

26.

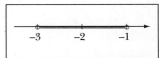

27.

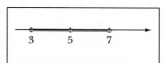

28.

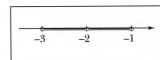

In Exercises 29 to 34, use the square root operation to solve.

29. $x^2 \leq 3$

30. $5x^2 > 10$

31. $3(x + 1)^2 > 9$

32. $\dfrac{1}{2}(x + 1)^2 < 1$

33. $|x^2 - 9| < 0.01$

34. $|1 - x^2| \geq \dfrac{1}{2}$

In Exercises 35 and 36, solve by squaring both sides first.

35. (a) $\sqrt{x^2 + 1} < 2$
    (b) $\sqrt{x^2 + 1} > 2$

36. (a) $\sqrt{x^2 + 9} < 5$
    (b) $\sqrt{x^2 + 9} > 5$

In Exercises 37 to 40, solve for x in terms of k. Assume $k > 0$.

37. $|3x + 4| < k$

38. $|2x - 1| < k$

39. $0 < |x + 4| < k$

40. $0 < |x - 2| < k$

In Exercises 41 to 43, complete the statement by filling in the blank with the appropriate real number.

41. If $|x - 1| < .2$, then $|10x - 10| <$ _____.

42. If $|x + 2| < .1$, then $|5x + 10| <$ _____.

43. If $|x - 5| < .01$, then $|(2x + 1) - 11| <$ _____.

1.6 CARTESIAN COORDINATES AND GRAPHS

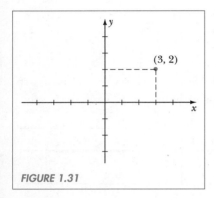

FIGURE 1.31

Two real lines in a plane, one horizontal and one vertical, intersecting at their origins establish a **rectangular Cartesian coordinate system** in the plane. See Figure 1.31. We call the horizontal line the **x-axis**, the vertical line the **y-axis**, and their point of intersection the **origin**. We refer to a plane with a rectangular Cartesian coordinate system as the *x, y-plane*, or *the plane* for short. Note that any letters could be chosen to name the axes; however, x and y are the ones most commonly used.

The collection of all points in the plane are in one-to-one correspondence with the set of ordered pairs (a, b) where $a, b \in \mathbf{R}$. The first number in the ordered pair is the **abscissa**, or **x-coordinate**; the second number is the **ordinate**, or **y-coordinate**. To locate the point $(3, 2)$, for example, start at the origin and move to the right 3 units in the positive x-direction and then up 2 units in the positive y-direction. The point in the plane corresponding to (a, b) is called the **graph** of (a, b). Figure 1.31 is the graph of $(3, 2)$.

Be careful not to confuse a point (a, b) with the interval (a, b). The meaning of (a, b) should be clear from the context. However, a point may be distinguished by placing a capital letter in front of the ordered pair, for example, $P(a, b)$.

The x- and y-axes divide the plane into four quadrants (see Figure 1.32).

1. Quadrant I = $\{(x, y): x > 0 \text{ and } y > 0\}$
2. Quadrant II = $\{(x, y): x < 0 \text{ and } y > 0\}$
3. Quadrant III = $\{(x, y): x < 0 \text{ and } y < 0\}$
4. Quadrant IV = $\{(x, y): x > 0 \text{ and } y < 0\}$

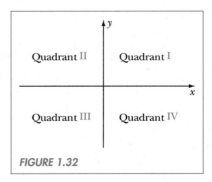

FIGURE 1.32

Example 1

Graph $A(-2, 1)$ and $B(6, 5)$. Use the graph to find the coordinates of the midpoint of the line segment joining A and B.

SOLUTION   Points A and B are shown in Figure 1.33. Let $M(\bar{x}, \bar{y})$ denote the midpoint of segment AB. As an aid to finding M, we drew a rectangle with diagonal AB and sides parallel to the axes. We see that M must have its x-coordinate halfway between the x-coordinates of A and B. It follows that $\bar{x} = 2$ because 2 is halfway between -2 and 6. Similarly, we find $\bar{y} = 3$. Thus the coordinates of M are $(2, 3)$. ∎

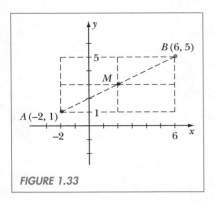

FIGURE 1.33

The midpoint of the line segment joining any two points in the plane can be found without resorting to a graph by using the following **midpoint formula**.

The Midpoint Formula
Let $M(\bar{x}, \bar{y})$ denote the midpoint of the line segment joining $P_1(x_1, y_1)$ and $P_2(x_2, y_2)$. Then

$$\bar{x} = \frac{x_1 + x_2}{2} \quad \text{and} \quad \bar{y} = \frac{y_1 + y_2}{2}$$

The formula says we simply average the coordinates of the two points to find the midpoint. Thus for $A(-2, 1)$ and $B(6, 5)$ in Example 1, the midpoint $M(\bar{x}, \bar{y})$ has coordinates

$$\bar{x} = \frac{x_1 + x_2}{2} = \frac{-2 + 6}{2} = 2 \quad \text{and} \quad \bar{y} = \frac{y_1 + y_2}{2} = \frac{1 + 5}{2} = 3$$

This of course agrees with our previous result.

The proof of the midpoint formula is in the exercises at the end of Section 1.7.

The following formula tells us how to compute the distance between two points in the plane.

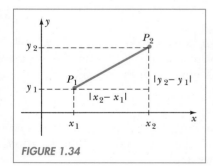

FIGURE 1.34

The Distance Formula
The distance between $P_1(x_1, y_1)$ and $P_2(x_2, y_2)$ is given by the formula

$$P_1 P_2 = \sqrt{(x_2 - x_1)^2 + (y_2 - y_1)^2}$$

Figure 1.34 shows how to prove the distance formula. We see that $P_1(x_1, y_1)$ and $P_2(x_2, y_2)$ determine a right triangle. Notice that the legs of the triangle have lengths $|x_2 - x_1|$ and $|y_2 - y_1|$. Now recall the Pythagorean Theorem, which states that the square of the hypotenuse in a right triangle is equal to the sum of the squares of the legs. Thus

$$(P_1P_2)^2 = (x_2 - x_1)^2 + (y_2 - y_1)^2$$

It follows that

$$P_1P_2 = \sqrt{(x_2 - x_1)^2 + (y_2 - y_1)^2}$$

Example 2

Find the distance between $P(a, b)$ and the (a) x-axis, (b) y-axis, (c) origin.

SOLUTION   The distance from a point to a line is measured by the length of the perpendicular line segment joining the point to the line. See Figure 1.35.

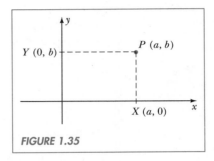

FIGURE 1.35

(a) The perpendicular from $P(a, b)$ to the x-axis joins the x-axis at $X(a, 0)$. Thus

$$XP = \sqrt{(a - a)^2 + (b - 0)^2} = \sqrt{b^2} = |b|$$

(b) The perpendicular from $P(a, b)$ to the y-axis joins the y-axis at $Y(0, b)$. Thus

$$YP = \sqrt{(a - 0)^2 + (b - b)^2} = \sqrt{a^2} = |a|$$

(c) The origin O has coordinates $(0, 0)$. Thus

$$OP = \sqrt{(a - 0)^2 + (b - 0)^2} = \sqrt{a^2 + b^2}$$ ∎

For a specific application of Example 2, consider the point $P(-3, -4)$. We find that P is $|-4| = 4$ units from the x-axis, $|-3| = 3$ units from the y-axis, and $\sqrt{(-3)^2 + (-4)^2} = \sqrt{25} = 5$ units from the origin.

Example 3

Consider the set S of all points $P(x, y)$ in the plane satisfying the condition that the distance between P and $C(3, 4)$ is 5 units. (a) Find an equation for the set S. (b) Find four specific points in S.

SOLUTION (a) We must translate "the distance between $P(x, y)$ and $C(3, 4)$ is 5 units" into an equation. The statement says that $CP = 5$. By the distance formula,

$$\sqrt{(x - 3)^2 + (y - 4)^2} = 5$$

Squaring both sides, we obtain the equivalent equation

$$(x - 3)^2 + (y - 4)^2 = 25$$

This is an equation for the set S because any point in S must satisfy this equation, and, conversely, any point satisfying this equation must be in S. Therefore

$$S = \{(x, y) : (x - 3)^2 + (y - 4)^2 = 25\}$$

Note S is a circle centered at $C(3, 4)$ with radius 5 units.

(b) We must find four ordered pairs (x, y) satisfying the equation $(x - 3)^2 + (y - 4)^2 = 25$. To find a point, we choose a value for x (or y), substitute it into the equation, and solve for the corresponding value of y (or x).

Choose $x = 3$:	*Choose $y = 0$:*
$(3 - 3)^2 + (y - 4)^2 = 25$	$(x - 3)^2 + (-4)^2 = 25$
$(y - 4)^2 = 25$	$(x - 3)^2 + 16 = 25$
$\lvert y - 4 \rvert = 5$	$(x - 3)^2 = 9$
	$\lvert x - 3 \rvert = 3$

$$y - 4 = 5 \quad or \quad y - 4 = -5 \qquad x - 3 = 3 \quad or \quad x - 3 = -3$$
$$y = 9 \qquad\qquad y = -1 \qquad\quad x = 6 \qquad\qquad x = 0$$

Therefore $(3, 9)$, $(3, -1)$, $(6, 0)$, and $(0, 0)$ are points in S. ∎

We can follow the same arguments presented in Example 3 to arrive at an equation for any circle.

Equation of a Circle
The circle with center $C(h, k)$ and radius r has equation

$$(x - h)^2 + (y - k)^2 = r^2$$

This is called the **standard form** for the equation of a circle.

An equation in two variables determines a **solution set** consisting of all the ordered pairs (x, y) that are solutions of the equation. By graphing all the points in the solution set, we obtain the **graph** of the equation. When referring to an equation, we use the words "graph" and "**curve**" interchangeably.

> **Definition of the Graph of an Equation**
> The graph of an x, y equation is obtained by graphing its solution set in the x, y-plane.

We will learn several techniques for graphing a variety of equations in later chapters. At this time we must rely on the following procedure:

1. Make an x, y table of ordered pairs for the solution set of the equation. Each ordered pair is found by choosing an appropriate value of x (or y), substituting it into the equation, and solving for the corresponding value of y (or x).
2. Graph the points from the x, y table. Continue to add more points until the general shape of the curve is recognizable.
3. Draw a curve through the points obtained in step 2.

Example 4

Graph $y = \dfrac{|x|}{x}$.

SOLUTION We make an x, y table of solutions by first choosing values for x and then finding the corresponding values for y.

x	-2	-1	$-1/2$	0	$1/2$	1	2	$3\ 1/2$
$y = \dfrac{|x|}{x}$	-1	-1	-1	Undefined	1	1	1	1

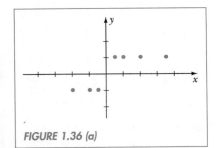

FIGURE 1.36 (a)

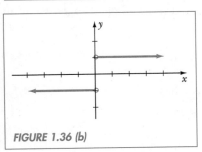

FIGURE 1.36 (b)

These points are graphed in Figure 1.36(a). We suspect that $y = 1$ whenever $x > 0$, and $y = -1$ whenever $x < 0$. This can be proven as follows.

$$\text{If } x > 0, \text{ then } \quad y = \frac{|x|}{x} = \frac{x}{x} = 1 \quad \text{since } |x| = x.$$

$$\text{If } x < 0, \text{ then } \quad y = \frac{|x|}{x} = \frac{-x}{x} = -1 \quad \text{since } |x| = -x.$$

The points are connected to obtain the final graph shown in Figure 1.36(b). Note the reason for the open circles: there is no point on the graph of $y = |x|/x$ with x-coordinate zero. ∎

When making a table for the graph of an x, y equation, include the points where the curve intersects the x-axis. These points are called the **x-intercepts** and are found by substituting $y = 0$ into the equation. Similarly, the points where the curve intersects the y-axis, called the **y-intercepts**, should also be included in the table. These points are found by substituting $x = 0$ into the equation.

Example 5

Find the x- and y-intercepts and graph $(x + 1)^2 + (y - 2)^2 = 5$.

SOLUTION For the x-intercepts, substitute $y = 0$.

$$(x + 1)^2 + (0 - 2)^2 = 5$$
$$(x + 1)^2 + 4 = 5$$
$$(x + 1)^2 = 1$$
$$\sqrt{(x + 1)^2} = \sqrt{1}$$
$$|x + 1| = 1$$
$$x + 1 = 1 \quad or \quad x + 1 = -1$$
$$x = 0 \quad \quad \quad x = -2$$

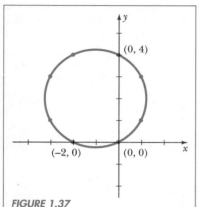

FIGURE 1.37

Therefore $(0, 0)$ and $(-2, 0)$ are the x-intercepts. Similarly, by substituting $x = 0$ and solving, we get $(0, 0)$ and $(0, 4)$ for the y-intercepts.

Notice that the equation represents a circle centered at $(-1, 2)$ with radius $\sqrt{5}$. Figure 1.37 is the graph. ■

Example 6

Graph $x = y^2$.

SOLUTION For this equation it is easiest to make a table of x, y values by choosing y first and then calculating the corresponding x. Figure 1.38 shows the table and the graph. The curve $x = y^2$ is called a *parabola*.

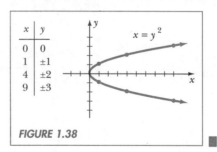

FIGURE 1.38

Example 7

Graph $y = \sqrt{x}$.

SOLUTION Figure 1.39 shows the table of x, y values and the graph. The table was constructed by choosing values of x first and then calculating the corresponding y. Note that x cannot be negative in the equation (the square root of a negative number is not real). Hence there is no graph for $x < 0$. Also, y is never negative because $\sqrt{x}$ means the nonnegative square root of x.

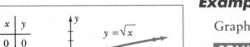

FIGURE 1.39

The shape of the curve appears to be half of the parabola that we graphed in Example 6. This is indeed the case because squaring both sides of $y = \sqrt{x}$ yields $y^2 = x$. Thus $y = \sqrt{x}$ is the upper half of the parabola $y^2 = x$. ∎

We now introduce a list of equations that are referred to so frequently in mathematics that we call them the **basic curves**. It is essential that you be able to graph any one of these curves quickly and accurately.

The Basic Curves

$$\text{Line:} \quad y = mx \qquad x = a \qquad y = a$$
$$\text{Parabola:} \quad y = x^2 \qquad x = y^2$$
$$\text{Square root:} \quad y = \sqrt{x} \qquad x = \sqrt{y}$$
$$\text{Cubic:} \quad y = x^3 \qquad x = y^3$$
$$\text{Absolute value:} \quad y = |x| \qquad x = |y|$$
$$\text{Circle:} \quad x^2 + y^2 = r^2$$
$$\text{Semicircle:} \quad y = \sqrt{r^2 - x^2} \qquad x = \sqrt{r^2 - y^2}$$

Figure 1.40 is a summary of the graphs for all the basic curves. Commit the curves to memory.

Remarks

1. The cubic $y = x^3$ flattens out and gets very close to the x-axis near the origin. This can be shown by making a detailed table of x, y values for x close to zero. Similarly, $x = y^3$ flattens out close to the y-axis near the origin.

2. By squaring both sides of the equation $y = \sqrt{r^2 - x^2}$, we obtain $y^2 = r^2 - x^2$, or $x^2 + y^2 = r^2$. Thus $y = \sqrt{r^2 - x^2}$ must be part of the circle $x^2 + y^2 = r^2$. Since $y \geq 0$ (y is a nonnegative square root), we conclude that it is the upper half of the circle. Similarly, one can argue that $x = \sqrt{r^2 - y^2}$ must be the right half of the circle $x^2 + y^2 = r^2$.

EXERCISES 1.6

In Exercises 1 to 4, graph the points A and B in the x, y-plane. Determine which quadrant contains A and which quadrant contains B. Find the midpoint of the line segment joining A and B. Find the distance between A and B.

1. $A(-3, 1)$, $B(5, -5)$ 2. $A(1, 2)$, $B(5, -4)$

3. $A(-2, 3)$, $B(-2, -1)$

4. $A(-3/4, -1/2)$, $B(1/4, 2/3)$

In Exercises 5 to 8, points A and B determine the diagonal AB of a rectangle whose sides are parallel to the x- and y-axes. Find the coordinates of the remaining two vertices of the rectangle. Find the area of the rectangle.

5. $A(-3, 0)$, $B(5, 3)$ 6. $A(2, 5)$, $B(-1, -4)$

7. $A(4, -2)$, $B(-7, 6)$

8. $A(\sqrt{2}, 2)$, $B(\pi, 0)$

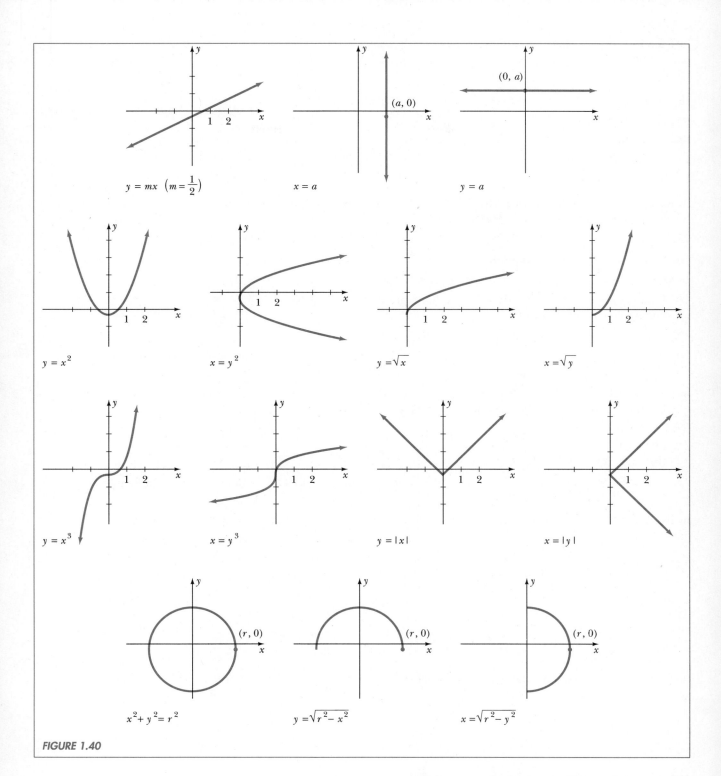

FIGURE 1.40

In Exercises 9 to 12, find the distance between the given point A and the (a) x-axis, (b) y-axis, (c) origin.

9. $A(-3, -5)$
10. $A(7, 0)$
11. $A(0, -12)$
12. $A(2, -\pi)$

In Exercises 13 and 14, use the distance formula to prove that the triangle with vertices A, B, and C is isosceles.

13. $A(0, 0)$, $B(-3, 4)$, $C(4, 3)$
14. $A(-4, 5)$, $B(1, 2)$, $C(4, 7)$

In Exercises 15 to 20, find the equation in standard form for the circle with the given description.

15. Center $(6, -3)$ and radius 2.

16. Center $(4, 3)$ and x-intercept $(4, 0)$.

17. Center $(-4, -5)$ and graph passing through the origin.

18. Radius 3 and center at the midpoint of the line segment joining $A(-2, -1)$ and $B(6, 7)$.

19. A diameter has endpoints $(-7, 5)$ and $(4, 1)$.

20. Radius 4, center in quadrant II, and the x- and y-axes are tangent to the circle.

In Exercises 21 and 22, find the center, radius, and x- and y-intercepts of the circle. Find four points on the circle but not on the x- or y-axis.

21. (a) $x^2 + y^2 = 16$
    (b) $(x + 2)^2 + (y - 3)^2 = 5$

22. (a) $(x + 1)^2 + y^2 = 2$
    (b) $(x - 2)^2 + (y - 5)^2 = 1$

In Exercises 23 and 24, make a table of x, y values and graph.

23. (a) $x = \dfrac{|y|}{y}$
    (b) $y = -\dfrac{|x|}{x}$
    (c) $y = -x$
    (d) $x = 3$
    (e) $y = 0$

24. (a) $y = \dfrac{2|x|}{x}$
    (b) $x = -\dfrac{2|y|}{y}$
    (c) $y = 2x$
    (d) $y = -2$
    (e) $x = 0$

In Exercises 25 to 34, find the x- and y-intercepts and graph.

25. $x^2 + y^2 = 9$
26. $(x - 4)^2 + (y + 3)^2 = 25$

27. $(x + 4)^2 + y^2 = 2$
28. $x^2 + (y + 2)^2 = 9$
29. $y = \sqrt{4 - x^2}$
30. $x = \sqrt{16 - y^2}$
31. $x + 2y = 3$
32. $x - y = 2$

In Exercises 33 and 34, make a table of x, y values for the two given equations at $x = 0$, $\pm 1/4$, $\pm 1/2$, $\pm 3/4$, ± 1, ± 2. Graph both equations on the same axes. Show detail near the origin. Where do the two curves intersect?

33. $y = x^2$, $y = x^3$
34. $y = \sqrt{|x|}$, $y = |x|$

In Exercises 35 to 38, graph the given curves on the same axes.

35. $y = x^2$, $y = 2x^2$
36. $y = x^3$, $y = 2x^3$
37. $y = |x|$, $y = \dfrac{1}{2}|x|$
38. $y = \sqrt{x}$, $y = \dfrac{1}{2}\sqrt{x}$

In Exercises 39 to 41, graph.

39. (a) $y = -x^2$
    (b) $y = -x^3$

40. (a) $y = -\sqrt{x}$
    (b) $y = -\sqrt{1 - x^2}$

41. (a) $y = -|x|$
    (b) $y = |-x|$

In Exercises 42 to 47, for the sets A and B, graph (a) $A \cup B$, (b) $A \cap B$.

42. $A = \{(x, y): y = |x|\}$, $B = \{(x, y): x = |y|\}$
43. $A = \{(x, y): y = \sqrt{x}\}$, $B = \{(x, y): x = \sqrt{y}\}$
44. $A = \{(x, y): y = \sqrt{9 - x^2}\}$,
    $B = \{(x, y): x = \sqrt{9 - y^2}\}$
45. $A = \{(x, y): y = \sqrt{9 - x^2}\}$,
    $B = \{(x, y): x = -\sqrt{9 - y^2}\}$
46. $A = \{(x, y): x = \sqrt{y}\}$, $B = \{(x, y): y = -x^3\}$
47. $A = \{(x, y): x = |y|\}$, $B = \{(x, y): x = y^3\}$

In Exercises 48 to 53, graph both curves on the same axes. From your graph, determine where the curves intersect.

48. $x = y^2$, $x = 4$
49. $y = 4x$, $y = x^3$
50. $y = x$, $x = y^3$
51. $y = x^2$, $x = \sqrt{y}$

52. $y = \frac{1}{2}x$, $y = \sqrt{x}$ 53. $y = -x^3$, $x^2 + y^2 = 2$

54. Establish a table of x, y values and graph $y = \sqrt[3]{x}$. How is this graph related to the curve $x = y^3$?

In Exercises 55–60, consider the set S of all points $P(x, y)$ satisfying the given conditions. (a) Find an equation for S; (b) find four specific points in S.

55. The distance between P and $Q(1, 2)$ is 3.

56. The distance between P and the x-axis is the same as the distance between P and the y-axis.

57. The distance between P and the x-axis is twice the distance between P and the y-axis.

58. The distance between P and $Q(0, 2)$ is the same as the distance between P and the x-axis.

59. The distance between P and the y-axis is the same as the distance between P and $Q(1, 0)$.

60. The distance between P and the origin is the same as the distance between P and the x-axis.

61. Consider points $A(0, 4)$ and $B(10, 6)$. Find the point $P(x, 0)$ on the x-axis such that the distances from P to A and from P to B are equal.

62. Repeat Exercise 61 if A and B have coordinates $A(0, c)$ and $B(a, b)$, a, b, and c are constants, and $a \neq 0$.

63. Given points $P(x, 1 + x)$ and $Q(0, 3)$. Find all x satisfying the given condition. (a) $PQ = 2$, (b) $PQ < 2$, (c) $PQ > 2$.

64. Repeat Exercise 63 for points $P(x, 2x)$ and $Q(1, 0)$.

65. Find the distance between the origin and the midpoint of $A(a, b)$ and $B(b, a)$.

1.7 SLOPE AND LINES

Slope. We assign a number to each nonvertical line in the x, y-plane that measures how much the line rises or falls per unit distance in the positive x-direction. We call this number the **slope** of the line and denote it with the letter m. Informally, the slope measures the steepness of a line.

> **Definition of Slope**
> Let $P_1(x_1, y_1)$ and $P_2(x_2, y_2)$ be two distinct points in the plane (see Figure 1.41). Then the slope of the line through P_1 and P_2 is given by
> $$m = \frac{y_2 - y_1}{x_2 - x_1}$$

The quantities $y_2 - y_1$ and $x_2 - x_1$ are called the *change in y* and the *change in x*, respectively, in going from P_1 to P_2.

Figure 1.41 shows P_1 on the left of P_2; however, this is not necessary for the computation of slope. We can reverse the roles of P_1 and P_2 and still come out with the same value for m.

$$\frac{y_1 - y_2}{x_1 - x_2} = \frac{-1(y_1 - y_2)}{-1(x_1 - x_2)} = \frac{y_2 - y_1}{x_2 - x_1}$$

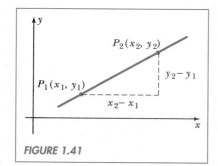

FIGURE 1.41

The computation for slope can also be made between any pair of distinct points on the line. Consider the points P_1, P_2 and P_1', P_2' shown in Figure 1.42. Since triangles P_1CP_2 and $P_1'C'P_2'$ are similar, the ratios of their corresponding sides are equal. Thus

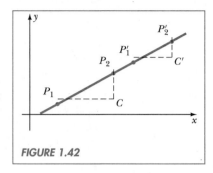

$$\frac{y_2 - y_1}{x_2 - x_1} = \frac{y_2' - y_1'}{x_2' - x_1'}$$

FIGURE 1.42

Example 1

Find the slope of the line passing through the points (a) $A(-5, -2)$, $B(-3, 2)$; (b) $A(-5, -3)$, $B(-7, -3)$; (c) $A(4, 1)$, $B(4, 8)$.

SOLUTION (a) Taking A for P_1 and B for P_2, we have

$$m = \frac{y_2 - y_1}{x_2 - x_1} = \frac{2 - (-2)}{-3 - (-5)} = \frac{4}{2} = 2$$

(b) Again, taking A for P_1 and B for P_2,

$$m = \frac{y_2 - y_1}{x_2 - x_1} = \frac{-3 - (-3)}{-7 - (-5)} = \frac{0}{-2} = 0$$

Note that the line passing through $A(-5, -3)$ and $B(-7, -3)$ will be horizontal.

In general, any two points P_1 and P_2 on a horizontal line will have the same y-coordinates, $y_1 = y_2$. It follows that the slope of a horizontal line must be *zero*.

(c) Taking A for P_1 and B for P_2, we have

$$m = \frac{y_2 - y_1}{x_2 - x_1} = \frac{8 - 1}{4 - 4} = \frac{7}{0}$$

which is undefined. Since division by zero is undefined, we say the slope of the line passing through $A(4, 1)$ and $B(4, 8)$ is *undefined*. Note that this line is vertical.

In general, if a line is vertical, then any two points P_1 and P_2 on the line will have their x-coordinates equal: $x_1 = x_2$. It follows that the slope of a vertical line must be undefined. (We avoid the phrase "no slope" for a vertical line because this term can be confused with *zero slope*. Horizontal lines have slope equal to zero.) ▪

Properties of Slope. The most important geometric properties of slope are illustrated in Figures 1.43 to 1.48. A line with positive slope will be *rising* upward (Figure 1.43) as we move along the line from left to right in the positive *x*-direction. A line with negative slope will be *falling* downward (Figure 1.44) as we move from left to right.

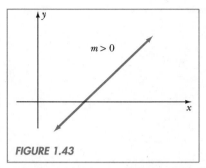

FIGURE 1.43

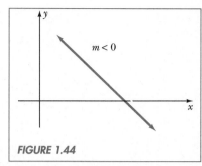

FIGURE 1.44

Horizontal lines (Figure 1.45) have zero slope. Vertical lines (Figure 1.46) have undefined slope.

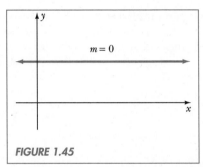

FIGURE 1.45

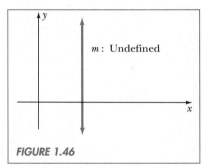

FIGURE 1.46

Two nonvertical lines are parallel (Figure 1.47) if and only if their slopes are equal. We shall prove this fact in Section 9.1 (Exercise 46). Two lines that are not vertical or horizontal (Figure 1.48) are perpendicular if and only if their slopes are negative reciprocals of each other. This fact will be proven in Section 7.4 (Exercise 44).

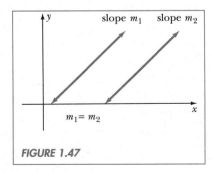

FIGURE 1.47

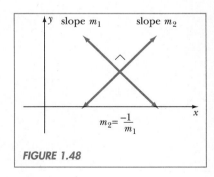

FIGURE 1.48

Equation of a Line. To find an equation for a line, no matter what the situation, two pieces of information are required: (1) a point on the line, $P_1(x_1, y_1)$, and (2) the slope of the line, m. The slope may be a real number or undefined. If m is a real number, let $P(x, y)$ be any point on the line other than P_1. By the definition of slope,

$$\frac{y - y_1}{x - x_1} = m$$

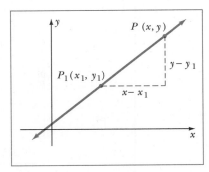

Rewriting this equation in the form

$$y - y_1 = m(x - x_1)$$

we obtain the **point-slope form** for the equation of a line. A point $P(x, y)$ is on the line if and only if it is a solution of this equation.

If m is undefined, we have a vertical line. Since the line passes through the point $P_1(x_1, y_1)$, every point $P(x, y)$ on the line must have its x-coordinate equal to x_1. Thus the equation for this vertical line is $x = x_1$. Following is a summary of our results.

Finding an Equation for a Line

Required information
1. Any point on the line, (x_1, y_1)
2. Slope of the line, m

Equation

$$\text{If } m \text{ is a real number:} \quad y - y_1 = m(x - x_1)$$

$$\text{If } m \text{ is undefined:} \quad x = x_1$$

Example 2

Find an equation for the line passing through the points $A(-3, 4)$ and $B(-6, -4)$.

SOLUTION We need a point on the line and its slope. For the point, we may use $A(-3, 4)$. For the slope, we have

$$m = \frac{y_2 - y_1}{x_2 - x_1} = \frac{-4 - 4}{-6 - (-3)} = \frac{-8}{-3} = \frac{8}{3}$$

Therefore the equation is $y - y_1 = m(x - x_1)$,

$$y - 4 = \frac{8}{3}(x + 3) \quad \blacksquare$$

Example 3

Find an equation for the line passing through the points $A(2, 3)$ and $B(2, -5)$.

SOLUTION For the point, we use $A(2, 3)$. For the slope, we have

$$m = \frac{y_2 - y_1}{x_2 - x_1} = \frac{-5 - 3}{2 - 2} = \frac{-8}{0} \quad \text{undefined}$$

Therefore the line is vertical and has equation $x = x_1$,

$$x = 2 \quad \blacksquare$$

The several useful ways for writing the equation of a line are now summarized.

Equations of Lines

Type	Equation	Comment
Point-slope form	$y - y_1 = m(x - x_1)$	Use this equation to find the equation of a line (unless m is undefined).
Vertical line	$x = x_1$	Slope is undefined.
Horizontal line	$y = y_1$	Slope is zero.
Slope-intercept form	$y = mx + b$	Use this equation to find the slope of a line (unless m is undefined).
General form	$ax + by = c$	The equation of any line can be written in this form.

The following theorem establishes the assertion that the coefficient of x in the slope-intercept form for the equation of a line is the slope.

The Slope of a Line from Its Equation

If the equation of a line can be written in the form $y = ax + b$, then $m = a$.

PROOF: The points $P_1(0, b)$ and $P_2(1, a + b)$ are on the line since they satisfy the equation. (You should verify this.) Hence

$$m = \frac{y_2 - y_1}{x_2 - x_1} = \frac{a + b - b}{1 - 0} = a. \ \blacksquare$$

Example 4

Find an equation of the line L that passes through the point $A(-4, 1)$ and is parallel to the line $3x - 2y = 6$. Write the equation in general form.

SOLUTION As usual, we must find two pieces of data for the equation: a point on L and its slope. The point $A(-4, 1)$ is given. For the slope, we know that parallel lines have equal slopes; therefore the slope of L is the same as the slope of $3x - 2y = 6$. We find the slope of $3x - 2y = 6$ by rewriting this equation in slope-intercept form.

$$3x - 2y = 6$$
$$-2y = -3x + 6$$
$$y = \frac{3}{2}x - 3$$

Therefore the value of the slope is 3/2. Now, using the point-slope form for the equation of L, we have

$$y - 1 = \frac{3}{2}(x + 4)$$

Multiplying both sides by 2 to clear the fraction, we can rewrite this equation in the general form as follows:

$$2(y - 1) = 3(x + 4)$$
$$2y - 2 = 3x + 12$$
$$-3x + 2y = 14 \ \blacksquare$$

Example 5

Find an equation of the tangent line to the circle $(x - 6)^2 + (y - 1)^2 = 25$ at the point $P(2, 4)$.

SOLUTION Let T be the tangent line. Since P is a point on T, we need only find the slope, m_T. From the equation of the circle, we see that its center is $C(6, 1)$. Recall that a tangent line to a circle intersects the circle in exactly one point and is perpendicular to the radius at the point of intersection. Thus T is perpendicular to the radius CP (see Figure 1.49). Hence

$$m_T = \frac{-1}{m_{CP}}$$

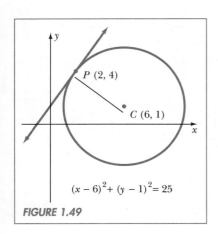

$(x - 6)^2 + (y - 1)^2 = 25$

FIGURE 1.49

We compute the slope of CP from the coordinates $C(6, 1)$ and $P(2, 4)$.

$$m_{CP} = \frac{y_2 - y_1}{x_2 - x_1} = \frac{4 - 1}{2 - 6} = -\frac{3}{4}$$

Therefore

$$m_T = \frac{-1}{m_{CP}} = \frac{-1}{-3/4} = \frac{4}{3}$$

Finally, we obtain the equation for T using P and m_T.

$$y - 4 = \frac{4}{3}(x - 2)$$

or, in general form,

$$-4x + 3y = 4 \quad \blacksquare$$

EXERCISES 1.7

In Exercises 1 to 6, compute the slope of the line that passes through the given points.

1. $A(-6, 1/3)$, $B(5, -5)$

2. $A(4, -1/2)$, $B(3/8, -3)$

3. $A(-4, -5)$, $B(-4, -10)$

4. $A(12, 3)$, $B(-12, 3)$

5. $A(a, ma + b)$, $B(c, mc + b)$ (assume $a \neq c$)

6. $A(a, b)$, $B(-b, -a)$ (assume $a \neq -b$)

In Exercises 7 to 10, find the value of k so that the slope of the line through A and B will equal m.

7. $A(1, 2)$, $B(3, k)$, $m = 6$

8. $A(1, 2)$, $B(k, 4)$, $m = 6$

9. $A(k, k)$, $B(-2, -1)$, $m = -1$

10. $A(k, 2k)$, $B(1, 0)$, $m = 3$

In Exercises 11 to 20, determine the slope, the x-intercept, the y-intercept, and graph each of the following. State whether the graph is rising or falling.

11. $x - y = 1$

12. $2x - y = 2$

13. $2x + 3y = 6$

14. $\dfrac{x}{2} + \dfrac{y}{4} = 1$

15. $\dfrac{4}{5}x - y = -2$

16. $\dfrac{2}{7}x + \dfrac{1}{7}y = -1$

17. $x + y = 0$

18. $x - 4y = 0$

19. $3y = -4$

20. $2x = 1$

In Exercises 21 to 24, find an equation for the line that passes through A and B. Write the equation in general form.

21. $A(-3, -2)$, $B(6, 0)$

22. $A(-7, 5)$, $B(-7, -3)$

23. $A(8, -2)$, $B(4, -2)$

24. $A(a, 0)$, $B(0, b)$ (assume $a, b \neq 0$)

In Exercises 25 to 27, find an equation for the line that passes through A with the given property. Write the equation in general form.

25. (a) $A(-3, -2)$, parallel to $3x - 2y = 11$
    (b) $A(2, 0)$, perpendicular to $2x + 3y = 44$

26. (a) $A(1, -5)$, parallel to $x + y = 6$
    (b) $A(-2, -5)$, perpendicular to $x = 1$

27. (a) $A(2, 1)$, parallel to the y-axis
    (b) $A(0, 0)$, perpendicular to $10x + 5y = 1$

In Exercises 28 to 31, find an equation in general form for the line shown.

28.

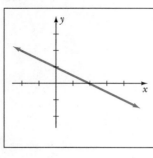

29.

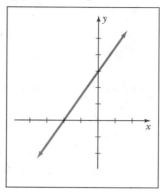

30.

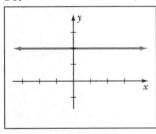

31.

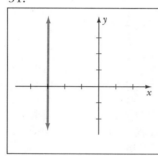

In Exercises 32 and 33, find an equation of the line passing through the midpoint of the line segment joining $A(-7, -4)$ and $B(3, 10)$ with the given property.

32. (a) Parallel to the x-axis
    (b) Passing through the origin

33. (a) Parallel to the y-axis
    (b) Passing through the x-intercept of $4x - 3y = 9$

In Exercises 34 to 36, find an equation for the perpendicular bisector of the line segment AB.

34. $A(5, 9), B(-1, 2)$

35. $A(-2, 1), B(-1, 2)$

36. $A(a, b), B(b, a)$ (assume $a \neq b$)

In Exercises 37 to 40, find an equation in general form for the tangent line to the given circle at the point P. Graph the circle and the tangent line.

37. $x^2 + y^2 = 13, P(3, 2)$

38. $(x + 2)^2 + (y - 2)^2 = 20, P(-6, 4)$

39. $(x - 3)^2 + (y - 2)^2 = 29$, P is an x-intercept of the circle. (There are two answers.)

40. $(x + 5)^2 + y^2 = 169$, (a) P is the positive x-intercept of the circle; (b) P is the positive y-intercept.

In Exercises 41 to 43, find an equation for the line described. (*Hint:* first draw a graph of the line.)

41. All points on the line are equidistant from $x - y = 1$ and $y - x = 1$.

42. The graph of the line has positive slope and all points on the line are equidistant to the lines $x = 3$ and $y = 2$.

43. Repeat Exercise 42, except that the graph of the line has negative slope.

In Exercises 44 to 47, a set of points is called **collinear** if a single line passes through them all. Determine whether or not the following sets are collinear:

44. $\{(22, 9), (0, 3), (-11, 0)\}$

45. $\{(-3, 7), (0, 0), (6, -15)\}$

46. $\{(1, 2), (3, 4), (5, 6), (7, 8)\}$

47. $\{(n, n + 1) : n \in \mathbf{Z}\}$

48. Consider the points $O(0, 0)$, $P(a, b)$, and $Q(-b, a)$. If $a, b \neq 0$, show that OP is perpendicular to OQ.

49. Prove that the triangle determined by the vertices $A(1/2, -5/4)$, $B(2, -2)$, and $C(3, 0)$ is a right triangle by showing that two of its sides are perpendicular. Which side is the hypotenuse?

In Exercises 50 and 51, find an equation for the set of all points $P(x, y)$ satisfying the given conditions.

50. The distance between P and $(5, 9)$ is the same as the distance between P and $(-1, 2)$.

51. The distance between P and $(-2, 1)$ is the same as the distance between P and $(-1, 2)$.

52. Let A and B represent the x- and y-intercept, respectively, of the line $4y - 3x = 12$. Let O be the origin.

(a) Find the perimeter of triangle AOB.

(b) Find the area of triangle AOB.

53. Repeat Exercise 52 for the line $2x - 3y = 30$.

54. This exercise proves the midpoint formula given in Section 1.6. Let $P_1(x_1, y_1)$ and $P_2(x_2, y_2)$ be any two points in the plane. Let $M(\overline{x}, \overline{y})$ denote the point with coordinates $\overline{x} = (x_1 + x_2)/2$ and $\overline{y} = (y_1 + y_2)/2$.

(a) If $x_1 \neq x_2$, compute the slope of the line through

P_1 and M, and through P_2 and M. Conclude that M is on the line through P_1 and P_2.

(b) If $x_1 = x_2$, show that M is on the line through P_1 and P_2.

(c) Use the distance formula to compute the distances MP_1 and MP_2.

(d) From the results of (a), (b), and (c), conclude that M is the midpoint of the line segment joining P_1 and P_2.

CHAPTER 1 / REVIEW

1. Graph the given number on the real line.

   (a) $\sqrt{2}$ (b) $-\sqrt{3}$ (c) π (d) -2.75

In Exercises 2 to 6, list the elements of the given set inside set braces. Write **R** if the set consists of all real numbers, $\varnothing$ if the set is empty.

2. $\left\{ x : x = \dfrac{1}{x} \right\}$

3. $\{ y : y + y = y \}$

4. $\{ t : t + 6 = t \}$

5. $\{ x : 3(x + 2) = 3x + 6 \}$

6. $\{ n : n \in \mathbf{N} \text{ and } -\pi < n < \pi \}$

In Exercises 7 and 8, use set builder notation to represent the set described.

7. Positive integers divisible by 3

8. Even integer multiples of $\sqrt{3}$

9. Translate into mathematical symbols using inequalities or set builder notation:

   (a) x is a positive real number.

   (b) The set of all positive real numbers.

10. Find a real number x satisfying $\sqrt{2} < x < 1.5$ such that (a) x is rational, (b) x is irrational.

11. Simplify the expression $\dfrac{h}{|h|}$ if (a) $h > 0$, (b) $h < 0$.

12. Simplify $\dfrac{|4x - x^2|}{|3x - 12|}$ (assume $x \neq 4$)

13. Answer true or false:

    (a) $|x^2| = x^2$

    (b) $|x - y||y - x| = x^2 + y^2 - 2xy$

    (c) $|a^2 + b^2| = a^2 + b^2$

    (d) $|a + b| = |a| + |b|$

In Exercises 14 and 15, simplify if possible.

14. (a) $\sqrt{(x + 1)^2}$ (b) $\sqrt{9a^2 + 36}$

15. (a) $\sqrt[3]{(x + 1)^3}$ (b) $\sqrt[3]{-8x}$

In Exercises 16 to 29, find all real solutions for the given equation.

16. $\left| 1 - \dfrac{x}{8} \right| = 7$

17. $|-x| = -3$

18. $\dfrac{1}{2 - \dfrac{3}{x}} = \dfrac{2}{3}$

19. $(x + 3)^2 = 625$

20. $x(x + 4) = 5$

21. $2x^2 = 3x + 4$

22. $x^2 + 9 = 0$

23. $\dfrac{x^2}{\sqrt{x^2 - 1}} + \sqrt{x^2 - 1} = 0$

24. $\dfrac{1}{3} + \dfrac{2}{2x - 1} = \dfrac{1}{x(2x - 1)}$

25. $\sqrt{x^2} = -x$

26. $2\sqrt{9x + 5} - 4 = 3x$

27. $3x^4 = x^2 + 4$

28. $4x^3 + 4x^2 = 3x$

29. $|\sqrt{x} - 5| = 8$

In Exercises 30 to 35, solve for the indicated variable.

30. $\dfrac{1}{a} + \dfrac{1}{y} = \dfrac{1}{b}$; y

31. $\dfrac{ax}{1 + bx} = c$; x

32. $x^2 + xy - y^2 = 0; y$ 33. $\sqrt{\dfrac{k}{m}} = 10; m$

34. $|x - a| = b; x$ 35. $\sqrt{x^2 + y^2} = r; y$

36. Consider the computer constructed by connecting the function machines A, C, and B in series, where A doubles the number, C takes the reciprocal, and B adds 1. See Figure 1.50.

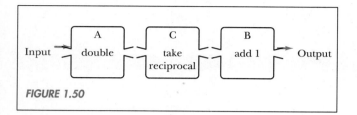

FIGURE 1.50

(a) Find the input number if the output is 13.
(b) Find the input number if the output is a. (Express the answer in terms of a.)
(c) Will this computer accept zero as an input? Explain.
(d) Is it possible for this computer to output a 1? Explain.

37. A wire is bent to form an isosceles right triangle. If the area of the triangle is 20 square units, find the length of the wire.

38. A capacitor with value C is connected in series with a capacitor of value 18. If this arrangement is equivalent to one capacitor of value 6, find C.

39. Graph the given set of real numbers and represent the set using interval notation.
(a) $\{x : x < 3 \text{ and } x \geq -2\}$
(b) $\{x : x < -3/2 \text{ or } x > 0\}$

In Exercises 40 to 45, solve each inequality. Write the solution set in interval notation.

40. $\dfrac{x - 3}{2} > 0$

41. $x^3 + x^2 > 0$

42. $\dfrac{1}{x + 1} < \dfrac{3}{4}$

43. $\sqrt{2x} < 3$

44. $-2x^2 > -18$

45. $\dfrac{1}{x - 3} \geq \dfrac{1}{x - 2}$

46. Use absolute value to write one inequality statement that represents the set of real numbers shown.

(a)

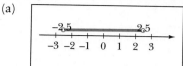

(b)

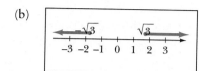

(c)

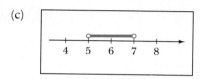

47. Graph $\{x : |x| > 1 \text{ and } |x| < 2\}$. Use interval notation to represent this set.

In Exercises 48 to 51, solve the inequality. Write the solution set in interval notation.

48. $\left| 4 - \dfrac{1}{3}x \right| \leq \dfrac{1}{2}$

49. $|(3x + 1) - 7| < .1$

50. $\left| \dfrac{1}{2x + 1} \right| > 1$

51. $4x^2 > 12$

52. A circle has a diameter with endpoints $A(4, -3)$ and $B(-2, 7)$. Find the
(a) center, (b) radius, (c) equation in standard form, (d) x- and y-intercepts

53. Answer true or false: the point (a, b) is on the graph of an x, y equation if and only if (a, b) is a solution of the equation.

In Exercises 54 to 60, graph and label x- and y-intercepts.

54. $y = \dfrac{|x|}{2x}$ 55. $x = -2$ 56. $y = 4$

57. $x = \sqrt{4 - y^2}$ 58. $y = \dfrac{|-x|}{2}$ 59. $y = \sqrt{4x}$

60. $2x - y = 6$

61. Graph $y = x^2$ and $y = x^3$ on the same axes.

62. Find an equation for the set of all points $P(x, y)$ such that the distance between P and $Q(0, 4)$ is the same as the distance between P and the x-axis.

63. Consider the points $A(3, 2)$ and $B(-4, 1)$.
    (a) Find the slope of the line passing through points A and B.
    (b) Find an equation in general form for the perpendicular bisector of the line segment AB.
    (c) Find an equation in general form for the line passing through A and parallel to $3x + 4y = 8$.

64. Find an equation for the line passing through the points $A(-3, 2)$ and $B(-3, -7)$.

CHAPTER 2 Functions

DEFINITION OF FUNCTION

A function represents a certain kind of relationship between two sets of objects. The objects can be almost anything: real numbers, words, names of people, and so on. However, most of the functions that will interest us deal with real numbers. Thus we shall formally define a **real function** as follows.

Definition of a Real Function
A real function consists of two things:
1. A set of real numbers called the *domain*.
2. A *rule* that assigns to each element in the domain exactly one real number.

To refer to a specific function, we will give the function a name by using a letter, such as F, G, or H, or several letters, such as SQ, ACE, or sin. Also, we shall write $\mathscr{D}_F$ to represent the domain of the function F. Unless otherwise noted, when we say function, it is understood that we mean real function.

Example 1

Determine whether the following defines a function A.
Domain for A: Z, the set of integers.
Rule for A: To each integer x, A assigns the number computed by adding 1 to the absolute value of x, $|x| + 1$.

SOLUTION A satisfies the definition of a function because it consists of two items: a domain and a rule that assigns one number to each element in the domain. Figure 2.1 illustrates examples of what A does to some of the elements in its domain. Arrows emphasize the correspondence established by A from an element in the domain to another real number. ∎

x	A assigns to x
2	⟶ 3
1	⟶ 2
0	⟶ 1
−1	⟶ 2
−2	⟶ 3
−7	⟶ 8

FIGURE 2.1

To represent the number that a given function assigns to an element in its domain, we use a special notation called **functional notation**. The format for this notation is

Name of function   element in domain
$$\downarrow \quad \downarrow$$
$$F \ (\Box)$$

As an example, consider the function A defined in Example 1. We write $A(1)$ to represent the number that A assigns to 1. This number is 2 (the absolute value of 1 plus 1). Thus

$$A(1) = 2.$$

We read this as *A of one equals two*. Other examples of functional notation with A are

$$A(2) = 3$$
$$A(-5) = 6$$
$$A(0) = 1$$

In fact, we can express the general rule for A as follows:

$$A(x) = |x| + 1$$

This equation says that A assigns to x the number $|x| + 1$. When we use the variable x in stating the rule for a function, it is understood that x represents *any* element in the domain of the function. In this situation, x is called an **independent variable**.

The collection of all real numbers (and only those numbers) that a function F assigns to the elements in its domain is called the **range** of F.

Definition of Range
If F is a real function with domain $\mathscr{D}_F$, then the range of F is the set of real numbers

$$\{y : y = F(x) \ and \ x \in \mathscr{D}_F\}$$

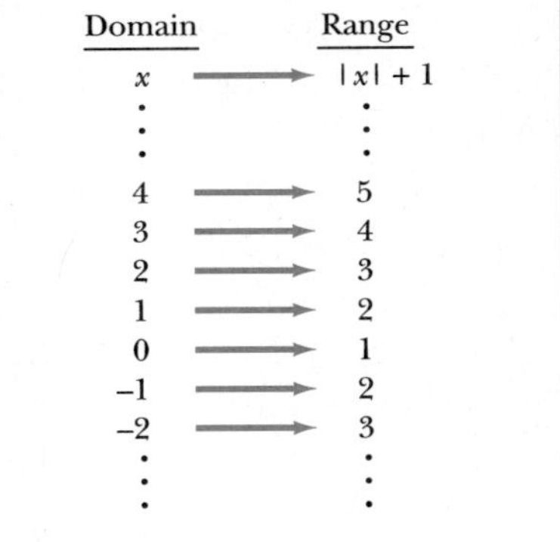

Domain   Range

Domain		Range
x	$\longrightarrow$	$|x| + 1$
.		.
.		.
.		.
4	$\longrightarrow$	5
3	$\longrightarrow$	4
2	$\longrightarrow$	3
1	$\longrightarrow$	2
0	$\longrightarrow$	1
-1	$\longrightarrow$	2
-2	$\longrightarrow$	3
.		.
.		.
.		.

FIGURE 2.2

Example 2

Find the range of the function A defined in Example 1.

SOLUTION   Figure 2.2 illustrates what A assigns to the elements in its domain. Evidently, the range of A is **N**, the set of all positive integers. ∎

Example 3

Consider the diagrams labeled F, G, and H in Figure 2.3. Which diagrams represent functions? For those diagrams that do represent functions, find the range.

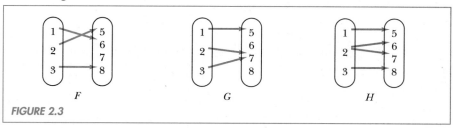

FIGURE 2.3

SOLUTION F satisfies the definition of a function because it has a set of numbers for the domain, $\{1, 2, 3\}$, and it assigns exactly one number to each element in the domain. The range of F is $\{F(1), F(2), F(3)\} = \{6, 5, 8\}$.

G also satisfies the definition of a function. Notice that it does assign exactly one number to each element in the domain. The fact that both 2 and 3 are assigned to 7 does not violate the definition of a function. The range of G is $\{5, 7\}$.

H does *not* satisfy the definition of a function. It violates the condition that the rule must assign exactly one number to each element in the domain. In this case, H assigns two different numbers to 2. ■

In Chapter 1 we introduced an object called a function machine in connection with certain word problems. Those machines do qualify as functions. Indeed, one may think of any function as a machine with an input slot and an output slot. The function accepts only a number in its domain through the input slot, operates on the number according to its rule, and then outputs the result. For each acceptable input number, we get exactly one output number.

Example 4

Consider the function machine G in Figure 2.4. Assume the domain is **R**.
(a) Write the rule for G by completing the statement $G(x) =$ _____.
(b) Find the range of G.

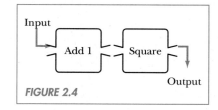

FIGURE 2.4

SOLUTION (a) Inputting x (Figure 2.5), we find

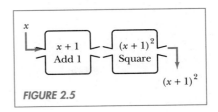

FIGURE 2.5

Thus the output is $(x + 1)^2$. Hence the rule for G is $G(x) = (x + 1)^2$.

(b) For any real number x, $(x + 1)^2 \geq 0$. Therefore the range of G is contained in $[0, +\infty)$. Now, is the range of G all of $[0, +\infty)$? Let $r \in [0, +\infty)$. To prove r is in the range of G, we must find an x in the domain such that $G(x) = r$. This means $(x + 1)^2 = r$. Solving for x, we find that $x = \sqrt{r} - 1$ or $x = -\sqrt{r} - 1$. Hence G will assign the real number $\sqrt{r} - 1$ (as well as $-\sqrt{r} - 1$) to r. We conclude that the range of G is exactly $[0, +\infty)$. ∎

Examples 5 to 11 develop essential functional notation skills.

Example 5

Let F be defined by the rule $F(x) = x^2$. Assume that -5, n, and $n + 1$ are in the domain of F. Find $F(-5)$, $F(n)$, and $F(n + 1)$.

SOLUTION Most functions are presented by expressing the rule using x as the independent variable. However, we must remember that x represents *any* number in the domain. Thus we interpret the statement $F(x) = x^2$ as meaning

$$F(\square) = \square^2$$

where any number or *expression* representing a number in the domain of F can be placed inside the box. Hence

$$F(-5) = (-5)^2 = 25$$

$$F(n) = n^2$$

$$F(n + 1) = (n + 1)^2 = n^2 + 2n + 1 \quad ∎$$

Example 6

Let G be defined by the rule $G(x) = 3x - 4$. Assuming that -6, $-x$, $n + 1$, and $x + h$ are all in the domain of G, find $G(-6)$, $G(-x)$, $G(n + 1)$, and $G(x + h)$.

SOLUTION We think of the rule for G as $G(\Box) = 3\Box - 4$. Thus

$$G(-6) = 3(-6) - 4 = -18 - 4 = -22$$

$$G(-x) = 3(-x) - 4 = -3x - 4$$

$$G(n + 1) = 3(n + 1) - 4 = 3n + 3 - 4 = 3n - 1$$

$$G(x + h) = 3(x + h) - 4 = 3x + 3h - 4 \; \blacksquare$$

If c is a constant, then we may define a function K to have domain **R** and rule that assigns to each $x \in \mathbf{R}$ the real number c, $K(x) = c$. This type of function is called a **constant function**.

Example 7

Consider the constant function K with domain **R** and rule $K(x) = 4$ for every real number x. Find $K(3)$, $K(-\pi)$, and $K(x + 1)$. What is the range of K?

SOLUTION We think of the rule for K as $K(\Box) = 4$. Thus

$$K(3) = 4$$

$$K(-\pi) = 4$$

$$K(x + 1) = 4$$

Since K assigns the number 4 to every element in its domain, we know that the range of K is {4}. $\blacksquare$

An expression involving functional notation must be carefully interpreted. Numbers appearing on the inside of the parentheses are to be distinguished from those on the outside.

Example 8

Let F be defined by the rule $F(x) = 2x$. Assuming x and $x + 1$ are in the domain of F, compare the expressions $F(x + 1)$ and $F(x) + 1$.

SOLUTION We think of the rule for F as $F(\Box) = 2\Box$. Thus

$$F(x + 1) = 2(x + 1) = 2x + 2.$$

This is the number that F assigns to $x + 1$. On the other hand, the expression $F(x) + 1$ means to add 1 to the number that F assigns to x. We have

$$F(x) + 1 = 2x + 1$$

Now, $2x + 2 \neq 2x + 1$ for any $x \in \mathbf{R}$. Hence the two expressions $F(x + 1)$ and $F(x) + 1$ are not the same for any real number x. $\blacksquare$

Example 9

Let L be the function defined by the rule $L(x) = x + 4$. Assuming that x and $2x$ are in the domain, compare $L(2x)$ with $2L(x)$.

SOLUTION We think of the rule for L as $L(\square) = \square + 4$. Thus

$$L(2x) = 2x + 4\cdot$$

Now, $2L(x)$ means to multiply $L(x)$ by 2. We have

$$2L(x) = 2(x + 4) = 2x + 8$$

Note that $2x + 4 \neq 2x + 8$ for any $x \in \mathbf{R}$, so the expressions $L(2x)$ and $2L(x)$ are not the same for any real number x. ■

Examples 5 to 9 illustrate three important points about functional notation. If F is any function, then

1. The symbols $F(\square)$ represent the number that F assigns to $\square$.
2. $F(kx)$ and $kF(x)$ have different meanings.
3. $F(x + k)$ and $F(x) + k$ have different meanings.

If F is a function, then the expression

$$\frac{F(x + h) - F(x)}{h}$$

is called a **difference quotient** for F. Certain computations made in calculus require the evaluation of difference quotients.

Example 10

Let F be the function defined by $F(x) = \dfrac{3}{x}$. Find $\dfrac{F(x + h) - F(x)}{h}$, assuming $h \neq 0$.

SOLUTION We think of the rule for F as $F(\square) = \dfrac{3}{\square}$. Thus

$$\frac{F(x + h) - F(x)}{h} = \frac{\dfrac{3}{x + h} - \dfrac{3}{x}}{h}$$

$$= \frac{3x - 3(x + h)}{hx(x + h)}$$

$$= \frac{-3h}{hx(x + h)}$$

$$= \frac{-3}{x(x + h)} \quad ■$$

Example 11

Let Q be the function defined by $Q(x) = x^2 - 2x + 1$.
Find $\dfrac{Q(x + h) - Q(x)}{h}$, assuming $h \neq 0$.

SOLUTION   We think of the rule for Q as $Q(\square) = \square^2 - 2\square + 1$.
Thus

$$\frac{Q(x + h) - Q(x)}{h} = \frac{(x + h)^2 - 2(x + h) + 1 - [x^2 - 2x + 1]}{h}$$

$$= \frac{x^2 + 2xh + h^2 - 2x - 2h + 1 - x^2 + 2x - 1}{h}$$

$$= \frac{2xh + h^2 - 2h}{h}$$

$$= 2x + h - 2$$

Notice that $-Q(x) = -[x^2 - 2x + 1] = -x^2 + 2x - 1$. The minus sign is distributed over the entire expression for $Q(x)$. ∎

The domain of a function F is part of its definition and therefore normally should be given along with the rule for F. However, mathematicians often write the rule for a function and leave it to the reader to decide what the domain should be. In this case, everyone uses the following assumption.

Finding the Domain of a Function
If the domain of a function is not given, then assume that it is the set of all possible real numbers for which the rule defining the function makes sense.

How do we find the numbers for which the rule makes sense? As we gain experience with different types of functions, we will be able to answer this question relatively easily. For now, we will concentrate on keeping two kinds of numbers *out* of the domain:

1. Numbers that cause us to divide by zero when the rule is applied to them
2. Numbers that cause us to take the square root of a negative number when the rule is applied to them

Example 12

Find the domain of the function H defined by the rule $H(x) = \sqrt{\dfrac{x}{x - 1}}$.

SOLUTION   The rule for H involves both a square root and division. We do not want to take the square root of a negative number, nor do we wish to divide by zero. Hence $H(x) = \sqrt{\dfrac{x}{x-1}}$ makes sense if and only if

$$\frac{x}{x-1} \geq 0 \quad \text{and} \quad x - 1 \neq 0$$

We shall solve $\dfrac{x}{x-1} \geq 0$ using the method of analysis of signs introduced in Section 1.4.

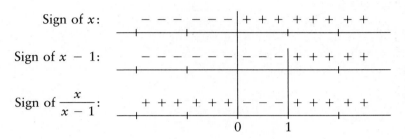

Therefore $x/(x-1) \geq 0$ when $x \in (-\infty, 0] \cup (1, +\infty)$. Notice that $x = 0$ works in the inequality and is included in the solution. However, $x = 1$ is not in the solution since its use results in division by zero. We conclude,

$$\mathcal{D}_H = (-\infty, 0] \cup (1, +\infty) \quad \blacksquare$$

We end this section with a discussion of what it means for two functions to be equal.

Definition of Equal Functions
Suppose F and G are functions. We write $F = G$ and say that F and G are **equal functions** if the following two conditions hold:
1. $\mathcal{D}_F = \mathcal{D}_G$
2. $F(x) = G(x)$ for every $x \in \mathcal{D}_F$ (or $\mathcal{D}_G$).

Example 13

Determine if F and G are equal functions:
(a) $F(x) = |x|$, $G(x) = \sqrt{x^2}$
(b) $F(x) = x + 1$, $G(x) = \dfrac{x^2 + x}{x}$

SOLUTION (a) The domains of F and G are both **R**. Furthermore, $F(x) = |x| = \sqrt{x^2} = G(x)$ for all $x \in \mathbf{R}$. Therefore $F = G$.

(b) The domains of F and G are not the same: $\mathcal{D}_F = \mathbf{R}$ and $\mathcal{D}_G = \{x : x \neq 0\}$. Therefore $F \neq G$. Notice that the rule for G reduces to the rule for F whenever $x \neq 0$. We have

$$\frac{x^2 + x}{x} = \frac{x(x + 1)}{x} = x + 1 \qquad \text{provided } x \neq 0$$

Thus F and G are almost the same function; it is just when $x = 0$ that they differ. ■

EXERCISES 2.1

In Exercises 1 to 4, determine if the diagram describes a function with domain $\{1, 2, 3\}$. Explain. If it does, find the range.

1.

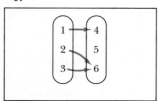

2.

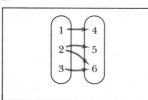

3.

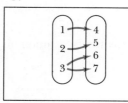

4.

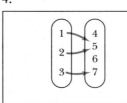

In Exercises 5 to 12, determine if the given rule defines a function F with the indicated domain. Explain. If it does, find the range of F.

5. Domain **R**, rule: $F(x) = x + 1$

6. Domain **R**, rule: $F(x) = |x|$

7. Domain **R**, rule: $F(x) = -1$ if $x \leq 0$, $F(x) = 1$ if $x \geq 0$

8. Domain **R**, rule: $F(x) = 0$ if $x < 1$, $F(x) = 1$ if $x \geq 1$

9. Domain **R**, rule: $F(x) = -2$

10. Domain **R**, rule: $F(x) = \dfrac{1}{x}$

11. Domain **N**, rule: $F(x) =$ all positive integer factors of x

12. Domain **Z**, rule: $F(x) = 1$ if x is even, $F(x) = -1$ if x is odd

In Exercises 13 to 18, invent a rule for a function F with domain $\mathcal{D}$ and range $\mathcal{R}$. Each rule can be expressed with a formula. Answers may vary.

13. $\mathcal{D} = \mathbf{R}, \mathcal{R} = [0, +\infty)$ 14. $\mathcal{D} = \mathbf{R}, \mathcal{R} = (-\infty, 0]$

15. $\mathcal{D} = \{x : x \neq 0\}, \mathcal{R} = \{1, -1\}$

16. $\mathcal{D} = \{x : x \neq 0\}, \mathcal{R} = \{2, -2\}$

17. $\mathcal{D} = \mathbf{Z}, \mathcal{R} = \mathbf{N}$ 18. $\mathcal{D} = [1, +\infty), \mathcal{R} = (0, 1]$

In Exercises 19 to 24, each of the following function machines describes a function G with domain **R**. (a) Write the rule for G by completing the statement $G(x) = $ _____. (b) Find the range of G.

19.

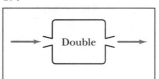

20.

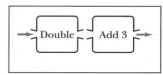

21.

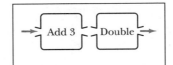

22.

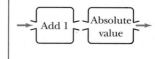

23.

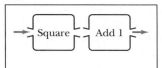

24.

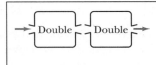

In Exercises 25 to 36, assume that -2, x, $-x$, a, $2x$, and $x + 1$ are in the domain of F. Find (a) $F(-2)$, (b) $F(-x)$, (c) $-F(a)$, (d) $F(2x)$, (e) $2F(x)$, (f) $F(x + 1)$, (g) $F(x) + 1$.

25. $F(x) = 3x$
26. $F(x) = x + 1$

27. $F(x) = 2x - 3$
28. $F(x) = 4 - 5x$

29. $F(x) = x^2 + x + 1$
30. $F(x) = x^2 + 16$

31. $F(x) = |x|$
32. $F(x) = |x| + x$

33. $F(x) = x^3$
34. $F(x) = 1/x$

35. $F(x) = \dfrac{x}{x + 1}$
36. $F(x) = \dfrac{x - 5}{2x + 1}$

37. Given $ACE(x) = 2x^2 + 1$, evaluate (a) $ACE(x - 1)$; (b) $ACE(x) - 1$; (c) $ACE(-x) - ACE(x)$; (d) $2\,ACE(1) - ACE(2)$; (e) $\dfrac{ACE(x) - ACE(a)}{x - a}$, $x \neq a$; (f) $\dfrac{ACE(x + h) - ACE(x)}{h}$, $h \neq 0$. Assume each number is in the domain of ACE. Write answers in simplest form.

38. Repeat Exercise 37 given $ACE(x) = \dfrac{x - 1}{x + 2}$.

In Exercises 39 and 40, using the given rule for R, evaluate (a) $R(1/3)$; (b) $R(1/x)$; (c) $R(x + 1) - R(x)$; (d) $\dfrac{R(x + h) - R(x)}{h}$, $h \neq 0$; (e) $\dfrac{R(x) - R(a)}{x - a}$, $x \neq a$. Assume each number is in the domain of R. Write answers in simplest form.

39. $R(x) = \dfrac{1}{x}$
40. $R(x) = -2$

In Exercises 41 and 42, using the given rule for SQ, evaluate (a) $SQ(8)$, (b) $SQ(x^2 - 1)$, (c) $SQ(0) + SQ(-1)$, (d) $-SQ(-x)$. Assume each number is in the domain of SQ. Write answers in simplest form.

41. $SQ(x) = \sqrt{x + 1}$
42. $SQ(x) = \sqrt{x^2 + 1}$

In Exercises 43 to 52, for each function F given below, find in simplest form (a) $F(-x) - F(x)$; (b) $\dfrac{F(k + 1)}{F(k)}$; (c) $\dfrac{F(x + h) - F(x)}{h}$, $h \neq 0$. Assume all numbers are in the domain of F.

43. $F(x) = 2 - x$
44. $F(x) = x + 2$

45. $F(x) = 2x + 1$
46. $F(x) = x^2$

47. $F(x) = 1 - x^2$
48. $F(x) = x - x^2$

49. $F(x) = 3$
50. $F(x) = \dfrac{2}{x - 1}$

51. $F(x) = \dfrac{1}{x^2}$
52. $F(x) = 3x^2 - 2x - 1$

In Exercises 53 to 66, find the domain of the given function.

53. (a) $F(x) = \sqrt{x - 5}$
54. (a) $F(x) = \sqrt{2 - x}$

(b) $F(x) = \dfrac{x}{x + 3}$
(b) $F(x) = \dfrac{2}{x^2 - 25}$

(c) $F(x) = \dfrac{x - 1}{2}$
(c) $F(x) = \dfrac{5x - 2}{x^2 + 4}$

55. $F(x) = \dfrac{\sqrt{4 - x}}{x}$
56. $F(x) = \sqrt{\dfrac{2x}{x + 5}}$

57. $P(x) = \sqrt{1 + \dfrac{1}{x}}$

58. $F(x) = \sqrt{x(x + 1)(x + 2)}$

59. $P(x) = \sqrt{1 + x^2}$
60. $F(x) = x\sqrt{1 - x}$

61. $Q(x) = \dfrac{1}{x^2 + x - 3}$
62. $S(t) = \sqrt{\dfrac{t}{t^2 - t - 2}}$

63. (a) $F(x) = \dfrac{-2}{|x|}$
64. (a) $V(x) = \dfrac{x}{|x|}$

(b) $H(x) = \dfrac{1}{1 + |x|}$
(b) $D(x) = \dfrac{1}{1 - |x|}$

65. $F(t) = \sqrt{\sqrt{t} - 1}$
66. $C(x) = \sqrt{2}$

In Exercises 67 to 72, find the domain and range for each of the following functions.

67. $F(x) = -x$
68. $SQR(x) = \sqrt{x}$

69. $PAR(t) = -t^2$
70. $C(x) = 7$

71. $F(x) = |x| + 1$
72. $F(x) = |x| - 2$

In Exercises 73 to 78, determine if F and G are equal functions. If $F \neq G$, explain why.

73. $F(x) = |x|^2$, $G(x) = x^2$

74. $F(x) = \sqrt{|x|}$, $G(x) = \sqrt{x}$

75. $F(x) = \dfrac{x + 2}{x^2 - 4}$, $G(x) = \dfrac{1}{x - 2}$

76. $F(x) = \dfrac{x^2 - 9}{x + 3}$, $G(x) = x - 3$

77. $F(x) = x^2 + 1$, $G(x) = (x + 1)^2$

78. $F(x) = \sqrt{(x - 2)^2}$, $G(x) = |x - 2|$

In Exercises 79 to 82, find the number of different functions possible with domain $\mathcal{D}$ and range $\mathcal{R}$.

79. $\mathcal{D} = \{1, 2\}$, $\mathcal{R} = \{3, 4\}$ 80. $\mathcal{D} = \{1, 2, 3\}$, $\mathcal{R} = \{3, 4\}$

81. $\mathcal{D} = \{1, 2, 3\}$, $\mathcal{R} = \{4\}$ 82. $\mathcal{D} = \{1, 2\}$, $\mathcal{R} = \{3, 4, 5\}$

2.2 THE GRAPH OF A FUNCTION

If F is any function, then it defines a set of points in the plane as follows:

$$\{(x, y): x \in \mathcal{D}_F \text{ and } y = F(x)\}$$

The **graph** of F is obtained by graphing in the plane all the ordered pairs in this set. Recall from Section 1.6 that we defined the graph of an x, y equation as the set of all points (x, y) in the plane that are solutions of the equation. It follows that the graph of the function F is just the graph of the equation

$$y = F(x)$$

subject to the restriction $x \in \mathcal{D}_F$. We call $y = F(x)$ the **equation for F**.

Example 1

Let F be the function with domain $\mathbf{Z}$ and rule $F(x) = 1 - |x|$. Graph F.

SOLUTION   We wish to graph the equation

$$y = 1 - |x|$$

subject to the restriction $x \in \mathbf{Z}$. We can make a table of values just as we do when graphing an x, y equation. However, we must remember that x may have only integer values since $\mathcal{D}_F = \mathbf{Z}$.

x	-3	-2	-1	0	1	2	3
$y = 1 - |x|$	-2	-1	0	1	0	-1	-2

Plotting these points gives us a portion of the graph of F in Figure 2.6. ■

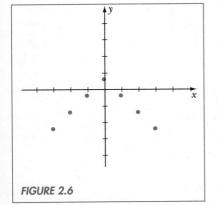

FIGURE 2.6

Example 2

Let SQR be the function defined by the rule $SQR(x) = \sqrt{x}$. Graph SQR.

SOLUTION   Note that the domain of SQR is the set of all real numbers that work in the rule $SQR(x) = \sqrt{x}$. This means that $x \in [0, +\infty)$. Thus the graph of *SQR* is obtained by graphing the equation

$$y = \sqrt{x}$$

where $x \in [0, +\infty)$. This is one of the familiar basic curves introduced in Section 1.6. Figure 2.7 shows a short table of x, y values and the graph.

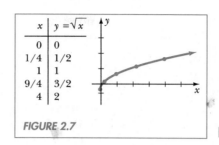

x	$y = \sqrt{x}$
0	0
1/4	1/2
1	1
9/4	3/2
4	2

FIGURE 2.7

At this time we wish to emphasize that the notation $F(a)$ for a given function F has two equivalent interpretations:

1. $F(a)$ is the number that F assigns to a,

$$a \longrightarrow F(a)$$

2. $F(a)$ is the y-coordinate of the point on the graph of F corresponding to $x = a$. See Figure 2.8.

We call interpretation 1 the *analytic interpretation* of $F(a)$ and interpretation 2 the *geometric interpretation*.

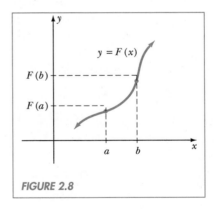

FIGURE 2.8

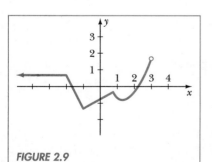

FIGURE 2.9

Example 3

For the function F whose graph is shown in Figure 2.9, find $F(0)$, $F(2)$, and $F(-5/2)$.

SOLUTION   To find $F(0)$, we go to $x = 0$ on the x-axis and look at the corresponding y-coordinate on the graph of F. We see that this is $-1/2$. Thus $F(0) = -1/2$.

To find $F(2)$, we go to $x = 2$ on the x-axis and find the corresponding y-coordinate on the graph of F. We see that this is zero. Hence $F(2) = 0$.

Finally, when $x = -5/2$, we see that the corresponding y-coordinate on the graph is 1. Therefore $F(-5/2) = 1$. ∎

A point (a, b) is on the graph of a function F if and only if a is in the domain of F and $F(a) = b$. Thus b is in the range of F. This gives us the following geometric interpretation for the domain and range of a function.

Geometric Interpretation of Domain and Range

The domain of F consists of the set of all x-coordinates belonging to points on the graph of F.

The range of F consists of the set of all y-coordinates belonging to points on the graph of F.

Example 4

For the function F whose graph is shown in Figure 2.9, find $\mathcal{D}_F$ and the range of F.

SOLUTION The x-coordinates belonging to the graph of F consist of all real numbers less than 3. Therefore $\mathcal{D}_F = \{x : x < 3\} = (-\infty, 3)$.

The y-coordinates belonging to the graph of F consist of all real numbers greater than or equal to -1 and strictly less than 2. Thus the range of F is $\{y : -1 \le y < 2\} = [-1, 2)$. ∎

Example 5

Find the range of the function F whose graph is shown in **Figure 2.6**.

SOLUTION The y-coordinates of the points on the graph in Figure 2.6 are $1, 0, -1, -2, -3, \ldots$. Therefore the range of F is $\{1, 0, -1, -2, -3, \ldots\}$. ∎

Example 6

Find the range of the function SQR whose graph is shown in Figure 2.7.

SOLUTION The y-coordinates of the points on the graph in Figure 2.7 consist of all nonnegative real numbers. Therefore the range of SQR is $\{y : y \ge 0\} = [0, +\infty)$. ∎

When graphing a curve in the plane, the location of the x- and y-intercepts is often desired for determining the position of the curve and analyzing its behavior. In the case of graphing a function, the x-intercepts are the ones of particular importance. We refer to the x-coordinates of these points as **zeros** of the function.

> **Definition of a Zero of a Function**
> Let F be a function and $c \in \mathcal{D}_F$. Then c is a *zero* of F if and only if $F(c) = 0$.

The graph of any function intersects the y-axis at most once. In fact, if $0 \in \mathcal{D}_F$, then $(0, F(0))$ is the y-intercept.

Example 7

Find the zeros of $F(x) = 2x^2 - x - 10$. Where does the graph intersect the y-axis?

SOLUTION   We know that x is a zero of F if and only if $F(x) = 0$. Thus

$$2x^2 - x - 10 = 0$$
$$(2x - 5)(x + 2) = 0$$
$$x = 5/2 \quad or \quad x = -2$$

Therefore the zeros of F are $5/2$ and -2. Since $F(0) = -10$, the graph intersects the y-axis at $(0, -10)$. ■

At this point we know that every function determines a graph in the plane. We now consider the converse of this statement: whether or not every curve in the plane is the graph of a function.

Let us examine the simple graph in Figure 2.10. If this is the graph of a function F, then there is only one element in the domain: 1. Furthermore, the rule would have to be $F(1) = \pm 2$.

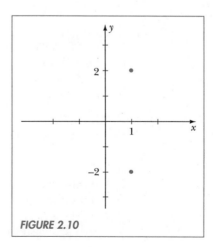

FIGURE 2.10

A function cannot assign two different numbers to an element in its domain. Therefore this graph is *not* the graph of a function.

Now that we know some graphs may not represent a function, the question becomes, Which ones do? A simple test can be used to answer this question. The main idea is this: if a vertical line intersects the graph in two (or more) points, then we have a situation similar to that in Figure 2.10: two y-coordinates correspond to one x-coordinate. This is an impossible situation for the graph of any function because it implies that the function would assign two numbers to one of the elements in its domain. Thus we have the following result.

The Vertical Line Test

If every vertical line in the plane intersects a curve no more than once, then the curve is the graph of a function.

Example 8

Determine if the given curve in Figures 2.11 and 2.12 is the graph of a function.

(a)
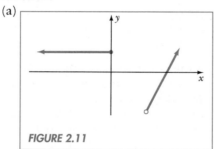
FIGURE 2.11

(b)
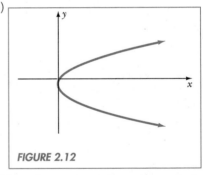
FIGURE 2.12

SOLUTION   (a) The curve in Figure 2.11 has the property that any vertical line will never intersect it more than once. Hence this curve is the graph of a function. We say that the graph *satisfies the vertical line test*.

(b) The curve in Figure 2.12 does not satisfy the vertical line test. Hence it is not the graph of a function. ∎

We end Section 2.2 with a definition concerning the behavior of a function that has a nice geometric interpretation.

Definition of Increasing or Decreasing Function

A function F is **increasing** if whenever $x_1, x_2 \in \mathcal{D}_F$ and $x_2 > x_1$, then $F(x_2) > F(x_1)$.

F is **decreasing** if whenever $x_1, x_2 \in \mathcal{D}_F$ and $x_2 > x_1$, then $F(x_2) < F(x_1)$.

In terms of the graph, F is increasing if the graph is rising as we move from left to right in the positive x-direction; F is decreasing if the graph is falling as we move from left to right.

Example 9

Determine whether the given function is increasing or decreasing: (a) $F(x) = x + 1$, (b) $G(x) = -x^3$.

SOLUTION   (a) Figure 2.13 shows that F is an increasing function. (b) Figure 2.14 shows that G is decreasing.

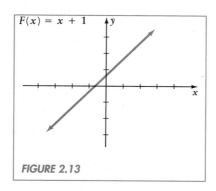

FIGURE 2.13

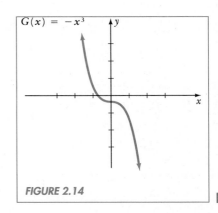

FIGURE 2.14

Example 10

Determine where $F(x) = 1 - x^2$ is increasing and where it is decreasing.

SOLUTION   The graph of F in Figure 2.15 shows that F does not qualify as an increasing or decreasing function over its entire domain. However, we can say that F is increasing on the interval $(-\infty, 0]$ and decreasing on the interval $[0, +\infty)$.

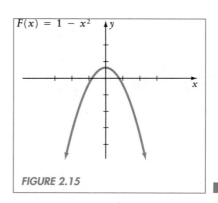

FIGURE 2.15

1. (a) How does a function of x determine a graph in the x, y-plane?
   (b) What graphs determine a function and how?

2. If G is a function and $a \in \mathcal{D}_G$, what does $G(a)$ mean? Give the analytic and geometric interpretations.

In Exercises 3 to 14, write the equation for the given function and graph. Find the range.

3. (a) $F(x) = |x|, \mathcal{D}_F = \mathbf{Z}$ (b) $G(x) = |x|$

4. (a) $F(x) = -|x|, \mathcal{D}_F = \mathbf{Z}$ (b) $G(x) = -|x|$

5. (a) $F(x) = -2x, \mathcal{D}_F = \mathbf{N}$ (b) $G(x) = -2x$

6. (a) $F(x) = \frac{1}{2}x, \mathcal{D}_F = \mathbf{N}$ (b) $G(x) = \frac{1}{2}x$

7. (a) $K(x) = -2, \mathcal{D}_K = \mathbf{Z}$ (b) $C(x) = -2$

8. (a) $K(x) = 0, \mathcal{D}_C = \mathbf{N}$ (b) $C(x) = 0$

9. (a) $S(x) = x^2, \mathcal{D}_S = \mathbf{Z}$
   (b) $SQ(x) = x^2, \mathcal{D}_S = [0, +\infty)$
   (c) $SQR(x) = x^2$

10. (a) $S(x) = \sqrt{x}, \mathcal{D}_S = \mathbf{N}$
    (b) $SQ(x) = \sqrt{x}, \mathcal{D}_{SQ} = [0, 4]$
    (c) $SQR(x) = \sqrt{x}$

11. (a) $KW(x) = x^3, \mathcal{D}_{KW} = (-\infty, 0]$
    (b) $KWB(x) = x^3$

12. (a) $KW(x) = x^3, \mathcal{D}_{KW} = [0, +\infty)$
    (b) $KWB(x) = x^3, \mathcal{D}_{KWB} = [-2, 2]$

13. (a) $HC(x) = \sqrt{16 - x^2}, \mathcal{D}_{HC} = [0, 4]$
    (b) $H(x) = \sqrt{16 - x^2}$

14. (a) $HC(x) = \sqrt{4 - x^2}, \mathcal{D}_{HC} = [-2, 0]$
    (b) $H(x) = \sqrt{4 - x^2}$

In Exercises 15 to 22, graph the function. State the domain, range, zeros, and y-intercept and where the function is increasing or decreasing.

15. $F(x) = \sqrt[3]{x}$ 16. $F(x) = -x$

17. $H(x) = \sqrt{9 - x^2}$ 18. $H(x) = \sqrt{25 - x^2}$

19. $G(x) = \dfrac{|x|}{x}$ 20. $G(x) = \dfrac{-|x|}{x}$

21. $R(x) = \sqrt{-x}$ 22. $R(x) = \sqrt[3]{-x}$

In Exercises 23 to 28, find the zeros and the y-intercept of the given function. Do not graph.

23. (a) $Q(x) = x^2 - 3x + 2$
    (b) $L(x) = 6x + 7$

24. (a) $Q(x) = 2x^2 - 5x - 3$
    (b) $L(x) = 4 - 3x$

25. (a) $F(x) = x^2 + x - 3$
    (b) $G(x) = |x| - 3$

26. (a) $F(x) = x^2 - 2x - 1$
    (b) $G(x) = |x + 1|$

27. (a) $R(x) = \dfrac{x + 2}{x^2 - 4}$ (b) $F(x) = \dfrac{x^2 + 1}{x - 3}$

28. (a) $R(x) = \dfrac{x^2 - 1}{x + 3}$ (b) $F(x) = \dfrac{1}{2x - 5}$

In Exercises 29 to 34, determine whether or not the given graph represents a function. If it is a function, state where it is increasing or decreasing.

29.

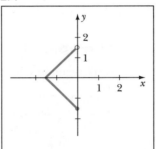

30.

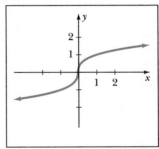

31.

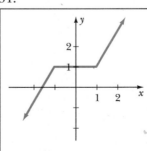

32.

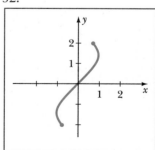

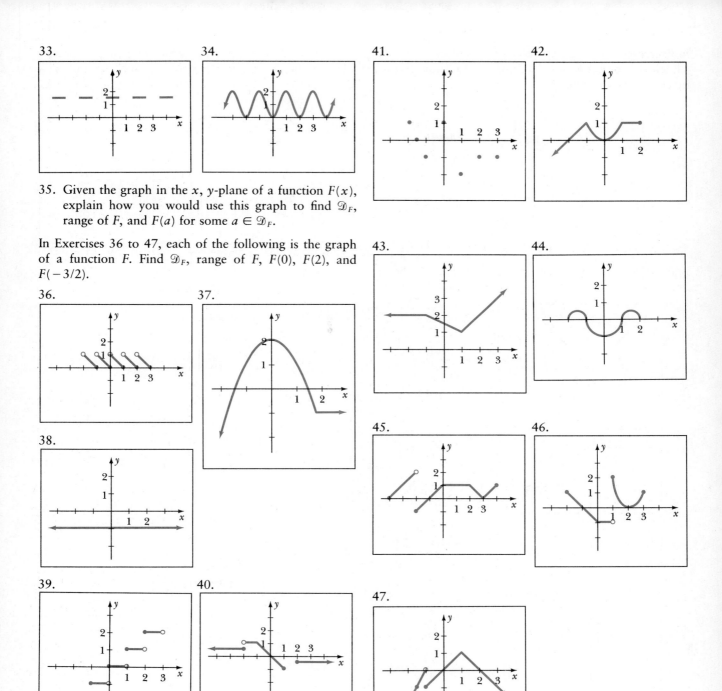

33.

34.

35. Given the graph in the x, y-plane of a function $F(x)$, explain how you would use this graph to find $\mathcal{D}_F$, range of F, and $F(a)$ for some $a \in \mathcal{D}_F$.

In Exercises 36 to 47, each of the following is the graph of a function F. Find $\mathcal{D}_F$, range of F, $F(0)$, $F(2)$, and $F(-3/2)$.

36.

37.

38.

39.

40.

41.

42.

43.

44.

45.

46.

47.

In Exercises 48 to 55, graph each of the following sets and tell whether or not each set represents a function. If it is a function, give the domain and range.

48. $\{(1, 1), (1.5, 2), (2, 1), (2.5, 2)\}$

49. $\{(1, 1), (2, 1), (3, 2), (3, 3)\}$

50. $\{(x, y) : x = 1\}$

51. $\{(x, y) : x = y^2 \text{ and } y \geq 0\}$

52. $\{(x, y) : x = y^3\}$

53. $\{(x, y) : x = \sqrt{y}\}$

54. $\{(x, y) : x^2 + y^2 = 4\}$

55. $\{(x, y) : x^2 + y^2 = 4 \text{ and } x \geq 0\}$

In Exercises 56 to 58, the graph of a function F is shown. Find a formula for its rule, $F(x)$.

56.
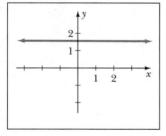

57.
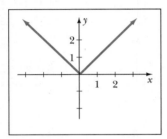

58.
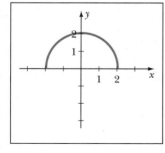

In Exercises 59 to 64, each of the following shows a portion of the graph of a function F with domain $\mathbf{N}$. Assuming the entire graph follows the obvious pattern, find a rule for $F(x)$.

59.
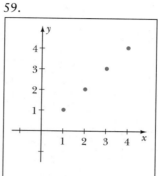

60.
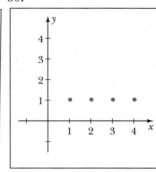

61.
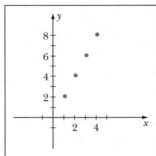

62.
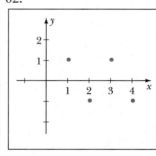

63.
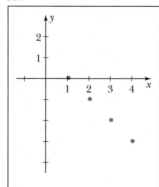

64.
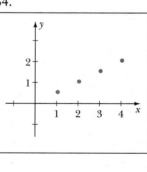

In Exercises 65 to 68, suppose that P_1 and P_2 are points on the graph of the function F with x-coordinates x_1 and

x_2, respectively. Find the slope of the line through P_1 and P_2. Assume that h is a nonzero constant.

65. $F(x) = x^2$
    (a) $x_1 = 1, x_2 = 2$
    (b) $x_1 = 1, x_2 = 1 + h$

66. $F(x) = x^3$
    (a) $x_1 = 1, x_2 = 2$
    (b) $x_1 = 1, x_2 = 1 + h$

67. $F(x) = \sqrt{x}$
    (a) $x_1 = 4, x_2 = 5$
    (b) $x_1 = 4, x_2 = 4 + h$

68. $F(x) = |x|$
    (a) $x_1 = 0, x_2 = 2$
    (b) $x_1 = 0, x_2 = -1$
    (c) $x_1 = 0, x_2 = h$, assume $h > 0$
    (d) $x_1 = 0, x_2 = h$, assume $h < 0$

69. Suppose F is a function and $\mathcal{D}_F = \{0, 1, 2, 3, 4, \ldots, 10\}$.
    (a) What is the maximum possible number of elements in the range of F?
    (b) What is the minimum possible number of elements in the range of F?

70. Prove that the graph of a function $F(x)$ cannot have more than one y-intercept.

2.3 SYMMETRY

We use the word "symmetry" to describe a certain type of balance in the structure of an object. The objects that interest us are functions, and the symmetry we look for will be displayed by the graph of a function. Let us begin by examining two types of symmetry that may occur in any graph in the plane: symmetry with respect to a line, and symmetry with respect to a point.

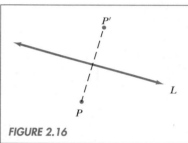

FIGURE 2.16

Symmetry with Respect to a Line
A graph in the plane is symmetric with respect to a line L if and only if every point P on the graph (not on L) has a partner P' on the graph such that L is the perpendicular bisector of PP'.

Figure 2.16 illustrates the definition. Every point P on the graph has its symmetric partner, or mirror image P', on the other side of the line. This means that the line of symmetry L divides the graph into two halves that will come together exactly on top of each other if the plane is folded along L. Figure 2.17 is an example.

FIGURE 2.17

Symmetry with Respect to a Point
A graph in the plane is symmetric with respect to a point Q if and only if every point P on the graph (different from Q) has a partner P' on the graph such that Q is the midpoint of PP'.

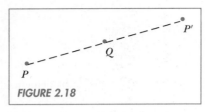

FIGURE 2.18

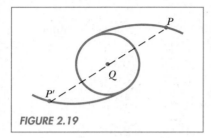

FIGURE 2.19

Figure 2.18 illustrates the definition. Figure 2.19 is an example of a graph displaying symmetry with respect to a point.

The following theorem describes the relationship between the coordinates of a point in the x, y-plane and its symmetric partner when there is symmetry with respect to the x-axis, the y-axis, the diagonal $y = x$, or the origin.

Symmetry Theorem

A graph displays the symmetry described in the first column if and only if the condition in the second column is true.

Symmetry with Respect to	Condition	
1. x-axis	(a, b) is on the graph implies $(a, -b)$ is on the graph.	
2. y-axis	(a, b) is on the graph implies $(-a, b)$ is on the graph.	
3. Diagonal $y = x$	(a, b) is on the graph implies (b, a) is on the graph.	
4. Origin	(a, b) is on the graph implies $(-a, -b)$ is on the graph.	

The results in this theorem follow from the definitions of symmetry and a few facts from elementary geometry. We omit the proof.

Example 1

Find the symmetric partner for $P(-2, 3)$ with respect to the (a) x-axis, (b) y-axis, (c) diagonal $y = x$, (d) origin.

SOLUTION

(a) $P_1(-2, -3)$
(b) $P_2(2, 3)$
(c) $P_3(3, -2)$
(d) $P_4(2, -3)$ See Figure 2.20.

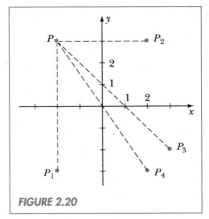

FIGURE 2.20

Suppose we have an x, y equation and wish to determine whether or not its graph will display symmetry with respect to either the x- or y-axis, the diagonal, or the origin. It would be particularly helpful if we could make this determination *before* attempting to draw the graph of the equation. The result we need is simply a restatement of the Symmetry Theorem in terms of x, y equations.

Symmetry in an Equation

The graph of an x, y equation displays the symmetry described in the first column if and only if an equivalent x, y equation is produced under the conditions of the second column.

Symmetry with Respect to	Condition
x-axis	y is replaced with $-y$.
y-axis	x is replaced with $-x$.
Diagonal $y = x$	x is replaced with y and y is replaced with x.
Origin	x is replaced with $-x$ and y is replaced with $-y$.

The examples in the following table illustrate how to apply this result.

Equation	Symmetry with Respect to	Reason
$x = y^2$	x-axis	Replacing y with $(-y)$, we get $x = (-y)^2$, which is equivalent to $x = y^2$.
$y = x^2$	y-axis	Replacing x with $(-x)$, we get $y = (-x)^2$, which is equivalent to $y = x^2$.
$x + xy + y = 0$	Diagonal	Replacing x with y and y with x, we get $y + yx + x = 0$, which is equivalent to $x + xy + y = 0$.
$y = x^3$	Origin	Replacing x with $(-x)$ and y with $(-y)$, we get $-y = (-x)^3$, which is equivalent to $y = x^3$.

Example 2

Show that $(x - y)^2 = x + y$ is symmetric with respect to the diagonal $y = x$. Graph. Does the equation define y as a function of x?

SOLUTION Replacing x with y and y with x, we have

$$(y - x)^2 = y + x$$

Note that $(y - x)^2 = [-1(x - y)]^2 = (x - y)^2$. Therefore this equation is equivalent to

$$(x - y)^2 = y + x$$

which is the same as the original equation. This proves symmetry with respect to the diagonal $y = x$.

To graph the curve, we make a table of x, y values. We shall exhibit one of the computations used to compile the table. Choosing $x = 3$, the corresponding y value is found as follows:

$$(3 - y)^2 = 3 + y$$

$$9 - 6y + y^2 = 3 + y$$

$$y^2 - 7y + 6 = 0$$

$$(y - 6)(y - 1) = 0$$

So $y = 6$ or $y = 1$. This gives us points $(3, 6)$ and $(3, 1)$ on the graph. Now, using the symmetry of the curve, we know that $(6, 3)$ and $(1, 3)$ are also on the graph. Thus, by choosing one value of x, we found four points on the curve. Working in this fashion, we obtain the following table of solutions for the equation:

x	0	0	1	1	3	3	6	6	10
y	0	1	0	3	1	6	3	10	6

This leads us to the graph in Figure 2.21. Notice that the graph fails the vertical line test. Therefore the equation $(x - y)^2 = x + y$ does *not* define y as a function of x.

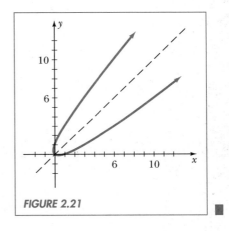

FIGURE 2.21

Example 3

Show that $xy = 1$ is symmetric with respect to the diagonal $y = x$ and the origin. Graph. Does the equation define y as a function of x?

SOLUTION Replacing x with y and y with x, we get $yx = 1$, which is the same as the original equation. This proves symmetry with respect to the diagonal $y = x$.

Replacing x with $-x$ and y with $-y$, we get $(-x)(-y) = 1$, which is equivalent to $xy = 1$. This proves symmetry with respect to the origin.

We compute a table of x, y values by choosing $x = 0, 1, 2, 3, 4$:

x	0	1	2	3	4
y	Undefined	1	1/2	1/3	1/4

Note that there is no point on this graph with x-coordinate zero. Now, using symmetry with respect to the diagonal, we have

x	1	1/2	1/3	1/4
y	1	2	3	4

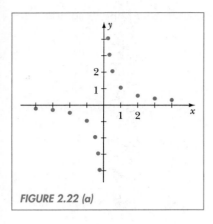

FIGURE 2.22 (a)

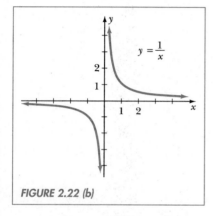

FIGURE 2.22 (b)

Finally, using symmetry with respect to the origin, we obtain

x	-4	-3	-2	-1	$-1/2$	$-1/3$	$-1/4$
y	$-1/4$	$-1/3$	$-1/2$	-1	-2	-3	-4

We graph the points from our tables in Figure 2.22(a). Next, we draw a curve through these points, as in Figure 2.22(b). Notice that the curve satisfies the vertical line test; hence the equation $xy = 1$ defines y as a function of x. In fact, $y = 1/x$, so if we call this function F, we have $F(x) = 1/x$. This is an important function that we study again in Chapter 3. ∎

To determine the symmetry properties of a function, we may apply the symmetry tests to the corresponding x, y equation of the function. However, for future reference, we rewrite two of the symmetry tests in terms of functional notation.

Symmetry in a Function
The graph of a function F displays the given symmetry if and only if the corresponding condition is satisfied:

Symmetry with Respect to	Condition
y-axis	$F(-x) = F(x)$ for all $x \in \mathcal{D}_F$.
Origin	$F(-x) = -F(x)$ for all $x \in \mathcal{D}_F$.

Notice that we do not discuss symmetry with respect to the x-axis for a function. By the vertical line test, the graph of any function cannot display symmetry with respect to the x-axis. Symmetry in a function with respect to the diagonal is discussed in Section 2.6.

Example 4

Without graphing the function, determine whether it is symmetric with respect to the y-axis or the origin. (a) $F(x) = \dfrac{1}{x^2 - 1}$, (b) $G(x) = x^3 + x$, (c) $H(x) = \dfrac{x + 1}{x^2}$.

SOLUTION (a) We note that $\mathcal{D}_F = \{x : x \neq 1, \ -1\}$. Now, for all $x \in \mathcal{D}_F$, we have

$$F(-x) = \frac{1}{(-x)^2 - 1} = \frac{1}{x^2 - 1}$$

Thus $F(-x) = F(x)$, which implies that F is symmetric with respect to the y-axis.

(b) In this case, $\mathcal{D}_G = \mathbf{R}$. For all $x \in \mathbf{R}$, we have

$$G(-x) = (-x)^3 + (-x) = -x^3 - x = -(x^3 + x)$$

Thus $G(-x) = -G(x)$, which implies that G is symmetric with respect to the origin.

(c) In this case, $\mathcal{D}_H = \{x : x \neq 0\}$. For all $x \in \mathcal{D}_H$, we have

$$H(-x) = \frac{(-x) + 1}{(-x)^2} = \frac{-x + 1}{x^2}$$

Unfortunately, $H(-x)$ is not the same as $H(x)$ or $-H(x)$. We conclude that H is not symmetric with respect to the y-axis or the origin. ∎

EXERCISES 2.3

1. Find the symmetric partner for $P(5, -2)$ with respect to the (a) x-axis, (b) y-axis, (c) diagonal $y = x$, (d) origin.

2. Repeat Exercise 1 for $P(-\pi, -\pi)$.

In Exercises 3 to 6, determine what additional points (if any) are necessary to make the graph of the given set symmetric with respect to the (a) x-axis, (b) y-axis, (c) diagonal $y = x$, (d) origin.

3. $\{(-2, 3), (3, -2)\}$

4. $\{(2, 0), (0, 2)\}$

5. $\{(-1, -2), (0, 0), (1, -2)\}$

6. $\{(-2, 0), (1, 1), (1, -1), (2, 0)\}$

In Exercises 7 to 20, determine whether the equation displays symmetry with respect to the x-axis, y-axis, the diagonal $y = x$, or the origin. Graph. Does the equation define a function $y = F(x)$? If so, state the rule for $F(x)$.

7. $x = \sqrt{1 - y^2}$
8. $x = \sqrt{y^2}$
9. $y = \sqrt{x^2}$
10. $y = x^2 + 1$
11. $x = y^3$
12. $y = 2x$
13. $x^2 + y^2 = 4$
14. $|y| = |x|$
15. $x = \dfrac{|y|}{y}$
16. $y = \dfrac{|x|}{x}$
17. $xy = 2$
18. $xy = -2$
19. $x^2 y = 1$
20. $x^2 y = -1$

In Exercises 21 to 27, verify that the equation possesses the given symmetry. Make an x, y table of values, and use the symmetry to graph the equation. Does the equation define y as a function of x?

21. $y^3 = x^2$; y-axis
22. $y^2 = x^3$; x-axis
23. $(x - y)^2 + x + y = 0$; diagonal $y = x$
24. $x^2 y^2 = 1$; x-axis, y-axis, diagonal $y = x$, and origin
25. $|x + y| = 1$; diagonal $y = x$ and origin
26. $(x - y)^2 = 1$; diagonal $y = x$ and origin (*Hint:* apply the square root operation.)
27. $4y^2 = x^2$; x-axis, y-axis, and origin (*Hint:* apply the square root operation.)

In Exercises 28 to 31, without graphing the function, determine whether it is symmetric with respect to the y-axis or the origin.

28. (a) $F(x) = \dfrac{1}{x^2}$ (b) $G(x) = \dfrac{x}{x^2 + 1}$

29. (a) $F(x) = \dfrac{1}{x^3}$ (b) $G(x) = 1 - 6x^2$

30. (a) $F(x) = |x - 1|$ (b) $G(x) = |x| - 1$

31. (a) $F(x) = \dfrac{1 + x}{1 - x}$ (b) $G(x) = |x - x^3|$

In Exercises 32 and 33, suppose that P is symmetric to P' with respect to the point Q. Find P'.

32. $P(5, -3)$, $Q(1, 2)$ 33. $P(-8, 3)$, $Q(-2, -1)$

34. Give an example of a curve that possesses an infinite number of lines of symmetry.

35. Consider the proposition: If a curve is symmetric with respect to the x- and y-axes, then the curve is symmetric with respect to the origin.
    (a) Prove this proposition is true by showing that if $P(a, b)$ is on the curve, then so is $P'(-a, -b)$.
    (b) State the converse of the proposition. Is the converse true or false?

36. Consider $\{(x, y) : |x| + |y| = 1\}$. (a) What symmetry does the set possess? (b) Graph. (*Hint:* consider two cases, $x \geq 0$ and $x < 0$.)

37. Prove that the graph of a nonzero function $F(x)$ cannot be symmetric with respect to the x-axis. Use the fact that $F(x) \neq 0$ for at least one x in its domain.

2.4 GRAPHING METHODS

In this section we discuss five types of graphing problems: (1) shifting, (2) vertical stretching and compressing, (3) reflecting, (4) functions defined by several rules, and (5) absolute value.

Shifting. Our first objective is to recognize when we have a shifted equation and how it is related to its unshifted counterpart. We then apply our results concerning equations to the problem of graphing functions.

Let us begin by considering the following five equations.

$$x^2 + y^2 = 1 \tag{1}$$

$$(x - 3)^2 + (y - 2)^2 = 1 \tag{2}$$

$$(x + 3)^2 + (y - 2)^2 = 1 \tag{3}$$

$$(x + 3)^2 + (y + 2)^2 = 1 \tag{4}$$

$$(x - 3)^2 + (y + 2)^2 = 1 \tag{5}$$

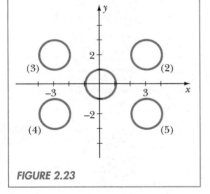

FIGURE 2.23

The graph of Equation (1) is a circle centered at the origin with radius 1. Equations (2) to (5) represent circles that also have radii 1 but different centers. (See Figure 2.23.) We can think of the graphs of these circles as various *shifts* of the original circle, $x^2 + y^2 = 1$. The results are summarized in the following table:

Unshifted Equation	Shift Graph in		Shifted Equation
	x-direction	*y*-direction	
$x^2 + y^2 = 1$	$+3$	$+2$	$(x - 3)^2 + (y - 2)^2 = 1$
$x^2 + y^2 = 1$	-3	$+2$	$(x + 3)^2 + (y - 2)^2 = 1$
$x^2 + y^2 = 1$	-3	-2	$(x + 3)^2 + (y + 2)^2 = 1$
$x^2 + y^2 = 1$	$+3$	-2	$(x - 3)^2 + (y + 2)^2 = 1$

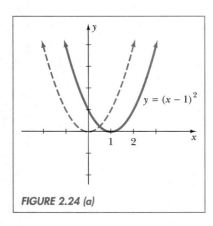

FIGURE 2.24 (a)

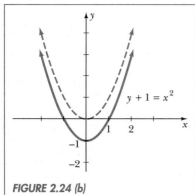

FIGURE 2.24 (b)

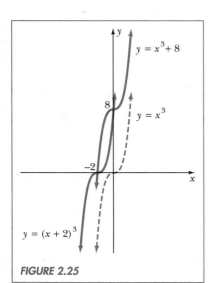

FIGURE 2.25

We are led to the following important fact.

> **Shifting Equations**
> Given an x, y equation, replacing x with $(x - h)$ and y with $(y - k)$ results in a shift of h units in the x-direction and k units in the y-direction.

Example 1

Find the equation for the graph obtained by shifting the curve $y = x^2$ (a) $+1$ unit in the x-direction and (b) -1 unit in the y-direction.

SOLUTION (a) To shift the graph of an equation $+1$ unit in the x-direction, we must replace x with $x - 1$. Thus

$$\text{Unshifted equation:} \quad y = x^2$$

$$\text{Shifted equation:} \quad y = (x - 1)^2$$

See Figure 2.24(a).

(b) To shift the equation -1 unit in the y-direction, we must replace y with $y - (-1) = y + 1$. Thus

$$\text{Unshifted equation:} \quad y = x^2$$

$$\text{Shifted equation:} \quad y + 1 = x^2$$

See Figure 2.24(b). ■

Example 2

Describe how the graphs of $y = (x + 2)^3$ and $y = x^3 + 8$ are related to $y = x^3$.

SOLUTION Both these equations are shifts of $y = x^3$. Compare

$$\text{Unshifted equation:} \quad y = x^3$$

$$\text{Shifted equation:} \quad y = (x + 2)^3$$

Evidently x has been replaced by $x + 2$, which means that the graph of $y = (x + 2)^3$ is obtained by shifting the graph of $y = x^3$ in the x-direction -2 units.

Next, consider the equation $y = x^3 + 8$, which is equivalent to $y - 8 = x^3$. Compare

$$\text{Unshifted equation:} \quad y = x^3$$

$$\text{Shifted equation:} \quad y - 8 = x^3$$

Thus y has been replaced by $y - 8$, which means that the graph of $y = x^3 + 8$ is obtained by shifting the graph of $y = x^3$ in the y-direction $+8$ units.

Figure 2.25 shows the graphs of $y = (x + 2)^3$ and $y = x^3 + 8$. Notice that these two curves are not the same. ■

Example 3

$(x + 2)^2 + y = 1$ is a shifted basic curve. Find the unshifted equation, determine the amount of the shift, and graph both curves, unshifted dashed and shifted solid, on the same axes.

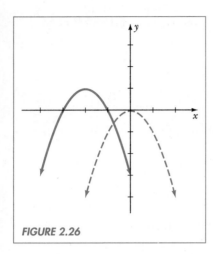

FIGURE 2.26

SOLUTION The equation contains x^2 and y. This reminds us of the basic curve $y = x^2$. With this in mind, we rearrange terms in the equation:

$$(x + 2)^2 + y = 1$$
$$y = 1 - (x + 2)^2$$
$$y - 1 = -(x + 2)^2$$

Note that we end up with a negative sign in front of the x^2 term. Therefore, compare

Unshifted equation: $y = -x^2$

Shifted equation: $y - 1 = -(x + 2)^2$

Thus the equation $y = -x^2$ has been shifted $+1$ unit in the y-direction and -2 units in the x-direction. Figure 2.26 is the graph. We confirm the accuracy of the shifted graph by calculating the x- and y-intercepts. x-intercepts: $(-3, 0)$ $(-1, 0)$; y-intercept: $(0, -3)$. ∎

Let us now apply our results on shifting equations to the problem of graphing functions. Recall that if G is a function, then its graph is determined by the equation

$$y = G(x)$$

subject to the restriction that x must be in the domain of G. Suppose we shift this equation by h units in the x-direction and k units in the y-direction. We have

Unshifted equation: $y = G(x)$

Shifted equation: $y - k = G(x - h)$

Solving the shifted equation for y yields $y = G(x - h) + k$. Therefore the function F defined by

$$F(x) = G(x - h) + k$$

has the same graph as G shifted h units in the x-direction and k units in the y-direction. We say that *F is the function G shifted*.

Example 4

Let $G(x) = |x|$. Find $F(x)$ if F is the function G shifted $+2$ units in the x-direction and -1 unit in the y-direction. Graph G dashed and F solid on the same axes.

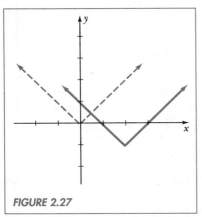

FIGURE 2.27

SOLUTION We need to shift the equation for G and then solve for y to obtain the formula for $F(x)$.

$$\text{Equation for } G: \qquad y = |x|$$
$$\text{Shifted equation:} \quad y - (-1) = |x - 2|$$
$$y + 1 = |x - 2|$$
$$y = |x - 2| - 1$$

Thus $F(x) = |x - 2| - 1$.

The graph of G (obtained by graphing $y = |x|$) is one of the basic curves that we studied in Chapter 1. By shifting this graph $+2$ units in the x-direction and -1 unit in the y-direction, we find the graph of F, which is shown in Figure 2.27. ■

Example 5

Figure 2.28 is the graph of a certain function S. Find the graph of $F(x) = S(x + 3) + 1$.

SOLUTION The graph of F is obtained by graphing the equation $y = S(x + 3) + 1$, or

$$y - 1 = S(x + 3)$$

This is S shifted -3 units in the x-direction and $+1$ unit in the y-direction. Figure 2.29 is the graph.

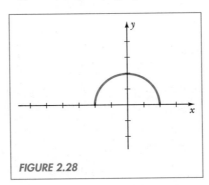

FIGURE 2.28

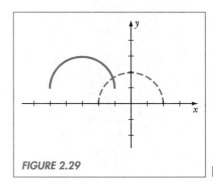

FIGURE 2.29

■

Vertical Stretching and Compressing. Multiplying a function F by the constant a forms a new function denoted by aF. The rule for aF is given as follows:

$$(aF)(x) = a \cdot F(x)$$

Thus, if $F(x) = x^2$, then

$$2F(x) = 2x^2 \quad \text{and} \quad \frac{1}{4}F(x) = \frac{1}{4}x^2$$

The domain of aF will be the same as the domain of F.

We wish to investigate the following question: How does multiplying a function by a constant affect its graph? Figure 2.30 compares the graphs of F and $2F$ on the same axes where $F(x) = x^2$. Notice that the y-coordinates on the graph of F are multiplied by 2 to obtain the graph of $2F$. Thus the graph of $2F$ is produced by *stretching* the graph of F vertically. Next, Figure 2.31 shows the graphs of F and $\frac{1}{4}F$ on the same axes. In this case we find that the graph of $\frac{1}{4}F$ is produced by *compressing* the graph of F vertically.

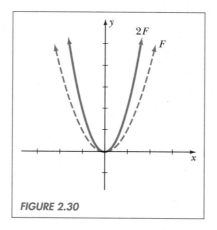

FIGURE 2.30

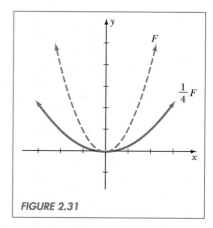

FIGURE 2.31

We can generalize the results in Figures 2.30 and 2.31 as follows.

> If F is any function and a is a positive constant, then the graph of aF is obtained from the graph of F by
>
> Stretching vertically if $a > 1$
>
> Compressing vertically if $0 < a < 1$

Example 6

Let $F(x) = x^3$. Graph $y = 2F(x)$ and $y = \frac{1}{4}F(x)$ by stretching or compressing the graph of F.

SOLUTION The graphs are shown in Figures 2.32 and 2.33.

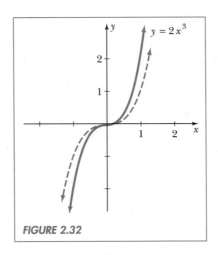

FIGURE 2.32

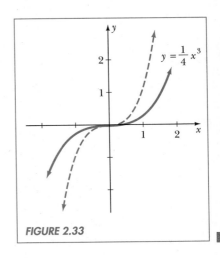

FIGURE 2.33

Reflecting. What happens when we multiply a function by -1? Figure 2.34 is the graph of a certain function $y = F(x)$ compared with the graph of $y = -F(x)$. The y-coordinates on the graph of $-F$ are obtained by changing the sign of the y-coordinates on the graph of F. In other words, the graph of $y = -F(x)$ is found by reflecting the graph of $y = F(x)$ across the x-axis.

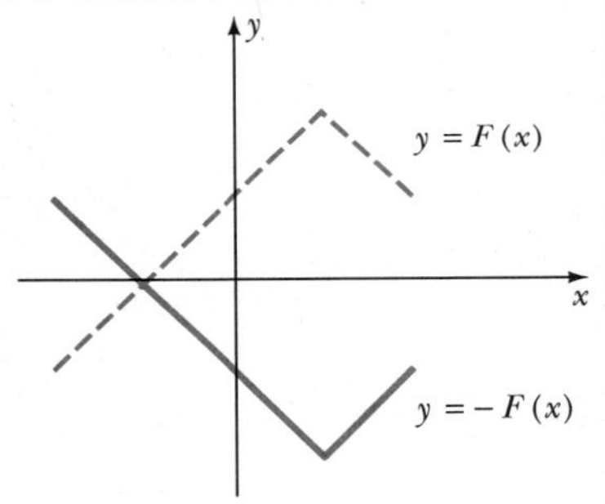

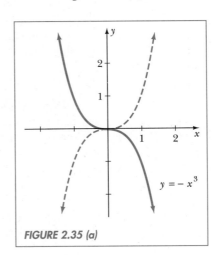

FIGURE 2.34

Example 7

Graph (a) $y = -x^3$, (b) $y = -(x - 1)^3 + 1$.

SOLUTION (a) We wish to graph $y = -F(x)$, where $F(x) = x^3$. Therefore we reflect $y = x^3$ across the x-axis to obtain the graph of $y = -x^3$, which is shown in Figure 2.35(a).

FIGURE 2.35 (a)

(b) We find the graph of $y = -(x - 1)^3 + 1$ by shifting $y = -x^3$. We have

$$y = -(x - 1)^3 + 1 \quad \text{is equivalent to} \quad y - 1 = -(x - 1)^3$$

Thus

Unshifted equation: $\quad y = -x^3$

Shifted equation: $\quad y - 1 = -(x - 1)^3$

Hence, shifting $y = -x^3$ by $+1$ unit in the x-direction and $+1$ unit in the y-direction, we obtain the desired graph, shown in Figure 2.35(b).

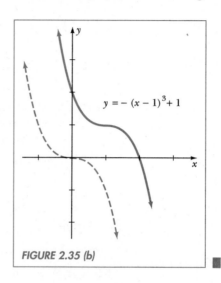

$$y = -(x - 1)^3 + 1$$

FIGURE 2.35 (b)

Remark

When an equation involves both reflection and shifting, as in Example 7(b), always reflect *first* and then shift.

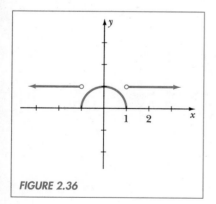

FIGURE 2.36

Functions Defined By Several Rules. The function F in Figure 2.36 appears to consist of two different functions: the semicircle $F(x) = \sqrt{1 - x^2}$ for x in the interval $[-1, 1]$ and the constant function $F(x) = 1$ for x outside $[-1, 1]$. In other words, $F(x)$ has two different rules, depending on the value of x:

$$F(x) = \sqrt{1 - x^2} \quad \text{for } x \in [-1, 1]$$

and

$$F(x) = 1 \quad \text{for } x \in (-\infty, -1) \cup (1, +\infty).$$

The formal way to define a function such as this is to use a brace followed by the rules, one per line, with the values of x for which the rule is valid. Thus

$$F(x) = \begin{cases} \sqrt{1 - x^2}, & |x| \leq 1 \\ 1, & |x| > 1 \end{cases}$$

Note that $|x| \leq 1$ if and only if $x \in [-1, 1]$, and $|x| > 1$ if and only if $x \in (-\infty, -1) \cup (1, +\infty)$.

The use of several rules to define a function is valid as long as we assign only one real number to each element in the domain.

Functions defined by several rules are often functions with graphs containing holes or breaks in the curve. Functions with holes or breaks in their graphs are said to be *discontinuous*.

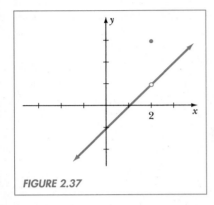

FIGURE 2.37

Example 8

Graph the function $F(x) = \begin{cases} x - 1, & x \neq 2 \\ 3, & x = 2 \end{cases}$

 **SOLUTION**   Figure 2.37 is the graph of F. Notice that the open circle at $(2, 1)$ indicates that this point is not on the graph. Because of this hole in the graph, we say that F is discontinuous. ■

Example 9

Graph the function $G(x) = \begin{cases} -x, & x < 0 \\ 1 - x, & x > 0 \end{cases}$

 **SOLUTION**   Figure 2.38 is the graph of G. Note that a break occurs in the graph when $x = 0$. Thus we say that G is discontinuous. In fact, notice that G is left undefined at $x = 0$. ■

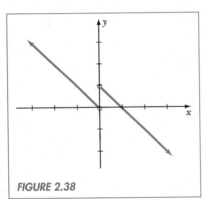

FIGURE 2.38

If the graph of a function contains no holes or breaks in it, then we say the function is *continuous*.

Example 10

Graph the function $H(x) = \begin{cases} x + 2, & x < -1 \\ x^2, & |x| \leq 1 \\ 1, & x > 1 \end{cases}$

Is H continuous?

 **SOLUTION**   As we can see in Figure 2.39, the graph of H is continuous. ■

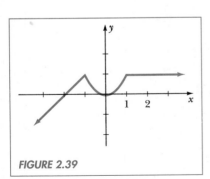

FIGURE 2.39

Absolute Value.   Suppose we wish to graph the absolute value of a function F, $y = |F(x)|$. By the definition of absolute value,

$$|F(x)| = \begin{cases} F(x) & \text{if } F(x) \geq 0 \\ -F(x) & \text{if } F(x) < 0 \end{cases}$$

Therefore we may obtain the graph of $y = |F(x)|$ from the graph of $y = F(x)$ by changing any point with a negative y-coordinate to one with the opposite, positive y-coordinate. This procedure amounts to reflecting across the x-axis any part of the curve $y = F(x)$ that is below the axis, and leaving any points on or above the x-axis alone.

Example 11

Graph $y = |x^2 - 4|$.

SOLUTION   We are graphing the absolute value of the function $F(x) = x^2 - 4$. We begin by graphing F. The equation for F is $y = x^2 - 4$, or

$$y + 4 = x^2$$

This is just the basic curve $y = x^2$ shifted -4 units in the y-direction. [See Figure 2.40(a).] Now, to obtain the graph of $y = |F(x)|$, we reflect across the x-axis any portion of the graph of F that is below the axis, as shown in Figure 2.40(b).

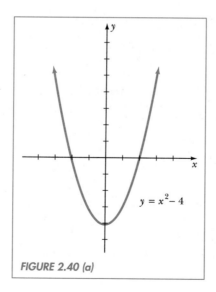

FIGURE 2.40 (a)

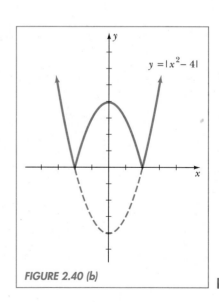

$y = |x^2 - 4|$

FIGURE 2.40 (b)

EXERCISES 2.4

1. Describe how the graphs of $y = \sqrt{x+1}$ and $y = \sqrt{x} + 1$ are related to $y = \sqrt{x}$. Graph all three equations on the same axes.

2. Describe how the graphs of $y = |x-1|$ and $y = |x| - 1$ are related to $y = |x|$. Graph all three equations on the same axes.

In Exercises 3 to 12, complete the table and graph the shifted equation.

Unshifted Equation	Shift Graph in x-direction	Shift Graph in y-direction	Shifted Equation
3. $y = x^2$	0	-3	?
4. $x = \dfrac{1}{2}y^2$	$+2$	-4	?
5. $y = -x^3$	-1	$+2$	?
6. $x = -y^3$	0	$+3$	?
7. $y = \sqrt{9 - x^2}$	$+3$	-3	?
8. $x^2 + y^2 = 9$	-3	-3	?
9. ?	?	?	$y - 2 = \dfrac{1}{4}(x - 1)^2$
10. ?	?	?	$y = \sqrt{x + 9}$
11. ?	?	?	$x = y^2 - 4$
12. ?	?	?	$y = \sqrt{1 - x^2} + 1$

In Exercises 13 to 22, find the unshifted equation and the amount of the shift for the given equation. Graph the unshifted equation dashed and the shifted equation solid on the same axes. Label x- and y-intercepts for the shifted curve.

13. $y = -x^2 + 2$

14. $(x + 2)^2 + y = 3$

15. $x + y^2 = 1$

16. $x = y^3 + 8$

17. $x = \sqrt{y + 1} - 2$

18. $y = \sqrt{x - 4} + 1$

19. $y = 1 - |x|$

20. $y + 1 = (x - 1)^3$

21. $x^2 = 4 - (y + 1)^2$

22. $(y - 4)^2 = 25 - (x + 5)^2$

In Exercises 23 to 26, find $F(x)$ if F is the function G shifted by the given amount shown in the table. Graph F. Label x- and y-intercepts.

Function $G(x)$	Shift in x-direction	Shift in y-direction
23. $G(x) = |x|$	$+3$	-2
24. $G(x) = x^2$	-2	$+1$
25. $G(x) = \sqrt{x}$	-4	-1
26. $G(x) = -x$	$+1$	-2

In Exercises 27 and 28, graph the function.

27. (a) $JMP(x) = \dfrac{|x|}{x} - 2$

    (b) $STP(x) = \dfrac{|x - 2|}{x - 2}$

28. (a) $S(x) = 1 - \sqrt{x}$

    (b) $SQ(x) = -\sqrt{x + 1}$

In Exercises 29 to 32, the graph of a function *JAG* is shown. Find the graph of each of the following:

(a) $y = JAG(x - 2) + 1$ (b) $y = -JAG(x)$

(c) $y = 2JAG(x)$ (d) $y = |JAG(x)|$

(e) $y = -\frac{1}{2}JAG(x + 1)$

29.

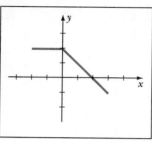

30.

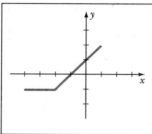

31.

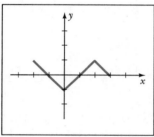

32.

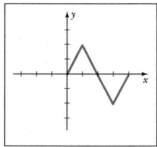

In Exercises 33 to 36, the graph of a function $y = F(x)$ is shown. Each graph is a shift of a basic curve. Find a formula for $F(x)$.

33.

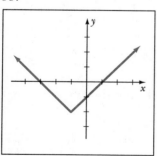

34.

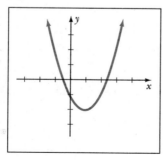

35.

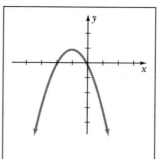

36.

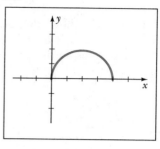

In Exercises 37 to 44, graph each function.

37. (a) $F(x) = 5|x|$ (b) $G(x) = -\frac{1}{5}|x|$

38. (a) $F(x) = \frac{|x|}{6}$ (b) $G(x) = -6|x|$

39. $F(x) = -\frac{1}{2}x^3 - 1$

40. $F(x) = -\frac{1}{4}(x - 2)^2$

41. (a) $F(x) = -\frac{|x|}{x}$ (b) $G(x) = \frac{2|x|}{x}$

42. (a) $F(x) = \frac{|x|}{2x}$ (b) $G(x) = -\frac{3|x|}{x}$

43. (a) $F(x) = -\frac{1}{4}\sqrt{x}$

(b) $G(x) = 4\sqrt{4 - x^2}$

44. (a) $F(x) = -\frac{1}{4}\sqrt{16 - x^2}$

(b) $G(x) = -2\sqrt{x}$

In Exercises 45 to 52, graph each of the following functions. Is the function continuous or discontinuous?

45. $F(x) = \begin{cases} 0, & x < 1 \\ 1, & x \geq 1 \end{cases}$

46. $F(x) = \begin{cases} 1 - x, & x \neq 1 \\ 1, & x = 1 \end{cases}$

47. $F(x) = \begin{cases} \sqrt{4 - x^2}, & |x| \le 2 \\ x, & |x| > 2 \end{cases}$

48. $F(x) = \begin{cases} 1 + x, & x < 0 \\ 1, & x \ge 0 \end{cases}$

49. $F(x) = \begin{cases} -x, & x < 0 \\ x^2, & 0 \le x \le 2 \\ x, & x > 2 \end{cases}$

50. $F(x) = \begin{cases} 0, & x < -1 \\ -\sqrt{1 - x^2}, & |x| \le 1 \\ x - 1, & x > 1 \end{cases}$

51. $F(x) = \begin{cases} x + 4, & x < 0 \\ (x + 2)^2, & x \ge 0 \end{cases}$

52. $F(x) = \begin{cases} x^2, & x < 0 \\ x^3, & x > 0 \end{cases}$

In Exercises 53 and 54, the graph of $F(x)$ is given. Use several rules to define $F(x)$.

53.

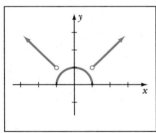

54.

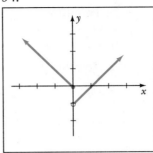

In Exercises 55 to 60, graph each equation.

55. $y = |x^2 - 9|$ 56. $y = |16 - x^2|$

57. $y = |\sqrt{x} - 2|$ 58. $y = |3x - 6|$

59. $y = |1 + x^3|$ 60. $y = ||x| - 1|$

In Exercises 61 and 62, the graph of a certain function is given. Write an equation for this function.

61.

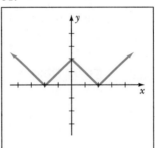

62.

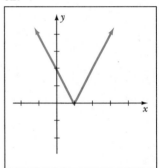

2.5   COMBINING FUNCTIONS

In this section we discuss several methods used to combine two functions to form a new function. Consider for a moment two real numbers a and b. What are the possible ways we can combine the numbers? The obvious arithmetic operations of addition, subtraction, multiplication, and division come to mind: $a + b$, $a - b$, $a \cdot b$, and a/b. These operations also make sense in the world of functions, as we state in the following definition.

Definitions of F + G, F − G, F · G, and F/G

Let F and G be functions. Then F and G can be combined using addition, subtraction, multiplication, or division according to the following rules:

Notation	Rule	Domain
$F + G$	$(F + G)(x) = F(x) + G(x)$	$\mathscr{D}_F \cap \mathscr{D}_G$
$F - G$	$(F - G)(x) = F(x) - G(x)$	$\mathscr{D}_F \cap \mathscr{D}_G$
$F \cdot G$	$(F \cdot G)(x) = F(x) \cdot G(x)$	$\mathscr{D}_F \cap \mathscr{D}_G$
$\dfrac{F}{G}$	$\left(\dfrac{F}{G}\right)(x) = \dfrac{F(x)}{G(x)}$	$\{x : x \in \mathscr{D}_F \cap \mathscr{D}_G \text{ and } G(x) \neq 0\}$

Note that the domains of $F + G$, $F - G$, and $F \cdot G$ consist of those x's that are in both $\mathscr{D}_F$ and $\mathscr{D}_G$ simultaneously. However, the domain of F/G has the additional restriction that $G(x) \neq 0$. This prevents us from dividing by zero (which is undefined).

Example 1

Given that $F(x) = x^2$ and $G(x) = 3x + 1$, find the rule and domain for $F + G$, $F - G$, $F \cdot G$, and $\dfrac{F}{G}$.

SOLUTION

Rule	Domain
$(F + G)(x) = x^2 + 3x + 1$	$\mathbf{R}$
$(F - G)(x) = x^2 - 3x - 1$	$\mathbf{R}$
$(F \cdot G)(x) = 3x^3 + x^2$	$\mathbf{R}$
$\left(\dfrac{F}{G}\right)(x) = \dfrac{x^2}{3x + 1}$	$\{x : x \neq -1/3\}$

It is interesting to see how the graph of $F + G$ is related to the individual graphs of F and G. If (x, y_1) is on the graph of F and (x, y_2) is on the graph of G, then $(x, y_1 + y_2)$ will be on the graph of $F + G$. In other words, by adding the y-coordinate of G at x to the y-coordinate of F at x, we obtain the y-coordinate for the graph of $F + G$ at x. In this manner we can obtain a rough graph of $F + G$ from the individual graphs of F and G. We call this procedure the **method of adding graphs**. The approach is visual rather than computational; instead of constructing a table of x, y values for $F + G$, we work directly from the graphs of F and G.

Example 2

Use the method of adding graphs to find the graph of $H(x) = \frac{1}{2}x^3 + x$.

SOLUTION   We think of H as the sum of two functions,

$$H(x) = F(x) + G(x) \qquad \text{where } F(x) = \frac{1}{2}x^3 \text{ and } G(x) = x$$

We begin by graphing F and G on the same axes in Figure 2.41(a). Next, select either one of the graphs, say $G(x) = x$, to add to the other. This means that for a given value of x, we measure the corresponding y-coordinate on the graph of G and add it to the graph of F at x. This procedure is illustrated in Figure 2.41(a) at four values of x: $x = -2, -1, 1,$ and 2. Notice that when $G(x)$ is positive, at $x = 1$ and 2, we add this amount *above* the graph of F, and when $G(x)$ is negative, at $x = -1$ and -2, we add this amount *below* the graph of F. Figure 2.41 (b) shows several more points obtained by adding the graphs. In Figure 2.41(c), we draw a curve through these points, giving us the graph of $H(x) = F(x) + G(x)$.

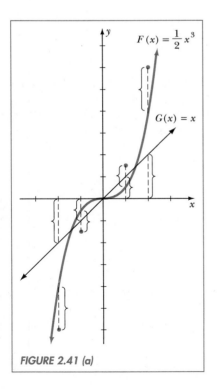

FIGURE 2.41 (a)

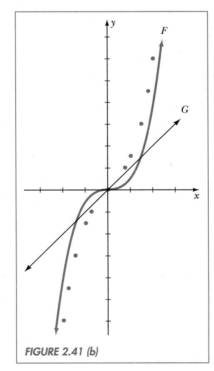

FIGURE 2.41 (b)

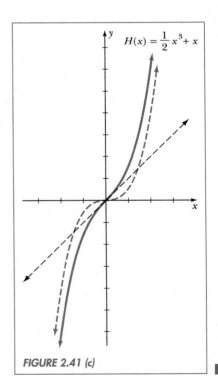

FIGURE 2.41 (c)

Remark

In Example 2, we selected one of the functions, $G(x) = x$, and always added it to the other function, $F(x) = \frac{1}{2}x^3$, to find a point on $F + G$. It does not matter which function is chosen to be added to the other because adding G to F is the same as adding F to G, $F(x) + G(x) = G(x) + F(x)$. However, we recommend being consistent: if you choose to add $G(x)$ to $F(x)$ for one point, then do the same for all points.

Example 3

Use the method of adding graphs to find the graph of $y = |x| - \frac{1}{2}x$.

SOLUTION We think of the formula for y as a sum of two functions, $y = |x| + \left(-\frac{1}{2}x \right)$. Thus

$$y = F(x) + G(x) \qquad \text{where } F(x) = |x| \text{ and } G(x) = -\frac{1}{2}x$$

We begin by graphing F and G on the same axes in Figure 2.42(a). Now we choose to always add $G(x) = -\frac{1}{2}x$ to $F(x)$. This is illustrated in Figure 2.42(a) at four values of x: $x = -4, -2, 2,$ and 4. Notice that when $G(x)$ is positive, we add it above the graph of F, and when $G(x)$ is negative, we add it below the graph of F. Figure 2.42(b) shows several more points for $F + G$; this gives us the desired graph of $F + G$ in Figure 2.42(c).

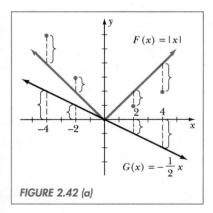

FIGURE 2.42 (a)

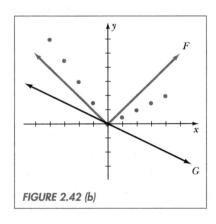

FIGURE 2.42 (b)

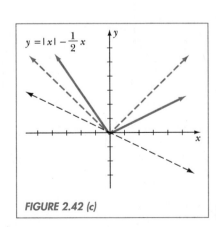

FIGURE 2.42 (c)

There are occasions, particularly in calculus, when it helps to recognize a function as a combination of simpler basic functions. Example 4 illustrates this idea.

Example 4

Let $K(x) = 2$ and $I(x) = x$. Write the given function as a combination of K and I.

(a) $F(x) = x + 2$ (b) $G(x) = x^2 + 2x$

SOLUTION   (a) $F = I + K$. To verify this, we have

$$(I + K)(x) = I(x) + K(x)$$
$$= x + 2$$

which is $F(x)$, as desired.

(b) $G = I \cdot I + K \cdot I$. To verify this, we have

$$(I \cdot I + K \cdot I)(x) = (I \cdot I)(x) + (K \cdot I)(x)$$
$$= I(x) \cdot I(x) + K(x) \cdot I(x)$$
$$= x^2 + 2x$$

which is $G(x)$, as desired. ■

We now introduce another important method for combining functions. Consider two functions represented by machines G and F, where G adds 1 to the input number (Figure 2.43) and F squares the input number (Figure 2.44).

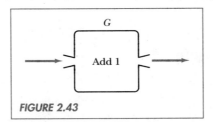

FIGURE 2.43

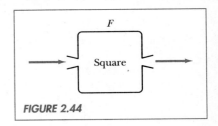

FIGURE 2.44

How can we combine these machines? The most obvious way is to hook them together, one after the other, so that the output from G automatically goes into F. See Figure 2.45.

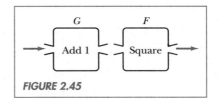

FIGURE 2.45

We call this setup the **composition of F with G**. Let us see what this new function does by first considering a specific input number, such as 2. If the number 2 is input, then we add 1 and square the result, obtaining the output 9. In general, if the input is x, then the output is $(x + 1)^2$. (See Figure 2.46).

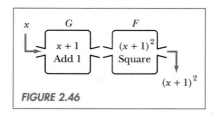

FIGURE 2.46

We now formally define the method of composition of functions.

> **Definition of F ∘ G**
>
> Let F and G be functions. The composition of F with G, denoted by $F \circ G$, is a new function given by
>
> Rule: $(F \circ G)(x) = F(G(x))$
>
> Domain: $\mathscr{D}_{F \circ G} = \{x : x \in \mathscr{D}_G \text{ and } G(x) \in \mathscr{D}_F\}$

We read $F \circ G$ as *F circle G* or *F composed with G*.

The rule for $F \circ G$, $(F \circ G)(x) = F(G(x))$ means to first apply the rule for G and then apply F to the result. We shall diagram this as follows:

$$x \xrightarrow{G} G(x) \xrightarrow{F} F(G(x))$$

Example 5

Let $F(x) = 2x$ and $G(x) = 3x - 1$. Find the rule for $F \circ G$.

SOLUTION Note that the rule for F can be stated $F(\square) = 2\,\square$, where any number can be placed in the box. With this in mind, we compute $(F \circ G)(x)$.

$$(F \circ G)(x) = F(G(x))$$
$$= F(3x - 1)$$
$$= 2(3x - 1)$$
$$= 6x - 2$$

Therefore $(F \circ G)(x) = 6x - 2$. We may diagram this result as follows:

$$x \xrightarrow{G} \boxed{3x - 1} \xrightarrow{F} 2(3x - 1) \ \blacksquare$$

Notice that the order in which we write the composition is important. $F \circ G$ means to apply G first, and $G \circ F$ means to apply F first. It follows that $F \circ G$ and $G \circ F$ could be very different functions.

Example 6

Given $F(x) = x^2 + 1$ and $G(x) = x + 2$. Compare the rules for $F \circ G$ and $G \circ F$.

SOLUTION   Remembering that $F(\square) = \square^2 + 1$, we have

$$(F \circ G)(x) = F(G(x))$$
$$= F(x + 2)$$
$$= (x + 2)^2 + 1$$
$$= x^2 + 4x + 5$$

Therefore $(F \circ G)(x) = x^2 + 4x + 5$. In terms of a diagram, we have

$$x \xrightarrow{\;G\;} \boxed{x + 2} \xrightarrow{\;F\;} (x + 2)^2 + 1$$

Remembering that $G(\square) = \square + 2$, we have

$$(G \circ F)(x) = G(F(x))$$
$$= G(x^2 + 1)$$
$$= (x^2 + 1) + 2$$
$$= x^2 + 3$$

Therefore $(G \circ F)(x) = x^2 + 3$. Diagramming this, we have

$$x \xrightarrow{\;F\;} \boxed{x^2 + 1} \xrightarrow{\;G\;} (x^2 + 1) + 2$$

We conclude that the rules for $F \circ G$ and $G \circ F$ are not the same. In fact, $(F \circ G)(x) \neq (G \circ F)(x)$ for all $x \neq -1/2$. ∎

Now that we know how to find the rule for the function $F \circ G$, let us discuss the domain. Since $(F \circ G)(x)$ is computed using two operations,

$$x \xrightarrow{\;G\;} G(x) \qquad \text{and} \qquad G(x) \xrightarrow{\;F\;} F(G(x))$$

there are two corresponding requirements for x to be in the domain of $F \circ G$. First, x must be acceptable to G for the first operation to work, which means that $x \in \mathcal{D}_G$. Second, $G(x)$ must be acceptable to F for the second operation to work, which means that $G(x) \in \mathcal{D}_F$. Finally, the domain consists of all x satisfying both conditions,

$$\mathcal{D}_{F \circ G} = \{x : x \in \mathcal{D}_G \text{ and } G(x) \in \mathcal{D}_F\}$$

Example 7

Let $F(x) = \dfrac{1}{x}$ and $G(x) = \dfrac{x-1}{x+2}$. Find the rule and domain for $F \circ G$.

SOLUTION   The following computation gives us the rule for $F \circ G$:

$$(F \circ G)(x) = F(G(x))$$
$$= F\left(\frac{x-1}{x+2}\right)$$
$$= \frac{1}{\left(\dfrac{x-1}{x+2}\right)}$$
$$= \frac{x+2}{x-1}$$

Thus $(F \circ G)(x) = \dfrac{x+2}{x-1}$.

Now, let us find the domain of $F \circ G$. From the diagram

$$x \xrightarrow{\ G\ } \boxed{\frac{x+2}{x-1}} \xrightarrow{\ F\ } \frac{1}{\left(\dfrac{x+2}{x-1}\right)}$$

we note that the first operation, $x \xrightarrow{\ G\ } \dfrac{x+2}{x-1}$, works for $x \neq 1$. (This is the domain of G.) Next, the second operation,

$$\frac{x-1}{x+2} \xrightarrow{\ F\ } \frac{1}{\dfrac{x-1}{x+2}}$$

makes sense whenever $\dfrac{x-1}{x+2}$ is a real number different from zero. This means that $x \neq 1$ and $x \neq -2$. Putting these two conditions together, we have

$$\mathcal{D}_{F \circ G} = \{x : x \neq 1, -2\}. \blacksquare$$

Example 8

Given $F(x) = x^2 + 2$ and $G(x) = \sqrt{x}$, find the rule and domain for $F \circ G$.

SOLUTION   We compute the rule for $F \circ G$ as follows:

$$(F \circ G)(x) = F(G(x))$$
$$= F(\sqrt{x})$$
$$= (\sqrt{x})^2 + 2$$
$$= x + 2$$

Therefore $(F \circ G)(x) = x + 2$.

To determine the domain of $F \circ G$, we consider the diagram

$$x \xrightarrow{G} \boxed{\sqrt{x}} \xrightarrow{F} (\sqrt{x})^2 + 2$$

The first operation,

$$x \xrightarrow{G} \sqrt{x}$$

works for $x \in [0, +\infty)$, the domain of G. Next, the second operation

$$\sqrt{x} \xrightarrow{F} (\sqrt{x})^2 + 2$$

works whenever $\sqrt{x}$ is a real number. This means, again, that $x \in [0, +\infty)$. Thus

$$\mathscr{D}_{F \circ G} = [0, +\infty).$$

Although the expression $x + 2$ is meaningful for all real numbers, the domain of $F \circ G$ is not $\mathbf{R}$. $F \circ G$ is composed of two operations, and the first one works only for nonnegative real numbers. ∎

Given two functions F and G, we can certainly compose them in any order that we please, either $F \circ G$ or $G \circ F$. Also, functions may be composed with themselves, such as $F \circ F$. We leave it to you to verify the rules for the following compositions of $F(x) = 1/(x + 1)$ and $G(x) = \sqrt{x}$:

$$(F \circ G)(x) = \frac{1}{\sqrt{x} + 1} \qquad (F \circ F)(x) = \frac{x + 1}{x + 2}$$

$$(G \circ F)(x) = \frac{1}{\sqrt{x + 1}} \qquad (G \circ G)(x) = \sqrt{\sqrt{x}}$$

In calculus, it is important to recognize certain functions as being compositions of simpler functions. We conclude this section with an example that illustrates this idea.

Example 9

Write the given function as a composition of two of the following functions:
$A(x) = x^2$, $B(x) = \sqrt{x}$, $C(x) = x + 3$.
(a) $F(x) = (x + 3)^2$ (b) $G(x) = \sqrt{x} + 3$

SOLUTION   (a) We want to find two functions that will complete the following diagram:

$$x \xrightarrow{?} \boxed{} \xrightarrow{?} (x + 3)^2$$

Evidently, we would like the first function to add 3 to x; then the second function should square the result. Therefore we try composing $A(x) = x^2$

with $C(x) = x + 3$ to obtain F. We have

$$(A \circ C)(x) = A(C(x))$$
$$= A(x + 3)$$
$$= (x + 3)^2$$

Thus $A \circ C = F$ as desired.

(b) Once again, we want two functions that will complete the following diagram:

$$x \xrightarrow{?} \boxed{} \xrightarrow{?} \sqrt{x} + 3$$

It appears that the first function should take the square root of x, and then the second function should add 3 to the result. Therefore we try composing $C(x) = x + 3$ with $B(x) = \sqrt{x}$ to obtain G. We have

$$(C \circ B)(x) = C(B(x))$$
$$= C(\sqrt{x})$$
$$= \sqrt{x} + 3$$

Thus $C \circ B = G$ as desired. ■

EXERCISES 2.5

In Exercises 1 to 4, find the rule and domain for (a) $F + G$, (b) $F - G$, (c) $F \cdot G$, (d) F/G.

1. $F(x) = 1 + x$, $G(x) = 1 - x$
2. $F(x) = x^2 + 1$, $G(x) = x + 1$
3. $F(x) = 4$, $G(x) = x^3$
4. $F(x) = x$, $G(x) = x^2 + 2$

In Exercises 5 to 8, the graphs of functions F and G are shown. Use the method of adding graphs to find the graph of $F + G$.

5.

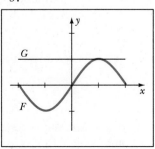

6.

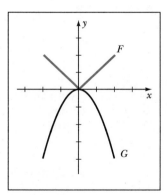

7.

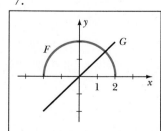

8.

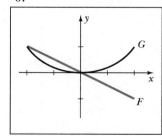

In Exercises 9 to 14, graph each function by the method of adding graphs.

9. $H(x) = \dfrac{1}{8}x^3 + x$ 10. $H(x) = \dfrac{|x|}{x} + x$

11. $H(x) = x^2 + \dfrac{|x|}{x}$ 12. $H(x) = |x| + x$

13. $H(x) = |x| - x$ 14. $H(x) = x^3 - x$

In Exercises 15 and 16, let $K(x) = 3$ and $I(x) = x$. Write the given function as a combination of K or I using addition, subtraction, multiplication, or division.

15. (a) $F(x) = x - 3$ 16. (a) $F(x) = \dfrac{1}{3}x$

(b) $F(x) = \dfrac{3}{x}$ (b) $F(x) = 3 + x$

(c) $F(x) = x^2 + x$ (c) $F(x) = x^2 - 3x$

(d) $F(x) = 3x^3$ (d) $F(x) = x^4$

In Exercises 17 to 34, find (a) $(F \circ G)(x)$, (b) $(G \circ F)(x)$, (c) $(F \circ F)(x)$, (d) $(G \circ G)(x)$.

17. $F(x) = 2x,\ G(x) = 3x$

18. $F(x) = x + 3,\ G(x) = x + 2$

19. $F(x) = 2x - 1,\ G(x) = 3x + 2$

20. $F(x) = 5 - 2x,\ G(x) = 1 - x$

21. $F(x) = (2x + 3)/4,\ G(x) = (4x - 3)/2$

22. $F(x) = (1 - 2x)/3,\ G(x) = (1 - 3x)/2$

23. $F(x) = 2x + 7,\ G(x) = 13$

24. $F(x) = 8,\ G(x) = x - 2$

25. $F(x) = x^2,\ G(x) = 5x - 2$

26. $F(x) = 4 - x^2,\ G(x) = 5x$

27. $F(x) = x^2 - 1,\ G(x) = x^2 + 1$

28. $F(x) = x^2 + 2x + 3,\ G(x) = 2x + 1$

29. $F(x) = |x| + 1,\ G(x) = x - 1$

30. $F(x) = |x + 2|,\ G(x) = x^2 - 2$

31. $F(x) = \dfrac{3}{x},\ G(x) = \dfrac{x + 1}{x - 2}$

32. $F(x) = \dfrac{1}{x + 1},\ G(x) = \dfrac{1}{x - 1}$

33. $F(x) = \dfrac{x}{x + 1},\ G(x) = \dfrac{x}{x - 1}$

34. $F(x) = x^3,\ G(x) = \sqrt[3]{x}$

35. Consider the functions $F(x) = 1/(x + 1)$ and $G(x) = \sqrt{x}$. Find the rule and domain for (a) $F \circ G$, (b) $G \circ F$, (c) $F \circ F$, (d) $G \circ G$.

In Exercises 36 to 41, find the rule and domain for $F \circ G$.

36. $F(x) = x^2,\ G(x) = \sqrt{x}$

37. $F(x) = 1/x,\ G(x) = (x + 3)/(x + 1)$

38. $F(x) = x/(x + 1),\ G(x) = 1/(x - 3)$

39. $F(x) = \sqrt{x},\ G(x) = (x + 1)^2$

40. $F(x) = 1/x,\ G(x) = x^3$

41. $F(x) = \sqrt[3]{x},\ G(x) = x$

In Exercises 42 to 47, evaluate the given expression using the graphs of F and G in Figures 2.47 and 2.48.

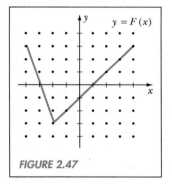

FIGURE 2.47

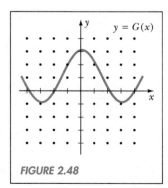

FIGURE 2.48

42. (a) $F(G(2))$ (b) $G(F(2))$ 43. (a) $G(F(3))$ (b) $F(G(3))$

44. $F(G(-2))$ 45. $G(F(-2))$

46. $F(F(1))$ 47. $G(G(-1))$

In Exercises 48 to 59, let $A(x) = x^2$, $B(x) = \sqrt{x}$, $C(x) = x + 1$, $D(x) = 1/x$, and $E(x) = 2x$. Write each of the following functions as a *composition* of two functions chosen from A, B, C, D or E. You may compose a function with itself.

48. $F(x) = \sqrt{x + 1}$ 49. $F(x) = \dfrac{2}{x}$

50. $F(x) = 2x + 1$ 51. $F(x) = \sqrt{x} + 1$

52. $F(x) = |x|$ 53. $F(x) = 4x^2$

54. $F(x) = x^2 + 2x + 1$ 55. $F(x) = x + 2$

56. $F(x) = \dfrac{1}{x^2}$ 57. $F(x) = x^4$

58. $F(x) = 4x$ 59. $F(x) = 2\sqrt{x}$

2.6   *INVERSE FUNCTIONS*

When we define a function F, we might ask whether there is another function that *undoes* what F does. For example, suppose F is the function that does the following: given a real number, it doubles the number and then adds 1.

$$F(x) = 2x + 1$$

To undo this operation, we should reverse the rule by first subtracting 1 and then dividing by 2. The function

$$G(x) = \frac{(x - 1)}{2}$$

ought to undo what F does. Let us see if this happens in the specific case when $x = 3$. First, apply F to 3.

$$F(3) = 2(3) + 1 = 7$$

Now, we apply G to this output, 7, hoping to undo what we just did.

$$G(7) = \frac{(7 - 1)}{2} = 3$$

Success! We have gotten back the number that we started with. The question now is whether G undoes what F does to *any* real number x. First, apply F.

$$F(x) = 2x + 1$$

Now, apply G.

$$G(2x + 1) = \frac{(2x + 1) - 1}{2} = x$$

Again, we are successful. Figure 2.49 illustrates the situation.

What we just showed is that the composite function $G \circ F$ always gives back the same number that we use for the input,

$$(G \circ F)(x) = x \qquad \text{for all } x \in \mathcal{D}_F$$

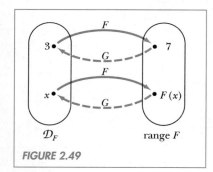

FIGURE 2.49

We also note that F has the same effect on G,

$$(F \circ G)(x) = x \qquad \text{for all } x \in \mathscr{D}_G$$

Whenever this situation occurs, F and G are called **inverse functions**.

Definition of Inverse Functions

Given a function F, we say that G is the inverse function for F provided the following two conditions hold:

1. $(G \circ F)(x) = x \qquad$ for all $x \in \mathscr{D}_F$
2. $(F \circ G)(x) = x \qquad$ for all $x \in \mathscr{D}_G$

Notice that if G is the inverse function for F, then F is also the inverse function for G. Therefore we sometimes just say F *and* G *are inverse functions.*

Example 1

Prove that $F(x) = \dfrac{1}{4}x + 3$ and $G(x) = 4x - 12$ are inverse functions.

SOLUTION   We must show that the two conditions in the definition are true.

1.   We wish to show that $(G \circ F)(x) = x$ for all $x \in \mathscr{D}_F = \mathbf{R}$. We have

$$(G \circ F)(x) = G(F(x))$$

$$= G\left(\frac{1}{4}x + 3\right)$$

$$= 4\left(\frac{1}{4}x + 3\right) - 12 \qquad \text{since } G(\square) = 4(\square) - 12$$

$$= x + 12 - 12$$

$$= x$$

2.   We must show that $(F \circ G)(x) = x$ for all $x \in \mathscr{D}_G = \mathbf{R}$. We have

$$(F \circ G)(x) = F(G(x))$$

$$= F(4x - 12)$$

$$= \frac{1}{4}(4x - 12) + 3 \qquad \text{since } F(\square) = \frac{1}{4}(\square) + 3$$

$$= x - 3 + 3$$

$$= x \quad \blacksquare$$

Example 2

Show that $F(x) = x^2$ and $G(x) = \sqrt{x}$ are not inverse functions.

SOLUTION If $x \in \mathscr{D}_F = \mathbf{R}$, then

$$(G \circ F)(x) = G(F(x))$$

$$= G(x^2)$$

$$= \sqrt{x^2}$$

$$= |x|$$

If x is negative, then $|x| \neq x$. Therefore it is *not* true that $(G \circ F)(x) = x$ for *all* $x \in \mathscr{D}_F$. Hence F and G are not inverse functions.

Note that if $x \in \mathscr{D}_G = [0, +\infty)$, then

$$(F \circ G)(x) = F(G(x))$$

$$= F(\sqrt{x})$$

$$= (\sqrt{x})^2$$

$$= x$$

Therefore $(F \circ G)(x) = x$. This example shows that it is possible for $(F \circ G)(x) = x$ for all $x \in \mathscr{D}_G$, and $(G \circ F)(x) \neq x$ for some $x \in \mathscr{D}_F$. ∎

A special notation is used when dealing with inverse functions. We write F^{-1} and say F *inverse* for the inverse function of F. Thus, if $F(x) = 2x + 1$, then

$$F^{-1}(x) = \frac{x - 1}{2}$$

Be careful not to confuse $F^{-1}(x)$ with the reciprocal function, $1/F(x)$; they are *not* the same.

Given a function F, two questions we want to answer are (1) Does F^{-1} exist? (2) If F^{-1} exists, then what is the rule for $F^{-1}(x)$?

Existence of F^{-1}. We begin with two simple functions A and B, illustrated in Figures 2.50 and 2.51. Does A have an inverse function? Yes. A^{-1} would map 4 back to 1, 5 back to 2, and 6 back to 3. Does B have an inverse function? No. The trouble occurs when trying to decide which number B^{-1} should assign to 5. The essential characteristic that distinguishes these two functions is that A has a certain *one-to-one* property, whereas B does not. It is this one-to-one property that guarantees the existence of an inverse function. Let us define this formally.

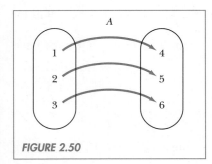

FIGURE 2.50

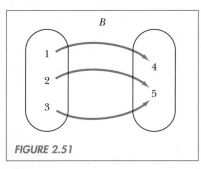

FIGURE 2.51

Definition of One-to-One Function
A function F is said to be **one-to-one** (abbreviated 1–1) if for any x_1, $x_2 \in \mathscr{D}_F$, we have

$$F(x_1) = F(x_2) \qquad \text{if and only if} \qquad x_1 = x_2$$

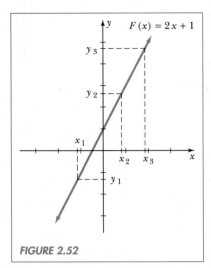

FIGURE 2.52

Example 3

Prove that $F(x) = 2x + 1$ is a 1–1 function.

SOLUTION   Let x_1, x_2 be any two numbers in the domain of F (which is **R**). We must show that $F(x_1) = F(x_2)$ if and only if $x_1 = x_2$. The proof is as follows:

$$F(x_1) = F(x_2) \qquad \text{if and only if} \qquad 2x_1 + 1 = 2x_2 + 1$$
$$2x_1 = 2x_2$$
$$x_1 = x_2 \quad \blacksquare$$

If a function is 1–1, then its graph has a distinctive characteristic. Consider the function $F(x) = 2x + 1$, which we just showed to be 1–1. Figure 2.52 illustrates three different y values on the graph, each one corresponding to exactly one x value in the domain. Indeed, any y value in the range of F cannot correspond to more than one x value; otherwise, the 1–1 property of the function would be violated. An equivalent interpretation of this property is that any horizontal line will intersect the graph of F at most once. In this case, we say that F satisfies the **horizontal line test**.

The Horizontal Line Test

A function is said to satisfy the horizontal line test if any horizontal line intersects the graph of the function at most once.

The following theorem summarizes the results on the existence of inverse functions:

Inverse Function Theorem

(Existence of Inverse Functions)

Let F be any function. Then the following statements are equivalent:
1. F has an inverse function.
2. F is 1–1.
3. F satisfies the horizontal line test.

Example 4

Determine whether or not the given function has an inverse function.
(a) $G(x) = \sqrt{1 - x^2}$ (b) $F(x) = \sqrt{x - 1}$

SOLUTION   If we know the graph of a function, then the easiest way to determine whether or not an inverse function exists is to apply the horizontal line test to the graph. Figures 2.53 and 2.54 on the next page show the graphs of G and F, respectively.

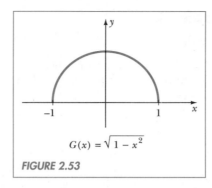

FIGURE 2.53

$$G(x) = \sqrt{1 - x^2}$$

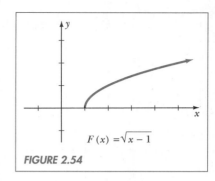

FIGURE 2.54

$$F(x) = \sqrt{x - 1}$$

(a) The graph of G clearly does not satisfy the horizontal line test. Therefore G is not 1–1 and does not have an inverse function.

(b) The graph of F is the upper half of a parabola. This graph satisfies the horizontal line test. Therefore F is 1–1 and has an inverse function. ∎

Finding $F^{-1}(x)$. Let us return to the function A illustrated in Figure 2.50. We know that this function is 1–1 and, therefore, has an inverse function, A^{-1}. Following are tables for A and A^{-1}; the corresponding graphs are in Figure 2.55.

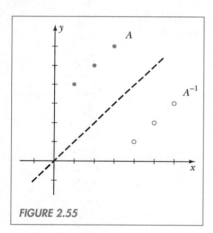

FIGURE 2.55

x	$A(x)$
1	4
2	5
3	6

x	$A^{-1}(x)$
4	1
5	2
6	3

Notice that (a, b) is on the graph of A if and only if (b, a) is on the graph of A^{-1}. This relationship between the points on the graph of a function and its inverse is true in general. For example, suppose (a, b) is on the graph of a 1–1 function F. This means that

$$F(a) = b$$

It follows that

$$F^{-1}(b) = a$$

Thus (b, a) must be on the graph of F^{-1}.

Now, recall the geometric connection between the points (a, b) and (b, a): they are symmetric with respect to the diagonal $y = x$. [See statement (3) in the Symmetry Theorem, Section 2.3.] Therefore, every point on the graph of F will have a symmetric partner across the diagonal on the graph of F^{-1}, and vice versa. This establishes the following result.

Example 5

Given the function $F(x) = \sqrt{x - 1}$. Graph F^{-1}. Find the domain and rule for F^{-1}.

SOLUTION Example 4 showed by the horizontal line test that F does have an inverse. We can obtain the graph of F^{-1} by reflecting the graph of F across the diagonal $y = x$, as shown in Figure 2.56. Several strategic points used to establish the graphs are

Point on $F(x) = \sqrt{x - 1}$	Point on Graph of F^{-1}
(1, 0)	(0, 1)
(2, 1)	(1, 2)
(5, 2)	(2, 5)

The graph of F^{-1} clearly shows that $\mathscr{D}_{F^{-1}} = [0, +\infty)$.

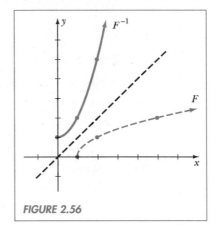

FIGURE 2.56

To determine the rule for $F^{-1}(x)$, we first write the equation for F.

$$y = \sqrt{x - 1}$$

Since the graph of F^{-1} is obtained by interchanging the x- and y-coordinates on the graph of F, the equation for F^{-1} is obtained by interchanging the roles of x and y in the equation for F.

$$x = \sqrt{y - 1}$$

Now, solve for y. Squaring both sides, we have

$$x^2 = y - 1$$

$$x^2 + 1 = y$$

This gives us the rule for $F^{-1}(x)$.

$$F^{-1}(x) = x^2 + 1 \blacksquare$$

In general, to find the rule for $F^{-1}(x)$, we execute the following steps when possible.

Finding $F^{-1}(x)$
1. Write the equation for F: $y = F(x)$
2.   Interchange x and y: $x = F(y)$
3.   Solve for y: $y = F^{-1}(x)$

Remarks

The third step of solving for y can sometimes be impossible to do algebraically. We consider these situations in later chapters.

Example 6

Find the domain and rule for the inverse function of $F(x) = 3x - 4$.

SOLUTION   Figure 2.57 is the graph of F. We note that F is 1–1 and therefore has an inverse function. The graph of F^{-1} is obtained in Figure 2.58 by taking the symmetric image of F across the diagonal $y = x$. We see from this graph that $\mathscr{D}_{F^{-1}} = \mathbf{R}$.

Now, let us find $F^{-1}(x)$.

$$\text{Equation for } F: \quad y = 3x - 4$$

$$\text{Interchange } x \text{ and } y: \quad x = 3y - 4$$

$$\text{Solve for } y: \quad x + 4 = 3y$$

$$\frac{x + 4}{3} = y$$

Therefore $F^{-1}(x) = \dfrac{x + 4}{3}$. You should verify that $(F^{-1} \circ F)(x) = x$ and $(F \circ F^{-1})(x) = x$. $\blacksquare$

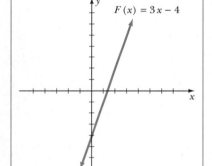

FIGURE 2.57

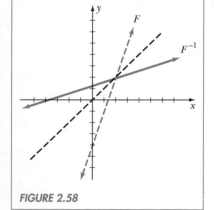

FIGURE 2.58

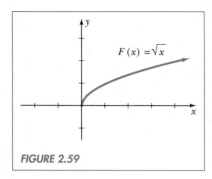

FIGURE 2.59

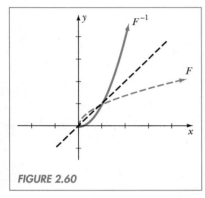

FIGURE 2.60

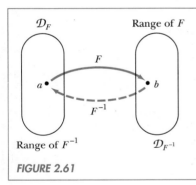

FIGURE 2.61

Example 7

Find the domain and rule for the inverse function of $F(x) = \sqrt{x}$.

SOLUTION Figure 2.59 is the graph of F. Note that F is 1–1 and therefore has an inverse function. The graph of F^{-1} is found in Figure 2.60 by reflecting the graph of F across the diagonal. From this graph we find that $\mathscr{D}_{F^{-1}} = [0, +\infty)$.

Now, let us find the rule for $F^{-1}(x)$.

$$\text{Equation for } F: \quad y = \sqrt{x}$$

$$\text{Interchange } x \text{ and } y: \quad x = \sqrt{y}$$

$$\text{Solve for } y: \quad x^2 = y$$

Therefore $F^{-1}(x) = x^2$.

Note that the domain of $F^{-1}(x) = x^2$ is restricted to $[0, +\infty)$. Without this restriction, x^2 and $\sqrt{x}$ would not be inverse functions. (This was illustrated in Example 2.) ∎

We close this section with a remark about the relationship between the domain and range of a function and its inverse. Suppose F is a function that has an inverse, F^{-1}. Let $a \in \mathscr{D}_F$ and $F(a) = b$. Then $b \in \mathscr{D}_{F^{-1}}$ and $F^{-1}(b) = a$. Figure 2.61 illustrates the situation. It follows that every number in the range of F must be in the domain of F^{-1}, and furthermore, F^{-1} must send each of these numbers back to an element in the domain of F. We conclude,

$$\mathscr{D}_{F^{-1}} = \text{Range of } F$$

$$\text{Range of } F^{-1} = \mathscr{D}_F$$

Note that we have no choice for the domain of F^{-1} or its rule; these are completely determined by the function F. This means that F^{-1} is *unique*; it is impossible for F to have more than one inverse function.

EXERCISES 2.6

In Exercises 1 to 6, prove that F and G are inverse functions.

1. $F(x) = 3x$, $G(x) = x/3$

2. $F(x) = 4x - 8$, $G(x) = x/4 + 2$

3. $F(x) = 1/x$, $G(x) = 1/x$

4. $F(x) = (x + 2)^3$, $G(x) = \sqrt[3]{x} - 2$

5. $F(x) = \sqrt{x} + 1$, $\mathscr{D}_F = [0, +\infty)$

$\quad G(x) = (x - 1)^2$, $\mathscr{D}_G = [1, +\infty)$

6. $F(x) = \dfrac{1}{3 - x}$, $\quad G(x) = 3 - \dfrac{1}{x}$

In Exercises 7 to 24, graph the given function. Determine if the inverse function exists. If the inverse does exist, graph it on the same axes and find its domain and rule.

7. $F(x) = 2x + 3$ 8. $G(x) = x - 2$

9. $L(x) = 4 - (2/3)x, \mathcal{D}_L = [-3, 3]$

10. $L(x) = 7 - 4x, \mathcal{D}_L = [1, 2]$

11. $S(x) = x^2, \mathcal{D}_S = (-\infty, 0]$

12. $Q(x) = \sqrt{1 - x^2}, \mathcal{D}_Q = [0, 1]$

13. $F(x) = (x - 1)^2, \mathcal{D}_F = [0, +\infty)$

14. $P(x) = (x + 2)^2, \mathcal{D}_P = (-\infty, 0]$

15. $F(x) = -\sqrt{9 - x^2}, \mathcal{D}_F = [0, 3]$

16. $F(x) = 1 - x^2, \mathcal{D}_F = [1, +\infty)$

17. $ABS(x) = |x|$ 18. $C(x) = 5$

19. $M(x) = -x$ 20. $SQ(x) = \sqrt{x + 1}$

21. $R(x) = \sqrt[3]{x}$ 22. $F(x) = \sqrt[3]{x - 1}$

23. $H(x) = -(x + 2)^2, \mathcal{D}_H = [-2, +\infty)$

24. $F(x) = (x + 1)^2 - 1, \mathcal{D}_F = [-1, +\infty)$

In Exercises 25 to 30, prove that the given function is one-to-one (see Example 3), then find the rule for its inverse function.

25. $F(x) = 4x + 3$ 26. $L(x) = 6 - 5x$

27. $R(x) = 1/(2x)$ 28. $F(x) = 1/(x + 2)$

29. $F(x) = \dfrac{x}{x + 1}$ 30. $F(x) = \dfrac{x - 2}{3x + 4}$

In Exercises 31 and 32, the graph of a function F is shown. (a) Explain why F has an inverse. (b) Find $F^{-1}(a)$ for $a = -2, 0, 1, 2, 3$. (c) Graph $y = F^{-1}(x)$.

31.

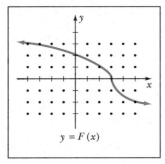

$y = F(x)$

32.

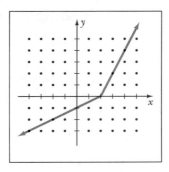

In Exercises 33 and 34, answer true or false.

33. If F has an inverse and $F(1) = 2$, then $2 \in \mathcal{D}_{F^{-1}}$ and $F^{-1}(2) = 1$.

34. If $\mathcal{D}_F = \{2, 3, 4\}$ and the range of F consists of three distinct real numbers, then F has an inverse.

In Exercises 35 to 39, we say that a function F is its own inverse if $(F \circ F)(x) = x$ for all $x \in \mathcal{D}_F$. Prove that each of the following functions is its own inverse.

35. $F(x) = \sqrt{4 - x^2}, \mathcal{D}_F = [0, 2]$

36. $F(x) = \dfrac{a}{x}, a$ a nonzero constant.

37. $F(x) = \dfrac{2x + 1}{3x - 2}$

38. $F(x) = a - x, a$ a constant.

39. $F(x) = \dfrac{x + a}{bx - 1}, a$ and b constants, $b \neq 0$.

40. This exercise refers to the definition given in the instructions to Exercises 35 to 39.
    (a) Complete the following statement: If a function F is its own inverse, then the graph of F must be symmetric with respect to _____.
    (b) Illustrate your answer to part (a) by graphing the function given in Exercise 35.

41. Find the range of $F(x) = \dfrac{3x - 2}{x + 4}$. (*Hint:* consider the domain of F^{-1}.)

42. Find the range of $F(x) = \dfrac{7 - 2x}{3x + 1}$.

43. If $F(x) = 1 - |x|, \mathcal{D}_F = (-\infty, 0]$, find $\mathcal{D}_{F^{-1}}$ and $F^{-1}(x)$.

1. State the definition of a real function.

2. Determine if the given rule defines a real function F with the indicated domain.
   (a) Domain $\mathbf{R}$, rule $F(x) = \sqrt{x}$
   (b) Domain $\mathbf{R}$, rule $F(x) = 1$ if $x \in \mathbf{Z}$; $F(x) = 0$ if $x \notin \mathbf{Z}$.

3. Invent a rule for a function F with
   (a) Domain $\mathbf{N}$ and range $\{-1, 1\}$.
   (b) Domain $\mathbf{Z}$ and range $\mathbf{N}$.

In Exercises 4 to 7, using the given rule for F, find $F(3)$, $F(0)$, $F(-1)$, $F(-x)$, $-F(x)$, $F(x + 1)$, $F(x) + F(1)$, $\dfrac{F(x + h) - F(x)}{h}$ (assume $h \neq 0$), and $\dfrac{F(k + 1)}{F(k)}$.

4. $F(x) = 1 - 2x$ 5. $F(x) = x - x^2$

6. $F(x) = \dfrac{1}{x + 2}$ 7. $F(x) = 7$

In Exercises 8 to 11, find the domain of F.

8. $F(x) = \dfrac{\sqrt{x}}{x - 1}$ 9. $F(x) = \sqrt{\dfrac{x - 1}{3x}}$

10. $F(x) = \dfrac{9 - x^2}{x + 2}$ 11. $F(x) = 2$

12. Determine if F and G are equal functions. Explain your answer.
    (a) $F(x) = x$, $G(x) = \dfrac{x^2 - x}{x - 1}$
    (b) $F(x) = |x + 1|$, $G(x) = \sqrt{(x + 1)^2}$

13. How many different functions are possible with domain $\{1, 2\}$ and range $\{0, 1\}$?

14. Assume that F is a function and $a \in \mathcal{D}_F$. Answer true or false: (a, b) is on the graph of F if and only if $F(a) = b$.

In Exercises 15 to 17, graph the given function. State the domain, range, zeros, y-intercept, and where the function is increasing or decreasing.

15. (a) $F(x) = -\dfrac{1}{2}x$, $\mathcal{D}_F = [-2, 4]$

    (b) $L(x) = -\dfrac{1}{2}x$

16. (a) $G(x) = \sqrt{36 - x^2}$, $\mathcal{D}_G = [-6, 0]$
    (b) $H(x) = \sqrt{36 - x^2}$

17. (a) $C(x) = \sqrt[3]{x}$; (b) $S(x) = x^3$

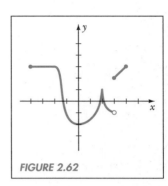

FIGURE 2.62

18. Consider the graph of F in Figure 2.62.
    (a) How do we know this is the graph of a function?
    (b) Find $\mathcal{D}_F$ and range of F.
    (c) Find $F(0)$.
    (d) Find $F(F(-2))$.

19. Let $F(x) = x^2$. Suppose P_1 and P_2 are points on the graph of F, with x-coordinates x_1 and x_2, respectively. Find the slope of the line through P_1 and P_2 if
    (a) $x_1 = 1$ and $x_2 = 4$.
    (b) $x_1 = 2$ and $x_2 = 2 + h$, $h \neq 0$.

In Exercises 20 to 22, determine whether the equation displays symmetry with respect to the x-axis, y-axis, the diagonal $y = x$, or the origin. Graph. Does the equation define a function $y = F(x)$? If so, state the rule for $F(x)$.

20. $xy^2 = 1$

21. $2xy = 1$

22. $x = -y^3$

In Exercises 23 to 25, without graphing, determine whether the function is symmetric with respect to the y-axis or the origin.

23. $F(x) = x^3 - 2x$

24. $F(x) = \sqrt{x^2 + 1}$

25. $F(x) = x^3 + 8$

In Exercises 26 to 29, complete the table. Graph the shifted equation.

Unshifted Equation	Shift Graph in x-direction	Shift Graph in y-direction	Shifted Equation
26. $y = \dfrac{\lvert x \rvert}{x}$	$+2$	0	?
27. $y = \dfrac{1}{2}x^2$	-2	-4	?
28. ?	?	?	$y + 1 = \sqrt{1 - x^2}$
29. ?	?	?	$x^2 + (y - 2)^2 = 4$

In Exercises 30 and 31, graph the given function. State the domain, range, zeros, and y-intercept.

30. $F(x) = 2 - \sqrt{x + 1}$ 31. $F(x) = \lvert 1 - x^3 \rvert$

In Exercises 32 and 33, graph.

32. $F(x) = \begin{cases} 2 - x^2, & \lvert x \rvert < 1 \\ 1, & \lvert x \rvert \geq 1 \end{cases}$

33. $F(x) = \begin{cases} \lvert x \rvert, & x < 0 \\ \sqrt{x}, & 0 \leq x \leq 4 \\ x, & x > 4 \end{cases}$

34. Figure 2.63 is the graph of a certain function J. Find the graph of each of the following:
 (a) $y = \lvert J(x) \rvert$ (b) $y = -2J(x)$
 (c) $y = J(x - 2)$ (d) $y = J(x + 1) - 1$

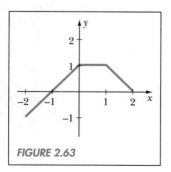

FIGURE 2.63

35. Graph $F(x) = x^2 - \lvert x \rvert$ by the method of adding graphs.

36. Let $K(x) = 4$ and $I(x) = x$. Write the given function as a combination of K or I using addition, subtraction, multiplication, or division.
 (a) $F(x) = 4 - x$
 (b) $F(x) = \dfrac{4}{x}$
 (c) $F(x) = x^2 + 4x$

37. Let $F(x) = \dfrac{1}{x} + 1$ and $G(x) = 3x - 2$. Find
 (a) $(F \circ G)(x)$ and $\mathscr{D}_{F \circ G}$
 (b) $(G \circ F)(x)$ and $\mathscr{D}_{G \circ F}$
 (c) $(F \circ F)(x)$ and $\mathscr{D}_{F \circ F}$
 (d) $(G \circ G)(x)$ and $\mathscr{D}_{G \circ G}$

In Exercises 38 to 41, graph the given function. Determine if the inverse function exists. If the inverse does exist, graph it on the same axes and find its domain and rule.

38. $F(x) = -2x + 4$

39. $F(x) = \sqrt{x + 4}$

40. $F(x) = x^2 + 1$

41. $F(x) = (x + 1)^2$, $\mathscr{D}_F = [-1, +\infty)$

42. Suppose a function F satisfies $F(1) = F(2)$. May we conclude that F cannot have an inverse? Explain.

CHAPTER 3 Polynomial and Rational Functions

3.1 LINEAR AND QUADRATIC FUNCTIONS

Consider the function L defined by the rule

$$L(x) = ax + b$$

where $a, b \in \mathbf{R}$ and $a \neq 0$. Since the equation for L, $y = ax + b$, is the equation of a straight line, we call L a **linear function**. Unless specifically restricted, the domain of L is $\mathbf{R}$.

Example 1

Graph $L(x) = \dfrac{3}{2}x - 3$. Find the range and zeros of L. Is L increasing or decreasing?

SOLUTION We recognize the equation for L, $y = \dfrac{3}{2}x - 3$, as the equation of the line with slope 3/2 and y-intercept $(0, -3)$. Using the table

x	-2	0	2
$L(x)$	-6	-3	0

we obtain the graph for L shown in Figure 3.1.

From the graph, we find that the range of L is $\mathbf{R}$, and the zero occurs at $x = 2$. Furthermore, as we move from left to right across the graph, we see that $L(x)$ is increasing. ■

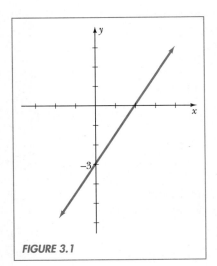

FIGURE 3.1

Given certain properties of a linear function L, suppose we wish to find its formula, $L(x)$. Since the graph of L is a line, we may write an equation for the line and then solve it for y, obtaining the desired formula $y = L(x)$. Recall that to write an equation for any line, we need a point on the line and its slope.

Example 2

Suppose $F(1) = 2$ and $F(3) = -4$ for the linear function F. Find $F(x)$.

SOLUTION $F(1)$ is the y-coordinate on the graph of F at $x = 1$. Hence $F(1) = 2$ means that $(1, 2)$ is a point on the graph of F. Similarly, $F(3) = -4$ means that $(3, -4)$ is also a point on the graph. The slope of the line passing through these two points is given by

$$m = \frac{y_2 - y_1}{x_2 - x_1} = \frac{-4 - 2}{3 - 1} = \frac{-6}{2} = -3$$

Using the point $(1, 2)$ and slope -3 in the point-slope formula $y - y_1 = m(x - x_1)$, we find the equation of the line

$$y - 2 = -3(x - 1) \qquad \text{or} \qquad y = -3x + 5$$

Therefore $F(x) = -3x + 5$. ∎

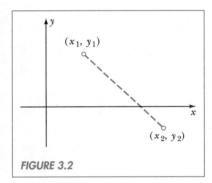

FIGURE 3.2

The problem of finding a linear function has important applications to the subject of approximating unknown functions. For example, suppose we run a laboratory experiment attempting to determine the behavior of a certain unknown function $U(x)$. Our equipment is limited, and we are able to determine the value of $U(x)$ at only two points, $x = x_1$ and $x = x_2$. Assume that we find $U(x_1) = y_1$ and $U(x_2) = y_2$. We call the points (x_1, y_1) and (x_2, y_2) **data points** for U. Now, the problem is to estimate the value of $U(x)$ for x between x_1 and x_2. The easiest thing to do is to connect the data points with a straight line segment (Figure 3.2); that is, we assume that U behaves like a linear function $L(x)$ between the data points. We call $L(x)$ the **linear approximation** for $U(x)$ between x_1 and x_2.

Let us find the formula for $L(x)$. The slope of the line segment connecting (x_1, y_1) and (x_2, y_2) is given by

$$\frac{y_2 - y_1}{x_2 - x_1}$$

Using the point-slope formula, we find the equation for $L(x)$.

$$y - y_1 = \frac{y_2 - y_1}{x_2 - x_1}(x - x_1) \qquad \text{or} \qquad y = y_1 + \frac{y_2 - y_1}{x_2 - x_1}(x - x_1)$$

This gives us the formula for $L(x)$. We conclude

The Linear Approximation Between Data Points

The linear approximation $L(x)$ for an unknown function between data points (x_1, y_1) and (x_2, y_2) is given by

$$L(x) = y_1 + \frac{y_2 - y_1}{x_2 - x_1}(x - x_1)$$

Example 3

An experiment is run that measures the temperature of an object at three instants of time. The results are

Time in seconds, t	1.5	1.75	2.0
Temperature, °F	−4.1	−0.6	1.4

(a) Find the linear approximation for the temperature as a function of time t between the first two data points.

(b) Find the linear approximation for the temperature function between the second two data points.

(c) Using these approximations, estimate the time for which the temperature is zero.

SOLUTION We see that the three data points in Figure 3.3 determine two linear approximations, one between the first two points and another between the second two points. We shall let $F(t)$ represent either one of these linear approximations for the temperature (in degrees Fahrenheit) as a function of time t (in seconds). Notice that we use the letter t for the independent variable instead of x because we are dealing with time.

(a) Using our formula for the linear approximation between the first two data points, $(1.5, -4.1)$ and $(1.75, -0.6)$, we have

$$F(t) = y_1 + \frac{y_2 - y_1}{x_2 - x_1}(t - x_1)$$

$$= -4.1 + \frac{-0.6 - (-4.1)}{1.75 - 1.5}(t - 1.5)$$

Simplifying, we obtain

$$F(t) = 14t - 25.1$$

(b) Using the second two data points, $(1.75, -0.6)$ and $(2.0, 1.4)$, we have

$$F(t) = y_1 + \frac{y_2 - y_1}{x_2 - x_1}(t - x_1)$$

$$= -0.6 + \frac{1.4 - (-0.6)}{2.0 - 1.75}(t - 1.75)$$

Simplifying, we obtain

$$F(t) = 8t - 14.6$$

(c) We wish to estimate when the temperature is zero, which means that we want to find t for which $F(t) = 0$. From Figure 3.3, we see that $F(t) = 0$ for some t in the interval $(1.75, 2.0)$. On this interval we have

$$F(t) = 8t - 14.6$$

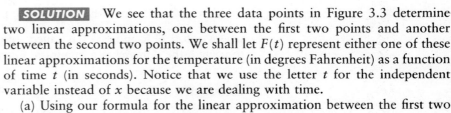

FIGURE 3.3

Setting this equal to zero and solving, we find

$$8t - 14.6 = 0$$
$$8t = 14.6$$
$$t = \frac{14.6}{8}$$
$$= 1.825$$

Thus, according to our linear approximations, the temperature will be zero when $t = 1.825$ seconds. ■

A function Q defined by the rule

$$Q(x) = ax^2 + bx + c$$

where $a, b, c \in \mathbf{R}$ and $a \neq 0$ is called a **quadratic function**. Unless specifically restricted, the domain of a quadratic function is $\mathbf{R}$.

Example 4

Given the quadratic function $Q(x) = x^2 - 4x + 3$. Find the (a) zeros and (b) y-intercept, and (c) graph. Determine the range of Q.

SOLUTION (a) The zeros occur when $Q(x) = 0$. We have

$$Q(x) = 0 \qquad \text{if and only if} \qquad x^2 - 4x + 3 = 0$$
$$(x - 1)(x - 3) = 0$$
$$x = 1 \qquad or \qquad x = 3$$

Therefore the zeros of Q are $x = 1$ and 3.

(b) The y-intercept occurs when $x = 0$. Since $Q(0) = 3$, the y-intercept is $(0, 3)$.

(c) To graph Q, we make a table as follows:

x	0	1	2	3	4
$Q(x)$	3	0	-1	0	3

The curve drawn through these points in Figure 3.4 appears to be a parabola. We can show that this is indeed the case by rewriting the equation for Q in a familiar form. We shall use the method of completing the square.

$$y = x^2 - 4x + 3$$
$$y - 3 = x^2 - 4x$$
$$y - 3 + 4 = x^2 - 4x + 4$$
$$y + 1 = (x - 2)^2$$

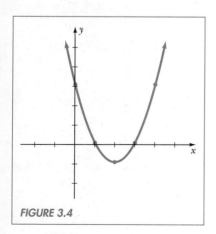

FIGURE 3.4

Now, $y + 1 = (x - 2)^2$ is the basic curve $y = x^2$ shifted $+2$ units in the x-direction and -1 unit in the y-direction. Hence the graph of Q is the parabola $y = x^2$ shifted.

From the graph, we find that the range of Q is $[-1, +\infty)$. ∎

By using the method of completing the square it is possible to show that the graph of any quadratic function $Q(x) = ax^2 + bx + c$ is the basic parabola $y = ax^2$ shifted. Beginning with the equation for Q,

$$y = ax^2 + bx + c$$

$$y - c = ax^2 + bx$$

$$= a\left(x^2 + \frac{b}{a}x\right)$$

$$y - c + \frac{b^2}{4a} = a\left(x^2 + \frac{b}{a}x + \frac{b^2}{4a^2}\right)$$

$$y - \left(c - \frac{b^2}{4a}\right) = a\left(x + \frac{b}{2a}\right)^2$$

This is the equation $y = ax^2$ shifted $-b/2a$ units in the x-direction and $c - b^2/4a$ units in the y-direction. Note that $c - b^2/4a = Q(-b/2a)$. Thus we may write the equation for Q in the form

$$y - k = a(x - h)^2$$

where $h = -b/2a$ and $k = Q(-b/2a)$.

Figure 3.5(a) is a typical graph for Q when $a > 0$. In this case the graph is a shifted upward curving parabola. Figure 3.5(b) illustrates the situation when $a < 0$.

We now determine some general properties for the quadratic function $Q(x) = ax^2 + bx + c$.

Vertex. The **vertex** of the basic parabola $y = ax^2$ is defined to be the origin. It follows that the vertex for the graph of Q is $(-b/2a, Q(-b/2a))$. Note that if $a > 0$, then the vertex is the lowest point on the graph of Q. We call this point a **minimum point**. If $a < 0$, then the vertex is the highest point on the curve, or the **maximum point**.

Symmetry. The basic parabola $y = ax^2$ is symmetric with respect to the y-axis because the equation remains unchanged if x is replaced with $-x$. Therefore the graph of Q must be symmetric with respect to the vertical line passing through its vertex, $x = -b/2a$. Each point on one side of this vertical line must have a symmetric partner on the other side.

Zeros. The zeros for any quadratic function can always be found by using the quadratic formula. We have

$$Q(x) = ax^2 + bx + c = 0 \qquad \text{if and only if} \qquad x = \frac{-b \pm \sqrt{b^2 - 4ac}}{2a}$$

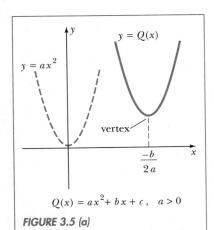

$Q(x) = ax^2 + bx + c, \quad a > 0$

FIGURE 3.5 (a)

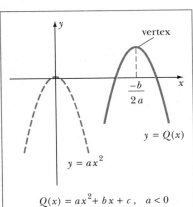

$Q(x) = ax^2 + bx + c, \quad a < 0$

FIGURE 3.5 (b)

It follows that

$$b^2 - 4ac < 0 \qquad \text{implies} \qquad Q \text{ has no zeros}$$

$$b^2 - 4ac = 0 \qquad \text{implies} \qquad Q \text{ has exactly one zero}$$

$$b^2 - 4ac > 0 \qquad \text{implies} \qquad Q \text{ has exactly two zeros}$$

These situations are illustrated in Figure 3.6(a), (b), and (c) in the case for $a > 0$. Let us summarize our discussion.

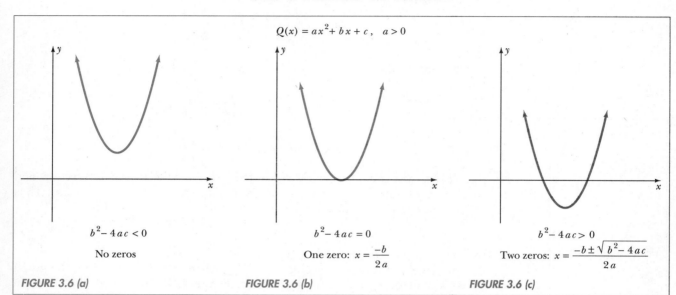

$$Q(x) = ax^2 + bx + c, \quad a > 0$$

$b^2 - 4ac < 0$

No zeros

FIGURE 3.6 (a)

$b^2 - 4ac = 0$

One zero: $x = \dfrac{-b}{2a}$

FIGURE 3.6 (b)

$b^2 - 4ac > 0$

Two zeros: $x = \dfrac{-b \pm \sqrt{b^2 - 4ac}}{2a}$

FIGURE 3.6 (c)

Properties of Quadratic Functions

Function: $\quad Q(x) = ax^2 + bx + c \qquad a \neq 0$

Equation: $\quad y = ax^2 + bx + c$ is $y = ax^2$ shifted,

$\qquad\qquad y - k = a(x - h)^2$

$\qquad\qquad$ where $h = -b/2a$ and $k = Q(-b/2a)$.

Vertex: $\quad \left(\dfrac{-b}{2a}, Q\left(\dfrac{-b}{2a}\right) \right)$

Zeros: $\quad x = \dfrac{-b \pm \sqrt{b^2 - 4ac}}{2a} \qquad$ provided that $b^2 - 4ac \geq 0$

Symmetry: with respect to the vertical line $x = \dfrac{-b}{2a}$.

Example 5

Given the quadratic function $Q(x) = -\dfrac{1}{2}x^2 - 2x + 1$. Find the (a) vertex, (b) zeros, (c) y-intercept, and (d) line of symmetry. Graph Q. Determine the range, where Q is increasing, and where it is decreasing.

SOLUTION (a) We have $Q(x) = ax^2 + bx + c$, where $a = -1/2$, $b = -2$, and $c = 1$. Therefore the vertex has coordinates

$$x = \frac{-b}{2a} = \frac{-(-2)}{2\left(-\dfrac{1}{2}\right)} = \frac{2}{-1} = -2$$

$$y = Q\left(\frac{-b}{2a}\right) = -\frac{1}{2}(-2)^2 - 2(-2) + 1 = -2 + 4 + 1 = 3$$

We write the coordinates of the vertex as an ordered pair, $(-2, 3)$.

(b) We must solve the equation $Q(x) = 0$ for x.

$$Q(x) = 0$$

$$-\frac{1}{2}x^2 - 2x + 1 = 0 \qquad \text{Multiply by } -2.$$

$$x^2 + 4x - 2 = 0$$

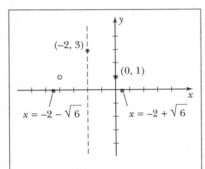

$x = -2 - \sqrt{6}$ $\quad$ $x = -2 + \sqrt{6}$

FIGURE 3.7 (a)

Applying the quadratic formula to this last equation, we obtain $x = -2 \pm \sqrt{6}$ for the zeros of Q. (See Example 5, Section 1.2.)

(c) $Q(0) = 1$, so the y-intercept is $(0, 1)$.

(d) The line of symmetry is the vertical line through the vertex, $x = -2$.

We know that $Q(x)$ is just $y = -\dfrac{1}{2}x^2$ shifted to the new vertex $(-2, 3)$.

Therefore the graph of Q will be a downward curving parabola. Since we know the general shape of Q, we need only a few points to determine the graph. We start by graphing the vertex, zeros, and y-intercept, as shown in Figure 3.7(a). Using the symmetry of the curve, we see that the y-intercept $(0, 1)$ has a symmetric partner at $(-4, 1)$. This gives us a total of five specific points on the graph of Q. Now we can draw the parabola passing through these points, as in Figure 3.7(b).

From the graph of Q, we see that the range is $(-\infty, 3]$. Q is *increasing* for $x \in (-\infty, -2]$; Q is *decreasing* for $x \in [-2, +\infty)$. ∎

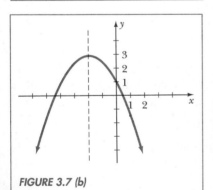

FIGURE 3.7 (b)

Given certain properties of a quadratic function Q, suppose we wish to find its formula, $Q(x)$. We have two ways to write the equation for Q:

General form: $\quad y = ax^2 + bx + c$

Standard form: $\quad y - k = a(x - h)^2$

If we know the location of the vertex (h, k), then we should use the standard form for the equation of Q. In this case, only the coefficient a is needed to determine $Q(x)$.

Example 6

Find the quadratic function Q whose graph passes through the point $(-1, 1)$ and has vertex $(-4, 7)$.

SOLUTION   Since we are given the vertex $(-4, 7)$ on the graph of Q, we shall use the standard form for the equation.

$$y - k = a(x - h)^2$$
$$y - 7 = a(x + 4)^2$$

To determine a, we use the fact that the graph passes through the point $(-1, 1)$, which means that $(-1, 1)$ is a solution of the equation for Q. Thus, substituting $x = -1$ and $y = 1$ into the equation, we have

$$1 - 7 = a(-1 + 4)^2$$
$$-6 = 9a$$
$$-\frac{2}{3} = a$$

Consequently, the equation for Q is $y - 7 = \dfrac{-2}{3}(x + 4)^2$. Solving for y gives us the formula for $Q(x)$.

$$Q(x) = -\frac{2}{3}(x + 4)^2 + 7 \ \blacksquare$$

When an inequality involves a quadratic function, it may be possible to find the solution quite easily from the graph of the function. For example, suppose we wish to solve

$$x^2 + x - 2 > 0$$

One way to find the solution is to factor the left side and apply the method of analysis of signs presented in Section 1.4. However, an alternate approach is to reinterpret the problem as follows:

$$\text{Solve} \quad Q(x) > 0 \quad \text{where} \quad Q(x) = x^2 + x - 2$$

In terms of the graph of Q, we wish to find all x for which $y = Q(x)$ is positive. In other words, we want to know where the graph of Q is above the x-axis. We shall make a quick sketch of the graph using only two pieces of data: the zeros of Q and the knowledge that the graph is a parabola curving upward.

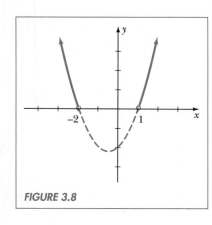

FIGURE 3.8

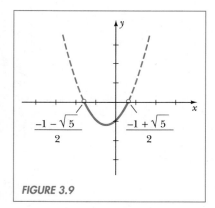

FIGURE 3.9

$Q(x) = x^2 + x - 2$ curves upward (since it is $y = x^2$ shifted)

Zeros: $\qquad x^2 + x - 2 = 0$

$$(x + 2)(x - 1) = 0$$

$$x = -2 \quad or \quad x = 1$$

Thus we have the graph shown in Figure 3.8. From the graph, we find that $Q(x) > 0$ (that is, the y-coordinates are positive) if and only if x is in the set $(-\infty, -2) \cup (1, +\infty)$.

Example 7

Solve $x^2 + x < 1$.

SOLUTION Begin by rewriting the inequality so that zero is on one side: $x^2 + x - 1 < 0$. We interpret the problem as follows.
Solve

$$Q(x) < 0 \qquad \text{where} \qquad Q(x) = x^2 + x - 1$$

Now, sketch the graph of Q.

$$Q(x) = x^2 + x - 1 \quad \text{curves upward}$$

Zeros: $\quad x^2 + x - 1 = 0$

$$x = \frac{-1 \pm \sqrt{5}}{2} \quad \text{(by the quadratic formula)}$$

The graph is shown in Figure 3.9. We conclude that $y = Q(x) < 0$ if and only if x is in the interval $\left(\dfrac{-1 - \sqrt{5}}{2}, \dfrac{-1 + \sqrt{5}}{2} \right)$. ∎

EXERCISES 3.1

In Exercises 1 to 4, (a) find the zero and y-intercept of L; (b) graph and determine whether L is increasing or decreasing.

1. $L(x) = 4x - 7$

2. $L(x) = -2x + 5$

3. $L(x) = -\dfrac{5}{3}x$

4. $L(x) = x + \dfrac{3}{2}$

In Exercises 5 to 8, find the linear function F with the given properties.

5. $F(1) = 3, F(0) = -2$

6. $F\left(-\dfrac{1}{2} \right) = \dfrac{5}{3}, F(4) = 3$

7. (a) $F(2) = -1, F(2x) = 2F(x)$
   (b) $F^{-1}(x) = (x + 6)/7$
   (c) $F(0) = 4, F(x + h) - F(x) = 3h$

8. (a) $F(1) = 4, F(x_1 + x_2) = F(x_1) + F(x_2)$ for $x_1, x_2 \in \mathbf{R}$
   (b) $F^{-1}(x) = 5 - 3x$
   (c) $F(1) = 2, F(x + h) - F(x) = -5h$

In Exercises 9 and 10, find the linear function $L(x)$ whose graph is shown.

9.

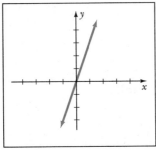

10.

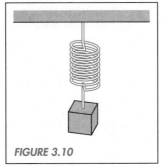

11. The force F exerted by a spring stretched to length x is a linear function of x, $F(x)$. Use the following information to find $F(x)$: the spring exerts a force of 2 pounds when stretched to 15 inches [$F(15) = 2$] and 10 pounds when stretched to 18 inches [$F(18) = 10$]. See Figure 3.10. What length corresponds to zero force?

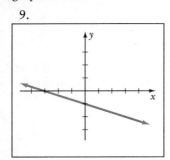

FIGURE 3.10

12. Repeat Exercise 11, assuming that the spring exerts a force of 5 pounds when stretched to 16 inches and 15 pounds when stretched to 20 inches.

In Exercises 13 and 14, assume that an unknown function has the given data points. Find the linear approximation L between the data points. Find the value of x when $L(x) = 0$.

13. $(1.6, -4.6)$, $(5, 2.2)$ 14. $(.5, 3.3)$, $(.75, -0.7)$

15. An experiment is run that measures the temperature of an object at three instants of time. The results are

t, time in seconds	0	.4	.8
y, temperature in °F	96.8	94.1	89.1

(a) Find the linear approximation for the temperature as a function of time t between the first two data points.
(b) Find the linear approximation for the temperature function between the second two data points.
(c) Using linear approximations, estimate the time at which the temperature is 90°F.

16. An experiment is run by placing a detector that counts the number of photons (light particles) arriving from a source some distance away along the x-axis. Let $U(x)$ be the number of counts per second the detector registers when it is positioned at x. (See Figure 3.11.) Suppose U has the following data points:

Position of detector, x	0	1	2	3
Counts per second, $U(x)$	8	12	18	15

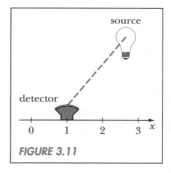

FIGURE 3.11

(a) Find linear approximations for U between each pair of consecutive data points. Graph.
(b) Estimate the values of x for which $U(x) = 16$.
(c) Estimate the value of $U(1/2.)$

In Exercises 17 and 18, for the points P and V given, suppose the graph of a quadratic function $Q(x)$ passes through P and has vertex V. Find the symmetric partner for P on the graph of $Q(x)$.

17. (a) $P(2, 3)$, $V(-1, -2)$
    (b) $P(-1, 2)$, $V(1, 5)$

18. (a) $P(2, 1)$, $V(-3, 3)$
    (b) $P(-4, -5)$, $V(2, -10)$

In Exercises 19 to 28, for each quadratic function Q, find the vertex, zeros, y-intercept, and the line of symmetry. Graph Q. Determine the range, where Q is increasing, and where it is decreasing.

19. $Q(x) = \frac{1}{2}x^2 + 2x + 2$

20. $Q(x) = -x^2 + 2x - 1$

21. $Q(x) = -4x^2 + 12x - 8$

22. $Q(x) = \frac{2}{3}x^2 - 4x$

23. $Q(x) = x^2 + 2x - 2$

24. $Q(x) = -\frac{1}{2}x^2 + 2x - 1$

25. $Q(x) = -\frac{1}{8}x^2 - 1$

26. $Q(x) = \frac{3}{4}x^2 + 3x + 6$

27. $Q(x) = -(x - 4)^2 + 4$

28. $Q(x) = (x + 3)^2 - 3$

In Exercises 29 to 32, find the quadratic function $Q(x)$ whose graph has vertex V and passes through the point P.

29. $V(1/2, 4)$, $P(5, 1)$ 30. $V(-3, -5)$, $P(0, 0)$

31. $V(2, 1)$, $P(-3, 2)$ 32. $V(-2, 1)$, $P(1, -3)$

In Exercises 33 to 35, find the quadratic function $Q(x)$ with the given properties.

33. Q has zeros at $x = -1$ and $x = 3$ and range $[-6, +\infty)$. (*Hint:* the x-coordinate of the vertex must be 1.)

34. Q has zeros at $x = 0$ and $x = 5$ and range $(-\infty, 2]$.

35. The graph of Q passes through $(1, 1)$ and $(7, 1)$ and the range is $[0, +\infty)$.

In Exercises 36 to 38, the graph of a quadratic function Q is shown. Find a formula for $Q(x)$.

36.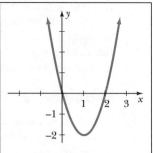

37.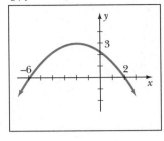

38.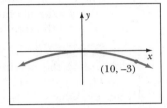

In Exercises 39 to 46, use the method presented in Example 7 to solve each inequality.

39. $x^2 + 2x - 3 > 0$ 40. $x^2 + 10x + 23 < 0$

41. $1 + 4x > x^2$ 42. $x^2 - 8x + 10 > 0$

43. $1 > x(1 - x)$ 44. $x^2 + 3 > 2x$

45. $x^2 + 8x + 19 < 0$ 46. $x^2 + x + 2 < 1$

In Exercises 47 and 48, find the domain of the function.

47. $F(x) = \sqrt{x^2 - 7}$ 48. $F(x) = \sqrt{3 - x^2}$

In Exercises 49 to 52, solve each inequality. (*Hint:* use the method presented in Example 7 and the analysis of signs.)

49. $(x + 4)(x^2 - x - 1) > 0$

50. $x(x^2 + 2x - 1) < 0$

51. $\dfrac{x + 2}{x^2 - 2} \le 0$

52. $\dfrac{x}{2x^2 - 2x - 1} \le 0$

If a ball is thrown into the air, we might wonder how high it will go and how long it will be in the air before hitting the ground. Under ideal conditions, classical physics tells us that the height of an object thrown into the air near the earth's surface will be a quadratic function of time. Given this function, we can answer questions about maximum height and time of flight.

Example 1

A ball is thrown up into the air from a point 21 feet above the ground. The height of the ball, H, measured in feet above the ground is given by the function

$$H(t) = -16t^2 + 80t + 21$$

where t is the time in seconds after releasing the ball.
(a) Find the maximum height the ball reaches.
(b) Find the total flight time before the ball hits the ground.

SOLUTION (a) Finding the maximum height the ball reaches is equivalent to finding the maximum value of $H(t)$. Figure 3.12 is a sketch of the graph of H, a parabola. Notice that the maximum value of $H(t)$ occurs at the vertex, where

$$t = -\frac{b}{2a} = -\frac{80}{2(-16)} = \frac{5}{2}$$

Thus the maximum height the ball reaches is $H(5/2)$,

$$H(5/2) = -16(5/2)^2 + 80(5/2) + 21$$
$$= -4(25) + 40(5) + 21$$
$$= 121 \text{ feet}$$

(b) The ball will hit the ground when the height H is zero. Therefore we want to find t so that $H(t) = 0$.

$$H(t) = 0$$
$$-16t^2 + 80t + 21 = 0$$
$$16t^2 - 80t - 21 = 0$$
$$(4t - 21)(4t + 1) = 0$$

$$4t - 21 = 0 \quad or \quad 4t + 1 = 0$$

$$t = \frac{21}{4} \qquad \qquad t = -\frac{1}{4}$$

We reject the negative value for t. Thus the ball will hit the ground after $21/4$ or $5\frac{1}{4}$ seconds. ∎

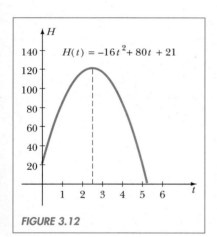

FIGURE 3.12

H

$H(t) = -16t^2 + 80t + 21$

The problem of finding the maximum height of the ball in Example 1 was relatively easy to solve for two reasons: (1) we knew the function for the height, and (2) we knew how to analyze this function. These two ingredients are essential for solving problems concerning the maximum value of a particular quantity. In Example 2 we solve another problem of this type, only this time we have to find the required function on our own.

Example 2

The sum of two real numbers is 10. What is the maximum possible value for the product of these two numbers? Prove your answer.

SOLUTION As a preliminary experiment, let us look at the products of a few pairs of numbers whose sum is 10.

Pair of Numbers	Product
5, 5	25
$5\frac{1}{2}$, $4\frac{1}{2}$	$24\frac{3}{4}$
6, 4	24
8, 2	16
11, -1	-11

We might assert (correctly) that the maximum product is 25 corresponding to the pair of numbers 5 and 5. Now we must prove this.

Our first step is to define specifically the quantity whose maximum value we seek. We let Q represent the quantity and define it in words:

Let Q = product of two real numbers.

Next, we need to write a formula for Q. Letting x and y represent two real numbers, we may write

$$Q = xy$$

Now, we must write Q as a function of just one variable. We know that the sum of x and y must be 10, $x + y = 10$, so $y = 10 - x$. Substituting this into the formula for Q, we have

$$Q = x(10 - x)$$

Therefore we find that $Q(x) = -x^2 + 10x$; this is a quadratic function that we know how to analyze. The graph is the parabola sketched in Figure 3.13. We see that the maximum value of $Q(x)$ occurs when

$$x = \frac{-b}{2a} = \frac{-10}{2(-1)} = 5$$

Thus $Q(5) = -(5)^2 + 10(5) = 25$ is the maximum value of Q, the product of two numbers whose sum is 10. ■

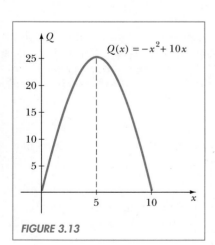

FIGURE 3.13

$Q(x) = -x^2 + 10x$

Examples 1 and 2 represent what we call **maximum–minimum problems**. In this type of problem, we wish to find the maximum or minimum value of a particular quantity given certain conditions. To solve the problem, we write the quantity under consideration as a function of one variable and then analyze this function. Let us outline the steps involved.

Steps for Solving Maximum–Minimum Problems

Step 1   Define in a word or phrase the quantity whose maximum or minimum is desired:

$$Q = \text{(quantity)}$$

Step 2   Write a formula for Q. When possible, it helps to draw and label a picture.

Step 3   Rewrite the formula for Q so that it becomes a function of just one variable, $Q(x)$.

Step 4   Find the maximum or minimum value of Q by sketching the graph of $Q(x)$.

Ideally, the function found in step 3 will be a quadratic function so that the analysis in step 4 can be carried out. If the function is not quadratic, then the solution may require techniques from calculus. We shall not consider such problems here.

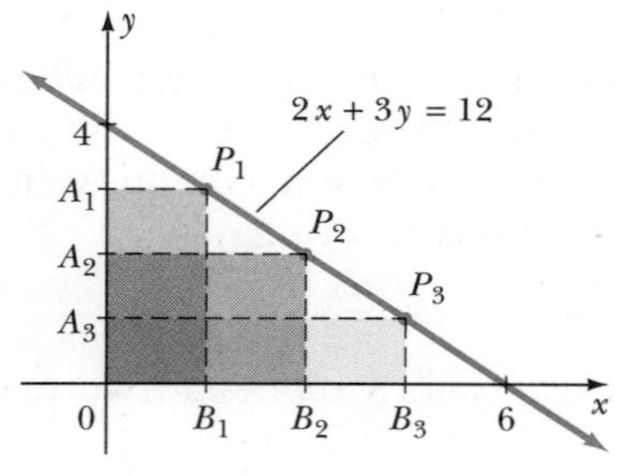

Example 3

Let $P(x, y)$ be any point on the line $2x + 3y = 12$ in the first quadrant. Consider the rectangle constructed with vertices at P, along the x- and y-axes, and at the origin. Find the maximum area possible for such a rectangle. What are the coordinates of the point P corresponding to the rectangle of maximum area?

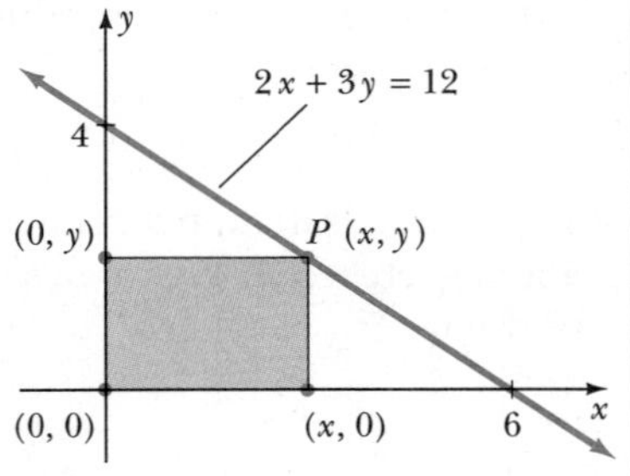

SOLUTION   Figure 3.14(a) shows three possible positions for P at P_1, P_2, and P_3 and the corresponding rectangles $A_1P_1B_1O$, $A_2P_2B_2O$, and $A_3P_3B_3O$. We see that each position of P along the line $2x + 3y = 12$ in the first quadrant determines a corresponding rectangle. Our task is to find the largest area among all the possible rectangles. We use the four-step procedure outlined above.

Step 1.   *Define the quantity whose maximum is desired.* We want to find the maximum area of a rectangle. Therefore

$$Q = \text{area of a rectangle}$$

Step 2.   *Write a formula for Q.* Figure 3.14(b) illustrates a typical rectangle corresponding to the point $P(x, y)$. The base of this rectangle is x units, and

FIGURE 3.14 (a)

FIGURE 3.14 (b)

the height is y units. Note that x and y are positive since $P(x, y)$ is in the first quadrant. It follows that

$$Q = xy$$

It is important to note that the symbols x and y appearing here represent the coordinates of P.

Step 3. *Rewrite Q as a function of one variable.* Since $P(x, y)$ is on the line $2x + 3y = 12$, the coordinates of P must satisfy the equation of the line. We have

$$2x + 3y = 12 \quad \text{implies} \quad y = \frac{12 - 2x}{3}$$

Substituting this into the formula for Q,

$$Q = xy = x\left(\frac{12 - 2x}{3}\right)$$

Simplifying, we have

$$Q(x) = -\frac{2}{3}x^2 + 4x$$

Step 4. *Find the maximum value of Q(x).* Figure 3.15 is a sketch of the graph of Q, a downward curving parabola. $Q(x)$ has a maximum value occurring at the vertex, where

$$x = \frac{-b}{2a} = \frac{-4}{2\left(-\frac{2}{3}\right)} = 3$$

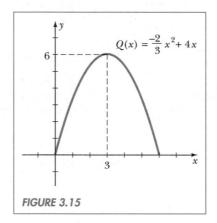

FIGURE 3.15

Therefore $Q(3) = -\frac{2}{3}(3)^2 + 4(3) = 6$ is the maximum area possible.

From our analysis in step 4, we know that the point P corresponding to the rectangle with maximum area must have x-coordinate 3. Since P is on the line $2x + 3y = 12$, we also know that its y-coordinate must satisfy

$$y = \frac{12 - 2x}{3}$$

Substituting $x = 3$ into this equation yields $y = 2$. Therefore the coordinates of P corresponding to the rectangle with maximum area are $(3, 2)$. ∎

All our examples so far have asked for maximum values of certain quantities. We conclude this section with a problem concerned with finding a minimum.

Example 4

Find the minimum distance between any point $P(x, y)$ on the curve $y = \sqrt{x}$ and the point $A(4, 0)$. What are the coordinates of the point P corresponding to this minimum distance?

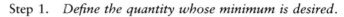 Figure 3.16(a) indicates three possible positions for P along the curve $y = \sqrt{x}$ and the corresponding distances to A, $P_1 A$, $P_2 A$, and $P_3 A$. Among all possible choices for P, we are looking for the one whose distance to A is minimum. Once again, we follow the four-step procedure for solving maximum–minimum problems.

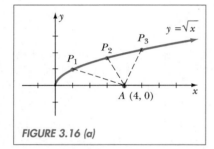

FIGURE 3.16 (a)

Step 1. Define the quantity whose minimum is desired.

$$Q = \text{distance } PA$$

Step 2. Write a formula for Q. We use the distance formula between $P(x, y)$ and $A(4, 0)$ [Figure 3.16(b)]:

$$Q = \sqrt{(x - 4)^2 + y^2}$$

Note that the symbols x and y appearing here represent the coordinates of P.

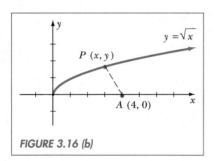

FIGURE 3.16 (b)

Step 3. Rewrite Q as a function of one variable. Since P is on the curve $y = \sqrt{x}$, the coordinates of P must satisfy this equation. Therefore we may substitute $y = \sqrt{x}$ into the formula for Q.

$$Q(x) = \sqrt{(x - 4)^2 + (\sqrt{x})^2}$$

Simplifying, we obtain

$$Q(x) = \sqrt{x^2 - 7x + 16}$$

Step 4. Find the minimum value of Q(x). $Q(x)$ is not a quadratic function because of the square root, so, how do we find its minimum? We shall use the following fact: If $F(x) \geq 0$, then the minimum value of $\sqrt{F(x)}$ will occur when $F(x)$ is minimum. Thus, to find the minimum value of $Q(x) = \sqrt{x^2 - 7x + 16}$, we look for the minimum value of $F(x) = x^2 - 7x + 16$. By sketching the graph of F in Figure 3.17, we see it has a minimum when

$$x = \frac{-b}{2a} = \frac{-(-7)}{2(1)} = \frac{7}{2}$$

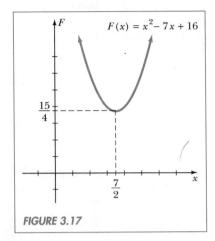

FIGURE 3.17

Therefore the minimum value of $F(x)$ is given by $F(7/2) = (7/2)^2 - 7(7/2) + 16 = 15/4$. It follows that the minimum value of $Q(x)$, that is, the minimum distance between any point P on the curve $y = \sqrt{x}$ and the point A, is $Q(7/2) = \sqrt{F(7/2)} = \sqrt{15/4} = \sqrt{15}/2$.

From our analysis in step 4 we know that the point P with minimum distance to A has x-coordinate 7/2. Since P is on the curve $y = \sqrt{x}$, its y-coordinate must be $y = \sqrt{7/2}$. Therefore the point P with the minimum distance to A has coordinates $(7/2, \sqrt{7/2})$. ∎

EXERCISES 3.2

In Exercises 1 to 4, assume that the given function $H(t)$ represents the height in feet above the ground of a ball t seconds after the ball has been thrown into the air. Find the (a) maximum height the ball reaches and (b) total flight time before the ball hits the ground.

1. $H(t) = -16t^2 + 40t + 8$

2. $H(t) = -16t^2 + 60t$

3. $H(t) = -16t^2 + 96t$

4. $H(t) = -16t^2 + 72t + 19$

5. The sum of two numbers is 15. What is the maximum possible value for the product of these two numbers? Prove your answer.

6. The sum of twice a number with another number is 10. What is the maximum possible value for the product of these two numbers? Prove your answer.

7. The sum of two numbers is 20. What is the minimum possible value for the sum of their squares? Prove your answer.

8. The difference of two numbers is 12. What is the minimum possible value for the product of these two numbers? Prove your answer.

In Exercises 9 to 13, the equation of a line is given. Let $P(x, y)$ be any point on the line in the first quadrant. Consider the rectangle constructed with vertices at P, along the x- and y-axes, and at the origin. Find the maximum area possible for such a rectangle. Also, find the coordinates of P corresponding to the rectangle of maximum area.

9. $2x + 5y = 10$ 10. $x + 2y = 4$

11. $16x + 5y = 80$ 12. $3x + 2y = 18$

13. $\dfrac{x}{a} + \dfrac{y}{b} = 1$, where a and b are positive constants. (Your answers will be in terms of a and b.)

In Exercises 14 to 16, the equation of a line is given. Let $P(x, y)$ be any point on the line in the first quadrant. Consider the triangle determined by the vertices P, $A(x, 0)$, and $O(0, 0)$. Find the maximum area possible for such a triangle. Also, find the coordinates of P corresponding to the triangle of maximum area.

14. $2x + y = 8$

15. $2x + 3y = 24$

16. $ax + by = ab$, where a and b are positive constants. (Your answers will be in terms of a and b.)

17. Let $P(x, y)$ be any point on the curve $y = 16 - x^2$ in the first quadrant. Consider the rectangle constructed with vertices at P, along the positive x- and y-axes, and at the origin. Find the maximum perimeter possible for such a rectangle.

18. Repeat Exercise 17 for the curve $y = 9 - x^2$.

19. A rectangle is to be constructed with its base on the x-axis and its top two vertices on the curve $y = 10 - x^2$ (in quadrants I and II). Find the maximum perimeter possible for such a rectangle. Find the coordinates of the top two vertices corresponding to this rectangle.

20. Repeat Exercise 19 for the curve $y = 12 - x^2$.

21. Find the minimum distance between any point $P(x, y)$ on the curve $y = \sqrt{x}$ and the point $A(19/2, 0)$. What are the coordinates of the point P corresponding to this minimum distance?

22. Find the point $P(x, y)$ on the curve $y = \sqrt{x + 3}$ that is closest to the point $A(3/2, 0)$.

23. Find the minimum distance between any point $P(x, y)$ on the curve $y = 3 - x^2$ and the origin. What are the coordinates of the points P corresponding to this minimum distance? (*Hint:* write the distance between P and the origin as a function of y.)

24. Find the minimum distance between any point $P(x, y)$ on the curve $y = x^2$ and the point $A(0, 2)$. What are the coordinates of the points P corresponding to this minimum distance? (*Hint:* write the distance between P and A as a function of y.)

25. A gardener wishes to use 200 feet of chain link fencing to enclose a rectangular area. The fencing will be used for only three sides of the rectangle, the fourth side being established by an existing wall. Find the maximum area possible for the rectangle. What are the dimensions of the rectangle with maximum area?

26. Repeat Exercise 25 without the wall so that the gardener must use the fencing for all four sides of the rectangle.

Exercises 27 and 28 refer to the following definition. Suppose F and G are functions and $F(x) \geq G(x)$. We define the **vertical distance** between F and G at x as $F(x) - G(x)$. This measures the length of the vertical line segment from $G(x)$ to $F(x)$. See Figure 3.18.

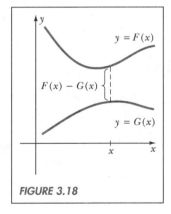

FIGURE 3.18

27. Graph $F(x) = -\dfrac{1}{3}(x - 2)^2 + 4$ and $G(x) = x^2$ on the same axes. Over what interval is $F(x) \geq G(x)$?

Find the maximum value of the vertical distance between F and G over this interval.

28. Graph $F(x) = x^2 + 2$ and $G(x) = -\dfrac{1}{2}(x - 2)^2 + 2$ on the same axes. Over what interval is $F(x) \geq G(x)$? Find the minimum value of the vertical distance between F and G over this interval.

29. A wire 4 feet in length is to be cut into two pieces. One piece will be bent into the shape of a square, the other piece into the shape of a rectangle whose length is twice its width. Where should the wire be cut so that the sum of the areas of the square and rectangle is minimum?

30. Consider the line $y = mx + b$, where m and b are real constants. Let P be the point on this line that is closest to the origin.
    (a) Use the techniques of this section to find the coordinates of P. $\left[\textit{Answer:} \left(\dfrac{-mb}{m^2 + 1}, \dfrac{b}{m^2 + 1} \right). \right]$
    (b) Show that the distance $OP = b/\sqrt{m^2 + 1}$.
    (c) Assuming that $m \neq 0$, show that the slope of OP is $-1/m$.

31. Let $P(x, 0)$ be a point on the x-axis such that the sum of the *squares* of the distances between P and the points A, B, and C is a minimum.
    (a) If the points have coordinates $A(2, 1)$, $B(4, 3)$, and $C(5, -6)$, show that $x = 11/3$.
    (b) If the points have coordinates $A(x_1, a)$, $B(x_2, b)$, and $C(x_3, c)$, show that $x = (x_1 + x_2 + x_3)/3$.
    (c) If instead of three points there are n points with x-coordinates $x_1, x_2, \ldots, x_n$, what do you think the value of x is?

32. A rectangle with perimeter 12 is to be constructed with a circle circumscribed about it. Find the dimensions for such a rectangle so that the area between the rectangle and the circle is a minimum.

3.3 GENERAL POLYNOMIAL FUNCTIONS

The linear and quadratic functions introduced in Section 3.1 are examples of a general category of functions called **polynomials**. A polynomial function assigns to any real number x a sum of terms, each term consisting of the product of a constant and a nonnegative integer power of x. This type of rule makes sense for any real number x so that the domain of a polynomial function is the set of all real numbers, **R**.

> **Definition of a Polynomial Function**
> Let n be a nonnegative integer and $a_0, a_1, a_2, \ldots, a_n$ real constants with $a_n \neq 0$. Any function F with rule
>
> $$F(x) = a_n x^n + a_{n-1}x^{n-1} + \cdots + a_2 x^2 + a_1 x + a_0$$
>
> is called a **polynomial function of degree n**. The constants $a_0, a_1, a_2, \ldots, a_n$ are called the **coefficients of F**. The **leading coefficient** is a_n, the **constant term** is a_0.

Following are typical examples of polynomial functions and their corresponding degree:

Function	Degree
$A(x) = -3x^4 + \frac{1}{2}x^2 + \sqrt{3}\,x + 1$	4
$B(x) = x(x + 1)(x + 2)$	3
$Q(x) = x^2 - 2x + 3$	2
$L(x) = 4x - 5$	1
$C(x) = 6$	0

Remarks

1. $A(x)$ has the form of a degree 4 polynomial function, $a_4x^4 + a_3x^3 + a_2x^2 + a_1x + a_0$, with leading coefficient -3 and constant term 1. Notice that the coefficient of x^3 in this polynomial is zero.
2. When the rule for function B is multiplied out, we get $B(x) = x^3 + 3x^2 + 2x$, a degree 3 polynomial with leading coefficient 1 and constant term zero.
3. Any quadratic function is a polynomial function of degree 2.
4. Any linear function is a polynomial function of degree 1.
5. Any nonzero constant function is a polynomial function of degree 0.

Now we give two examples of functions that do not qualify as polynomials. First, consider

$$F(x) = 1 + \frac{1}{x}$$

The rule for F contains a negative power of x, x^{-1}, which violates the definition of a polynomial function. Next, consider

$$G(x) = \sqrt{x^2 + x + 1}$$

The square root appearing in $G(x)$ prevents us from rewriting this formula as a sum of nonnegative integer powers of x. Hence G is not a polynomial function.

Next, let us turn our attention to the problem of graphing polynomial functions. We begin with two geometric properties that serve as general guidelines.

Geometric Properties of Polynomial Functions

Let F be a polynomial function.

1. The graph of F is smooth and continuous.
2. For $|x|$ sufficiently large, $F(x)$ behaves like its highest-degree term.

Discussion of Property 1. The first property says that the graph of any polynomial function must be *smooth* and *continuous*. If we imagine that the graph represents a road we are driving along, then the drive is not interrupted by any holes or breaks in the roadway; the road is continuous. Furthermore, there are no sharp corners to negotiate; the road is smooth. To get an idea of how such a curve might look, consider Figure 3.19(a), which is the graph of a degree 5 polynomial function. This graph is continuous and smooth, consisting of a series of mountains and valleys. This is typical of a high-degree polynomial. On the other hand, Figures 3.19(b) and (c) illustrate curves that cannot represent polynomial functions. The graph in Figure 3.19(b) exhibits a break at $x = x_0$; hence it is not continuous. The curve in Figure 3.19(c) possesses a sharp corner at $x = x_0$, and thus it is not smooth. The fact that the graph of any polynomial function must be continuous and smooth requires calculus techniques to prove; therefore we accept its validity without proof.

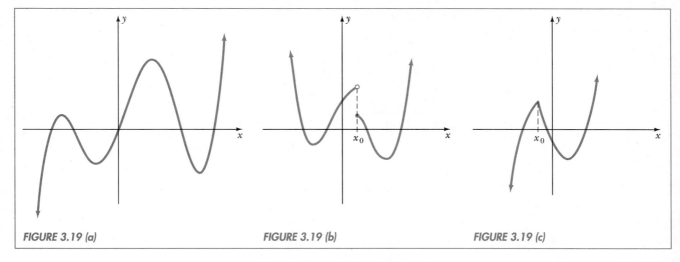

FIGURE 3.19 (a)   FIGURE 3.19 (b)   FIGURE 3.19 (c)

Discussion of Property 2. The second geometric property of polynomial functions tells us about the behavior of the function when $|x|$ is large. Suppose

$$F(x) = a_n x^n + a_{n-1} x^{n-1} + \cdots + a_1 x + a_0$$

Assuming $|x|$ is large (and hence $x \neq 0$), we may factor out x^n as follows:

$$F(x) = x^n \left(a_n + \frac{a_{n-1}}{x} + \frac{a_{n-2}}{x^2} + \cdots + \frac{a_1}{x^{n-1}} + \frac{a_0}{x^n} \right)$$

Now, if $|x|$ is very large, then each term in the sum

$$\frac{a_{n-1}}{x} + \frac{a_{n-2}}{x^2} + \cdots + \frac{a_1}{x^{n-1}} + \frac{a_0}{x^n}$$

will be so small that the sum itself is very small. Thus

$$F(x) = x^n(a_n + (\text{very small number}))$$

Therefore we may make the approximation $F(x) \approx a_n x^n$. In other words, when $|x|$ is sufficiently large, $F(x)$ behaves like its highest-degree term.

$$|x| \text{ sufficiently large} \quad \text{implies} \quad F(x) \approx a_n x^n$$

Geometric property 2 of polynomial functions lets us determine the behavior of their graphs far away from the origin in either the positive or negative x-direction. To make this discussion easier, we write $x \to +\infty$, read "x tends toward positive infinity," to mean x increases, taking on positive values of arbitrarily large magnitude, such as 10^6, 10^{10}, 10^{100}, etc. In using this terminology, we understand that x will never reach $+\infty$ (or even get close to it) because the positive x-axis never ends. Similarly, we write $x \to -\infty$, read "x tends toward negative infinity," to mean x decreases, taking on negative values of arbitrarily large magnitude.

Example 1

Discuss the behavior of the graph of the given function as $x \to +\infty$ and as $x \to -\infty$.
(a) $F(x) = 2x^3 - x^2 + x - 1$
(b) $G(x) = x^4 - 3x^3 - x - 10$

SOLUTION (a) The highest-degree term of $F(x) = 2x^3 - x^2 + x - 1$ is $2x^3$. Thus

$$|x| \text{ large} \quad \text{implies} \quad F(x) \approx 2x^3$$

Now, as $x \to +\infty$, $2x^3$ tends toward $+\infty$. For example, consider the following brief table of values:

x	10	100	1000
$2x^3$	2000	2000000	2000000000

Thus

$$x \to +\infty \quad \text{implies} \quad F(x) \to +\infty$$

This means that as we move away from the origin toward the right, the graph of F will curve upward, its y-coordinates tending toward $+\infty$.

Next, as $x \to -\infty$, $2x^3$ will tend toward $-\infty$ (a negative number cubed will be negative). Thus

$$x \to -\infty \quad \text{implies} \quad F(x) \to -\infty$$

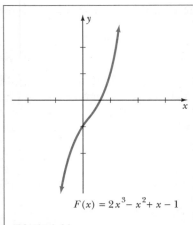

$$F(x) = 2x^3 - x^2 + x - 1$$

FIGURE 3.20

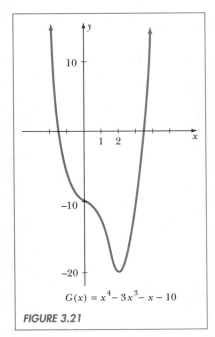

$$G(x) = x^4 - 3x^3 - x - 10$$

FIGURE 3.21

This means that as we move away from the origin toward the left, the graph of F will curve downward, its y-coordinates tending toward $-\infty$.

Compare this description with the graph of F in Figure 3.20.

(b) The highest-degree term of $G(x) = x^4 - 3x^3 - x - 10$ is x^4. Thus

$$|x| \text{ large } \quad \text{implies} \quad G(x) \approx x^4$$

Since x^4 will be positive whether x is positive or negative, it follows that

$$x \to +\infty \quad \text{implies} \quad G(x) \to +\infty$$

$$x \to -\infty \quad \text{implies} \quad G(x) \to +\infty$$

This says that as we move away from the origin toward the right or left, the graph of G will curve upward, its y-coordinates tending toward positive infinity.

Compare this description with the graph of G in Figure 3.21. ∎

We now summarize the general consequences of the second geometric property of polynomial functions.

Behavior of a Polynomial Function for $|x|$ Large

Let $F(x)$ be a polynomial function of degree $n > 0$ with leading coefficient a.

Then

$$|x| \text{ large } \quad \text{implies} \quad F(x) \approx ax^n$$

Furthermore

n even
$$\begin{cases} a > 0 & \begin{cases} x \to +\infty & \text{implies} & F(x) \to +\infty \\ x \to -\infty & \text{implies} & F(x) \to +\infty \end{cases} \\ a < 0 & \begin{cases} x \to +\infty & \text{implies} & F(x) \to -\infty \\ x \to -\infty & \text{implies} & F(x) \to -\infty \end{cases} \end{cases}$$

n odd
$$\begin{cases} a > 0 & \begin{cases} x \to +\infty & \text{implies} & F(x) \to +\infty \\ x \to -\infty & \text{implies} & F(x) \to -\infty \end{cases} \\ a < 0 & \begin{cases} x \to +\infty & \text{implies} & F(x) \to -\infty \\ x \to -\infty & \text{implies} & F(x) \to +\infty \end{cases} \end{cases}$$

Thus the y-coordinates on the graph of F always tend toward positive infinity or negative infinity as we move away from the origin toward the right or left. When F has even degree, the y-coordinates will go in the same direction as we move toward the right or left. On the other hand, when the degree of F

is odd, the y-coordinates will go in opposite directions as we move toward the right or left.

Example 2

Determine whether or not the given graph (Figures 3.22 to 3.25) could represent a polynomial function. If it can, determine the sign of the leading coefficient and whether the degree is even or odd. If the graph does not represent a polynomial function, explain why.

(a)

(b)

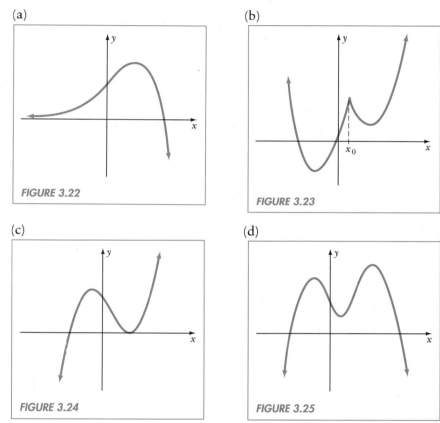

FIGURE 3.22

FIGURE 3.23

(c)

(d)

FIGURE 3.24

FIGURE 3.25

SOLUTION (a) If the graph in Figure 3.22 represents a polynomial function, then its degree would be at least 1 since it is not a constant function. However, the y-coordinates on the graph of any polynomial function of positive degree must tend toward $+\infty$ or $-\infty$ as we move away from the origin in either direction. The y-coordinates on this graph do not do this as we move toward the left. Therefore the graph in Figure 3.22 does not represent a polynomial function.

(b) The graph in Figure 3.23 violates the first geometric property of polynomials: it is not smooth. A sharp corner occurs when $x = x_0$. Therefore the graph in Figure 3.23 does not represent a polynomial function.

(c) The graph in Figure 3.24 is smooth and continuous, and its y-coordinates tend toward $+\infty$ and $-\infty$ as we move away from the origin. Therefore the graph in Figure 3.24 could represent a polynomial function, say F. In addition, we have

$$x \to +\infty \quad \text{implies} \quad F(x) \to +\infty$$

$$x \to -\infty \quad \text{implies} \quad F(x) \to -\infty$$

Looking at our summary of results concerning the behavior of a polynomial function for $|x|$ large, we see that these statements require that F have odd degree and leading coefficient positive.

(d) The graph in Figure 3.25 is also smooth and continuous, and its y-coordinates tend toward $-\infty$ as we move away from the origin. Therefore the graph in Figure 3.25 could represent a polynomial function. Calling this function G, we note that

$$x \to +\infty \quad \text{implies} \quad G(x) \to -\infty$$

$$x \to -\infty \quad \text{implies} \quad G(x) \to -\infty$$

We find that these statements require that G have even degree and leading coefficient negative. ∎

Now we are ready to do some graphing. We will not concern ourselves with polynomial functions of degree 0, 1, or 2. A degree 0 polynomial is just a constant function whose graph consists of a horizontal line. Degree 1 and 2 polynomials are the linear and quadratic functions already covered in Section 3.1. This brings us to the degree 3 polynomials.

An arbitrary degree 3 polynomial function with leading coefficient positive has a graph similar to one of those in Figure 3.26. By reflecting these graphs across the x-axis, we get the corresponding graphs for a degree 3 polynomial function with leading coefficient negative.

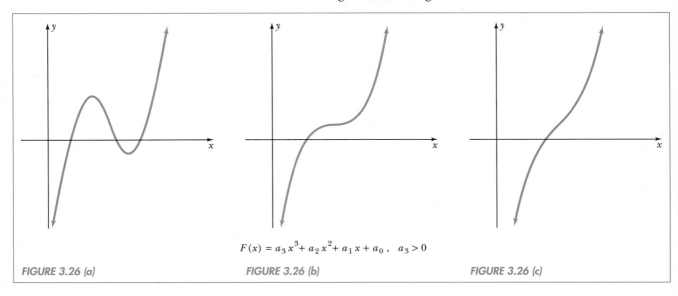

$$F(x) = a_3 x^3 + a_2 x^2 + a_1 x + a_0, \quad a_3 > 0$$

FIGURE 3.26 (a)   FIGURE 3.26 (b)   FIGURE 3.26 (c)

Example 3

Graph $F(x) = x^3 - 2x^2 + x$.

SOLUTION Our approach is to establish a table of strategic points for the curve. When possible, it helps to write the function in factored form, to make the zeros easier to find. We have

$$F(x) = x^3 - 2x^2 + x$$
$$= x(x^2 - 2x + 1)$$
$$= x(x - 1)^2$$

Thus F has zeros at $x = 0$ and $x = 1$, so we construct the following table:

x	-1	0	$1/2$	1	2
$F(x)$	-4	0	$1/8$	0	2

Before graphing these points, let us look at the behavior of $F(x)$ for $|x|$ large. We know that when $|x|$ is large, $F(x) \approx x^3$. Therefore

$$x \to +\infty \quad \text{implies} \quad F(x) \to +\infty$$

$$x \to -\infty \quad \text{implies} \quad F(x) \to -\infty$$

This means that the graph of F should go downward as $x \to -\infty$ and upward as $x \to +\infty$.

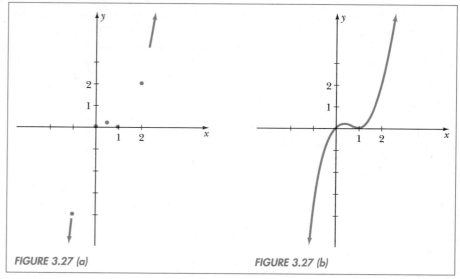

FIGURE 3.27 (a)   FIGURE 3.27 (b)

Figure 3.27(a) shows the points from our table and the information concerning the behavior of $F(x)$ for $|x|$ large. The next step is to fill in the graph with a smooth, continuous curve, as in Figure 3.27(b). Notice how the final graph behaves near $x = 1$: it does not cross over the x-axis but instead just touches it. If we were to (incorrectly) draw the curve crossing over at $x = 1$, then it would have to come back up through the x-axis crossing at a third point. This would contradict the fact that F has real zeros occurring only at $x = 0$ and $x = 1$. ∎

Two questions we might want to ask about the graph of F are, How do we know where it turns between $x = 0$ and $x = 1$? And, Shouldn't this happen when $x = 1/2$? Unfortunately, it is difficult to answer these questions without using some calculus. In this case, the graph actually turns when $x = 1/3$; that is, $F(x)$ increases from $x = 0$ to $x = 1/3$ and then decreases from $x = 1/3$ to $x = 1$. Since calculus is involved, we will not worry about finding turning points when graphing polynomials. Instead, we shall use an adequate number of points to find a reasonable approximation to the correct graph.

Example 4

Graph $G(x) = x^3 + x + 5$.

SOLUTION To graph this function, we use a slightly different approach from that in Example 3. By dropping the constant term $+5$ from $G(x)$, we obtain the function with equation $y = x^3 + x$. This curve is quite easy to graph by adding the functions x^3 and x, as shown in Figure 3.28(a).

Now, to get the graph of $G(x) = x^3 + x + 5$, we simply shift the graph of $y = x^3 + x$ by $+5$ units in the y-direction, as is done in Figure 3.28(b). ∎

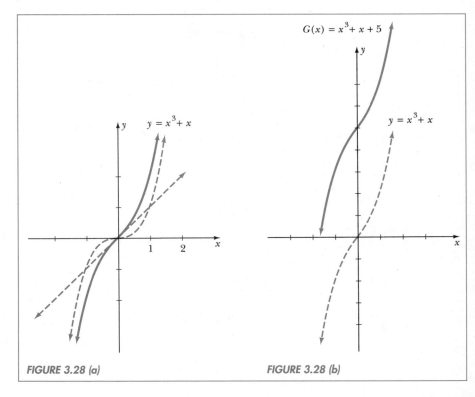

FIGURE 3.28 (a)   FIGURE 3.28 (b)

To graph higher-degree polynomials, we use the following techniques.

Graphing Polynomials

Basic Techniques
1. Make a table of strategic points. Include zeros, y-intercept, and points between zeros. When possible, factoring the function may help locate zeros.
2. Determine the behavior when $|x|$ is large.
3. Using the information from steps 1 and 2, draw a smooth, continuous graph for the function.

Additional Techniques
4. Check for symmetry:

$$F(-x) = F(x) \qquad \text{implies symmetry with respect to } y\text{-axis.}$$

$$F(-x) = -F(x) \quad \text{implies symmetry with respect to origin.}$$

5. If dropping the constant term from the function results in a curve easier to graph (as in Example 4), do so. Then shift the curve in the y-direction by the constant term to obtain the desired graph.

Example 5

Graph $F(x) = -2x^4 + 9x^2 - 4$.

SOLUTION   First, note that $F(-x) = F(x)$. Therefore the graph of F will be symmetric to the y-axis. Also, we can factor $F(x)$ as follows:

$$F(x) = -2x^4 + 9x^2 - 4$$
$$= -(2x^4 - 9x^2 + 4)$$
$$= -(2x^2 - 1)(x^2 - 4)$$

Hence F has zeros at $x = \pm 1/\sqrt{2}$ and $x = \pm 2$. Now, we construct a table of values:

x	0	$\pm 1/\sqrt{2} \approx \pm .7$	± 1	$\pm 3/2$	± 2	± 3
$F(x)$	-4	0	3	$\dfrac{49}{8}$	0	-85

Next, we determine the behavior of $F(x)$ when $|x|$ is large. We know that when $|x|$ is large, $F(x) \approx -2x^4$. Thus

$$x \to +\infty \quad \text{implies} \quad F(x) \to -\infty$$

$$x \to -\infty \quad \text{implies} \quad F(x) \to -\infty$$

This means that the y-coordinates on the graph of F go down toward negative infinity as we move away from the origin toward the right or left. The points from our table and the information about the behavior of $F(x)$ when $|x|$ is large are in Figure 3.29(a). By connecting the points with a smooth, continuous curve, we obtain the final graph of F, shown in Figure 3.29(b). ■

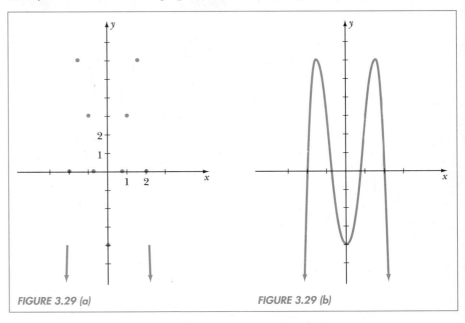

FIGURE 3.29 (a)   FIGURE 3.29 (b)

EXERCISES 3.3

In Exercises 1 and 2, determine whether the given function qualifies as a polynomial. If it does, find the degree, the leading coefficient, and the constant term.

1. (a) $F(x) = \sqrt{3}x + 1$
    (b) $G(x) = -x + x^{-1}$
    (c) $H(x) = 1 + \dfrac{1}{x} + \dfrac{1}{x^2}$
    (d) $K(x) = \pi$
    (e) $P(x) = (2x - 1)^2(x + 2)$

2. (a) $F(x) = \sqrt{3x} + 1$
    (b) $G(x) = (1 - x)^2$
    (c) $H(x) = x^4 + \dfrac{x}{2} + \dfrac{1}{4}$
    (d) $K(x) = \sqrt{x^4 + x^2}$
    (e) $P(x) = (x^3 + 3x^2 + 3x + 1)^{1/3}$

In Exercises 3 and 4, determine whether the graph could possibly represent a polynomial function. If it can, determine the sign of the leading coefficient and whether the degree must be even or odd. If it is not a polynomial function, explain why.

3. (a)   (b)

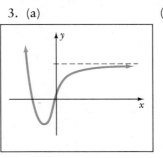
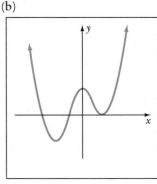

(c)

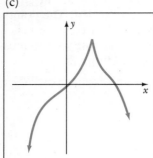

(d)

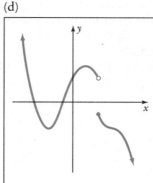

(d)

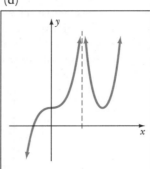

(e)

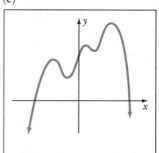

(e)

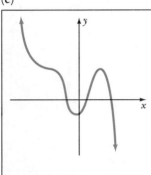

5. Graph the functions $F(x) = x^n$, for $n = 1, 2, 3, 4, 5,$ and 6 over the interval $[-1, 1]$, all on the same axes. For each function, complete the statements $x \to +\infty$ implies $F(x) \to$ _____, and $x \to -\infty$ implies $F(x) \to$ _____.

In Exercises 6 to 32, graph the given polynomial function. Follow the steps outlined in this section.

6. $F(x) = x^3 + x$ 7. $F(x) = x^3 - x$

8. $F(x) = x^3 - x + 2$ 9. $F(x) = x^3 + x - 3$

10. $F(x) = 2 - \dfrac{1}{8}x^3$ 11. $F(x) = \dfrac{1}{8}(x - 2)^3 + 2$

12. $F(x) = (x - 1)^2(x - 2)$ 13. $F(x) = x^3 - 4x^2 + 4x$

14. $F(x) = 2x - x^2 - x^3$

15. $F(x) = 4 + 8x - x^2 - 2x^3$

16. $F(x) = 4x - x^3$

17. $F(x) = 9x - x^3$

18. $F(x) = x^3 + 2x^2 - 3x + 1$ (*Hint:* graph $y = x^3 + 2x^2 - 3x$ first.)

19. $F(x) = x^4 + 2$

20. $F(x) = 16 - x^4$

21. $F(x) = x^4 - x^2$

22. $F(x) = x^4 - 11x^2 + 18$

23. $F(x) = (x - 1)(x - 2)(x + 1)^2$

24. $F(x) = x(x - 2)(x - 3)^2$

25. $F(x) = x - x^4$

26. $F(x) = x^4 - x^3$

4. (a)

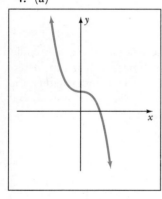

(b)

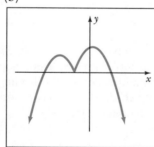

(c)

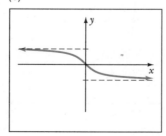

27. $F(x) = (x + 5)^5$ 28. $F(x) = x^5 - 1$

29. $F(x) = x^{5} - 5x^3 + 4x$

30. $F(x) = x^5 - x^4 - 2x^3$

31. $F(x) = x - \dfrac{1}{16}x^5$

32. $F(x) = -2x + \dfrac{5}{8}x^3 - \dfrac{1}{32}x^5$

In Exercises 33 and 34, answer true or false.

33. An odd degree polynomial function must have at least one real zero.

34. An odd degree polynomial function cannot be symmetric to the y-axis.

In Exercises 35 and 36, graph the given function.

35. $A(x) = |x^3 - x^5|$ 36. $A(x) = |x^4 - 2x^2 - 1|$

3.4 RECIPROCAL FUNCTIONS

Suppose we know the graph of a function G and want to find the graph of its reciprocal function,

$$y = \frac{1}{G(x)}$$

Two approaches we may take are

1. *Numerical method:* Construct a table of x, y values for the function $y = 1/G(x)$ and plot points.

2. *Geometric method:* Obtain the graph of $y = 1/G(x)$ visually from the graph of $y = G(x)$.

In this section we use both methods to draw the graph and analyze the behavior of reciprocal functions. We especially emphasize the geometric method because, as part of our goal, we want to understand how the graph of a function is related to its reciprocal. Along the way we introduce the important concepts of asymptote and limit.

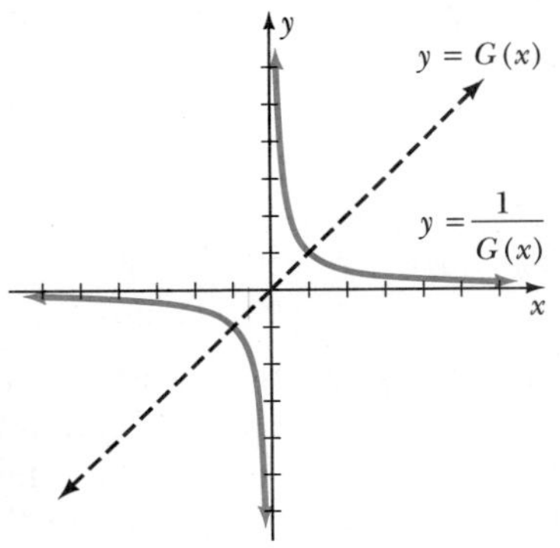

FIGURE 3.30

Example 1

Consider the function $G(x) = x$. Graph G and $1/G$ on the same axes. Discuss the relationship between the graphs of G and $1/G$.

SOLUTION   The equation for G is $y = x$. We recognize the graph of $y = x$ as the diagonal through the first and third quadrants. Next, consider the equation for $1/G$, $y = 1/x$. Recall that this equation was graphed in Example 3, Section 2.3. (See Figures 2.22(a) and (b).) Thus we obtain the graphs of G and $1/G$ in Figure 3.30.

Let us examine the relationship between the graphs of $G(x) = x$ and its reciprocal. First, we notice that when the y-coordinates on the graph of G are large in magnitude, then the corresponding y-coordinates on $1/G$ are small in magnitude. For example, $(10, 10)$ is on the graph of G, whereas $(10, .1)$ is on the graph of $1/G$. On the other hand, when the y-coordinates on G are small in magnitude, then the corresponding y-coordinates on $1/G$ are large in magnitude. Thus $(.01, .01)$ is on the graph of G, and $(.01, 100)$ is on the graph of $1/G$. Finally, we note that when $G(x) = 0$ at $x = 0$, the function $1/G$ is undefined. ■

The relationships we described between the graph of G and its reciprocal in Figure 3.30 turn out to be true for any function. Thus we have the following general results.

Relationship Between Graphs of G and 1/G

point on graph of G		point on graph of $1/G$
(a, b)	$\longleftrightarrow$	$\left(a, \dfrac{1}{b}\right)$

By taking the reciprocal of each y-coordinate on the graph of G, we obtain the corresponding y-coordinates for the graph of $1/G$. Furthermore,

(a, b) far from x-axis   implies $\left(a, \dfrac{1}{b}\right)$ close to x-axis

(a, b) close to x-axis   implies $\left(a, \dfrac{1}{b}\right)$ far from x-axis

Finally, if $G(a) = 0$, then $1/G(x)$ is undefined at $x = a$.

As a consequence of these results, we may obtain the graph of a reciprocal function, $1/G$, without plotting many points, provided we know the graph of G. Example 2 illustrates this technique.

Example 2

Graph $F(x) = \dfrac{1}{\sqrt{4 - x^2}}$.

SOLUTION   We begin by observing that $F(x)$ is the reciprocal of $G(x) = \sqrt{4 - x^2}$. We recognize the graph of G as a semicircle of radius 2 centered at the origin, which is shown in Figure 3.31(a). We can take a few specific points on the graph of G and find their corresponding points on the graph of $F = 1/G$ as follows:

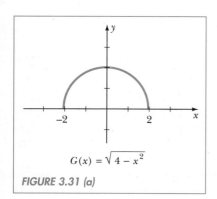

$G(x) = \sqrt{4 - x^2}$

FIGURE 3.31 (a)

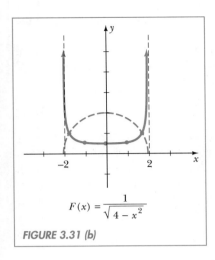

$$F(x) = \frac{1}{\sqrt{4 - x^2}}$$

FIGURE 3.31 (b)

Point on Graph of G		Point on Graph of $F = 1/G$
$(0, 2)$	$\longleftrightarrow$	$\left(0, \dfrac{1}{2}\right)$
$(\pm 1, \sqrt{3})$	$\longleftrightarrow$	$\left(\pm 1, \dfrac{1}{\sqrt{3}}\right) \approx (\pm 1, .6)$
$(\pm \sqrt{3}, 1)$	$\longleftrightarrow$	$(\pm \sqrt{3}, 1)$
$(\pm 2, 0)$	$\longleftrightarrow$	Undefined

Using these points and the fact that when $G(x)$ is close to the x-axis $F(x) = 1/G(x)$ is far from the x-axis, we obtain the graph of F in Figure 3.31(b). ∎

Remark

Notice that the reciprocal function $1/G$ in Figure 3.31(b) is undefined when $x = 2$ or $x = -2$. The dashed vertical lines at these points help illustrate the behavior of $1/G$ near $x = 2$ and $x = -2$. We call these lines *vertical asymptotes* of the function $1/G$. Informally speaking, a vertical asymptote occurs where the graph of a function blows up, taking on y-values of larger and larger magnitude as the curve approaches the asymptote. We provide a more formal definition of vertical asymptotes after Example 3.

Example 3

Graph $F(x) = \dfrac{1}{x - 2}$. Discuss the behavior of $F(x)$ as $x \to +\infty$, as $x \to -\infty$, and when x is near 2.

SOLUTION   We notice that F is the reciprocal of the linear function $G(x) = x - 2$. From the graph of G in Figure 3.32(a), we take the reciprocal of several of its y-coordinates to find the corresponding y-coordinates on the graph of $F = 1/G$. These points are verified in the following table:

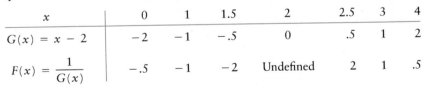

x	0	1	1.5	2	2.5	3	4
$G(x) = x - 2$	-2	-1	$-.5$	0	$.5$	1	2
$F(x) = \dfrac{1}{G(x)}$	$-.5$	-1	-2	Undefined	2	1	$.5$

Using these points and our knowledge of how the graph of a function is related to its reciprocal, we obtain the curve for F in Figure 3.32(b).

Another way to obtain the graph of F is to note that $y = 1/(x - 2)$ is the same as $y = 1/x$ shifted $+2$ units in the x-direction. Our graph in

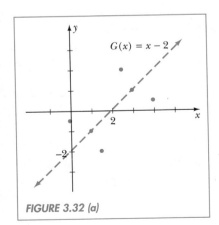

$G(x) = x - 2$

FIGURE 3.32 (a)

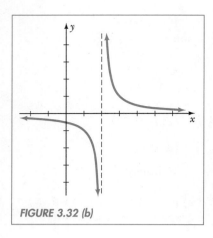

FIGURE 3.32 (b)

Figure 3.32(b) certainly agrees with this observation. (See Figure 3.30 for the graph of $y = 1/x$ given in Example 1.)

Now we wish to discuss the behavior of $F(x) = 1/(x - 2)$ in three critical areas: (1) as x tends toward positive infinity, $x \to +\infty$, (2) as x tends toward negative infinity, $x \to -\infty$, and (3) when x is near 2.

1. Behavior of $F(x) = \dfrac{1}{x - 2}$ as $x \to +\infty$. We recall that the notation $x \to +\infty$ means that x takes on positive values of arbitrarily larger and larger magnitude. What we wish to know is, As $x \to +\infty$, what happens to $F(x)$? We can answer this question by looking at the graph of F in Figure 3.32(b). Since $F(x)$ is the y-coordinate on the graph of F at x, we find that $y = F(x)$ gets closer and closer to zero as $x \to +\infty$. We may verify this with some specific numbers:

x	102	1002	10002	$\longrightarrow$	$+\infty$
$F(x) = \dfrac{1}{x - 2}$	.01	.001	.0001	$\longrightarrow$	0

In summary, we have

$$x \to +\infty \quad \text{implies} \quad F(x) \to 0$$

Note that we can only say x *tends toward* $+\infty$; we do not say x equals $+\infty$. Also, it happens that $F(x) = 1/(x - 2)$ *approaches* zero but never actually equals zero, hence the notation $F(x) \to 0$. We are now dealing with the concept of a **limit**. We shall rewrite our results in the following way:

$$\lim_{x \to +\infty} F(x) = 0$$

This is read "the limit of $F(x)$ as x tends toward positive infinity is equal to zero." Again, even though this notation ends with "$= 0$," it does not mean that $F(x)$ must actually equal zero, only that it gets closer and closer to zero as x becomes larger and larger.

One final note concerning the geometric interpretation of $\lim_{x \to +\infty} F(x) = 0$. Since $y = F(x)$ gets arbitrarily close to zero as $x \to +\infty$, we can say that the graph of F behaves like $y = 0$ for large positive x. In other words, the graph of F and the horizontal line $y = 0$ are almost indistinguishable as $x \to +\infty$. When this happens, we say that the line $y = 0$ is a *horizontal asymptote* of F.

2. Behavior of $F(x) = \dfrac{1}{x - 2}$ as $x \to -\infty$. We want to know what happens to $F(x)$ as $x \to -\infty$. From the graph of F in Figure 3.32(b), we find that $F(x)$ gets closer and closer to zero as $x \to -\infty$. Using limit notation, we write

$$\lim_{x \to -\infty} F(x) = 0$$

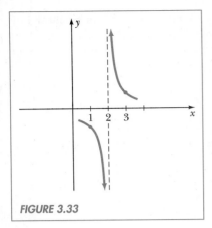

FIGURE 3.33

This is read, "the limit of $F(x)$ as x tends toward $-\infty$ is equal to zero." In geometric terms, this means that the graph of F behaves like the horizontal line $y = 0$ when x is far to the left of the origin.

3. Behavior of $F(x) = \dfrac{1}{x-2}$ when x is near 2. A detail of the graph of F near $x = 2$ is in Figure 3.33. Notice that the description of the behavior of F in this region depends on whether we look on the right or left side of $x = 2$.

On the right side of 2, $F(x)$ becomes larger and larger as x gets closer and closer to 2. We verify this with the following table:

x	2.1	2.01	2.001	$\longrightarrow$	2^+
$F(x) = \dfrac{1}{x-2}$	10	100	1000	$\longrightarrow$	$+\infty$

We write $x \to 2^+$ to mean that x approaches 2 from the right side of 2. We have

$$x \to 2^+ \quad \text{implies} \quad F(x) \to +\infty$$

or, using limit notation,

$$\lim_{x \to 2^+} F(x) = +\infty$$

This is read, "the limit of $F(x)$ as x approaches 2 from the right is equal to positive infinity." For obvious reasons, we call this a **right-hand limit**. Similarly, we have a **left-hand limit** to describe the behavior of $F(x)$ as x approaches 2 from the left side. Thus

$$\lim_{x \to 2^-} F(x) = -\infty \quad \blacksquare$$

In Example 3 we introduced the concept of left- and right-hand limits and limits at infinity. Our purpose now is to develop an intuitive geometric understanding of these limits, leaving their rigorous definitions for study in calculus. With this in mind, let us summarize our understanding of limits and limit notation. In what follows, assume that F is a function and N represents $+\infty$, $-\infty$, or a (finite) real number.

Limits at Infinity

$\displaystyle\lim_{x \to +\infty} F(x) = N$ As x tends toward $+\infty$, $F(x)$ approaches N.

$\displaystyle\lim_{x \to -\infty} F(x) = N$ As x tends toward $-\infty$, $F(x)$ approaches N.

If N is a real number in either of these limits, then the horizontal line $y = N$ is called a **horizontal asymptote** for F. Geometrically, this means that the graph of F gets closer and closer to the line $y = N$ as we move farther and farther out from the origin in the appropriate direction.

Left- and Right-Hand Limits

$$\lim_{x \to a^+} F(x) = N \qquad \text{As } x \text{ approaches } a \text{ from the right side of } a, \; F(x) \text{ approaches } N.$$

$$\lim_{x \to a^-} F(x) = N \qquad \text{As } x \text{ approaches } a \text{ from the left side of } a, \; F(x) \text{ approaches } N.$$

If N is $+\infty$ or $-\infty$ in either of these limits, then the vertical line $x = a$ is called a **vertical asymptote** for F. In terms of the graph of F, any one of the following four situations (see Figures 3.34 to 3.37) implies the existence of a vertical asymptote at $x = a$.

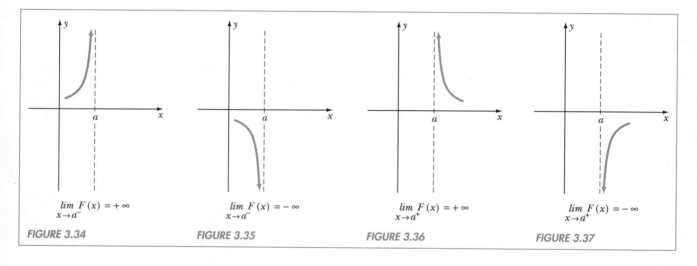

$$\lim_{x \to a^-} F(x) = +\infty$$

FIGURE 3.34

$$\lim_{x \to a^-} F(x) = -\infty$$

FIGURE 3.35

$$\lim_{x \to a^+} F(x) = +\infty$$

FIGURE 3.36

$$\lim_{x \to a^+} F(x) = -\infty$$

FIGURE 3.37

Example 4

Graph $F(x) = \dfrac{1}{x^2}$. Use limits to describe the behavior of $F(x)$ as $x \to +\infty$, as $x \to -\infty$, and when x is near zero.

SOLUTION Notice that $F(x)$ is the reciprocal of $G(x) = x^2$, the familiar parabola graphed in Figure 3.38(a). Let us take a few specific points on the graph of G and find their corresponding points on the graph of $F = 1/G$.

Point on Graph of G		Point on Graph of $F = 1/G$
$\left(\pm\dfrac{1}{2}, \dfrac{1}{4}\right)$	$\longleftrightarrow$	$\left(\pm\dfrac{1}{2}, 4\right)$
$(\pm 1, 1)$	$\longleftrightarrow$	$(\pm 1, 1)$
$(\pm 2, 4)$	$\longleftrightarrow$	$\left(\pm 2, \dfrac{1}{4}\right)$

Using these points and our knowledge of how the graph of a function is related to its reciprocal, we obtain the graph of F in Figure 3.38(b). From the graph, we make the following conclusions about the behavior of $F(x)$:

$$\lim_{x \to -\infty} F(x) = 0 \qquad \lim_{x \to +\infty} F(x) = 0$$

$$\lim_{x \to 0^-} F(x) = +\infty \qquad \lim_{x \to 0^+} F(x) = +\infty$$

It follows that F has horizontal asymptote $y = 0$ and vertical asymptote $x = 0$. Note that F is symmetric with respect to the y-axis.

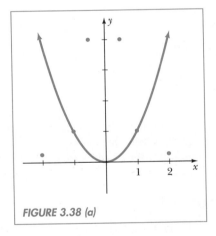

FIGURE 3.38 (a)

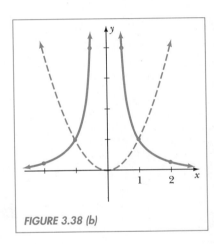

FIGURE 3.38 (b)

The graph of a general quadratic function $Q(x) = ax^2 + bx + c$ with $a > 0$ will be an upward curving parabola with its vertex situated above, on, or below the x-axis. Figures 3.39(a) to (d) on page 160 illustrate typical upward curving quadratic functions and the graphs of their reciprocals. When the graph of Q gets closer to the x-axis, as in Figure 3.39(a) and (b), we find the graph of $1/Q$ rising. [The point $(a, Q(a))$ is close to the x-axis implies the point $(a, 1/Q(a))$ will be far from the x-axis.] When the graph of Q just touches the x-axis, as in Figure 3.39(c), then the graph of $1/Q$ blows up at

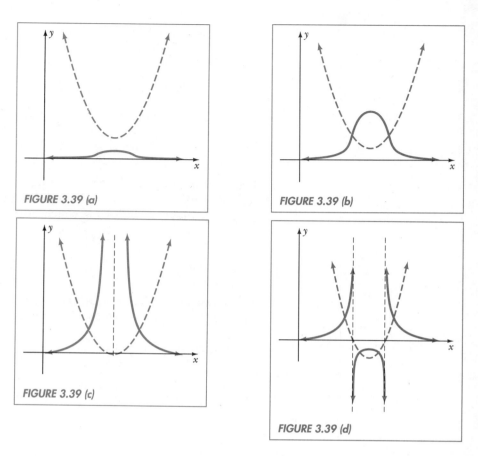

FIGURE 3.39 (a)

FIGURE 3.39 (b)

FIGURE 3.39 (c)

FIGURE 3.39 (d)

this point. Example 4 illustrated this particular situation. Finally, when the vertex of Q is below the x-axis, as in Figure 3.39(d), we find the graph of $1/Q$ blowing up at both zeros of Q. Notice the symmetry each graph displays with respect to a vertical line through the vertex of Q.

Example 5

Graph the given quadratic function Q dashed and its reciprocal $F = 1/Q$ with a solid curve on the same axes. Find all horizontal and vertical asymptotes for F.

(a) $Q(x) = (x - 1)^2 + \dfrac{1}{2}$

(b) $Q(x) = (x - 1)^2 - 1$

SOLUTION   (a) The equation for Q is $y = (x - 1)^2 + \dfrac{1}{2}$ or $y - \dfrac{1}{2} = (x - 1)^2$. This is the curve $y = x^2$ shifted to the new vertex $(1, 1/2)$. We can graph Q and then obtain some corresponding points on the graph of

$F = 1/Q$ as follows:

$$\text{Point on } Q: \quad \left(-1, \frac{9}{2}\right) \quad \left(0, \frac{3}{2}\right) \quad \left(1, \frac{1}{2}\right) \quad \left(2, \frac{3}{2}\right) \quad \left(3, \frac{9}{2}\right)$$

$$\text{Point on } F = \frac{1}{Q}: \quad \left(-1, \frac{2}{9}\right) \quad \left(0, \frac{2}{3}\right) \quad (1, 2) \quad \left(2, \frac{2}{3}\right) \quad \left(3, \frac{2}{9}\right)$$

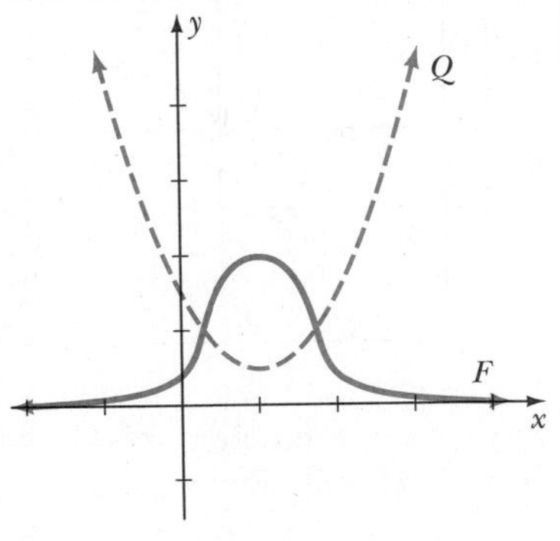

These points help us draw the graph of F, which is shown with the graph of Q in Figure 3.40. Notice the symmetry with respect to the vertical line $x = 1$.

From the graph of F we make the following observations:

Horizontal asymptote for F: $y = 0$.

Note that

$$\lim_{x \to -\infty} F(x) = 0 \quad \text{and} \quad \lim_{x \to +\infty} F(x) = 0$$

Vertical asymptote for F: None.

(b) The equation for Q is $y = (x - 1)^2 - 1$ or $y + 1 = (x - 1)^2$. Once again we have the curve $y = x^2$ shifted to a new vertex at $(1, -1)$. We graph Q accordingly and then obtain some corresponding points on the graph of $1/Q$ as follows.

$$\text{Point on } Q: \quad (-1, 3) \quad (0, 0) \quad (1, -1) \quad (2, 0) \quad (3, 3)$$

$$\text{Point on } \frac{1}{Q}: \quad \left(-1, \frac{1}{3}\right) \quad \text{Undefined} \quad (1, -1) \quad \text{Undefined} \quad \left(3, \frac{1}{3}\right)$$

Figure 3.41 shows the graphs of Q and $F = 1/Q$. In obtaining the graph of F, we relied heavily on the knowledge that $(a, Q(a))$ is close to the x-axis implies $(a, 1/Q(a))$ is far from the x-axis. Also, we know that the graphs are symmetric with respect to the vertical line $x = 1$.

From the graph of F we make the following observations:

Horizontal asymptote for F: $y = 0$.

Note that

$$\lim_{x \to -\infty} F(x) = 0 \quad \text{and} \quad \lim_{x \to +\infty} F(x) = 0$$

Vertical asymptote for F: $x = 0$ and $x = 2$.

Note that near $x = 0$,

$$\lim_{x \to 0^-} F(x) = +\infty \quad \text{and} \quad \lim_{x \to 0^+} F(x) = -\infty$$

and near $x = 2$,

$$\lim_{x \to 2^-} F(x) = -\infty \quad \text{and} \quad \lim_{x \to 2^+} F(x) = +\infty \quad \blacksquare$$

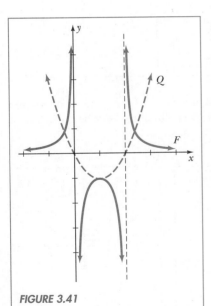

FIGURE 3.40

FIGURE 3.41

EXERCISES 3.4

In Exercises 1 and 2, the graph of G is shown. Graph $1/G$.

1. (a)   (b)

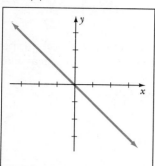   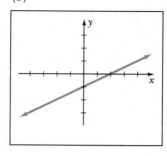

2. (a)   (b)

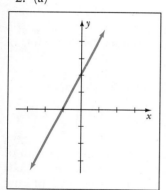   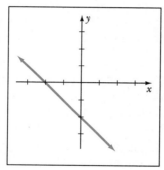

In Exercises 3 to 12, graph the given function. Determine any symmetry, and find all horizontal and vertical asymptotes. (*Hint:* each function has the form $F(x) = 1/G(x)$. Graph G first.)

3. $F(x) = \dfrac{1}{\sqrt{1 - x^2}}$ 4. $F(x) = \dfrac{1}{\sqrt{9 - x^2}}$

5. $F(x) = \dfrac{1}{\sqrt{x}}$ 6. $F(x) = \dfrac{1}{\sqrt{x + 1}}$

7. $F(x) = \dfrac{1}{|x|}$ 8. $F(x) = \dfrac{1}{|x| - 1}$

9. $F(x) = \dfrac{1}{x^3}$ 10. $F(x) = \dfrac{1}{x^3 + 1}$

11. $F(x) = \dfrac{1}{|x| + 1}$ 12. $F(x) = \dfrac{1}{\sqrt[3]{x}}$

In Exercises 13 and 14, the graph of a function F is shown. Evaluate $\lim\limits_{x \to +\infty} F(x)$ and $\lim\limits_{x \to -\infty} F(x)$.

13.   14.

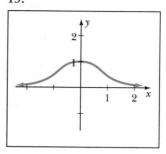   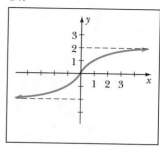

In Exercises 15 and 16, the graph of a function F is shown near $x = -3$. Evaluate $\lim\limits_{x \to -3^+} F(x)$ and $\lim\limits_{x \to -3^-} F(x)$.

15.   16.

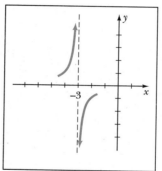   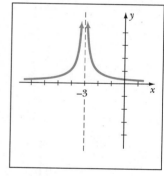

In Exercises 17 and 18, a portion of the graph of a function F is given. Use an appropriate limit to describe the behavior of F.

17. (a)   (b)

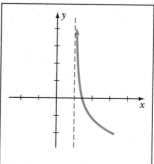   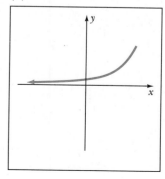

18. (a)

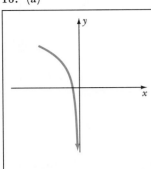

(b)

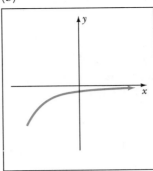

32.

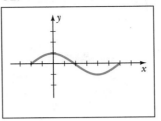

In Exercises 19 to 24, graph the function. Determine the horizontal and vertical asymptotes. Use limits to describe the behavior of $F(x)$ as $x \to +\infty$, as $x \to -\infty$, and for x near the vertical asymptote. (Each function has the form $F(x) = \dfrac{1}{G(x)}$. Graph G first.)

19. $F(x) = \dfrac{1}{2x}$

20. $F(x) = \dfrac{1}{-3x}$

21. $F(x) = \dfrac{1}{x-3}$

22. $F(x) = \dfrac{1}{3-x}$

23. $F(x) = \dfrac{1}{2-4x}$

24. $F(x) = \dfrac{1}{3x-2}$

In Exercises 25 to 30, graph the quadratic function Q dashed and its reciprocal $F = 1/Q$ with a solid curve on the same axes. Find all horizontal and vertical asymptotes for F.

25. (a) $Q(x) = 2x^2 + 1$
    (b) $Q(x) = 2x^2$
    (c) $Q(x) = 2x^2 - 2$

26. (a) $Q(x) = -x^2 + 1$
    (b) $Q(x) = -x^2$
    (c) $Q(x) = -x^2 - 2$

27. $Q(x) = -(x-2)^2$

28. $Q(x) = \dfrac{1}{2}(x+4)^2$

29. $Q(x) = -x^2 + 6x - 8$

30. $Q(x) = x^2 - 4x + 2$

In Exercises 31 to 36, the graph of a certain function G is shown. Graph $F = 1/G$.

31.

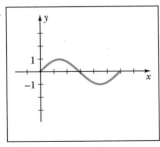

33.

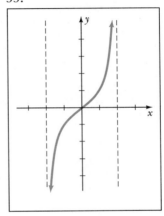

34.

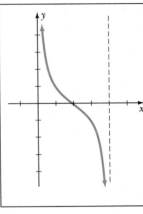

35.

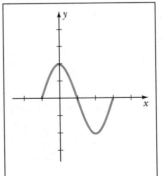

36.

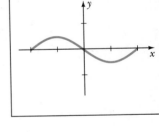

37. Let P_1 and P_2 be points on the graph of $F(x) = 1/x$, where P_1 has x-coordinate 2 and P_2 has x-coordinate $x, x > 0$. Let m be the slope of the line passing through P_1 and P_2. Find m as a function of x. Evaluate $\lim_{x \to +\infty} m(x)$ and $\lim_{x \to 0^+} m(x)$. Can you explain your answers for these limits geometrically, in terms of the points P_1 and P_2?

38. Repeat Exercise 41 if P_1 has x-coordinate -1.

3.5 RATIONAL FUNCTIONS: PRELIMINARIES

If we form the quotient p/q of two nonzero integers p and q, the result is another real number, which we call a rational number. Similarly, in the world of functions, if we take the quotient $P(x)/Q(x)$ of two nonzero polynomial functions $P(x)$ and $Q(x)$, we get what is called a **rational function**.

> **Definition of Rational Function**
> Let $P(x)$ and $Q(x)$ be two polynomial functions (neither of which is the constant zero function). Then
>
> $$F(x) = \frac{P(x)}{Q(x)}$$
>
> is called a rational function.

The formula $P(x)/Q(x)$ makes sense for all real numbers x except when $Q(x)$ is zero (division by zero is undefined). Therefore the domain of a rational function $F(x) = P(x)/Q(x)$ consists of all real numbers except the zeros of Q. Following are examples of rational functions and their domains:

$$F(x) = \frac{2x^3 + x - 1}{x^2 + x - 2} \qquad \mathcal{D}_F = \{x : x \neq 1, -2\}$$

$$G(x) = \frac{x}{x^2 + 4} \qquad \mathcal{D}_G = \mathbf{R}$$

$$S(x) = 1 + \frac{1}{x + 1} \qquad \mathcal{D}_S = \{x : x \neq -1\}$$

Note that $S(x)$, by simple addition, is equivalent to $\dfrac{x + 2}{x + 1}$, which satisfies the definition of a rational function.

Technically, any polynomial function $P(x)$ qualifies as a rational function because we can write it in the form

$$P(x) = \frac{P(x)}{1}$$

Polynomials were discussed in Section 3.3, so we will avoid this special case. Thus, for the remainder of the chapter, we consider only rational functions with polynomial denominator $Q(x)$ of degree at least 1.

Occasionally, the numerator and denominator of a rational function may contain a common factor. In this case, the formula for the function can be simplified by reducing the quotient. However, in doing so we must stipulate a restriction on the domain for the new formula.

Example 1

Simplify the formula for the given function. State the domain and graph.

(a) $F(x) = \dfrac{x^2 + x}{x + 1}$ (b) $G(x) = \dfrac{x + 2}{x^2 - 4}$

SOLUTION   (a) We may simplify the formula for $F(x)$ as follows:

$$F(x) = \frac{x^2 + x}{x + 1} = \frac{x(x + 1)}{x + 1} = x \qquad \text{provided } x \neq -1$$

Notice that the original formula for $F(x)$ is undefined when $x = -1$; therefore we must stipulate that x cannot equal -1 when we take out the common factor $x + 1$. The original formula for $F(x)$ always determines its domain. Thus we write

$$F(x) = x \qquad \mathcal{D}_F = \{x : x \neq -1\}$$

The graph of F is just the graph of the line $y = x$ with a hole in it when $x = -1$ (since F is undefined there). This is shown in Figure 3.42.

   (b) We may simplify the formula for $G(x)$ as follows:

$$G(x) = \frac{x + 2}{x^2 - 4} = \frac{x + 2}{(x + 2)(x - 2)} = \frac{1}{x - 2} \qquad \text{provided } x \neq -2$$

Thus

$$G(x) = \frac{1}{x - 2} \qquad \mathcal{D}_G = \{x : x \neq -2, 2\}$$

To graph G, we graph the reciprocal function $y = 1/(x - 2)$ with a hole in it when $x = -2$ (since G is undefined there). The graph of $y = 1/(x - 2)$ was constructed in Example 3, Section 3.4. (See Figure 3.32(b).) Therefore we obtain the graph for G in Figure 3.43.

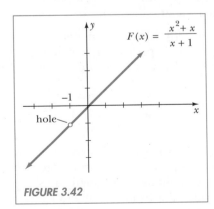

FIGURE 3.42

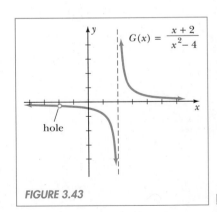

FIGURE 3.43

For the remainder of the section we shall concentrate on the following two key steps required in graphing a typical rational function:

1. Determine all vertical asymptotes and the behavior of the function near each asymptote.
2. Determine the behavior of the function when $|x|$ is large.

In Section 3.6 we shall use the information from these two steps to help us draw the complete graph of a given rational function.

Recall that a function F is said to have a vertical asymptote at $x = c$ if its graph behaves like any one of the curves in Figure 3.44 near the vertical line $x = c$. If F is the rational function $F(x) = P(x)/Q(x)$, then F may have a vertical asymptote at $x = c$, where c is a zero of Q, $Q(c) = 0$, because when x is near c, $Q(x)$ gets close to zero, which may cause the quotient $P(x)/Q(x)$ to become quite large in magnitude (since we are dividing by a very small number).

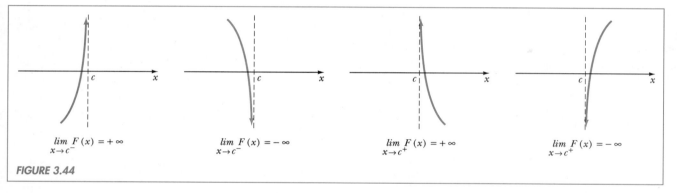

$$\lim_{x \to c^-} F(x) = +\infty \qquad \lim_{x \to c^-} F(x) = -\infty \qquad \lim_{x \to c^+} F(x) = +\infty \qquad \lim_{x \to c^+} F(x) = -\infty$$

FIGURE 3.44

Example 2

Given $F(x) = \dfrac{3x - 3}{x - 2}$. Determine the vertical asymptote for F. Analyze the behavior of F near the asymptote, and use limits to describe the results. Sketch the graph of F in the vicinity of the asymptote.

SOLUTION   The denominator in the formula for $F(x)$ has a zero at $x = 2$, so we suspect that F may have a vertical asymptote there. Let us begin our analysis by making a short table of values for $F(x)$ when x is near 2.

x	1	$1\frac{1}{2}$	2	$2\frac{1}{2}$	3
$F(x)$	0	-3	undefined	9	6

The table does not contain enough information to guarantee that $x = 2$ is a vertical asymptote. One solution is to compute more points. However, instead of more computation, we shall use some *thought analysis*. For this analysis, we consider the behavior of $F(x)$ in two cases: x approaching 2 from the right, $x \to 2^+$, and x approaching 2 from the left, $x \to 2^-$.

Behavior of F(x) as x → 2⁺

$$x \to 2^+ \qquad \text{implies} \qquad F(x) = \frac{3x - 3}{x - 2} \to \frac{3}{0^+} \to +\infty$$

This says that when x approaches 2 from the right, the quantity $(3x - 3)/(x - 2)$ approaches 3 in the numerator and zero in the denominator. These numbers are found by substituting 2 for x in the formula $(3x - 3)/(x - 2)$. Furthermore, we write 0^+ in the denominator to emphasize that $x - 2$ approaches zero through small positive numbers. (Think of x as taking on values such as 2.1, 2.01, and 2.001; then $x - 2$ takes on values .1, .01, and .001, all positive.) This is important because when a number close to 3 is divided by a small *positive* number, we get a large *positive* number. Thus we write $3/0^+ \to +\infty$. Using limit notation, we summarize our result by writing

$$\lim_{x \to 2^+} F(x) = +\infty$$

This means that the y-coordinates on the graph of F go up toward positive infinity as we approach the right side of $x = 2$. (See Figure 3.45.)

Behavior of F(x) as x → 2⁻

$$x \to 2^- \qquad \text{implies} \qquad F(x) = \frac{3x - 3}{x - 2} \to \frac{3}{0^-} \to -\infty$$

Notice the use of 0^- in the denominator to indicate that $x - 2$ approaches zero through small *negative* numbers. (Think of x as taking on values such as 1.9, 1.99, and 1.999; then $x - 2$ takes on values $-.1$, $-.01$, and $-.001$, all negative.) We conclude

$$\lim_{x \to 2^-} F(x) = -\infty$$

Thus the y-coordinates on the graph of F go down toward negative infinity as we approach the left side of $x = 2$.

Using the points from our table and our thought analysis near $x = 2$, we sketch the portion of the graph of F in Figure 3.45. We see that $x = 2$ is definitely a vertical asymptote for F. ■

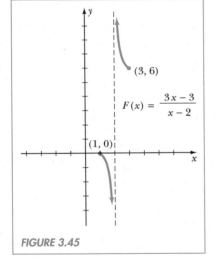

FIGURE 3.45

$F(x) = \dfrac{3x - 3}{x - 2}$

The technique of thought analysis introduced in Example 2 requires some skill with mental computation and estimation. These are important skills that can be developed with practice.

Example 3

Given $G(x) = \dfrac{1 - x}{x^2 - x - 2}$. Determine where G has vertical asymptotes. Analyze the behavior of G near each asymptote, and use limits to describe the results. Sketch the graph of G in the vicinity of each asymptote.

SOLUTION Factoring the denominator of $G(x)$, we have

$$G(x) = \frac{1 - x}{(x + 1)(x - 2)}$$

Vertical asymptotes should occur when the denominator is zero, at $x = -1$ and $x = 2$. We shall make a short table of values of $G(x)$ when x is near -1 and 2 and then analyze the behavior of $G(x)$ in the vicinity of these points.

x	-2	-1	0	1	2	3
$G(x)$	$\frac{3}{4}$	undefined	$-\frac{1}{2}$	0	undefined	$-\frac{1}{2}$

Behavior of $G(x)$ as $x \to -1^+$

When we say that $x \to -1^+$, we are thinking that x takes on values such as $-.9$, $-.99$, $-.999$, and so on.

$$x \to -1^+ \text{ implies } G(x) = \frac{1 - x}{(x + 1)(x - 2)} \to \frac{2}{(0^+)(-3)} \to \frac{2}{0^-} \to -\infty$$

Note that we replaced $(0^+)(-3)$ with 0^- because a small positive number multiplied by -3 will be a small negative number. We conclude,

$$\lim_{x \to -1^+} G(x) = -\infty$$

Behavior of $G(x)$ as $x \to -1^-$

In this case, x takes on values such as -1.1, -1.01, -1.001, and so on.

$$x \to -1^- \text{ implies } G(x) = \frac{1 - x}{(x + 1)(x - 2)} \to \frac{2}{(0^-)(-3)} \to \frac{2}{0^+} \to +\infty$$

Therefore

$$\lim_{x \to -1^-} G(x) = +\infty$$

Behavior of $G(x)$ as $x \to 2^+$

$$x \to 2^+ \text{ implies } G(x) = \frac{1 - x}{(x + 1)(x - 2)} \to \frac{-1}{(3)(0^+)} \to \frac{-1}{0^+} \to -\infty$$

Therefore

$$\lim_{x \to 2^+} G(x) = -\infty$$

Behavior of $G(x)$ as $x \to 2^-$

$$x \to 2^- \text{ implies } G(x) = \frac{1 - x}{(x + 1)(x - 2)} \to \frac{-1}{(3)(0^-)} \to \frac{-1}{0^-} \to +\infty$$

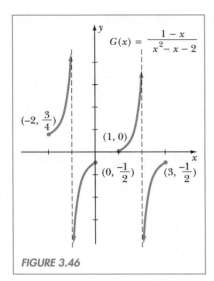

$$G(x) = \frac{1-x}{x^2-x-2}$$

$\left(-2, \frac{3}{4}\right)$

$(1, 0)$

$\left(0, \frac{-1}{2}\right)$ $\left(3, \frac{-1}{2}\right)$

FIGURE 3.46

Therefore

$$\lim_{x \to 2^-} G(x) = +\infty$$

Using the points from our table and the results of our thought analysis, we sketch the portion of the graph of G in Figure 3.46. ■

Now we consider how to determine the behavior of a rational function $F(x)$ when $|x|$ is large. This means that we look at what $F(x)$ does when x tends toward positive or negative infinity. Often we find that the graph of F behaves like a straight line when we look far away from the origin. In other words, the distance between the graph $y = F(x)$ and a fixed line approaches zero as x tends toward $+\infty$ or $-\infty$. If the line is horizontal, as we saw for reciprocal functions in Section 3.4, we call it a horizontal asymptote. However, for rational functions the line may have nonzero slope, in which case we call it a **slant asymptote**.

Figure 3.47 shows two graphs, one with a horizontal asymptote and the other with a slant asymptote. In Figure 3.47(a), the function F has a horizontal asymptote at $y = 1$. In Figure 3.47(b), G has a slant asymptote at $y = x$. Note that in both cases the function also has a vertical asymptote at $x = 0$.

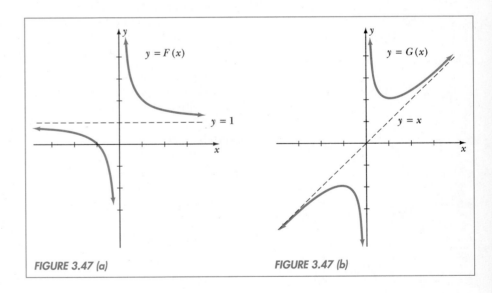

FIGURE 3.47 (a)   **FIGURE 3.47 (b)**

Example 4

Determine the behavior of $G(x) = \dfrac{1-x}{x^2-x-2}$ when $|x|$ is large.

SOLUTION Recall from Section 3.3 that a polynomial function behaves like its highest-degree term when $|x|$ is large. Thus $1 - x \approx -x$ and $x^2 - x - 2 \approx x^2$ for $|x|$ large. It follows that

$$|x| \text{ large} \quad \text{implies} \quad G(x) = \frac{1-x}{x^2-x-2} \approx \frac{-x}{x^2} = \frac{-1}{x}$$

When $|x|$ is large, we know $-1/x$ will be very close to zero. Thus

$$|x| \text{ large} \quad \text{implies} \quad G(x) \approx 0$$

Geometrically, this says that $y = G(x)$ approaches the line $y = 0$ when $|x|$ is large. Therefore $y = 0$ is a horizontal asymptote for G. ∎

Example 5

Determine the behavior of $F(x) = \dfrac{3x-3}{x-2}$ when $|x|$ is large.

SOLUTION Using the fact that $3x - 3 \approx 3x$ and $x - 2 \approx x$ when $|x|$ is large, we have

$$|x| \text{ large} \quad \text{implies} \quad F(x) = \frac{3x-3}{x-2} \approx \frac{3x}{x} = 3$$

Thus $F(x)$ approaches 3 when $|x|$ is large. In geometric terms, the graph $y = F(x)$ approaches the line $y = 3$ when $|x|$ is large. Therefore $y = 3$ is a horizontal asymptote for F. ∎

In Examples 4 and 5 we had degree $P \leq$ degree Q for the given function $P(x)/Q(x)$. Next, we consider a function with degree $P >$ degree Q. This calls for a different method of analysis.

Example 6

Determine the behavior of $H(x) = \dfrac{x^2+x+1}{x-1}$ when $|x|$ is large.

SOLUTION When the degree of the numerator is greater than the degree of the denominator, we use longhand division, dividing the denominator into the numerator, to rewrite the formula for the function. In this case we have

$$
\begin{array}{r}
x + 2 \\
x - 1 \overline{\smash{)}\ x^2 + x + 1} \\
\underline{x^2 - x } \\
2x + 1 \\
\underline{2x - 2} \\
3
\end{array}
$$

This allows us to write

$$H(x) = x + 2 + \frac{3}{x - 1}$$

Now, when $|x|$ is large, the term $3/(x - 1)$ approaches zero. Therefore

$$|x| \text{ large} \qquad \text{implies} \qquad H(x) \approx x + 2$$

This says that the graph of $y = H(x)$ gets close to the line $y = x + 2$ as x tends toward positive or negative infinity. We conclude that $y = x + 2$ is a slant asymptote for H. ∎

The arguments presented in Examples 4 to 6 can be generalized to establish the following results.

Behavior of $F(x) = P(x)/Q(x)$ When $|x|$ Is Large

Condition	Behavior When $|x|$ Is Large	Conclusion
Degree P < degree Q	$F(x) \approx 0$	Horizontal asymptote at $y = 0$.
Degree P = degree Q	$F(x) \approx k$, k is the ratio of the leading coefficients of P and Q.	Horizontal asymptote at $y = k$.
Degree P > degree Q	$F(x) \approx$ the quotient obtained by dividing $Q(x)$ into $P(x)$.	A slant asymptote occurs when degree P = degree $Q + 1$.

To further illustrate these results, we offer the following examples.

Function	Behavior When $|x|$ Is Large	Conclusion
$F(x) = \dfrac{2x + 1}{x^2 - 1}$	≈ 0	Horizontal asymptote $y = 0$.
$F(x) = \dfrac{10 - 4x}{5x + 8}$	$\approx -\dfrac{4}{5}$	Horizontal asymptote $y = -\dfrac{4}{5}$.
$F(x) = \dfrac{2x^2}{x - 1}$	$\approx 2x + 2$	Slant asymptote $y = 2x + 2$.
$F(x) = \dfrac{x^3 + 1}{x}$	$\approx x^2$	No horizontal or slant asymptote.

EXERCISES 3.5

In Exercises 1 to 6, simplify the formula for the given function. State the domain and graph.

1. (a) $F(x) = \dfrac{x^2 - x}{x - 1}$ (b) $F(x) = \dfrac{x - 1}{x^2 - x}$

2. (a) $F(x) = \dfrac{x - 1}{x^2 - 1}$ (b) $F(x) = \dfrac{x^2 - 1}{x - 1}$

3. $F(x) = \dfrac{9 - x^2}{x + 3}$ 4. $F(x) = \dfrac{25 - x^2}{x + 5}$

5. $F(x) = \dfrac{x + 1}{x^2 - 4x - 5}$ 6. $F(x) = \dfrac{x + 2}{x^2 - x - 6}$

In Exercises 7 to 10, analyze the behavior of $F(x)$ near $x = c$ and use limits to describe the results. Sketch the graph of F in the vicinity of $x = c$.

7. $F(x) = \dfrac{x + 2}{x - 3}, c = 3$ 8. $F(x) = \dfrac{x - 1}{x + 4}, c = -4$

9. $F(x) = \dfrac{x}{(x - 1)(x + 2)}$ 10. $F(x) = \dfrac{x + 3}{x^2 - 1}$
     (a) $c = 1$ (a) $c = 1$
     (b) $c = -2$ (b) $c = -1$

In Exercises 11 to 16, find the horizontal or slant asymptote for F.

11. $F(x) = \dfrac{6x - 1}{4x + 5}$ 12. $F(x) = \dfrac{3x + 2}{x^2 - 5}$

13. $F(x) = \dfrac{8x}{(x + 1)(x + 2)}$ 14. $F(x) = \dfrac{x}{3x + 4}$

15. $F(x) = \dfrac{2x^2 + x + 2}{x + 1}$ 16. $F(x) = \dfrac{(x + 1)^2}{x + 3}$

In Exercises 17 to 24, (a) if possible, find the vertical asymptotes for F. Analyze the behavior of F near each asymptote, and use limits to describe the results. Sketch the graph of F in the vicinity of each vertical asymptote. (b) If possible, find the horizontal or slant asymptote for F.

17. $F(x) = \dfrac{1}{x^3 - 2x^2}$ 18. $F(x) = \dfrac{1}{x^3 - 6x^2 + 9x}$

19. $F(x) = \dfrac{-x^2}{x^2 + 4}$ 20. $F(x) = \dfrac{x^2}{2x^2 + 1}$

21. $F(x) = \dfrac{x^2}{x + 1}$ 22. $F(x) = \dfrac{(x + 1)^2}{x}$

23. $F(x) = \dfrac{2x^2}{(x - 2)^2}$ 24. $F(x) = \dfrac{-x^2}{(x + 4)^2}$

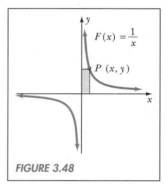

FIGURE 3.48

25. Let $P(x, y)$ be any point on the graph of $F(x) = 1/x$ in the first quadrant. Consider the rectangle constructed with vertices at P, along the x- and y-axes, and at the origin. See Figure 3.48.
     (a) Find $A(x)$, the area of the rectangle as a function of x. Complete the statement: $x \to +\infty$ implies $A(x) \to$?.
     (b) Find $B(x)$, the perimeter of the rectangle as a function of x. Complete the statement: $x \to +\infty$ implies $B(x) \to$?.

26. Repeat Exercise 25 in the case when $F(x) = 1/x^2$.

3.6   RATIONAL FUNCTIONS: GRAPHING

In Section 3.5 we discussed three factors concerning rational functions: (1) how to simplify the function when the numerator and denominator share a common factor, (2) how to determine the behavior of the function near its vertical asymptotes, and (3) how to determine the behavior of the function far away from the origin when $|x|$ is large. These items of analysis provide us with the first three steps in the following general procedure for graphing rational functions.

> **Graphing a Rational Function $F(x) = P(x)/Q(x)$**
> 1. *If possible, reduce the formula for $F(x)$.*
>   If $P(x)$ and $Q(x)$ share a common factor, reduce the quotient $P(x)/Q(x)$. Note any restrictions on $\mathcal{D}_F$.
> 2. *Find vertical asymptotes.*
>   Check the zeros of Q. If $Q(c) = 0$, then determine the behavior of $F(x)$ near $x = c$ by evaluating
>
> $$\lim_{x \to c^-} F(x) \quad \text{and} \quad \lim_{x \to c^+} F(x)$$
>
> 3. *Determine the behavior of $F(x)$ when $|x|$ is large.*
>   Conclude whether F has a horizontal or slant asymptote. If degree $P >$ degree Q, divide $Q(x)$ into $P(x)$ first before doing the analysis.
> 4. *Make a table of x, y values.*
>   Include the y-intercept, x-intercepts, and a few other strategic points. Check F for symmetry.
> 5. *Draw the graph.*
>   Use the information from steps 1 to 4.
>
> *Additional techniques:*
> Is $F(x)$ the reciprocal of a simple function?
> Is $F(x)$ a shift of a simple function?
> Is $F(x)$ the sum of two simple functions?

We shall demonstrate this procedure with several examples. For simplicity, our examples do not include a function where step 1 applies. This does not imply that step 1 is irrelevant. In fact, you are urged to reexamine Examples 1 and 2, Section 3.5, to see the importance of step 1.

Example 1

Graph $F(x) = \dfrac{x - 1}{x - 2}$.

SOLUTION Step 1. There are no common factors in the formula for $F(x)$.

Step 2. A vertical asymptote occurs when $x = 2$. To determine the behavior of $F(x)$ near $x = 2$, we do the following analysis:

$$x \to 2^- \quad \text{implies} \quad F(x) = \frac{x - 1}{x - 2} \to \frac{1}{0^-} \to -\infty$$

$$x \to 2^+ \quad \text{implies} \quad F(x) = \frac{x - 1}{x - 2} \to \frac{1}{0^+} \to +\infty$$

Therefore

$$\lim_{x \to 2^-} F(x) = -\infty \qquad \text{and} \qquad \lim_{x \to 2^+} F(x) = +\infty$$

We use this information to draw the vertical asymptote $x = 2$ and the graph of F near this asymptote, shown in Figure 3.49(a). Figure 3.49(a) also includes points from the table established in step 4.

Step 3.   When $|x|$ is large, we have

$$F(x) = \frac{x - 1}{x - 2} \approx \frac{x}{x} = 1$$

Therefore $y = 1$ is a horizontal asymptote for F. This is drawn as a dashed line in Figure 3.49(a).

Step 4.

x	−1	0	1	2	3	4
$F(x)$	$\frac{2}{3}$	$\frac{1}{2}$	0	undefined	2	$\frac{3}{2}$

Step 5.   The information from steps 1 to 4 is in Figure 3.49(a). We draw the complete graph of F in Figure 3.49(b). ∎

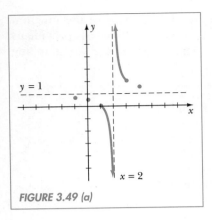

FIGURE 3.49 (a)

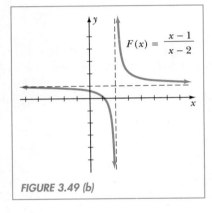

FIGURE 3.49 (b)

Remark

An alternate approach to graphing $F(x) = (x - 1)/(x - 2)$ is to notice that it is a shift of a simple reciprocal function. By dividing the denominator into the numerator of this function, we find

$$x - 2 \overline{\smash{\big)}\, x - 1} \qquad \text{implies} \qquad F(x) = 1 + \frac{1}{x - 2}$$
$$\underline{x - 2}$$
$$1$$

Therefore the equation for F is equivalent to

$$y - 1 = \frac{1}{x - 2}$$

We recognize this as a shift of the curve $y = 1/x$ by $+2$ units in the x-direction and $+1$ unit in the y-direction. This result agrees with our graph in Figure 3.49(b).

Example 2

Graph $G(x) = \dfrac{1 - x}{x^2 + x - 6}$

SOLUTION   Step 1.   We may factor the denominator of $G(x)$, obtaining

$$G(x) = \frac{1 - x}{(x + 3)(x - 2)}$$

There are no common factors in this formula.

Step 2. Vertical asymptotes occur at $x = -3$ and $x = 2$. For the analysis near $x = -3$, we have

$$x \to -3^- \quad \text{implies} \quad G(x) = \frac{1-x}{(x+3)(x-2)} \to \frac{4}{(0^-)(-5)} \to \frac{4}{0^+} \to +\infty$$

$$x \to -3^+ \quad \text{implies} \quad G(x) = \frac{1-x}{(x+3)(x-2)} \to \frac{4}{(0^+)(-5)} \to \frac{4}{0^-} \to -\infty$$

Therefore

$$\lim_{x \to -3^-} G(x) = +\infty \quad \text{and} \quad \lim_{x \to -3^+} G(x) = -\infty$$

For the analysis near $x = 2$, we have

$$x \to 2^- \quad \text{implies} \quad G(x) = \frac{1-x}{(x+3)(x-2)} \to \frac{-1}{(5)(0^-)} \to \frac{-1}{0^-} \to +\infty$$

$$x \to 2^+ \quad \text{implies} \quad G(x) = \frac{1-x}{(x+3)(x-2)} \to \frac{-1}{(5)(0^+)} \to \frac{-1}{0^+} \to -\infty$$

Therefore

$$\lim_{x \to 2^-} G(x) = +\infty \quad \text{and} \quad \lim_{x \to 2^+} G(x) = -\infty$$

We use these results to draw the vertical asymptotes $x = -3$ and $x = 2$ and the graph of G near these asymptotes shown in Figure 3.50(a). Figure 3.50(a) also includes points from the table established in step 4.

Step 3. When $|x|$ is large, we have

$$G(x) = \frac{1-x}{x^2 + x - 6} \approx \frac{-x}{x^2} = \frac{-1}{x}$$

Since $-1/x \approx 0$ when $|x|$ is large, we conclude

$$|x| \text{ large} \quad \text{implies} \quad G(x) \approx 0$$

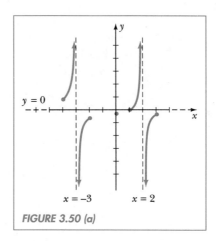

FIGURE 3.50 (a)

Therefore G has a horizontal asymptote at $y = 0$. We indicate this with a dashed line in Figure 3.50(a).

Step 4.

x	-4	-3	-2	0	1	2	3
$G(x)$	$\dfrac{5}{6}$	undefined	$-\dfrac{3}{4}$	$-\dfrac{1}{6}$	0	undefined	$-\dfrac{1}{3}$

Step 5.   Figure 3.50(a) contains the information from steps 1 to 4. The complete graph of G is in Figure 3.50(b).

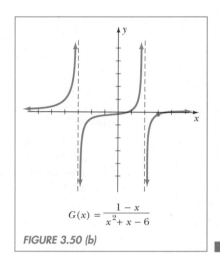

$$G(x) = \frac{1 - x}{x^2 + x - 6}$$

FIGURE 3.50 (b)

Remark

Example 2 shows that the graph of a function may intersect its horizontal asymptote. Remember that a horizontal asymptote is used to describe the behavior of a function far away from the origin (when $|x|$ is large) and has no bearing on the graph in the vicinity of the origin.

Example 3

Graph $H(x) = \dfrac{x^2 + 2}{x - 1}$

SOLUTION   Step 1.   The formula for $H(x)$ does not reduce.

Step 2.   A vertical asymptote occurs when $x = 1$. For the analysis near $x = 1$, we have

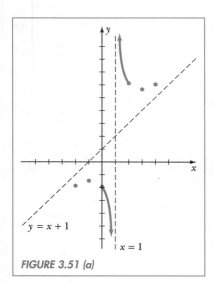

FIGURE 3.51 (a)

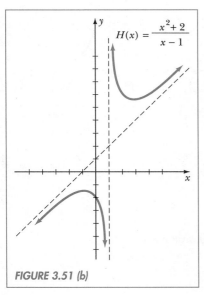

FIGURE 3.51 (b)

$$x \to 1^- \quad \text{implies} \quad H(x) = \frac{x^2 + 2}{x - 1} \to \frac{3}{0^-} \to -\infty$$

$$x \to 1^+ \quad \text{implies} \quad H(x) = \frac{x^2 + 2}{x - 1} \to \frac{3}{0^+} \to +\infty$$

Therefore

$$\lim_{x \to 1^-} H(x) = -\infty \quad \text{and} \quad \lim_{x \to 1^+} H(x) = +\infty$$

We use this information to draw the vertical asymptote $x = 1$ and the graph of H near this asymptote shown in Figure 3.51(a). Figure 3.51(a) also includes points from the table established in step 4.

Step 3. Since the degree of the numerator in the formula for $H(x)$ is greater than the degree of the denominator, we divide first to determine the behavior of $H(x)$ when $|x|$ is large.

$$
\begin{array}{r}
x + 1 \\
x - 1 \overline{\smash{\big)}\ x^2 \qquad + 2} \\
\underline{x^2 - x} \\
x + 2 \\
\underline{x - 1} \\
3
\end{array}
\quad \text{implies} \quad H(x) = x + 1 + \frac{3}{x - 1}
$$

Now, when $|x|$ is large, $3/(x - 1) \approx 0$; hence $H(x) \approx x + 1$. We conclude that $y = x + 1$ is a slant asymptote for H. We draw this with a dashed line in Figure 3.51(a).

Step 4.

x	-2	-1	0	1	2	3	4
$H(x)$	-2	$-\dfrac{3}{2}$	-2	undefined	6	$\dfrac{11}{2}$	6

Step 5. Figure 3.51(a) contains the information from steps 1 to 4. Figure 3.51(b) is the complete graph of H. ∎

Example 4

Graph $F(x) = \dfrac{x}{(x - 1)^2}$

SOLUTION Step 1. The formula for $F(x)$ does not reduce.

Step 2. A vertical asymptote occurs when $x = 1$. We leave it to you to show

$$\lim_{x \to 1^-} F(x) = +\infty \quad \text{and} \quad \lim_{x \to 1^+} F(x) = +\infty$$

Step 3.   You can verify that when $|x|$ is large, $F(x) \approx 0$. Therefore F has a horizontal asymptote at $y = 0$.

Step 4.

x	-2	-1	0	$\frac{1}{2}$	1	$\frac{3}{2}$	2	3
$F(x)$	$-\dfrac{2}{9}$	$-\dfrac{1}{4}$	0	2	undefined	6	2	$\dfrac{3}{4}$

Step 5.   We sketch the information from steps 1 to 4 in Figure 3.52(a). The graph for F is completed in Figure 3.52(b).

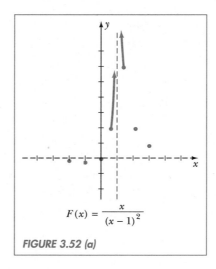

$$F(x) = \frac{x}{(x-1)^2}$$

FIGURE 3.52 (a)

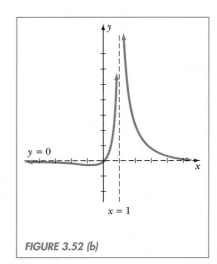

$$x = 1$$

FIGURE 3.52 (b)

Example 5

Graph $F(x) = \dfrac{x^2}{x^2 + 1}$ by shifting a simpler function.

SOLUTION   By performing the division

$$x^2 + 1 \overline{\big)\ \begin{array}{r} 1 \\ \hline x^2 \\ \underline{x^2 + 1} \\ -1 \end{array}}$$

we may rewrite the formula for $F(x)$ as

$$F(x) = 1 + \frac{-1}{x^2 + 1}$$

Therefore the equation for F is equivalent to

$$y - 1 = \frac{-1}{x^2 + 1}$$

We graph $y = 1/(x^2 + 1)$ in Figure 3.53(a). Next, we reflect this across the x-axis to obtain the graph of $y = -1/(x^2 + 1)$ in Figure 3.53(b). Finally, we shift $+1$ unit in the y-direction to get the desired graph of F in Figure 3.53(c). Note that $F(-x) = F(x)$, so the graph of F is symmetric with respect to the y-axis. Furthermore, F has a horizontal asymptote at $y = 1$ and no vertical asymptotes. ∎

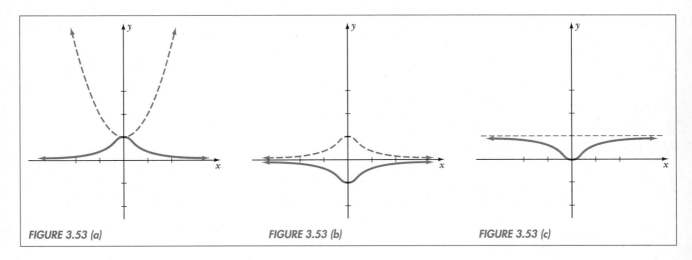

FIGURE 3.53 (a)   FIGURE 3.53 (b)   FIGURE 3.53 (c)

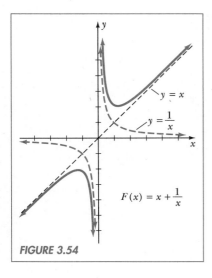

FIGURE 3.54

Example 6

Graph $F(x) = x + \dfrac{1}{x}$ by adding graphs.

SOLUTION   Since $F(x)$ is the sum of two simple functions, $y = x$ and $y = 1/x$, we add their graphs to obtain the graph of F. The result is shown in Figure 3.54. Note that $F(-x) = -F(x)$, so the graph of F is symmetric with respect to the origin. Furthermore, we see that F has a vertical asymptote at $x = 0$ and the slant asymptote $y = x$. ∎

EXERCISES 3.6

In Exercises 1 to 30, use the steps presented in this section to graph the given rational function. Label x- and y-intercepts when possible.

1. $F(x) = \dfrac{x - 1}{x + 1}$

2. $F(x) = \dfrac{-x}{x - 3}$

3. $F(x) = \dfrac{1 - 2x}{x - 1}$

4. $F(x) = \dfrac{2x + 1}{x - 1}$

5. $F(x) = \dfrac{x + 1}{x^2 + x - 2}$

6. $F(x) = \dfrac{2 - x}{x^2 - 2x - 3}$

7. $F(x) = \dfrac{4}{x^2 - x - 2}$

8. $F(x) = \dfrac{2x}{x^2 + x - 6}$

9. $F(x) = \dfrac{x}{1 - x^2}$

10. $F(x) = \dfrac{x^2 + 1}{x^2 - 1}$

11. $F(x) = \dfrac{x^2}{4 - x^2}$

12. $F(x) = \dfrac{2x}{4 - x^2}$

13. $F(x) = \dfrac{x^2 + 1}{x + 1}$

14. $F(x) = \dfrac{x^2 + 1}{x - 2}$

15. $F(x) = \dfrac{x^2 - 4}{x}$

16. $F(x) = \dfrac{1 - x^2}{x}$

17. $F(x) = \dfrac{x}{x^2 + 1}$

18. $F(x) = \dfrac{-3x}{x^2 + 2}$

19. $F(x) = \dfrac{10}{x^3 - 9x}$

20. $F(x) = \dfrac{5}{x^3 - 4x}$

21. $F(x) = \dfrac{x^2 - 1}{x^3}$

22. $F(x) = \dfrac{4 - x^2}{x^3}$

23. $F(x) = \dfrac{x}{(x + 2)^2}$

24. $F(x) = \dfrac{x}{(2x - 3)^2}$

25. $F(x) = \dfrac{2x + 1}{(x - 2)^2}$

26. $F(x) = \dfrac{3x - 1}{(x + 1)^2}$

27. $F(x) = \dfrac{x^2}{(x + 1)^2}$

28. $F(x) = \dfrac{-x^2}{(x - 2)^2}$

29. $F(x) = \dfrac{x^3 + 3x^2 + 2x}{x^2 + 3x + 2}$

30. $F(x) = \dfrac{x^2 - 16}{x^3 - 16x}$

31. Consider the function $F(x) = \dfrac{1}{x - 1} + 2$.
    (a) Describe how to shift $G(x) = 1/x$ to obtain F.
    (b) Use the result in (a) to graph F.
    (c) State the horizontal and vertical asymptotes for F.

32. Graph $F(x) = \dfrac{1}{x + 3} - 2$ by shifting the graph of $G(x) = 1/x$. State the horizontal and vertical asymptote for F.

In Exercises 33 and 34, graph each function. To graph H, write $H(x)$ as a shift of $G(x)$ by dividing the denominator of $H(x)$ into the numerator. State all asymptotes for each function.

33. (a) $F(x) = \dfrac{1}{x^2 + 2}$
    (b) $G(x) = \dfrac{-1}{x^2 + 2}$
    (c) $H(x) = \dfrac{x^2 + 1}{x^2 + 2}$

34. (a) $F(x) = \dfrac{1}{x^2 + 1}$
    (b) $G(x) = \dfrac{-1}{x^2 + 1}$
    (c) $H(x) = \dfrac{2x^2 + 1}{x^2 + 1}$

In Exercises 35 to 37, use the method of adding graphs to find the graph of F. State all asymptotes.

35. $F(x) = x - \dfrac{1}{x}$

36. $F(x) = x + \dfrac{1}{x^2}$

37. $F(x) = \dfrac{1}{x^2} - x$

In Exercises 38 and 39, graph the given equation. (*Hint:* write y as a function of x.)

38. $x + xy + y = 0$

39. $x - 2xy + y = 0$

40. Let P_1 and P_2 be points on the graph of $F(x) = 1/x^2$, where P_1 has x-coordinate 2 and P_2 has x-coordinate x. Let m be the slope of the line passing through P_1 and P_2.
    (a) Find m as a function of x.
    (b) Evaluate $\lim\limits_{x \to +\infty} m(x)$ and $\lim\limits_{x \to 0^+} m(x)$.
    (c) Graph $m(x)$.

41. Repeat Exercise 41 if P_1 has x-coordinate -1.

1. Graph $L(x) = 7 - 2x$. Find the zero and y-intercept of L. Is L increasing or decreasing?

2. Find a linear function F such that $F(3) = 5$ and $F(-2) = 0$.

3. The force F exerted by a spring compressed to x units in length is a linear function of x, $F(x)$. If the spring exerts a force of 3 pounds when compressed to 8 inches and 9 pounds when compressed 6 inches, find $F(x)$. What length corresponds to zero force?

4. The temperature of an object measures 18.4°C (degrees Centigrade) after 2 seconds and 19.2°C after 3.6 seconds. Find the linear approximation for the temperature T as a function of time t between these two data points. Find when $T = 20$°C.

In Exercises 5 and 6, for the given quadratic function Q, find the vertex, zeros, y-intercept, and line of symmetry. Graph Q. Determine the range, where Q is increasing and where Q is decreasing.

5. $Q(x) = -\frac{1}{2}x^2 + 2x + 6$

6. $Q(x) = 2x^2 + 4x - 2$

7. Find the quadratic function $Q(x)$ whose graph has vertex $V(3, 1)$ and passes through $P(1, 3)$.

8. Find a formula for the linear function $L(x)$ in Figure 3.55.

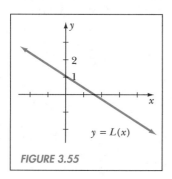

FIGURE 3.55

9. Find a formula for the quadratic function $Q(x)$ in Figure 3.56.

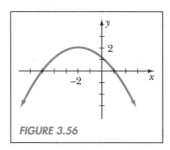

FIGURE 3.56

In Exercises 10 and 11, solve the given inequality.

10. $x^2 - x - 1 > 0$

11. $(x + 1)(x - 2) < 3$

12. The sum of two numbers is 4. Let S denote the sum of twice the first number and the square of the second number. Find the numbers that make the value of S a minimum.

13. A rectangle is to be constructed in the first quadrant, with vertices located at the origin, along the positive x- and y-axes and at the point $P(x, y)$ on the line $2x + 6y = 12$. Write the area of the rectangle as a function of x, the x-coordinate of P. Find the maximum area for such a rectangle.

14. Find the minimum distance between any point $P(x, y)$ on the curve $y = \sqrt{x + 4}$ and the point $A(11/2, 0)$. Find the coordinates of the point corresponding to this minimum distance.

15. A rectangle is to be constructed with its base on the x-axis and its top two vertices on the curve $y = 8 - \frac{1}{2}x^2$ (in quadrants I and II). Find the maximum perimeter possible for such a rectangle. Find the coordinates of the top two vertices corresponding to this rectangle.

In Exercises 16 and 17, the graph of a certain polynomial function is shown. Determine the sign of the leading coefficient and whether the degree is even or odd.

16.

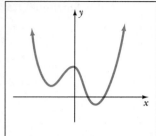

17.

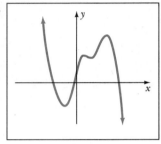

In Exercises 18 and 19, explain why the graph shown cannot represent a polynomial function.

18.

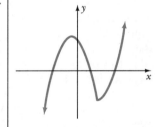

19.

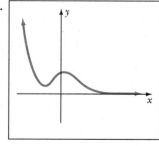

In Exercises 20 to 22, graph the given polynomial function.

20. $F(x) = x^3 + x^2 - x - 1$

21. $F(x) = \frac{1}{2}x(x + 2)(x - 2)^2$

22. $F(x) = \frac{1}{10}(x^5 + x^4 - 9x^3 - 9x^2)$

In Exercises 23 and 24, the graph of G is shown. Sketch the graph of $1/G$.

23.

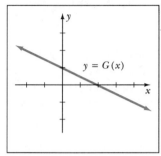

24.

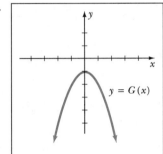

In Exercises 25 to 27, graph the given function. State any symmetry. Find all horizontal and vertical asymptotes. (*Hint:* each function has the form $F(x) = 1/G(x)$. Graph G first.)

25. $F(x) = \dfrac{1}{x^4}$

26. $F(x) = \dfrac{1}{\sqrt{x + 1}}$

27. $F(x) = \dfrac{1}{x^2 - 4x + 3}$

28. Graph $F(x) = \dfrac{1}{x^3 - 1}$. Evaluate $\lim\limits_{x \to +\infty} F(x)$, $\lim\limits_{x \to -\infty} F(x)$, $\lim\limits_{x \to 1^+} F(x)$, and $\lim\limits_{x \to 1^-} F(x)$.

29. Consider $F(x) = \dfrac{x - 1}{x^2 - x - 6}$.

   (a) Find the vertical asymptotes for F. Analyze the behavior of F near each asymptote and use limits to describe the results. Sketch the graph of F in the vicinity of each asymptote.

   (b) Find the horizontal asymptote for F.

In Exercises 30 to 35, follow the procedure given in Section 3.6 to graph each of the following rational functions. State all vertical, horizontal, and slant asymptotes. Label x- and y-intercepts when possible.

30. $F(x) = \dfrac{2x - 3}{x - 1}$

31. $F(x) = \dfrac{x}{x^2 - 2x}$

32. $F(x) = \dfrac{x}{x^2 - 4}$

33. $F(x) = \dfrac{x^2 + 2x + 1}{x + 2}$

34. $F(x) = \dfrac{x^2 - 4x + 3}{1 - x}$

35. $F(x) = \dfrac{x}{(x - 1)^2}$

36. (a) Graph $G(x) = \dfrac{1}{x^2 + \frac{1}{2}}$.

   (b) Graph $F(x) = \dfrac{-2x^2}{x^2 + \frac{1}{2}}$ by shifting the graph of G in part (a). State the horizontal asymptote for F.

In Exercises 37 and 38, use the method of adding graphs to find the graph of F. State all vertical and slant asymptotes.

37. $F(x) = \dfrac{1}{x^3} + x$

38. $F(x) = \dfrac{1}{x^3} - x$

CHAPTER 4

Zeros of Polynomial Functions

4.1 THE FACTOR THEOREM

This chapter is devoted to the search for the zeros of polynomial functions, a subject of interest for three reasons. First, as noted in Section 3.3, the zeros of a polynomial function provide significant points for the construction of its graph. Second, by finding when $F(x) = 0$, we are solving a polynomial equation and thus enhancing our overall ability to solve equations. Finally, the zeros of a polynomial function tell us a great deal about its structure. The key result in this direction says the following: if $x = c$ is a zero of a polynomial function $F(x)$, then $(x - c)$ is a factor of $F(x)$. This is the result we want to obtain in this section, but first we must discuss how to factor terms out of a polynomial, a procedure that involves the division algorithm for polynomials.

An algorithm is a method for solving a particular problem by the repeated application of a basic process. For example, the division algorithm for integers is as follows. Suppose we wish to divide 951 by 4. The longhand division involves three steps:

$$
\text{Step 1: } 4\overline{)951} \qquad \text{Step 2: } 4\overline{)951} \qquad \text{Step 3: } 4\overline{)951}
$$

Each step repeats the basic process of divide, multiply, subtract. The whole procedure is summarized by writing

$$951 = 4 \cdot 237 + 3$$

We call 951 the *dividend*, 4 the *divisor*, 237 the *quotient*, and 3 the *remainder*. In general, when a positive integer m is divided by another positive integer p, the result is

$$m = p \cdot q + r$$

where q and r are unique integers, and $0 \le r < p$. This is called the division algorithm for integers.

Now we extend the division algorithm to polynomial functions. We state the algorithm without proof.

> **Division Algorithm for Polynomials**
> Let $F(x)$ and $P(x)$ be polynomial functions. Then there exist unique polynomials $Q(x)$ and $R(x)$ such that
>
> $$F(x) = P(x) \cdot Q(x) + R(x)$$
>
> where $R(x) = 0$ or $0 \le$ degree of $R <$ degree of P.

Example 1

State the result of the division algorithm when $F(x) = x^5 + x^4 - 2x^2 + 4x$ is divided by $P(x) = x^2$.

SOLUTION   We may use longhand division as follows.

Step 1:
$$
\begin{array}{r}
x^3 \\
x^2 \overline{)\, x^5 + x^4 - 2x^2 + 4x} \\
\underline{x^5 } \\
x^4 - 2x^2 + 4x
\end{array}
$$

Step 2:
$$
\begin{array}{r}
x^3 + x^2 \\
x^2 \overline{)\, x^5 + x^4 - 2x^2 + 4x} \\
\underline{x^5 } \\
x^4 - 2x^2 + 4x \\
\underline{x^4 } \\
-2x^2 + 4x
\end{array}
$$

Step 3:
$$
\begin{array}{r}
x^3 + x^2 - 2 \\
x^2 \overline{)\, x^5 + x^4 - 2x^2 + 4x} \\
\underline{x^5 } \\
x^4 - 2x^2 + 4x \\
\underline{x^4 } \\
-2x^2 + 4x \\
\underline{-2x^2 } \\
4x
\end{array}
$$

Thus we have

$$
\begin{aligned}
F(x) &= P(x) \cdot Q(x) \quad\; + R(x) \\
&= x^2(x^3 + x^2 - 2) + 4x
\end{aligned}
$$

The quotient is $x^3 + x^2 - 2$; the remainder is $4x$. Notice that the degree of the remainder, 1, is less than the degree of the divisor, 2. ∎

Example 2

State the result of the division algorithm when $F(x) = x^4 + 2x^3 + 5x + 1$ is divided by $P(x) = x^2 - x + 1$.

SOLUTION   Working the division out longhand, we have

$$
\begin{array}{r}
x^2 + 3x + 2 \\
x^2 - x + 1 \overline{\smash{\big)}\ x^4 + 2x^3 + 0 + 5x + 1} \\
\underline{x^4 - x^3 + x^2} \\
3x^3 - x^2 + 5x + 1 \\
\underline{3x^3 - 3x^2 + 3x} \\
2x^2 + 2x + 1 \\
\underline{2x^2 - 2x + 2} \\
4x - 1
\end{array}
$$

We write the result in the form $F(x) = P(x) \cdot Q(x) + R(x)$.

$$F(x) = (x^2 - x + 1) \cdot (x^2 + 3x + 2) + 4x - 1$$

The quotient is $x^2 + 3x + 2$; the remainder is $4x - 1$. Once again, note that the degree of the remainder is less than the degree of the divisor. ∎

When the divisor happens to be a linear polynomial of the form $x - c$, c a constant, the process of division is particularly easy. Suppose, for example, we wish to divide $F(x) = 2x^4 - 19x^2 + 5x - 15$ by $x - 3$. Using longhand division, we have

$$
\begin{array}{r}
2x^3 + 6x^2 - x + 2 \\
x - 3 \overline{\smash{\big)}\ 2x^4 + 0 - 19x^2 + 5x - 15} \\
\underline{2x^4 - 6x^3} \\
6x^3 - 19x^2 + 5x - 15 \\
\underline{6x^3 - 18x^2} \\
-x^2 + 5x - 15 \\
\underline{-x^2 + 3x} \\
2x - 15 \\
\underline{2x - 6} \\
-9
\end{array}
\qquad
\begin{array}{r}
2 \quad 6 \quad -1 \quad 2 \\
x - 3 \overline{\smash{\big)}\ 2 \quad 0 \quad -19 \quad 5 \quad -15} \\
\underline{2 \quad -6} \\
6 \\
\underline{6 \quad -18} \\
-1 \\
\underline{-1 \quad 3} \\
2 \\
\underline{2 \quad -6} \\
-9
\end{array}
$$

The left side shows the ordinary longhand division. Notice how the powers of x serve as place holders, and the computations depend only on the coefficients of the polynomials. Thus, on the right, we drop out all the x's and eliminate some details to emphasize the essential calculations. It is possible to eliminate more, reducing the division process to a minimum number of computations, giving us a method called **synthetic division**.

Let us illustrate how to divide $F(x) = 2x^4 - 19x^2 + 5x - 15$ by $x - 3$ once again, but this time using synthetic division. We begin by writing down the constant term c of the divisor $x - c$; in our case, this is 3 because the divisor is $x - 3$. Next, we draw a vertical bar followed by the coefficients of the dividend $F(x)$ in descending order.

$$3 \quad | \quad 2 \quad 0 \quad -19 \quad 5 \quad -15$$

Note that we are careful to write a zero for any missing powers of x in $F(x)$. Next, we draw a horizontal line below this data and bring down the first coefficient, which is a 2.

Multiply this coefficient by 3 and put the result under the next coefficient. Thus

$$
\begin{array}{r|rrrrr}
3 & 2 & 0 & -19 & 5 & -15 \\
  &   & 6 &   &   &   \\
\hline
  & 2 &   &   &   &   \\
\end{array}
$$

Now, we add this new column, obtaining

$$
\begin{array}{r|rrrrr}
3 & 2 & 0 & -19 & 5 & -15 \\
  &   & 6 &   &   &   \\
\hline
  & 2 & 6 &   &   &   \\
\end{array}
$$

We repeat the process of multiplying by 3, putting the result under the next coefficient and adding until there are no further coefficients to work with.

$$
\begin{array}{r|rrrrr}
3 & 2 & 0 & -19 & 5 & -15 \\
  &   & 6 & 18 & -3 & 6 \\
\hline
  & 2 & 6 & -1 & 2 & -9 \\
\end{array}
$$

We interpret this result as follows. The last number in the bottom row, -9, is the remainder. The other numbers in the bottom row, 2, 6, -1, and 2, are the coefficients of the quotient polynomial, $Q(x)$, in descending order. Thus

$$Q(x) = 2x^3 + 6x^2 - x + 2$$

We conclude that

$$F(x) = (x - 3)(2x^3 + 6x^2 - x + 2) - 9$$

which agrees with our previous computation using longhand division.

Example 3

Use synthetic division to find the result of the division algorithm when $F(x) = x^3 + 5x^2 - 2x + 1$ is divided by $x + 2$.

SOLUTION The divisor is $x + 2$, which we can write as $x - (-2)$. Therefore, to set up the synthetic division, we write -2 followed by the coefficients of $F(x)$ in descending order. We have

$$
\begin{array}{r|rrrr}
-2 & 1 & 5 & -2 & 1 \\
   &   & -2 & -6 & 16 \\
\hline
   & 1 & 3 & -8 & 17 \\
\end{array}
$$

which tells us that the remainder is 17 and the coefficients of the quotient polynomial, in descending order, are 1, 3, and -8. Thus

$$F(x) = (x + 2)(x^2 + 3x - 8) + 17 \blacksquare$$

When we divide a polynomial function by another polynomial, the division algorithm tells us that the remainder will either be zero or have degree less than the degree of the divisor. Therefore, if the divisor has degree 1, such as when we divide by $Q(x) = x - c$, then the remainder will either be 0 or have degree 0. In either case, the remainder is a constant. As we shall see, this constant has another important interpretation.

As an example, let us find the remainder when the function $F(x) = 2x^3 - 3x^2 - 6$ is divided by $x - 3$. Using snythetic division,

$$
\begin{array}{r|rrrr}
3 & 2 & -3 & 0 & -6 \\
  &   & 6 & 9 & 27 \\
\hline
  & 2 & 3 & 9 & 21
\end{array}
$$

we see that the remainder is 21. Now consider the value of the function when $x = 3$.

$$F(3) = 2(3)^3 - 3(3)^2 - 6 = 21$$

Again, we get the number 21. This is not just a lucky coincidence. Whenever any nonconstant polynomial function is divided by a linear term of the form $x - c$, the remainder will always be the value of the function at $x = c$. We call this result the Remainder Theorem.

Remainder Theorem

If $F(x)$ is a polynomial function of degree $n > 0$ and c is any number, then

$$F(x) = (x - c)Q(x) + r$$

where $r = F(c)$ and $Q(x)$ is a polynomial function of degree $n - 1$.

PROOF: Dividing $F(x)$ by $(x - c)$, the division algorithm for polynomials says

$$F(x) = (x - c)Q(x) + R(x)$$

where $R(x)$ is either 0 or $0 \le$ degree of $R <$ degree of $(x - c)$. Since the degree of $(x - c)$ is 1, it follows that $R(x)$ is either 0 or its degree is 0. In either case, $R(x)$ is a constant function, say r, and we may write

$$F(x) = (x - c)Q(x) + r$$

Now, evaluate this expression at $x = c$.

$$F(c) = (c - c)Q(c) + r$$
$$= 0 \cdot Q(c) + r$$
$$= r$$

Thus $r = F(c)$, as we wished to show. Finally, note that the degree of $Q(x)$ must be $n - 1$ since the degree of $F(x)$ is n. This completes the proof. ■

Example 4

Without performing the division, find the remainder when $F(x) = x^4 + x^3 + x^2 + x + 1$ is divided by $x + 1$.

SOLUTION The Remainder Theorem says

$$F(x) = (x + 1)Q(x) + r$$

where $r = F(-1)$. [Since $x + 1 = x - (-1)$, we take $c = -1$]. Therefore the remainder is given by

$$r = F(-1)$$
$$= (-1)^4 + (-1)^3 + (-1)^2 + (-1) + 1$$
$$= 1 \ ▪$$

The Remainder Theorem provides us with the link between the zeros of a polynomial function and its structure. The key result, which we call the Factor Theorem, can be summarized by saying *zeros factor out*.

Factor Theorem

Let $F(x)$ be a polynomial function of degree at least 1. Then $x = c$ is a zero of F if and only if $x - c$ is a factor of $F(x)$.

$$F(c) = 0 \quad \text{if and only if} \quad F(x) = (x - c)Q(x)$$

PROOF: By the Remainder Theorem,

$$F(x) = (x - c)Q(x) + r$$

where $r = F(c)$. Hence $F(c) = 0$ if and only if $F(x) = (x - c)Q(x)$. This completes the proof. ■

Example 5

Consider $F(x) = 2x^3 + 5x^2 - 10x - 3$. Evaluate $F(c)$ if $c = 1$ or $c = 3/2$. If c is a zero of F, express $F(x)$ with this zero factored out.

SOLUTION We have two methods for evaluating $F(c)$. First, we can determine $F(c)$ directly, using the formula for F. Second, we can use synthetic division: dividing by $x - c$, the remainder will be $F(c)$.

For $c = 1$, we shall evaluate $F(c)$ directly.

$$F(1) = 2(1)^3 + 5(1)^2 - 10(1) - 3 = -6$$

Since $F(1) = -6$, $c = 1$ is not a zero of F.

For $c = 3/2$, we shall use synthetic division.

$$
\begin{array}{r|rrrr}
\frac{3}{2} & 2 & 5 & -10 & -3 \\
 & & 3 & 12 & 3 \\
\hline
 & 2 & 8 & 2 & 0
\end{array}
$$

Thus $F(3/2) = 0$. Furthermore, we may write

$$F(x) = (x - 3/2)(2x^2 + 8x + 2) \ \blacksquare$$

EXERCISES 4.1

In Exercises 1 to 10, state the result of the division algorithm when $F(x)$ is divided by $P(x)$. Use longhand division.

1. $F(x) = 4x^4 - 2x^2 + x + 10, P(x) = x^2 - x + 1$
2. $F(x) = x^4 + 2x^2 + 1, P(x) = x^2 + x + 5$
3. $F(x) = x^3 - 5x^2 + 2x - 3, P(x) = x^2 - 2$
4. $F(x) = x^3 + 1, P(x) = x$
5. $F(x) = x^6, P(x) = x^2 - 1$
6. $F(x) = x^5, P(x) = x^3 + x - 1$
7. $F(x) = 6x^6 - 9x^5 - 4x^4 + 6x^3 + 2x^2 - 7x + 6$, $P(x) = 2x - 3$
8. $F(x) = 9x^9 - 1, P(x) = 3x^3 - 1$
9. $F(x) = -2x, P(x) = 3x + 4$
10. $F(x) = 5x + 4, P(x) = 2x - 1$

In Exercises 11 to 26, use synthetic division to find the result of the division algorithm when $F(x)$ is divided by $P(x)$.

11. $F(x) = x^3 + x^2 + x + 1, P(x) = x + 1$
12. $F(x) = 3x^3 + 4x^2 - 5x + 4, P(x) = x + 2$
13. $F(x) = 3x^3 + 4x^2 - 5x - 2, P(x) = x + 2$
14. $F(x) = 2x^3 - 11x^2 + 12x + 8, P(x) = x + 1$
15. $F(x) = 3x^3 + 8x - 21, P(x) = x - 2$
16. $F(x) = x^4 + 3x^3 - 2x^2 + 5x - 12, P(x) = x + 4$
17. $F(x) = x^3 + 3x^2 + 2x, P(x) = x - 3$
18. $F(x) = x^4 + 2x^2, P(x) = x - 1$
19. $F(x) = 4x^4 + 4x^3 + 3x^2 - x - 1, P(x) = x - 1/2$
20. $F(x) = 2x^3 - x^2 - 7x + 6, P(x) = x - 3/2$
21. $F(x) = 4x^3 - 3x^2 + 11x + 5, P(x) = x + 1/4$
22. $F(x) = 6x^4 - 13x^3 + 6x^2 + 12x - 7, P(x) = x - 2/3$
23. $F(x) = x^6, P(x) = x - 1$
24. $F(x) = x^{21} + 1, P(x) = x + 1$
25. $F(x) = x^{100}, P(x) = x + 1$
26. $F(x) = x^n, P(x) = x - 1$ (assume n is a positive integer)

In Exercises 27 to 37, evaluate $F(c)$ for each value of c given. If c is a zero of F, then express $F(x)$ with this zero factored out; that is, write $F(x)$ in the form $F(x) = (x - c)Q(x)$.

27. $F(x) = 4x^2 + 15x - 4$
    (a) $c = -3$ (b) $c = 1/4$

28. $F(x) = x^3 - 4x^2 + 3x + 2$
   (a) $c = 2$ (b) $c = -1$

29. $F(x) = 6x^3 - 5x^2 + 3x - 1$
   (a) $c = 1$ (b) $c = 1/2$

30. $F(x) = 2x^4 + 5x^3 - 2x^2 + 4x + 3$
   (a) $c = -3$ (b) $c = 0$

31. $F(x) = 4x^3 - x$
   (a) $c = 0$ (b) $c = 1/2$

32. $F(x) = 2x^3 + x^2 - 6x + 1$
   (a) $c = 3/2$ (b) $c = -1$

33. $F(x) = x^4 + x^3 - x^2 - 2x - 2$
   (a) $c = -\sqrt{2}$ (b) $c = -2$

34. $F(x) = 3x^3 + 5x^2 - 2x$
   (a) $c = 1/3$ (b) $c = 0$

35. $F(x) = x^4 + x^3 - 5x^2 + 25x$
   (a) $c = 5$ (b) $c = -5$

36. $F(x) = x^{100} + 2x - 3$
   (a) $c = 1$ (b) $c = -1$

37. $F(x) = x^{40} + 2x^{25} + 3x^{10} + 4x^5 + 5x^2$
   (a) $c = -1$ (b) $c = 0$

In Exercises 38 to 44, write the result of the division algorithm $(F(x) = P(x)Q(x) + R(x))$ using the given functions. Solve for the unknown function by adding, subtracting, multiplying, or dividing the appropriate polynomials.

38. If the polynomial $F(x)$ is divided by $x^2 + x + 1$, then the quotient is $x^2 - x + 2$ and the remainder is $x - 3$. Find $F(x)$.

39. If the polynomial $F(x)$ is divided by $x^2 + x + 1$, then the quotient is $x^3 - x^2 + x$ and the remainder is $2x - 1$. Find $F(x)$.

40. If $F(x) = x^4 + 16$ is divided by the polynomial $P(x)$, then the quotient is $x^3 + 2x^2 + 4x + 8$ and the remainder is 32. Find the divisor, $P(x)$.

41. If $F(x) = 4x^5 - 4x^4 + x^3 - 8x^2 + 1$ is divided by the polynomial $P(x)$, then the quotient is $2x^3 - x^2 - x - 4$ and the remainder is $-3x + 5$. Find the divisor, $P(x)$.

42. If the polynomial $F(x)$ is divided by $2x - 3$, then the quotient is $4x^3 - x + 3$ and the remainder is 9. Find $F(x)$.

43. If $F(x) = x^6 + 1$ is divided by $P(x)$, then the quotient is $x^5 + 2x^4 + 4x^3 + 8x^2 + 16x + 32$ and the remainder is 65. Find the divisor, $P(x)$.

44. When the polynomial $F(x)$ is divided by $x - 1$, the quotient is $3x^3 + 5x^2 + 4x + 4$ plus a remainder. If $F(1) = 9$, find $F(x)$.

45. Prove or disprove: If $F(x) = x^2 + 1$ is divided by $x - c$ for any real number c, then the remainder is always positive.

46. Prove or disprove: If $F(x) = x^4 + x^2 + 1$ is divided by $x - c$ for any real number c, then the remainder is always positive.

47. Prove that $n + 1$ is an integer factor of $n^{2n-1} + 1$, where n is any positive integer. For example, if $n = 3$, then $n + 1 = 4$ is a factor of $n^{2n-1} + 1 = 3^5 + 1 = 244$. [*Hint:* consider $F(x) = x^{2n-1} + 1$.]

4.2 COMPLEX NUMBERS

The problem of finding the zeros of a polynomial function gets quite difficult, even in the simplest cases. Consider, for instance, the function $F(x) = x^2 + 1$. To find a zero for this function, we have to solve the equation

$$x^2 + 1 = 0$$

But this is equivalent to saying $x^2 = -1$, and there is no real number whose square is -1. At this point, we may conclude that F has no *real* zeros.

However, mathematicians have created a different type of number, appropriately called an *imaginary* number, that will solve the equation $x^2 + 1 = 0$.

> **Definition of i**
> The symbol i represents an imaginary number with the property $i^2 = -1$.

Now our function $F(x) = x^2 + 1$ has a zero since

$$F(i) = i^2 + 1 = -1 + 1 = 0$$

Using this new number i, we can expand the set of real numbers **R** into a larger set called the **complex numbers**, denoted by **C**. As we shall see in Section 4.3, all the zeros for any nonconstant polynomial function can be found in **C**.

> **Definition of Complex Numbers**
> The set of complex numbers is given by
> $$\mathbf{C} = \{z : z = a + bi \quad and \quad a, b \in \mathbf{R}\}.$$

Thus we have the following examples of complex numbers:

$$1 + 2i \qquad 3 - \sqrt{2}i \qquad 4 \qquad \frac{1}{6}i$$

A typical complex number, say, $z = a + bi$, consists of two parts: the **real part** of z, represented by a, and the **imaginary part** of z, represented by b. Whenever we say that two complex numbers are equal, we mean that both their real and imaginary parts are identical.

If we consider the collection of all complex numbers whose imaginary part is zero, we just get the set of real numbers. We demonstrate this as follows:

$$\{z : z = a + bi \quad and \quad a \in \mathbf{R}, b = 0\}$$
$$= \{z : z = a \quad and \quad a \in \mathbf{R}\}$$
$$= \mathbf{R}$$

Hence the real numbers are a subset of the complex numbers. Thus we shall remember when talking about an arbitrary complex number z that it is possible that z could be an ordinary real number.

To say that **C** is a valid number system, we must be able to do arithmetic in **C**, which means we should be able to add, subtract, multiply, and divide complex numbers. We accomplish this by following the usual rules of algebra, treating i as a variable symbol subject to the property that $i^2 = -1$.

> **Arithmetic with Complex Numbers**
>
> Let $z_1 = a + bi$ and $z_2 = c + di$.
>
> 1. $z_1 + z_2 = (a + c) + (b + d)i$
> 2. $z_1 - z_2 = (a - c) + (b - d)i$
> 3. $z_1 \cdot z_2 = (ac - bd) + (ad + bc)i$
> 4. $\dfrac{z_1}{z_2} = \dfrac{ac + bd}{c^2 + d^2} + \left(\dfrac{bc - ad}{c^2 + d^2} \right)i$

Rules 1 and 2 are understandable, but 3 and 4 are not so obvious. Let us examine the reasoning behind rule 3. We begin by writing

$$z_1 \cdot z_2 = (a + bi)(c + di)$$

Now, multiply these two binomials just as you do in an ordinary algebra expression of this type.

$$= ac + adi + bci + bdi^2$$

Next, use the fact that $i^2 = -1$.

$$= ac + adi + bci - bd$$

Finally, combine like terms.

$$= (ac - bd) + (ad + bc)i$$

This establishes rule 3.

Rule 4 is a bit trickier to see. We begin by writing the fraction

$$\frac{z_1}{z_2} = \frac{a + bi}{c + di}$$

Now, multiply the numerator and denominator of this fraction by the number $c - di$.

$$\frac{z_1}{z_2} = \frac{(a + bi)(c - di)}{(c + di)(c - di)}$$

$$= \frac{(ac + bd) + (bc - ad)i}{c^2 + d^2}$$

$$= \frac{ac + bd}{c^2 + d^2} + \left(\frac{bc - ad}{c^2 + d^2} \right)i$$

This establishes rule 4.

If $z = a + bi$ is any complex number, then its **conjugate**, denoted by $\bar{z}$, is given by $\bar{z} = a - bi$. In establishing rule 4, we multiplied both the numerator and denominator of a fraction by the conjugate of the denominator.

Example 1

Given $z_1 = 1 + 2i$ and $z_2 = 3 - 4i$, find (a) $z_1 - 3z_2$, (b) $z_1 \cdot z_2$, (c) z_1/z_2, (d) $z_1\bar{z}_1$.

SOLUTION (a)
$$z_1 - 3z_2 = 1 + 2i - 3(3 - 4i)$$
$$= 1 + 2i - 9 + 12i$$
$$= -8 + 14i$$

(b) Rather than apply rule 3 for $z_1 \cdot z_2$ directly from memory, we shall perform the actual multiplication.

$$z_1 \cdot z_2 = (1 + 2i)(3 - 4i)$$
$$= 3 - 4i + 6i - 8i^2$$
$$= 3 - 4i + 6i + 8$$
$$= 11 + 2i$$

(c) Again, rather than apply rule 4 for z_1/z_2, we shall use the method employed in deriving the rule: multiply the numerator and denominator by the conjugate of the denominator, $\bar{z}_2 = 3 + 4i$.

$$\frac{z_1}{z_2} = \frac{1 + 2i}{3 - 4i}$$

$$= \frac{(1 + 2i)(3 + 4i)}{(3 - 4i)(3 + 4i)}$$

$$= \frac{3 + 4i + 6i + 8i^2}{9 + 12i - 12i - 16i^2}$$

$$= \frac{-5 + 10i}{25}$$

$$= \frac{-1}{5} + \frac{2}{5}i$$

(d)
$$z_1\bar{z}_1 = (1 + 2i)(1 - 2i)$$
$$= 1 - 2i + 2i - 4i^2$$
$$= 1 + 4$$
$$= 5 \ \blacksquare$$

Since i has the property that $i^2 = -1$, we can define $\sqrt{-1}$ as equal to i. We generalize this definition as follows.

If $r > 0$, then $\sqrt{-r} = \sqrt{r}\,i$.

Example 2

Find the zeros of $F(x) = x^2 - 2x + 4$ among the complex numbers.

SOLUTION We must solve the equation $F(x) = 0$,

$$x^2 - 2x + 4 = 0$$

Applying the quadratic formula, we get

$$x = \frac{-(-2) \pm \sqrt{(-2)^2 - 4(1)(4)}}{2(1)}$$

$$= \frac{2 \pm \sqrt{-12}}{2}$$

Now, $\sqrt{-12} = \sqrt{12}i = 2\sqrt{3}i$. Hence

$$x = \frac{2 \pm 2\sqrt{3}i}{2} = 1 \pm \sqrt{3}i$$

Therefore the zeros of F are $1 + \sqrt{3}i$ and $1 - \sqrt{3}i$. ■

We began this section by introducing the complex number i so that the polynomial function $F(x) = x^2 + 1$ would have a zero.

$$F(i) = i^2 + 1 = -1 + 1 = 0$$

As it happens, $-i$ is also a zero of this function.

$$F(-i) = (-i)^2 + 1 = i^2 + 1 = -1 + 1 = 0$$

So, we have two complex zeros for F, $\pm i$. Now, recall from Example 2, where we found two complex zeros for the function $F(x) = x^2 - 2x + 4$, that these zeros were $1 \pm \sqrt{3}i$. At least in these two examples, we find that complex zeros of polynomials come in conjugate pairs: if c is a zero, then so is its complex conjugate $\bar{c}$. This result, which turns out to be true in general, is called the Conjugate Root Theorem.

Conjugate Root Theorem

If F is a polynomial function with real coefficients and $c = a + bi$ is a zero of F, then $\bar{c} = a - bi$ is also a zero of F.

To prove this theorem, we need to know the following three elementary properties of complex conjugates:

1. $(\overline{z_1 + z_2}) = \overline{z_1} + \overline{z_2}$

2. If $a \in \mathbf{R}$, then $(\overline{az}) = a\bar{z}$

3. $(\overline{z^n}) = (\bar{z})^n$

The proofs of properties 1 and 2 are left for the exercises; property 3 will be established in Section 11.4 (Exercise 23).

> **PROOF:** To prove the Conjugate Root Theorem, let $F(x) = a_n x^n + a_{n-1} x^{n-1} + \cdots + a_1 x + a_0$, where $a_0, a_1, \ldots, a_n \in \mathbf{R}$. Suppose $c = a + bi$ is a zero of F, $F(c) = 0$. We begin the proof with the statement that the complex conjugate of zero is zero. Hence
>
> $$0 = \overline{0}$$
> $$= \overline{F(c)}$$
> $$= \overline{(a_n c^n + a_{n-1} c^{n-1} + \cdots + a_1 c + a_0)}$$

Using property 1 of complex conjugates, we have

$$0 = (\overline{a_n c^n}) + (\overline{a_{n-1} c^{n-1}}) + \cdots + (\overline{a_1 c}) + \overline{a_0}$$

Using property 2,

$$0 = a_n(\overline{c^n}) + a_{n-1}(\overline{c^{n-1}}) + \cdots + a_1 \overline{c} + a_0$$

Finally, using property 3,

$$0 = a_n(\overline{c})^n + a_{n-1}(\overline{c})^{n-1} + \cdots + a_1 \overline{c} + a_0$$
$$= F(\overline{c})$$

Thus we have shown that $0 = F(\overline{c})$, so $x = \overline{c}$ is also a zero of F. This completes the proof. ∎

Before applying the Conjugate Root Theorem to an example, we recall the division algorithm for polynomial functions given in Section 4.1. Although the algorithm was stated for polynomials with real coefficients (this was part of our definition of a polynomial function), the algorithm remains valid even if the polynomial has complex numbers for coefficients. It follows that the Remainder and the Factor Theorems also remain valid when complex numbers are allowed. Thus for a polynomial function $F(x)$ with either real or imaginary coefficients, if $c \in \mathbf{C}$ is a zero of F, then $x - c$ is a factor of $F(x)$, $F(x) = (x - c)Q(x)$.

Example 3

Consider the function $F(x) = x^3 - 5x^2 + 8x - 6$. Given that $c = 1 + i$ is a zero of F, find the remaining zeros among the complex numbers.

SOLUTION Since $c = 1 + i$ is a zero of F and $F(x)$ has real coefficients, the Conjugate Root Theorem tells us that $\overline{c} = 1 - i$ is also a zero. Using the Factor Theorem, we know that $(x - (1 + i))$ and $(x - (1 - i))$ are factors of $F(x)$. This means that

$$F(x) = (x - (1 + i))(x - (1 - i))Q(x)$$

We wish to find $Q(x)$. First, we multiply $(x - (1 + i))(x - (1 - i))$, obtaining $x^2 - 2x + 2$. We leave it to you to verify this as an exercise. Now,

$$F(x) = (x^2 - 2x + 2)Q(x)$$

Dividing both sides of this equation by $x^2 - 2x + 2$, we have

$$\frac{F(x)}{x^2 - 2x + 2} = Q(x)$$

Since $F(x) = x^3 - 5x^2 + 8x - 6$, we can divide out the left side of this equation longhand. Doing so, we obtain $Q(x) = x - 3$. Therefore

$$F(x) = (x - (1 + i))(x - (1 - i))(x - 3)$$

The zeros of F are $1 \pm i$ and 3. ∎

EXERCISES 4.2

In Exercises 1 to 8, find (a) $z_1 - 2z_2$, (b) $z_1 \cdot z_2$, (c) $\frac{z_1}{z_2}$, (d) $(z_1)^2$ for the given z_1 and z_2.

1. $z_1 = 3 + i, z_2 = 1 + 2i$
2. $z_1 = 4 - 3i, z_2 = 2 + i$
3. $z_1 = 1 + \sqrt{2}i, z_2 = 1 - \sqrt{2}i$
4. $z_1 = 3, z_2 = i$
5. $z_1 = 2i, z_2 = 4$
6. $z_1 = \sqrt{3}i, z_2 = \sqrt{3}i$
7. $z_1 = 5 + 2i, z_2 = 5 - 2i$
8. $z_1 = \frac{\sqrt{3}}{2} - \frac{1}{2}i, z_2 = \frac{1}{2} + i$

In Exercises 9 to 13, find $\bar{z}$, the conjugate of the given number z, and then compute $z\bar{z}$.

9. $z = 3 + 6i$
10. $z = \sqrt{3} - \sqrt{2}i$
11. $z = 4i$
12. $z = 7$
13. $z = i + \frac{1}{i}$ (simplify z first)

In Exercises 14 to 17, determine whether the given number c is a zero of the function F.

14. $F(x) = x^4 + 2x^2 + 1, c = -i$
15. $F(x) = x^2 - 2x + 3, c = 1 + \sqrt{2}i$
16. $F(x) = x^7 + x^6 + x^5 + x^4 + x^3 + x^2 + x + 1, c = i$
17. $F(x) = x^3 - 4x^2 + 5x, c = 2 - i$

In Exercises 18 to 23, find the zeros of the given function among the complex numbers.

18. $F(x) = x^2 - 2x + 5$
19. $F(x) = x^2 - 4x + 5$
20. $F(x) = x^2 + x + 1$
21. $F(x) = 3x^2 + 2x + 1$
22. $F(x) = 6 + x^2$
23. $F(x) = 3 - x + \frac{1}{2}x^2$

In Exercises 24 to 30, multiply and simplify.

24. $(x - \sqrt{2}i)(x + \sqrt{2}i)$
25. $(x - \sqrt{3}i)(x + \sqrt{3}i)$
26. $(x - (1 + i))(x - (1 - i))$
27. $(x - (2 + i))(x - (2 - i))$
28. $(x + (1 - 3i))(x + (1 + 3i))$
29. $(x + (3 - 2i))(x + (3 + 2i))$
30. $(x - (a + bi))(x - (a - bi))$, a and b constants.

In Exercises 31 to 37, a polynomial function $F(x)$ is given along with one of its zeros, c. Find the remaining zeros of F among the complex numbers.

31. $F(x) = x^3 - 5x^2 + 9x - 5, c = 2 + i$
32. $F(x) = x^3 + 2x^2 - 5x + 12, c = 1 - \sqrt{2}i$
33. $F(x) = 2x^3 + x^2 + 2x + 1, c = i$
34. $F(x) = x^4 - 2x^3 + 6x^2 + 8x - 40, c = 1 - 3i$

35. $F(x) = 2x^4 + x^3 + 5x^2 + 3x - 3$, $c = -\sqrt{3}i$

36. $F(x) = x^6 - 2x^5 + 2x^4 + x^2 - 2x + 2$, $c = 1 + i$

37. $F(x) = x^4 - 5x^3 + 18x^2 + 31x - 25$, $c = 3 + 4i$

38. Prove $\overline{(z_1 + z_2)} = \bar{z}_1 + \bar{z}_2$ for any $z_1, z_2 \in \mathbf{C}$.

39. Prove if $a \in \mathbf{R}$, then $\overline{(az)} = a\bar{z}$ for any $z \in \mathbf{C}$.

40. Prove: for any $z \in \mathbf{C}$, $z\bar{z}$ is a nonnegative real number.

In Exercises 41 and 42, answer true or false.

41. Any nonconstant polynomial function with real coefficients will have only an even number of complex zeros with nonzero imaginary parts.

42. Any odd degree polynomial function with real coefficients will have only an odd number of real zeros.

∘4.3   *THE FUNDAMENTAL THEOREM OF ALGEBRA*

We are ready to state a theorem first proven by Carl Friedrich Gauss (1777–1855). Gauss, who many consider one of the greatest mathematicians, made significant contributions in both physics and mathematics. The theorem we are about to state reveals the basic structure of all polynomial functions; we call it the Fundamental Theorem of Algebra.

> **Fundamental Theorem of Algebra**
> If $F(x)$ is a polynomial function of degree at least 1, then there exists a number $c \in \mathbf{C}$ such that $F(c) = 0$.

The Fundamental Theorem tells us that any nonconstant polynomial function has at least one zero among the complex numbers, but it does not say how to find the zero. The question of how to go about finding zeros of polynomials is considered in Section 4.4.

Recall that the definition of a polynomial function given in Section 3.3 required that the coefficients in the polynomial be real numbers. It turns out that the Fundamental Theorem remains true even if the polynomial has complex coefficients (with nonzero imaginary parts). In any case, the proof of this theorem is difficult, and so we omit it.

As mentioned, the Fundamental Theorem leads us to the basic structure of all polynomial functions. We state the result as follows.

> **Structure Theorem for Polynomials**
> If $F(x)$ is a polynomial function of degree $n > 0$, then
> $$F(x) = a(x - c_1)(x - c_2) \cdots (x - c_n)$$
> where a is a constant and $c_1, c_2, \ldots c_n \in \mathbf{C}$.

PROOF: By the Fundamental Theorem, there exists $c_1 \in \mathbf{C}$ such that $F(c_1) = 0$. By the Factor Theorem, we can write

$$F(x) = (x - c_1)Q_1(x)$$

where $Q_1(x)$ is a polynomial (possibly with imaginary coefficients) of degree $n - 1$.

Next, if $n - 1 > 0$, then we can apply the Fundamental Theorem to $Q_1(x)$. It follows that there exists $c_2 \in \mathbf{C}$ such that $Q_1(c_2) = 0$. Again using the Factor Theorem, we have

$$Q_1(x) = (x - c_2)Q_2(x)$$

where $Q_2(x)$ is a polynomial of degree $n - 2$. Now, if $n - 2 > 0$, we apply the Fundamental Theorem to $Q_2(x)$, and so on. The process is repeated until we reach $Q_n(x)$, which has degree $n - n = 0$, which means that $Q_n(x)$ must be a constant, a. Putting all these results together, we have

$$
\begin{aligned}
F(x) &= (x - c_1)Q_1(x) \\
&= (x - c_1)(x - c_2)Q_2(x) \\
&\qquad \vdots \\
&= (x - c_1)(x - c_2)\cdots(x - c_n)Q_n(x) \\
&= (x - c_1)(x - c_2)\cdots(x - c_n)a
\end{aligned}
$$

which is the desired result. ■

Remarks

1. Suppose we have a polynomial function F of degree $n > 0$ and we write it in factored form as indicated by the Structure Theorem.

$$F(x) = a(x - c_1)(x - c_2)\cdots(x - c_n)$$

By multiplying this out, we obtain

$$F(x) = a_n x^n + a_{n-1}x^{n-1} + \cdots + a_1 x + a_0$$

Notice that the leading coefficient, a_n, must be the same as the constant term a that appears in the factored form of $F(x)$.

2. The numbers $c_1, c_2, \ldots, c_n$ in the Structure Theorem are precisely the zeros of the polynomial function. This list of numbers may have repeats in it. For example, consider the function

$$F(x) = 2(x - 4)(x - 4)(x - i)(x + i)$$

The zeros of this function are 4, i, and $-i$; however, the factor $x - 4$ occurs twice, so we call 4 a zero of **multiplicity** 2.

Example 1

A certain polynomial function F has degree 3 and zeros at $x = 1, -2$, and 3. Furthermore, $F(0) = 5$. Find $F(x)$.

SOLUTION According to the Structure Theorem, F has the form

$$F(x) = a(x - c_1)(x - c_2)(x - c_3)$$

Since we know that the zeros of F are 1, -2, and 3, we have

$$F(x) = a(x - 1)(x + 2)(x - 3)$$

Now, we must find the leading coefficient a. We are told that $F(0) = 5$; hence

$$5 = F(0) = a(0 - 1)(0 + 2)(0 - 3)$$

$$5 = a(-1)(2)(-3)$$

$$5 = 6a$$

$$\frac{5}{6} = a$$

Therefore $F(x) = \frac{5}{6}(x - 1)(x + 2)(x - 3)$. If we multiply this out, we find

that $F(x) = \frac{5}{6}x^3 - \frac{5}{3}x^2 - \frac{25}{6}x + 5$. ■

Example 2

Find the polynomial function $P(x)$ with real coefficients such that the degree of P is 4, -1 is a zero of multiplicity 2, $P(i) = 0$, and $P(-2) = 10$.

SOLUTION Since the degree of P is 4, $P(x)$ must look like

$$P(x) = a(x - c_1)(x - c_2)(x - c_3)(x - c_4)$$

Now, what are the zeros of P? We know that -1 is a zero of multiplicity 2. Furthermore, $P(i) = 0$, so that i is another zero. Now we need just one more zero. Recall that if i is a zero of a polynomial with real coefficients, then by the Conjugate Root Theorem its complex conjugate $-i$ is also a zero. Therefore

$$P(x) = a(x + 1)(x + 1)(x - i)(x + i)$$

Finally, to get the leading coefficient a, we use the fact that $P(-2) = 10$.

$$10 = P(-2) = a(-2 + 1)(-2 + 1)(-2 - i)(-2 + i)$$

$$10 = a(-1)^2(-2 - i)(-2 + i)$$

$$10 = a(4 - i^2)$$

$$10 = 5a$$

$$2 = a$$

Thus $P(x) = 2(x + 1)(x + 1)(x - i)(x + i)$. If we multiply this out, we find that $P(x) = 2x^4 + 4x^3 + 4x^2 + 4x + 2$. ■

Examples 1 and 2 illustrate once again that if we know a zero of a polynomial function, say, $x = c$, then we know that $x - c$ is a linear factor of the function. Of course, the converse of this is also true: if we know a linear factor, then we know a zero. It follows that the processes of factoring a polynomial function and finding its zeros are intimately connected.

Example 3

Factor $F(x)$ into a product of linear terms, as in the Structure Theorem. State the zeros of F and their multiplicity.

(a) $F(x) = x^4 - 18x^2 + 81$

(b) $F(x) = 4x^3 - 4x^2 - x$

(c) $F(x) = x^3 - x^2 + 2$ (*Hint:* $x = -1$ is a zero.)

SOLUTION (a) We factor $F(x)$ as follows:

$$x^4 - 18x^2 + 81$$

$$(x^2 - 9)(x^2 - 9)$$

$$(x + 3)(x - 3)(x + 3)(x - 3)$$

Thus $F(x) = (x + 3)(x - 3)(x + 3)(x - 3)$. So F has zeros -3 and 3, each of multiplicity 2.

(b) We can factor out an x in $F(x) = 4x^3 - 4x^2 - x$.

$$F(x) = x(4x^2 - 4x - 1)$$

Now, $4x^2 - 4x - 1$ does not factor readily. However, we can find its zeros using the quadratic formula.

$$x = \frac{4 \pm \sqrt{16 - 4(4)(-1)}}{2(4)} = \frac{1 \pm \sqrt{2}}{2}$$

Thus, by the Structure Theorem, we have

$$F(x) = 4x^3 - 4x^2 - x = ax\left(x - \frac{1 + \sqrt{2}}{2}\right)\left(x - \frac{1 - \sqrt{2}}{2}\right)$$

Note that a must equal the leading coefficient 4 (see Remark 1 under the Structure Theorem). We conclude that

$$F(x) = 4x\left(x - \frac{1 + \sqrt{2}}{2}\right)\left(x - \frac{1 - \sqrt{2}}{2}\right)$$

The zeros of F are 0, $\dfrac{1 + \sqrt{2}}{2}$, and $\dfrac{1 - \sqrt{2}}{2}$, each of multiplicity 1.

(c) Since we are told that -1 is a zero, we know $x - (-1) = x + 1$ is a factor of $F(x) = x^3 - x^2 + 2$. Using synthetic division,

$$
\begin{array}{r|rrrr}
-1 & 1 & -1 & 0 & 2 \\
   &   & -1 & 2 & -2 \\
\hline
   & 1 & -2 & 2 & 0 \\
\end{array}
$$

we have

$$F(x) = (x + 1)(x^2 - 2x + 2)$$

Next, applying the quadratic formula to $x^2 - 2x + 2$, we find that its zeros are $x = 1 \pm i$. It follows that

$$F(x) = (x + 1)(x - (1 + i))(x - (1 - i))$$

The zeros of F are -1, $1 + i$, and $1 - i$, each of multiplicity 1. ∎

The following result is an important consequence of the Structure Theorem.

Corollary to the Structure Theorem

A polynomial function of degree $n > 0$ has exactly n zeros, counting multiplicities, among the complex numbers.

PROOF: Suppose F is a polynomial function of degree $n > 0$. Then the Structure Theorem says that

$$F(x) = a(x - c_1)(x - c_2) \cdots (x - c_n)$$

Thus F has the n zeros $c_1, c_2, \ldots, c_n$. But could F have any other zeros? We wish to show that this is not possible. To do this, let c be any complex number different from $c_1, c_2, \ldots, c_n$. We must demonstrate that $F(c) \neq 0$. We have

$$F(c) = a(c - c_1)(c - c_2) \cdots (c - c_n)$$

which is a product of $n + 1$ *nonzero* complex numbers. Although it may seem obvious, it must be *proved* that a finite product of nonzero complex numbers is not zero (this is proved in Section 11.4 using mathematical induction). Accepting this fact, we have $F(c) \neq 0$ as desired. Thus the only zeros of F are precisely $c_1, c_2, \ldots, c_n$. This completes the proof. ∎

The corollary says that any polynomial function of positive degree n has exactly n *complex* zeros, counting multiplicities. How many of these zeros are *real*? Consider the following examples, which show that none, some, or all of the zeros can be real:

Function	Zeros Complex	Zeros Real
$F(x) = x^2 + 1$	$i, -i$	None
$G(x) = (x^2 + 1)(x - 1)$	$i, -i$	1
$H(x) = x(x^2 - 1)$	None	$0, 1, -1$

In any case, the number of real zeros of a nonconstant polynomial function cannot exceed the degree of the function. Examples 4 and 5 illustrate applications of this result.

Example 4

Prove that if F is a polynomial function of degree $n > 1$, then any straight line cannot intersect the graph of F more than n times.

SOLUTION The graph of $y = F(x)$ intersects the graph of the line $y = mx + b$ whenever x satisfies

$$F(x) = mx + b$$

This is equivalent to

$$F(x) - mx - b = 0$$

But $F(x) - mx - b$ is a polynomial function of degree n and therefore cannot have more than n real zeros. Thus the graph of F cannot intersect the line $y = mx + b$ more than n times. In the special case when the line is vertical, the graph of F will intersect it only once (by the definition of a function). ■

For a specific case of Example 4, consider Figure 4.1, the graph of $F(x) = x^3 - 6x^2 + 8x + 4$, a degree 3 polynomial function, with three straight lines ℓ_1, ℓ_2, and ℓ_3 that intersect the graph of F once, twice, and three times, respectively. It does not appear that any line will intersect the graph of F more than three times, but this is exactly what we should expect. Example 4 says that no line may intersect the graph of a degree 3 polynomial function more than three times. On the other hand, the function G in Figure 4.2 has a line that intersects the graph five times, which means that the degree of G must be at least 5.

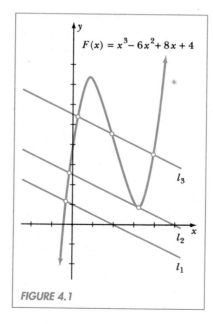

$F(x) = x^3 - 6x^2 + 8x + 4$

FIGURE 4.1

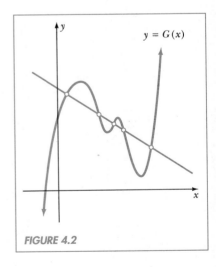

$y = G(x)$

FIGURE 4.2

Example 5

Prove that if two polynomial functions are equal, then their corresponding coefficients are equal.

SOLUTION Before starting the proof, let us see what this result says in a specific case. For instance, suppose

$$F(x) = a_3 x^3 + a_2 x^2 + a_1 x + a_0 \quad \text{and} \quad G(x) = b_3 x^3 + b_2 x^2 + b_1 x + b_0$$

are equal functions, so that $F(x) = G(x)$ for all x. The conclusion we are going to prove is that the corresponding coefficients of $F(x)$ and $G(x)$ are equal. Thus

$$a_3 = b_3 \qquad \text{equating } x^3\text{-coefficients}$$
$$a_2 = b_2 \qquad \text{equating } x^2\text{-coefficients}$$
$$a_1 = b_1 \qquad \text{equating } x\text{-coefficients}$$
$$a_0 = b_0 \qquad \text{equating constant terms}$$

Now let us begin the proof.

Let F and G be any two polynomial functions with the coefficients of x^k given by a_k in $F(x)$ and b_k in $G(x)$. Now suppose that F and G are equal functions, so that $F(x) = G(x)$ for all x. We wish to show that $a_k = b_k$ for each integer $k \geq 0$. To do so, consider the polynomial function $F(x) - G(x)$. Note that the coefficient of x^k in $F(x) - G(x)$ is $a_k - b_k$. Furthermore,

$$F(x) - G(x) = 0 \qquad \text{for all } x$$

since $F(x) = G(x)$ for all x. Therefore $F(x) - G(x)$ has infinitely many zeros. However, we know that the number of zeros for any polynomial function of positive degree n cannot exceed n, so it follows that $F(x) - G(x)$ must have degree 0. This means that the coefficient of x^k in $F(x) - G(x)$ is zero for each $k \geq 1$. Thus

$$a_k - b_k = 0 \qquad \text{for each } k \geq 1$$

This implies that $a_k = b_k$ for each $k \geq 1$, which leaves us with $F(x) - G(x) = a_0 - b_0$ for all x. But $F(x) - G(x) = 0$ from above, so $a_0 - b_0 = 0$, or $a_0 = b_0$. Thus we have shown that $a_k = b_k$ for each integer $k \geq 0$. ∎

Example 6

Suppose $Ax^2 + Bx + C = 2x$ for all x. Find A, B, and C.

SOLUTION We are given two equal polynomial functions, $Ax^2 + Bx + C$ and $2x$. Note that we may write

$$Ax^2 + Bx + C = 0x^2 + 2x + 0$$

By example 5, the corresponding coefficients must be equal. Thus we have

$$A = 0 \qquad \text{equating } x^2\text{-coefficients}$$

$$B = 2 \qquad \text{equating } x\text{-coefficients}$$

$$C = 0 \qquad \text{equating constant terms} \quad \blacksquare$$

Example 7

Suppose $A(x + 2) + Bx = 1$ for all x. Find A and B.

SOLUTION The left side of the equation is a polynomial function of degree 1. We rewrite it in descending powers of x as follows:

$$A(x + 2) + Bx$$

$$Ax + 2A + Bx$$

$$Ax + Bx + 2A$$

$$(A + B)x + 2A$$

We know that this polynomial is equal to the constant function 1, which we may write as $0x + 1$. Thus

$$(A + B)x + 2A = 0x + 1$$

It follows that

$$A + B = 0 \qquad \text{equating } x\text{-coefficients}$$

$$2A = 1 \qquad \text{equating constant terms}$$

From the second equation, $A = 1/2$. Substituting this into the first equation, we find that $1/2 + B = 0$, or $B = -1/2$. $\quad \blacksquare$

EXERCISES 4.3

In Exercises 1 to 6, list the zeros of the given function. State the multiplicity if it is greater than 1.

1. $F(x) = 4(x - 2)(x + 5)(x + 5)$

2. $F(x) = -2(x + 1)^3(x - 2)$

3. $F(x) = -3x(x^2 - 2)$

4. $F(x) = 5x^2(x + 2)$

5. $F(x) = (x^2 + 1)(x^2 - 9)$

6. $F(x) = (x^2 - 3)(x^2 + 3)$

In Exercises 7 to 10, find a polynomial function G with real coefficients satisfying the given conditions. (Answers are not unique.)

7. G has exactly three zeros at 0, 1/2, and -3.

8. The degree of G is 4, G has zeros at 1, 2, and $2i$.

9. The degree of G is 7, G has zeros at $1 + i$, $2 + i$, and a zero of multiplicity 3 at $x = 3$.

10. The degree of G is 6, 1/4 is a zero of multiplicity 3, $G(6) = 0$, and $G(\sqrt{2}i) = 0$.

In Exercises 11 to 16, find the polynomial function P with real coefficients satisfying the given conditions.

11. P has degree 3, zeros at $x = -1, 2, 3$, and $P(0) = -12$.

12. P has degree 4, zeros at $x = 0, 1, 2, 3$, and $P(-2) = 3$.

13. P has degree 2, $P(\sqrt{3}i) = 0$, and $P(2) = 21$.

14. P has degree 3, $P(0) = 0$, $P(-1 - i) = 0$, and $P(1) = 1$.

15. P has degree 4, $P(i + 4) = 0$, -3 is a zero of multiplicity 2, and $P(1) = 4$.

16. P has degree 3, $P(-4) = 0$, $P\left(\dfrac{1}{2}i\right) = 0$, and $P(-1) = 15$.

In Exercises 17 to 32, factor $F(x)$ into a product of linear terms, as in the Structure Theorem. Find the zeros of F. State the multiplicity if it is greater than 1.

17. $F(x) = x^3 - x^2 + x - 1$

18. $F(x) = 4x^3 - x^2 + 4x - 1$

19. $F(x) = x^3 + 1$

20. $F(x) = x^3 + 8$

21. $F(x) = 3x^3 + 2x^2 - x$

22. $F(x) = x^3 - 6x^2 + x$

23. $F(x) = x^3 + x + 10$ (*Hint:* $x = -2$ is a zero.)

24. $F(x) = x^3 - 6x^2 + 13x - 10$ (*Hint:* $x = 2$ is a zero.)

25. $F(x) = x^4 - 81$

26. $F(x) = x^4 - 16$

27. $F(x) = x^4 + 8x^3 + 16x^2$

28. $F(x) = x^4 + x^2 - 6$

29. $F(x) = 12x^5 - 13x^3 - 4x$

30. $F(x) = x^5 - x^3 + x^2 - 1$

31. $F(x) = x^6 - 7x^3 - 8$

32. $F(x) = 2x^6 + 52x^3 - 54$

In Exercises 33 to 38, find the constants A and B, or A, B, and C, so that the equation holds for all x.

33. $Ax^2 + Bx + C = 2x^2 - 1$

34. $Ax^2 + Bx + C = 3x^2 - x$

35. $Ax^2 + Bx + C = 5x + 4$

36. $Ax^2 + Bx + C = 1$

37. $A(x + 1) + Bx = 1$

38. $Ax + B(x - 3) = 1$

In Exercises 39 to 42, find the degree 3 polynomial function $F(x)$ with the given graph.

39.

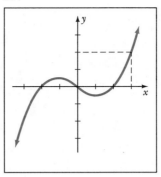

40.

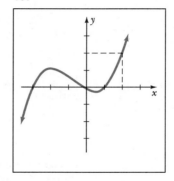

41. (*Hint:* first find $y = F(x) - 4$.)

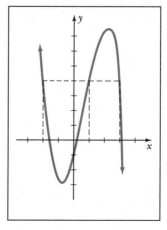

42. (*Hint:* first find $y = F(x) - 2$.)

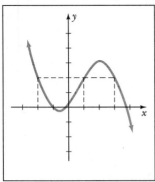

In Exercises 43 and 44, find the degree 4 polynomial function $G(x)$ with the given graph.

43.

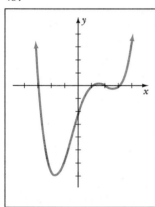

44.

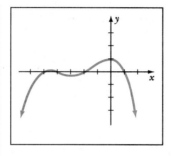

45. Determine the maximum number of times the graphs of two polynomial functions F and G may intersect, assuming F and G are not equal and (a) degree F = degree G = 2; (b) degree F = degree G = 3; (c) n = degree $F \geq$ degree $G \geq 1$. [*Hint:* the functions intersect when $F(x) = G(x)$, or equivalently, when $F(x) - G(x) = 0$.]

46. Prove that any line in the plane must intersect the graph of a degree 3 polynomial function at least once. [*Hint:* The line $y = mx + b$ intersects F when $F(x) - (mx + b) = 0$. A vertical line automatically intersects F once.]

47. **The graph** of a polynomial function is shown. Determine the minimum degree possible for the function.

(a)

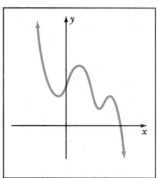

(b)

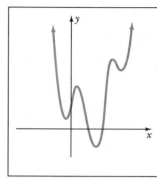

(c)

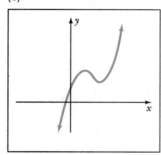

In Exercises 48 to 51, answer true or false.

48. **It is possible for a degree 3 polynomial (with real coefficients) to have no real zeros.**

49. **It is possible for a degree 4 polynomial (with real coefficients) to have no real zeros.**

50. **For any positive integer n, there exists a polynomial function whose graph intersects the x-axis at exactly n distinct points.**

51. **There exists a non-constant polynomial function whose graph intersects the x-axis an infinite number of times.**

In this section we examine the problem of finding the zeros for a general polynomial function,

$$F(x) = a_n x^n + a_{n-1} x^{n-1} + \cdots + a_1 x + a_0$$

If the degree of F is 1, then the solution is easy since we have

$$a_1 x + a_0 = 0 \quad \text{implies} \quad x = \frac{-a_0}{a_1}$$

If the degree of F is 2, then we must solve the equation

$$a_2 x^2 + a_1 x + a_0 = 0$$

which is accomplished by using the quadratic formula. When the degree of F is 3 or 4, the problem becomes more difficult, although formulas have been found. For example, the zeros of the degree 3 polynomial function $F(x) = x^3 + ax + b$ can be found by using the following formula:

$$x = \sqrt[3]{-\frac{b}{2} + \sqrt{\frac{a^3}{27} + \frac{b^2}{4}}} + \sqrt[3]{-\frac{b}{2} - \sqrt{\frac{a^3}{27} + \frac{b^2}{4}}}$$

We also have the formula

$$x = \pm \sqrt{\frac{-a \pm \sqrt{a^2 - 4b}}{2}}$$

which gives us the zeros of the degree 4 polynomial function $G(x) = x^4 + ax^2 + b$. Of course, all degree 3 and 4 polynomials do not have the same form as the functions $F(x)$ and $G(x)$ above for which these formulas were written. This problem can be overcome, but only at the cost of making the formulas much more complicated. Finally, with degree 5 polynomials, the situation becomes hopeless since it has been shown that no general formula for the zeros exists. Thus, for polynomial functions of degree 3 or more, to find the zeros we shall use a method of careful guessing rather than formulas. Our objective in this section is to see how one makes the appropriate guesses.

Example 1

Find the zeros of $F(x) = x^4 + x^3 - 7x^2 + 5x$.

SOLUTION We notice that $x = 0$ is a zero of F. Factoring $x - 0$ out of F, we have

$$F(x) = x(x^3 + x^2 - 7x + 5)$$

Now our problem reduces to finding the zeros of $Q(x) = x^3 + x^2 - 7x + 5$. By guessing, we discover that $x = 1$ is a zero of Q. Using synthetic division, we factor $x - 1$ out of $Q(x)$.

$$
\begin{array}{r|rrrr}
1 & 1 & 1 & -7 & 5 \\
  &   & 1 & 2 & -5 \\
\hline
  & 1 & 2 & -5 & 0
\end{array}
$$

Therefore $Q(x) = (x - 1)(x^2 + 2x - 5)$. It follows that

$$F(x) = x(x - 1)(x^2 + 2x - 5)$$

Finally, we must find the zeros of $x^2 + 2x - 5$. Applying the quadratic formula, we find that the zeros are $-1 \pm \sqrt{6}$. Therefore we conclude that

$$F(x) = x(x - 1)(x - (-1 + \sqrt{6}))(x - (-1 - \sqrt{6}))$$

and the zeros of F are 0, 1, $-1 + \sqrt{6}$, and $-1 - \sqrt{6}$. ∎

Each time we find a zero of a polynomial function, we factor it out and look at the quotient $Q(x)$ for the remaining zeros. Since $Q(x)$ has a smaller degree, the search for the remaining zeros usually becomes easier. Therefore, the question we must ask is, How do we make the initial guesses for the zeros? The following result, known as the Rational Root Theorem, gives us a method for finding real zeros that happen to be rational numbers.

Rational Root Theorem

Let $F(x) = a_n x^n + a_{n-1} x^{n-1} + \cdots + a_1 x + a_0$ be a polynomial function of degree $n > 0$ with integer coefficients. If $x = r/s$ is a rational zero of F (r and s integers, r/s reduced to lowest terms), then r must be an integer factor of a_0 and s must be an integer factor of a_n.

Warning

This theorem works only for polynomial functions with integer coefficients. Furthermore, it only tells us what a rational zero of the function must look like. It does *not* say that the function should have any rational zeros; indeed, there may be none at all.

We outline a proof of the Rational Root Theorem in the exercises.

Example 2

Use the Rational Root Theorem to list all the possible rational zeros for the given function.

(a) $F(x) = 6x^2 + x - 2$ (b) $G(x) = x^4 - x^2 - 2$.

SOLUTION (a) According to the theorem, any rational zero of F must look like

$$\pm \frac{\text{factor of 2}}{\text{factor of 6}}$$

(We are taking into account all the possible signs involved with the $\pm$ symbol in front.) The positive integer factors of 2 are 1 and 2, and the positive integer factors of 6 are 1, 2, 3, and 6. We use the notation

$$\pm \frac{1, 2}{1, 2, 3, 6}$$

to summarize the possibilities. Listing them, we have

$$\pm 1 \qquad \pm \frac{1}{2} \qquad \pm \frac{1}{3} \qquad \pm \frac{1}{6} \qquad \pm 2 \qquad \pm \frac{2}{3}$$

Note that a simple application of the quadratic formula to F gives us the zeros directly, $x = 1/2$ and $x = -2/3$. These are of course in our list of possibilities.

(b) Any rational zero of G must look like

$$\pm \frac{\text{factor of 2}}{\text{factor of 1}} \qquad \text{or} \qquad \pm \frac{1, 2}{1}$$

This gives us the possibilities ± 1, ± 2. It is interesting to note here that none of these possibilities are zeros for G, as can be verified by substituting each of them into the formula for $G(x)$. This process tells us that if there are any real zeros of G, then they must be irrational numbers because all the rational possibilities do not check out. ∎

Example 3

Find the zeros of $F(x) = 6x^4 + 5x^3 - 9x^2 - x + 2$.

SOLUTION By the Rational Root Theorem, any rational zero of F (if there are any) must look like

$$\pm \frac{\text{factor of 2}}{\text{factor of 6}} \qquad \text{or} \qquad \pm \frac{1, 2}{1, 2, 3, 6}$$

The possibilities are ± 1, $\pm \frac{1}{2}$, $\pm \frac{1}{3}$, $\pm \frac{1}{6}$, ± 2, $\pm \frac{2}{3}$. By trial and error, using synthetic division, we find that ± 1 and $1/2$ do not work. Next, trying

$x = -1/2$, we have

$$-\frac{1}{2} \;\Big|\; \begin{array}{rrrrr} 6 & 5 & -9 & -1 & 2 \\ & -3 & -1 & 5 & -2 \\ \hline 6 & 2 & -10 & 4 & 0 \end{array}$$

Therefore $-1/2$ is a zero and we may write

$$F(x) = \left(x + \frac{1}{2}\right)(6x^3 + 2x^2 - 10x + 4)$$

It is advisable to factor out any common factors when possible because doing so makes the quotient function easier to work with. Thus we write

$$F(x) = 2\left(x + \frac{1}{2}\right)(3x^3 + x^2 - 5x + 2)$$

Our next step is to look for the zeros of $Q(x) = 3x^3 - x^2 - 5x + 2$. Applying the Rational Root Theorem to Q, we find that the possible rational zeros are $\pm 1, \ \pm \frac{1}{3}, \ \pm 2, \ \pm \frac{2}{3}$. Since ± 1 did not work before, we do not bother to consider them again. We are left with the possibilities $\pm \frac{1}{3}, \ \pm 2,$ $\pm \frac{2}{3}$. By trial and error, we find that $x = 2/3$ works.

$$\frac{2}{3} \;\Big|\; \begin{array}{rrrr} 3 & 1 & -5 & 2 \\ & 2 & 2 & -2 \\ \hline 3 & 3 & -3 & 0 \end{array}$$

Thus

$$F(x) = 2\left(x + \frac{1}{2}\right)\left(x - \frac{2}{3}\right)(3x^2 + 3x - 3)$$

or, factoring out the common factor of 3 from the last term,

$$F(x) = 6\left(x + \frac{1}{2}\right)\left(x - \frac{2}{3}\right)(x^2 + x - 1)$$

Now, we apply the quadratic formula to $x^2 + x - 1$, which yields zeros at $x = (-1 \pm \sqrt{5})/2$. Hence

$$F(x) = 6\left(x + \frac{1}{2}\right)\left(x - \frac{2}{3}\right)\left(x - \left(\frac{-1 + \sqrt{5}}{2}\right)\right)\left(x - \left(\frac{-1 - \sqrt{5}}{2}\right)\right)$$

We conclude that the zeros of F are $-\frac{1}{2}, \frac{2}{3}, \frac{-1 + \sqrt{5}}{2}$, and $\frac{-1 - \sqrt{5}}{2}$. ∎

Example 4

Find the zeros of

$$F(x) = \frac{1}{3}x^3 + x^2 + \frac{5}{12}x - \frac{1}{2}.$$

SOLUTION   The Rational Root Theorem does not apply to F because it does not have integer coefficients. However, we notice the following fact:

c is a zero of F if and only if c is a zero of $k \cdot F$, where k is a nonzero constant.

This is obvious because $F(c) = 0$ if and only if $k \cdot F(c) = 0$, $k \neq 0$. Therefore we shall consider the function G, where

$$G(x) = 12 \cdot F(x)$$
$$= 4x^3 + 12x^2 + 5x - 6$$

The zeros of F are precisely the zeros of G. Now, applying the Rational Root Theorem to G, we find that the possible rational zeros look like

$$\pm \frac{\text{factor of } 6}{\text{factor of } 4} \quad \text{or} \quad \pm \frac{1, 2, 3, 6}{1, 2, 4}$$

which gives us the possibilities ± 1, $\pm \frac{1}{2}$, $\pm \frac{1}{4}$, ± 2, ± 3, $\pm \frac{3}{2}$, $\pm \frac{3}{4}$, and ± 6.

By trial and error, we find that $x = 1/2$ works.

$$
\begin{array}{r|rrrr}
\frac{1}{2} & 4 & 12 & 5 & -6 \\
 & & 2 & 7 & 6 \\
\hline
 & 4 & 14 & 12 & 0 \\
\end{array}
$$

Thus

$$G(x) = \left(x - \frac{1}{2}\right)(4x^2 + 14x + 12)$$

$$= 2\left(x - \frac{1}{2}\right)(2x^2 + 7x + 6)$$

Now, $2x^2 + 7x + 6$ factors as $(2x + 3)(x + 2)$. Therefore

$$G(x) = 2\left(x - \frac{1}{2}\right)(2x + 3)(x + 2)$$

$$= 4\left(x - \frac{1}{2}\right)\left(x + \frac{3}{2}\right)(x + 2)$$

We conclude that the zeros of G and F are $\frac{1}{2}$, $-\frac{3}{2}$, and -2. ∎

In the examples, we were fortunate to find at least one rational zero from the list of possibilities given by the Rational Root Theorem. However, what happens when none of the possibilities work? In this case, if a real zero exists, it must be irrational. We discuss how to approximate these irrational zeros in Section 4.6.

EXERCISES 4.4

In Exercises 1 and 2, list the possible rational zeros for the function as given by the Rational Root Theorem.

1. $F(x) = 3x^5 - 4x^3 + x - 6$
2. $F(x) = 4x^4 - 2x^3 + x^2 - 10$

In Exercises 3 to 8, find the zeros of the given function. Write the function in the form $F(x) = a(x - c_1)(x - c_2) \cdots (x - c_n)$.

3. $F(x) = x^3 + 8x^2 - 9$
4. $F(x) = x^4 + 2x^3 - 3x^2 - 4x$
5. $F(x) = x^4 - x^3 - 6x^2 - 4x$
6. $F(x) = 6x^4 - 31x^3 + 41x^2 - 18x + 2$
7. $F(x) = 12x^4 - 8x^3 - 19x^2 + 4x + 3$
8. $F(x) = 6x^5 + 17x^4 + 15x^3 + 9x^2 - 2x$

In Exercises 9 to 12, prove that the only real zeros of the given function must be irrational.

9. $G(x) = x^3 + x - 1$
10. $G(x) = x^5 - x^4 + x^3 - x^2 + x - 2$
11. $G(x) = 2x^4 - x + 3$
12. $G(x) = 3x^3 + x^2 - 5$

In Exercises 13 and 14, find all points of intersection of the graphs of F and G.

13. $F(x) = 3x^3 + x^2 - 7x + 4, G(x) = -x^3 + 2x^2 - 6x + 9$
14. $F(x) = 24x^3 - 5x^2, G(x) = 21x^2 - 9x + 1$

In Exercises 15 to 18, find the zeros of the given function.

15. $F(x) = x^3 + \frac{1}{10}x^2 - \frac{7}{10}x + \frac{1}{5}$
16. $F(x) = 3x^3 - \frac{25}{4}x^2 + \frac{3}{8}x + \frac{1}{4}$
17. $F(x) = 3x^3 + x^2 - \frac{25}{4}x - 3$
18. $F(x) = \frac{1}{3}x^3 + \frac{41}{36}x^2 + \frac{1}{3}x - \frac{1}{4}$

In Exercises 19 to 24, graph the given polynomial function. Label the zeros.

19. $F(x) = 3x^3 - 6x^2 + x + \frac{10}{9}$
20. $F(x) = x^3 - x^2 - \frac{11}{4}x + \frac{3}{2}$
21. $F(x) = x^4 - x^3 - \frac{19}{4}x^2 + x + \frac{15}{4}$
22. $F(x) = x^4 + x^3 - \frac{5}{4}x^2 - \frac{3}{4}x$

23. The product of three consecutive integers is three more than their sum. Find the integers.

24. Three consecutive integers have the following property: the sum of the squares of the first two equals the cube of the third. Find the integers.

25. A frame for a box with square base is to be made from 36 inches of wire (see Figure 4.3). If the volume of the box is to be 20 cubic inches, find the dimensions of the box.

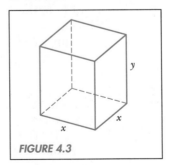

FIGURE 4.3

26. An isosceles triangle (Figure 4.4) has perimeter 18 and area 12 square units. Find the lengths of the sides.

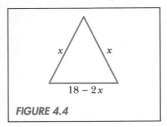

FIGURE 4.4

In Exercises 27 to 31, use the Rational Root Theorem to prove each statement.

27. The square root of 2 is irrational. [*Hint:* consider the polynomial function $F(x) = x^2 - 2$.]

28. The square root of 3 is irrational.

29. The square root of a prime number is irrational.

30. The cube root of a prime number is irrational.

31. It is not possible to find three consecutive integers such that the sum of the cubes of the first two equals the cube of the third.

32. This exercise outlines a proof for the Rational Root Theorem. Let $F(x) = a_n x^n + a_{n-1} x^{n-1} + \cdots + a_1 x + a_0$ be a polynomial function of degree $n > 0$ with integer coefficients. Suppose $x = r/s$ is a rational zero of F. Assume r/s is reduced to lowest terms (this means r and s have no common factors). Since $F(r/s) = 0$, we have

$$a_n \left(\frac{r}{s}\right)^n + a_{n-1} \left(\frac{r}{s}\right)^{n-1} + \cdots + a_1 \left(\frac{r}{s}\right) + a_0 = 0$$

Multiplying this equation by s^n, we get

$$a_n r^n + a_{n-1} r^{n-1} s + \cdots + a_1 r s^{n-1} + a_0 s^n = 0 \quad (*)$$

We wish to show that s is a factor of a_n and r is a factor of a_0.

(a) Solve Equation $(*)$ for $a_n r^n$. Show that s is a factor of $a_n r^n$.

(b) It is known that if b and c are integers that do not have any common factors (other than ± 1) and c is a factor of ab^n, then c must be a factor of a. Use this fact together with the result of part (a) to establish that s is a factor of a_n.

(c) Solve Equation $(*)$ for $a_0 s^n$. Show that r is a factor of $a_0 s^n$. Use the fact stated in part (b) to establish that r must be a factor of a_0.

4.5 UPPER AND LOWER BOUNDS

In this section we attempt to locate all the real zeros of a polynomial function F within some interval, say, $[a, b]$. This means that if c is any real zero of F, then c must be in the interval $[a, b]$. Thus $a \le c$ and $c \le b$ for every real zero c of F. In this case, we call a a **lower bound** for the zeros of F and b an **upper bound**.

Upper and lower bounds are not unique. For example, the function

$$F(x) = (x + 2)\left(x - \frac{3}{2}\right)(x - 1)(x - \pi)$$

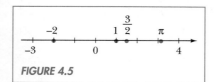

FIGURE 4.5

has four real zeros at -2, $3/2$, 1, and π; these points are graphed on the real line in Figure 4.5. Note that all four zeros are located inside the interval $[-3, 4]$. Therefore we can say that $a = -3$ is a lower bound for the zeros of F and $b = 4$ is an upper bound. However, there are other choices. For instance, the zeros are all within the interval $[-2, 3.5]$. Thus $a = -2$ and

$b = 3.5$ also qualify as lower and upper bounds, respectively, for the zeros of F. Notice that any number larger than -2 does not qualify as a lower bound. In this sense, $a = -2$ is the *best* lower bound for the zeros of F. The best upper bound is $b = \pi$.

For our purposes, we shall try to find the best *integer* upper and lower bounds for the real zeros of a given polynomial function F. The actual value of each zero will not concern us initially. Instead, we will be satisfied with simply locating the zeros within an interval $[a, b]$, a and b integers.

Example 1

Find integer upper and lower bounds for the real zeros of $F(x) = 2x^3 - 4x^2 + 1$. Prove that your choices are correct.

SOLUTION   Our first step is to guess at the upper and lower bounds by examining a rough graph of the function. We have

x	-2	-1	0	1	2	3
$F(x)$	-31	-5	1	-1	1	19

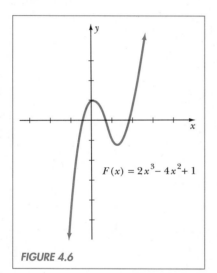

$F(x) = 2x^3 - 4x^2 + 1$

FIGURE 4.6

Figure 4.6 is a sketch of the graph of F using this table. It appears that all the real zeros of F are located within the interval $[-1, 2]$. Thus we claim that $b = 2$ is an upper bound for the real zeros, and $a = -1$ is a lower bound. Our next step is to *prove* our claim.

Proof that $b = 2$ is an upper bound. The graph in Figure 4.6 indicates that $F(x)$ approaches $+\infty$ as we move away from the origin in the positive x-direction. As a consequence, $F(x)$ is always positive as we move in this direction. Therefore our strategy for the proof will be to show

$$x > 2 \quad \text{implies} \quad F(x) > 0 \qquad (*)$$

It follows that $F(x) \neq 0$ for $x > 2$, so that all the real zeros of F must be less than or equal to 2. This would establish $b = 2$ as an upper bound for the zeros of F. Hence we focus our attention on proving Statement $(*)$.

We began by rewriting $F(x)$ with $x - 2$ factored out. Using synthetic division,

$$
\begin{array}{r|rrrr}
2 & 2 & -4 & 0 & 1 \\
  &   & 4 & 0 & 0 \\
\hline
  & 2 & 0 & 0 & 1
\end{array}
$$

we find

$$F(x) = (x - 2)(2x^2) + 1$$

With $F(x)$ written in this form, Statement $(*)$ is relatively easy to establish. We have $x > 2$ implies $x - 2 > 0$. Also, $2x^2 > 0$ for any nonzero x. Thus

$$x > 2 \quad \text{implies} \quad x - 2 > 0 \quad \text{and} \quad 2x^2 > 0$$

Since the product of two positive numbers is positive, we have

$$(x - 2)(2x^2) > 0$$

Next, by adding 1 to a positive number, the result remains positive.

$$(x - 2)(2x^2) + 1 > 0$$

Now we have the formula for $F(x)$ on the left side of this inequality. Thus we conclude

$$x > 2 \quad \text{implies} \quad F(x) > 0$$

This proves Statement (∗) and, by the previous remarks, completes the proof that $b = 2$ is an upper bound for the zeros of F.

   Proof that $a = -1$ is a lower bound. The graph of F indicates that $F(x)$ approaches $-\infty$ as we move away from the origin in the negative x-direction. Therefore our strategy for the proof will be to show

$$x < -1 \quad \text{implies} \quad F(x) < 0 \qquad\qquad (\ast\ast)$$

It follows that $F(x) \neq 0$ for $x < -1$, so that all the real zeros of F must be greater than or equal to -1. This would establish $a = -1$ as a lower bound for the zeros of F. Hence we focus our attention on proving Statement (∗∗).

   We begin by rewriting $F(x)$ with $x + 1$ factored out. Using synthetic division.

$$
\begin{array}{r|rrrr}
-1 & 2 & -4 & 0 & 1 \\
   &   & -2 & 6 & -6 \\
\hline
   & 2 & -6 & 6 & -5
\end{array}
$$

we find

$$F(x) = (x + 1)(2x^2 - 6x + 6) - 5$$

Now, $x < -1$ implies $x + 1 < 0$. Also, when x is negative, $2x^2 - 6x + 6$ is positive because it is a sum of three positive terms. Thus

$$x < -1 \quad \text{implies} \quad x + 1 < 0 \quad \text{and} \quad 2x^2 - 6x + 6 > 0$$

Since the product of a negative number and a positive number is negative, we have

$$(x + 1)(2x^2 - 6x + 6) < 0$$

Next, by adding -5 to a negative number, the result remains negative.

$$(x + 1)(2x^2 - 6x + 6) - 5 < 0$$

Now we have the formula for $F(x)$ on the left side of this inequality. Thus we conclude

$$x < -1 \quad \text{implies} \quad F(x) < 0$$

This proves Statement (∗∗) and, by our remarks above, completes the proof that $a = -1$ is a lower bound for the zeros of F. ∎

The proof for the upper bound b in Example 1 depended on rewriting $F(x)$ in the form

$$F(x) = (x - b)Q(x) + r$$

This was done using synthetic division, which we may represent as follows:

$$b \quad | \quad F$$
$$\overline{\quad q_1 \quad q_2 \quad \cdots \quad q_n \quad r}$$

The numbers in the bottom row of the process, represented here by $q_1, q_2, \cdots q_n, r$, determine the quotient function $Q(x)$ and remainder r when $F(x)$ is divided by $(x - b)$. It is reasonable to believe that we can determine whether $(x - b)Q(x) + r$ is always positive (or negative) when $x > b$ by examining the signs of the numbers $q_1, q_2, \cdots q_n, r$. Similar statements hold for a lower bound a. This introduces us to the following result.

A Boundary Test

Let $F(x)$ be a polynomial function of degree at least 1. Suppose $b \geq 0$ and the synthetic division

$$b \quad | \quad F$$
$$\overline{\quad q_1 \quad q_2 \quad \cdots \quad q_n \quad r}$$

yields the bottom row with all its nonzero numbers having the same sign. Then b is an upper bound for the real zeros of F.

Suppose $a \leq 0$ and the synthetic division

$$a \quad | \quad F$$
$$\overline{\quad q_1 \quad q_2 \quad \cdots \quad q_n \quad r}$$

yields the bottom row of numbers having alternating signs, counting zeros for either sign required. Then a is a lower bound for the real zeros of F.

This theorem does not always yield the best upper and lower bounds. One drawback is that it works only when we wish to establish an upper bound that is positive and a lower bound that is negative. However, what we give up in accuracy can be gained back in simplicity.

As an example, let us apply the Boundary Test to the function $F(x) = 2x^3 - 4x^2 + 1$ given in Example 1. The candidate for an upper bound was $b = 2$. We have

$$
\begin{array}{r|rrrr}
2 & 2 & -4 & 0 & 1 \\
  &   & 4 & 0 & 0 \\
\hline
  & 2 & 0 & 0 & 1
\end{array}
$$

The bottom row of nonzero numbers all have the same sign. Therefore, by the theorem, $b = 2$ is an upper bound. Now, for the lower bound candidate, $a = -1$, we have

$$
\begin{array}{r|rrrr}
-1 & 2 & -4 & 0 & 1 \\
   &   & -2 & 6 & -6 \\
\hline
   & 2 & -6 & 6 & -5 \\
\end{array}
$$

The bottom row of numbers have alternating signs: $+, -, +, -$. Hence $a = -1$ is a lower bound for the zeros. In this case, the Boundary Test establishes an upper and lower bound for the zeros of F in a much simpler way than the direct arguments used in Example 1.

We leave the proof of the Boundary Test for the exercises. However, Example 2 includes remarks that should help you understand why the Boundary Test works.

Example 2

Find integer upper and lower bounds for the zeros of $F(x) = 2x^4 + 8x^3 + x^2 - 10x - 4$. Prove that your choices are correct by applying the Boundary Test.

SOLUTION First, we must find candidates for the upper and lower bounds of the zeros. Therefore we shall make a table of values for $F(x)$ and draw a rough graph.

x	-4	-3	-2	-1	0	1	2
$F(x)$	52	-19	-12	1	-4	-3	76

From our graph in Figure 4.7, it appears that the real zeros for F must all be inside the interval $[-4, 2]$. Thus we claim that $b = 2$ is an upper bound and $a = -4$ is a lower bound for the zeros of F.

Proof that $b = 2$ is an upper bound. We have

$$
\begin{array}{r|rrrrr}
2 & 2 & 8 & 1 & -10 & -4 \\
  &   & 4 & 24 & 50 & 80 \\
\hline
  & 2 & 12 & 25 & 40 & 76 \\
\end{array}
$$

Since all the numbers in the bottom row have the same sign, this establishes that $b = 2$ is an upper bound.

Remark. Why is this true? The synthetic division tells us that $F(x) = (x - 2)Q(x) + 76$, where $Q(x) = 2x^3 + 12x^2 + 25x + 40$. When $x > 2$, every term in $Q(x)$ has the same sign: positive. Thus when $x > 2$, $F(x) = (x - 2)Q(x) + 76$ is a product and sum of positive numbers. Hence, $F(x) > 0$ for $x > 2$. In particular, $F(x)$ is never zero for $x > 2$.

Proof that $a = -4$ is a lower bound. We have

$$
\begin{array}{r|rrrrr}
-4 & 2 & 8 & 1 & -10 & -4 \\
   &   & -8 & 0 & -4 & 56 \\
\hline
   & 2 & 0 & 1 & -14 & 52 \\
\end{array}
$$

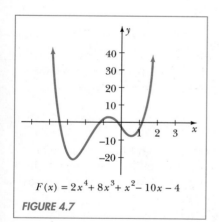

$F(x) = 2x^4 + 8x^3 + x^2 - 10x - 4$

FIGURE 4.7

Since the numbers in the bottom row have alternating signs, provided we count zero as a negative, this establishes that $a = -4$ is a lower bound.

Remark. Why is this true? The synthetic division tells us that $F(x) = (x + 4)Q(x) + 52$, where $Q(x) = 2x^3 + x - 14$. When $x < -4$, every term in $Q(x)$ will be negative. (Note that a negative number to an odd power is negative.) Thus, when $x < -4$, $(x + 4)Q(x)$ will be a product of two negative numbers and hence positive. It follows that $F(x) = (x + 4)Q(x) + 52 > 0$ for $x < -4$. In particular, $F(x)$ is never zero for $x < -4$. ■

Example 3

Use the Boundary Test to prove that $b = 2$ and $a = 0$ are upper and lower bounds, respectively, for the real zeros of $F(x) = -9x^3 + 12x^2 - 7x + 2$. Find the zeros of F.

SOLUTION *Proof that $b = 2$ is an upper bound.* We have

$$
\begin{array}{r|rrrr}
2 & -9 & 12 & -7 & 2 \\
  &   & -18 & -12 & -38 \\
\hline
  & -9 & -6 & -19 & -36
\end{array}
$$

Since all the numbers in the bottom row have the same sign, this establishes that $b = 2$ is an upper bound for the real zeros of F.

Proof that $a = 0$ is a lower bound. We have

$$
\begin{array}{r|rrrr}
0 & -9 & 12 & -7 & 2 \\
  &   & 0 & 0 & 0 \\
\hline
  & -9 & 12 & -7 & 2
\end{array}
$$

Since the numbers in the bottom row have alternating signs, this establishes that $a = 0$ is a lower bound for the real zeros of F.

By the Rational Root Theorem, the possible rational zeros for F are given by

$$
\pm \frac{\text{factor of } 2}{\text{factor of } -9} \quad \text{or} \quad \pm \frac{1, 2}{1, 3, 9}
$$

Thus the list of possibilities is ± 1, $\pm \frac{1}{3}$, $\pm \frac{1}{9}$, ± 2, $\pm \frac{2}{3}$, and $\pm \frac{2}{9}$. However, from our work above we know that all the real zeros of F must occur in the interval $[0, 2]$, which reduces our list of possible rational zeros to 1, $\frac{1}{3}$, $\frac{1}{9}$, 2, $\frac{2}{3}$, and $\frac{2}{9}$. By trial and error, we find that $x = 2/3$ works.

$$
\begin{array}{r|rrrr}
\frac{2}{3} & -9 & 12 & -7 & 2 \\
            &   & -6 & 4 & -2 \\
\hline
            & -9 & 6 & -3 & 0
\end{array}
$$

Thus

$$F(x) = \left(x - \frac{2}{3}\right)(-9x^2 + 6x - 3)$$

$$= -3\left(x - \frac{2}{3}\right)(3x^2 - 2x + 1)$$

Using the quadratic formula on $3x^2 - 2x + 1$, we find that the remaining zeros are $(1 \pm i\sqrt{2})/3$. ∎

In Example 3 we showed that $b = 2$ is an upper bound for the real zeros of F. However, this is not the *best* integer upper bound. We explore this situation further in Exercise 11, where we find by direct arguments that $b = 1$ is an upper bound for the real zeros even though the Boundary Test does not apply.

EXERCISES 4.5

In Exercises 1 to 4, find the best integer upper and lower bounds for the real zeros of the given function.

1. $F(x) = (x^2 - 3)(x^2 + 1)(2x + 5)$

2. $F(x) = (x^2 - 4)(x + \pi)$

3. $F(x) = x(x^2 - 9)(x - \pi)$

4. $F(x) = x(3x + 1)(x^2 + 3)$

In Exercises 5 to 8, use the Boundary Test to prove that b is an upper bound and a is a lower bound for the real zeros of the given function.

5. $F(x) = 9x^3 - 18x^2 + 10;\ a = -1,\ b = 2$

6. $F(x) = x^4 - x^3 + x^2 - x - 3;\ a = -1,\ b = 2$

7. $F(x) = 5x^4 - 3x^3 - 14x^2 + 9x + 6;\ a = -2,$
   $b = 2$

8. $F(x) = 2x^6 - 5x^5 - 6x^4 + 10x^3 - x^2 - x - 15;$
   $a = -2,\ b = 4$

9. Consider $F(x) = x^3 - x^2 - 1$.
   (a) Use the Boundary Test to prove that $b = 2$ is an upper bound for the real zeros of F.
   (b) Show that $a = 1$ is a lower bound for the real zeros of F as follows. Write $F(x)$ in the form $(x - 1)Q(x) + r$. Explain why $x < 1$ implies $(x - 1)Q(x) + r < 0$.

10. Consider $F(x) = x^3 + x + 3$.
    (a) Use the Boundary Test to prove that $a = -2$ is a lower bound for the real zeros of F.
    (b) Show that $b = -1$ is an upper bound for the real zeros of F as follows. Write $F(x)$ in the form $(x + 1)(Q(x) + r$. Explain why $Q(x) > 0$ for all x. Use this fact to show that $x > -1$ implies $(x + 1)Q(x) + 1 > 0$.

11. Consider $F(x) = -9x^3 + 12x^2 - 7x + 2$. (See Example 3.)
    (a) Show that the Boundary Test fails to prove that $b = 1$ is an upper bound for the real zeros of F.
    (b) Show that $b = 1$ is an upper bound as follows. Write $F(x)$ in the form $(x - 1)Q(x) + r$. Explain why $Q(x) < 0$ for all x. Use this fact to show that $x > 1$ implies $(x - 1)Q(x) + r < 0$.

In Exercises 12 to 18, find integer upper and lower bounds for the real zeros of the given function. Prove that your choices are correct.

12. $F(x) = x^3 - 5x^2 + 3x + 1$

13. $F(x) = x^3 + x^2 - x$

14. $F(x) = 2x^4 - 4x^3 - 3x^2 + 2x + 1$

15. $F(x) = 4x^4 - 9x^2 + 2$

16. $F(x) = x^5 - 2x^3 - 15x$

17. $F(x) = x^3 - 3x^2 + 3x - 2$. (*Hint:* the lower bound is positive. Use a direct argument as in Example 1 to prove the lower bound.)

18. $F(x) = -3x^3 - 12x^2 - 19x - 11$ (*Hint:* the best upper bound is negative. Use a direct argument to prove the upper bound.)

In Exercises 19 to 22, find the zeros of the given function. Start by determining upper and lower bounds for the real zeros, and then apply the Rational Root Theorem.

19. $F(x) = 2x^3 - 5x^2 + 12x - 30$

20. $F(x) = 2x^3 - 3x^2 + 24x - 36$

21. $F(x) = 4x^4 - 8x^3 + 7x^2 - 8x + 3$

22. This exercise outlines a proof of the Boundary Test. Let F be a polynomial function of degree at least 1. Let c be a constant. Suppose $F(x) = (x - c)Q(x) +$

r, where $Q(x)$ and r are determined by the synthetic division

$$
\begin{array}{c|cccc}
c & F & & & \\
\hline
& q_1 & q_2 \;\cdots\; q_n & r
\end{array}
$$

(a) If $c \geq 0$ and $q_1, q_2, \cdots q_n, r$ are all positive or zero, explain why $x > c$ implies $(x - c)Q(x) + r > 0$.

(b) If $c \geq 0$ and $q_1, q_2, \cdots q_n, r$ are all negative or zero, explain why $x > c$ implies $(x - c)Q(x) + r < 0$.

(c) If $c \leq 0$ and $q_1, q_2, \cdots q_n, r$ have alternating signs with r negative, explain why $x < c$ implies $(x - c)Q(x) + r < 0$.

(d) If $c \leq 0$ and $q_1, q_2, \cdots q_n, r$ have alternating signs with r positive, explain why $x < c$ implies $(x - c)Q(x) + r > 0$.

4.6 APPROXIMATING REAL ZEROS

In searching for the zeros of a polynomial function, we used three methods: an application of the Rational Root Theorem, factoring, and the quadratic formula. So far, we have been able to find the zeros of a given polynomial using at least one of these methods. Now we consider the question of what to do when all these methods fail to produce a zero. In this case, if the function has at least one real zero, then we can approximate its value numerically using certain algorithms (procedures). We shall describe one such algorithm, which we call the *midpoint algorithm*.

Let us begin by considering the function

$$F(x) = x^3 + 2x - 1$$

Notice that the Rational Root Theorem says the only possible rational zeros of F are ± 1. However, as you can check, neither number works. We conclude that the only possible real zeros for F must be irrational numbers.

Let us make a table of values for F and sketch its graph:

x	-2	-1	0	1	2
$F(x)$	-13	-4	-1	2	11

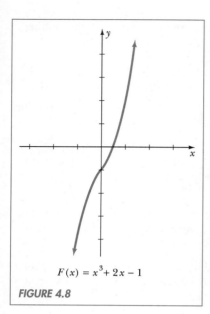

$F(x) = x^3 + 2x - 1$

FIGURE 4.8

The graph in Figure 4.8 indicates that F must have a real zero somewhere inside the interval $[0, 1]$. In a moment we will approximate this zero to three decimal places, but first we describe how the approximation method works.

The method of approximation that we shall use is based on the following result:

If F is a continuous function on $[a, b]$ (the graph of F has no holes or breaks in it) and $F(a)$ and $F(b)$ have opposite signs, then F must have a zero inside the interval $[a, b]$.

If a function has a zero somewhere inside an interval $[a, b]$, then the most obvious guess to make for this zero is the midpoint of the interval, $(a + b)/2$. This is exactly what the midpoint algorithm does.

Midpoint Algorithm

Step 1   Locate an interval $[a, b]$, where $F(a)$ and $F(b)$ have opposite signs.

Step 2   Let $m = \dfrac{a + b}{2}$. If $F(m)$ and $F(a)$ have opposite signs, go to step 3a. If $F(m)$ and $F(b)$ have opposite signs, go to step 3b.

Step 3a Set $b = m$. Go to step 2.

Step 3b Set $a = m$. Go to step 2.

In step 1 we locate an interval $[a, b]$ where a zero can be found. In step 2 we test the value of the function at the midpoint of the interval. Finally, in step 3, we create a new interval $[a, b]$ where the zero must be found. Since the new interval is half the length of the old one, our new interval is squeezing in on the location of the zero. We can repeat the process over and over, each time narrowing the length of the interval where the zero occurs.

Example 1

Use the midpoint algorithm to help find an approximation to three decimal places for the zero of $F(x) = x^3 + 2x - 1$.

SOLUTION   Step 1. Since $F(0) = -1$ and $F(1) = 2$ have opposite signs, we start with the interval $[0, 1]$.

Step 2. Test the midpoint $m = \dfrac{1}{2}$ by comparing the sign of $F\left(\dfrac{1}{2}\right)$ with the sign of $F(0)$ and $F(1)$:

x	0	$\dfrac{1}{2}$	1
$F(x)$	-1	$+\dfrac{1}{8}$	$+2$

Step 3. Since $F\left(\dfrac{1}{2}\right)$ and $F(0)$ have opposite signs, we set $b = \dfrac{1}{2}$, forming the new interval $\left[0, \dfrac{1}{2}\right]$. The zero that we are after must be in this new interval. Figure 4.9 illustrates what has happened so far.

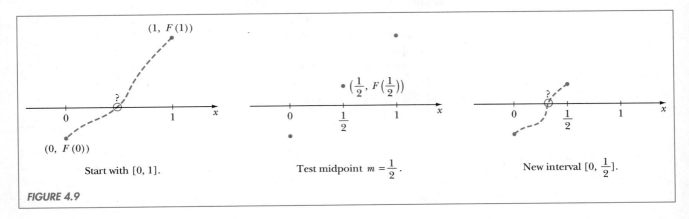

Start with $[0, 1]$.   Test midpoint $m = \dfrac{1}{2}$.   New interval $[0, \dfrac{1}{2}]$.

FIGURE 4.9

Next we return to step 2 and repeat the process, this time using the interval $\left[0, \dfrac{1}{2}\right]$.

Step 2. Test the midpoint $m = \dfrac{1}{4}$ by comparing the sign of $F\left(\dfrac{1}{4}\right)$ with the sign of $F(0)$ and $F\left(\dfrac{1}{2}\right)$:

x	0	$\dfrac{1}{4}$	$\dfrac{1}{2}$
$F(x)$	-1	$-\dfrac{31}{64}$	$+\dfrac{1}{8}$

Step 3. Since $F\left(\dfrac{1}{4}\right)$ and $F\left(\dfrac{1}{2}\right)$ have opposite signs, we set $a = \dfrac{1}{4}$, forming the new interval $\left[\dfrac{1}{4}, \dfrac{1}{2}\right]$.

Now we know that the zero for F is located in the interval $\left[\dfrac{1}{4}, \dfrac{1}{2}\right]$. From here we can continue to repeat the algorithm, each time narrowing the size of the interval where the zero is located. To make things easier, we may program a machine (such as a programmable calculator or personal computer) to do the calculations for us. After running through the algorithm ten times for the function F, we obtain the following results:

[a, b]	Midpoint m	Sign of F(a)	Approximate Value of F(m)	Sign of F(b)
[0, 1]	.5	−	+.1250	+
[0, .5]	.25	−	−.4844	+
[.25, .5]	.375	−	−.1973	+
[.375, .5]	.4375	−	−.0413	+
[.4375, .5]	.46875	−	+.0405	+
[.4375, .46875]	.453125	−	−.0007	+
[.453125, .46875]	.460938	−	+.0198	+
[.453125, .460938]	.457031	−	+.0095	+
[.453125, .457031]	.455078	−	+.0044	+
[.453125, .455078]	.454182	−	+.0018	+
[.453125, .454182]				

From the last line we conclude that the zero for F is somewhere in the interval [.453125, .454182]. If we wish to write an approximation to three decimal places for the zero, then we have a choice between .453 and .454. Comparing the value of F at these two points and at the midpoint .4535, we have

x	.453	.4535	.454
$F(x)$	−.0010	+.0002	+.0016

Therefore the zero of F is in the interval (.453, .4535). Hence its approximate value to three decimal places is .453. ∎

Notice that when applying the midpoint algorithm it is not necessary for F to be a polynomial function. The only requirement is that $F(a)$ and $F(b)$ have opposite signs and that $F(x)$ be continuous on $[a, b]$.

Example 2

Graph $G(x) = 1/x$ and $H(x) = \sqrt{x} - 1$ on the same axes. Use the midpoint algorithm to help find an approximation to two decimal places for the x-coordinate of the intersection point of the two graphs.

SOLUTION   The graphs of G and H are in Figure 4.10. We see that the intersection point has the x-coordinate very close to 2. To find this point, we shall consider the function $F(x) = G(x) - H(x) = 1/x - (\sqrt{x} - 1) = 1/x - \sqrt{x} + 1$. Notice that

$$G(x) = H(x) \qquad \text{if and only if} \qquad F(x) = 0$$

Thus the problem of finding where the graphs of G and H intersect is equivalent

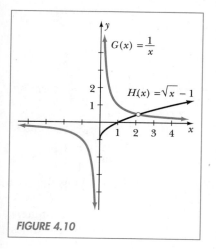

FIGURE 4.10

to finding the zero of the function $F(x) = 1/x - \sqrt{x} + 1$. Using our calculator, we find that

$$F(2) \approx +.0858 \qquad F(3) = -.3987$$

Since $F(2)$ and $F(3)$ have opposite signs, the zero for F must be in the interval $[2, 3]$. (This agrees with the graphs in Figure 4.10.) Applying the midpoint algorithm to $F(x)$ three times, we obtain the following results:

$[a, b]$	Midpoint m	Sign of $F(a)$	Approximate Value of $F(m)$	Sign of $F(b)$
$[2, 3]$	2.5	+	$-.1811$	$-$
$[2, 2.5]$	2.25	+	$-.0556$	$-$
$[2, 2.25]$	2.125	+	$+.0129$	$-$
$[2.125, 2.25]$				

At this stage we shall make our own choices for the point m rather than use the midpoint, to make the numbers a little easier to enter and compute on our calculator. Thus, instead of taking $m = 2.1875$ (the midpoint of the interval $[2.125, 2.25]$), we choose $m = 2.15$.

$[a, b]$	Our Choice for m	Sign of $F(a)$	Approximate Value of $F(m)$	Sign of $F(b)$
$[2.125, 2.25]$	2.15	+	$-.0012$	$-$
$[2.125, 2.15]$	2.14	+	$+.0044$	$-$
$[2.14, 2.15]$	2.145	+	$+.0016$	$-$
$[2.145, 2.15]$				

We see that the zero for F must be in the interval $[2.145, 2.15]$, so an approximation to two decimal places is given by 2.15. We conclude that the graphs of G and H intersect when $x \approx 2.15$. ∎

Remark

Example 2 illustrates that if we must use our calculator to do the computations, then at some stage of the process we may wish to make our own choice for the point m (instead of using the midpoint). This will simplify the work somewhat. The midpoint algorithm remains important, however, because it gives us a definite procedure to follow when beginning the search for a zero. Most important, however, is the fact that the algorithm can be used to program a machine to do the work for us.

EXERCISES 4.6

In Exercises 1 and 2, complete the table, repeating the midpoint algorithm three times for the given function starting with the interval [0, 1].

[a, b]	Midpoint m	Sign of F(a)	F(m)	Sign of F(b)
[0, 1]				

1. $F(x) = 2x^3 - 1$ 2. $F(x) = 3x^3 - 2$

In Exercises 3 and 4, complete the table, repeating the midpoint algorithm three times for the given function starting with the interval [1, 2].

[a, b]	Midpoint m	Sign of F(a)	F(m)	Sign of F(b)
[1, 2]				

3. $F(x) = x^3 - 4$ 4. $F(x) = x^4 - 2$

In Exercises 5 to 10, show that any real zero of F must be irrational. Use the midpoint algorithm to help find an approximation to two decimal places for the real zero of F.

5. $F(x) = 2x^3 + x - 1$
6. $F(x) = x^3 + x + 1$
7. $F(x) = x^3 + x - 3$
8. $F(x) = 1 - x - x^3$
9. $F(x) = x^5 - 2$
10. $F(x) = x^5 - 3$

In Exercises 11 to 14, graph the functions G and H on the same axes. Find an approximation to two decimal places for the x-coordinate of the point of intersection. [*Hint:* consider the function $F(x) = G(x) - H(x)$.]

11. $G(x) = -x^3, H(x) = x - 1$
12. $G(x) = x^5, H(x) = x + 1$
13. $G(x) = \dfrac{2}{x}, H(x) = x^3 - 1$
14. $G(x) = \sqrt{x}, H(x) = x^3 - 1$

In Exercises 15 and 16, find an approximate solution to the given equation. Use accuracy to two decimal places.

15. $\sqrt{x} = \dfrac{1}{x^2 + 1}$

16. $x^2 = \dfrac{1}{x + 1}$

17. The **linear interpolation** algorithm is another method for approximating zeros of a function F. The steps are the same as in the midpoint algorithm, except for the choice of m. Using linear interpolation, we take m to be the point where the line segment joining $(a, F(a))$ and $(b, F(b))$ intersects the x-axis; Figure 4.11 illustrates the situation.

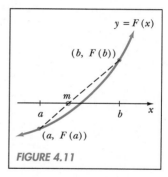

FIGURE 4.11

(a) Show that $m = \dfrac{aF(b) - bF(a)}{F(b) - F(a)}$. [Note that we are making a linear approximation for $F(x)$ between the data points $(a, F(a))$ and $(b, F(b))$. See Section 3.1.]

(b) Apply the linear interpolation algorithm to the function $F(x) = x^3 + 2x - 1$. Start with the interval [0, 1] and apply the algorithm twice, showing that $m = 1/3$ the first time and $m \approx .4194$ the second time.

18. Apply the linear interpolation algorithm to the function $F(x) = 1/x - \sqrt{x} + 1$ (see Exercise 17). Start with the interval [2, 3] and apply the algorithm twice. Compare your second value of m with the zero of F found in Example 2.

In Exercises 1 and 2, state the result of the division algorithm when $F(x)$ is divided by $P(x)$.

1. $F(x) = 2x^5 - 3x^3 + 4x^2 - 2x + 3$, $P(x) = x^2 + 1$

2. $F(x) = 3x^5 - 7x^4 + 5x^3 + 8x^2 - 6x + 2$, $P(x) = x - 1/3$

In Exercises 3 and 4, evaluate $F(c)$ for each value of c given. If c is a zero of F, then express $F(x)$ with this zero factored out, $F(x) = (x - c)Q(x)$.

3. $F(x) = x^4 - 2x^3 + 3x^2 - 4x - 4$
   (a) $c = 2$
   (b) $c = 5$

4. $F(x) = 2x^4 + x^3 - 5x^2 + x + 6$
   (a) $c = -4$
   (b) $c = -3/2$

5. If a polynomial function $F(x)$ is divided by $2x^2 - x + 3$, then the quotient is $x^3 - 5$ and the remainder is $x + 2$. Find $F(x)$.

6. Given that $z_1 = 1 + i$ and $z_2 = 3 - 2i$, find
   (a) $z_1 + 2z_2$, (b) $z_1 z_2$, (c) z_1/z_2, (d) $(z_1)^2$, (e) $\overline{z_1}$,
   (f) $z_2 \overline{z_2}$.

7. Find the zeros of $F(x) = x^2 - 2x + 3$ among the complex numbers.

8. Given that $c = -1 + \sqrt{3}i$ is a zero of $F(x) = x^4 + 6x^3 + 15x^2 + 22x + 12$. Find the remaining zeros of F among the complex numbers. Write $F(x)$ as a product of linear factors.

9. Find the polynomial function $F(x)$ with real coefficients such that the degree of F is 5, 1 is a zero of multiplicity 3, $F(2 - i) = 0$, and $F(0) = 3$.

10. Factor $F(x)$ into a product of linear terms and state the zeros of F.
    (a) $F(x) = 2x^4 + 3x^3 - 2x^2 - 3x$
    (b) $F(x) = x^6 - 4x^4 - x^2 + 4$

11. Find the constants A and B or A, B, and C so that the equation holds for all x.
    (a) $A + B(x - 2) = x + 1$
    (b) $Ax^2 + Bx + C = 6x^2 - 1$

12. Figure 4.12 is the graph of a polynomial function $G(x)$.
    (a) State the minimum possible degree of G.
    (b) Find a formula for $G(x)$ given that its degree is the value stated in part (a).

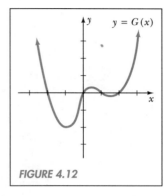

FIGURE 4.12

13. Explain why any straight line in the plane must intersect the graph of a polynomial function of degree n, where n is odd and $n \geq 3$.

In Exercises 14 and 15, for the given polynomial function F, (a) list all possible rational zeros given by the Rational Root Theorem; (b) determine integer upper and lower bounds for the real zeros of F. Use the Boundary Test to establish each answer; (c) find the zeros of F; (d) express $F(x)$ as a product of linear factors; (e) graph F.

14. $F(x) = 12x^3 - 20x^2 - x + 6$

15. $F(x) = 25x^4 + 30x^3 - 11x^2 - 12x + 4$

16. Prove that $\sqrt[3]{7}$ is irrational.

17. Show that if $F(x) = 2x^3 - 10x^2 + 18x - 11$, then $x > 2$ implies $F(x) > 0$.

18. Show that any real zero of $F(x) = x^3 + 2x - 2$ must be irrational. Use the midpoint algorithm to help find an approximation to two decimal places for the zero of F.

CHAPTER 5 Exponential and Logarithmic Functions

5.1 EXPONENTIAL EXPRESSIONS

In this chapter we explore how to use expressions of the form a^r, where a and r are real numbers. We call a^r an **exponential expression**, read "a raised to the rth power," or just "a to the r"; r is the **exponent** and a is the **base** of the exponent. We assume that you have had experience working with exponential expressions that involve integer or rational exponents. Appendix A.4 is a review of this material.

The emphasis in this section is on understanding what an exponential expression means. Of course, a^r represents a number, but what number is it? We shall answer this question in three stages: first, when r is an integer; second, when r is rational; and third, when r is irrational.

The Meaning of a^r when r is an Integer. If n is a positive integer, then we have

$$a^n = \overbrace{a \cdot a \cdot a \cdots a}^{n \text{ factors}}$$

Thus $a^1 = a$, $a^2 = a \cdot a$, $a^3 = a \cdot a \cdot a$, and so forth. We extend the definition to allow zero or a negative integer in the exponent as follows:

$$\text{If } a \neq 0, \text{ then} \qquad a^0 = 1 \qquad \text{and} \qquad a^{-n} = \frac{1}{a^n}.$$

Note that we do not allow the base to be zero when using negative exponents to avoid expressions resulting in division by zero. Also, the expression 0^0 is left undefined.

Using these definitions, we obtain the following results for real numbers a and b and integers m and n.

> **Rules for Exponents**
>
> $$a^n a^m = a^{n+m} \qquad (ab)^n = a^n b^n$$
>
> $$(a^n)^m = a^{nm} \qquad \left(\frac{a}{b}\right)^n = \frac{a^n}{b^n}$$
>
> $$\frac{a^n}{a^m} = a^{n-m} \qquad \left(\frac{a}{b}\right)^{-n} = \left(\frac{b}{a}\right)^n$$

In writing these rules, we assume that the base is nonzero whenever it appears in a denominator or when it has an exponent that is zero or negative.

Example 1

Simplify $\left(\dfrac{a^2 a^3}{b}\right)^{-4}$ to positive exponents.

SOLUTION
$$\left(\frac{a^2 a^3}{b}\right)^{-4} = \left(\frac{a^5}{b}\right)^{-4} = \left(\frac{b}{a^5}\right)^4 = \frac{b^4}{(a^5)^4}$$
$$= \frac{b^4}{a^{20}} \ \blacksquare$$

Example 2

Simplify $\left(\dfrac{4b^2}{b^5}\right)^3$ so that b appears in the numerator only.

SOLUTION
$$\left(\frac{4b^2}{b^5}\right)^3 = (4b^{-3})^3 = 4^3(b^{-3})^3$$
$$= 64b^{-9} \ \blacksquare$$

The Meaning of a^r when r is Rational. To define a^r when r is rational, we first look at the special case when $r = 1/n$, n a positive integer. Intuitively, we have

$$a^{1/n} = x \quad \text{means} \quad x^n = a$$

In other words, $a^{1/n}$ is the *nth root* of a, a number that, when raised to the nth power, equals a. For instance,

$$8^{1/3} = 2 \quad \text{since} \quad 2^3 = 8$$

Unfortunately, this definition of $a^{1/n}$ is not precise enough. In certain cases, $a^{1/n}$ does not exist as a real number; in other cases, $a^{1/n}$ has two possible values. Consider the following four examples;

1. n odd, $a < 0$: $(-32)^{1/5}$. If $(-32)^{1/5} = x$, then $x^5 = -32$. There is a unique real number that satisfies this equation, $x = -2$. Thus $(-32)^{1/5} = -2$.

2. n odd, $a > 0$: $(32)^{1/5}$. If $(32)^{1/5} = x$, then $x^5 = 32$. There is a unique real number that satisfies this equation, $x = 2$. Thus $(32)^{1/5} = 2$.

3. n even, $a < 0$: $(-16)^{1/4}$. If $(-16)^{1/4} = x$, then $x^4 = -16$. But there are no real numbers that satisfy this equation because an even power of any real number cannot be negative. Hence $(-16)^{1/4}$ does not exist as a real number.

4. n even, $a > 0$: $(81)^{1/4}$. If $(81)^{1/4} = x$, then $x^4 = 81$. There are two real numbers that satisfy this equation, $x = 3$ and $x = -3$. Which number should we choose for the value of $(81)^{1/4}$?

Examples 1 and 2 are typical when n is odd: $a^{1/n}$ exists and is unique regardless of the base a. However, the third example shows that when n is even, the base a cannot be negative. Furthermore, if n is even and a is positive, as in the fourth example, then we have two choices for $a^{1/n}$. To resolve this difficulty, we define $a^{1/n}$ to be the positive root. Thus we are led to the following definition.

Definition of $a^{1/n}$

If a is a real number and n is a positive integer, then

$$a^{1/n} = x \quad \text{means} \quad x^n = a$$

subject to the condition that when n is even, then $a \geq 0$ and $a^{1/n} \geq 0$.

We may also write the radical symbol $\sqrt[n]{a}$ to represent $a^{1/n}$.

Example 3

Find (a) $(-8)^{1/3}$, (b) $(-8)^{1/4}$, (c) $0^{1/6}$, (d) $(81)^{1/4}$.

SOLUTION (a) $(-8)^{1/3} = -2$ Explanation: $(-8)^{1/3} = x$ means $x^3 = -8$. Thus $x = -2$.

(b) $(-8)^{1/4}$ is not a real number. Explanation: if $(-8)^{1/4} = x$, then $x^4 = -8$. But an even power of any real number cannot be negative.

(c) $0^{1/6} = 0$ Explanation: $0^{1/6} = x$ means $x^6 = 0$. Thus $x = 0$.

(d) $(81)^{1/4} = 3$ Explanation: $(81)^{1/4} = x$ means $x^4 = 81$. Thus $x = 3$ or -3. By definition, we are to take the positive value, so $x = 3$. ■

Example 4

Find $2^{1/3}$.

SOLUTION The difficulty here is that $2^{1/3}$ is not a number that can be guessed easily. By definition, $2^{1/3}$ is the real number satisfying $x^3 = 2$, or $x^3 - 2 = 0$. Therefore, finding $2^{1/3}$ is equivalent to finding the real zero of the polynomial function $F(x) = x^3 - 2$. From the graph of F in Figure 5.1, we see that this zero occurs somewhere in the interval $[1, 2]$. Note that by the Rational Root Theorem, the only possible rational zeros of F are ± 1 and ± 2. But none of these work, so we know that $2^{1/3}$ must be an irrational number.

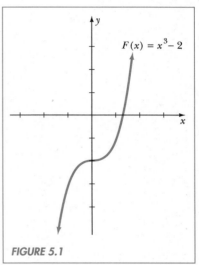

FIGURE 5.1

How do we calculate $2^{1/3}$? In Section 4.6 we developed the midpoint algorithm to solve just such a problem. Given enough time and patience, we could use this algorithm to work out the value of $2^{1/3}$ with pencil and paper to any number of decimal places required. Of course, we do not need to do this because our calculator is conveniently programmed to do all these calculations for us in a fraction of a second. Using a standard algebraic entry calculator, we key in the following:

$$2 \quad \boxed{y^x} \quad \boxed{(} \quad 1 \quad \boxed{\div} \quad 3 \quad \boxed{)} \quad \boxed{=}$$

or

$$2 \quad \boxed{y^x} \quad 3 \quad \boxed{1/x} \quad \boxed{=}$$

The result is 1.25992, accurate to five decimal places.

This does not solve the problem "find $2^{1/3}$." We know this number exists, but since it is irrational we cannot write a decimal of finite length that will represent its exact value. Therefore we must be satisfied by saying that $2^{1/3}$ is *approximately* equal to 1.25992. ∎

Example 4 shows how we can restate the definition of $a^{1/n}$ in terms of the real zeros of a polynomial function. We have

$$a^{1/n} \quad \text{is a real zero of} \quad F(x) = x^n - a$$

Figure 5.2 shows the various possibilities. Notice that when n is odd, a real zero for F exists and is unique; when n is even, a real zero for F does not exist if $a < 0$, and there are two real zeros if $a > 0$. In the latter case, we take the positive zero for the value of $a^{1/n}$.

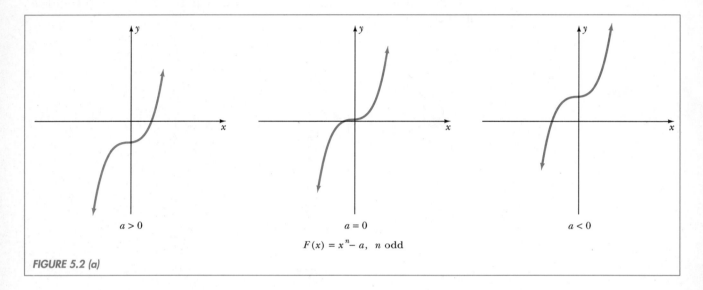

$a > 0$ $a = 0$ $a < 0$

$$F(x) = x^n - a, \quad n \text{ odd}$$

FIGURE 5.2 (a)

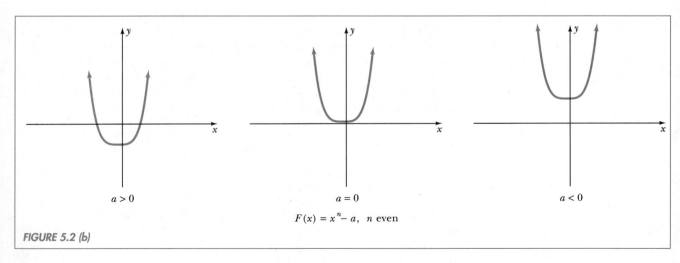

$a > 0$ $a = 0$ $a < 0$

$$F(x) = x^n - a, \quad n \text{ even}$$

FIGURE 5.2 (b)

Example 5

Use the calculator to approximate $(-10)^{1/5}$ to five decimal places.

SOLUTION Since we are taking an odd root, we know that $(-10)^{1/5}$ exists and is unique even though the base in this expression is negative. However, using a negative base with the $\boxed{y^x}$ key on the calculator may cause an error display. To get around this problem, we note that $(-10)^{1/5}$ and $10^{1/5}$ differ only in sign.

$$(-10)^{1/5} = -(10^{1/5})$$

(See remark 1 below.) Thus, to compute $(-10)^{1/5}$, we use the positive base 10, compute $10^{1/5}$, and then change the sign of the answer. By executing the following keystrokes,

$$10 \quad \boxed{y^x} \quad 5 \quad \boxed{1/x} \quad \boxed{=}$$

we obtain $10^{1/5} = 1.58489$. Therefore $(-10)^{1/5} = -1.58489$. ∎

Remarks

1. To see why $(-10)^{1/5} = -(10^{1/5})$, let $x = -(10^{1/5})$. We must show that $x^5 = -10$.

$$
\begin{aligned}
x^5 &= (-(10^{1/5})^5 \\
&= (-1)^5(10^{1/5})^5 \\
&= -10
\end{aligned}
$$

   Since $x^5 = -10$, it follows by definition that $x = (-10)^{1/5}$ as desired.
2. In general, whenever n is *odd*, we have

$$(-a)^{1/n} = -(a^{1/n})$$

   The proof is left for the exercises. Please note, however, that this is definitely *not* true when n is even (unless $a = 0$). An even root of a negative base does not exist as a real number.
3. Parentheses are often dropped when a negative sign is not part of the base of an exponential expression. Thus we write $-10^{1/5}$ instead of $-(10^{1/5})$. In general, $-a^r$ means $-(a^r)$.

Now that we know the definition of $a^{1/n}$, we can define a^r for any rational number r. Recall that if r is rational, then it can be written in the form m/n, where m and n are integers and n is positive.

> **Definition of $a^{m/n}$**
>
> If a is a real number and m and n are integers, n positive, then
>
> $$a^{m/n} = (a^{1/n})^m = (a^m)^{1/n}$$
>
> subject to the condition that when $a < 0$, then m/n must be reduced and n must be odd.

With this definition, the rules for exponents that we quoted earlier remain valid for rational numbers.

Example 6

Evaluate (a) $(81)^{3/4}$, (b) $((32)^6)^{1/5}$, (c) $(16)^{-5/4}$.

SOLUTION (a) $(81)^{3/4} = ((81)^{1/4})^3 = (3)^3 = 27$

(b) $((32)^6)^{1/5} = ((32)^{1/5})^6 = (2)^6 = 64$

In this case, changing the order of the exponents proved quite helpful.

(c) $(16)^{-5/4} = ((16)^{1/4})^{-5} = (2)^{-5} = \dfrac{1}{2^5} = \dfrac{1}{32}$ ■

Note the condition that $a^{m/n}$ does not exist when a is negative unless m and n are such that m/n is reduced and n is odd. Without this restriction, the rules for exponents would lead to contradictions. For example,

$$-1 = (-1)^1 = (-1)^{2/2} = ((-1)^2)^{1/2} = (1)^{1/2} = 1$$

The problem in this argument occurs when using the exponent 2/2 with a negative base. The exponent 2/2 is not reduced, so $(-1)^{2/2}$ is not defined unless we reduce the exponent. Thus -1 is not equal to 1, as this argument incorrectly asserts.

Example 7

Evaluate (a) $(-8)^{2/3}$, (b) $(-32)^{.6}$.

SOLUTION (a) Notice that the base in this expression is negative. However, the exponent $m/n = 2/3$ is reduced and its denominator $n = 3$ is odd as required. Thus

$$(-8)^{2/3} = ((-8)^{1/3})^2 = (-2)^2 = 4$$

(b) The base in this case is negative, so the exponent, as required by definition, must be reduced. Since $.6 = 6/10 = 3/5$, we have

$$(-32)^{.6} = (-32)^{3/5} = ((-32)^{1/5})^3 = (-2)^3 = -8 ■$$

Example 8

Simplify the expression $x^{-1/3} y^{1/3} - x^{2/3} y^{-2/3}$ to one fraction with no negative exponents.

SOLUTION We use the rules for exponents to change each term so that the exponents are positive.

$$x^{-1/3} y^{1/3} - x^{2/3} y^{-2/3}$$

$$\frac{y^{1/3}}{x^{1/3}} - \frac{x^{2/3}}{y^{2/3}}$$

Multiplying the first fraction by $y^{2/3}/y^{2/3}$ and the second fraction by $x^{1/3}/x^{1/3}$, we convert both fractions to the common denominator $x^{1/3} y^{2/3}$.

$$\frac{y^{1/3} y^{2/3}}{x^{1/3} y^{2/3}} - \frac{x^{1/3} x^{2/3}}{x^{1/3} y^{2/3}}$$

Simplifying, then adding, we obtain

$$\frac{y}{x^{1/3} y^{2/3}} - \frac{x}{x^{1/3} y^{2/3}}$$

$$\frac{y - x}{x^{1/3} y^{2/3}} \quad \blacksquare$$

The Meaning of a^r when r is Irrational. Unfortunately, we cannot give a rigorous definition of a^r when r is irrational. To do so properly, we would need to use some methods from calculus to formulate the definition and show that the rules for exponents remain valid. Instead, we shall give an intuitive definition and just accept the fact that the rules, with proper restrictions, still work for irrational exponents. To avoid undefined expressions, we must restrict the base a to be positive whenever the exponent is irrational. We do allow zero in the base, $0^r = 0$, but only when $r > 0$.

So, let us consider an irrational number r. Here is the key to the argument: we let x approach r through *rational numbers only*. It is reasonable to assume that

$$x \text{ approaches } r \quad \text{implies} \quad a^x \text{ approaches } a^r$$

Thus we find the value of a^r by examining the value of a^x for rational numbers x getting closer and closer to r. Since x is a rational number, we can calculate each a^x. The closer x is to r, the nearer a^x is to a^r.

Example 9

Estimate the value of $2^{\sqrt{2}}$ to three decimal places in the following two ways:
(a) By examining 2^x for rational numbers x approaching $\sqrt{2}$.
(b) By computing $2^{\sqrt{2}}$ directly on the calculator.

SOLUTION (a) We need a sequence of rational numbers getting closer and closer to $\sqrt{2}$. We know that $\sqrt{2}$ has an infinite decimal representation that begins with 1.41421356. Therefore we can use the following list of rational numbers approaching $\sqrt{2}$:

$$1, 1.4, 1.41, 1.414, 1.4142, 1.41421, \text{ and so on.}$$

For each x in this list, we evaluate 2^x using our calculator. The following table shows the results for the first six values of x rounded off to five decimal places:

x	1	1.4	1.41	1.414	1.4142	1.41421	$\rightarrow$	$\sqrt{2}$
2^x	2	2.63902	2.65737	2.66475	2.66512	2.66514	$\rightarrow$	?

By definition, the numbers in the 2^x row should approach the value of $2^{\sqrt{2}}$. At this point, we estimate that $2^{\sqrt{2}} \approx 2.665$.

(b) The direct computation of $2^{\sqrt{2}}$ on an algebraic entry calculator is accomplished by executing the following keystrokes:

$$2 \quad \boxed{y^x} \quad \boxed{(} \quad 2 \quad \boxed{\sqrt{x}} \quad \boxed{)} \quad \boxed{=}$$

We obtain $2^{\sqrt{2}} \approx 2.66514414$. Rounded off to three decimal places, we get 2.665, which agrees with the result given in part (a). ■

EXERCISES 5.1

In Exercises 1 to 6, evaluate if possible. Do not use a calculator. Note that some expressions may be undefined.

1. (a) $(-32)^{-.4}$
   (b) $(-27)^{2/6}$
   (c) $(-64)^{1/6}$

2. (a) $(-1)^{3/7}$
   (b) $(2 - \sqrt{4})^{-1/8}$
   (c) $(49)^{1.5}$

3. (a) $\left(-\dfrac{4}{5}\right)^{-1}$
   (b) $((64)^4)^{1/3}$
   (c) $(-9)^{2.5}$

4. (a) $(-8)^{-5/3}$
   (b) $(81)^{-5/4}$
   (c) $0^{5/6}$

5. (a) $(256)^{-3/4}$
   (b) $(-8)^{1/6}(-8)^{1/6}$
   (c) $1^{-1.5}$

6. (a) $(-4^{2/5})^5$
   (b) -2^{-2}
   (c) $(-\sqrt{2})^0$

In Exercises 7 to 10, use your calculator to find an approximation to five decimal places for the given exponential expression.

7. (a) $5^{1/4}$
   (b) $8^{.3}$

8. (a) $9^{1/5}$
   (b) $(10)^{.6}$

9. (a) $(2.5)^{-3/4}$
   (b) $(-3)^{1/3}$

10. (a) $(6.1)^{-1.6}$
    (b) $(-16)^{1/5}$

In Exercises 11 to 20, use the Rules for Exponents to simplify the given expression. Assume that any symbol in the base of an exponential expression represents a positive constant.

11. (a) $2(2^x)$
    (b) $\left(\dfrac{a}{b^3 b^4}\right)^{-3}$

12. (a) $\dfrac{3^x}{3}$
    (b) $\left(\dfrac{x^4 x^2}{y^3}\right)^{-5}$

13. (a) $(e^{-x})^2$
    (b) $e(e^{x^2})$

14. (a) $(e^x)^2$
    (b) $e^x e^{-x}$

15. (a) $\dfrac{x^{2r}}{x^r}$
    (b) $\dfrac{x}{x^r}$

16. (a) $\dfrac{x^{r+s}}{x^{r-s}}$
    (b) $(e^{x-1})^{-1}$

17. (a) $\left(\dfrac{e^x + e^{-x}}{2}\right)^2$ 18. (a) $\left(\dfrac{e^x - e^{-x}}{2}\right)^2$

(b) $e^{-\sqrt{2}}e^{\sqrt{2}}$ (b) $(xe^{\sqrt{2}})^2$

19. (a) $(x^\pi)^{1/\pi}$ 20. (a) $\dfrac{2^x 3^x}{6^x}$

(b) $(2^{1/x})2^x$ (b) $(x^{.1}y^{.2})^{10}$

In Exercises 21 and 22, simplify so that all symbols appear in the numerator.

21. (a) $\left(\dfrac{2x^2}{x^6}\right)^3$

(b) $\left(\dfrac{a^{1/2}b^{1/3}}{a^2 b}\right)^{-4}$

22. (a) $\left(\dfrac{y^4}{xy}\right)^{-3}$

(b) $\left(\dfrac{a^{3/2}b}{3a^{1/2}b}\right)^2$

In Exercises 23 to 28, simplify the given expression to one fraction with no negative exponents.

23. $\dfrac{4}{3}x^{1/3} - \dfrac{1}{3}x^{-4/3}$

24. $x^{-1/2} - \dfrac{1}{2}x^{-3/2}$

25. $(1 + x)^{-2/3} - \dfrac{2}{3}x(1 + x)^{-5/3}$

26. $\dfrac{1}{3}x(x + 4)^{-2/3} + (x + 4)^{1/3}$

27. $\dfrac{3}{5}(1 + x)^{-2/5}(2 - x)^{2/5} - \dfrac{2}{5}(1 + x)^{3/5}(2 - x)^{-3/5}$

28. $(1 + x)^{-1/3}(1 - x)^{1/3} - (1 + x)^{2/3}(1 - x)^{-2/3}$

In Exercises 29 and 30, graph the function. Use your calculator to approximate the real zero of F to five decimal places.

29. (a) $F(x) = x^3 - 10$

(b) $F(x) = x^3 + 10$

30. (a) $F(x) = x^5 - 5$

(b) $F(x) = x^5 + 5$

In Exercises 31 and 32, graph the function. Use your calculator to approximate the positive real zero of F to five decimal places.

31. $F(x) = x^4 - 20$

32. $F(x) = x^6 - 20$

In Exercises 33 and 34, use your calculator to complete the table. Approximate all answers to four decimal places.

x	3.1	3.14	3.141	3.1415	3.14159	π
33. 2^x						
34. 3^x						

35. Use your calculator to compute
(a) $10(10^{.4})(10^{.01})(10^{.004})(10^{.0002})$
(b) $10^{\sqrt{2}}$ [Compare this answer with part (a).]

36. Use your calculator to compute
(a) $10^3(10^{.1})(10^{.04})(10^{.001})(10^{.0006})$
(b) 10^π [Compare this answer with part (a).]

37. Use your calculator to compute (a) π^π, (b) $-2^{1/\pi}$

In Exercises 38 to 50, answer true or false.

38. $2^{1/3} \cdot 3^{1/2} = 6^{5/6}$

39. $(-1)^0 = 1$

40. If $a > 0$, then $(a^{\sqrt{2}})^{\sqrt{2}} = a^2$

41. $(3^x)^2 = 9^x$

42. If $a > 0$, then $(a^r)^2 = a^{r^2}$

43. $\sqrt[3]{x}\sqrt[3]{x} = x^{2/3}$

44. $2^{\sqrt{3}} = 8^{1/\sqrt{3}}$

45. $(a + b)^{1/3}(a - b)^{1/3} = (a^2 - b^2)^{1/3}$

46. $1^x = 1$ for all $x \in \mathbf{R}$

47. $2^{-x} = \left(\dfrac{1}{2}\right)^x$ for all $x \in \mathbf{R}$

48. $(a^2 + b^2)^{1/2} = a + b$

49. $(-6)^6 = -6^6$

50. $3 \cdot 4^x = 12^x$

51. Prove: if n is odd, then $(-a)^{1/n} = -a^{1/n}$. (*Hint:* let $x = -a^{1/n}$. Show that $x^n = -a$.)

5.2 EXPONENTIAL FUNCTIONS

In Section 5.1 we found that when a is a positive real number, then the exponential expression a^r makes sense for every real number r regardless of whether r is rational, irrational, positive, or negative. This information allows us to define a new type of function.

Definition of Exponential Function
Let a be a positive real number, $a \neq 1$. The rule

$$F(x) = a^x$$

defines a function with domain **R**. We call F an **exponential function** with base a.

The base $a = 1$ is excluded because the function $F(x) = 1^x$ is just a constant function, $1^x = 1$ for all x.

Example 1

Graph $F(x) = 2^x$.

SOLUTION   We wish to graph the equation

$$y = 2^x$$

We begin by making a table of values:

x	-2	-1	0	1	2
$y = 2^x$	$\frac{1}{4}$	$\frac{1}{2}$	1	2	4

These points are graphed in Figure 5.3(a). Using our calculator, we may fill in more points as follows:

x	-1.5	$-.5$	$.5$	1.5
$y = 2^x$	.35	.71	1.41	2.83

The values for 2^x are rounded off to two decimal places. Figure 5.3(b) shows these points added into the graph. This gives us a pretty good idea of the shape of the curve close to the origin. Finally, in Figure 5.3(c) we connect the points with a continuous, smooth curve to obtain the graph of F. ∎

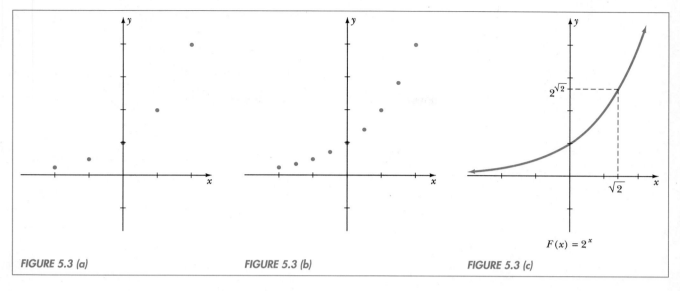

FIGURE 5.3 (a)   FIGURE 5.3 (b)   FIGURE 5.3 (c)

Remarks

Several remarks are in order here concerning the geometric behavior of $F(x) = 2^x$.

1. When we draw the continuous curve in Figure 5.3(c), we are saying (among other things) that 2^x exists for every x; in other words, the domain of 2^x is **R**. This agrees with our discussion in Section 5.1. As an example, the graph may be used to estimate the value of $2^{\sqrt{2}}$. As shown in the figure, we find that $2^{\sqrt{2}} \approx 2.7$. This number agrees with the result given in Example 9, Section 5.1.

2. As we move from left to right across the graph in Figure 5.3(c), we see that the curve is increasing. We have

$$x_2 > x_1 \quad \text{implies} \quad 2^{x_2} > 2^{x_1}$$

Thus 2^x is increasing.

3. Recall that the range of a function consists of the set of all y-coordinates appearing on the graph of the function. From the graph in Figure 5.3(c), we find

$$\text{Range of } 2^x = \{y : y > 0\} = (0, +\infty)$$

Thus 2^x is never negative or zero. We may write

$$2^x > 0 \qquad \text{for all } x \in \mathbf{R}.$$

4. Finally, we consider the behavior of 2^x when $|x|$ is large. As x increases toward positive infinity, 2^x gets bigger and bigger. For example,

$2^{10} = 1{,}024$, $2^{20} = 1{,}048{,}576$, $2^{30} = 1{,}073{,}741{,}824$, and so on. Therefore

$$\lim_{x \to +\infty} 2^x = +\infty$$

On the other hand, as x tends toward negative infinity, 2^x gets closer and closer to zero. For example, $2^{-10} = 1/2^{10} \approx .00098$, $2^{-20} = 1/2^{20} \approx .00000095$, $2^{-30} = 1/2^{30} \approx .00000000093$, and so on. Thus

$$\lim_{x \to -\infty} 2^x = 0$$

This tells us that the graph of $y = 2^x$ has a horizontal asymptote at $y = 0$ (the x-axis).

We now summarize these remarks.

Properties of $F(x) = 2^x$
1. The domain of 2^x is $\mathbf{R}$.
2. 2^x is increasing.
3. $2^x > 0$ for all $x \in \mathbf{R}$; its range is $(0, +\infty)$.
4. $\lim\limits_{x \to -\infty} 2^x = 0 \qquad \lim\limits_{x \to +\infty} 2^x = +\infty$

Horizontal asymptote at $y = 0$.

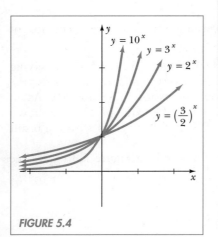

FIGURE 5.4

All these results are contained in the graph in Figure 5.3(c). This is the beauty of graphing a function: several important results are summarized in a single visual picture.

The exponential functions a^x with base $a > 1$ all share the same properties given for $y = 2^x$ in Example 1. For instance, Figure 5.4 shows the graphs of $y = a^x$ for $a = 3/2, 2, 3$, and 10, all on the same axes. We note that all these graphs behave more or less the same way; only their curvatures are slightly different.

Example 2

Graph $G(x) = \left(\dfrac{1}{2}\right)^x$. Find the range of G, the behavior of $G(x)$ for $|x|$ large, asymptotes, and whether $G(x)$ is increasing or decreasing.

SOLUTION We wish to graph the equation

$$y = \left(\frac{1}{2}\right)^x$$

We begin with a short table of values:

x	-2	-1	0	1	2
$y = \left(\dfrac{1}{2}\right)^x$	4	2	1	$\dfrac{1}{2}$	$\dfrac{1}{4}$

After graphing these points in Figure 5.5(a), we connect them with the continuous, smooth curve shown in Figure 5.5(b).

From the graph of G, we deduce the following properties that answer the questions concerning the behavior of $G(x)$.

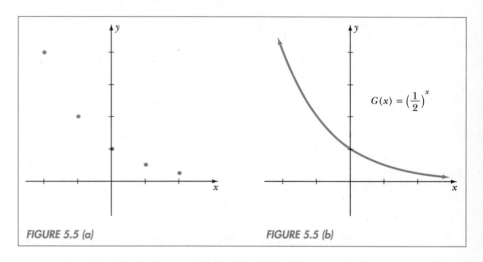

FIGURE 5.5 (a)   FIGURE 5.5 (b)

Properties of $G(x) = \left(\dfrac{1}{2}\right)^x$

1. The domain of $\left(\dfrac{1}{2}\right)^x$ is $\mathbf{R}$.

2. $\left(\dfrac{1}{2}\right)^x$ is decreasing.

3. $\left(\dfrac{1}{2}\right)^x > 0$ for all $x \in \mathbf{R}$; its range is $(0, +\infty)$.

4. $\displaystyle\lim_{x \to -\infty} \left(\dfrac{1}{2}\right)^x = +\infty \qquad \lim_{x \to +\infty} \left(\dfrac{1}{2}\right)^x = 0$

   Horizontal asymptote at $y = 0$. ■

The graph of any exponential function $F(x) = a^x$ behaves like the graph of $y = 2^x$ if $a > 1$ or $y = (1/2)^x$ if $0 < a < 1$. Thus we can draw two typical curves for $y = a^x$ as shown in Figure 5.6; Figure 5.6(a) corresponds to $a > 1$, and Figure 5.6(b) corresponds to $0 < a < 1$. To locate specific points on the graph, we can make a table of values containing $x = 1, 0$, and -1. These values are easy to evaluate because

$$a^1 = a \qquad a^0 = 1 \qquad a^{-1} = \frac{1}{a}$$

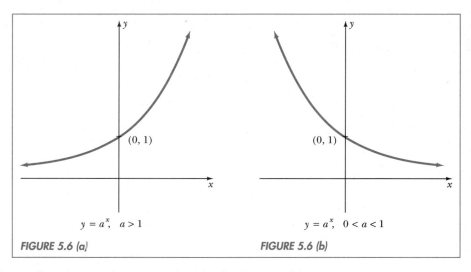

$y = a^x, \quad a > 1$

$y = a^x, \quad 0 < a < 1$

FIGURE 5.6 (a)   **FIGURE 5.6 (b)**

Finally, observe the following properties for the graph:

1. The domain of a^x is **R**.
2. a^x is $\begin{cases} \text{increasing if } a > 1 \\ \text{decreasing if } 0 < a < 1 \end{cases}$
3. $a^x > 0$ for all $x \in \mathbf{R}$; its range is $(0, +\infty)$.
4. $y = a^x$ has a horizontal asymptote at $y = 0$.

Example 3

Evaluate (a) $\lim\limits_{x \to +\infty} \left(\dfrac{6}{5}\right)^x$, (b) $\lim\limits_{x \to +\infty} \left(\dfrac{3}{4}\right)^x$.

SOLUTION   (a) Consider the exponential function $F(x) = \left(\dfrac{6}{5}\right)^x$. We want to determine what $F(x)$ does as x increases toward positive infinity. Since the base $6/5$ is a number greater than 1, the graph of F looks like the curve in Figure 5.6(a). Thus from the graph we observe that

$$\lim_{x \to +\infty} \left(\frac{6}{5}\right)^x = +\infty$$

(b) Now we consider the exponential function $G(x) = \left(\dfrac{3}{4}\right)^x$. Since the base of this function is between 0 and 1, the graph of G looks like the curve in Figure 5.6(b). Thus we see that $G(x)$ is decreasing toward zero as x tends toward positive infinity.

$$\lim_{x \to +\infty} \left(\frac{3}{4}\right)^x = 0 \ \blacksquare$$

Example 4

Graph the given function. (a) $H(x) = \left(\dfrac{2}{3}\right)^x$, (b) $G(x) = -\left(\dfrac{2}{3}\right)^x$, (c) $F(x) = 1 - \left(\dfrac{2}{3}\right)^{x+2}$. For the function in part (c), determine if $F(x)$ is increasing or decreasing, find its range and asymptotes, evaluate $\lim\limits_{x \to +\infty} F(x)$ and $\lim\limits_{x \to -\infty} F(x)$.

SOLUTION (a) We wish to graph the equation

$$y = \left(\frac{2}{3}\right)^x$$

We know the curve looks like Figure 5.6(b). To obtain a more precise graph, we make a short table of values:

x	-2	-1	0	1	2
$y = \left(\dfrac{2}{3}\right)^x$	$\dfrac{9}{4}$	$\dfrac{3}{2}$	1	$\dfrac{2}{3}$	$\dfrac{4}{9}$

This gives us the graph in Figure 5.7(a) on page 244.

(b) We wish to graph the equation

$$y = -\left(\frac{2}{3}\right)^x$$

Note that this is just the reflection across the x-axis of the curve $y = \left(\dfrac{2}{3}\right)^x$, which we graphed in part (a). Hence we get the curve in Figure 5.7(b).

(c) We wish to graph the equation

$$y = 1 - \left(\frac{2}{3}\right)^{x+2}$$

which is equivalent to

$$y - 1 = -\left(\frac{2}{3}\right)^{x+2}$$

We recognize this as a shift of the curve found in part (b). By comparing

$$\text{Unshifted equation:} \quad y = -\left(\frac{2}{3}\right)^x$$

$$\text{Shifted equation:} \quad y - 1 = -\left(\frac{2}{3}\right)^{x+2}$$

we conclude that the curve in part (b) is shifted by -2 units in the x-direction and $+1$ unit in the y-direction. Thus we obtain the graph in Figure 5.7(c). From the graph, we find that

1. $F(x)$ is increasing.
2. Range: $(-\infty, 1)$.
3. Horizontal asymptote: $y = 1$.
4. $\lim\limits_{x \to -\infty} F(x) = -\infty \qquad \lim\limits_{x \to +\infty} F(x) = 1$ ■

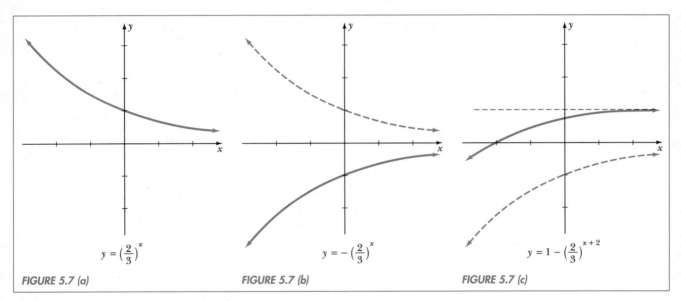

$$y = \left(\frac{2}{3}\right)^x$$

FIGURE 5.7 (a)

$$y = -\left(\frac{2}{3}\right)^x$$

FIGURE 5.7 (b)

$$y = 1 - \left(\frac{2}{3}\right)^{x+2}$$

FIGURE 5.7 (c)

EXERCISES 5.2

In Exercises 1 to 10, graph the given function F. Then find (a) whether $F(x)$ is increasing or decreasing, (b) the domain and range of F, (c) $\lim\limits_{x \to +\infty} F(x)$, (d) $\lim\limits_{x \to -\infty} F(x)$, (e) asymptotes.

$\left(\text{Note that } a^{-x} = \left(\frac{1}{a}\right)^x.\right)$

1. $F(x) = \left(\frac{1}{3}\right)^x$

2. $F(x) = 3^x$

3. $F(x) = \left(\frac{2}{3}\right)^{-x}$

4. $F(x) = 4^{-x}$

5. $F(x) = 2^{2x}$

6. $F(x) = (\sqrt{3})^{-2x}$

7. $F(x) = 2\left(\frac{1}{2}\right)^x$

8. $F(x) = \frac{1}{4}(2^x)$

9. $F(x) = -2^x$

10. $F(x) = -2^{-x}$

In Exercises 11 and 12, graph all three equations on the same axes.

11. $y = 2^x$, $y = 3^x$, $y = 4^x$

12. $y = \left(\dfrac{1}{2}\right)^x$, $y = \left(\dfrac{1}{3}\right)^x$, $y = \left(\dfrac{1}{4}\right)^x$

In Exercises 13 to 16, evaluate. (*Hint:* think of the appropriate graph.)

13. $\displaystyle\lim_{x \to +\infty} \left(\dfrac{4}{5}\right)^x$

14. $\displaystyle\lim_{x \to +\infty} \left(\dfrac{8}{7}\right)^x$

15. $\displaystyle\lim_{x \to +\infty} (1.01)^x$

16. $\displaystyle\lim_{x \to +\infty} (.98)^x$

In Exercises 17 to 22, each function is a shift of $y = a^x$ or $y = -a^x$. Find the unshifted equation and the amount of the shift. Graph F. Find (a) whether $F(x)$ is increasing or decreasing, (b) range of F, (c) $\displaystyle\lim_{x \to +\infty} F(x)$, (d) values of x for which $F(x) > 0$.

17. $F(x) = 2^x - 1$

18. $F(x) = 2^{x-1}$

19. $F(x) = 2^{x+1} + 1$

20. $F(x) = 2^{1-x} - 2$

21. $F(x) = 3 - 3^{-x}$

22. $F(x) = 4 - 4^x$

In Exercises 23 and 24, use the graph in Figure 5.8 to approximate the given number to one decimal place.

23. (a) $a^{1/2}$
    (b) $a^{\sqrt{2}}$

24. (a) a
    (b) $a^{-1/2}$

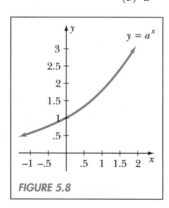

FIGURE 5.8

In Exercises 25 to 28, for the given function F, find $F(k + 1)/F(k)$ in simplest terms.

25. $F(x) = \dfrac{2^{x-1}}{3^x}$

26. $F(x) = 4^{-x}$

27. $F(x) = x5^{-x}$

28. $F(x) = \dfrac{3^{x+1}}{5^{x-1}}$

29. A pile of pennies is created by starting with one penny and doubling the pile's size after each day. Thus, after the first day there are two pennies, after 2 days, four pennies, after 3 days, eight pennies, and so on. Let $F(n)$ be the value in dollars of the pile of pennies after n days.
    (a) Find a formula for $F(n)$.
    (b) In how many days will the value of the pennies exceed \$1,000,000?

30. Suppose through unfortunate circumstances you are marooned on a deserted island with only 1 gallon of fresh water. You decide to limit the amount of water you drink each day to only one-fourth of the amount available at the beginning of the day. Thus, during the first day you drink 1/4 gallon, leaving 3/4 gallon at the end of the day. During the second day you drink one-fourth of the 3/4 gallon available, leaving 9/16 gallon at the end of the day.
    (a) Let $F(n)$ be the amount in ounces of water left after n days. Find a formula for $F(n)$. Note that there are 128 ounces in 1 gallon.
    (b) In how many days will less than 1 ounce of water be left?

In Exercises 31 to 34, use the technique of adding graphs to find the graph of the given function. Then determine the (a) range of F, (b) $\displaystyle\lim_{x \to +\infty} F(x)$.

31. $F(x) = 2^x + 2^{-x}$

32. $F(x) = 2^x - 2^{-x}$

33. $F(x) = 3^{-x} - 3^x$

34. $F(x) = \left(\dfrac{3}{2}\right)^x + \left(\dfrac{3}{2}\right)^{-x}$

In Exercises 35 to 38, determine what type of symmetry the given function possesses. Graph. Find the range.

35. $F(x) = 2^{-x^2}$

36. $F(x) = 3^{x^2}$

37. $F(x) = 1 - 2^{x^2}$

38. $F(x) = 1 - 3^{-x^2}$

39. (a) Graph $y = 2^x$ and $y = \left(\dfrac{1}{2}\right)^x$ on the same axes. What symmetry relationship exists between these two curves?
    (b) What symmetry relationship exists between the curves $y = a^x$ and $y = \left(\dfrac{1}{a}\right)^x$? Assume that $a > 0$ and $a \neq 1$.

Interest rates play an important role in such business transactions as loans or investments. In this section we concentrate on how to calculate the amount of money that can be earned on an investment at a given rate of interest. If we begin with P dollars invested at the interest rate r, our question is, After time t, how much money will we have? We wish to find a function $A(t)$ that tells us the amount of money in our investment at time t. As it turns out, $A(t)$ is an exponential function.

Example 1

Suppose $100 is deposited into a savings account that pays an annual interest rate of 6 percent. How much money is in the account after 1 year? After 10 years?

SOLUTION At the end of the first year, the interest earned on the account is 6 percent of $100:

$$\text{Interest earned} = 100(.06) = 6$$

So, we would have a total of $106 in the account, which can be expressed as follows:

$$\text{Beginning amount} + \text{interest earned} = \text{total}$$
$$100 \quad + \quad 100(.06) \quad = 100(1 + .06)$$

Notice that we wrote the total in an unusual way, $100(1 + .06)$, instead of 106. The reason for this will become clear as we do more calculations.

Now, for the second year, we begin with $106.

$$\text{Beginning amount} + \text{interest earned} = \text{total}$$
$$106 \quad + \quad 106(.06) \quad = 106 + 106(.06)$$
$$= 106(1 + .06)$$
$$= 100(1 + .06)(1 + .06)$$
$$= 100(1 + .06)^2$$

The total comes out to $112.36. More importantly, note how we wrote this total: $100(1 + .06)^2$.

Moving on to the third year, we begin with $112.36.

$$\text{Beginning amount} + \text{interest earned} = \text{total}$$
$$112.36 \quad + \quad 112.36(.06) \quad = 112.36 + 112.36(.06)$$
$$= 112.36(1 + .06)$$
$$= 100(1 + .06)^2(1 + .06)$$
$$= 100(1 + .06)^3$$

So, the total after 3 years is $100(1 + .06)^3$.

At this point we see an obvious pattern emerging. After each year, the total amount in the account grows by a factor of $(1 + .06)$. So, after 4 years we would have $100(1 + .06)^4$, after 5 years $100(1 + .06)^5$, and so on. Finally, after 10 years the total amount in the account would be

$$100(1 + .06)^{10}$$

This number can be computed on a standard algebraic entry calculator by pressing the following sequence of keys:

$$100 \quad \boxed{x} \quad 1.06 \quad \boxed{y^x} \quad 10 \quad \boxed{=}$$

Rounded off to the nearest penny (two decimal places), we obtain \$179.08. ∎

We can easily generalize our results from Example 1. First, if the length of time involved were t years instead of 10, the total amount would be

$$100(1 + .06)^t$$

Furthermore, we could have started with P dollars instead of 100 at the interest rate r (expressed as a decimal) instead of .06. The total would be

$$P(1 + r)^t$$

Example 2

An investment plan pays a yearly interest rate of 7.5 percent. If we start with \$2000, find how much money would be in the account after 5 years?

SOLUTION We know that $A(t) = P(1 + r)^t = 2000(1 + .075)^t$. Therefore

$$A(5) = 2000(1 + .075)^5$$

Using the calculator, this comes out to \$2871.26. ∎

When interest is paid into an account once a year, as in Examples 1 and 2, we say that the interest is **compounded** annually. Most investment opportunities, however, compound the interest more often than just once a year. For example, a bank may offer an annual interest rate of 6 percent compounded quarterly, which means that the 6 percent rate is divided into four payments per year, each payment at the rate of .06/4.

Example 3

A bank advertises savings deposits that pay 6 percent compounded quarterly. Suppose we invest \$100. How much money will be in the account after 1 year? After 2 years? After t years?

SOLUTION When a bank quotes an interest rate, it is referring to a *yearly* rate unless stated otherwise. So 6 percent compounded quarterly means that the bank will pay interest into the account four times per year, each time at the rate of .06/4. One might expect that after four payments the total interest earned would be a simple 6 percent. However, this is *not* the case, as the following calculations demonstrate.

After 3 months the account receives its first interest payment.

$$\text{Beginning amount } + \text{ interest earned } = \text{ total}$$

$$100 \quad + \quad 100\left(\frac{.06}{4}\right) \quad = 100\left(1 + \frac{.06}{4}\right)$$

This gives us $100\left(1 + \frac{.06}{4}\right)$, or $101.50. From 3 months to 6 months we earn interest on this new amount. Thus for the second payment we get

$$\text{Beginning amount } + \text{ interest earned } = \text{ total}$$

$$101.50 \quad + \quad 101.50\left(\frac{.06}{4}\right) \quad = 101.50 + 101.50\left(\frac{.06}{4}\right)$$

$$= 101.50\left(1 + \frac{.06}{4}\right)$$

$$= 100\left(1 + \frac{.06}{4}\right)\left(1 + \frac{.06}{4}\right)$$

$$= 100\left(1 + \frac{.06}{4}\right)^2$$

It is clear that each time a payment is made, the total amount in our investment grows by a factor $\left(1 + \frac{.06}{4}\right)$. Thus after the third payment (9 months), we have

$$100\left(1 + \frac{.06}{4}\right)^3$$

And after the fourth payment (1 year), we have

$$100\left(1 + \frac{.06}{4}\right)^4$$

This comes out to approximately $106.14. Note that this is 14 cents more than the total would be for the same investment at 6 percent compounded annually.

During the second year we will have four more interest payments, each one increasing the account total by the factor $\left(1 + \frac{.06}{4}\right)$. Thus, starting with the total of $100\left(1 + \frac{.06}{4}\right)^4$ after the first year, we have

$$100\left(1 + \frac{.06}{4}\right)^4 \left(1 + \frac{.06}{4}\right)^4 = 100\left(1 + \frac{.06}{4}\right)^8$$

at the end of the second year. This comes out to approximately \$112.65.

For each year that passes there will be four interest payments, each one increasing the account total by the factor $\left(1 + \frac{.06}{4}\right)$. So, after 3 years the total will be $100\left(1 + \frac{.06}{4}\right)^{12}$, after 4 years, $100\left(1 + \frac{.06}{4}\right)^{16}$, and so on. We conclude that after t years the total amount is given by the formula

$$100\left(1 + \frac{.06}{4}\right)^{4t} \quad \blacksquare$$

We may generalize Example 3 as follows. If an interest rate r is compounded n times per year, then each payment increases the total investment by the factor $1 + \frac{r}{n}$. Thus we have the following result.

Compound Interest
An investment of P dollars at the annual interest rate r (expressed as a decimal) compounded n times per year will yield after t years the amount $A(t)$ given by

$$A(t) = P\left(1 + \frac{r}{n}\right)^{nt}$$

Example 4

Find the amount yielded by investing \$2000 at 7.5 percent compounded daily for 10 years.

SOLUTION   Compounded daily means that there are 365 payments per year. Therefore the amount after t years is given by

$$A(t) = P\left(1 + \frac{r}{n}\right)^{nt} = 2000\left(1 + \frac{.075}{365}\right)^{365t}$$

So, after 10 years we have

$$A(10) = 2000\left(1 + \frac{.075}{365}\right)^{365(10)}$$

$$= 2000\left(\frac{365.075}{365}\right)^{3650}$$

Computing this on our calculator, we get $A(10) \approx \$4233.67.$ $\blacksquare$

Figure 5.9

Let us imagine that we are standing at the beginning of an infinitely long road that has nothing but banks down one side: bank 1, bank 2, bank 3, and so on, in line, one after the other (Figure 5.9). Our goal is to establish our own bank, which will pay a better interest rate than any of the banks along this road. To do this, we will need to determine the rates which all these banks are paying and then set our own rate just higher.

We begin our task by walking down the street and asking what each bank will pay on a $1 investment for 1 year. Bank 1 surprises us with an offer of 100 percent interest, which means that our $1 investment would yield $2 at the end of a year. Bank 2 offers to pay 100 percent compounded twice a year, which is a better offer than bank 1 because our $1 would yield

$$\left(1 + \frac{1}{2}\right)^2$$

or $2.25. Now we visit bank 3, which pays 100 percent compounded three times a year. Hence our $1 would yield

$$\left(1 + \frac{1}{3}\right)^3$$

or $2.37037037. . . . We now see a pattern forming. Bank 4 pays $\left(1 + \frac{1}{4}\right)^4$. Bank 5 pays $\left(1 + \frac{1}{5}\right)^5$, and so on. At bank n, the interest of 100 percent is compounded n times per year, yielding

$$\left(1 + \frac{1}{n}\right)^n$$

How much should we offer to return on a $1 investment for 1 year to do better than all these banks? To answer this question, we need to find out what happens to the quantity $\left(1 + \frac{1}{n}\right)^n$ as n gets bigger and bigger. In other words, we want to evaluate

$$\lim_{n \to +\infty} \left(1 + \frac{1}{n}\right)^n$$

Using our calculator, we obtain the following table of values:

n	$\left(1 + \dfrac{1}{n}\right)^n$	Approximate Value
10	$(1 + 1/10)^{10}$	2.59374246
100	$(1 + 1/100)^{100}$	2.704813815
1000	$(1 + 1/1000)^{1000}$	2.716923842
10,000	$(1 + 1/10,000)^{10,000}$	2.718145918
100,000	$(1 + 1/100,000)^{100,000}$	2.718254646
1,000,000	$(1 + 1/1,000,000)^{1,000,000}$	2.718281828

It appears that the limit is approximately 2.718 to three decimal places. As it turns out, evaluating this limit is a well-known problem in calculus. The answer is an irrational number denoted by the letter e in honor of the eighteenth-century mathematician Leonard Euler. Thus

$$\lim_{n \to +\infty} \left(1 + \frac{1}{n}\right)^n = e$$

The value of e rounded off to fifteen decimal places is 2.718281828459045.

Now, to conclude our original objective, we need to establish a bank that pays $\$e$, or about $2.72, for every dollar invested per year in order to offer a better return than any of the banks along the road.

We can think of e as the amount yielded on a $1 investment at the rate of 100 percent *compounded continuously* for 1 year. It follows that an investment of P dollars at 100 percent compounded continuously would yield after t years the amount $A(t)$ given by

$$A(t) = Pe^t$$

Example 5

Find the amount of money yielded by investing $1 at 100 percent compounded continuously for 5 years.

SOLUTION   We have $A(t) = Pe^t$, where $P = 1$ and $t = 5$. Thus

$$A(5) = e^5$$

Most calculators have an $\boxed{e^x}$ key, which allows us to compute this quantity by pressing

$$5 \quad \boxed{e^x}$$

The answer, rounded off to two decimal places, is 148.41. Thus our $1 investment has grown to $148.41. If your calculator does not have an $\boxed{e^x}$ key, in its place use the following two keystrokes: $\boxed{\text{INV}}$ $\boxed{\ln}$. ∎

Remark

For calculators that display each keystroke, to evaluate e^5, press $\boxed{e^x}$ first and then 5.

Of course, most banks are not going to offer 100 percent interest rates, but many do offer interests rates compounded continuously. To compute the yield on these types of accounts, we give the following result from calculus.

> **Interest Compounded Continuously**
> An investment of P dollars at an annual interest rate of r (expressed as a decimal) compounded continuously will yield after t years the amount $A(t)$ given by
> $$A(t) = Pe^{rt}$$

Example 6

A bank offers to pay 6 percent compounded continuously on savings deposits. If we invest $100, how much is yielded after 1 year? After 10 years?

SOLUTION   According to the formula, after 1 year we have

$$A(1) = 100e^{.06} \approx \$106.18$$

After 10 years, we get

$$A(10) = 100e^{(.06)10} = 100e^{.6} \approx \$182.21$$

Compare these results with Example 1. ∎

In Examples 5 and 6 we used an exponential expression with base e to solve a problem. Although e seems like a very awkward number to work with, in calculus it turns out to be of fundamental importance. As a result, we call the exponential function with base e

$$F(x) = e^x$$

the exponential function. Since the base e satisfies $e > 1$, the graph of F is increasing. Figure 5.10 is the graph of F on the same axes with $y = 2^x$ and $y = 3^x$. We know that $2 < e < 3$, so it is natural to expect the graph of $F(x) = e^x$ to be located between the two curves $y = 2^x$ and $y = 3^x$. A rough graph of $y = e^x$ can be drawn quickly by hand if we plot the following three points:

$$x = -1, \ y = e^{-1} = \frac{1}{e} \approx \frac{1}{3}$$

$$x = 0, \ y = e^0 = 1$$

$$x = 1, \ y = e^1 \approx 2.7$$

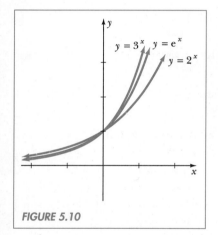

FIGURE 5.10

EXERCISES 5.3

In Exercises 1 to 6, find the amount yielded by an investment of $100 at the given rate for the time period of (a) 1 year, (b) 10 years.

1. 10 percent compounded annually

2. $5\frac{1}{2}$ percent compounded annually

3. 10 percent compounded monthly

4. $5\frac{1}{2}$ percent compounded monthly

5. 10 percent compounded continuously

6. $5\frac{1}{2}$ percent compounded continuously

In Exercises 7 to 10 you wish to invest $2000 for 20 years. Two possible investment opportunities are described. Determine which opportunity is best and the amount it will yield on your $2000.

7. 10 percent compounded daily or $10\frac{1}{2}$ percent compounded annually.

8. 7 percent compounded quarterly or 6.9 percent compounded continuously.

9. 10 percent compounded annually for the first 10 years, then 5 percent compounded annually for the last 10 years, or $7\frac{1}{2}$ percent compounded annually.

10. 6 percent compounded annually for the first 8 years, then 8 percent compounded annually for the last 12 years, or 7 percent compounded quarterly.

11. An investment offers $5e$ percent compounded continuously. Determine the yield after t years on an investment of P dollars (in terms of P and t).

12. An investment offers $4e$ percent compounded daily. Determine the yield after t years on an investment of P dollars (in terms of P and t).

13. Find the amount yielded by a \$1 investment after 1 year in a bank that pays 100 percent interest compounded (a) monthly, (b) daily, (c) hourly, (d) every 15 minutes.

14. Find (a) $\lim\limits_{x \to +\infty} e^x$, (b) $\lim\limits_{x \to -\infty} e^x$

In Exercises 15 and 16, without a calculator, determine which number is larger. (*Hint:* think of the graph of $y = e^x$.)

15. (a) $e^{1.3}, e^{1.4}$
    (b) $e^{-.01}, e^{-.09}$

16. (a) $e^{\sqrt{2}}, e^{\sqrt{3}}$
    (b) $e^{-\pi}, e^{-3}$

In Exercises 17 to 20, graph each function. State the domain, range, zero (if it exists), and y-intercept.

17. $F(x) = e^{x+1}$

18. $F(x) = -e^{x-1}$

19. $F(x) = e - e^x$

20. $F(x) = e^x - e$

21. Graph $y = e^{-x}$. Find (a) $\lim\limits_{x \to +\infty} e^{-x}$, (b) $\lim\limits_{x \to -\infty} e^{-x}$.

In Exercises 22 and 23, use the method of adding graphs to graph the function. Determine the range from your graph.

22. $S(x) = \dfrac{e^x - e^{-x}}{2}$

23. $C(x) = \dfrac{e^x + e^{-x}}{2}$

24. Let $F(x) = e^x$.
    (a) Show that $\dfrac{F(x + h) - F(x)}{h} = e^x \left(\dfrac{e^h - 1}{h} \right)$.
    (b) Use the calculator to find the value of $\dfrac{e^h - 1}{h}$ for $h = .1, .01, .001, .0001$.
    (c) Use part (b) to guess the value of $\lim\limits_{h \to 0^+} \dfrac{e^h - 1}{h}$.

In Exercises 25 to 27, answer true or false.

25. e^x cannot be negative or zero for any x.

26. e^{-x} could be negative for some x.

27. $e^x > 2^x$ for all x.

28. Suppose two investment plans are available, one paying the rate r (expressed as a decimal) compounded annually, the other paying rate r compounded twice annually. If P dollars are invested in each plan, show that the difference in yields after 1 year is $Pr^2/4$.

29. Suppose bank A pays interest rate r (expressed as a decimal) compounded twice annually, and bank B pays rate R compounded annually. Show that at the end of any number of years the yield on an investment of P dollars will be the same in either bank if $R = r + r^2/4$.

30. Let r_1, r_2, and r_3 be interest rates expressed as decimals. Suppose an investment earns the rate r_1 for the first year, r_2 for the second year, and r_3 for the third year (each compounded annually). If P dollars were originally invested, find the amount yielded after 3 years.

5.4 LOGARITHMIC FUNCTIONS

Let us consider once again an arbitrary exponential function

$$F(x) = a^x$$

Throughout this section we assume that $a > 0$ and $a \neq 1$. Recall Figures 5.6(a) and (b), which show that the graph of F is either increasing (when $a > 1$) or decreasing (when $0 < a < 1$). In either case, we have

1. The graph of $y = a^x$ satisfies the Horizontal Line Test; that is, any horizontal line intersects the graph at most once.
2. a^x is one-to-one; that is, $a^{x_1} = a^{x_2}$ if and only if $x_1 = x_2$.

Both these statements are equivalent. Furthermore, by the Inverse Function Theorem of Section 2.6, both statements say that F has an inverse function, F^{-1}. Instead of writing the general symbol F^{-1}, we give the inverse of the exponential function a special name and notation; we call it the **logarithmic function base a**, denoted by $\log_a$. In this section we explore the properties of this new function, $\log_a(x)$.

The first objective is to find the rule for $F^{-1}(x) = \log_a(x)$. The usual procedure is as follows:

$$\text{Equation for } F: \quad y = a^x$$

$$\text{Interchange } x \text{ and } y: \quad x = a^y$$

$$\text{Solve for } y: \quad ?$$

Since it is not possible for us to find y directly, we use an indirect approach.

Definition of $\log_a(x)$

Symbols: $\log_a(x) = y$ means $a^y = x$.

Words: $\log_a(x)$ is the power to which we raise a to get x.

Before considering the domain and the graph of $y = \log_a(x)$, let us evaluate the function in some specific cases.

Example 1

Evaluate (a) $\log_3(81)$,   (b) $\log_2\left(\dfrac{1}{8}\right)$,   (c) $\log_{10}(1)$,   (d) $\log_3(0)$, (e) $\log_2(-1)$.

SOLUTION   (a) Using the word definition, $\log_3(81)$ is the power to which we raise 3 to get 81. In symbols,

$$\log_3(81) = y \quad \text{means} \quad 3^y = 81$$

We know that $3^4 = 81$; therefore $\log_3(81) = 4$.

(b) The word definition says $\log_2\left(\dfrac{1}{8}\right)$ is the power to which we raise 2 to get 1/8. In symbols,

$$\log_2\left(\dfrac{1}{8}\right) = y \quad \text{means} \quad 2^y = \dfrac{1}{8}$$

We know that $2^{-3} = \dfrac{1}{8}$; therefore $\log_2\left(\dfrac{1}{8}\right) = -3$.

(c) $\log_{10}(1)$ is the power to which we raise 10 to get 1. In symbols,

$$\log_{10}(1) = y \quad \text{means} \quad 10^y = 1$$

We know that $10^0 = 1$; therefore $\log_{10}(1) = 0$.

(d) $\log_3(0)$ is the power to which we raise 3 to get 0. In symbols,

$$\log_3(0) = y \quad \text{means} \quad 3^y = 0$$

But 3^y is *never* zero for any value of y. Therefore $\log_3(0)$ does not exist. Evidently, zero is not in the domain of $\log_3(x)$.

(e) $\log_2(-1)$ is the power to which we raise 2 to get -1. In symbols,

$$\log_2(-1) = y \quad \text{means} \quad 2^y = -1$$

But 2^y is *never* negative for any value of y. Therefore $\log_2(-1)$ does not exist. In other words, -1 is not in the domain of $\log_2(x)$. ∎

Examples 1(d) and (e) illustrate that not all real numbers are contained in the domain of a logarithmic function. Since the logarithmic and exponential functions are inverses, the domain of $\log_a(x)$ is the same as the range of a^x. It follows that the domain of $\log_a(x)$ is the set of all positive real numbers, $(0, +\infty)$.

Example 2

Graph $F(x) = \log_2(x)$. Discuss the domain, range, and behavior of $\log_2(x)$.

SOLUTION We wish to graph the equation

$$y = \log_2(x)$$

We may follow two approaches. First, recall that the graphs of inverse functions are symmetric images of each other with respect to the diagonal $y = x$. Therefore we can graph $y = 2^x$ and then reflect it across the diagonal to obtain the graph of $y = \log_2(x)$. This is what we have done in Figure 5.11.

Of course, the second approach to graphing the curve is to make a table and plot points. We shall construct a table here to check the accuracy of our graph:

x	$\dfrac{1}{8}$	$\dfrac{1}{4}$	$\dfrac{1}{2}$	1	2	4
$y = \log_2(x)$	-3	-2	-1	0	1	2

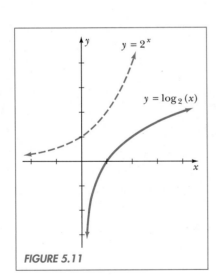

FIGURE 5.11

The reader can verify that the graph in Figure 5.11 does contain these points. From the graph, we make the following observations.

> **Properties of $F(x) = log_2(x)$**
> 1. The domain of $\log_2(x)$ is $\{x : x > 0\}$.
> 2. $\log_2(x)$ is increasing.
> 3. The range of $\log_2(x)$ is $(-\infty, +\infty)$.
> 4. $\lim\limits_{x \to 0^+} \log_2(x) = -\infty$ $\lim\limits_{x \to +\infty} \log_2(x) = +\infty$
>
>   Vertical asymptote at $x = 0$. ■

The logarithmic functions with base $a > 1$ all share the same properties listed above for $\log_2(x)$. For example, Figure 5.12 shows the graphs of $y = \log_a(x)$ for $a = 3/2$, 2, 3, and 10, on the same axes. Although these graphs have slightly different curvatures, we see that they all have the same domain and range, the same vertical asymptote at $x = 0$, and they are all increasing.

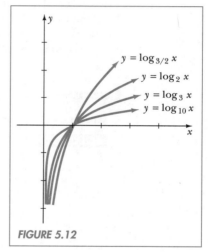

FIGURE 5.12

Example 3

Graph $G(x) = \log_{1/2}(x)$. Discuss the domain, range, and behavior of $\log_{1/2}(x)$.

SOLUTION   We wish to graph the equation

$$y = \log_{1/2}(x)$$

Since $\log_{1/2}(x)$ and $\left(\dfrac{1}{2}\right)^x$ are inverse functions, we reflect the graph of

$y = \left(\dfrac{1}{2}\right)^x$ across the diagonal to obtain the desired curve, as shown in Figure 5.13.

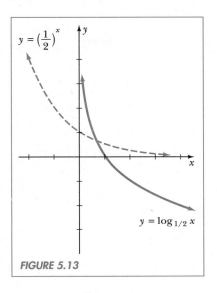

$y = \left(\dfrac{1}{2}\right)^x$

$y = \log_{1/2} x$

FIGURE 5.13

From the graph, we make the following observations.

Properties of $G(x) = \log_{1/2}(x)$
1. The domain of $\log_{1/2}(x)$ is $\{x : x > 0\}$.
2. $\log_{1/2}(x)$ is decreasing.
3. The range of $\log_{1/2}(x)$ is $(-\infty, +\infty)$.

4. $\displaystyle\lim_{x \to 0^+} \log_{1/2}(x) = +\infty$ $\displaystyle\lim_{x \to +\infty} \log_{1/2}(x) = -\infty$

  Vertical asymptote at $x = 0$. ■

The graph of any logarithmic function $F(x) = \log_a(x)$ behaves like the graph of $y = \log_2(x)$ if $a > 1$ or $y = \log_{1/2}(x)$ if $0 < a < 1$. Thus we can draw two typical curves for $y = \log_a(x)$, as shown in Figure 5.14 on page 258; Figure 5.14(a) corresponds to $a > 1$, and Figure 5.14(b) corresponds to $0 < a < 1$. To locate specific points on either graph, make a table of values containing $x = a$, 1, and $1/a$, which are easy to evaluate using the definition of $\log_a(x)$.

$$\log_a(a) = 1 \qquad \log_a(1) = 0 \qquad \log_a\!\left(\frac{1}{a}\right) = -1$$

Finally, observe the following properties for the graph:

1. The domain of $\log_a(x)$ is $\{x : x > 0.\}$
2. $\log_a(x)$ is $\begin{cases} \text{increasing if } a > 1 \\ \text{decreasing if } 0 < a < 1 \end{cases}$
3. The range of $\log_a(x)$ is $(-\infty, +\infty)$.
4. $y = \log_a(x)$ has a vertical asymptote at $x = 0$.

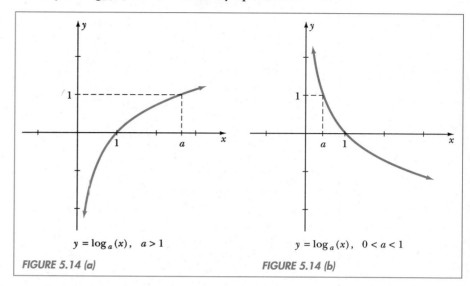

$$y = \log_a(x), \quad a > 1$$

FIGURE 5.14 (a)

$$y = \log_a(x), \quad 0 < a < 1$$

FIGURE 5.14 (b)

Recall the special number e introduced in Section 5.3. Since $e \approx 2.718$, we can use it for the base of a logarithmic function. In this case, we have a special name and notation for the logarithm.

Definition of ln(x)

$\log_e(x)$ is called the **natural logarithm** of x and is denoted by $\ln(x)$.

As we mentioned, e is a very important number in calculus, so it is not surprising to learn that $\ln(x)$ is a very important function as well. Note that e^x and $\ln(x)$ are inverse functions.

Example 4

Evaluate (a) $\ln(e)$, (b) $\ln(2)$.

SOLUTION (a) By the word definition of a logarithm, $\ln(e)$ is the power to which we raise e to get e. In symbols,

$$\ln(e) = y \quad \text{means} \quad e^y = e$$

Of course, $e^1 = e$, so $\ln(e) = 1$.

(b) $\ln(2)$ is the power to which we raise e to get 2. In symbols,

$$\ln(2) = y \quad \text{means} \quad e^y = 2$$

The value of y is not obvious in this case, so we shall have to use our calculator. Fortunately, the calculator does have an $\boxed{\ln}$ key. We press 2 $\boxed{\ln}$ to obtain $\ln(2) \approx .69315$ rounded off to five decimal places. ■

Remark

For calculators that display each keystroke, evaluate $\ln(2)$ by pressing $\boxed{\ln}$ first and then 2.

EXERCISES 5.4

In Exercises 1 to 15, evaluate each logarithm without using a calculator.

1. $\log_2(8)$
2. $\log_2(1)$
3. $\log_{1/2}(1/2)$
4. $\log_{1/2}(2)$
5. $\log_3(1/9)$
6. $\log_3(27)$
7. $\log_\pi\left(\dfrac{1}{\pi}\right)$
8. $\log_{10}(10)$
9. $\log_{1/3}(9)$
10. $\log_{1/3}(3)$
11. $\log_{10}(.001)$
12. $\log_{10}(.1)$
13. $\ln(e^2)$
14. $\ln(1/e)$
15. $\ln(1)$

In Exercises 16 to 24, the value of the given logarithm is between what two consecutive integers? Do not use a calculator.

16. $\log_2(15)$
17. $\log_2(3)$
18. $\log_3(15)$
19. $\log_3(25)$
20. $\log_8(16)$
21. $\log_{1/2}(5)$
22. $\ln(3)$
23. $\ln(2.5)$
24. $\ln(8)$

In Exercises 25 and 26, graph on the same axes.

25. $y = \log_4(x)$, $y = \log_6(x)$
26. $y = \log_{1/4}(x)$, $y = \log_{1/6}(x)$

In Exercises 27 to 40, graph the function F. Is the function increasing or decreasing? Find its domain, range, zeros, and evaluate $\lim_{x \to 0^+} F(x)$ and $\lim_{x \to +\infty} F(x)$.

27. $F(x) = \log_{1/3}(x)$
28. $F(x) = \log_3(x)$
29. $F(x) = \log_{10}(x)$
30. $F(x) = \log_{1/10}(x)$
31. $F(x) = \ln(x)$
32. $F(x) = -\ln(x)$

33. $F(x) = |\log_2(x)|$
34. $F(x) = |\log_3 x|$
35. $F(x) = \dfrac{1}{\log_2(x)}$
36. $F(x) = \dfrac{1}{\log_{1/2}(x)}$
37. $F(x) = \log_3|x|$
38. $F(x) = \log_2|x|$
39. $F(x) = -\log_{3/2}(x)$
40. $F(x) = \log_4(\sqrt{x})$

In Exercises 41 to 44, find $F^{-1}(x)$. Graph F and F^{-1} on the same axes.

41. $F(x) = \log_4(x)$
42. $F(x) = e^x$
43. $F(x) = \left(\dfrac{1}{3}\right)^x$
44. $F(x) = \log_{1/4}(x)$

In Exercises 45 to 50, recall that $y = F(x - h) + k$ is the same as $y = F(x)$ shifted h units in the x-direction and k units in the y-direction. (a) For each function $G(x)$ given, find the unshifted function $F(x) = \log_a(x)$ and tell how to shift F to get G. (b) Graph G. Find (c) the domain of G, (d) the vertical asymptote, (e) all x such that $G(x) > 0$.

45. $G(x) = \log_2(x - 2)$
46. $G(x) = \log_2(x + 1)$
47. $G(x) = 1 - \log_2(x)$
48. $G(x) = 2 - \log_{1/2}(x)$
49. $G(x) = \log_2(x + 4) - 2$
50. $G(x) = 1 - \ln(x + 1)$

In Exercises 51 to 53, evaluate.

51. (a) $\lim_{x \to +\infty} \log_{\sqrt{2}}(x)$
(b) $\lim_{x \to 0^+} \log_{\sqrt{2}}(x)$

52. (a) $\lim\limits_{x \to +\infty} \log_{1/\pi}(x)$

   (b) $\lim\limits_{x \to 0^+} \log_{1/\pi}(x)$

53. (a) $\lim\limits_{x \to +\infty} \ln(x - 1)$

   (b) $\lim\limits_{x \to 1^+} \ln(x - 1)$

54. Graph $F(x) = \dfrac{1}{3} \log_2(x)$ and $G(x) = \log_8(x)$. How do these functions compare?

55. (a) Graph $F(x) = -\log_3(x)$ and $G(x) = \log_{1/3}(x)$. How do these functions compare?

   (b) How do you think $-\log_a(x)$ and $\log_{1/a}(x)$ compare?

In Exercises 56 to 58, answer true or false.

56. (r, s) is on the graph of $F(x) = \log_a(x)$ if and only if $s = a^r$.

57. a^x is increasing implies $\log_a(x)$ is decreasing.

58. a^x is decreasing implies $\log_a(x)$ is decreasing.

59. Suppose $F(x) = \log_a(x)$ and $G(x) = a^x$. Find (a) $F(a^3)$, (b) $F(G(x))$, (c) $G(\log_a(7))$, (d) $G(F(x))$.

60. Prove $F(x) = \log_{1/a}(x)$ and $G(x) = -\log_a(x)$ are equal functions.

5.5 ALGEBRA OF LOGARITHMS

In this section we concentrate on manipulating and simplifying expressions involving logarithms. As usual, whenever a appears as a base, we assume that $a > 0$ and $a \neq 1$.

Recall the definition of the logarithm;

$$y = \log_a(x) \quad \text{is equivalent to} \quad a^y = x$$

On the left is an equation in *logarithmic form*, and on the right is an equivalent equation in *exponential form*. Whenever we are working with one form of these equations, it is possible to replace it with the other form. This can be helpful in situations where it may be easier to work with exponents rather than logarithms, or vice versa.

Example 1

Write the equation $\log_4(x + 1) = 3$ in its equivalent exponential form.

SOLUTION $\log_4(x + 1) = 3$ is equivalent to $4^3 = x + 1$ ■

Occasionally, we shall write a logarithm without indicating the base. Whenever we do so, the base is understood to be 10. Thus $\log(x)$ means $\log_{10}(x)$. We refer to the logarithm with base 10 as the **common logarithm**.

Example 2

Write the equation $10^x = 42$ in its equivalent logarithmic form.

SOLUTION $\qquad 10^x = 42 \quad$ is equivalent to $\quad x = \log(42)$ ∎

Example 3

Find $F^{-1}(x)$ if $F(x) = \log_2(x - 1) + 3$.

SOLUTION Note that $F(x)$ is just the function $\log_2(x)$ shifted. We know that the graph of $y = \log_2(x)$ satisfies the Horizontal Line Test; therefore, so does the graph of $y = F(x)$. Hence F must have an inverse. Now, let us find the rule for $F^{-1}(x)$.

$$\text{Equation for } F\!: \quad y = \log_2(x - 1) + 3$$

$$\text{Interchange } x \text{ and } y\!: \quad x = \log_2(y - 1) + 3$$

$$\text{Solve for } y\!: \quad x - 3 = \log_2(y - 1)$$

At this point we have an equation in logarithmic form, which we can rewrite in its equivalent exponential form. Thus

$$x - 3 = \log_2(y - 1) \quad \text{is equivalent to} \quad 2^{x-3} = y - 1$$

It follows that

$$y = 1 + 2^{x-3}$$

Therefore $F^{-1}(x) = 1 + 2^{x-3}$. ∎

We now list five algebraic properties of logarithms. These properties play an important role in simplifying expressions and solving equations that contain exponents or logarithms.

Properties of Logarithms

Assume that a, x, and y are positive real numbers and $a \neq 1$:

1. $\log_a(xy) = \log_a(x) + \log_a(y)$.

2. $\log_a\!\left(\dfrac{x}{y}\right) = \log_a(x) - \log_a(y)$.

3. $\log_a(x^r) = r \log_a(x),\ r \in \mathbf{R}$.

4. $a^{\log_a(x)} = x$.

5. One-to-one property of logarithms: for positive real numbers x_1 and x_2,

$$x_1 = x_2 \quad \text{if and only if} \quad \log_a(x_1) = \log_a(x_2)$$

PROOF OF 1: Let $n = \log_a(x)$ and $m = \log_a(y)$. By the definition of the logarithm, these equations are equivalent to

$$a^n = x \qquad \text{and} \qquad a^m = y$$

This implies

$$a^n a^m = xy$$

$$a^{n+m} = xy \qquad \text{by the rules for exponents}$$

$$\log_a(xy) = n + m \qquad \text{by the definition of logarithm}$$

Now, substituting for n and m, we have

$$\log_a(xy) = \log_a(x) + \log_a(y)$$

This completes the proof.

PROOF OF 2: We leave this for an exercise.

PROOF OF 3: Let $n = \log_a(x)$. By the definition of the logarithm, this is equivalent to

$$a^n = x$$

Since $x > 0$, x^r is defined for any $r \in \mathbf{R}$. Therefore

$$x^r = (a^n)^r = a^{rn}$$

by the rules for exponents. Now,

$$x^r = a^{rn} \quad \text{is equivalent to} \quad \log_a(x^r) = rn$$

Substituting for n in this last equation, we have

$$\log_a(x^r) = r \log_a(x)$$

This completes the proof.

PROOF OF 4: By definition, $\log_a(x)$ is the power to which we raise a to get x. Thus, if we raise a to the power $\log_a(x)$,

$$a^{\log_a(x)}$$

we must get x,

$$a^{\log_a(x)} = x$$

This completes the proof.

PROOF OF 5: The function $\log_a(x)$ has an inverse, a^x. It follows from the Inverse Function Theorem given in Section 2.6 that $\log_a(x)$ is one-to-one. By definition of the one-to-one function (also given in Section 2.6), we have

$$x_1 = x_2 \qquad \text{if and only if} \qquad \log_a(x_1) = \log_a(x_2)$$

for any x_1 and x_2 in the domain of $\log_a(x)$. Finally, we note that the domain of $\log_a(x)$ is the set of all positive real numbers. This completes the proof. ∎

Examples 4 to 7 illustrate applications of the properties of logarithms.

Example 4

Let $A = \log_a(x)$ and $B = \log_a(y)$. Find $\log_a\left(x\sqrt{\dfrac{x}{y}}\right)$ in terms of A and B.

SOLUTION We can change the expression $x\sqrt{\dfrac{x}{y}}$ to exponents,

$$x\sqrt{\frac{x}{y}} = x\left(\frac{x}{y}\right)^{1/2} = x\frac{x^{1/2}}{y^{1/2}} = \frac{x^{3/2}}{y^{1/2}}$$

Thus

$$\log_a\left(x\frac{x}{y}\right) = \log_a\left(\frac{x^{3/2}}{y^{1/2}}\right)$$

$$= \log_a(x^{3/2}) - \log_a(y^{1/2}) \qquad \text{by property 2}$$

$$= \frac{3}{2}\log_a(x) - \frac{1}{2}\log_a(y) \qquad \text{by property 3}$$

$$= \frac{3}{2}A - \frac{1}{2}B \quad ∎$$

Example 5

Assume that $\log 2 = A$ and $\log 3 = B$. Find $\log\left(\dfrac{1}{54}\right)$ in terms of A and B.

SOLUTION We know that $54 = 2(27) = 2(3^3)$. Therefore

$$\log\left(\frac{1}{54}\right) = \log 1 - \log 54$$

$$= 0 - \log(2(3^3))$$

$$= -(\log 2 + \log(3^3))$$

$$= -(\log 2 + 3\log 3)$$

$$= -(A + 3B)$$

$$= -A - 3B \quad ∎$$

Example 6

Combine the expression $2 \ln(3) + \ln(y) - \frac{1}{2} \ln(x)$ into one logarithm.

SOLUTION

$$2 \ln(3) + \ln(y) - \frac{1}{2} \ln(x) = \ln(3^2) + \ln(y) + \ln(x^{-1/2}) \qquad \text{by property 3}$$

$$= \ln(3^2 y) + \ln(x^{-1/2}) \qquad \text{by property 1}$$

$$= \ln(3^2 y x^{-1/2}) \qquad \text{by property 1}$$

$$= \ln\left(\frac{9y}{\sqrt{x}}\right) \ \blacksquare$$

Example 7

Simplify (a) $2^{\log_2(\pi)}$, (b) $e^{-\ln x}$, (c) $e^{\log_2(8)}$.

SOLUTION (a) $2^{\log_2(\pi)} = \pi$ by property 4.

(b) Before simplifying this expression, we point out that we wrote $\ln x$ instead of $\ln(x)$; the parentheses are often dropped when the meaning is clear. Now, by property 3, we have

$$-\ln x = \ln(x^{-1})$$

Therefore

$$e^{-\ln x} = e^{\ln(x^{-1})}$$

$$= x^{-1} \qquad \text{by property 4}$$

$$= \frac{1}{x}$$

(c) Note that property 4 does *not* apply to $e^{\log_2(8)}$ because the base of the exponent, e, is not the same as the base of the logarithm, 2. However, we do know that $\log_2(8) = 3$. Thus

$$e^{\log_2(8)} = e^3 \ \blacksquare$$

The one-to-one property of logarithms, stated as property 5 above, gives us a new operation that can be applied to both sides of an equation. In the past we used the operations of addition, multiplication, squaring, and square root to solve equations. Now, if both sides of an equation are positive, we may take the logarithm of each side and still have a valid equation.

As an application of this property, we shall prove that every logarithm can be written in terms of the natural logarithm.

Changing to Base e

$$\log_a(x) = \frac{\ln x}{\ln a}$$

PROOF: We note that $a > 0$, $a \neq 1$, and $x > 0$. Let $n = \log_a(x)$. By the definition of the logarithm, this equation is equivalent to

$$a^n = x$$

Since both sides of this equation are positive, we may take the natural logarithm of each side (property 5), obtaining

$$\ln(a^n) = \ln x$$

By Property 3, we have

$$n \ln a = \ln x$$

So,

$$n = \frac{\ln x}{\ln a}$$

Now, substituting for n, we have

$$\log_a(x) = \frac{\ln x}{\ln a}$$

This completes the proof. ■

The Changing to Base e Theorem says there is only one logarithmic function necessary, the natural logarithm. All other logarithmic functions can be expressed in terms of $\ln x$.

Example 8

Use the calculator to approximate $\log_2(3)$ to four decimal places.

SOLUTION The calculator does not have a $\log_2$ key; it does have a common logarithm key, $\boxed{\log}$, and a natural logarithm key, $\boxed{\ln}$. In view of the Changing to Base e Theorem, we shall use the natural logarithm to compute $\log_2(3)$. We have

$$\log_2(3) = \frac{\ln 3}{\ln 2}$$

by pressing

$$3 \quad \boxed{\ln} \quad \boxed{\div} \quad 2 \quad \boxed{\ln} \quad \boxed{=}$$

we obtain $\log_2(3) \approx 1.5850$. ■

Remark

It is possible to use the log key to find $\log_2(3)$ as follows:

$$\log_2(3) = \frac{\log 3}{\log 2}$$

You may enter this into the calculator to verify that it produces the same result for $\log_2(3)$ found in Example 8. This is a special case of the general result,

$$\log_a(x) = \frac{\log_b(x)}{\log_b(a)}$$

The proof for this is similar to the Changing to Base e Theorem and is left as an exercise.

EXERCISES 5.5

In Exercises 1 to 6, write the equation in its equivalent exponential form.

1. (a) $\log_4(64) = 3$
   (b) $\log x = 4$
   (c) $\ln 3 = y$

2. (a) $\log_2\left(\dfrac{1}{16}\right) = -4$
   (b) $\log 5 = y$
   (c) $\ln x = 2$

3. $-2 \log x = 3$

4. $-\ln x = 1$

5. $1 + \ln(x + 1) = 5$

6. $4 + \log(x - 1) = 6$

In Exercises 7–10, write the equation in its equivalent logarithmic form.

7. (a) $4^{1/2} = 2$
   (b) $e^x = 2.5$
   (c) $10^b = a$

8. (a) $8^{1/3} = 2$
   (b) $10^x = 1.7$
   (c) $e^a = b$

9. $x + y = e^3$

10. $x + y = 10^4$

In Exercises 11 to 16, find $F^{-1}(x)$.

11. (a) $F(x) = \log_2(x + 1)$
    (b) $F(x) = \log_2(x) + 1$

12. (a) $F(x) = \ln x - 2$
    (b) $F(x) = \ln(x - 2)$

13. (a) $F(x) = 1 - \ln x$
    (b) $F(x) = \log(x - 5) + 10$

14. (a) $F(x) = 3 - \ln x$
    (b) $F(x) = \log(x + 1) - 5$

15. $F(x) = 2 \ln\left(\dfrac{1}{x}\right)$

16. $F(x) = -3 \ln x$

In Exercises 17 to 22, simplify.

17. (a) $10^{\log \pi}$
    (b) $e^{\ln 10}$
    (c) $2^{\log 10}$

18. (a) $e^{\ln \pi}$
    (b) $10^{\log e}$
    (c) $2^{\ln e}$

19. (a) $e^{2 \ln x}$
    (b) $10^{-\log 4}$
    (c) $2^{\log_4(25)}$

20. (a) $10^{2 \log 3}$
    (b) $e^{-\ln 2}$
    (c) $3^{\log_9(4)}$

21. (a) $\ln(e^x)$
    (b) $e^{-x \ln 2}$

22. (a) $\log\left(\dfrac{1}{10^x}\right)$
    (b) $\log(100^x)$

In Exercises 23 to 26, let $A = \ln x$, $B = \ln y$, and $C = \ln c$. Write the given expression in terms of A, B, and C.

23. (a) $\ln(xy)$
    (b) $\ln(x^3)$
    (c) $\ln(c\sqrt{xy})$
    (d) $\ln\left(\dfrac{1}{c}\right)$
    (e) $\ln\left(\dfrac{c^2 y}{x^3}\right)$

24. (a) $\ln(cxy)$
    (b) $\ln(y^2)$
    (c) $\ln\left(c\sqrt{\dfrac{x}{y}}\right)$
    (d) $\ln\left(\dfrac{1}{xy}\right)$
    (e) $\ln\left(\dfrac{c}{y^2\sqrt{x}}\right)$
    (b) $\ln\left(\dfrac{c}{e^2}\right)$

25. (a) $\ln(ex)$

26. (a) $\ln\left(\dfrac{y}{e}\right)$ (b) $\ln(ce^3)$

In Exercises 27 and 28, assume that $\log_2(x) = 2.4$ and $\log_2(y) = 0.6$. Evaluate the given expression.

27. (a) $\log_2(\sqrt{x})$
   (b) $\log_2\left(\dfrac{1}{xy}\right)$
   (c) $\log_2(4x^3)$
   (d) $\log_2\left(\dfrac{16}{y}\right)$

28. (a) $\log_2(x\sqrt{y})$
   (b) $\log_2\left(\dfrac{x}{y}\right)$
   (c) $\log_2(32y^2)$
   (d) $\log_2\left(\dfrac{1}{2x}\right)$

In Exercises 29 and 30, assume that $\log_2(3) = A$ and $\log_2(5) = B$. Evaluate the given expression in terms of A and B.

29. (a) $\log_2\left(\dfrac{16}{25}\right)$
   (b) $\log_2\left(\dfrac{5}{6}\right)$
   (c) $\log_2\left(\dfrac{15}{8}\right)$

30. (a) $\log_2\left(\dfrac{1}{12}\right)$
   (b) $\log_2\left(\dfrac{3}{40}\right)$
   (c) $\log_2(270)$

In Exercises 31 to 34, combine the given expression into one logarithm.

31. (a) $\log x - 2 \log y$
   (b) $\ln y - \dfrac{1}{3}\ln x + 2 \ln 10$

32. (a) $3 \log x + \log\left(\dfrac{y}{2}\right)$
   (b) $\ln x - \ln(x + 1) + \ln 2$

33. (a) $1 + \ln x$ (*Hint:* $\ln e = 1$.)
   (b) $2 - \ln x$

34. (a) $3 + \log x$ (*Hint:* $\log 1000 = 3$.)
   (b) $1 - \log x$

In Exercises 35 to 38, use the $\boxed{\ln}$ key on the calculator to approximate the given logarithm to four decimal places.

35. (a) $\ln 10$
   (b) $\log_2 10$
   (c) $\log \pi$
   (d) $\log_{1/3} 6$

36. (a) $\ln 2$
   (b) $\log_3 10$
   (c) $\log e$
   (d) $\log_{1/2} 6$

37. $\log_\pi \sqrt{2}$

38. $\log_{\sqrt{2}}(\pi)$

In Exercises 39 to 52, answer true or false.

39. $\dfrac{\log a}{\log b} = \log a - \log b$

40. $\log_a(a) = 0$

41. $\log_a(a^2) - \log_a(1) = 2$

42. $\log 5 + \log 3 = \log 15$

43. $-\log x = \dfrac{1}{\log x}$

44. $\sqrt{\log x} = \dfrac{1}{2}\log x$

45. $(\ln x)^2 = 2 \ln x$

46. $\log\left(\dfrac{2}{x}\right) = -\log(2x)$

47. $\ln(xy^3) = 3 \ln(xy)$

48. $\log e = \dfrac{1}{\ln 10}$

49. $a^{\ln x} = x$

50. $(\ln x)(\ln x) = \ln(x^2)$

51. $\log_4(x) = 2 \log_2(x)$

52. $\log(a + b) = (\log a)(\log b)$

In Exercises 53 to 56, graph the given equation. (*Hint:* simplify the equation first.)

53. $y = \log_2\left(\dfrac{1}{x}\right)$

54. $y = \log_2\left(\dfrac{2}{x}\right)$

55. $y = \log_2(\sqrt{x})$

56. $y = \log_2(x^2)$

57. Prove property 2 of logarithms:

$$\log_a\left(\frac{x}{y}\right) = \log_a(x) - \log_a(y).$$ [*Hint:* let $n = \log_a(x)$ and $m = \log_a(y)$. Then $a^n = x$ and $a^m = y$.]

58. Assume that a and b are positive and not equal to 1.
    (a) Prove the Changing to Base 10 Theorem.

    $$\log_a(x) = \frac{\log x}{\log a}$$

    (*Hint:* follow the proof of the Changing to Base e Theorem.)
    (b) Prove

    $$\log_a(x) = \frac{\log_b(x)}{\log_b(a)}$$

59. Suppose $\log_a(x) = 2 \log_2(x)$ for all $x > 0$. Find a. (*Hint:* use the result from Exercise 58.)

60. (a) Graph $y = \log_{1/2}(x)$ and $y = \log_2\left(\frac{1}{x}\right)$. How do these curves compare?

    (b) How do you think $\log_{1/a}(x)$ and $\log_a\left(\frac{1}{x}\right)$ compare? Prove your answer.

61. (a) Use your calculator to compare the values of $2^{\log 3}$ and $3^{\log 2}$.
    (b) Use your calculator to compare the values of $e^{\log \pi}$ and $\pi^{\log e}$.
    (c) Prove for all positive real numbers x and y,

    $$y^{\log x} = x^{\log y}$$

    [*Hint:* start with the fact that $(\log x)(\log y) = (\log y)(\log x)$ and use property (3) of logarithms.]

62. When an earthquake occurs, seismologists may assign it a number M, called the Richter scale magnitude, as follows: $M = \log(E/E_0)$, where E is a measure of the ground strain caused by the earthquake and E_0 is a constant. Find the magnitude of an earthquake that produces a ground strain (a) 100 times that of a magnitude 5 earthquake, (b) 1000 times that of a magnitude 5 earthquake.

5.6 EQUATIONS WITH EXPONENTS AND LOGARITHMS

In this section we solve equations that have three basic forms: $a^x = k$, $\log_a(x) = k$, and $x^a = k$, where a and k are constants. We assume throughout the discussion that $a > 0$ and $a \neq 1$ when it is the base of an exponential expression or logarithm.

Equations of the Form $a^x = k$. In this case, k must be positive; otherwise there is no solution. So, assuming $k > 0$, we may take the natural logarithm of both sides (by property 5 of logarithms).

$$\ln(a^x) = \ln(k)$$

Now, we apply property 3 of logarithms.

$$x \ln(a) = \ln(k)$$

Hence

$$x = \frac{\ln k}{\ln a}$$

Example 1

Solve for x, $6^x = 12.29$.

SOLUTION Applying the natural logarithm to both sides of this equation, we have

$$\ln(6^x) = \ln(12.29)$$

$$x \ln 6 = \ln(12.29)$$

$$x = \frac{\ln(12.29)}{\ln 6} \approx 1.4002 \quad \blacksquare$$

An alternate method for solving this equation is to write the equivalent logarithmic equation:

$$6^x = 12.29 \quad \text{is equivalent to} \quad x = \log_6(12.29)$$

However, if we want an approximate numerical value for x, we must change this answer to natural logarithms (or common logarithms) to use our calculator. Thus

$$x = \log_6(12.29) = \frac{\ln(12.29)}{\ln 6}$$

which is, of course, the same result we obtained in Example 1.

Example 2

Solve for x, $2 + 3^{2x} = 17$.

SOLUTION We begin by isolating the exponential expression.

$$2 + 3^{2x} = 17$$

$$3^{2x} = 15$$

Now, apply the natural logarithm.

$$\ln(3^{2x}) = \ln(15)$$

$$2x \ln 3 = \ln 15$$

$$x = \frac{\ln 15}{2 \ln 3} \approx 1.2325 \quad \blacksquare$$

Example 3

Solve for x, $\dfrac{e^x}{e^x - 1} = 2$.

SOLUTION We manipulate this equation so that it looks like $a^x = k$. We start by multiplying both sides by $e^x - 1$ to eliminate the fraction.

$$(e^x - 1)\left(\frac{e^x}{e^x - 1}\right) = 2(e^x - 1)$$

$$e^x = 2e^x - 2$$

$$-e^x = -2$$

$$e^x = 2$$

Now, apply the natural logarithm.

$$\ln e^x = \ln 2$$

$$x = \ln 2 \approx .6931 \ \blacksquare$$

Equations of the Form $\log_a(x) = k$. In this case, we write the equivalent exponential statement.

$$\log_a(x) = k \quad \text{is equivalent to} \quad x = a^k$$

It is necessary to check answers when solving these types of equations.

Example 4

Solve for x, $\quad \log(x + 1) = 1.4$.

SOLUTION We have

$$\log(x + 1) = 1.4 \quad \text{is equivalent to} \quad x + 1 = 10^{1.4}$$

It follows that $x = 10^{1.4} - 1 \approx 24.1189$. You may check this answer in the original equation. $\blacksquare$

Example 5

Solve for x, $\quad \frac{1}{2} \log x = 2 \log 3 - \log 6 + 2$.

SOLUTION We collect all the logarithms on one side and then use the laws of logarithms to combine them into one term.

$$\frac{1}{2} \log x = 2 \log 3 - \log 6 + 2$$

$$\frac{1}{2} \log x - 2 \log 3 + \log 6 = 2$$

$$\log(x^{1/2}) + \log(3^{-2}) + \log 6 = 2$$

$$\log(x^{1/2} \cdot 3^{-2} \cdot 6) = 2$$

$$\log\left(\frac{2x^{1/2}}{3}\right) = 2$$

This is equivalent to the exponential equation

$$\frac{2x^{1/2}}{3} = 10^2$$

$$x^{1/2} = \frac{3}{2}(100)$$

$$x^{1/2} = 150$$

$$x = (150)^2 = 22,500$$

You may check this answer in the original equation. ∎

Example 6

Solve for x, $\ln x + \ln(x + 3) = 2 \ln 2$.

SOLUTION We combine the logarithms.

$$\ln x + \ln(x + 3) = 2 \ln 2$$

$$\ln(x(x + 3)) = \ln(2^2)$$

$$\ln(x^2 + 3x) = \ln(4)$$

At this point we could proceed as in Example 5, moving all the logarithms to the left side of the equation, combining them, and then converting to an equivalent exponential statement. However, another option is to apply property 5, the one-to-one property of logarithms.

$$\ln(x^2 + 3x) = \ln(4) \quad \text{implies} \quad x^2 + 3x = 4$$

Thus we must solve

$$x^2 + 3x = 4$$

$$x^2 + 3x - 4 = 0$$

$$(x + 4)(x - 1) = 0$$

$$x = -4 \quad or \quad x = 1$$

Now, we check these answers in the original equation.

$$\text{Checking } x = -4: \quad \ln(-4) + \ln(-4 + 3) \overset{?}{=} 2 \ln 2$$

This does not work since the first two logarithms are undefined.

$$\text{Checking } x = 1: \quad \ln(1) + \ln(1 + 3) \overset{?}{=} 2 \ln 2$$

$$0 + \ln 4 \quad \bigg| \quad \ln(2^2)$$

$$\text{Yes}$$

Therefore the final solution is $x = 1$. ∎

Warning

When attempting to remove logarithms from an equation, we may apply the one-to-one property of logarithms only to equations of the form

$$\log_a(x_1) = \log_a(x_2)$$

From this equation we may conclude that $x_1 = x_2$. On the other hand, equations of the form

$$\log_a(x_1) = \log_a(x_2) + \log_a(y) \quad \text{or} \quad \log_a(x_1) = \log_a(x_2) + y$$

cannot be simplified to $x_1 = x_2 + y$. Instead, we have

$$\log_a(x_1) = \log_a(x_2) + \log_a(y) \qquad\qquad \log_a(x_1) = \log_a(x_2) + y$$

$$\log_a(x_1) = \log_a(x_2 y) \qquad\qquad \log_a(x_1) - \log_a(x_2) = y$$

$$x_1 = x_2 y \qquad\qquad \log_a\!\left(\frac{x_1}{x_2}\right) = y$$

$$\frac{x_1}{x_2} = a^y$$

$$x_1 = x_2 a^y$$

Equations of the Form $x^a = k$.   There are two cases to consider, depending on whether the exponent a is irrational or rational. When a *is irrational*, we must assume that x and k are positive. In this case we use the fact that

$$(x^a)^{1/a} = x$$

Example 7

Solve for x, $x^\pi = 5$.

SOLUTION   We raise both sides of this equation to the power $1/\pi$.

$$(x^\pi)^{1/\pi} = (5)^{1/\pi}$$

$$x = 5^{1/\pi} \approx 1.6691 \quad\blacksquare$$

When a *is rational*, $(x^a)^{1/a}$ does not always simplify to x. For example, when x is unknown, we must write $(x^2)^{1/2} = |x|$ because $(x^2)^{1/2}$ means the nonnegative square root of x^2. Similarly, $(x^4)^{1/4} = |x|$. On the other hand, cube roots are unique, so $(x^3)^{1/3} = x$. In general, we have the following result.

> If n is a positive integer, then
>
> $$(x^n)^{1/n} = \begin{cases} |x| & \text{if } n \text{ is even} \\ x & \text{if } n \text{ is odd} \end{cases}$$

Example 8

Solve for x, $\quad x^{4/3} = 16$.

SOLUTION We first cube both sides to eliminate the fraction in the exponent.

$$(x^{4/3})^3 = 16^3$$
$$x^4 = 16^3$$

Now we raise both sides to the power 1/4.

$$(x^4)^{1/4} = (16^3)^{1/4}$$
$$|x| = 16^{3/4}$$
$$|x| = 8$$
$$x = 8 \quad or \quad x = -8$$

So, we get two solutions for x. You may check that both solutions work in the original equation. ∎

Example 9

Solve for x, $\quad x^{.7} = 3$.

SOLUTION We write .7 as 7/10. Thus

$$(x^{7/10})^{10} = (3)^{10}$$
$$x^7 = 3^{10}$$
$$(x^7)^{1/7} = (3^{10})^{1/7}$$
$$x = 3^{10/7} \approx 4.8040 \quad ∎$$

EXERCISES 5.6

In Exercises 1 to 50, solve each equation for x. Write the exact answer and then use the calculator (if necessary) to obtain an approximation accurate to four decimal places.

1. $2^x = 100$

2. $7^{2x} = 42$

3. $3^{x/2} = 6$

4. $2^x 3^x = 9$

5. $2^x e^x = 10$

6. $10^x - 1 = 54$

7. $\dfrac{2}{3} = e^{2x}$

8. $\dfrac{1}{2} = e^{150x}$

9. $35 = 70 e^{x \ln(3/4)}$

10. $12 = 30 e^{x \ln(2/3)}$

11. $2^{1/x} = 3$ 12. $2^x = 2(3^x)$

13. $\dfrac{2^x}{2^x - 1} = 5$ 14. $\dfrac{3^x}{3^x - 2} = 6$

15. $\dfrac{e^x + e^{-x}}{e^x - e^{-x}} = 2$ 16. $\dfrac{e^x - e^{-x}}{e^x + e^{-x}} = \dfrac{1}{3}$

17. $2^x = 2^{2x}$ 18. $2^{x/3} = 3^{x/2}$

19. $\ln x = 6$ 20. $\ln x = e + 2$

21. $\log x + \log(x - 5) = \log 6$

22. $\log x + \log(x + 1) = 0$

23. $\log x = 1 - \log(x - 3)$

24. $\log x = \log(2x - 1) + 2$

25. $\ln(x^2 + 3) = \ln 4 + \ln x$

26. $2 \log(x) = \log\left(\dfrac{3x}{4}\right) - 1$

27. $\ln(x + 1) + \ln(x - 1) = \ln(15x) - 2 \ln 2$

28. $\ln(2x - 3) = \ln 7 - \ln(x - 4)$

29. $\ln x = \dfrac{1}{2} \ln x$ 30. $\dfrac{2}{\log x} + 3 = \dfrac{4}{\log x}$

31. $\log_2(x) = \dfrac{1}{\log_3(x)}$ (*Hint:* change both logarithms to ln.)

32. $\log x = \ln x$

33. $x^{\sqrt{2}} = 10$ 34. $x^{\sqrt{3}} = 8$

35. $x^{2/3} = 4$ 36. $x^{4/3} = 81$

37. $x^{.9} = 2$ 38. $x^{1.3} = 15$

39. $(x + 1)^{2/3} = 1$ 40. $(2x - 1)^{4/5} = 1$

41. $x^{-.4} = 9$ 42. $x^{-1.2} = 64$

43. $\dfrac{4}{3} x^{1/3} - \dfrac{1}{3} x^{-4/3} = 0$ 44. $\dfrac{5}{4} x^{1/4} - \dfrac{1}{4} x^{-5/4} = 0$

45. $\dfrac{1}{3} x(x + 2)^{-2/3} + (x + 2)^{1/3} = 0$

46. $\dfrac{1}{5} x(2x - 1)^{-4/5} + (2x - 1)^{1/5} = 0$

47. $x - x^9 = 0$ 48. $x + x^4 = 0$

49. $x2^x + 2^x = 0$ 50. $x^2 e^x - 9e^x = 0$

In Exercises 51 to 58, solve the given equation graphically.

51. $2^x = 3^x$ 52. $e^x = e^{-x}$

53. $e^x = \ln x$ 54. $2^x = \log_2(x)$

55. $\log_2(x) = \dfrac{1}{4} x^2 - x + 2$ 56. $2^x = \dfrac{1}{2} x^2$

57. $\log_3 x = \log_{1/3}(x)$ 58. $2x - x^2 - x^3 = \ln x$

59. Suppose your calculator has only three function keys: 10^x, $\ln x$, and $\div$. How would you compute e^x?

60. Suppose your calculator has only three function keys: e^x, $\ln x$, and $\times$. How would you compute 10^x?

5.7 EXPONENTIAL GROWTH AND DECAY

Exponents and logarithms are helpful for describing how certain physical, quantities vary with respect to time. To say that a quantity varies with time simply means that it is a function of time.

Let $Q(t)$ be the amount of a substance at time t. Suppose further that the formula for $Q(t)$ happens to be

$$Q(t) = Q_0 e^{kt}$$

where Q_0 and k are constants. When this situation occurs, we say that Q *varies exponentially with time*. The number Q_0 is the amount of the substance at time $t = 0$ because

$$Q(0) = Q_0 e^{k(0)} = Q_0 e^0 = Q_0$$

The number k determines whether Q will increase (grow) or decrease (decay) with the passage of time (assuming Q_0 is positive). As the graphs in Figure 5.15 indicate, Q will *grow* if k is positive, and Q will *decay* if k is negative.

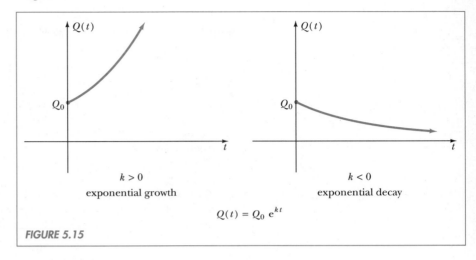

$$Q(t) = Q_0\, e^{kt}$$

FIGURE 5.15

Example 1

It is known that the population of an uninhibited rabbit colony grows exponentially with time. The colony begins with 20 rabbits and, after 1 month, grows to 30. How many rabbits will be in this colony at the end of 1 year?

SOLUTION   Let $Q(t) =$ the number of rabbits present in the colony at time t. We are told that $Q(t)$ grows exponentially. This means that

$$Q(t) = Q_0 e^{kt}$$

We know there are 20 rabbits when $t = 0$. Hence

$$20 = Q(0) = Q_0$$

Therefore

$$Q(t) = 20 e^{kt}$$

If we let t be measured in units of months, then we want to find $Q(12)$. But first we must find k. It is given that there are 30 rabbits when $t = 1$ month. Hence

$$30 = Q(1) = 20 e^{k(1)}$$

$$\frac{30}{20} = e^{k}$$

$$k = \ln\left(\frac{3}{2}\right)$$

The formula for $Q(t)$ is therefore

$$Q(t) = 20e^{\ln(3/2) \cdot t}$$

Now, to see how many rabbits are in the colony after 1 year, we evaluate $Q(t)$ when $t = 12$ months.

$$Q(12) = 20e^{\ln(3/2)12} \approx 2595 \ \blacksquare$$

Remarks

1. We often write an exponent such as $\ln(3/2) \cdot t$ as $\ln(3/2)t$ without indicating the multiplication. Thus to compute $\ln(3/2)12$ we find $\ln(3/2)$ and then multiply it by 12.
2. We did not use the calculator until the *last* step in Example 1. By using exact expressions for all quantities involved until the last step, we avoid making round-off errors. For example, suppose we used the calculator to express $k = \ln(3/2)$ as .405, its approximate value accurate to three decimal places. Then, we would have written $Q(t)$ as

$$20e^{(.405)t}$$

Evaluating this expression when $t = 12$ yields 2580 rabbits, a difference of 15 from the correct answer. Therefore, always write exact expressions for quantities when possible and refrain from using the calculator until the last step in the computation of an answer.
3. If desired, the formula for $Q(t)$ may be simplified by using the fact that $e^{\ln(3/2)} = 3/2$. Thus

$$Q(t) = 20e^{\ln(3/2)t} = 20\left(\frac{3}{2}\right)^t$$

Example 2

Carbon 14 is a radioactive element that decays exponentially with time. It is known that the amount of carbon 14 present in any given specimen will be reduced by 50 percent after 5750 years. How much carbon 14 would be left from an original 10-gram specimen after 10,000 years?

SOLUTION Let $Q(t) = $ the amount of carbon 14 present in the specimen at time t. We will measure t in units of years. Since $Q(t)$ decays exponentially, we have

$$Q(t) = Q_0 e^{kt}$$

Our objective is to find $Q(10,000)$, the amount of carbon 14 present when $t = 10,000$ years. But first we must find Q_0 and k. Since there are 10 grams in the specimen at time $t = 0$, we have

$$10 = Q(0) = Q_0$$

Hence $Q(t) = 10e^{kt}$. After 5750 years, the original 10 grams of carbon 14 will be reduced by 50 percent to 5 grams. Therefore

$$5 = Q(5750) = 10e^{k(5750)}$$

$$\frac{5}{10} = e^{5750k}$$

$$\ln\left(\frac{1}{2}\right) = 5750k$$

$$k = \frac{\ln(1/2)}{5750}$$

Hence

$$Q(t) = 10e^{\frac{\ln(1/2)}{5750}t}$$

Finally, the amount of carbon 14 present when $t = 10,000$ years is given by

$$Q(10,000) = 10e^{\frac{\ln(1/2)}{5750}10,000} = 2.9955 \text{ grams} \quad \blacksquare$$

Remark

Once again, we may simplify the formula for $Q(t)$ if desired. Note that

$$e^{\frac{\ln(1/2)}{5750}t} = \left(e^{\ln(1/2)}\right)^{\frac{t}{5750}} = \left(\frac{1}{2}\right)^{\frac{t}{5750}}$$

Therefore

$$Q(t) = 10e^{\frac{\ln(1/2)}{5750}t} = 10\left(\frac{1}{2}\right)^{\frac{t}{5750}}$$

When the formula is written in this form, it is evident that every 5750 years the amount of carbon 14 present will be reduced by the factor 1/2. Thus, after $2(5750) = 11,500$ years, there will be $(1/2)^2 = 1/4$ the original amount present, after $3(5750) = 17,250$ years, there will be $(1/2)^3 = 1/8$ the original amount, and so on.

If a given substance decays exponentially, then the amount of time that it takes to decrease by 50 percent is called the **half-life**. For example, the half-life of radium 226 is about 1620 years. Example 2 dealt with the half-life of carbon 14, which is 5750 years.

Another example of exponential growth is when an investment pays an interest rate that is compounded continuously. In Section 5.3 we introduced the formula

$$A(t) = Pe^{rt}$$

which gives the amount of money yielded by an investment of P dollars after t years at the interest rate r (expressed as a decimal) compounded continuously.

Example 3

Suppose \$2000 is invested at $5\frac{1}{2}$ percent compounded continuously. How long will it take for the amount to grow to \$3000?

SOLUTION We have

$$A(t) = 2000e^{.055t}$$

We wish to find t so that

$$A(t) = 3000$$

This means that

$$2000e^{.055t} = 3000$$

$$e^{.055t} = \frac{3}{2}$$

$$.055t = \ln\left(\frac{3}{2}\right)$$

$$t = \frac{\ln(3/2)}{.055} \approx 7.37 \text{ years} \quad \blacksquare$$

In Examples 1 to 3, the original amount of the quantity under consideration was given: 20 rabbits in Example 1, 10 grams of carbon 14 in Example 2, and \$2000 in Example 3. Now we examine how to find $Q(t)$ in the case when Q_0 is not given directly.

Example 4

Suppose $Q(t) = Q_0e^{kt}$ and $Q(1) = 4$ and $Q(3) = 12$. Find Q_0 and k.

SOLUTION Using the formula $Q(t) = Q_0e^{kt}$, we have

$$Q(1) = 4 \quad \text{implies} \quad Q_0e^k = 4 \tag{1}$$

$$Q(3) = 12 \quad \text{implies} \quad Q_0e^{3k} = 12 \tag{2}$$

This gives us two equations for the two unknowns Q_0 and k. Using Equation (1),

$$Q_0e^k = 4 \quad \text{implies} \quad Q_0 = \frac{4}{e^k} = 4e^{-k} \tag{3}$$

Substituting this into Equation (2) for Q_0,

$$4e^{-k}e^{3k} = 12$$

$$e^{2k} = 3$$

$$2k = \ln 3$$

$$k = \frac{1}{2}\ln 3 = \ln\sqrt{3}$$

Substituting this into Equation (3) for k,

$$Q_0 = 4e^{-k} = 4e^{-\ln\sqrt{3}} = 4e^{\ln(1/\sqrt{3})}$$

$$= \frac{4}{\sqrt{3}}$$

Thus $Q_0 = 4/\sqrt{3}$ and $k = \ln\sqrt{3}$. ∎

Example 5

The voltage measured across two plates in an electrical experiment decays exponentially with time. One second after the start of the experiment the voltage measures 160 volts, and 3 seconds after the start it measures 90 volts.

(a) Find the original voltage at the start of the experiment.
(b) When will the voltage be decreased to 50 percent of its original amount?

SOLUTION Let $Q(t)$ represent the voltage t seconds after the start of the experiment. Since the voltage decays exponentially with time, we have

$$Q(t) = Q_0 e^{kt}$$

Our plan is to find Q_0 and k from the given information, and then any questions about the voltage may be answered using the formula for $Q(t)$.

We are given that $Q(1) = 160$ and $Q(3) = 90$. Using the formula $Q(t) = Q_0 e^{kt}$, we have

$$Q(1) = 160 \quad \text{implies} \quad Q_0 e^k = 160 \tag{1}$$

$$Q(3) = 90 \quad \text{implies} \quad Q_0 e^{3k} = 90 \tag{2}$$

From Equation (1),

$$Q_0 e^k = 160 \quad \text{implies} \quad Q_0 = \frac{160}{e^k} = 160e^{-k} \tag{3}$$

Substituting this into Equation (2) for Q_0,

$$160e^{-k}e^{3k} = 90$$

$$e^{2k} = \frac{9}{16}$$

$$2k = \ln\left(\frac{9}{16}\right)$$

$$k = \frac{1}{2}\ln\left(\frac{9}{16}\right) = \ln\left(\frac{3}{4}\right)$$

Substituting this into Equation (3) for k,

$$Q_0 = 160e^{-k} = 160e^{-\ln(3/4)} = 160e^{\ln(4/3)}$$

$$= 160\left(\frac{4}{3}\right)$$

$$= \frac{640}{3}$$

Thus $Q_0 = 640/3$ and $k = \ln(3/4)$, so that we may write $Q(t)$ as

$$Q(t) = \frac{640}{3}e^{\ln(3/4)t}$$

(a) To find the original voltage at the start of the experiment, we evaluate $Q(t)$ when $t = 0$.

$$Q(0) = \frac{640}{3}e^0 = \frac{640}{3}$$

Thus the original voltage is $640/3$, or $213\frac{1}{3}$ volts. Of course, this is just Q_0.

(b) If the voltage is 50 percent of its original amount, then it equals

$$\frac{1}{2}Q_0 = \frac{1}{2}\left(\frac{640}{3}\right) = \frac{320}{3} \text{ volts.}$$

To find when the voltage equals $320/3$ volts, we must solve the equation

$$Q(t) = \frac{320}{3}$$

Using our formula for $Q(t)$, we have

$$\frac{640}{3}e^{\ln(3/4)t} = \frac{320}{3}$$

$$e^{\ln(3/4)t} = \frac{3}{640}\left(\frac{320}{3}\right)$$

$$e^{\ln(3/4)t} = \frac{1}{2}$$

Taking the natural logarithm of both sides,

$$\ln\left(\frac{3}{4}\right)t = \ln\left(\frac{1}{2}\right)$$

$$t = \frac{\ln(1/2)}{\ln(3/4)} \approx 2.41 \text{ seconds} \blacksquare$$

EXERCISES 5.7

1. The population of a rabbit colony grows exponentially with time. Suppose the colony originally has 8 rabbits. If, after 2 months, the colony has grown to 20 rabbits, then how many rabbits will there be after (a) t months, (b) 6 months?

2. The population of a certain country grows exponentially with time. Suppose the population was 2 million in 1980. In 1982, after 2 years, it has increased by 20,000 to 2.02 million. Find what the population will be in the year 2000.

3. How much carbon 14 would be left from an original 10-gram specimen after 100,000 years?

4. Suppose an original specimen of radium 226 is 10 grams. How much would be left after (a) t years, (b) 100 years? (The half-life of radium 226 is 1620 years.)

5. Suppose an original 5-gram sample of material X decays to 4.1 grams after 4 days. Assuming X decays exponentially with time, find the half-life of X.

6. The half-life of plutonium 239 is 24,360 years. How long would it take a sample of plutonium 239 to decay to 1/100 of its original amount?

7. The temperature of an object is found to decay exponentially with time. At $t = 0$ seconds, the temperature is 100°C, and at $t = 4$ seconds it is 98°C. Find the temperature at (a) t seconds, (b) 10 seconds.

8. The voltage across two metal plates in an electrical circuit decays exponentially with time. The beginning voltage is 500 volts. If the voltage measures 450 volts after 1 second, how long will it take the voltage to reach 250 volts?

9. Suppose $2000 is invested at $5\frac{1}{2}$ percent compounded continuously. How long will it take the amount to grow to $4000?

10. If $1000 is invested at 8 percent compounded continuously, how long will it take to grow to $1500?

11. An investment is to be made that pays 6 percent compounded continuously. How much should be invested so that the amount is $5000 after 5 years?

12. If an investment pays 7 percent compounded continuously, how much should we invest so that after 4 years the total amount is $6000?

13. What rate is necessary for an investment of $10,000 to grow to $12,000 after 3 years? The rate is to be compounded continuously.

14. We want to invest $50,000 into an account that pays an annual interest rate compounded continuously. What rate should we find so that our investment grows to $60,000 in 5 years?

In Exercises 15 to 18, suppose $Q(t) = Q_0 e^{kt}$. Find Q_0 and k from the given information. Do not approximate.

15. $Q(1) = 2$, $Q(4) = 16$ 16. $Q(1) = 3$, $Q(2) = 9$

17. $Q(2) = 9$, $Q(4) = 6$ 18. $Q(2) = 40$, $Q(5) = 10$

19. The temperature of an object is measured during an experiment. At $t = 1$ hour, the temperature is 500°C, and at $t = 5$ hours, the temperature is 410°C. If it is known that the temperature decays exponentially with time, then what was the temperature of the object at the start of the experiment (when $t = 0$)?

20. At the start of an experiment we inject a solution containing ϕX cells into a test tube. We do not know how many ϕX cells are present initially, but we do know that in our laboratory conditions they will grow exponentially with time. Six hours later we are able to measure the density of ϕX at 4.2 million cells per milliliter. After 8 hours, we find that there are 6.6 million cells per milliliter. Find the density of ϕX

cells in millions per milliliter t hours after the start of the experiment. What was the density of ϕX at the start of the experiment?

21. A certain material called ZZ lays dormant in our laboratory test tube at the density of 1.5 million cells per milliliter. It is known that at some time these ZZ cells will start to grow exponentially with time. After leaving the cells overnight in the lab, we arrive the next day and find that they are growing. At 8 A.M., the density of the ZZ cells is 2.4 million per milliliter, and at 9 A.M. the density is 3.0 million per milliliter. At what time (to the nearest minute) did the ZZ cells start to grow? (*Hint:* Use 8 A.M. as the time $t = 0$.)

22. The voltage across two plates in our laboratory is held constant at 400 volts. We expect that a particular circuit in the apparatus may fail, and when it does, the voltage will decay exponentially with time. After going out to lunch, we come back to find that the circuit failed while we were gone and the voltage is decaying. At 2 P.M. the voltage measures 320 volts, and at 3 P.M. it is down to 240 volts. When did the circuit fail?

23. Suppose $Q(t_1) = a$ and $Q(t_2) = b$. Assuming $Q(t) = Q_0 e^{kt}$, show that

(a) $k = \dfrac{\ln(b/a)}{t_2 - t_1}$ (b) $Q_0 = a\left(\dfrac{a}{b}\right)^{t_1/(t_2 - t_1)}$

CHAPTER 5 / REVIEW

1. Evaluate if possible. Do not use a calculator.
   (a) $(-8)^{-4/3}$ (b) $(-1)^{1.5}$ (c) $(128^4)^{1/7}$

2. Use the calculator to approximate the exponential expression to five decimal places.
   (a) $(-6)^{4/5}$ (b) $2^{\pi/2}$

3. Simplify so that all symbols appear in the numerator.
   (a) $\left(\dfrac{1 + e^{2x}}{e^x}\right)^2$ (b) $\left(\dfrac{x^{1/2} y^{1/3}}{2 y^{-2}}\right)^{-3}$

4. If $F(x) = e^{-2x}$, find $F(k + 1)/F(k)$ in simplest terms.

5. Simplify to one fraction with no negative exponents,
   $$\frac{2}{3}x(x^2 + 1)^{-2/3}(x^2 - 1)^{2/3} + \frac{4}{3}x(x^2 + 1)^{1/3}(x^2 - 1)^{-1/3}.$$

In Exercises 6 to 8, graph the given function F. Determine whether F is increasing or decreasing. Find the domain, range and asymptotes. Evaluate $\lim\limits_{x \to +\infty} F(x)$ and $\lim\limits_{x \to -\infty} F(x)$.

6. $F(x) = 2^{-x}$

7. $F(x) = 1 - 2^x$

8. $F(x) = e^{x+1} - e$

9. Evaluate (a) $\lim\limits_{x \to +\infty} \left(\dfrac{5}{6}\right)^x$ (b) $\lim\limits_{x \to +\infty} (\sqrt{2})^x$

10. Use the technique of adding graphs to find the graph of F. Determine the range of F.
    (a) $F(x) = 2^x + 2^{-x}$ (b) $F(x) = 2^x - 2^{-x}$

11. Find the amount yielded by an investment of $10,000 for 5 years at the rate of 7.8 percent compounded (a) annually, (b) daily, (c) continuously.

12. Suppose bank A pays interest rate r (expressed as a decimal) compounded three times annually, and bank B pays rate R compounded annually. Show that at the end of any number of years the yield on an investment of P dollars will be the same in either bank if $R = r + r^2/3 + r^3/27$.

13. Evaluate each expression without using a calculator.
    (a) $\log\left(\dfrac{1}{100}\right)$ (b) $\ln \sqrt{e}$ (c) $\log_{1/2}(1)$

14. The value of $\ln(4)$ is between what two consecutive integers? Do not use a calculator.

In Exercises 15 to 18, graph the function. Is the function increasing or decreasing? Find the domain, range, zero, and vertical asymptote.

15. $F(x) = \log_2(x + 2)$ 16. $F(x) = 1 + \ln x$

17. $F(x) = \log_{1/4}(x)$

18. $F(x) = \ln\left(\dfrac{1}{x}\right)$

19. Evaluate (a) $\lim\limits_{x \to +\infty} \log_\pi(x)$, (b) $\lim\limits_{x \to 0^+} \log_\pi(x)$.

20. Write $\ln(x + 1) = 4$ in its equivalent exponential form.

21. Write $10^P = r - 1$ in its equivalent logarithmic form.

In Exercises 22 and 23, find $F^{-1}(x)$.

22. $F(x) = 3^x$

23. $F(x) = 1 + \ln(x + 1)$

24. Simplify $3^{\ln e} + \log(1) + e^{-2 \ln 2}$

25. Let $\log_b(2) = A$ and $\log_b(3) = B$. Evaluate the given expression in terms of A and B.

(a) $\log_b\left(\dfrac{b}{12}\right)$

(b) $\log_b(18\sqrt{b})$

26. Combine the given expression into one logarithm.

$$\ln 4 + 3 \ln x - \frac{\ln y}{2}$$

27. Use the calculator to approximate $\log_\pi(10)$ to four decimal places.

In Exercises 28 to 34, solve each equation for x. Write the exact answer, and then use the calculator (if necessary) to obtain an approximation accurate to four decimal places.

28. $100 = 25 e^{\ln(4/5)x}$

29. $\dfrac{3^x + 3^{-x}}{3^x - 3^{-x}} = 3$

30. $2 \log x = \log(x + 2) + 1$

31. $(2x + 1)^{2/3} = 4$

32. $x^{\sqrt{2}} = 2$

33. $(2x + 1)e^x = 0$

34. $e^x = e^{1/x}$

35. The population of country X is growing exponentially with time. In 1980, the population was 13 million. In 1985, it was 14.6 million. Find what the population will be in the year 2000. In what year will the population exceed 26 million?

36. Find the interest rate that, if compounded continuously, will cause an investment of $5000 to grow to $100,000 after 25 years.

37. The temperature of an object decays exponentially with time. At $t = 0$ seconds the temperature is $135°C$, and at $t = 10$ seconds it is $90°C$. Find the temperature at (a) t seconds, (b) 30 seconds.

In Exercises 38 to 43, answer true or false.

38. The functions $F(x) = 2^{x+1}$ and $G(x) = 4^x$ are equal.

39. There is no real solution for x in the equation $2e^{-x} + 1 = 0$.

40. $\dfrac{\log x}{\log \sqrt{yz}} = \log x - \dfrac{1}{2}(\log y + \log z)$

41. $\dfrac{\ln(3/4)}{\ln(5/4)} = \ln\left(\dfrac{3}{5}\right)$

42. $e^{\log_3(1)} = 1$

43. $\log_a(a^x) = a^{\log_a x}$

CHAPTER 6 Trigonometric Functions

6.1 ANGLES, RADIANS, AND DEGREES

In this section we discuss five topics related to angles: the definition of an angle, radian measure, assigning a measure to an angle, degrees, and coterminal angles.

Angles. An **angle** is a geometric object in the plane consisting of two rays (referred to as the sides of the angle) emanating from a common point called the **vertex.** For purposes of measurement, the angle can be placed in **standard position** as follows: one of the rays, designated as the **initial side**, is positioned along the positive x-axis with the vertex at the origin. The second ray is then referred to as the **terminal side**. Figure 6.1 shows a typical angle in standard position.

The measure of an angle is related to the amount of rotation required to get from the initial side to the terminal side. An angle has a **positive** measure if we designate a **counterclockwise** rotation from the initial side to the terminal side. An angle has a **negative** measure if the rotation is **clockwise.** It follows that an angle may be given either a positive or a negative measure. See Figure 6.2.

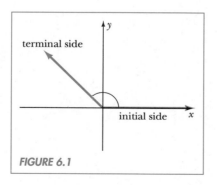

FIGURE 6.1

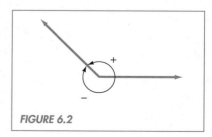

FIGURE 6.2

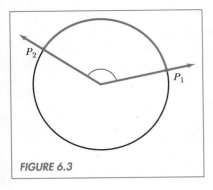

FIGURE 6.3

Figure 6.3 shows an angle with its vertex located at the center of a circle and its sides cutting off an arc on the circle with endpoints P_1 and P_2. We refer to such an angle as the **central angle** determined by the arc P_1P_2.

Radian Measure. We wish to establish the following important result.

> To each real number t there corresponds a unique angle whose measure we define as t.

To establish the correspondence between real numbers and angles, we start with a circle of radius $r > 0$, $x^2 + y^2 = r^2$. Let us define the angle corresponding to $t = 1$. From the point $(r, 0)$, move along the circle in the counterclockwise direction until the length of the arc traveled equals *one radius*, r. The arc on the circle obtained by this motion determines a central angle that we associate with the real number 1. To emphasize the method of construction, we say this angle has measure **1 radian**. See Figure 6.4.

Next, we construct the angle corresponding to the real number $t = 2$. Again, starting at the point $(r, 0)$, we move along the circle in the counterclockwise direction until the length of the arc traveled equals *two* radii, $2r$. The central angle determined by this arc has measure equal to 2 radians. See Figure 6.5.

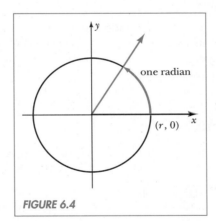

FIGURE 6.4

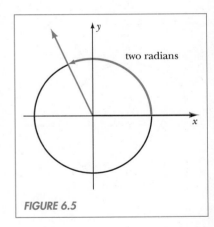

FIGURE 6.5

It is easy to see how to generalize this method of construction to any real number t. If t is positive, then an angle of measure t radians is constructed by moving along the circle in the counterclockwise direction until the length of the arc traveled equals t radii, or tr units. In case t is negative, we move along the circle in the clockwise direction until the length of arc traveled equals $|t|$ radii, or $|t|r$ units. We may summarize this discussion as follows.

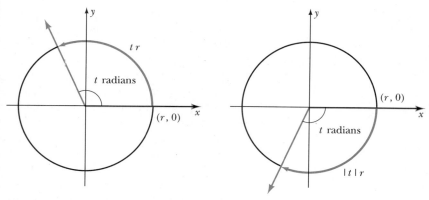

Constructing an Angle of Measure t Radians

Let $t \in \mathbf{R}$. From the point $(r, 0)$, $r > 0$, move along the circle $x^2 + y^2 = r^2$ until the length of arc traveled equals $|t|r$ units,

Counterclockwise if $t \geq 0$ *Clockwise if $t < 0$*

The central angle determined by this arc has measure t radians.

Remark

This construction is independent of the radius of the circle. In other words, if we construct an angle of t radians twice, once using a circle of radius r_1 and again using a different circle of radius r_2, then both angles will be congruent. See Figure 6.6.

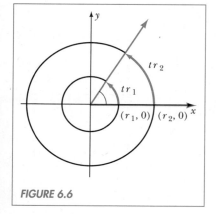

FIGURE 6.6

Example 1

Find the length of arc required to travel along the circle $x^2 + y^2 = 4$ in order to construct an angle with measure (a) 1.5 radians, (b) -4 radians.

SOLUTION   (a) The circle $x^2 + y^2 = 4$ has radius $r = 2$. For $t = 1.5$ radians, we need to travel along the circle so that the length of arc equals

$$tr = 1.5(2) = 3 \text{ units}$$

Figure 6.7 shows an angle of measure 1.5 radians constructed in the circle $x^2 + y^2 = 4$.

(b) For $t = -4$ radians, we need to travel along the circle so that the length of arc equals

$$|t|r = |-4|(2) = 8 \text{ units}$$

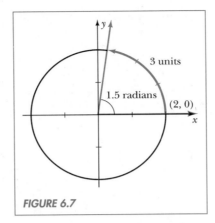

FIGURE 6.7

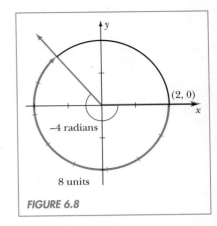

FIGURE 6.8

Figure 6.8 shows an angle of measure -4 radians constructed in the circle $x^2 + y^2 = 4$. Note that the movement is clockwise since the measure is negative. ▪

Recall the relationship between the circumference of a circle C and its radius r:

$$C = 2\pi r$$

Now, let us construct an angle of measure $t = 2\pi$ radians. To do this, we must move in the counterclockwise direction along a circle of radius r until the length of the arc traveled equals $tr = 2\pi r$ units. Of course, this takes us exactly once around the circle, with us ending up at the point where we started. Thus a revolution around the circle corresponds to 2π radians. It follows that a half revolution corresponds to π radians and a quarter revolution corresponds to $\pi/2$ radians. These are important angles to remember; Figure 6.9 illustrates them with their decimal approximations.

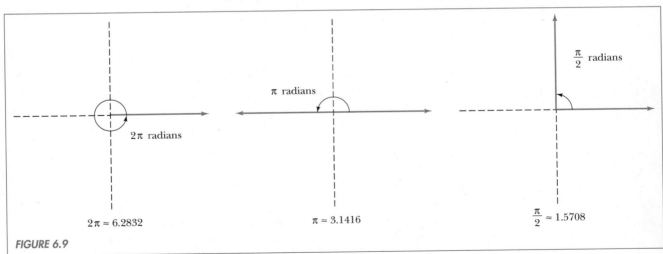

FIGURE 6.9

Another important angle corresponds to three-quarters of a revolution, or $\pi + \pi/2 = 3\pi/2$ radians; this angle is shown in Figure 6.10.

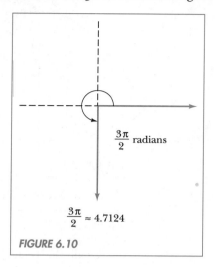

$\frac{3\pi}{2}$ radians

$\frac{3\pi}{2} \approx 4.7124$

FIGURE 6.10

Example 2

Draw a reasonably accurate picture of an angle with the given measure in standard position: (a) 3 radians, (b) 7 radians, (c) $7 + \pi$ radians.

SOLUTION (a) We know that an angle of measure π radians in standard position has its terminal side along the negative x-axis. Since 3 is slightly less than π, an angle of measure 3 radians must have its terminal side just above the negative x-axis in quadrant II. Figure 6.11 shows an angle of 3 radians in standard position.

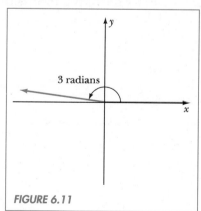

3 radians

FIGURE 6.11

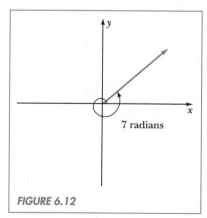

7 radians

FIGURE 6.12

(b) We know that $7 > 2\pi \approx 6.28$. Therefore, to construct an angle of measure 7 radians requires more than one revolution around a circle. Also, $7 < 2\pi + \pi/2 \approx 7.85$, so it is less than one and a quarter revolutions.

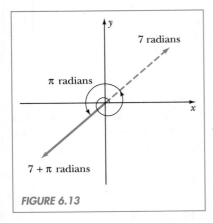

FIGURE 6.13

It follows that an angle of measure 7 radians in standard position must have its terminal side in quadrant I, as illustrated in Figure 6.12. Note that we drew the terminal side of the angle about halfway between the positive x- and y-axes since 7 is roughly halfway between $2\pi \approx 6.28$ and $2\pi + \pi/2 \approx 7.85$.

(c) To draw an angle of measure $7 + \pi$ radians in standard position, we start by drawing an angle of measure 7 radians (see Figure 6.12) and then rotate an additional π radians (a half revolution), as demonstrated in Figure 6.13. The terminal side of the angle of measure $7 + \pi$ radians will be in quadrant III. ■

Remark

It is cumbersome to always make a distinction between an angle that is a geometric object and its measure, which is a real number. Therefore, when referring to "an angle of measure t," we may just say "the angle t." Furthermore, when the context is clear, we may assume that the angle is in standard position without actually saying so. Thus we can shorten the statement "the angle of measure 3 radians in standard position has its terminal side in quadrant II" to "3 radians has its terminal side in quadrant II."

Assigning a Measure to an Angle. Given an angle in the plane, we may position it so that the vertex is at the center of a circle of radius r. The sides of the angle then cut off an arc of length s on the circle. If we wish to assign a positive measure of t radians to this angle, then the arc length s must equal tr units,

$$tr = s$$

Solving this for t gives us the following result.

If a central angle in a circle of radius r cuts off an arc of length s, then a positive radian measure for this angle is given by $t = \dfrac{s}{r}$.

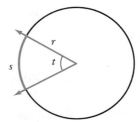

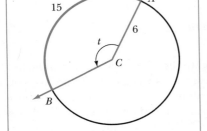

FIGURE 6.14

Example 3

Find a positive measure t for the central angle ACB in the circle in Figure 6.14. Assume that CA is the initial side.

We have

$$t = \frac{s}{r} = \frac{15}{6} = \frac{5}{2}$$

Thus $t = 2.5$ radians. ■

Degrees. Recall that an angle of 360 degrees sweeps out an arc that completes exactly one revolution around a circle. Furthermore, such an arc also corresponds to an angle with measure 2π radians. Thus

$$2\pi \text{ radians is equivalent to } 360 \text{ degrees}$$

or, dividing by 2, we have

$$\pi \text{ radians is equivalent to } 180 \text{ degrees}$$

This leads us to the following proposition, which we use to convert radians to degrees or degrees to radians.

Relationship Between Radians and Degrees
x radians is equivalent to y degrees if and only if

$$\frac{x \text{ (radians)}}{\pi} = \frac{y \text{ (degrees)}}{180}$$

Example 4

Convert 1 radian to its equivalent degree measure.

SOLUTION Let y represent the equivalent degree measure for 1 radian. Then

$$\frac{1}{\pi} = \frac{y}{180}$$

Solving for y, we find

$$y = \frac{180}{\pi}$$

Thus 1 radian is equivalent to $180/\pi$ degrees or approximately 57.3 degrees. ■

Recall that the symbol ° written above and to the right of a number indicates degrees. If no symbol is used when the measure of an angle is referred to, you should assume that the dimensions are radians unless the context implies otherwise. Thus the measure 1 by itself means 1 radian, whereas 1° means 1 degree. By Example 4, 1 radian is about 57.3°, so an angle of measure 1 is very different from an angle of measure 1°.

Example 5

Convert $-160°$ to radian measure.

SOLUTION Let x represent the number of radians in $-160°$. We have

$$\frac{x}{\pi} = \frac{-160}{180}$$

Solving for x,

$$x = \frac{-160}{180}(\pi) = -\frac{8\pi}{9}$$

Thus $-160°$ is equivalent to $-8\pi/9$ radians. Unless a decimal approximation is required, we leave the symbol π in the answer when dealing with radians. ∎

Coterminal Angles. We may assign many different measures to any given geometric angle in the plane. For example, Figure 6.15(a) shows an angle with measure $3\pi/4$ radians in standard position. Figures 6.15(b) to (d) show how we might assign other measures to this angle.

> **Coterminal Angles**
> If the terminal sides of two angles with assigned measures coincide when the angles are placed in standard position, then the angles are **coterminal**.

Thus all four angles in Figure 6.15 are coterminal.

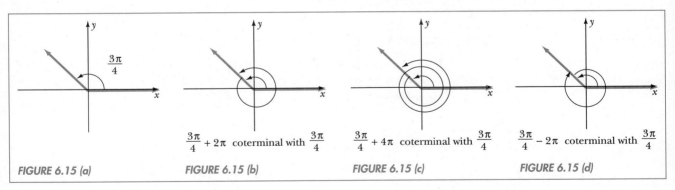

$\frac{3\pi}{4}$

FIGURE 6.15 (a)

$\frac{3\pi}{4} + 2\pi$ coterminal with $\frac{3\pi}{4}$

FIGURE 6.15 (b)

$\frac{3\pi}{4} + 4\pi$ coterminal with $\frac{3\pi}{4}$

FIGURE 6.15 (c)

$\frac{3\pi}{4} - 2\pi$ coterminal with $\frac{3\pi}{4}$

FIGURE 6.15 (d)

It is apparent that any angle coterminal with one of measure $3\pi/4$ must have measure

$$\frac{3\pi}{4} + 2\pi k$$

where k is an integer. We may generalize this result as follows.

> Let θ be the measure of an angle. All angles coterminal with θ have measure
>
> $$\theta + 2\pi k \qquad k \in \mathbf{Z}$$
>
> in radians, or
>
> $$\theta + 360°k \qquad k \in \mathbf{Z}$$
>
> in degrees.

Example 6

Find all angles coterminal with $-10\pi/3$. Name one of measure θ such that $0 \leq \theta < 2\pi$.

SOLUTION All angles coterminal with $-10\pi/3$ are represented by the expression

$$\frac{-10\pi}{3} + 2\pi k \qquad k \in \mathbf{Z}$$

To find an angle whose measure is between 0 and 2π, we evaluate the expression for various k until we get the right number. Trying $k = 1, 2,$ and 3, we have

k	1	2	3
$\dfrac{-10\pi}{3} + 2\pi k$	$\dfrac{-4\pi}{3}$	$\dfrac{2\pi}{3}$	$\dfrac{8\pi}{3}$

Thus we find that $\theta = 2\pi/3$ is coterminal with $-10\pi/3$ and $0 \leq \theta < 2\pi$. ∎

Example 7

Find the measures in degrees of all possible angles coterminal with 1000°. Name one of measure θ such that $0 \leq \theta < 360°$.

SOLUTION All angles coterminal with 1000° must have degree measures

$$1000° + 360°k \qquad k \in \mathbf{Z}$$

Setting $k = -2$, we find

$$1000 + 360(-2) = 1000 - 720$$
$$= 280$$

Thus $\theta = 280°$ is coterminal with 1000° and $0 \leq \theta < 360°$. ∎

We close this section with an example that will play an important role later when solving trigonometric equations.

Example 8

Find all angles x such that $3x$ is coterminal with $\pi/2$ and $0 \leq x < 2\pi$.

SOLUTION First, we find all $(3x)$ such that $(3x)$ is coterminal with $\pi/2$. We have

$$3x = \frac{\pi}{2} + 2\pi k \qquad k \in \mathbf{Z}$$

Now, multiply this equation by 1/3

$$x = \frac{\pi}{6} + \frac{2\pi k}{3} \qquad k \in \mathbf{Z}$$

This gives us *all* x such that $3x$ is coterminal with $\pi/2$. To find those values satisfying $0 \leq x < 2\pi$, we substitute in values of k:

k	-1	0	1	2	3
$x = \dfrac{\pi}{6} + \dfrac{2\pi k}{3}$	$-\dfrac{\pi}{2}$	$\dfrac{\pi}{6}$	$\dfrac{5\pi}{6}$	$\dfrac{3\pi}{2}$	$\dfrac{13\pi}{6}$

Thus the values of x satisfying $3x$ coterminal with $\pi/2$ and $0 \leq x < 2\pi$ are $\dfrac{\pi}{6}, \dfrac{5\pi}{6},$ and $\dfrac{3\pi}{2}$. ∎

EXERCISES 6.1

In Exercises 1 and 2, find the length of arc required to travel along the circle $x^2 + y^2 = 100$ in order to construct an angle with the given measure.

1. (a) 1/2 radian
   (b) π radians
   (c) -2.5 radians

2. (a) 10 radians
   (b) $\pi/2$ radians
   (c) -4.5 radians

In Exercises 3 to 8, draw a reasonably accurate picture of an angle in standard position with the given radian measure.

3. (a) $\pi/4$
   (b) $-\pi/2$
   (c) 1.5
   (d) -3.5

4. (a) $5\pi/4$
   (b) $-\pi/8$
   (c) 6
   (d) -4.8

5. (a) 3π
   (b) $1 - \pi$
   (c) $1 + 2\pi$

6. (a) $-7\pi/2$
   (b) $2 + \pi$
   (c) $2 - 2\pi$

7. (a) 8
   (b) $8 + \pi/2$
   (c) $8 + \pi$
   (d) $8 + 3\pi/2$

8. (a) -5
   (b) $-5 + \pi/2$
   (c) $-5 + \pi$
   (d) $-5 + 3\pi/2$

In Exercises 9 and 10, find a positive measure for the central angle ACB in the given circle. Assume that CA is the initial side.

9. (a)

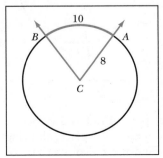

(b)

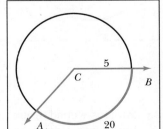

10. (a)

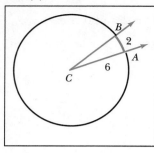

(b)

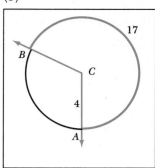

In Exercises 11 and 12, for the central angle *ACB* with initial side *CA*, find a positive and a negative measure for the angle.

11. (a)

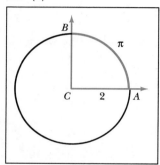

(b)

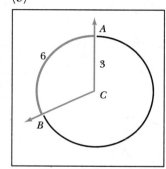

12. (a)

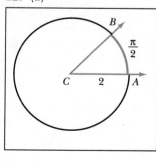

(b)

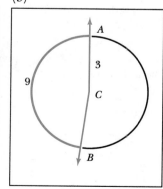

13. String is wound around a spool of radius 3 inches. If the string is pulled, the spool rotates in place, allowing us to unwind the string. Find the radian measure of the angle through which the spool rotates if the length of string pulled off is (a) 6 inches, (b) 24 inches.

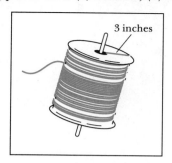

14. A wheel of diameter 12 inches rolls along the ground. Find the angle through which the wheel rotates if the distance the wheel travels is (a) 3 feet, (b) 12 feet.

In Exercises 15 and 16, draw the angle whose measure is given in standard position. Find its equivalent degree measure.

15. (a) $\pi/3$
    (b) $2\pi/3$
    (c) $-2\pi/3$

16. (a) $\pi/6$
    (b) $5\pi/6$
    (c) $-5\pi/6$

In Exercises 17 and 18, draw the angle whose measure is given in standard position. Find its equivalent radian measure.

17. (a) 30°
    (b) $-210°$
    (c) 330°

18. (a) 60°
    (b) $-240°$
    (c) 300°

19. Complete the following table. Draw the terminal side of each angle in standard position. (You may use the same axes for all angles.)

(a)

Radians	$\pi/4$	$\pi/2$	$3\pi/4$	π	$5\pi/4$	$3\pi/2$	$7\pi/4$	2π
Degrees								

(b)

Radians	$\pi/6$	$\pi/3$	$2\pi/3$	$5\pi/6$	$7\pi/6$	$4\pi/3$	$5\pi/3$	$11\pi/6$
Degrees								

In Exercises 20 and 21, convert degrees to radians. Give exact answers.

20. (a) 100°
    (c) $-7.2°$
    (b) $-36°$
    (d) 157.5°

21. (a) $-12°$ (b) $1800°$
    (c) $195°$ (d) $-22.5°$

In Exercises 22 and 23, convert radians to degrees. Give exact answers.

22. (a) $\pi/5$ 23. (a) $\pi/10$
    (b) $-\pi/9$ (b) $-\pi/20$
    (c) $7\pi/30$ (c) $-6\pi/5$
    (d) $-2/45$ (d) $5/72$

In Exercises 24 and 25, find the length of arc cut off by a central angle of measure θ in a circle of radius r.

24. (a) $\theta = 2\pi/9, r = 6$ 25. (a) $\theta = 5\pi/9, r = 18$
    (b) $\theta = 200°, r = 9$ (b) $\theta = 24°, r = 10$

26. How many radians are in π degrees?

In Exercises 27 and 28, find the measures of all possible angles coterminal with the angle of given measure. Use radians if the angle is given in radians, degrees if the angle is given in degrees.

27. (a) $\pi/4$ 28. (a) $-2\pi/3$
    (b) -2.5 (b) 4
    (c) $100°$ (c) $15°$

In Exercises 29 to 32, find the measure of an angle θ such that θ is coterminal with the given angle and $0 \leq \theta < 2\pi$ or $0° \leq \theta < 360°$.

29. (a) 3π
    (b) $810°$
    (c) $-55°$
30. (a) $7\pi/2$
    (b) $375°$
    (c) $-75°$

31. (a) $11\pi/4$
    (b) $2800°$
    (c) $-5\pi/6$
32. (a) $7\pi/3$
    (b) $720°$
    (c) $-5\pi/4$

In Exercises 33 to 36, find the measures of all angles x such that $2x$ is coterminal with the given angle and $0 \leq x < 2\pi$ or $0° \leq \theta < 360°$.

33. (a) $\pi/2$ 34. (a) $\pi/3$
    (b) π (b) $5\pi/6$
    (c) $330°$ (c) $270°$

35. (a) $\pi/6$ 36. (a) $\pi/4$
    (b) 0 (b) $-\pi/6$
    (c) $-60°$ (c) $-90°$

In Exercises 37 and 38, find the measures of all angles x such that $3x$ is coterminal with the given angle and $0 \leq x < 2\pi$.

37. (a) $\pi/3$ 38. (a) π
    (b) $-\pi/4$ (b) $-\pi/2$
    (c) $90°$ (c) $0°$

In Exercises 39 and 40, draw the angle with measure t whose terminal side is on the given line in the given quadrant. Then find a point on the terminal side of t.

39. (a) $y = 3x$, quadrant I; (b) $y = -\frac{1}{2}x$, quadrant II

40. (a) $y = 2x$, quadrant III; (b) $y = -\frac{3}{2}x$, quadrant IV

6.2 SINE, COSINE, AND TANGENT

In this section we introduce three real functions that we call the trigonometric functions **sine, cosine**, and **tangent**. Recall that a real function consists of two things: a set of numbers for the domain, and a rule that assigns one number to each element in the domain. Thus we need to state the rule and domain for each of our new functions. We do this as follows (the symbols sin, cos, and tan designate sine, cosine, and tangent, respectively).

Suppose $t \in \mathbf{R}$.

Step 1 Draw the angle of measure t radians in standard position.

Step 2 Choose any point other than the origin on the terminal side of t, $P(x, y)$.

Step 3 Set $r = \sqrt{x^2 + y^2}$, the distance OP.

Step 4 Define

$$\sin(t) = \frac{y}{r}$$

$$\cos(t) = \frac{x}{r}$$

$$\tan(t) = \frac{y}{x}$$

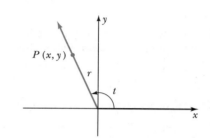

According to the rules just described, $\sin(t) = y/r$ and $\cos(t) = x/r$ make sense for any value of t since r is never zero. Thus the domain for sine or cosine is the set of all real numbers. However, the rule for tangent, $\tan(t) = y/x$, does not make sense if $x = 0$. This corresponds to the point P on the terminal side of t having its x-coordinate equal to zero. This in turn implies that the terminal side of t is along the y-axis, coterminal with $\pi/2$ or $-\pi/2$. Thus, for $\tan(t)$ to make sense, we must have $t \neq \pm\pi/2$, $\pm 3\pi/2$, $\pm 5\pi/2, \ldots$.

Domain of Sine, Cosine, and Tangent

$$\mathcal{D}_{\sin} = \mathbf{R}$$

$$\mathcal{D}_{\cos} = \mathbf{R}$$

$$\mathcal{D}_{\tan} = \{t : t \neq \pm\pi/2,\ \pm 3\pi/2,\ \pm 5\pi/2, \ldots\}$$

Do the rules for sine, cosine, and tangent actually define functions? To answer this question, we must check that each rule assigns exactly one number to each element t in the domain. The only difficulty that might occur is in step 2 of the rules, where we are allowed to choose *any* point P (other than the origin) on the terminal side of t. If we use two different points, do we still get the same result?

Figure 6.16 shows a first quadrant angle t with two points $P_1(x_1, y_1)$ and $P_2(x_2, y_2)$ on the terminal side. Notice that triangles OAP_1 and OBP_2 are similar, so the ratios of their corresponding sides are equal. Letting $r_1 = OP_1$

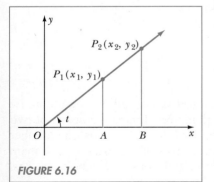

FIGURE 6.16

and $r_2 = OP_2$, we get the following equations for the ratios corresponding to $\sin(t)$, $\cos(t)$, and $\tan(t)$:

$$
\begin{array}{ccc}
sin(t) & cos(t) & tan(t) \\[4pt]
\dfrac{y_1}{r_1} = \dfrac{y_2}{r_2} & \dfrac{x_1}{r_1} = \dfrac{x_2}{r_2} & \dfrac{y_1}{x_1} = \dfrac{y_2}{x_2}
\end{array}
$$

This shows that the values for $\sin(t)$, $\cos(t)$, and $\tan(t)$ are independent of the choice for P. If t is in quadrant II, III, or IV, a similar argument holds. Finally, if t has its terminal side on the x- or y-axis, direct computations show that the choice of P does not affect the value of $\sin(t)$, $\cos(t)$, or $\tan(t)$. (See the remark after Example 3.) We conclude that our rules for sine, cosine, and tangent do define functions.

Example 1

Suppose the angle t in standard position has the point $P(2, 1)$ on its terminal side. Find $\sin(t)$, $\cos(t)$, and $\tan(t)$.

SOLUTION   We draw the angle t with point $P(2, 1)$ on its terminal side in Figure 6.17. We have $x = 2$, $y = 1$, and $r = \sqrt{x^2 + y^2} = \sqrt{2^2 + 1^2} = \sqrt{5}$. It follows that

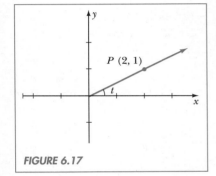

FIGURE 6.17

$$
\sin(t) = \frac{y}{r} = \frac{1}{\sqrt{5}}
$$

$$
\cos(t) = \frac{x}{r} = \frac{2}{\sqrt{5}}
$$

$$
\tan(t) = \frac{y}{x} = \frac{1}{2} \ \blacksquare
$$

Note in Example 1 that if we take the value of $\sin(t)$ and divide it by $\cos(t)$, we get $\tan(t)$: $(1/\sqrt{5}) \div (2/\sqrt{5}) = 1/2$. This relationship between sine, cosine, and tangent is true in general.

For all $t \in \mathscr{D}_{\tan}$, we have

$$
\tan(t) = \frac{\sin(t)}{\cos(t)}
$$

PROOF: Let $P(x, y)$ be on the terminal side of t and $r = \sqrt{x^2 + y^2}$. By definition, we have

$$
\frac{\sin(t)}{\cos(t)} = \frac{y/r}{x/r} = \frac{y}{x} = \tan(t)
$$

This completes the proof. $\blacksquare$

Whenever we compute values for $\sin(t)$, $\cos(t)$, and $\tan(t)$, we should check that $\sin(t)/\cos(t)$ is the same as $\tan(t)$.

Example 2

If $P(2, 1)$ is on the terminal side of t, find $\sin(\theta)$, $\cos(\theta)$, and $\tan(\theta)$ when (a) $\theta = t + \pi$, (b) $\theta = -t$.

SOLUTION (a) We see in Figure 6.18 that the terminal side of $t + \pi$ is the symmetric image of t with respect to the origin. Therefore the symmetric image of $P(2, 1)$ with respect to the origin, $P_1(-2, -1)$, is on the terminal side of $t + \pi$. We have $x = -2$, $y = -1$, and $r = \sqrt{x^2 + y^2} = \sqrt{5}$. It follows that

$$\sin(t + \pi) = \frac{y}{r} = \frac{-1}{\sqrt{5}}$$

$$\cos(t + \pi) = \frac{x}{r} = \frac{-2}{\sqrt{5}}$$

$$\tan(t + \pi) = \frac{y}{x} = \frac{-1}{-2} = \frac{1}{2}$$

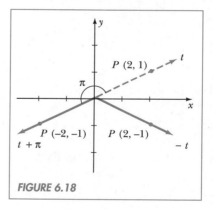

FIGURE 6.18

(b) Figure 6.18 also shows that $-t$ is the symmetric image of t with respect to the x-axis. Taking the symmetric image of $P(2, 1)$ with respect to the x-axis, we find that $P_2(2, -1)$ is on the terminal side of $-t$. We have $x = 2$, $y = -1$, and $r = \sqrt{x^2 + y^2} = \sqrt{5}$. It follows that

$$\sin(t) = \frac{y}{r} = \frac{-1}{\sqrt{5}}$$

$$\cos(t) = \frac{x}{r} = \frac{2}{\sqrt{5}}$$

$$\tan(t) = \frac{y}{x} = \frac{-1}{2}$$

Example 3

Find $\sin(t)$, $\cos(t)$, and $\tan(t)$ when $t = 0$, $\pi/2$, π, $3\pi/2$, and 2π.

SOLUTION We shall draw each angle in standard position, choose a point $P(x, y)$ on the terminal side, and then compute the sine, cosine, and tangent.

$t = 0$

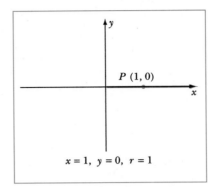

$$\sin(0) = \frac{y}{r} = \frac{0}{1} = 0$$

$$\cos(0) = \frac{x}{r} = \frac{1}{1} = 1$$

$$\tan(0) = \frac{y}{x} = \frac{0}{1} = 0$$

$t = \pi/2$

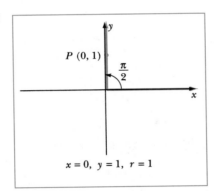

$$\sin\left(\frac{\pi}{2}\right) = \frac{y}{r} = \frac{1}{1} = 1$$

$$\cos\left(\frac{\pi}{2}\right) = \frac{x}{r} = \frac{0}{1} = 0$$

$$\tan\left(\frac{\pi}{2}\right) \quad \text{Undefined}$$

$t = \pi$

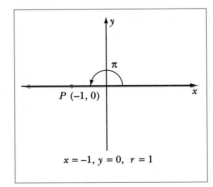

$$\sin(\pi) = \frac{y}{r} = \frac{0}{1} = 0$$

$$\cos(\pi) = \frac{x}{r} = \frac{-1}{1} = -1$$

$$\tan(\pi) = \frac{y}{x} = \frac{0}{-1} = 0$$

$t = 3\pi/2$

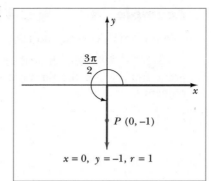

$x = 0,\ y = -1,\ r = 1$

$$\sin\left(\frac{3\pi}{2}\right) = \frac{y}{r} = \frac{-1}{1} = -1$$

$$\cos\left(\frac{3\pi}{2}\right) = \frac{x}{r} = \frac{0}{1} = 0$$

$$\tan\left(\frac{3\pi}{2}\right) \quad \text{Undefined}$$

$t = 2\pi$

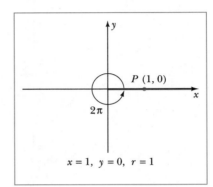

$x = 1,\ y = 0,\ r = 1$

This is the same as $t = 0$.

$\sin(2\pi) = 0$

$\cos(2\pi) = 1$

$\tan(2\pi) = 0$ ■

Remark

It is easy to see that any choice for P on the terminal side of t gives us the same results for $\sin(t)$, $\cos(t)$, and $\tan(t)$ found above. For example, when $t = 0$, we could have chosen the point $P(7, 0)$ instead of $(1, 0)$. In this case, $x = 7$, $y = 0$, and $r = \sqrt{x^2 + y^2} = 7$. It follows that

$$\sin(t) = \frac{y}{r} = \frac{0}{7} = 0$$

$$\cos(t) = \frac{x}{r} = \frac{7}{7} = 1$$

$$\tan(t) = \frac{y}{x} = \frac{0}{7} = 0$$

These are the same values we found when using the point $(1, 0)$.

Example 4

Suppose the terminal side of angle t is in the second quadrant on the line $y = -\frac{3}{2}x$. Find $\sin(t)$, $\cos(t)$, and $\tan(t)$.

SOLUTION Figure 6.19 shows t in standard position. We need a point P on the line $y = -\frac{3}{2}x$ in quadrant II. Letting $x = -2$, we find

$$y = -\frac{3}{2}x = -\frac{3}{2}(-2) = 3$$

Thus $P(-2, 3)$ is on the terminal side of t. We have $x = -2$, $y = 3$, and $r = \sqrt{x^2 + y^2} = \sqrt{(-2)^2 + 3^2} = \sqrt{13}$. It follows that

$$\sin(t) = \frac{y}{r} = \frac{3}{\sqrt{13}}$$

$$\cos(t) = \frac{x}{r} = \frac{-2}{\sqrt{13}}$$

$$\tan(t) = \frac{y}{x} = \frac{3}{-2} = -\frac{3}{2}$$

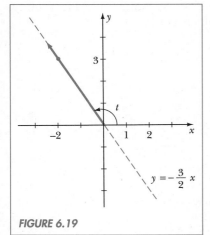

FIGURE 6.19

The values for $\sin(t)$, $\cos(t)$, and $\tan(t)$ can be positive, negative, or zero, depending on where we find the terminal side of t. For instance, we see that $\sin(t) = y/r$ is positive when $y > 0$ and negative when $y < 0$. Also, $\cos(t) = x/r$ is positive when $x > 0$ and negative when $x < 0$. Finally, $\tan(t) = y/x$ is positive when x and y have the same sign and negative when x and y have opposite signs. In summary, we have

| Sign of $\sin(t)$ | Sign of $\cos(t)$ | Sign of $\tan(t)$ |

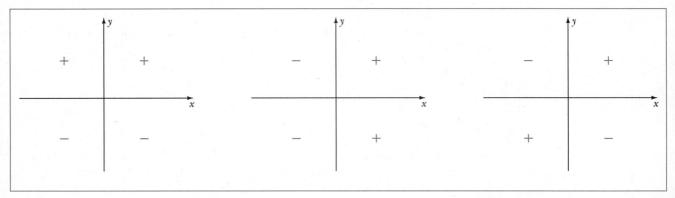

Example 5

Suppose $\sin(t) < 0$ and $\cos(t) > 0$. In what quadrant is the terminal side of t?

SOLUTION We know that $\sin(t) < 0$ implies t is in quadrant III or IV, and $\cos(t) > 0$ implies t is in quadrant I or IV. As indicated in Figure 6.20, both conditions hold if and only if t is in quadrant IV.

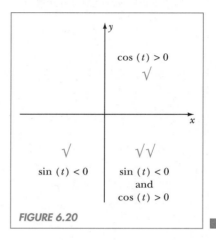

FIGURE 6.20

Example 6

Suppose point P is on the terminal side of the angle with measure t. If t is in quadrant IV, $\cos t = 2/3$, and $OP = 6$, find the coordinates of P.

SOLUTION It is very helpful to start a problem like this with a picture of the situation; this is shown in Figure 6.21. Let P have coordinates (x, y). We have $r = \sqrt{x^2 + y^2} = OP = 6$. By the definition of $\cos(t)$,

$$\frac{x}{r} = \cos(t)$$

Solving for x and substituting the known quantities $r = 6$ and $\cos(t) = 2/3$, we have

$$x = r\cos(t)$$
$$= 6\left(\frac{2}{3}\right)$$
$$= 4$$

Now that we know x and r, we can find y. Figure 6.21 shows that P determines a right triangle with sides x, $|y|$, and r. Therefore, by the Pythagorean Theorem,

$$x^2 + y^2 = r^2$$
$$4^2 + y^2 = 6^2$$
$$y^2 = 20$$
$$|y| = 2\sqrt{5}$$
$$y = 2\sqrt{5} \quad or \quad y = -2\sqrt{5}$$

Since P is in the fourth quadrant, y must be negative. We conclude that the coordinates of P are $(4, -2\sqrt{5})$.

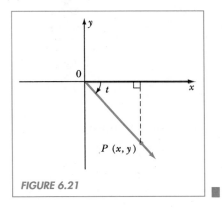

FIGURE 6.21

Example 7

Suppose point $P(x, y)$ moves along the circle $x^2 + y^2 = 100$. Let t be the measure of the angle in standard position whose terminal side passes through P.
(a) Find the y-coordinate of P as a function of t.
(b) Find the x-coordinate of P as a function of t.

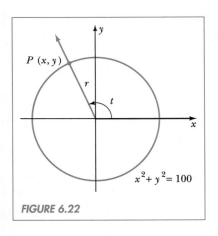

FIGURE 6.22

SOLUTION (a) Figure 6.22 shows a typical position for the point P on the circle $x^2 + y^2 = 100$. Note that the radius of the circle is $r = 10$. By definition of $\sin(t)$, we have

$$\frac{y}{r} = \sin(t)$$

This implies that $y = r\sin(t)$. Substituting $r = 10$, we get

$$y = 10\sin(t)$$

Thus we have expressed y as a function of t.
(b) By definition of $\cos(t)$, we have

$$\frac{x}{r} = \cos(t)$$

This implies that $x = r\cos(t)$. Substituting $r = 10$, we get

$$x = 10\cos(t)$$

This expresses x as a function of t. ∎

Example 8

Suppose $P(x, y)$ moves along the vertical line $x = -3$. Let t be the measure of the angle in standard position whose terminal side passes through P. Find the y-coordinate of P as a function of t.

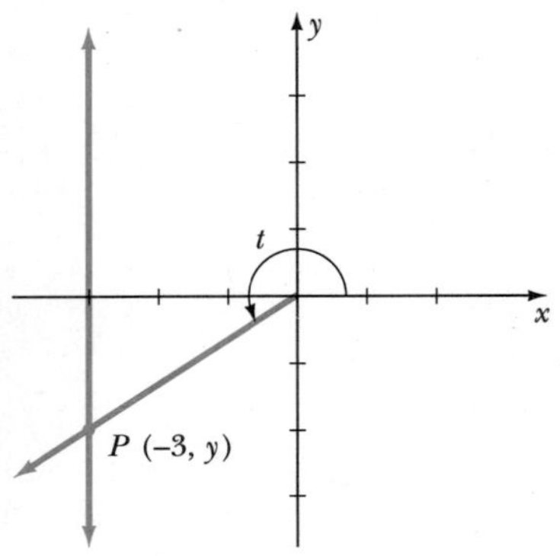

SOLUTION Figure 6.23 shows a typical position for the point $P(-3, y)$ on the vertical line $x = -3$. Note that the x-coordinate of P must always be -3. By definition of $\tan(t)$, we have

$$\frac{y}{-3} = \tan(t)$$

Solving for y,

$$y = -3\tan(t)$$

This expresses the y-coordinate of P as a function of t. ∎

FIGURE 6.23

The domains of the trigonometric functions are sets of real numbers, not degrees. However, to evaluate a trigonometric function at a real number t, we think of t as defining a certain angle. Since degrees are often used when discussing angles, it is common practice to write expressions such as $\sin(90°)$ instead of $\sin(\pi/2)$. We formally define such expressions as follows.

Trigonometric Functions of Degree Measures

Suppose F is a trigonometric function. Let θ be the measure of an angle expressed in degrees. We define

$$F(\theta) = F(t)$$

where t is the equivalent radian measure of θ.

For example, we have $\sin(180°) = \sin(\pi) = 0$, $\cos(180°) = \cos(\pi) = -1$, and $\tan(180°) = \tan(\pi) = 0$.

We close this section with a brief remark concerning functional notation. Expressions such as $\sin(t)$ occur so frequently in trigonometry that the parentheses are sometimes dropped when the meaning is clear. Thus we may write $\sin t$ instead of the more formal $\sin(t)$, $\cos t$ instead of $\cos(t)$, and $\tan t$ instead of $\tan(t)$.

EXERCISES 6.2

1. Given $P_1(1, 2)$ and $P_2(3, 6)$, both points on the terminal side of the angle with measure t in standard position. Compute $\sin(t)$ two ways, first using P_1 and then using P_2. Verify that both answers are the same.

2. Repeat Exercise 1 for $\cos(t)$.

In Exercises 3 to 8, suppose that the given point P is on the terminal side of the angle with measure t in standard position. Find $\sin(\theta)$, $\cos(\theta)$, and $\tan(\theta)$ if (a) $\theta = t$, (b) $\theta = t + \pi$, (c) $\theta = -t$.

3. $P(-4, 3)$ 4. $P(6, 8)$

5. $P(3, -1)$ 6. $P(5, -12)$

7. $P(0, 5)$ 8. $P(-9, 0)$

In Exercises 9 to 14, follow steps 1 to 4 of the rules for sine, cosine, and tangent to compute $\sin(t)$, $\cos(t)$, and $\tan(t)$ for each value of t. Note that $\tan(t)$ will be undefined for some values of t.

9. $t = 3\pi, -4\pi, 100\pi$ 10. $t = 6\pi, -\pi, 115\pi$

11. $t = 5\pi/2, -\pi/2, 99\pi/2$

12. $t = 7\pi/2, -3\pi/2, 101\pi/2$

13. $t = \pi/4, -\pi/4$ 14. $t = 3\pi/4, -3\pi/4$

In Exercises 15 to 18, suppose that the terminal side of angle t is on the line whose equation is given and also in the designated quadrant. Find $\sin(t)$, $\cos(t)$, and $\tan(t)$.

15. $y = -3x$; (a) quadrant II, (b) quadrant IV

16. $y = 5x/12$; (a) quadrant I, (b) quadrant III

17. $y = \pi x$; (a) quadrant I, (b) quadrant III

18. $y = -\sqrt{2}x$; (a) quadrant II, (b) quadrant IV

In Exercises 19 to 22, suppose that P is on the terminal side of the angle with measure t. Find a point on the terminal side of $t + \pi/2$ and use it to compute $\sin(t + \pi/2)$, $\cos(t + \pi/2)$, and $\tan(t + \pi/2)$. (*Hint*: The terminal sides of t and $t + \pi/2$ are perpendicular.)

19. $P(2, 1)$ 20. $P(-3, 2)$

21. $P(-4, -4)$ 22. $P(a, b)$, $a, b \neq 0$

In Exercises 23 to 26, find the quadrant in which the terminal side of t lies from the given conditions.

23. $\sin t > 0$, $\cos t < 0$ 24. $\sin t < 0$, $\cos t < 0$

25. $\tan t > 0$, $\sin t < 0$ 26. $\tan t < 0$, $\cos t > 0$

In Exercises 27 to 30, suppose that point P is on the terminal side of the angle with measure t. Find the x- and y-coordinates of P from the given information. OP designates the distance from the origin to P.

27. t is in quadrant I, $\sin(t) = 3/5$, $OP = 15$.

28. t is in quadrant II, $\sin(t) = 3/4$, $OP = 8$.

29. $\cos(t) = 1/5$, $OP = 1$, and $\sin(t) < 0$.

30. $\tan(t) = 2$, $OP = 4$, and $\cos(t) < 0$.

31. Suppose that point $P(x, y)$ moves along the circle $x^2 + y^2 = 64$. Let t be the measure of the angle in standard position whose terminal side passes through P.
    (a) Find the x-coordinate of P as a function of t.
    (b) Find the y-coordinate of P as a function of t.

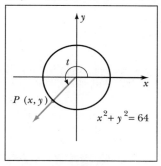

32. A point P moves along the circle $x^2 + (y - 1)^2 = 1$. Consider points $A(1, 1)$ and $C(0, 1)$. Let t be the measure of angle ACP with initial side CA.
    (a) Write the y-coordinate of P as a function of t.
    (b) Write the x-coordinate of P as a function of t.

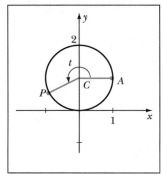

33. Suppose a point P moves along the circle $(x - 2)^2 + y^2 = 4$ with center $C(2, 0)$. Let t be the measure of angle ACP (with initial side CA), where A has coordinates $(4, 0)$. Find as a function of t the (a) y-coordinate of P, (b) x-coordinate of P.

34. Suppose the point P moves along the vertical line $x = 3$. Let t be the measure of the angle in standard position whose terminal side passes through P. Find the y-coordinate of P as a function of t.

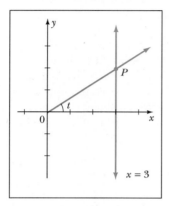

35. Repeat Exercise 34 if P moves along the vertical line $x = 1$.

36. Repeat Exercise 34 if P moves along the vertical line $x = -10$.

In Exercises 37 to 40, from the given information find $\tan(t)$.

37. $\sin(t) = .8$, $\cos(t) = -.6$

38. $\sin(t) = -2/\sqrt{5}$, $\cos(t) = 1/\sqrt{5}$

39. $\sin(t) = 0$

40. $\sin(t) = a$, $\cos(t) = b$, a and b constants, $b \neq 0$.

In Exercises 41 and 42, simplify the given expression. Assume the expression is defined for t.

41. $\tan(t) \cos(t)$

42. $\dfrac{\sin(t)}{\tan(t)}$

43. Suppose F is the sine, cosine, or tangent function. Explain why $F(t + 2\pi k) = F(t)$ for any $k \in \mathbf{Z}$.

6.3 REFERENCE TRIANGLES

In this section we describe how to construct a certain right triangle associated with a given real number. This will allow us to interpret the trigonometric functions as ratios of sides in a right triangle. To make the discussion simpler, we consider only real numbers t such that the terminal side of the angle of measure t in standard position is not on the x- or y-axis. In other words, t is not an integer multiple of $\pi/2$. Under this assumption we describe the construction as follows.

Given a real number t, draw the angle of measure t in standard position and choose any point $P(x, y)$ (other than the origin) on the terminal side of t. Now draw a perpendicular from P to the x-axis forming the line segment PQ. In this way we obtain a right triangle PQO, which we call a **reference triangle** for t. Figure 6.24 shows reference triangles for several values of t. Note that a reference triangle is not unique since the point P is arbitrary.

The angle POQ in a reference triangle is called the **reference angle** for t. The reference angle for each value of t in Figure 6.24 is labeled with the

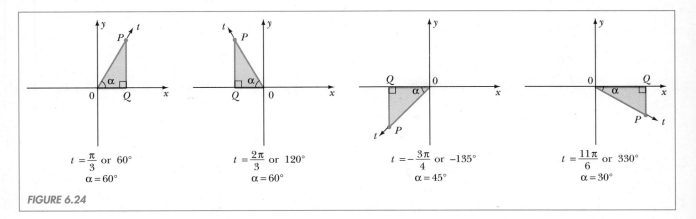

FIGURE 6.24

symbol α. Note that α is just the smallest angle between the terminal side of t and the x-axis. Even though α may not be in standard position, we always assign to it the positive measure satisfying $0 \leq \alpha \leq \pi/2$. With this definition, the reference angle for any real number t is unique.

Referring to a typical reference triangle in Figure 6.25, we define

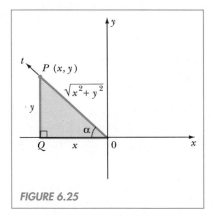

FIGURE 6.25

Opposite leg $= y$	directed length of the leg opposite α in $\triangle PQO$
Adjacent leg $= x$	directed length of the leg adjacent α in $\triangle PQO$
Hypotenuse $= \sqrt{x^2 + y^2}$	length of the hypotenuse in $\triangle PQO$

In these definitions we identify the opposite and adjacent legs of the reference triangle with the y- and x-coordinate of P, respectively. Thus we say they are *directed* lengths since they can be positive or negative. Note, however, that

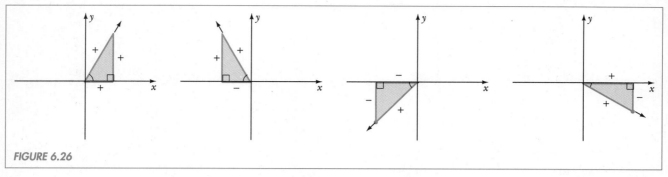

FIGURE 6.26

the hypotenuse is always assigned a positive value. Figure 6.26 illustrates the signs for the sides of a reference triangle in each quadrant.

Example 1

Suppose the terminal side of t is in quadrant II on the line $y = -3x$. Draw and label a reference triangle for t.

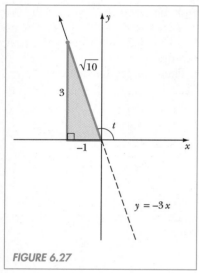

FIGURE 6.27

SOLUTION We choose the point $P(-1, 3)$ on the line $y = -3x$ in the second quadrant. This determines the reference triangle for t shown in Figure 6.27. We have

$$\text{Opposite leg} = 3$$
$$\text{Adjacent leg} = -1$$
$$\text{Hypotenuse} = \sqrt{3^2 + (-1)^2} = \sqrt{10}$$

Other possible reference triangles exist, each corresponding to a different choice of P on the terminal side of t. However, they all would be similar (in the geometric sense) to the one in Figure 6.27. ■

As an immediate consequence of the definitions of the trigonometric functions, we have the following:

> ### Reference Triangle Interpretation
> In a reference triangle for t,
>
> $$\sin(t) = \frac{y}{r} = \frac{\text{opposite leg}}{\text{hypotenuse}}$$
>
> $$\cos(t) = \frac{x}{r} = \frac{\text{adjacent leg}}{\text{hypotenuse}}$$
>
> $$\tan(t) = \frac{y}{x} = \frac{\text{opposite leg}}{\text{adjacent leg}}$$

Example 2

Use the reference triangle interpretation to compute $\sin(t)$, $\cos(t)$, and $\tan(t)$ for the angle t given in Example 1.

SOLUTION   Using the reference triangle for t in Figure 6.27, we have

$$\sin(t) = \frac{\text{opposite leg}}{\text{hypotenuse}} = \frac{3}{\sqrt{10}}$$

$$\cos(t) = \frac{\text{adjacent leg}}{\text{hypotenuse}} = \frac{-1}{\sqrt{10}}$$

$$\tan(t) = \frac{\text{opposite leg}}{\text{adjacent leg}} = \frac{3}{-1} = -3 \quad \blacksquare$$

Example 3

Suppose $\tan(t) = 4$ and $\pi < t < 3\pi/2$. Draw and label a reference triangle for t and use it to compute $\sin(t)$ and $\cos(t)$.

SOLUTION   To construct a reference triangle for t, we begin by drawing a ray in the third quadrant to represent the terminal side of t (since $\pi < t < 3\pi/2$). Next, draw a perpendicular from a point on the terminal side of t to the x-axis, forming the sides of a reference triangle as shown in Figure 6.28(a). The required signs for the legs and the hypotenuse are indicated.

Our next step is to label the sides of the reference triangle. We know

$$\tan(t) = \frac{\text{opposite leg}}{\text{adjacent leg}} = 4$$

Therefore we may let opposite leg $= -4$ and adjacent leg $= -1$. Note that both are negative and have the required ratio equal to 4. Now, by the Pythagorean Theorem,

$$\text{Hypotenuse} = \sqrt{4^2 + 1^2} = \sqrt{17}$$

Thus we may label the reference triangle for t as shown in Figure 6.28(b). It follows that

$$\sin(t) = \frac{\text{opposite leg}}{\text{hypotenuse}} = \frac{-4}{\sqrt{17}}$$

$$\cos(t) = \frac{\text{adjacent leg}}{\text{hypotenuse}} = \frac{-1}{\sqrt{17}} \quad \blacksquare$$

We could have evaluated the trigonometric functions in Example 3 without resorting to a reference triangle by simply observing that $P(-1, -4)$ must be on the terminal side of t. We then could proceed as in Section 6.2 to evaluate $\sin(t)$ and $\cos(t)$ directly from their definitions in terms of the coordinates of P. Similar remarks hold for Example 2. However, the reference triangle inter-

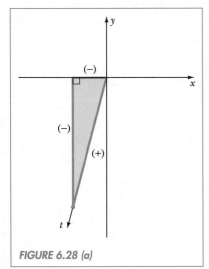

FIGURE 6.28 (a)

FIGURE 6.28 (b)

pretation is helpful when we evaluate the trigonometric functions at certain *standard* angles. Let us describe these angles next.

Consider the equilateral triangle with sides of length 2 in Figure 6.29. By drawing an altitude a, we obtain a 30–60–90 right triangle. The base of this triangle is 1, and its hypotenuse is 2. It follows from the Pythagorean Theorem that

$$a = \sqrt{2^2 - 1^2} = \sqrt{3}$$

Hence the lengths of the sides of this 30–60–90 right triangle are 1, 2, and $\sqrt{3}$.

Now, consider the isosceles right triangle with legs of length 1 in Figure 6.30. This is a 45–45–90 right triangle. It follows from the Pythagorean Theorem that the hypotenuse h is given by

$$h = \sqrt{1^2 + 1^2} = \sqrt{2}$$

Hence the lengths of the sides of this 45–45–90 right triangle are 1, 1, and $\sqrt{2}$.

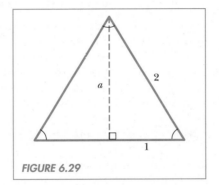

FIGURE 6.29

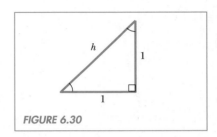

FIGURE 6.30

We have thus established the following *standard triangles*.

The Standard Triangles

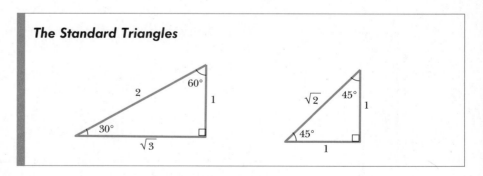

We call t a **standard angle** if it has a reference angle equal to $\pi/6$ (30°), $\pi/4$ (45°), or $\pi/3$ (60°). If t is a standard angle, then it will have a reference triangle congruent to one of the standard triangles displayed above. Figure 6.31 illustrates all the standard angles between 0 and 2π.

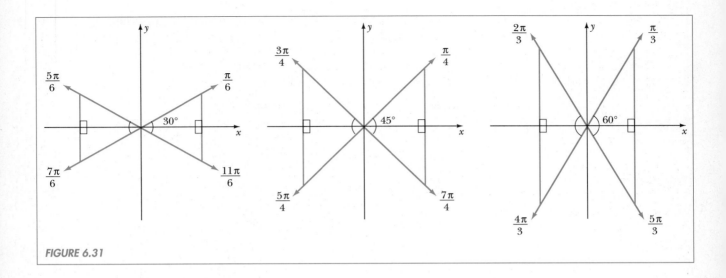

FIGURE 6.31

It is important to be able to draw and label from memory a reference triangle for any standard angle t. This allows us to compute the exact values of $\sin(t)$, $\cos(t)$, and $\tan(t)$ whenever t is a standard angle.

Example 4

For the angle with given measure t, find the reference angle, draw and label a reference triangle, and compute $\sin(t)$, $\cos(t)$, and $\tan(t)$: (a) $t = 150°$, (b) $t = 7\pi/4$, (c) $t = -11\pi/3$.

SOLUTION (a) Drawing $t = 150°$ in standard position, we find its reference angle $\alpha = 180° - 150° = 30°$. See Figure 6.32. It follows that a reference triangle for $150°$ is the 30–60–90 right triangle shown. We label the sides 1, 2, $\sqrt{3}$; however, the adjacent leg must be given a negative sign in quadrant II. Referring to Figure 6.32, it follows that

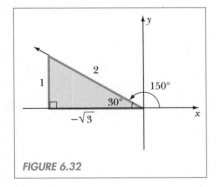

FIGURE 6.32

$$\sin(150°) = \frac{\text{opp.}}{\text{hyp.}} = \frac{1}{2}$$

$$\cos(150°) = \frac{\text{adj.}}{\text{hyp.}} = \frac{-\sqrt{3}}{2}$$

$$\tan(150°) = \frac{\text{opp.}}{\text{adj.}} = \frac{1}{-\sqrt{3}} = \frac{-1}{\sqrt{3}}$$

(opp. means opposite leg, adj. means adjacent leg, and hyp. means hypotenuse.)

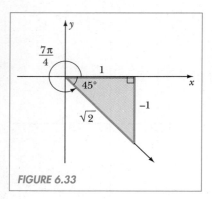

FIGURE 6.33

(b) Figure 6.33 shows $t = 7\pi/4$ drawn in standard position. We see that the reference angle $\alpha = 2\pi - 7\pi/4 = \pi/4$, or 45°. Thus a reference triangle for $7\pi/4$ is the 45−45−90 right triangle shown. We label the sides 1, 1, $\sqrt{2}$; however, the opposite leg must be given a negative sign in quadrant IV. Referring to Figure 6.33, we find that

$$\sin\left(\frac{7\pi}{4}\right) = \frac{\text{opp.}}{\text{hyp.}} = \frac{-1}{\sqrt{2}}$$

$$\cos\left(\frac{7\pi}{4}\right) = \frac{\text{adj.}}{\text{hyp.}} = \frac{1}{\sqrt{2}}$$

$$\tan\left(\frac{7\pi}{4}\right) = \frac{\text{opp.}}{\text{adj.}} = \frac{-1}{1} = -1$$

(c) To locate the terminal side of $t = -11\pi/3$, it is easiest to count as follows:

$$-\frac{11\pi}{3} = -6\left(\frac{\pi}{3}\right) \quad - \quad 3\left(\frac{\pi}{3}\right) \quad - \quad 2\left(\frac{\pi}{3}\right)$$

One
revolution   Half
revolution

As shown in Figure 6.34, this places the terminal side of $-11\pi/3$ in quadrant I with the reference angle $\alpha = 4\pi - 11\pi/3 = \pi/3$, or 60°. Therefore a reference triangle for $-11\pi/3$ is the 30−60−90 right triangle shown. We label the sides 1, 2, $\sqrt{3}$ (all positive in the first quadrant). Referring to Figure 6.34, we have

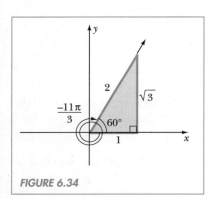

FIGURE 6.34

$$\sin\left(-\frac{11\pi}{3}\right) = \frac{\sqrt{3}}{2}$$

$$\cos\left(-\frac{11\pi}{3}\right) = \frac{1}{2}$$

$$\tan\left(-\frac{11\pi}{3}\right) = \sqrt{3} \quad\blacksquare$$

Example 5

Given the right triangle with acute angle of measure θ in Figure 6.35(a), find $\sin(\theta)$, $\cos(\theta)$, and $\tan(\theta)$.

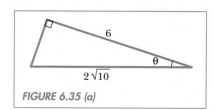

FIGURE 6.35 (a)

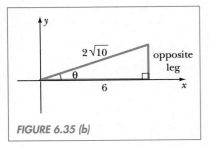

FIGURE 6.35 (b)

SOLUTION   We place the right triangle so that angle θ is in standard position, as in Figure 6.35(b). Our right triangle is now a reference triangle for θ with adjacent leg 6 and hypotenuse $2\sqrt{10}$. By the Pythagorean Theorem,

$$\text{Opposite leg} = \sqrt{(2\sqrt{10})^2 - 6^2} = \sqrt{40 - 36} = 2$$

It follows that

$$\sin(\theta) = \frac{\text{opp.}}{\text{hyp.}} = \frac{2}{2\sqrt{10}} = \frac{1}{\sqrt{10}}$$

$$\cos(\theta) = \frac{\text{adj.}}{\text{hyp.}} = \frac{6}{2\sqrt{10}} = \frac{3}{\sqrt{10}}$$

$$\tan(\theta) = \frac{\text{opp.}}{\text{adj.}} = \frac{2}{6} = \frac{1}{3} \ \blacksquare$$

Example 5 illustrates the following general result: *If θ is an acute angle in a right triangle, then the triangle itself is a reference triangle for θ.* Thus we may always use the length of the side opposite θ for the opposite leg and the length of the side adjacent θ for the adjacent leg in a reference triangle for θ.

Example 6

In the right triangle of Figure 6.36 with sides of length a, b, and c, determine the opposite leg, adjacent leg, and hypotenuse in a reference triangle for α. Compute $\sin(\alpha)$, $\cos(\alpha)$, and $\tan(\alpha)$. Do the same for β.

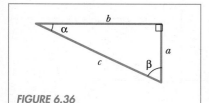

FIGURE 6.36

SOLUTION   For angle α, opposite $= a$, adjacent $= b$, and hypotenuse $= c$. Thus,

$$\sin(\alpha) = \frac{a}{c}$$

$$\cos(\alpha) = \frac{b}{c}$$

$$\tan(\alpha) = \frac{a}{b}$$

For angle β, opposite $= b$, adjacent $= a$, and hypotenuse $= c$. Thus,

$$\sin(\beta) = \frac{b}{c}$$

$$\cos(\beta) = \frac{a}{c}$$

$$\tan(\beta) = \frac{b}{a} \ \blacksquare$$

EXERCISES 6.3

In Exercises 1 and 2, suppose the terminal side of angle t in standard position is on the line whose equation is given and also in the designated quadrant. Draw and label a reference triangle for t.

1. $y = \dfrac{3}{4}x$; (a) quadrant I, (b) quadrant III

2. $y = -\dfrac{1}{2}x$; (a) quadrant II, (b) quadrant IV

In Exercises 3 to 10, from the given information, draw and label a reference triangle for t. Use the reference triangle to compute $\sin(t)$, $\cos(t)$, and $\tan(t)$.

3. $\tan(t) = 3$, $0 < t < \pi/2$
4. $\tan(t) = 2$, $\pi < t < 3\pi/2$
5. $\tan(t) = -1/2$, $\pi/2 < t < \pi$
6. $\tan(t) = -1/3$, $3\pi/2 < t < 2\pi$
7. $\sin(t) = -7/25$, $\pi < t < 3\pi/2$
8. $\sin(t) = 1/4$, $\pi/2 < t < \pi$
9. $\cos(t) = 4/5$, $3\pi/2 < t < 2\pi$
10. $\cos(t) = \sin(t)$, $0 < t < \pi/2$

In Exercises 11 and 12, draw and label a reference triangle for each value of t. Complete the table.

11.

t	$\pi/6$	$\pi/4$	$\pi/3$	$2\pi/3$	$3\pi/4$	$5\pi/6$
$\sin t$						
$\cos t$						
$\tan t$						

12.

t	$-\pi/6$	$-\pi/4$	$-\pi/3$	$-2\pi/3$	$-3\pi/4$	$-5\pi/6$
$\sin t$						
$\cos t$						
$\tan t$						

In Exercises 13 to 18, draw and label a reference triangle for each value of t. Find $\sin(t)$, $\cos(t)$, and $\tan(t)$.

13. (a) $4\pi/3$
    (b) $-240°$
    (c) $11\pi/3$

14. (a) $5\pi/3$
    (b) $480°$
    (c) $-8\pi/3$

15. (a) $-5\pi/4$
    (b) $315°$

16. (a) $11\pi/4$
    (b) $405°$

17. (a) $13\pi/6$
    (b) $-750°$

18. (a) $11\pi/6$
    (b) $-510°$

In Exercises 19 to 26, find $\sin\theta$, $\cos\theta$, and $\tan\theta$.

19.

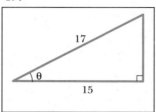

20.

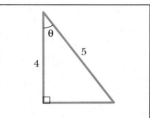

21.

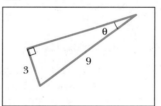

22.

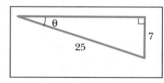

23.

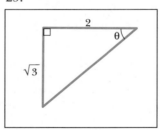

24.

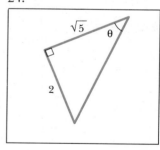

25.

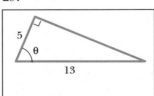

26.

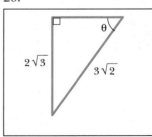

In Exercises 27 to 30, in the given right triangle, determine the opposite leg, adjacent leg, and hypotenuse in a reference triangle for α. Compute sin(α), cos(α), and tan(α). Do the same for β.

27.

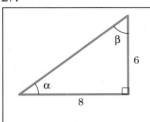

28.

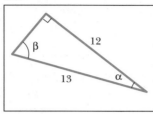

29.

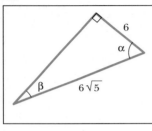

30.

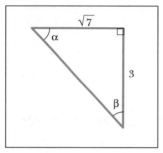

In Exercises 31 and 32, find d in terms of θ.

31. (a)

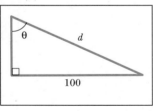

(b)

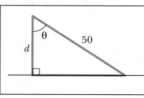

(c)

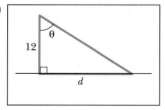

32. (a)

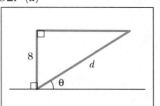

(b)

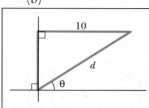

(c)

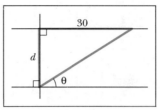

In Exercises 33 to 36, refer to Figure 6.37.

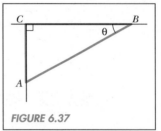

FIGURE 6.37

33. Suppose $AB = 4$. Find in terms of θ: (a) length AC, (b) length BC.

34. Repeat Exercise 33 if $AB = 1$.

35. Suppose $BC = 6$. Find in terms of θ: (a) length AC, (b) length AB.

36. Repeat Exercise 35 if $BC = 1$.

In Exercises 37 and 38, find x in terms of θ.

37.

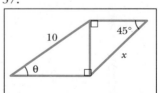

38.

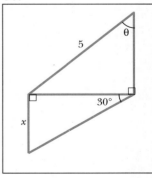

In Exercises 39 and 40, find x and y in terms of θ.

39.

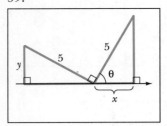

40.

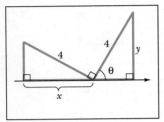

41. Find the area of the triangle in terms of θ. [Use $\sin(\theta)$ to find the height of the triangle.]

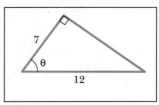

42. Express the area of right triangle ABC in terms of (a) θ if $AB = 7$, (b) θ and a if $AB = a$.

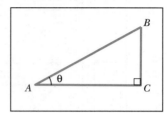

6.4 **PROPERTIES OF SINE AND COSINE**

Recall that the definitions for $\sin(t)$ and $\cos(t)$ given in Section 6.2 required that we select any point $P(x, y)$ on the terminal side of t; then $\sin(t) = y/r$ and $\cos(t) = x/r$, where $r = \sqrt{x^2 + y^2}$. If we choose to select P so that its distance from the origin is 1 unit, then we say that P is on the **unit circle**, the circle $x^2 + y^2 = 1$ of radius 1 centered at the origin. Choosing P on the unit circle gives the sine and cosine an important interpretation that deserves special emphasis.

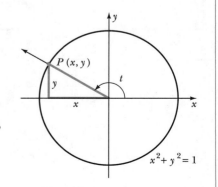

The Unit Circle Interpretation
Let $P(x, y)$ be the point where the terminal side of t intersects the unit circle. Then

$$\sin(t) = y \qquad y\text{-coordinate of } P$$

$$\cos(t) = x \qquad x\text{-coordinate of } P$$

PROOF: Since $P(x, y)$ is on the unit circle, $r = \sqrt{x^2 + y^2} = 1$. Thus

$$\sin(t) = \frac{y}{r} = \frac{y}{1} = y \quad \text{and} \quad \cos(t) = \frac{x}{r} = \frac{x}{1} = x$$

This completes the proof. ■

Example 1

Use the unit circle interpretation of sine to determine which value in the set is largest and which is smallest: $\{\sin(1), \sin(2), \sin(3)\}$.

SOLUTION We make a rough drawing of the angles with measures 1, 2, and 3 in standard position, as shown with the unit circle in Figure 6.38(a).

The y-coordinates where the terminal sides of 1, 2, and 3 intersect the unit circle are labeled y_1, y_2, and y_3, respectively. By the unit circle interpretation, we have

$$\sin(1) = y_1 \qquad \sin(2) = y_2 \qquad \sin(3) = y_3$$

Clearly, the smallest value is $y_3 = \sin(3)$. Now, the largest value is either y_1 or y_2. In Figure 6.38(b), we note that α has positive measure 1 and β has positive measure $\pi - 2 \approx 1.14$. Since $\beta > \alpha$, it follows that $y_2 > y_1$. Thus the largest value is $\sin(2)$.

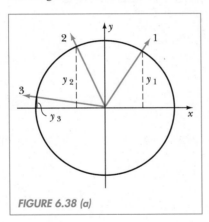

FIGURE 6.38 (a)

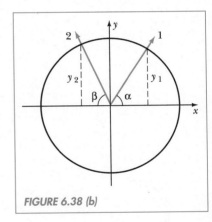

FIGURE 6.38 (b)

Example 2

Find the exact coordinates of the point where the terminal side of $\pi/3$ intersects the unit circle.

SOLUTION Let $P(x, y)$ be the point of intersection. By the unit circle interpretation of sine and cosine, we have $x = \cos(\pi/3)$ and $y = \sin(\pi/3)$. Since $\pi/3 = 60°$ has the 30–60–90 reference triangle shown in Figure 6.39, we find

$$\cos\left(\frac{\pi}{3}\right) = \frac{1}{2} \qquad \sin\left(\frac{\pi}{3}\right) = \frac{\sqrt{3}}{2}$$

Thus the coordinates of P are $(1/2, \sqrt{3}/2)$. ■

FIGURE 6.39

Now we wish to draw the graph of the sine function. To do this, we must graph the equation

$$y = \sin(t)$$

in the x, y-plane. This means that t will be the x-coordinate of each point and $\sin(t)$ the corresponding y-coordinate. In other words, we want to graph the set of points

$$\{(t, \sin(t)) : t \in \mathbf{R}\}$$

We begin by recalling the values for $\sin(t)$ when $t = 0$, $\pi/2$, π, $3\pi/2$, and 2π. This can be done quickly using the unit circle interpretation: $\sin(t)$ is the y-coordinate of the point where the terminal side of t intersects the unit circle. We have

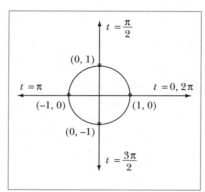

t	0	$\pi/2$	π	$3\pi/2$	2π
$\sin(t)$	0	1	0	-1	0

Thus we start the graph with the following points.

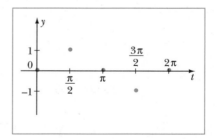

As we will show, these points establish one cycle of the sine curve. (A cycle is a portion of the curve that can be repeated to obtain the entire graph. We have more to say about this in Section 6.6.) At this time note how the dimensions on the graph are consistent; 1 unit along the y-axis is the same length as 1 unit along the x-axis. (Since $\pi/2 \approx 1.57$, 1 unit should be about two-thirds the distance from $x = 0$ to $x = \pi/2$.) Unless required to do otherwise, set up graphs so that the dimensions are consistent.

Now we use the unit circle interpretation to help us fill in the details of the graph over the four intervals $[0, \pi/2]$, $[\pi/2, \pi]$, $[\pi, 3\pi/2]$, and $[3\pi/2, 2\pi]$.

Interval [0, π/2]. We see on the unit circle that as t increases from 0 to $\pi/2$, $\sin(t)$ increases from 0 to 1. Using the standard angles, we compute some specific values of $\sin(t)$ for the graph.

t	0	$\pi/6$	$\pi/4$	$\pi/3$	$\pi/2$
$\sin(t)$	0	.5	$\dfrac{1}{\sqrt{2}} \approx .71$	$\dfrac{\sqrt{3}}{2} \approx .87$	1

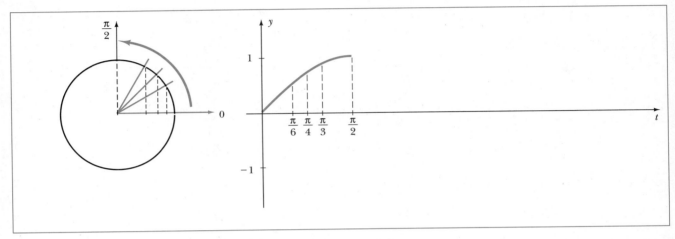

Interval [π/2, π]. As t increases from $\pi/2$ to π, $\sin(t)$ decreases from 1 to 0.

t	$\pi/2$	$2\pi/3$	$3\pi/4$	$5\pi/6$	π
$\sin(t)$	1	$\dfrac{\sqrt{3}}{2} \approx .87$	$\dfrac{1}{\sqrt{2}} \approx .71$	.5	0

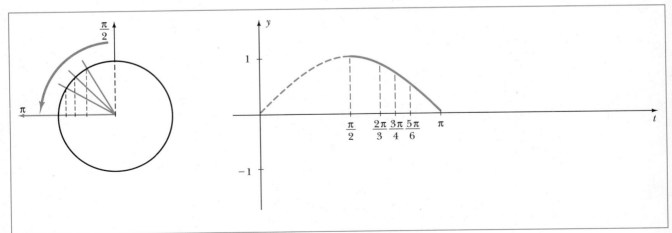

Interval [π, 3π/2]. As t increases from π to $3\pi/2$, $\sin(t)$ decreases from 0 to -1.

t	π	$7\pi/6$	$5\pi/4$	$4\pi/3$	$3\pi/2$
$\sin(t)$	0	$-.5$	$\dfrac{-1}{\sqrt{2}} \approx -.71$	$\dfrac{-\sqrt{3}}{2} \approx -.87$	-1

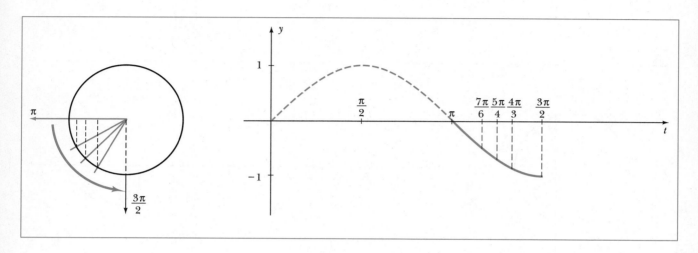

Interval [3π/2, 2π]. As t increases from $3\pi/2$ to 2π, $\sin(t)$ increases from -1 to 0.

t	$3\pi/2$	$5\pi/3$	$7\pi/4$	$11\pi/6$	2π
$\sin(t)$	-1	$-\dfrac{\sqrt{3}}{2} \approx -.87$	$\dfrac{-1}{\sqrt{2}} \approx -.71$	$-.5$	0

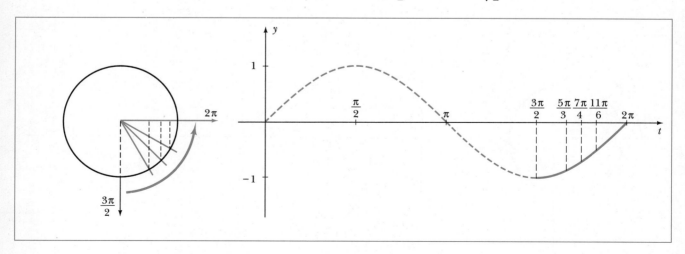

Putting all four parts together, we obtain the graph of the sine function over the interval $[0, 2\pi]$.

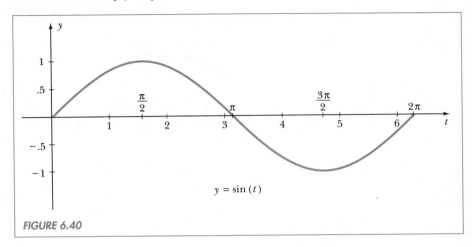

FIGURE 6.40

What happens over the interval $[2\pi, 4\pi]$? Since $t + 2\pi$ is coterminal with t, $\sin(t + 2\pi) = \sin(t)$. It follows that the graph of sine over the interval $[2\pi, 4\pi]$ will be exactly the same as the graph over $[0, 2\pi]$. Indeed, by continually repeating the curve in Figure 6.40 toward the right and the left, we can find the graph of the sine function over any desired interval.

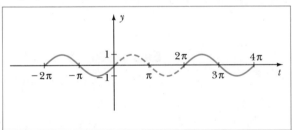

From the graph of the sine function we notice three important properties. First, the range of the function is limited: $-1 \le \sin(t) \le 1$. Second, the value of $\sin(t)$ repeats itself every 2π units. Functions that behave this way are said to be **periodic**; we say more about periodic functions in Section 6.6. Finally, the graph of $\sin(t)$ is symmetric with respect to the origin: $\sin(-t) = -\sin(t)$. We may formally establish these properties as follows.

Properties of Sine

Range: $[-1, 1]$
Periodicity: $\sin(t + 2\pi k) = \sin(t)$ for all $k \in \mathbf{Z}$
Symmetry: $\sin(-t) = -\sin(t)$

PROOF: By the unit circle interpretation, the range of $\sin(t)$ corresponds to all the possible y-coordinates appearing on the unit circle, $x^2 + y^2 = 1$. These y-coordinates take on all values from -1 to 1. Hence the range of $\sin(t)$ is $[-1, 1]$.

We know that $t + 2\pi k$ is coterminal with t for any integer k. Therefore $t + 2\pi k$ and t intersect the unit circle at the same point, say, $P(a, b)$. It follows that $\sin(t + 2\pi k) = b = \sin(t)$. This establishes the periodicity property.

Figure 6.41 indicates that if t intersects the unit circle at $P(a, b)$, then $-t$ will intersect the unit circle at $P(a, -b)$. Hence $\sin(-t) = -b = -\sin(t)$. This completes the proof. ■

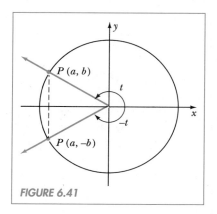

FIGURE 6.41

Example 3

Use the periodicity and symmetry properties of the sine function to simplify

$$\sin(\alpha + 10\pi) - \sin(10\pi - \alpha)$$

SOLUTION Observe that 10π is a multiple of 2π, $10\pi = (2\pi)5$, so that $\alpha + 10\pi$ is coterminal with α and $10\pi - \alpha$ is coterminal with $-\alpha$. Thus

$$\sin(\alpha + 10\pi) - \sin(10\pi - \alpha) = \sin\alpha - \sin(-\alpha) \qquad \text{by periodicity}$$
$$= \sin\alpha - (-\sin\alpha) \qquad \text{by symmetry}$$
$$= \sin\alpha + \sin\alpha$$
$$= 2\sin\alpha \quad ■$$

Example 4

Is there a real number t such that $\sin(t) = 1.5$?

SOLUTION No! The range of $\sin(t)$ is $[-1, 1]$, so $\sin(t)$ cannot be greater than 1. ■

Our next objective is to draw the graph of the cosine curve. This means that we wish to graph the points $(t, \cos(t))$ in the x, y-plane. We begin by considering the values of $\cos(t)$ corresponding to $t = 0, \pi/2, \pi, 3\pi/2$, and 2π. Once again, the unit circle interpretation proves helpful: $\cos(t)$ is the x-coordinate of the point where the terminal side of t intersects the unit circle.

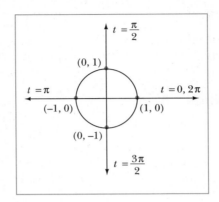

t	0	$\pi/2$	π	$3\pi/2$	2π
$\cos(t)$	1	0	-1	0	1

Graphing these points, we have

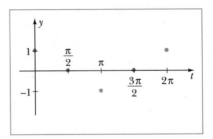

These points are enough to guide us in drawing one cycle of the cosine curve as shown in Figure 6.42. The details of how we obtain this graph are similar to the arguments given for the sine curve and so we omit them. The graph for $t < 0$ or $t > 2\pi$ is obtained by repeating the cycle in Figure 6.42 toward the left and right.

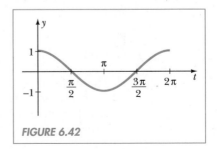

FIGURE 6.42

Notice that the only difference from the properties of sine is the symmetry property: $\cos(-t) = \cos(t)$. This says that the graph is symmetric with respect to the y-axis. The proofs of these properties are left for the exercises.

Example 5

Graph the function $F(t) = \cos(t)$ for t in the interval $[-4\pi, -2\pi]$.

SOLUTION We know that the cosine graph repeats itself every 2π units, so we repeat the portion of the graph of cosine shown in Figure 6.43 toward the left until we cover the interval $[-4\pi, -2\pi]$. The results are shown in Figure 6.43.

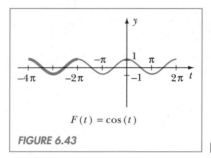

$$F(t) = \cos(t)$$

FIGURE 6.43

You may notice that the cosine curve looks identical to the sine curve except that it is shifted to the left a bit. This is indeed the case, as we now state.

Cosine Is Sine Shifted

$$\cos(t) = \sin\left(t + \frac{\pi}{2}\right) \qquad \text{for all } t \in \mathbf{R}$$

The proof is in Chapter 7. In terms of the graphs, this result says that if we shift the sine curve to the left $\pi/2$ units, we get the cosine curve.

Another relationship between the sine and cosine functions is the following result. We write $\sin^2 t$ to mean $(\sin t)^2$ and $\cos^2 t$ to mean $(\cos t)^2$.

> **The Fundamental Identity**
>
> $$\sin^2 t + \cos^2 t = 1 \qquad \text{for all } t \in \mathbf{R}$$

PROOF: Let $P(x, y)$ be on the terminal side of t and $r = \sqrt{x^2 + y^2}$. Then

$$\sin^2 t + \cos^2 t = \left(\frac{y}{r}\right)^2 + \left(\frac{x}{r}\right)^2 = \frac{y^2}{r^2} + \frac{x^2}{r^2}$$

$$= \frac{y^2 + x^2}{r^2} = \frac{r^2}{r^2}$$

$$= 1$$

Alternate forms of the Fundamental Identity are

$$1 - \cos^2 t = \sin^2 t \qquad 1 - \sin^2 t = \cos^2 t \quad \blacksquare$$

Example 6

Simplify $\dfrac{\cos^2 \theta}{1 - \sin^2 \theta}$

SOLUTION   Assuming $1 - \sin^2 \theta \neq 0$, we have

$$\frac{\cos^2 \theta}{1 - \sin^2 \theta} = \frac{\cos^2 \theta}{\cos^2 \theta} = 1 \quad \blacksquare$$

We close this section with a word about calculators. Up to this point we have evaluated the trigonometric functions only in situations where a calculator was not necessary: we knew a point on the terminal side of t, we were given a reference triangle for t, or t was a standard angle. Now, suppose we want to evaluate the sine of an angle that does not fit into any of these categories, such as 1 radian. From the graph of the sine function in Figure 6.40, we would estimate $\sin(1) \approx .8$. To obtain a better approximation, we can use the calculator. The calculator is programmed with formulas derived from calculus that will give us values of the trigonometric functions accurate to several decimal places.

Example 7

Use the calculator to find an approximate value of $\sin(1)$ accurate to five decimal places.

SOLUTION   A standard scientific calculator has three modes: *degree*, *radian*, and *grad*. Since we wish to find the sine of 1 radian, we must switch the calculator into radian mode before making our computation. Next, we enter the following keystrokes:

$$1 \boxed{\text{sin}}$$

This is valid for most standard calculators (algebraic entry or reverse Polish notation). The display should read .84147098. Rounding off to five decimal places, we have

$$\sin(1) \approx .84147 \ \blacksquare$$

Remark

For calculators that display each keystroke as if the expression is being written by hand, enter the function first and then the angle. Thus for sin(1) enter

$$\boxed{\text{sin}} \ 1 \quad \text{then press} \quad \boxed{=} \ \text{or} \ \boxed{\text{EXEC}}$$

Example 8

Find an approximate value accurate to five decimal places for the x-coordinate of the point where the terminal side of 1000° intersects the unit circle.

SOLUTION By the unit circle interpretation, the x-coordinate of the point of intersection is given by cos(1000°). Making certain the calculator is in degree mode, we key in

$$1000 \boxed{\text{cos}}$$

This gives us cos(1000°) $\approx$ .17365. $\blacksquare$

EXERCISES 6.4

In Exercises 1 and 2, use the unit circle interpretation to determine which value in the given set is largest and which is smallest. (Do not use a calculator.)

1. (a) {cos(5), cos(6), cos(7)}
   (b) {sin(4), sin(5), sin(6)}

2. (a) {cos(2), cos(3), cos(4)}
   (b) {sin(1.5), sin(1.6), sin(1.7)}

In Exercises 3 and 4, suppose $0 < \alpha < \beta < \pi/2$. Determine which number in the given set is largest.

3. (a) {sin α, sin β}
   (b) {cos α, cos β}

4. (a) {sin α, sin($\pi - \beta$)}
   (b) {cos β, cos($\pi - \alpha$)}

In Exercises 5 and 6, find the exact coordinates of the point where the terminal side of the given angle t intersects the unit circle. (Do not use a calculator.)

5. (a) $t = \pi/6$
   (b) $t = 3\pi/4$
   (c) $t = -\pi/3$

6. (a) $t = 5\pi/6$
   (b) $t = -\pi/4$
   (c) $t = 2\pi/3$

In Exercises 7 to 12, graph the given function over the designated interval.

7. $F(t) = \sin t$, $[-3\pi, 0]$

8. $F(t) = \sin t$, $[4\pi, 6\pi]$

9. $F(t) = \cos t$, $[3\pi, 5\pi]$

10. $F(t) = \cos t$, $[-3\pi, 0]$

11. $F(t) = \sin t$, $[9\pi/2, 11\pi/2]$

12. $F(t) = \cos t$, $[-7\pi/2, -5\pi/2]$

In Exercises 13 to 20, graph each function over the interval $[-2\pi, 2\pi]$.

13. $F(t) = \sin t - 1$

14. $F(t) = \cos t + 1$

15. $F(t) = \cos(t - \pi/2) + 1$

16. $F(t) = \sin(t - \pi/2) - 1$

17. $F(t) = |\sin t|$

18. $F(t) = |\cos t|$

19. $F(t) = \sin(-t)$

20. $F(t) = \cos(-t)$

In Exercises 21 to 28, use the periodicity and symmetry properties to simplify the given expression.

21. $\sin(\theta) - \sin(-\theta)$

22. $\cos(\theta) - \cos(-\theta)$

23. $\cos(2 + 2\pi)$

24. $\sin(1 + 4\pi)$

25. $\sin(\theta + 4\pi) - \sin(\theta - 2\pi)$

26. $\cos(\theta + 2\pi) - \cos(2\pi - \theta)$

27. $\cos(14\pi - \theta) + \cos(\theta)$

28. $\sin(4 + 4\pi) + \sin(4 - 4\pi)$

In Exercises 29 to 36, simplify the given expression.

29. $\cos^2 (5) + \sin^2 (5)$

30. $\sin^2 (12°) + \cos^2 (12°)$

31. $\dfrac{1 - \sin^2 \theta}{\cos \theta}$

32. $\dfrac{1 - \cos^2 \theta}{\sin \theta}$

33. $\sqrt{4 - 4\cos^2 \theta}$

34. $\sqrt{2 - 2\sin^2 \theta}$

35. $\sin(\theta + \pi/2) - \cos \theta$

36. $\sin(\theta + \pi/2) + \cos \theta$

37. Is there a real number θ such that $\sin \theta = -1.07$? Explain.

38. Is there a real number θ such that $\cos \theta = 2$? Explain.

In Exercises 39 and 40, use the calculator to find an approximation to five decimal places for the given expression.

39. (a) $\sin 4$
    (b) $\sin(\pi/12)$
    (c) $\cos(-400°)$

40. (a) $\cos 4$
    (b) $\cos(\pi/15)$
    (c) $\sin(-600°)$

41. The terminal side of $\theta = 2$ radians intersects the unit circle at point P. Find the coordinates of P accurate to four decimal places.

42. Repeat Exercise 41 if $\theta = 130°$.

43. A point P is on the terminal side of $\theta = 290°$. If the distance from the origin to P is 10, find the coordinates of P accurate to four decimal places.

44. Repeat Exercise 43 if $\theta = 5.5$ radians.

45. Use the calculator to complete the following table:

t	1	.5	.25	.1	.01	.001
$\dfrac{\sin(t)}{t}$						

Guess the value of $\lim\limits_{t \to 0^+} \dfrac{\sin(t)}{t}$.

46. Use the calculator to complete the following table:

t	.2	.1	.01	.001	.0001
$\dfrac{1 - \cos(t)}{t}$					

Guess the value of $\lim\limits_{t \to 0^+} \dfrac{1 - \cos(t)}{t}$.

47. Let $F(t) = \cos(t)$. Using the calculator (in radian mode), compute the following accurate to three decimal places: $F(1)$, $(F \circ F)(1)$, $(F \circ F \circ F)(1)$, $(F \circ F \circ F \circ F)(1)$, $(F \circ F \circ F \circ F \circ F)(1)$. What is the value if we compose F with itself ten times?

48. Prove that the arch of the sine curve between 0 and π cannot be a semicircle.

6.5 PROPERTIES OF TANGENT AND RECIPROCAL FUNCTIONS

This section deals with the tangent function and three new trigonometric functions related to the sine, cosine, and tangent. We start by presenting certain properties of tangent and establishing its graph.

Tangent and Slope

Suppose the terminal side of angle t in standard position coincides with a nonvertical line of slope m. Then

$$m = \tan(t)$$

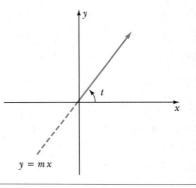

PROOF: Let $P(a, b)$ be a point on the terminal side of t (other than the origin). By the definition of $\tan(t)$,

$$\tan(t) = \frac{b}{a}$$

Now, $O(0, 0)$ and $P(a, b)$ are two points on the line coinciding with the terminal side of t. Therefore the slope of this line is given by

$$m = \frac{y_2 - y_1}{x_2 - x_1} = \frac{b - 0}{a - 0} = \frac{b}{a}$$

Thus $m = \tan(t)$. This completes the proof. ∎

Figure 6.44(a) illustrates several lines corresponding to the terminal sides of angles between 0 and $\pi/2$. We see that as angle t increases toward $\pi/2$ $\left(\text{in symbols, } t \to \dfrac{\pi^-}{2}\right)$, the slopes of the corresponding lines increase toward $+\infty$. Since $\tan(t)$ is the slope of the line corresponding to t, we conclude that

$$\lim_{t \to \frac{\pi^-}{2}} \tan(t) = +\infty$$

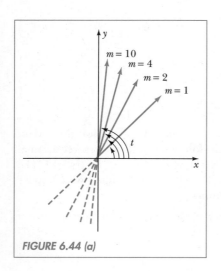

FIGURE 6.44 (a)

Figure 6.44(b) shows that as t decreases toward $-\pi/2$ $\left(\text{in symbols, } t \to \dfrac{-\pi}{2}^+\right)$, the corresponding slopes tend toward $-\infty$. Thus

$$\lim_{t \to \frac{-\pi}{2}^+} \tan(t) = -\infty$$

The limits established above imply that the graph of the tangent function will have vertical asymptotes at $t = \pi/2$ and $t = -\pi/2$. This information, together with the following short table of values, helps us obtain the graph of $\tan(t)$ in Figure 6.45:

t	$-\pi/4$	0	$\pi/4$
$\tan(t)$	-1	0	1

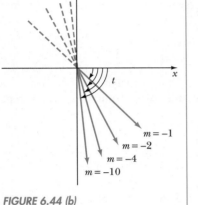

FIGURE 6.44 (b)

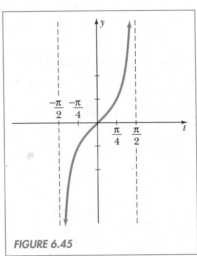

FIGURE 6.45

Properties of Tangent

Range: $(-\infty, +\infty)$

Periodicity: $\tan(t + \pi k) = \tan(t)$ for all $k \in \mathbf{Z}$

Symmetry: $\tan(-t) = -\tan(t)$

PROOF: Let $r \in \mathbf{R}$ and let t be the angle whose terminal side passes through the point $P(1, r)$. By the definition of tangent,

$$\tan(t) = \frac{r}{1} = r$$

Thus every real number is in the range of the tangent function.

To establish the periodicity property, we note that if $P(x, y)$ is on the terminal side of angle t, then $P_1(-x, -y)$ is on the terminal side of $t + \pi$.

It follows that

$$\tan(t + \pi) = \frac{-y}{-x} = \frac{y}{x} = \tan(t)$$

(assuming $t \in \mathscr{D}_{\tan}$). Now, $t + \pi k$ is coterminal with either $t + \pi$ or t for all $k \in \mathbf{Z}$. Thus $\tan(t + \pi k)$ equals $\tan(t + \pi)$ or $\tan(t)$, but these are the same as we have shown above.

Finally, to prove the symmetry property for tangent, we use the symmetry properties for sine and cosine.

$$\tan(-t) = \frac{\sin(-t)}{\cos(-t)} = \frac{-\sin(t)}{\cos(t)} = -\tan(t)$$

Notice that the periodicity property of tangent says that it repeats itself every π units. This differs from the sine and cosine functions, which repeat every 2π units. ∎

Example 1

Simplify $\tan(57\pi - \theta)$

SOLUTION $\tan(57\pi - \theta) = \tan(-\theta + 57\pi)$

$$= \tan(-\theta) \qquad \text{by periodicity}$$

$$= -\tan\theta \qquad \text{by symmetry} \ \blacksquare$$

Example 2

Graph $F(t) = \tan(t)$ for t in the interval $(-3\pi/2, \ 3\pi/2)$.

SOLUTION   By the periodicity property of tangent, we may repeat the graph in Figure 6.45 to obtain the desired graph in Figure 6.46. ∎

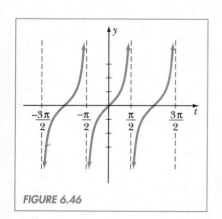

FIGURE 6.46

Next, we introduce the reciprocal functions cosecant, abbreviated csc, and secant, abbreviated sec.

Definition of Cosecant and Secant

$$\csc(t) = \frac{1}{\sin(t)} \qquad \mathscr{D}_{\csc} = \{t : t \neq 0, \ \pm\pi, \ \pm2\pi, \ \pm3\pi, \ldots\}$$

$$\sec(t) = \frac{1}{\cos(t)} \qquad \mathscr{D}_{\sec} = \left\{t : t \neq \pm\frac{\pi}{2}, \ \pm\frac{3\pi}{2}, \ \pm\frac{5\pi}{2}, \ldots\right\}$$

The domains of these functions are restricted so that division by zero does not occur. By taking the reciprocal of the sin(t) graph, we obtain the graph of csc(t) in Figure 6.47. Similarly, we take the reciprocal of the cos(t) graph to obtain the graph of sec(t) in Figure 6.48.

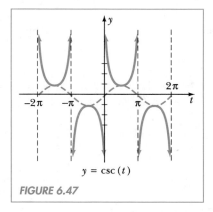

$y = \csc(t)$

FIGURE 6.47

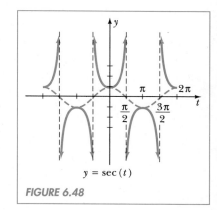

$y = \sec(t)$

FIGURE 6.48

Properties of csc(t) and sec(t)

	csc(t)	sec(t)
Range:	$(-\infty, -1] \cup [1, +\infty)$	$(-\infty, -1] \cup [1, +\infty)$
Periodicity:	$\csc(t + 2\pi k) = \csc(t)$	$\sec(t + 2\pi k) = \sec(t)$
Symmetry:	$\csc(-t) = -\csc(t)$	$\sec(-t) = \sec(t)$

PROOF: The range for these functions can be seen from the graphs. Periodicity and symmetry properties follow from the corresponding properties for sin(t) and cos(t). We omit the details. ∎

Example 3

Is there a real number t such that (a) $\sec(t) = 2$, (b) $\sec(t) = .9$?

SOLUTION (a) Yes; we know 2 is in the range of sec(t). In fact, $\cos(\pi/3) = 1/2$, so $\sec(\pi/3) = 1/\cos(\pi/3) = 1/(1/2) = 2$.
(b) No; .9 is not in the range of sec(t). ∎

When working with expressions involving the cosecant or secant functions, it is often more comfortable to rewrite these expressions in terms of sines and cosines.

Example 4

Simplify $1 - \dfrac{1}{\csc^2 \theta}$.

SOLUTION $1 - \dfrac{1}{\csc^2 \theta} = 1 - \dfrac{1}{\left(\dfrac{1}{\sin \theta}\right)^2}$ by definition

$$= 1 - \sin^2 \theta$$

$$= \cos^2 \theta \quad \text{from the Fundamental Identity} \quad \blacksquare$$

Example 5

Use the calculator to find an approximation to five decimal places for (a) $\csc(21.4°)$, (b) $\sec(7\pi/12)$.

SOLUTION (a) The calculator does not have a cosecant key. However, we note that

$$\csc(21.4°) = \dfrac{1}{\sin(21.4°)}$$

Therefore we first compute $\sin(21.4°)$ and then press the $\boxed{1/x}$ key to obtain $\csc(21.4°) \approx 2.74065$.

(b) Note that

$$\sec(7\pi/12) = \dfrac{1}{\cos(7\pi/12)}$$

Therefore we put the calculator in radian mode, compute $\cos(7\pi/12)$, and then press the $\boxed{1/x}$ key to obtain $\sec(7\pi/12) \approx -3.86370$. $\blacksquare$

We may interpret $\csc(t)$ and $\sec(t)$ in terms of the coordinates of a point $P(x, y)$ on the terminal side of t or in terms of a reference triangle determined by t. We have

$$\csc(t) = \dfrac{r}{y} = \dfrac{\text{hypotenuse}}{\text{opposite leg}} \qquad \sec(t) = \dfrac{r}{x} = \dfrac{\text{hypotenuse}}{\text{adjacent leg}}$$

where $r = \sqrt{x^2 + y^2}$. These are just the reciprocals of the formulas for $\sin(t)$ and $\cos(t)$.

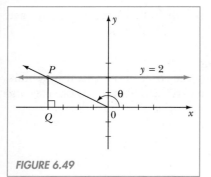

FIGURE 6.49

Example 6

A point P moves along the line $y = 2$. Let θ be an angle in standard position whose terminal side passes through P. Find the distance from the origin to P as a function of θ.

SOLUTION The situation appears in Figure 6.49. We drew a reference triangle for θ, triangle PQO. We want to find OP, the distance from the origin to P, which is the hypotenuse in the reference triangle. Note that opposite leg = $PQ = 2$ since P is on the line $y = 2$. Thus

$$\frac{OP}{2} = \frac{\text{hypotenuse}}{\text{opposite leg}} = \csc(\theta)$$

Hence $OP = 2 \csc(\theta)$. ∎

Finally, we introduce the cotangent function, abbreviated cot.

Definition of Cotangent
Let $P(x, y)$ be a point (other than the origin) on the terminal side of angle t in standard position. Then

$$\cot(t) = \frac{x}{y}$$

or, in a reference triangle for t,

$$\cot(t) = \frac{\text{adjacent leg}}{\text{opposite leg}}$$

$$\mathcal{D}_{\cot} = \{t : t \neq 0, \pm\pi, \pm 2\pi, \pm 3\pi, \ldots\}$$

The domain of the cotangent is restricted to avoid division by zero. This occurs when the point $P(x, y)$ has $y = 0$, which means that t is coterminal with 0 or π.

Example 7

Evaluate $\cot(t)$ if t is an odd multiple of $\pi/2$, $t = \pm\pi/2, \pm 3\pi/2$, etc.

SOLUTION If t is an odd multiple of $\pi/2$, then it must be coterminal with $\pi/2$ or $-\pi/2$. Therefore a point on the terminal side of t will have coordinates $(0, y)$, $y \neq 0$. It follows that

$$\cot(t) = \frac{x}{y} = \frac{0}{y} = 0 \quad ∎$$

The following relationships are valid whenever all the expressions involved are defined:

$$\cot(t) = \frac{\cos(t)}{\sin(t)} \qquad \cot(t) = \frac{1}{\tan(t)} \qquad \tan(t) = \frac{1}{\cot(t)}$$

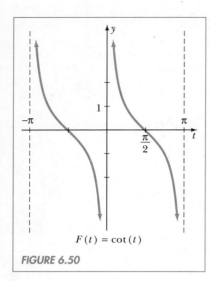

$F(t) = \cot(t)$

FIGURE 6.50

Since $\cot(t) = 1/\tan(t)$ [except when $\cot(t) = 0$], we can take the reciprocal of the tangent graph to help find the cotangent graph. Adding in the points where $\cot(t) = 0$ ($t = \pm\pi/2, \pm3\pi/2, \ldots$), we obtain the graph for $\cot(t)$ in Figure 6.50. Note that the graph repeats itself every π units.

We end this section with a simple application involving the cotangent and tangent functions.

Example 8

Find the length l of the rectangle in Figure 6.51 in terms of θ.

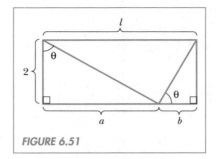

FIGURE 6.51

SOLUTION   We wish to find a and b since $l = a + b$. In the right triangle on the left, we have

$$\frac{a}{2} = \tan\theta \quad \text{implies} \quad a = 2\tan\theta$$

Note that the right side of the rectangle also has length 2. Therefore, in the right triangle on the right,

$$\frac{b}{2} = \cot\theta \quad \text{implies} \quad b = 2\cot\theta$$

Thus $l = a + b = 2\tan\theta + 2\cot\theta$. ∎

EXERCISES 6.5

In Exercises 1 to 4, graph the given function over the designated interval.

1. $F(t) = \tan(t), [2\pi, 3\pi]$ 2. $F(t) = \tan(-t), [0, 2\pi]$

3. $F(t) = \csc(t), [3\pi, 5\pi]$ 4. $F(t) = \sec(t), [3\pi, 5\pi]$

In Exercises 5 to 10, graph each function over the interval $[-2\pi, 2\pi]$.

5. $F(t) = |\tan t|$ 6. $F(t) = |\csc t|$

7. $F(t) = |\sec t|$ 8. $F(t) = \tan(t - \pi)$

9. $F(t) = \csc(t + \pi)$ 10. $F(t) = \sec(t + \pi)$

11. Complete the following table. Write *undefined* where appropriate.

t	0	$\pi/2$	π	$3\pi/2$	2π
$\csc(t)$					
$\sec(t)$					
$\cot(t)$					

In Exercises 12 and 13, complete the table.

12.

t	$\pi/6$	$\pi/4$	$\pi/3$	$\pi/2$	$2\pi/3$	$3\pi/4$	$5\pi/6$
$\csc(t)$							
$\sec(t)$							
$\cot(t)$							

13.

t	$-\pi/6$	$-\pi/4$	$-\pi/3$	$-\pi/2$	$-2\pi/3$	$-3\pi/4$	$-5\pi/6$
$\csc(t)$							
$\sec(t)$							
$\cot(t)$							

In Exercises 14 to 17, find $\csc(\theta)$, $\sec(\theta)$, $\cot(\theta)$.

14.

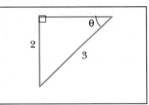

15.

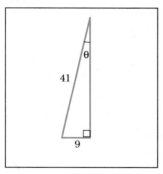

16.

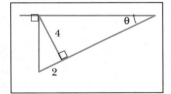

17.

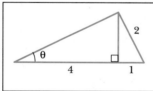

In Exercises 18 to 23, from the given information, find the value of all six trigonometric functions at t: $\sin t$, $\cos t$, $\tan t$, $\csc t$, $\sec t$, $\cot t$.

18. $\csc(t) = -4$, $\cos(t) > 0$

19. $\sec(t) = \sqrt{5}$, $\tan(t) > 0$

20. $\cot(t) = 24/7$, $\sec(t) > 0$

21. $\cot(t) = 2/3$, $\csc(t) < 0$

22. The terminal side of t is on the line $y = -5x$ and $\csc(t) > 0$.

23. The slope of the terminal side of t is -7 and $\sec(t) < 0$.

In Exercises 24 and 25, use the calculator to find an approximation to five decimal places for the given expression.

24. (a) $\cot(-100°)$
(b) $\csc(\pi/7)$
(c) $\sec(15)$

25. (a) $\cot(14.8°)$
(b) $\csc(11\pi/12)$
(c) $\sec(33)$

26. Complete the following table:

t	.1	.01	.001	.0001
$\dfrac{\tan(t)}{t}$				

Guess the value of $\displaystyle\lim_{t \to 0^+} \frac{\tan(t)}{t}$.

27. Complete the following table:

t	.1	.01	.001	.0001
$\dfrac{\sec(t)}{t}$				

Guess the value of $\displaystyle\lim_{t \to 0^+} \frac{\sec(t)}{t}$.

28. Suppose the point P moves along the line $y = 6$. Let θ be the angle in standard position whose terminal side passes through P.
(a) Find the distance OP from the origin to P as a function of θ.
(b) Find the x-coordinate of P as a function of θ.

29. Repeat Exercise 28 if P moves along the line $y = -3$.

30. The point P moves along the vertical line $x = 4$. Let θ be in standard position with terminal side OP. Find as a function of θ the (a) distance OP and (b) y-coordinate of P.

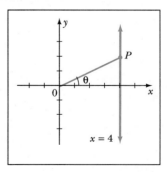

31. Repeat Exercise 30 if P is on the vertical line $x = -4$.

32. Refer to the rectangle $ABCD$ in Figure 6.52.
    (a) Find the height of the rectangle as a function of θ.
    (b) Find the sum of the lengths AE and DE as a function of θ.

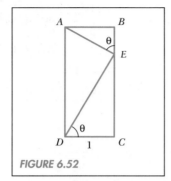

FIGURE 6.52

33. Refer to the right triangle ABC in Figure 6.53. Using the given information, find the area of the triangle as a function of θ: (a) $BC = 9$, (b) $AC = 4$, (c) $AB = 15$.

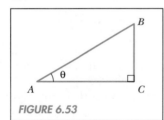

FIGURE 6.53

34. Consider the rectangle in Figure 6.54. Find the length of the diagonal in terms of (a) θ and a; (b) θ and b.

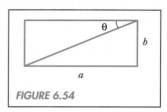

FIGURE 6.54

In Exercises 35 to 44, simplify. Assume all expressions are defined for θ.

35. $\tan(3 + \pi) - \tan(3 - \pi)$

36. $\tan(1 + \pi) + \tan(1 + 2\pi)$

37. $\tan(9\pi - \theta)$ 38. $\tan(\theta - 11\pi)$

39. $\sin\theta \csc\theta$ 40. $\cos\theta \sec\theta$

41. $\cot\theta \tan\theta$ 42. $\dfrac{\csc\theta}{\sec\theta}$

43. $\dfrac{2}{\csc^2\theta} + \dfrac{2}{\sec^2\theta}$ 44. $\tan(-\theta)\cot\theta$

45. Answer true or false.
    (a) $\csc(-t) = \csc(t)$
    (b) $\sec(-t) = \sec(t)$
    (c) $\cot(-t) = \cot(t)$

6.6   GRAPHING $a\sin(bx)$, $a\cos(bx)$, $a\tan(bx)$

By examining the graph of the sine function over a large interval, we find a curve that continually repeats itself. Figure 6.55(a) shows how the entire sine curve can be obtained by repeating the portion of the graph over the interval $[0, 2\pi]$. Figure 6.55(b) shows how the curve can also be obtained by repeating the portion over the interval $[-\pi/2, 3\pi/2]$. Indeed, any interval of length 2π may be used to produce the entire sine curve. When the graph of a function behaves in this repetitive way, we call the function **periodic**. Let us formally define this concept.

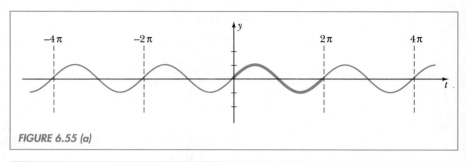

FIGURE 6.55 (a)

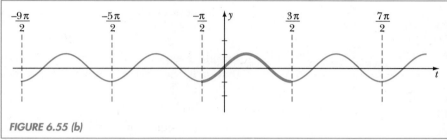

FIGURE 6.55 (b)

Definition of Periodic Function

Let F be any nonconstant function. If there is a number $p > 0$ such that

$$F(x + p) = F(x) \qquad \text{for all } x \in \mathcal{D}_F$$

then we say F is **periodic**. The smallest p for which this is true is called the **period** of F.

Figure 6.56(a) shows a periodic function F with period p and domain **R**. Thus

$$F(x + p) = F(x)$$

for all $x \in \mathbf{R}$. Figure 6.56(b) shows the same graph divided into intervals of length p. Notice that the graph of F is the same over each interval. Any portion of the graph over an interval of length p is called one **cycle** of the curve. Figure 6.56(c) shows a typical example of one cycle of F.

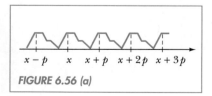

FIGURE 6.56 (a)

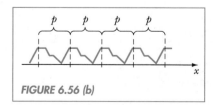

FIGURE 6.56 (b)

FIGURE 6.56 (c)

Example 1

Find the period of the sine, cosine, and tangent functions.

SOLUTION By the periodicity properties of sine and cosine given in Section 6.4, we know that

$$\sin(x + 2\pi) = \sin x \qquad \cos(x + 2\pi) = \cos x$$

for every $x \in \mathbf{R}$. Furthermore, examination of the graphs of $\sin(x)$ (Figure 6.40) and $\cos(x)$ (Figure 6.42) shows that 2π is the smallest number for which this is true. Therefore both $\sin(x)$ and $\cos(x)$ have period 2π.

The periodicity property of tangent given in Section 6.5 implies

$$\tan(x + \pi) = \tan(x)$$

for all $x \in \mathcal{D}_{\tan}$. Furthermore, the graph of $\tan(x)$ (Figure 6.45) shows that π is the smallest number for which this is true. Therefore $\tan(x)$ has period π. ∎

If a periodic function has a finite maximum value and a finite minimum value, then we assign to it a number called the **amplitude** of the function.

Definition of Amplitude

Let F be a periodic function. If $Y_{\max}$ = maximum value of $F(x)$ and $Y_{\min}$ = minimum value of $F(x)$, then

$$\text{Amplitude of } F = \frac{Y_{\max} - Y_{\min}}{2}$$

In words, the amplitude measures half the vertical spread in the graph of the function.

Example 2

Find the amplitude of $\sin(x)$ and $\cos(x)$.

SOLUTION We know that $Y_{\max} = 1$ and $Y_{\min} = -1$ for either $\sin(x)$ or $\cos(x)$. Thus

$$\text{Amplitude} = \frac{Y_{\max} - Y_{\min}}{2} = \frac{1 - (-1)}{2} = \frac{2}{2} = 1$$

for both $\sin(x)$ and $\cos(x)$. ∎

Example 3

Find the amplitude of $\tan(x)$.

SOLUTION We know that $\lim\limits_{x \to \frac{\pi}{2}^-} \tan(x) = +\infty$ and $\lim\limits_{x \to \frac{-\pi}{2}^+} \tan(x) = -\infty$.

Therefore $\tan(x)$ does not have a maximum value or a minimum value. We must say that the amplitude for the tangent function is *undefined*. ∎

Example 4 demonstrates that the value of the amplitude is not always the same as the maximum value of the function.

Example 4

Find the amplitude of $F(x) = |\sin(x)|$.

SOLUTION The graph of F is shown in Figure 6.57. We have $Y_{\max} = 1$ and $Y_{\min} = 0$. Thus

$$\text{Amplitude} = \frac{Y_{\max} - Y_{\min}}{2} = \frac{1 - 0}{2} = \frac{1}{2} \quad ∎$$

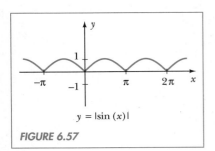

$y = |\sin(x)|$

FIGURE 6.57

Recall that any function F can be multiplied by a constant a, forming a new function aF. The graph of aF is obtained by stretching or compressing the graph of F in the vertical direction and possibly reflecting across the x-axis if a is negative. In any case, we expect that aF will have a different amplitude than F, but the period should remain the same.

Example 5

Graph and find the period and amplitude of (a) $y = 3 \sin x$; (b) $y = -\frac{1}{2} \cos x$.

SOLUTION To find the graph of $y = 3 \sin x$, multiply the y-coordinates on the graph of $y = \sin x$ by 3; that is, we stretch the graph of $y = \sin x$ vertically by the factor 3. Thus we obtain the graph in Figure 6.58. As we can see, the period is still 2π, but the amplitude is now 3.

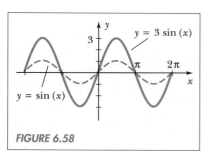

FIGURE 6.58

(b) To obtain the graph of $y = -\frac{1}{2} \cos x$, first multiply the y-coordinates on the graph of $y = \cos x$ by 1/2 (that is, compressing $y = \cos x$ by the factor 1/2) and then reflect this across the x-axis. The result is shown in Figure 6.59. From the graph, we find that the period is still 2π, but the amplitude is now 1/2. ∎

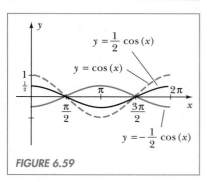

FIGURE 6.59

In general, multiplying the sine or cosine function on the *outside* by a nonzero constant a changes its amplitude but not its period. In summary,

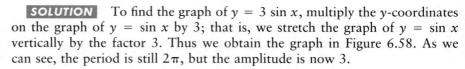

Function	Period	Amplitude
$a \sin x$	2π	$|a|$
$a \cos x$	2π	$|a|$

Now we would like to see what happens to sin x and cos x when we introduce a constant factor, say, b, on the *inside* of the function. By this, we mean to consider sin(bx) and cos(bx).

Example 6

Graph $y = \sin(2x)$. Determine the period and amplitude.

SOLUTION   We note that $\sin(2x)$ is the composition of two functions: multiply by 2 and then apply sine.

$$x \rightarrow 2x \rightarrow \sin(2x)$$

Therefore we shall construct a table of x, y values with three columns: x, $2x$, and $y = \sin(2x)$.

x	$2x$	$y = \sin(2x)$
0	0	0
$\pi/6$	$\pi/3$	$\sqrt{3}/2$
$\pi/4$	$\pi/2$	1
$\pi/3$	$2\pi/3$	$\sqrt{3}/2$
$\pi/2$	π	0
$2\pi/3$	$4\pi/3$	$-\sqrt{3}/2$
$3\pi/4$	$3\pi/2$	-1
$5\pi/6$	$5\pi/3$	$-\sqrt{3}/2$
π	2π	0

To graph the equation $y = \sin(2x)$, we use the ordered pairs $(x, \sin(2x))$ from the first and last columns of our table. This leads us to the curve in Figure 6.60.

We observe from the graph that $\sin(2x)$ has period π. Let us establish this in more detail. We know that $\sin(\square)$ will complete one cycle when $\square$ varies from 0 to 2π. Thus we can reason as follows:

$\sin(\square)$ completes one cycle when $0 \le \square \le 2\pi$

$\sin(2x)$ completes one cycle when $0 \le 2x \le 2\pi$

if and only if $0 \le x \le \pi$

We obtain the last inequality statement by dividing the previous one by 2. Therefore the period of $\sin(2x)$ is π.

Finally, the amplitude of $\sin(2x)$ is 1 since $Y_{\max} = 1$ and $Y_{\min} = -1$. (Note that $-1 \le \sin\square \le 1$ for any value of $\square$.) ■

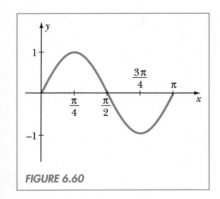

FIGURE 6.60

In Figure 6.61 we compare the graph of $y = \sin(2x)$ with the ordinary sine curve, $y = \sin x$. Notice that the graph of $y = \sin(2x)$ has the same shape as $y = \sin x$, but compressed horizontally. Since the period of $\sin(2x)$ is one-half the period of $\sin(x)$, we can say that the factor of 2 on the inside of the sine function changes its period by the factor 1/2.

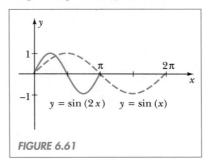

FIGURE 6.61

Let us generalize these observations by considering the following results:

Function	Completes One Cycle When	Equivalent Statement	Period
$\sin(3x)$	$0 \le 3x \le 2\pi$	$0 \le x \le \dfrac{2\pi}{3}$	$\dfrac{2\pi}{3}$
$\sin(4x)$	$0 \le 4x \le 2\pi$	$0 \le x \le \dfrac{2\pi}{4}$	$\dfrac{2\pi}{4} = \dfrac{\pi}{2}$
$\sin(bx), b > 0$	$0 \le bx \le 2\pi$	$0 \le x \le \dfrac{2\pi}{b}$	$\dfrac{2\pi}{b}$

Similar results hold for the cosine function. This leads us to the following conclusion.

Assume a and b are nonzero constants, $b > 0$.

Function	Period	Amplitude
$a \sin(bx)$	$\dfrac{2\pi}{b}$	$|a|$
$a \cos(bx)$	$\dfrac{2\pi}{b}$	$|a|$

Let us consider the basic function $F(x) = a \sin(bx)$, with a and b positive. We know the graph of F will complete one cycle over the interval $[0, 2\pi/b]$.

6.6 / Graphing a sin(bx), a cos(bx), a tan(bx)   ▪   **341**

Thus the graph of F should look like the curve in Figure 6.62. This is similar to the straight sine curve, $y = \sin x$, except that it may be stretched or compressed vertically or horizontally, depending on the values of a and b. Notice that by dividing the interval $[0, 2\pi/b]$ into fourths we obtain the most important points for the purpose of defining the curve: where $F(x)$ is maximum, minimum, and zero.

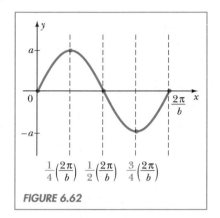

FIGURE 6.62

Similar remarks hold for the function $F(x) = a \cos(bx)$. This brings us to the following graphing procedure.

Graphing $y = a \sin(bx)$ or $y = a \cos(bx)$, $b > 0$

Step 1   Determine the period $p = 2\pi/b$.

Step 2   Along the x-axis divide the interval $[0, p]$ into fourths. Label $x = 0, p/4, p/2, 3p/4, p$. Along the y-axis label $y = -a$ and a.

Step 3   Graph the points corresponding to $x = 0, p/4, p/2, 3p/4$, and p.

x	0	$\dfrac{p}{4}$	$\dfrac{p}{2}$	$\dfrac{3p}{4}$	p
$a \, \sin(bx)$	0	a	0	$-a$	0
$a \, \cos(bx)$	a	0	$-a$	0	a

Step 4   Draw a smooth curve through the points found in step 3.

Step 5   Repeat the portion of the curve obtained in step 4 as needed.

In case $b < 0$, we simply remove the negative sign from inside the function by using the symmetry properties:

$$\sin(-\theta) = -\sin\theta \qquad \cos(-\theta) = \cos\theta$$

This is done before step 1 in the graphing procedure.

Example 7

Graph two cycles of the function $F(x) = 2\cos(-x/3)$, one to the right and one to the left of the origin. Label zeros. List maximum and minimum points.

SOLUTION First, we remove the negative sign from the inside of the function. By symmetry, $2\cos(-x/3) = 2\cos(x/3)$. Therefore we wish to graph $F(x) = 2\cos(x/3)$. We follow our graphing procedure outlined above.

Step 1. The period is $\dfrac{2\pi}{(1/3)} = 6\pi$.

Steps 2, 3. We divide the interval $[0, 6\pi]$ into fourths, label the axes, and plot the following points in Figure 6.63(a).

x	0	$\dfrac{3\pi}{2}$	3π	$\dfrac{9\pi}{2}$	6π
$F(x) = 2\cos\left(\dfrac{x}{3}\right)$	2	0	-2	0	2

Steps 4, 5. We draw a smooth curve through the points established in step 3 and then repeat one cycle of the curve to the left of the origin. The results are shown in Figure 6.63(b). Maximum points are $(\pm\, 6\pi,\, 2)$ and $(0, 2)$; minimum points are $(\pm\, 3\pi,\, -2)$. ∎

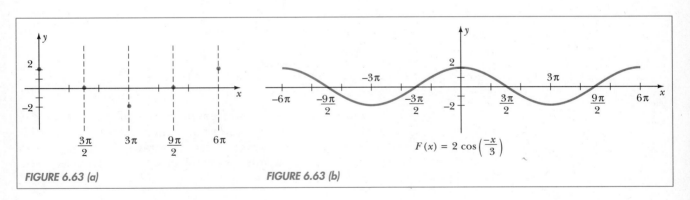

FIGURE 6.63 (a)

$$F(x) = 2\cos\left(\frac{-x}{3}\right)$$

FIGURE 6.63 (b)

Example 8

Find a function of the form $F(x) = a\sin(bx)$, $b > 0$, that has the given properties (a) amplitude 4, period 3π; (b) $F(5) = -3$, $F(10) = 0$.

SOLUTION (a) We want $F(x) = a\sin(bx)$ to have amplitude 4 and period 3π. Therefore we choose $a = 4$, and b so that

$$\frac{2\pi}{b} = 3\pi$$

This implies $b = 2/3$. Thus $F(x) = 4 \sin\left(\dfrac{2}{3}x\right)$ is the desired function. Note that the solution is not unique since we could have chosen a to be negative.

(b) We want $F(x) = a \sin(bx)$ so that $F(5) = -3$ and $F(10) = 0$. This means that the graph of F should pass through the points $(5, -3)$ and $(10, 0)$. Consider Figure 6.64, which shows a graph for one possible choice of F. Since the amplitude of this curve is 3, we shall need $|a| = 3$. Furthermore, the period is 20. Hence

$$\frac{2\pi}{b} = 20 \quad \text{implies} \quad b = \frac{\pi}{10}$$

Now, if we take $a = -3$ and $b = \pi/10$, we have

$$F(x) = -3 \sin\left(\frac{\pi x}{10}\right)$$

You can verify that $F(5) = -3$ and $F(10) = 0$ as desired. We note once again that this answer is not unique. For example, we could have taken $a = -3$ and $b = \pi/2$. In fact, there are infinitely many choices for $F(x)$.

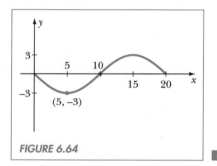

FIGURE 6.64

Now we turn our attention to the problem of graphing $F(x) = a \tan(bx)$, $b > 0$. First, recall the graph of $y = \tan(x)$ in Section 6.5 (see Figure 6.45). We see that one basic cycle of $\tan(x)$ can be drawn over the interval $(-\pi/2, \pi/2)$, with vertical asymptotes occurring at $x = -\pi/2$ and $x = \pi/2$. Thus

$$\tan(\square) \quad \text{completes one cycle when} \quad \frac{-\pi}{2} < \square < \frac{\pi}{2}$$

It follows that

$$a \tan(bx) \quad \text{completes one cycle when} \quad \frac{-\pi}{2} < bx < \frac{\pi}{2}$$

$$\text{if and only if} \quad \frac{-\pi}{2b} < x < \frac{\pi}{2b}$$

We conclude that $a \tan(bx)$ will have period $\pi/2b - (-\pi/2b) = \pi/b$. To graph the function, we may use the following procedure.

Graphing $y = a \tan(bx)$, $b > 0$

Step 1 Determine the period $p = \pi/b$.

Step 2 Along the x-axis draw vertical asymptotes at $x = -p/2$ and $x = p/2$. Label $x = -p/4$ and $p/4$. Along the y-axis label $y = -a$ and a.

Step 3 Graph the points corresponding to $x = -p/4, 0, p/4$.

x	$-\dfrac{p}{4}$	0	$\dfrac{p}{4}$
$a \tan(bx)$	$-a$	0	a

Step 4 Draw a smooth curve through the points found in step 3 so that the y-coordinates tend toward $+\infty$ or $-\infty$ as we approach the vertical asymptotes at $x = p/2$ and $x = -p/2$.

Step 5 Repeat the portion of the curve obtained in step 4 as needed.

In case $a \tan(bx)$ has $b < 0$, we remove the negative sign from inside the function by using the symmetry property.

$$\tan(-\theta) = -\tan \theta$$

We do this before beginning step 1 in the graphing procedure.

Example 9

Graph two cycles of the function (a) $F(x) = \dfrac{1}{2} \tan\left(\dfrac{x}{2}\right)$, (b) $G(x) = \dfrac{1}{2} \tan\left(-\dfrac{x}{2}\right)$.

SOLUTION   (a) We follow our procedure outlined above.

Step 1. The period is $\dfrac{\pi}{(1/2)} = 2\pi$.

Steps 2, 3. We draw vertical asymptotes at $x = -\pi$ and $x = \pi$. Next, we label the axes and plot the following points. See Figure 6.65(a).

x	$-\dfrac{\pi}{2}$	0	$\dfrac{\pi}{2}$
$F(x) = \dfrac{1}{2} \tan\left(\dfrac{x}{2}\right)$	$-\dfrac{1}{2}$	0	$\dfrac{1}{2}$

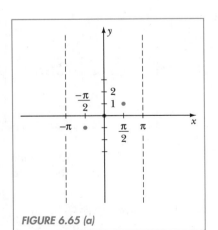

FIGURE 6.65 (a)

Steps 4, 5.   We draw a smooth curve through the points established in step 3 and then repeat one cycle to the right. See Figure 6.65(b).

(b)   We have $G(x) = \dfrac{1}{2}\tan\left(-\dfrac{x}{2}\right) = -\dfrac{1}{2}\tan\left(\dfrac{x}{2}\right)$ by symmetry. Therefore $G(x) = -F(x)$, so we may find the graph of G by reflecting the graph of F [given in Figure 6.65(b)] across the x-axis. The results are shown in Figure 6.66.

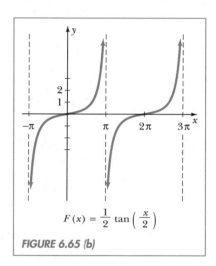

$$F(x) = \frac{1}{2}\tan\left(\frac{x}{2}\right)$$

FIGURE 6.65 (b)

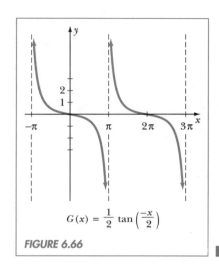

$$G(x) = \frac{1}{2}\tan\left(\frac{-x}{2}\right)$$

FIGURE 6.66

EXERCISES 6.6

In Exercises 1 to 4, determine if the graph represents a periodic function. If it does, draw one cycle of the curve. Assume each graph continues in both directions with the obvious pattern.

1.

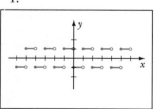

2.

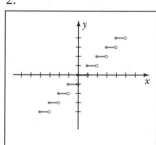

3.

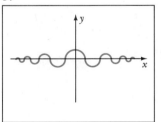

4.

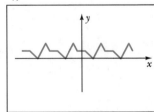

In Exercises 5 to 10, graph the given function. Determine whether the function is periodic. If it is, state the period and amplitude.

5. $F(x) = \cos(x) - 2$ 6. $F(x) = |\sec(x)|$

7. $F(x) = |\tan(x)|$ 8. $F(x) = \sin(x - \pi)$

9. $F(x) = |\cos x|$ 　　　 10. $F(x) = \sin(x) + 1$

In Exercises 11 and 12, using the given graph of F, find the graph of (a) $2F$, (b) $-\dfrac{1}{2}F$. State the period and amplitude in each case.

11. 　　　　　　　　　　　12.

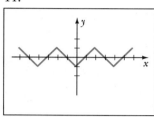 　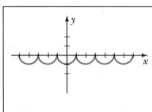

In Exercises 13 to 18, graph the function over the interval $[-2\pi, 2\pi]$. State the period and amplitude.

13. $F(x) = 3\cos(x)$ 　　　 14. $F(x) = -4\sin(x)$

15. $F(x) = \dfrac{1}{2}\tan(x)$ 　　 16. $F(x) = \dfrac{\cos(x)}{4}$

17. $F(x) = -\pi\sin(x)$ 　　 18. $F(x) = 2\tan(-x)$

In Exercises 19 to 32, follow the appropriate procedure outlined in this section to graph two cycles of the given function, one to the left and one to the right of the origin. Label zeros. List maximum and minimum points.

19. $F(x) = \sin(3x)$

20. $F(x) = \cos(4x)$

21. $F(x) = 3\cos\left(\dfrac{x}{2}\right)$

22. $F(x) = 4\sin\left(\dfrac{x}{4}\right)$

23. $F(x) = \tan(\pi x)$

24. $F(x) = \tan\left(-\dfrac{x}{2}\right)$

25. $F(x) = \dfrac{1}{2}\sin(-2x)$

26. $F(x) = 2\sin(\pi x)$

27. $F(x) = -\dfrac{1}{2}\cos\left(\dfrac{\pi}{2}x\right)$

28. $F(x) = \dfrac{1}{2}\cos\left(-\dfrac{3}{2}x\right)$

29. $F(x) = \pi\sin\left(-\dfrac{x}{3}\right)$

30. $F(x) = \pi\cos(2\pi x)$

31. $F(x) = 2\tan\left(\dfrac{x}{4}\right)$

32. $F(x) = \dfrac{1}{2}\tan(2x)$

In Exercises 33 to 36, find a function F of the form $F(x) = a\sin(bx)$ that has the given properties.

33. (a) Amplitude 5, period π
　　(b) Amplitude 1, graph passes through $(3\pi/4, 0)$
　　(c) Y_{max} occurs at $(\pi/5, 3)$

34. (a) Amplitude 1/6, period $\pi/6$
　　(b) Amplitude 8, $F(2) = 0$
　　(c) Y_{min} occurs at $(-2, -\pi)$

35. (a) $F(1) = 0$, amplitude 10
　　(b) Y_{min} occurs at $(\pi/3, -4)$
　　(c) $F(\pi) = -2$, $F(2\pi) = -2$

36. (a) $F(26) = 0$, $F(13) = -7$
　　(b) Y_{max} occurs at $(1/2, 1/2)$
　　(c) $F(12) = 4$, $F(18) = 0$

In Exercises 37 to 40, find a function G of the form $G(x) = a\cos(bx)$ that has the given properties.

37. (a) Amplitude 4, period 4π
　　(b) Amplitude 2, graph passes through $(\pi/4, 0)$
　　(c) Y_{max} occurs at $(5\pi, 3)$

38. (a) Amplitude 1/3, period $\pi/3$
　　(b) Amplitude 10, $G(2) = 0$
　　(c) Y_{min} occurs at $(-2, -\pi)$

39. (a) $G(\pi/2) = 5$, amplitude 5
　　(b) Y_{max} occurs at $(4, 1/3)$
　　(c) $G(12) = 0$, amplitude 4

40. (a) $G(21) = 0$, $G(14) = -7$
　　(b) Y_{min} occurs at $(-1, -2)$
　　(c) $G(-\pi/6) = 0$, $G(0) = -8$

6.7 FURTHER GRAPHING

In this section we use the graphing techniques of taking the reciprocal, shifting, and adding to obtain the graphs of a variety of trigonometric functions. We begin with the general cosecant and secant functions.

Suppose we wish to graph $y = a \csc(bx)$. We note that

$$a \csc(bx) = a \left(\frac{1}{\sin(bx)} \right)$$

Therefore we graph $y = \sin(bx)$, take its reciprocal, and then multiply the result by a to obtain the graph of $y = a \csc(bx)$. Similar remarks hold for $y = a \sec(bx)$. We now outline the procedure.

Graphing $y = a \csc(bx)$, $b > 0$

Step 1 Graph $y = \sin(bx)$.

Step 2 Draw the reciprocal of the graph in step 1, $y = \dfrac{1}{\sin(bx)}$.

Step 3 Multiply the y-coordinates on the graph in step 2 by a to obtain $y = a\left(\dfrac{1}{\sin(bx)}\right) = a \csc(bx)$.

Graphing $y = a \sec(bx)$, $b > 0$

Follow the above procedure, except use $\cos(bx)$ in place of $\sin(bx)$.

Remark

If $b < 0$, we remove the negative sign from inside the function by using the properties $\csc(-\theta) = -\csc\theta$ and $\sec(-\theta) = \sec\theta$.

Example 1

Graph two cycles of $y = \dfrac{1}{2} \csc(\pi x)$.

SOLUTION. Since $y = \dfrac{1}{2} \csc(\pi x) = \dfrac{1}{2}\left(\dfrac{1}{\sin(\pi x)}\right)$, our procedure for graphing this function is as follows.

Step 1. Graph two cycles of $y = \sin(\pi x)$. See Figure 6.67(a).

Step 2. Take the reciprocal of the graph in step 1 to obtain $y = 1/\sin(\pi x)$. See Figure 6.67(b).

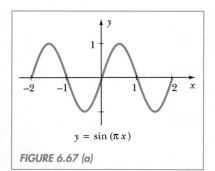

$y = \sin(\pi x)$

FIGURE 6.67 (a)

Step 3. Multiply the y-coordinates on the graph in step 2 by 1/2 to obtain $y = \dfrac{1}{2}\left(\dfrac{1}{\sin(\pi x)}\right) = \dfrac{1}{2}\csc(\pi x)$. Thus we have the desired graph in Figure 6.67(c).

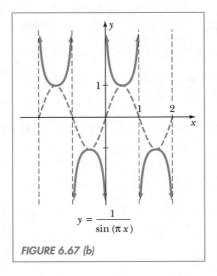

$$y = \frac{1}{\sin(\pi x)}$$

FIGURE 6.67 (b)

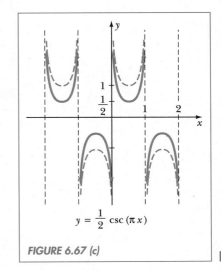

$$y = \frac{1}{2}\csc(\pi x)$$

FIGURE 6.67 (c)

Now we consider functions of the form $aF(bx + c) + d$, where a, b, c, and d are constants and F is one of the trigonometric functions. The equation for this function,

$$y = aF(bx + c) + d$$

can be rewritten in the form

$$y - d = aF\left(b\left(x + \frac{c}{b}\right)\right)$$

This is just the function $y = aF(bx)$ shifted $-c/b$ units in the x-direction and d units in the y-direction. For example, if we wish to graph

$$y = \frac{1}{2}\sin\left(2x - \frac{\pi}{2}\right) + 1$$

we rewrite this equation in the form

$$y - 1 = \frac{1}{2}\sin\left(2\left(x - \frac{\pi}{4}\right)\right)$$

It follows that the graph of this equation is just the graph of $y = \dfrac{1}{2}\sin(2x)$ (which we know how to draw) shifted $\pi/4$ units in the x-direction and 1 unit in the y-direction. We now outline the general procedure for graphing this type of function.

> **Graphing y = aF(bx + c) + d, b > 0**
>
> Step 1   Rewrite the equation in the form $y - d = aF\left(b\left(x + \dfrac{c}{b}\right)\right)$.
>
> Step 2   Graph one cycle of the unshifted function $y = aF(bx)$.
>
> Step 3   Shift the graph found in step 2 by $-c/b$ units in the x-direction and d units in the y-direction.
>
> Step 4   Repeat the portion of the curve obtained in step 3 as needed.

Example 2

Graph one cycle of $G(x) = \dfrac{1}{2}\sin\left(2x - \dfrac{\pi}{2}\right) + 1$. List maximum and minimum points. State the amplitude and period.

SOLUTION   The equation for G is $y = \dfrac{1}{2}\sin(2x - \pi/2) + 1$. To find the graph of this equation, we follow the steps outlined above.

Step 1.   We rewrite the equation in the form

$$y - 1 = \frac{1}{2}\sin\left(2\left(x - \frac{\pi}{4}\right)\right)$$

Step 2.   The unshifted function is $y = \dfrac{1}{2}\sin(2x)$. This has period π and amplitude 1/2. We draw one cycle of the curve in Figure 6.68(a).

Step 3.   From Step 1 we see that shifting the graph of $y = \dfrac{1}{2}\sin(2x)$ by $+\pi/4$ units in the x-direction and $+1$ unit in the y-direction gives us the graph of G(x). This is shown in Figure 6.68(b). A maximum point is $(\pi/2, 3/2)$; a minimum point is $(\pi, 1/2)$.

The shifted function has the same amplitude and period as the unshifted function: period π and amplitude 1/2.

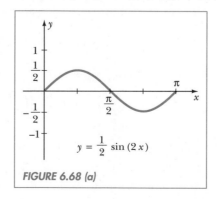

FIGURE 6.68 (a)

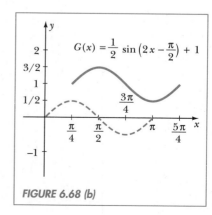

FIGURE 6.68 (b)

In Section 2.5 we discussed the method of adding graphs to construct the graph of a function that is the sum of two other functions. We review the procedure here with the trigonometric functions.

Suppose H is a function that can be written as the sum of two functions F and G.

$$H(x) = F(x) + G(x)$$

If (x, y_1) is on the graph of F and (x, y_2) is on the graph of G, then $(x, y_1 + y_2)$ is on the graph of H. This means that the graph of H can be obtained by graphing F and G on the same axes and then adding the y-coordinates over each x.

Example 3

Find the graph of $H(x) = \sin(x) + \cos(x)$ by adding graphs.

SOLUTION   Graph $y = \sin(x)$ and $y = \cos(x)$ on the same axes. Next, add the y-coordinates at various positions along the x-axis and plot these points. We do this directly on the graph using a straightedge placed vertically to help measure the y-coordinates. The results are shown in Figure 6.69(a).

After plotting enough points to determine the graph of H, connect them with a smooth curve; Figure 6.69(b) shows the results.

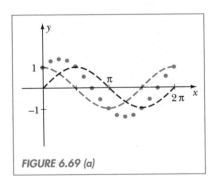

FIGURE 6.69 (a)

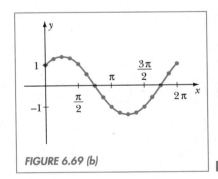

FIGURE 6.69 (b)

As with any graph drawn by hand, the results are only approximate. Even so, if we are careful, the final graph can give us useful information about the behavior of the function. For example, Figure 6.69(b) shows that the only zeros of H in the interval $[0, 2\pi]$ occur between $\pi/2$ and π and also between $3\pi/2$ and 2π, where sine and cosine have exact opposite values. We know this happens when $x = 3\pi/4$ and $7\pi/4$,

$$\sin\left(\frac{3\pi}{4}\right) = \frac{1}{\sqrt{2}} = -\cos\left(\frac{3\pi}{4}\right) \qquad \sin\left(\frac{7\pi}{4}\right) = -\frac{1}{\sqrt{2}} = -\cos\left(\frac{7\pi}{4}\right)$$

Thus the zeros of $H(x)$ in the interval $[0, 2\pi]$ are $x = 3\pi/4$ and $7\pi/4$. Another interesting aspect that the graph makes apparent is that H is periodic with period 2π.

Example 4

Let $F(x) = \cos(x)$ and $G(x) = \cos(2x)$. Graph one cycle of $H(x) = F(x) + G(x)$. What is the period of H?

SOLUTION Figure 6.70(a) is the graph of F and G over the interval $[0, 2\pi]$. For several values of x we took the y-coordinate on the graph of G and added it to the corresponding y-coordinate on the graph of F, to obtain the points shown in the figure for the graph of $F + G$. Notice that the graphs of F and G are symmetric with respect to the vertical line $x = \pi$. It follows that $F + G$ will display the same symmetry. In Figure 6.70(b) we show the graph of $F + G$ obtained by connecting our points with a smooth curve.

It is clear that the situation in Figure 6.70(a) repeats itself over the intervals $[2\pi, 4\pi]$, $[-2\pi, 0]$, and so on. Therefore we conclude that the period of H is 2π.

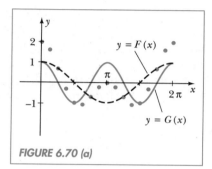

FIGURE 6.70 (a)

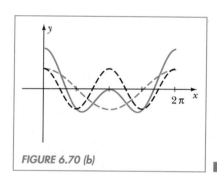

FIGURE 6.70 (b)

EXERCISES 6.7

In Exercises 1 to 4, graph two cycles of $F(x)$ by following the procedure given in this section for graphing $y = a\,\csc(bx)$ or $y = a\,\sec(bx)$.

1. $F(x) = 2\,\csc\!\left(\dfrac{1}{2}x\right)$

2. $F(x) = \dfrac{1}{2}\,\csc(2x)$

3. $F(x) = \dfrac{1}{4}\,\sec(\pi x)$

4. $F(x) = -\sec\!\left(\dfrac{x}{3}\right)$

In Exercises 5 to 14, graph one cycle of $G(x)$ using the procedure given in this section for graphing $y = F(ax + c) + d$. List maximum and minimum points. State the amplitude and period.

5. $G(x) = 2\,\sin\!\left(x - \dfrac{\pi}{4}\right)$

6. $G(x) = -2\,\cos\!\left(x + \dfrac{\pi}{4}\right)$

7. $G(x) = \cos\!\left(\dfrac{x}{2} - \dfrac{\pi}{4}\right)$

8. $G(x) = \sin(2x + \pi)$

9. $G(x) = \dfrac{1}{2}\,\sin\!\left(-x + \dfrac{\pi}{2}\right) + 1$

10. $G(x) = -\dfrac{1}{2}\,\cos\!\left(3x + \dfrac{\pi}{2}\right) - 1$

11. $G(x) = 2\,\cos\!\left(\dfrac{\pi}{4}x\right) - 2$

12. $G(x) = 2\,\sin\!\left(\dfrac{\pi}{4}x\right) + 2$

13. $G(x) = \tan(2x + \pi)$

14. $G(x) = \tan\!\left(-\dfrac{x}{2}\right) + 1$

In Exercises 15 and 16, find a function F of the form $F(x) = a \sin(bx) + d$ with the given properties.

15. The graph has a maximum point at $(\pi/8, 3)$ and a minimum point at $(3\pi/8, -1)$.

16. $F(0) = 2$, $F(2) = 2$, amplitude 3.

In Exercises 17 and 18, find a function G of the form $G(x) = a \cos(bx) + d$ with the given properties.

17. $G(0) = -5$, $G(6) = -5$, amplitude 4.

18. The graph has a maximum point at $(0, 3)$ and a minimum point at $(\pi/4, -1)$.

In Exercises 19 to 24, graph one cycle of the function by adding graphs. Use your graph to determine the period of H.

19. $H(x) = \sin(x) + \sin(2x)$

20. $H(x) = 2 \sin(x) + \sin(2x)$

21. $H(x) = \cos(2x) - \cos(x)$

22. $H(x) = \sin(2x) + \cos(x)$

23. $H(x) = \cos(2x) + \sin(x)$

24. $H(x) = \cos\left(\frac{\pi}{2}x\right) + \sin\left(\frac{\pi}{4}x\right)$

In Exercises 25 to 32, graph the function by adding graphs.

25. $H(x) = \frac{1}{2} \cos\left(\frac{x}{2}\right) + \cos(x)$

26. $H(x) = \cos(x) - \frac{1}{2} \cos\left(\frac{x}{2}\right)$

27. $H(x) = \sin\left(x + \frac{\pi}{2}\right) + \cos(x)$

28. $H(x) = \cos\left(x - \frac{\pi}{2}\right) + \sin(x)$

29. $H(x) = \sin(x + \pi) + \sin(x)$

30. $H(x) = \cos(x + \pi) + \sin(x)$

31. $H(x) = x + \sin(\pi x)$

32. $H(x) = x + \cos(x)$

33. Graph $y = \sin(x) + \csc(x)$.

34. Graph $y = \sin(x + \pi/4) + \sin(x - \pi/4)$.

35. Suppose P moves along the circle $(x - 6)^2 + (y - 4)^2 = 9$. Let A and C be the points $A(9, 4)$ and $C(6, 4)$. Let θ be the measure of angle PCA with initial side CA.
   (a) Find the y-coordinate of P as a function of θ.
   (b) Find the x-coordinate of P as a function of θ.

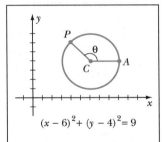

$(x - 6)^2 + (y - 4)^2 = 9$

CHAPTER 6 / REVIEW

1. Draw a reasonably accurate picture of the angle with measure t in standard position: (a) $t = 4$; (b) $t = -.5$; (c) $t = \pi - .5$.

2. Consider the circle of radius 6 with central angle t shown in Figure 6.71.
   (a) Find the length of the arc AB if $t = 7\pi/12$.
   (b) Find a positive measure for t if arc AB has length 10.

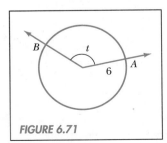

FIGURE 6.71

3. Find all angles θ coterminal with the given angle. State the value of θ satisfying $0 \le \theta < 2\pi$ or $0 \le \theta < 360°$.
   (a) $17\pi/3$ (b) -9 (c) $3000°$

4. Find the measures of all angles x such that $2x$ is coterminal with the given angle and $0 \le x \le 2\pi$.
   (a) $\pi/3$ (b) $-45°$

5. Suppose $P(15, -8)$ is on the terminal side of angle t. Find $\sin(\theta)$, $\cos(\theta)$, $\tan(\theta)$, $\csc(\theta)$, $\sec(\theta)$, and $\cot(\theta)$ if (a) $\theta = t$, (b) $\theta = t + \pi$, (c) $\theta = -t$.

6. Find the coordinates of the point P on the terminal side of angle t if $\sin(t) = .7$, $OP = 20$, and $\cos(t) < 0$.

7. Evaluate all six trigonometric functions at $t = 0$, $\pi/2$, π, $3\pi/2$, and 2π.

8. Draw and label a reference triangle for each angle t. State the equivalent degree measure for t. Find the exact values of $\sin(t)$, $\cos(t)$, and $\tan(t)$.
   (a) $t = \pi/6$ (b) $t = \pi/4$ (c) $t = \pi/3$

9. Find the exact value of each expression or write "undefined." Do not use a calculator.
   (a) $\cos(900°)$ (b) $\sin(-450°)$ (c) $\tan(120°)$
   (d) $\csc(-5\pi/4)$ (e) $\sec(5\pi/3)$ (f) $\cot(7\pi/6)$
   (g) $\tan(\pi/2)$ (h) $\cos(0)$ (i) $\sin(\pi)$

In Exercises 10 to 13, refer to the right triangle ABC in Figure 6.72.

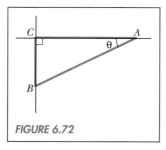

FIGURE 6.72

10. Find AC and BC in terms of θ if $AB = 6$.

11. Find BC and AB in terms of θ if $AC = 5$.

12. Find AC and AB in terms of θ if $BC = 4$.

13. Find the area of triangle ABC in terms of θ if $AB = 10$.

14. Find the exact coordinates of the point where the terminal side of $4\pi/3$ intersects the unit circle.

15. Suppose point P is on the terminal side of the given angle θ. If the distance from the origin to P is 8, use the calculator to find an approximation to four decimal places for the coordinates of P.
    (a) $\theta = 160°$ (b) $\theta = -\pi/5$

16. The terminal side of angle t is parallel to the line $y = 6x - 7$. What is the value of $\tan(t)$?

In Exercises 17 and 18, graph each function over the interval $[-2\pi, 2\pi]$. State the period and amplitude.

17. $F(x) = \sin(-x) - 1$ 18. $F(x) = |\cos x|$

In Exercises 19 and 20, graph two cycles of the function.

19. $F(x) = -\tan(\pi x)$ 20. $F(x) = \csc\left(\dfrac{\pi x}{2}\right)$

21. Is there a real number t such that (a) $\cos(t) = -\sqrt{3}$, (b) $\tan(t) = -\sqrt{3}$, (c) $\sec t = 0$?

In Exercises 22 to 27, simplify the given expression. Assume the expression is defined for θ.

22. $1 - \sin^2(\theta + 4\pi)$ 23. $\tan\left(\dfrac{\pi}{4} + 13\pi\right)$

24. $\cos\theta + \cos(-\theta)$ 25. $\csc(-\theta)\sin\theta$

26. $\sin\theta\cot\theta$ 27. $\dfrac{\sec\theta}{\csc\theta}$

28. Evaluate $\sin^2(10°) + \cos^2(370°)$ without using a calculator.

29. Refer to Figure 6.73, which shows the circle $x^2 + y^2 = 81$ with line l tangent to the circle at point P. Write as a function of θ:
    (a) x-coordinate of P (b) y-coordinate of P
    (c) slope of OP (d) slope of l
    (e) x-intercept of l (f) y-intercept of l

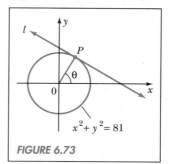

FIGURE 6.73

30. Suppose the terminal side of angle θ intersects the vertical line $x = -10$ at the point P. Find in terms of θ the (a) y-coordinate of P, (b) distance from the origin to P.

31. Suppose P moves along the horizontal line $y = 7$ and the terminal side of angle θ passes through P. Find as a function of θ the (a) x-coordinate of P, (b) distance from the origin to P.

In Exercises 32 to 35, graph one cycle of $F(x)$. List maximum and minimum points. State the amplitude and period.

32. $F(x) = 2 \sin\left(-\dfrac{x}{4}\right)$

33. $F(x) = \dfrac{1}{2} \cos\left(2x + \dfrac{\pi}{2}\right) + 1$

34. $F(x) = \tan\left(x - \dfrac{\pi}{2}\right)$

35. $F(x) = 3 \sec\left(\dfrac{x}{2}\right)$

36. Graph one cycle of the function $H(x) = \cos(\pi x) + 2 \sin(\pi x/2)$ by adding graphs. State the period of H.

37. Find a function of the form $F(x) = a \sin(bx)$ such that $F(3) = 4$ and $F(9) = -4$.

38. Find a function of the form $G(x) = a \cos(bx)$ such that the graph passes through $(1, 2)$ and the amplitude is 2.

CHAPTER

7 Trigonometric Equations and Identities

7.1 INVERSE TRIGONOMETRIC FUNCTIONS

When dealing with the concept of an inverse function for a function F, two important questions to answer are:

1. Does F^{-1} exist?
2. If F^{-1} exists, then what is the rule for $F^{-1}(x)$?

In this section we answer these questions for the trigonometric functions sine, cosine, and tangent. First, however, we consider a simple polynomial function that illustrates the type of thinking required when we define the inverse trigonometric functions.

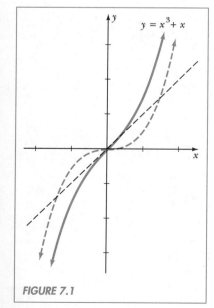

FIGURE 7.1

Consider $F(x) = x^3 + x$. We recall from Section 2.6 that an inverse function for F exists if and only if F is one-to-one or, equivalently, the graph of F satisfies the horizontal line test (HLT). By adding the graphs of $y = x^3$ and $y = x$, we obtain the graph of F in Figure 7.1. It is clear that the graph satisfies the HLT; hence F^{-1} exists.

Now that we know F^{-1} exists, what is the rule for $F^{-1}(x)$? Following the usual procedure, we have

$$\text{Equation for } F: \quad y = x^3 + x$$

$$\text{Interchange } x \text{ and } y: \quad x = y^3 + y$$

$$\text{Solve for } y: \quad ?$$

Solving $y^3 + y = x$ for y is, unfortunately, beyond our present capabilities. As a result, we do not define $F^{-1}(x)$ directly; instead, we resort to an indirect definition as follows.

Definition of $F^{-1}(x)$.

Symbols: $y = F^{-1}(x)$ means $F(y) = x$ or $y^3 + y = x$
Words: $F^{-1}(x)$ represents the number whose sum with its cube equals x.

Let us evaluate $F^{-1}(x)$ for some specific values of x.

Evaluate $F^{-1}(2)$. In words, $F^{-1}(2)$ represents the number whose sum with its cube equals 2. In symbols,

$$y = F^{-1}(2) \quad \text{means} \quad F(y) = 2 \quad \text{or} \quad y^3 + y = 2$$

At this point, we must use some guesswork to find y. Of course, $y = 1$ since $1^3 + 1 = 2$. Therefore $F^{-1}(2) = 1$.

Evaluate $F^{-1}(10)$. In words, $F^{-1}(10)$ represents the number whose sum with its cube equals 10. In symbols,

$$y = F^{-1}(10) \quad \text{means} \quad F(y) = 10 \quad \text{or} \quad y^3 + y = 10$$

Again, we use some guesswork to find y. Since $2^3 + 2 = 10$, we see that $y = 2$. Therefore $F^{-1}(10) = 2$.

Of course, we can only evaluate $F^{-1}(x)$ in this manner for choices of x that make the equation $y^3 + y = x$ easy to solve by guessing. In this sense, the indirect method for defining $F^{-1}(x)$ is rather limited. However, in the case of the trigonometric functions, this limitation can be removed by using the calculator.

We now focus our attention on the sine function. By examining the normal sine curve in Figure 7.2(a), we find that it does not satisfy the horizontal line test. (A horizontal line such as $y = 1/2$ will intersect the graph an infinite number of times!) Thus the sine function does not have an inverse *unless* we restrict its domain. As Figure 7.2(b) shows, the portion of the sine curve between $-\pi/2$ and $\pi/2$ does satisfy the horizontal line test. Therefore the *restricted sine function*, whose domain is $[-\pi/2, \pi/2]$, does possess an inverse. We write $\sin^{-1}(x)$ and say *the inverse sine function*, but it will be understood that it represents the inverse of the restricted sine function.

Now, what about the rule for $\sin^{-1}(x)$? Following the usual procedure, we have

$$\text{Equation for sine:} \quad y = \sin(x)$$
$$\text{Interchange } x \text{ and } y: \quad x = \sin(y)$$
$$\text{Solve for } y: \quad ?$$

The solution for y is possible to obtain, but the result is quite complicated and requires calculus to derive. Therefore we settle for the following description.

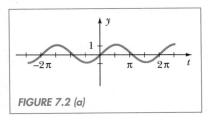

FIGURE 7.2 (a)

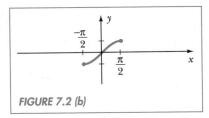

FIGURE 7.2 (b)

Definition of $\sin^{-1}(x)$

Symbols: $y = \sin^{-1}(x)$ means $\sin(y) = x$ and $-\dfrac{\pi}{2} \le y \le \dfrac{\pi}{2}$.

Words: $\sin^{-1}(x)$ is the measure of the angle between $-\dfrac{\pi}{2}$ and $\dfrac{\pi}{2}$ whose sine equals x.

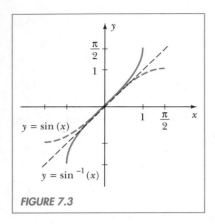

FIGURE 7.3

There are two concepts to point out concerning this definition. First, $\sin^{-1}(x)$ is a real number; we think of this number as the radian measure of an angle whose sine equals x. Second, $\sin^{-1}(x)$ must be a number between $-\pi/2$ and $\pi/2$ because we are talking about the inverse of the restricted sine function whose domain is $[-\pi/2, \pi/2]$.

Let us examine the graph of $y = \sin^{-1}(x)$. Recall that the graphs of inverse functions are symmetric images of each other across the diagonal $y = x$. Therefore, to obtain the graph of $y = \sin^{-1}(x)$, we can reflect the graph of the restricted sine function across the diagonal, as is done in Figure 7.3.

Notice that the domain of $\sin^{-1}(x)$ is $[-1, 1]$ and the range is $[-\pi/2, \pi/2]$. In other words,

$$\sin^{-1}(x) \text{ is defined only for } x \in [-1, 1]$$

$$-\frac{\pi}{2} \leq \sin^{-1}(x) \leq \frac{\pi}{2}$$

Example 1

Evaluate (a) $\sin^{-1}\left(\dfrac{1}{2}\right)$, (b) $\sin^{-1}(-1)$.

SOLUTION (a) Using the word definition for the inverse sine, we have $\sin^{-1}\left(\dfrac{1}{2}\right)$ is the measure of the angle between $-\dfrac{\pi}{2}$ and $\dfrac{\pi}{2}$ whose sine equals $\dfrac{1}{2}$. What angle between $-\pi/2$ and $\pi/2$ has its sine equal to 1/2? Since $\sin(\pi/6) = 1/2$, the answer is $\pi/6$. We write

$$\sin^{-1}\left(\frac{1}{2}\right) = \frac{\pi}{6}$$

Note that $\pi/6$ is the *only* possible answer; there is no other angle with measure between $-\pi/2$ and $\pi/2$ whose sine equals 1/2.

Warning: In certain situations in calculus it is incorrect to write 30° for the answer to this problem. By definition, $\sin^{-1}(x)$ is a real number, not a degree measure. Therefore, unless stated otherwise, we usually refrain from using degrees when evaluating inverse trigonometric functions.

(b) Again, we state the word definition: $\sin^{-1}(-1)$ is the measure of the angle between $-\dfrac{\pi}{2}$ and $\dfrac{\pi}{2}$ whose sine equals -1. What angle between $-\pi/2$ and $\pi/2$ has its sine equal to -1? Since $\sin\left(-\dfrac{\pi}{2}\right) = -1$, the answer is $-\pi/2$.

We write

$$\sin^{-1}(-1) = -\frac{\pi}{2}$$

Warning: It is incorrect to use $3\pi/2$ as the answer to this problem, even though $\sin(3\pi/2) = -1$. By definition, the value of $\sin^{-1}(-1)$ must be a number in the interval $[-\pi/2, \pi/2]$, and $3\pi/2$ does not meet this requirement. ▪

In Example 1, $\sin^{-1}(x)$ turned out to be a standard angle that we could easily guess. But what do we do in the case when $\sin^{-1}(x)$ is not a standard angle? We can use the calculator, which has been programmed with formulas derived from calculus, to evaluate the inverse trigonometric functions.

Example 2

Evaluate $\sin^{-1}\left(\dfrac{3}{5}\right)$.

SOLUTION By definition, $\sin^{-1}\left(\dfrac{3}{5}\right)$ is the measure of the angle between $-\dfrac{\pi}{2}$ and $\dfrac{\pi}{2}$ whose sine equals 3/5. Therefore we put the calculator in *radian mode* and key in

$$3 \quad \boxed{\div} \quad 5 \quad \boxed{\text{INV}} \quad \boxed{\sin}$$

We obtain $\sin^{-1}(3/5) = .64350$, accurate to five decimal places. ▪

If we wish to find $\sin^{-1}(3/5)$ in terms of degrees, we simply put the calculator in degree mode before entering the same keystrokes above. We find $\sin^{-1}(3/5) \approx 36.87°$. However, as mentioned before, this is not usually done in calculus.

Example 3

Evaluate $\sin^{-1}(1.5)$.

SOLUTION By definition, $\sin^{-1}(1.5)$ is the measure of the angle between $-\dfrac{\pi}{2}$ and $\dfrac{\pi}{2}$ whose sine equals 1.5. What angle between $-\pi/2$ and $\pi/2$ has its sine equal to 1.5? None. In fact, the sine of any angle must be between -1 and 1. Hence, $\sin^{-1}(1.5)$ is not defined. We write

$$\sin^{-1}(1.5) \quad \text{undefined}$$

This agrees with our previous observation that the domain of $\sin^{-1}(x)$ is $[-1, 1]$. ▪

An examination of the cosine graph (Figure 7.4) reveals that it will satisfy the horizontal line test if we restrict the curve to the interval $[0, \pi]$. Therefore, the *restricted cosine function* (Figure 7.5) with domain $[0, \pi]$ has an inverse, which we describe as follows.

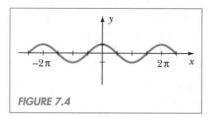

FIGURE 7.4

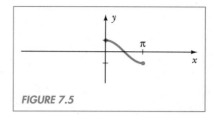

FIGURE 7.5

> **Definition of $\cos^{-1}(x)$**
> Symbols: $y = \cos^{-1}(x)$ means $\cos(y) = x$ and $0 \le y \le \pi$.
> Words: $\cos^{-1}(x)$ is the measure of the angle between 0 and π whose cosine equals x.

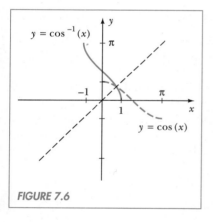

FIGURE 7.6

Note that $\cos^{-1}(x)$ is a number between 0 and π; this differs from the definition of $\sin^{-1}(x)$.

Figure 7.6 is the graph of $\cos^{-1}(x)$. Notice that the domain is $[-1, 1]$ and the range is $[0, \pi]$. Thus,

> $\cos^{-1}(x)$ is defined only for $x \in [-1, 1]$
>
> $$0 \le \cos^{-1}(x) \le \pi$$

Example 4

Evaluate $\cos^{-1}(-\sqrt{2}/2)$.

SOLUTION From the word definition, $\cos^{-1}(-\sqrt{2}/2)$ is the measure of the angle between 0 and π whose cosine equals $-\dfrac{\sqrt{2}}{2}$. What angle between 0 and π has its cosine equal to $-\sqrt{2}/2$? We know $\cos(3\pi/4) = -\sqrt{2}/2$, so the answer is $3\pi/4$. We write

$$\cos^{-1}\left(-\frac{\sqrt{2}}{2}\right) = \frac{3\pi}{4} \quad \blacksquare$$

We define one more inverse trigonometric function, the inverse tangent. We observe from the graph of tangent (Figure 7.7) that it does satisfy the horizontal line test over the interval $(-\pi/2, \pi/2)$. Hence the *restricted tangent function* (Figure 7.8) with domain $(-\pi/2, \pi/2)$ has an inverse, which we describe as follows.

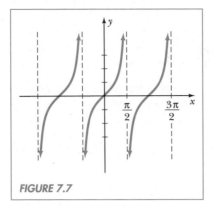

FIGURE 7.7

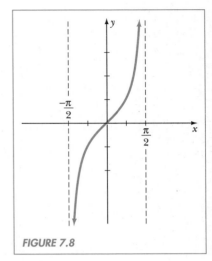

FIGURE 7.8

Definition of tan⁻¹(x)

Symbols: $y = \tan^{-1}(x)$ means $\tan(y) = x$ and $-\dfrac{\pi}{2} < y < \dfrac{\pi}{2}$.

Words: $\tan^{-1}(x)$ is the measure of the angle between $-\dfrac{\pi}{2}$ and $\dfrac{\pi}{2}$ whose tangent equals x.

Figure 7.9 is the graph of $\tan^{-1}(x)$. Note that the domain is $(-\infty, +\infty)$ and the range is $(-\pi/2, \pi/2)$. Therefore,

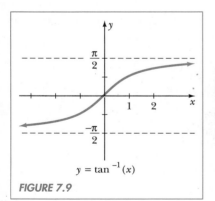

$y = \tan^{-1}(x)$

FIGURE 7.9

$$\tan^{-1}(x) \text{ is defined for all real } x$$

$$-\frac{\pi}{2} < \tan^{-1}(x) < \frac{\pi}{2}$$

Example 5

Evaluate (a) $\tan^{-1}(-1)$, (b) $\tan^{-1}(2)$.

SOLUTION   (a) By the word definition, $\tan^{-1}(-1)$ is the measure of the angle between $-\dfrac{\pi}{2}$ and $\dfrac{\pi}{2}$ whose tangent equals -1. What angle between

$-\pi/2$ and $\pi/2$ has its tangent equal to -1? Since $\tan\left(-\dfrac{\pi}{4}\right) = -1$, the answer is $-\pi/4$. We write

$$\tan^{-1}(-1) = -\frac{\pi}{4}$$

Note that $3\pi/4$ is incorrect, even though $\tan\left(\dfrac{3\pi}{4}\right) = -1$.

(b) $\tan^{-1}(2)$ is not a standard angle, so we use the calculator. Setting the calculator to radian mode and entering

$$2 \quad \boxed{\text{INV}} \quad \boxed{\text{tan}}$$

we find $\tan^{-1}(2) \approx 1.10715$ accurate to five decimal places. ∎

Examples 6 and 7 close this section; they evaluate expressions involving the inverse trigonometric functions.

Example 6

Let $\theta = \cos^{-1}\left(-\dfrac{2}{3}\right)$. (a) Draw a reference triangle for θ. (b) Without using the calculator, evaluate $\sin(\theta)$.

SOLUTION (a) By definition, $\theta = \cos^{-1}\left(-\dfrac{2}{3}\right)$ is the measure of the angle between 0 and π whose cosine equals $-2/3$. This angle must be in quadrant II since the cosine is negative there. Furthermore, the reference triangle for θ must satisfy

$$\frac{\text{Adjacent leg}}{\text{Hypotenuse}} = \frac{-2}{3}$$

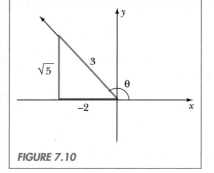

FIGURE 7.10

This gives us the reference triangle in Figure 7.10. By the Pythagorean Theorem,

$$\text{Opposite leg} = \sqrt{3^2 - 2^2} = \sqrt{5}$$

(b) From the reference triangle obtained in part (a),

$$\sin(\theta) = \frac{\text{opposite leg}}{\text{hypotenuse}} = \frac{\sqrt{5}}{3} \quad ∎$$

Example 7

Evaluate in terms of x, $\cos(\tan^{-1}(x))$.

SOLUTION We shall evaluate the expression by drawing a reference triangle for $\tan^{-1}(x)$. By definition, $\tan^{-1}(x)$ is the measure of the angle

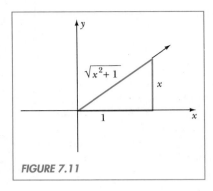

FIGURE 7.11

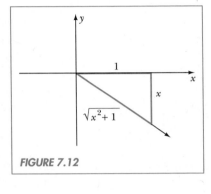

FIGURE 7.12

between $-\dfrac{\pi}{2}$ and $\dfrac{\pi}{2}$ whose tangent equals x. Therefore the reference triangle for $\tan^{-1}(x)$ must satisfy

$$\frac{\text{Opposite leg}}{\text{Adjacent leg}} = \frac{x}{1}$$

There are two cases to consider: $x \geq 0$ and $x < 0$.

Case 1: $x \geq 0$ (Figure 7.11) $\mathrm{Tan}^{-1}(x)$ must be in quadrant I (or possibly 0). Labeling the opposite leg x and the adjacent leg 1, we find by the Pythagorean Theorem

$$\text{Hypotenuse} = \sqrt{x^2 + 1}$$

It follows that

$$\cos(\tan^{-1}(x)) = \frac{\text{adjacent leg}}{\text{hypotenuse}} = \frac{1}{\sqrt{x^2 + 1}}$$

Case 2: $x < 0$ (Figure 7.12) $\mathrm{Tan}^{-1}(x)$ must be in quadrant IV. The opposite leg is again x (which is now a negative number); the adjacent leg is 1. We find that the hypotenuse is $\sqrt{x^2 + 1}$ by the Pythagorean Theorem. Thus

$$\cos(\tan^{-1}(x)) = \frac{\text{adjacent leg}}{\text{hypotenuse}} = \frac{1}{\sqrt{x^2 + 1}}$$

Therefore we obtain the same answer in either case. We conclude that for any x,

$$\cos(\tan^{-1}(x)) = \frac{1}{\sqrt{x^2 + 1}} \quad \blacksquare$$

EXERCISES 7.1

1. Sketch the graph of $F(x) = x^3 + 2x$. Does it satisfy the horizontal line test? Evaluate $F^{-1}(0)$, $F^{-1}(3)$, $F^{-1}(-12)$.

2. Sketch the graph of $F(x) = x^3 + x + 1$. Is F one-to-one? Evaluate $F^{-1}(-1)$, $F^{-1}(11)$, $F^{-1}(1)$.

In Exercises 3 to 23, evaluate each expression without a calculator. Some expressions may be undefined.

3. $\sin^{-1}(0)$

4. $\cos^{-1}(0)$

5. $\tan^{-1}(0)$

6. $\cos^{-1}(-1)$

7. $\cos^{-1}(-1/2)$

8. $\sin^{-1}(-\sqrt{3}/2)$

9. $\sin^{-1}(-1/\sqrt{2})$

10. $\tan^{-1}(-\sqrt{3})$

11. $\sin^{-1}(\sqrt{2})$

12. $\tan^{-1}(1) - \tan^{-1}(-1)$

13. $\sin^{-1}(1) - \sin^{-1}(-1)$

14. $\cos^{-1}(1) - \cos^{-1}(-1)$

15. $\tan^{-1}(\sqrt{3}) - \tan^{-1}\left(\dfrac{1}{\sqrt{3}}\right)$

16. $\sin^{-1}\left(\dfrac{\pi}{2}\right)$

17. $\cos^{-1}(\pi)$

18. $\tan^{-1}(\tan \pi)$

19. $\sin^{-1}(\sin \pi)$

20. $\cos^{-1}(\cos(-1))$

21. $\tan^{-1}(\tan(-1))$

22. $\tan^{-1}\left(\dfrac{4}{3}\right) - \tan^{-1}\left(-\dfrac{3}{4}\right)$

23. $\tan^{-1}\left(\dfrac{1}{2}\right) - \tan^{-1}(-2)$

In Exercises 24 to 31, use the calculator to approximate the given expression to five decimal places.

24. $\tan^{-1}(3)$
25. $\tan^{-1}(4)$
26. $\sin^{-1}(-.7229)$
27. $\cos^{-1}(-.4)$
28. $\cos^{-1}\left(\dfrac{\pi}{4}\right)$
29. $\sin^{-1}\left(-\dfrac{\pi}{4}\right)$
30. $\tan^{-1}\left(\dfrac{1 + \sqrt{5}}{2}\right)$
31. $\sin^{-1}\left(\dfrac{1 - \sqrt{5}}{2}\right)$

In Exercises 32 to 35, (a) draw a reference triangle for θ; (b) evaluate $\sin \theta$, $\cos \theta$, and $\tan \theta$.

32. $\theta = \cos^{-1}(1/3)$
33. $\theta = \sin^{-1}(1/4)$
34. $\theta = \tan^{-1}(3)$
35. $\theta = \tan^{-1}(-.75)$

In Exercises 36 to 41, without using a calculator, evaluate the given expression.

36. $\sin(\sin^{-1}(.55))$
37. $\cos(\cos^{-1}(-.3))$
38. $\cos(\tan^{-1}(-2))$
39. $\sin\left(\tan^{-1}\left(\dfrac{-1}{2}\right)\right)$
40. $\tan\left(\cos^{-1}\left(\dfrac{5}{13}\right)\right)$
41. $\tan(\sin^{-1}(.6))$

In Exercises 42 to 45, evaluate the given expression in terms of x.

42. $\tan(\sin^{-1} x)$
43. $\tan(\cos^{-1} x)$
44. $\sin(\tan^{-1} x)$
45. $\csc(\sin^{-1} x)$

46. Find the exact value of the given expression. Do not use a calculator.
   (a) $\cos^{-1}(\cos 2)$
   (b) $\sin^{-1}(\sin 2)$
   (c) $\tan^{-1}(\tan 2)$

47. A point $P(x, 4)$ moves along the horizontal line $y = 4$. (See Figure 7.13.) Let θ be the angle in standard position whose terminal side passes through P. Write θ as a function of x. (*Hint:* since θ will vary between 0 and π, try using the inverse cosine function.)

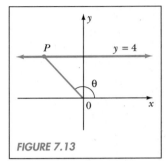

FIGURE 7.13

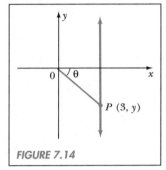

FIGURE 7.14

48. A point $P(3, y)$ moves along the vertical line $x = 3$. (See Figure 7.14.) Let θ be the angle in standard position whose terminal side passes through P. Write θ as a function of y.

49. If $x \in [-1, 1]$, find the value of $\sin^{-1}(x) + \cos^{-1}(x)$.

50. Find the value of $\tan^{-1}(x) + \tan^{-1}\left(\dfrac{1}{x}\right)$ if (a) $x > 0$, (b) $x < 0$.

7.2 BASIC TRIGONOMETRIC EQUATIONS

A trigonometric equation is an equation containing one or more trigonometric functions, with the unknown variable appearing on the inside of the functions. In this section we concentrate on solving trigonometric equations of the form

$$\sin(bx) = r \qquad \cos(bx) = r \qquad \tan(bx) = r$$

where b and r are constants. These are rather simple equations to solve, but, as we shall see later, they provide the key to solving more general types of trigonometric equations. Before considering specific examples, we state the following useful result.

> **Reference Angle Fact**
> Suppose F is a trigonometric function. If $F(\theta_1) = F(\theta_2)$, then θ_1 and θ_2 have the same reference angle.

IDEA OF THE PROOF: If $F(\theta_1) = F(\theta_2)$, then the ratios of two corresponding sides in the reference triangles for θ_1 and θ_2 are the same. Since both triangles are right triangles, it follows that all corresponding sides are proportional, which means that both reference triangles are similar, and hence their reference angles are equal. ∎

Example 1

We know that angle $\theta = 2\pi/3$ is one solution to the equation $\cos\theta = -1/2$. What other angles between 0 and 2π satisfy this equation?

SOLUTION Since $\cos(2\pi/3) = -1/2$, the reference angle fact tells us that any other θ satisfying $\cos(\theta) = -1/2$ must have the same reference angle as $2\pi/3$. Since the reference angle for $2\pi/3$ is 60°, the possible angles to consider are $\pi/3, 4\pi/3$, and $5\pi/3$. See Figure 7.15. Of course, $\cos(4\pi/3) = -1/2$, and $\cos(\pi/3)$ and $\cos(5\pi/3)$ are both positive $1/2$. Therefore, the only angle between 0 and 2π other than $2\pi/3$ satisfying $\cos\theta = -1/2$ is $\theta = 4\pi/3$. ∎

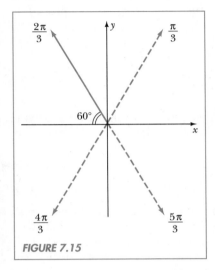

FIGURE 7.15

Example 2

Find all θ satisfying $\sin(\theta) = 1/2$.

SOLUTION Step 1. First, we find *one* value of θ that satisfies the equation. We have

$$\sin(\theta) = \frac{1}{2} \quad \text{implies} \quad \theta = \sin^{-1}\left(\frac{1}{2}\right) = \frac{\pi}{6}$$

We call this particular solution θ_o.

Step 2. Draw θ_o found in step 1 in standard position and look for all *other* solutions in $[0, 2\pi)$ that are not coterminal with θ_o. By the reference angle fact, any other solution must have the same reference angle as θ_o, which is 30°. Therefore the possible angles to consider are $5\pi/6, 7\pi/6$, and $11\pi/6$. See Figure 7.16. Note that $\sin(5\pi/6) = 1/2$, but $\sin(7\pi/6)$ and $\sin(11\pi/6)$ are both $-1/2$. Therefore the only other angle in $[0, 2\pi]$ not coterminal with θ_o satisfying $\sin\theta = 1/2$ is $\theta = 5\pi/6$.

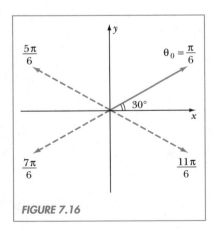

FIGURE 7.16

Step 3. All solutions of $\sin(\theta) = 1/2$ must be coterminal with the angles found in steps 1 and 2. Therefore the solutions are given by

$$\theta = \begin{cases} \dfrac{\pi}{6} + 2\pi k \\[2mm] \dfrac{5\pi}{6} + 2\pi k \end{cases} \quad k \in \mathbf{Z} \; \blacksquare$$

Now we consider a trigonometric equation where the argument inside the function has the form bx. This will add one more step to the procedure outlined in Example 2.

Example 3

Find all x satisfying $\cos(2x) = -\dfrac{1}{\sqrt{2}}$. Determine which solutions are in the interval $[0, 2\pi)$.

SOLUTION Step 1. We let $\theta = 2x$ and proceed to find all θ satisfying $\cos(\theta) = -1/\sqrt{2}$. Begin by finding one value of θ that satisfies this equation. We have

$$\theta_o = \cos^{-1}\left(-\frac{1}{\sqrt{2}}\right) = \frac{3\pi}{4}$$

Step 2. Draw θ_o found in step 1 in standard position. We look for all other solutions to $\cos \theta = -1/\sqrt{2}$ in $[0, 2\pi)$ that are not coterminal with θ_o. These angles must have the same reference angle as θ_o, which is 45°.

Figure 7.17 shows that the angle with measure $\theta = 5\pi/4$ satisfies our equation. Furthermore, no other angle with measure between 0 and 2π other than $3\pi/4$ and $5\pi/4$ will satisfy the equation.

Step 3. All solutions of $\cos(\theta) = -1/\sqrt{2}$ must be coterminal with the angles found in steps 1 and 2. Therefore

$$\theta = \begin{cases} \dfrac{3\pi}{4} + 2\pi k \\[2mm] \dfrac{5\pi}{4} + 2\pi k \end{cases} \quad k \in \mathbf{Z}$$

Step 4. Since $\theta = 2x$, we have $x = \theta/2$. Thus

$$x = \frac{\theta}{2} = \begin{cases} \dfrac{3\pi}{8} + \pi k \\[2mm] \dfrac{5\pi}{8} + \pi k \end{cases} \quad k \in \mathbf{Z}$$

To determine which solutions are in the interval $[0, 2\pi)$, we evaluate the expressions for x at several values of k.

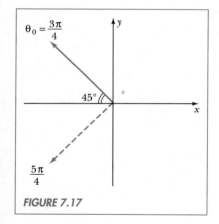

FIGURE 7.17

k	-1	0	1	2
$x = \dfrac{3\pi}{8} + \pi k$	$-\dfrac{5\pi}{8}$	$\dfrac{3\pi}{8}$	$\dfrac{11\pi}{8}$	$\dfrac{3\pi}{8} + 2\pi$
$x = \dfrac{5\pi}{8} + \pi k$	$-\dfrac{3\pi}{8}$	$\dfrac{5\pi}{8}$	$\dfrac{13\pi}{8}$	$\dfrac{5\pi}{8} + 2\pi$

Now we can see that the only solutions in $[0, 2\pi)$ are $3\pi/8$, $11\pi/8$, $5\pi/8$, and $13\pi/8$. ■

Here is a general outline of the procedure followed in Examples 2 and 3.

Solving Basic Trigonometric Equations
Suppose F is the sine, cosine, or tangent function. If b and r are constants, we solve

$$F(bx) = r$$

as follows.

Step 1   Let $\theta = bx$. Find one value of θ satisfying $F(\theta) = r$: $\theta_o = F^{-1}(r)$.

Step 2   Find all other θ in $[0, 2\pi)$ not coterminal with θ_o satisfying $F(\theta) = r$. (These θ will have the same reference angle as θ_o.)

Step 3   Add $2\pi k$, $k \in \mathbf{Z}$, to each value of θ found in steps 1 and 2.

Step 4   Since $x = \theta/b$, we find x by multiplying each expression for θ given in step 3 by $1/b$.

We note that if $b = 1$, as in Example 2, then step 4 is not necessary.

Example 4

Find all x satisfying $\tan\left(-\dfrac{x}{2}\right) = .6745$. Use degrees to represent the solutions. Which solutions are between 0 and 360°?

SOLUTION   Before beginning the steps for solving the equation, we choose to remove the negative sign from inside the function. Recall that $\tan(-\theta) = -\tan(\theta)$. Thus

$$\tan\left(-\frac{x}{2}\right) = .6745 \quad \text{if and only if} \quad -\tan\left(\frac{x}{2}\right) = .6745$$

$$\text{if and only if} \quad \tan\left(\frac{x}{2}\right) = -.6745$$

Now we solve $\tan(x/2) = -.6745$.

Step 1. Let $\theta = x/2$. We wish to find one value of θ satisfying $\tan\theta = -.6745$. Therefore

$$\theta_o = \tan^{-1}(-.6745) \approx -34°$$

As requested, we express θ_o in degrees. This was accomplished on our calculator by setting it in degree mode before pressing the inverse tangent key.

Step 2. As shown in Figure 7.18, the only other angle θ in $[0, 2\pi)$ (between 0° and 360°) not coterminal with $-34°$ that satisfies our equation is $\theta = 146°$.

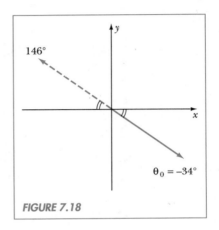

FIGURE 7.18

Step 3. We have

$$\theta = \begin{cases} -34° + 360k° \\ 146° + 360k° \end{cases} \quad k \in \mathbf{Z}$$

Step 4. Since $\theta = x/2$, we have

$$x = 2\theta = \begin{cases} -68° + 720k° \\ 292° + 720k° \end{cases} \quad k \in \mathbf{Z}$$

To find which solutions are between 0 and 360°, we begin by listing some of the values given in step 4. Substituting $k = -1, 0, 1$, and 2, we have

k	-1	0	1	2
$x = -68° + 720k°$	$-788°$	$-68°$	$652°$	$1372°$
$x = 292° + 720k°$	$-428°$	$292°$	$1012°$	$1732°$

Therefore, we see that $x = 292°$ is the only solution between 0 and 360°. ∎

Example 5

A rocket is to be launched from a position 5000 feet on level ground from our camera. At what angle should we point the camera to photograph the rocket when it reaches an altitude of 1 mile? Assume the rocket travels straight up.

SOLUTION Let θ be the desired angle. The situation is illustrated in Figure 7.19. Note that 1 mile is equivalent to 5280 feet. Thus

$$\tan \theta = \frac{5280}{5000}$$

It follows that

$$\theta = \tan^{-1}\left(\frac{5280}{5000}\right) \approx 46.56°$$

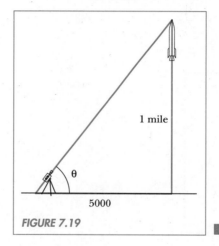

FIGURE 7.19

EXERCISES 7.2

In Exercises 1 to 24, find all x satisfying the given equation. Give exact answers when possible. If it is necessary to use the calculator, use degrees and approximate to two decimal places. Determine which solutions are in the interval $[0, 2\pi)$.

1. $\sin(x) = -1$
2. $\cos(x) = 0$
3. $2 \cos(x) = 1$
4. $\tan(x) = -\sqrt{3}$
5. $\tan(-x) = 1.95$
6. $\sin(-x) = .6$
7. $\sin^2(x) = .25$
8. $\tan^2(x) = 1$
9. $\sin(x - \pi/4) = 0$
10. $\cos(x + \pi/3) = -1$
11. $\cos(2x) = .5$
12. $\sin(2x) = .766$
13. $\tan(2x) = -1$
14. $\sin\left(-\frac{x}{2}\right) = \frac{1}{\sqrt{2}}$
15. $\cos\left(-\frac{2}{3}x\right) = \frac{\sqrt{3}}{2}$
16. $\tan(3x) = \sqrt{3}$
17. $\sin(3x) = .454$
18. $\sec(x) = -2$
19. $\cos(4x) + 1 = 0$
20. $\sin(4x) + 3 = 4$

21. $\cos(x) = \dfrac{-4 - \sqrt{10}}{2}$ 22. $\sin(10x) = \dfrac{5}{4}$

23. $\sec\left(\dfrac{x}{2}\right) = \sqrt{2}$ 24. $2\cos(3x) = -1$

In Exercises 25 to 27, write an equation for the angle x and then solve.

25.

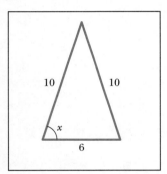

26.

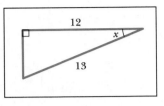

27.

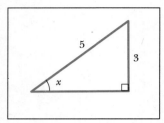

In Exercises 28 to 31, draw and label the right triangle ABC described. Assume angle C is 90°. Find angles A and B (in degrees) and the third side of the triangle. Approximate answers to two decimal places.

28. $AB = 28$; $AC = 20$

29. $AC = 30$; $BC = 40$

30. $AC = 2\sqrt{15}$; $BC = 14$

31. $AB = 23.34$; $BC = 15$

32. In Example 5, at what angle should we point the camera to photograph the rocket when it reaches an altitude of 2 miles?

33. Figure 7.20 shows a telescope aimed directly at object O. The line CO is called the *optic axis* of the telescope. On the perpendicular to the optic axis at O we find points A and B farthest from the optic axis that can be viewed with the telescope. We call the angle ACB the **angular field of view** of the telescope.

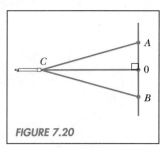

FIGURE 7.20

Find the angular field of view necessary for a telescope to view the entire surface of the moon from the earth. Assume that the distance to the moon is 238,000 miles and its diameter is 2160 miles.

34. Find the angular field of view necessary for a camera to photograph a rectangular painting from a distance of 20 feet. The painting has length 10 feet and height 6 feet. (Angular field of view is defined in Exercise 33.)

In Exercises 35 and 36, find all sides and angles for the triangle with the given vertices.

35. $A(1, 2)$, $B(3, 4)$, $C(-2, 7)$

36. $A(-2, 3)$, $B(4, 5)$, $C(3, -2)$

37. The position y of an object undergoing **simple harmonic motion** is given by the general formula

$$y = a \sin\left(\dfrac{2\pi}{T} t + \alpha\right)$$

where $|a|$ is the amplitude, T is the period, and α is the **phase shift**.

(a) If $y = 0$ when $t = 0$, show that α must be an integer multiple of π.

(b) Find α if $y = \dfrac{1}{2}a$ when $t = 0$.

(c) Find α if $y = a$ when $t = 0$.

7.3 FUNDAMENTAL IDENTITIES

A trigonometric equation in one variable, say, θ, is called an **identity** if it is true for all values of θ for which the expressions involved are defined. The following equations are examples of identities. Because each statement can be established from the definitions of sine, cosine, or tangent, we call these the **fundamental identities**.

Fundamental Identities

1. $\csc \theta = \dfrac{1}{\sin \theta}$

2. $\sec \theta = \dfrac{1}{\cos \theta}$

3. (a) $\tan \theta = \dfrac{\sin \theta}{\cos \theta}$

   (b) $\tan \theta = \dfrac{1}{\cot \theta}$

4. (a) $\cot \theta = \dfrac{\cos \theta}{\sin \theta}$

   (b) $\cot \theta = \dfrac{1}{\tan \theta}$

5. (a) $\sin^2 \theta + \cos^2 \theta = 1$
   (b) $\quad\quad\sin^2 \theta = 1 - \cos^2 \theta$
   (c) $\quad\quad\cos^2 \theta = 1 - \sin^2 \theta$

6. $1 + \tan^2 \theta = \sec^2 \theta$

7. $1 + \cot^2 \theta = \csc^2 \theta$

8. (a) $\sin(-\theta) = -\sin \theta$
   (b) $\cos(-\theta) = \cos \theta$
   (c) $\tan(-\theta) = -\tan \theta$

9. For all $k \in \mathbf{Z}$,
   (a) $\sin(\theta + 2\pi k) = \sin(\theta)$
   (b) $\cos(\theta + 2\pi k) = \cos(\theta)$
   (c) $\tan(\theta + \pi k) = \tan(\theta)$

Let us review why these statements are true.

Identities 1 to 4 follow immediately from the definitions of tangent, cotangent, and the reciprocal functions.

Identity 5(a) is *the* fundamental identity established in Section 6.4. Clearly, statements 5(b) and (c) are equivalent to 5(a).

To prove identity 6, we have

$$1 + \tan^2 \theta = 1 + \frac{\sin^2 \theta}{\cos^2 \theta} \qquad \text{by identity 3(a)}$$

$$= \frac{\cos^2 \theta + \sin^2 \theta}{\cos^2 \theta}$$

$$= \frac{1}{\cos^2 \theta} \qquad \text{by identity 5(a)}$$

$$= \sec^2 \theta \qquad \text{by identity 2}$$

The proof of identity 7 is left as an exercise.

Finally, identities 8 and 9 state the symmetry and periodicity properties of sine, cosine, and tangent, which we proved in Section 6.4 and 6.5.

Having established the fundamental identities, we wish to gain some experience using them to simplify and rewrite trigonometric expressions. The exercises give us this opportunity by asking us to prove a variety of trigonometric equations are identities.

For example, suppose we are asked to prove that $\csc \theta \tan \theta - \cos \theta = \sin \theta \tan \theta$. We begin by writing

$$\csc \theta \tan \theta - \cos \theta \overset{?}{=} \sin \theta \tan \theta$$

A question mark appears over the equal sign, indicating that we wish to establish the validity of the statement. This is accomplished by working on each side of the equation using known identities or algebraic manipulations so that the last expression on the left side is identical to the last expression on the right side. A vertical line down the middle emphasizes that we must work on each side independently.

Following is a complete proof:

$$\csc \theta \tan \theta - \cos \theta \overset{?}{=} \sin \theta \tan \theta$$

$$\frac{1}{\sin \theta}\frac{\sin \theta}{\cos \theta} - \cos \theta \qquad\Bigg|\qquad \sin \theta \, \frac{\sin \theta}{\cos \theta}$$

$$\frac{1}{\cos \theta} - \cos \theta \qquad\Bigg|\qquad \frac{\sin^2 \theta}{\cos \theta}$$

$$\frac{1}{\cos \theta} - \frac{\cos^2 \theta}{\cos \theta}$$

$$\frac{1 - \cos^2 \theta}{\cos \theta}$$

$$\frac{\sin^2 \theta}{\cos \theta}$$

Note that each line on the left side follows from the one above it by either a fundamental identity or an algebraic manipulation. The same is true for the right side of the proof. We have thus established the validity of the original statement.

How do we decide what to do on each side of the proof? Unfortunately, there is no general procedure that always works. However, we can make some recommendations that sometimes prove helpful.

Hints for Proving Identities

1. Rewrite all functions in terms of sines and cosines.
2. Add fractions and reduce; simplify compound fractions. Factoring may help.
3. If the trigonometric function F appears on one side but not the other, try to replace it by using a fundamental identity or eliminate it with an algebra manipulation.
4. Do not be afraid to *try* something; it is not always clear whether something will work unless we try it.

Example 1

Prove that $\csc^2 \theta - 1 = \cot^2 \theta$.

SOLUTION
$$\csc^2 \theta - 1 \overset{?}{=} \cot^2 \theta$$

Change to sines and cosines:	$\dfrac{1}{\sin^2 \theta} - 1$	$\dfrac{\cos^2 \theta}{\sin^2 \theta}$
Add fractions:	$\dfrac{1}{\sin^2 \theta} - \dfrac{\sin^2 \theta}{\sin^2 \theta}$	
	$\dfrac{1 - \sin^2 \theta}{\sin^2 \theta}$	
5(c) implies:	$\dfrac{\cos^2 \theta}{\sin^2 \theta}$	

Example 2

Prove $\dfrac{\sin^2 \theta}{1 + \cos \theta} = 1 - \cos \theta$.

SOLUTION Before beginning the proof, notice that $\sin \theta$ appears on the left side but not on the right. We shall try replacing it with cosine by using identity 5(b).

$$\frac{\sin^2 \theta}{1 + \cos \theta} \overset{?}{=} 1 - \cos \theta$$

5(b) implies: $\dfrac{1 - \cos^2 \theta}{1 + \cos \theta}$

Factor: $\dfrac{(1 + \cos \theta)(1 - \cos \theta)}{(1 + \cos \theta)}$

Reduce: $1 - \cos \theta$

Example 3

Prove $\dfrac{\cot \theta - \tan \theta}{\csc \theta + \sec \theta} = \cos \theta - \sin \theta$

SOLUTION

$$\frac{\cot \theta - \tan \theta}{\csc \theta + \sec \theta} \qquad\overset{?}{=}\qquad \cos \theta - \sin \theta$$

$$\frac{\dfrac{\cos \theta}{\sin \theta} - \dfrac{\sin \theta}{\cos \theta}}{\dfrac{1}{\sin \theta} + \dfrac{1}{\cos \theta}}$$

$$\frac{\sin \theta \cos \theta \left(\dfrac{\cos \theta}{\sin \theta} - \dfrac{\sin \theta}{\cos \theta} \right)}{\sin \theta \cos \theta \left(\dfrac{1}{\sin \theta} + \dfrac{1}{\cos \theta} \right)}$$

$$\frac{\cos^2 \theta - \sin^2 \theta}{\cos \theta + \sin \theta}$$

$$\frac{(\cos \theta + \sin \theta)(\cos \theta - \sin \theta)}{\cos \theta + \sin \theta}$$

$$\cos \theta - \sin \theta$$

When reducing fractions, we refrain from crossing out items by drawing lines through them. This makes the presentation of the proof clearer and more readable.

Remark

Our purpose for proving identities is to gain experience in manipulating and simplifying various trigonometric expressions. Our method requires the manipulation of each side independently of the other. It is for this reason that we do not approach the identity as an equation to be solved. Therefore, we do *not* apply the operations of addition, multiplication, squaring, or square root in our proofs.

Remember that an identity is an equation that is true *whenever the expressions involved are defined*. If one or both sides of the equation are *undefined* for some value of the variable, this fact does not invalidate the identity. For example, consider identity 4(b),

$$\cot \theta = \frac{1}{\tan \theta}$$

When $\theta = \pi/2$, the left side reads $\cot(\pi/2)$, which equals zero, and the right side reads $1/\tan(\pi/2)$, which is undefined! In fact, whenever θ is an integer multiple of $\pi/2$ $(0, \pm \pi/2, \pm \pi, \pm 3\pi/2, \ldots)$, either the right side or both sides are undefined. However, if θ takes on any value other than an integer multiple of $\pi/2$, then both $\cot \theta$ and $1/\tan \theta$ are defined and equal. This makes the equation an identity.

Do not think that every trigonometric equation is an identity. We can show that a given equation is not an identity by exhibiting a **counterexample**. In other words, we present a particular value of the variable that, when substituted into the equation, yields a real number on one side and a *different* real number on the other side.

Example 4

Prove or disprove with a counterexample that $(\sin \theta + \cos \theta)^2 = 1$.

SOLUTION

$$(\sin \theta + \cos \theta)^2 \overset{?}{=} 1$$

$$(\sin \theta + \cos \theta)(\sin \theta + \cos \theta)$$

$$\sin^2 \theta + 2 \sin \theta \cos \theta + \cos^2 \theta$$

Using 5(a): $\qquad 1 + 2 \sin \theta \cos \theta$

We observe that the left side is definitely not equal to 1 for all θ. In this case, we present a counterexample as follows:

$$(\sin \theta + \cos \theta)^2 \overset{?}{=} 1$$

$\theta = \dfrac{\pi}{4} \quad$ implies $\quad \left(\sin\dfrac{\pi}{4} + \cos\dfrac{\pi}{4}\right)^2$

$$\left(\frac{1}{\sqrt{2}} + \frac{1}{\sqrt{2}}\right)^2$$

$$\left(\frac{2}{\sqrt{2}}\right)^2$$

$$2$$

No

Therefore the statement $(\sin \theta + \cos \theta)^2 = 1$ is not an identity since it is not true for all θ which makes sense in the given expressions. ∎

EXERCISES 7.3

In Exercises 1 to 20, a basic algebra manipulation that will be useful for proving trigonometric identities is reviewed. We present the problem in two ways: (a) with letters representing variables, and (b) with trigonometric functions as the variables. For example, simplify:

(a) $\dfrac{1 - x^2}{1 + x}$; (b) $\dfrac{1 - \cos^2 \theta}{1 + \cos \theta}$

Solution:

(a) $\dfrac{1 - x^2}{1 + x} = \dfrac{(1 + x)(1 - x)}{1 + x} = 1 - x$

(b) $\dfrac{1 - \cos^2 \theta}{1 + \cos \theta} = \dfrac{(1 + \cos \theta)(1 - \cos \theta)}{1 + \cos \theta} = 1 - \cos \theta$

1. Factor (a) $1 - y^2$, (b) $1 - \sin^2 \theta$
2. Factor (a) $y^2 - x^2$, (b) $\sin^2 \theta - \cos^2 \theta$
3. Factor (a) $y^4 - x^4$, (b) $\sin^4 \theta - \cos^4 \theta$
4. Factor (a) $2y^4 - 1 - y^2$, (b) $2 \sin^4 \theta - 1 - \sin^2 \theta$
5. Factor (a) $y^3 + x^3$, (b) $\sin^3 \theta + \cos^3 \theta$
6. Factor (a) $y^3 - x^3$, (b) $\sin^3 \theta - \cos^3 \theta$

7. Reduce (a) $\dfrac{y - 1}{y^2 - 1}$, (b) $\dfrac{\sin \theta - 1}{\sin^2 \theta - 1}$

8. Reduce (a) $\dfrac{x^2 - y^2}{x^2 y - xy^2}$, (b) $\dfrac{\cos^2 \theta - \sin^2 \theta}{\cos^2 \theta \sin \theta - \cos \theta \sin^2 \theta}$

9. Simplify (a) $\dfrac{\frac{1}{y}}{\frac{1}{x}}$, (b) $\dfrac{\frac{\sin \theta}{1}}{\frac{1}{\cos \theta}}$

10. Simplify (a) $\dfrac{1}{\frac{y}{x} + \frac{x}{y}}$, (b) $\dfrac{1}{\frac{\sin \theta}{\cos \theta} + \frac{\cos \theta}{\sin \theta}}$

11. Simplify (a) $\dfrac{1 + \frac{1}{x}}{y + \frac{y}{x}}$, (b) $\dfrac{1 + \frac{1}{\cos \theta}}{\sin \theta + \frac{\sin \theta}{\cos \theta}}$

12. Simplify (a) $\dfrac{\frac{1}{yx}}{\frac{1}{y} + \frac{1}{x}}$, (b) $\dfrac{\frac{1}{\sin \theta \cos \theta}}{\frac{1}{\sin \theta} + \frac{1}{\cos \theta}}$

13. Add (a) $\dfrac{y}{x} + \dfrac{x}{y}$, (b) $\dfrac{\sin \theta}{\cos \theta} + \dfrac{\cos \theta}{\sin \theta}$

14. Add (a) $\dfrac{y^2}{x^2} + 1$, (b) $\dfrac{\sin^2 \theta}{\cos^2 \theta} + 1$

15. Add (a) $\dfrac{1}{y^2} - 1$, (b) $\dfrac{1}{\sin^2 \theta} - 1$

16. Add (a) $\dfrac{1}{1 - x} + \dfrac{1}{1 + x}$, (b) $\dfrac{1}{1 - \cos \theta} + \dfrac{1}{1 + \cos \theta}$

17. Multiply (a) $(y + x)^2$, (b) $(\sin \theta + \cos \theta)^2$
18. Multiply (a) $(1 + x)(1 - x)$,
    (b) $(1 + \cos \theta)(1 - \cos \theta)$

19. Multiply (a) $(y + x)\left(\dfrac{y}{x} + \dfrac{x}{y}\right)$,

    (b) $(\sin \theta + \cos \theta)\left(\dfrac{\sin \theta}{\cos \theta} + \dfrac{\cos \theta}{\sin \theta}\right)$

20. Multiply (a) $\left(\dfrac{1}{x} + \dfrac{1}{y}\right)\left(\dfrac{y}{x} + \dfrac{x}{y}\right)$,

    (b) $\left(\dfrac{1}{\cos \theta} + \dfrac{1}{\sin \theta}\right)\left(\dfrac{\sin \theta}{\cos \theta} + \dfrac{\cos \theta}{\sin \theta}\right)$

In Exercises 21 to 32, simplify the expression given in the left-hand column. Find the correct answer among the choices in the right-hand column.

21. $1 + \cot^2 \theta$ (A) $\sin \theta + \cos \theta$
22. $\sec^2 \theta - 1$ (B) $2 \sec^2 \theta$
23. $\sqrt{1 - \sin^2 \theta}$ (C) $1 + 2 \sin \theta \cos \theta$
24. $\sqrt{1 - \cos^2 \theta}$ (D) $\csc \theta \sec \theta$
25. $(\sin \theta - \cos \theta)^2$ (E) $\tan^2 \theta$
26. $(\sin \theta + \cos \theta)^2$ (F) $\sin^2 \theta - \cos^2 \theta$
27. $\tan \theta + \cot \theta$ (G) $|\sin \theta|$
28. $\sec^2 \theta + \csc^2 \theta$ (H) $1 - 2 \sin \theta \cos \theta$

29. $\dfrac{1}{1 - \sin \theta} + \dfrac{1}{1 + \sin \theta}$ (I) $\csc^2 \theta$

     (J) $\sec^2 \theta \csc^2 \theta$

     (K) $|\cos \theta|$

30. $\dfrac{1}{1 - \cos \theta} + \dfrac{1}{1 + \cos \theta}$ (L) $2 \csc^2 \theta$

31. $\sin^3 \theta + \sin^2 \theta \cos \theta + \sin \theta \cos^2 \theta + \cos^3 \theta$
32. $\sin^4 \theta - \cos^4 \theta$

In Exercises 33 to 78, prove that the given equation is an identity.

33. $\csc \theta = \csc(\theta + 4\pi)$

34. $\sec(\theta - 2\pi) = \sec \theta$

35. $\cos \theta + \cos(-\theta) = 2 \cos \theta$

36. $\sin \theta + \sin(-\theta) = 0$

37. $\dfrac{\tan \theta}{\sec \theta} = \sin \theta$

38. $\dfrac{\csc \theta}{\sec \theta} = \cot \theta$

39. $\csc \theta \tan \theta = \sec \theta$

40. $\sec \theta \cot \theta = \csc \theta$

41. $\cos \theta(\sec \theta - \cos \theta) = \sin^2 \theta$

42. $\sec \theta(\sec \theta - \cos \theta) = \tan^2 \theta$

43. $\sin^2 \theta + \tan^2 \theta + \cos^2 \theta = \sec^2 \theta$

44. $2 \cos^2 \theta - 1 = 1 - 2 \sin^2 \theta$

45. $(\cos \theta - \sin \theta)(\cos \theta + \sin \theta) = 2 \cos^2 \theta - 1$

46. $(\sin \theta + \cos \theta)^2 - (\sin \theta - \cos \theta)^2 = 4 \sin \theta \cos \theta$

47. $\dfrac{1}{\sqrt{1 + \tan^2 \theta}} = |\cos \theta|$

48. $\dfrac{1}{\sqrt{1 + \cot^2 \theta}} = |\sin \theta|$

49. $\cot^2 \theta - \cos^2 \theta = \cos^2 \theta \cot^2 \theta$

50. $\tan^2 \theta - \sin^2 \theta = \tan^2 \theta \sin^2 \theta$

51. $\sec^2 \theta - \csc^2 \theta = \tan^2 \theta - \cot^2 \theta$

52. $(1 + \tan^2 \theta)\cot^2 \theta = \csc^2 \theta$

53. $\dfrac{1 + \tan^2 \theta}{\tan^2 \theta} = \csc^2 \theta$

54. $\dfrac{1 + \csc \theta}{\sec \theta} - \cot \theta = \cos \theta$

55. $1 - \sin \theta = \dfrac{\cos^2 \theta}{1 + \sin \theta}$

56. $1 - \dfrac{\sin^2 \theta}{1 + \cos \theta} = \cos \theta$

57. $\sqrt{\dfrac{1 + \cos \theta}{1 - \cos \theta}} = \dfrac{|\sin \theta|}{1 - \cos \theta}$

(*Note:* $1 - \cos \theta \geq 0$ for all θ.)

58. $\sqrt{\dfrac{1 - \sin \theta}{1 + \sin \theta}} = \dfrac{|\cos \theta|}{1 + \sin \theta}$

(*Note:* $1 + \sin \theta \geq 0$ for all θ.)

59. $\dfrac{1}{1 + \sin \theta} + \dfrac{1}{1 - \sin \theta} = 2 \sec^2 \theta$

60. $\dfrac{1}{1 - \cos \theta} - \dfrac{1}{1 + \cos \theta} = 2 \csc \theta \cot \theta$

61. $\dfrac{1}{\tan \theta + \cot \theta} = \sin \theta \cos \theta$

62. $\dfrac{1 + \sec \theta}{\sin \theta + \tan \theta} = \csc \theta$

63. $\left(\dfrac{1 + \sin \theta}{1 + \cos \theta}\right)\left(\dfrac{1 + \sec \theta}{1 + \csc \theta}\right) = \tan \theta$

64. $\cot \theta(\cot \theta + \csc \theta) = \dfrac{\cos \theta}{1 - \cos \theta}$

65. $\dfrac{2 \sin^2 \theta - 1}{\sin \theta + \cos \theta} = \sin \theta - \cos \theta$

66. $\dfrac{\sin \theta \cot \theta}{1 + \sin \theta} = \sec \theta - \tan \theta$

67. $\sec \theta + \csc \theta = \dfrac{\tan \theta}{\sin \theta - \cos \theta} + \dfrac{\cot \theta}{\cos \theta - \sin \theta}$

68. $\dfrac{\tan \theta}{1 + \cos \theta} = \csc \theta(\sec \theta - 1)$

69. $\dfrac{\sin \theta}{1 - \cos \theta} = \dfrac{1 + \cos \theta}{\sin \theta}$

70. $\sec \theta + \tan \theta = \dfrac{\cos \theta}{1 - \sin \theta}$

71. $\sec \theta + \csc \theta = (\sin \theta + \cos \theta)(\tan \theta + \cot \theta)$

72. $(\sec \theta + \csc \theta)(\cot \theta + \tan \theta) = \dfrac{\sin \theta + \cos \theta}{\sin^2 \theta \cos^2 \theta}$

73. $\sin \theta + \cos \theta = \dfrac{\sin \theta}{1 - \cot \theta} + \dfrac{\cos \theta}{1 - \tan \theta}$

74. $\dfrac{\csc \theta}{2 - \csc^2 \theta} + \dfrac{\sec \theta}{2 - \sec^2 \theta} = \dfrac{1}{\sin \theta + \cos \theta}$

75. $\sin^4 \theta + \cos^2 \theta = \sin^2 \theta + \cos^4 \theta$

76. $\sin^4 \theta - 2 \sin^2 \theta + 1 = \cos^4 \theta$

77. $\cos^4 \theta - 2 \cos^2 \theta + 1 = \sin^4 \theta$

78. $\sin^4 \theta + \cos^4 \theta = 1 - 2 \sin^2 \theta \cos^2 \theta$

In Exercises 79 to 88, prove or disprove with a counter-example.

79. $\tan^2 \theta - \cot^2 \theta = \cos^2 \theta - \sin^2 \theta$

80. $\csc^2 \theta + \sec^2 \theta = 1$

81. $\sin^2 \theta - \cos^2 \theta = -1$

82. $\sqrt{\tan(\theta^2)} = |\tan \theta|$ 83. $\sqrt{1 - \cos^2 \theta} = \sin \theta$

84. $\tan(\theta + \pi) + \tan(\theta + 2\pi) = \tan(\theta + 3\pi)$

85. $\tan(\pi\theta) = \tan \theta$

86. $\sin(2\pi\theta) = \sin \theta$

87. $\tan^2 \theta + \cot^2 \theta - \sec^2 \theta - \csc^2 \theta = -2$

88. $\cot(\theta - \pi) \tan(\pi - \theta) = \csc(\theta - 2\pi) \sin(2\pi - \theta)$

89. Prove identity 7.

7.4 ADDITION FORMULAS

In this section we discuss another set of identities, the **addition formulas**. The first such formula provides the key to all the rest; we state it as follows:

10. $\cos(\alpha - \beta) = \cos \alpha \cos \beta + \sin \alpha \sin \beta$

The symbols α and β represent arbitrary real numbers. As we shall see, the fundamental identities together with identity 10 are enough to establish all the remaining trigonometric identities that will be used in this text.

PROOF OF IDENTITY 10:   First, if $\alpha = \beta$, then the result is obvious since

$\cos(\alpha - \beta) = \cos(0)$ 　　 and 　　 $\cos \alpha \cos \beta + \sin \alpha \sin \beta = \cos^2 \alpha + \sin^2 \alpha$

$= 1$ 　　　　　　　　　　　　　　 $= 1$

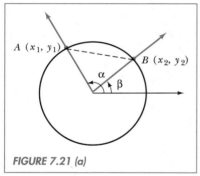

FIGURE 7.21 (a)

Now, suppose $\alpha \neq \beta$. Figure 7.21(a) is a typical example of two angles with radian measures α and β drawn in standard position. Although the figure indicates that both α and β are positive with $\alpha > \beta$, this is not essential. The terminal sides of α and β intersect the unit circle at points $A(x_1, y_1)$ and $B(x_2, y_2)$, respectively. Figure 7.21(b) shows the angle $\alpha - \beta$ in standard position, with its terminal side intersecting the unit circle at point $C(x_3, y_3)$.

We notice that arc AB and arc CD have the same central angle, $\alpha - \beta$. It follows that the line segments AB and CD are equal in length.

$$AB = CD$$

Using the distance formula, we have

$$\sqrt{(x_2 - x_1)^2 + (y_2 - y_1)^2} = \sqrt{(x_3 - 1)^2 + y_3^2}$$

Squaring and then multiplying, we find

$$x_2^2 - 2x_1 x_2 + x_1^2 + y_2^2 - 2y_1 y_2 + y_1^2 = x_3^2 - 2x_3 + 1 + y_3^2$$

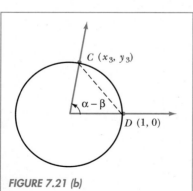

FIGURE 7.21 (b)

Now (x_1, y_1), (x_2, y_2), and (x_3, y_3) are points on the unit circle, which means that they satisfy the equation $x^2 + y^2 = 1$. Therefore we may replace each of the expressions $x_1^2 + y_1^2$, $x_2^2 + y_2^2$, and $x_3^2 + y_3^2$

with 1. Our equation now becomes

$$2 - 2x_1x_2 - 2y_1y_2 = 2 - 2x_3$$

Further simplification yields

$$x_1x_2 + y_1y_2 = x_3$$

Finally, by the unit circle interpretation, $x_1 = \cos \alpha$, $x_2 = \cos \beta$, $y_1 = \sin \alpha$, $y_2 = \sin \beta$, and $x_3 = \cos(\alpha - \beta)$. Thus

$$\cos \alpha \cos \beta + \sin \alpha \sin \beta = \cos(\alpha - \beta)$$

This completes the proof. ∎

Example 1

Prove that the sine function is the cosine function shifted by $+\pi/2$ units,

$$\sin(x) = \cos\left(x - \frac{\pi}{2}\right)$$

SOLUTION   We use identity 10 with $\alpha = x$ and $\beta = \pi/2$ to show that $\cos(x - \pi/2)$ is the same as $\sin(x)$. We have

$$\cos\left(x - \frac{\pi}{2}\right) = \cos x \cos\frac{\pi}{2} + \sin x \sin\frac{\pi}{2}$$

$$= \cos x \cdot 0 + \sin x \cdot 1$$

$$= \sin x \quad ∎$$

By Example 1, we may write

$$\sin\left(x + \frac{\pi}{2}\right) = \cos\left(\left(x + \frac{\pi}{2}\right) - \frac{\pi}{2}\right)$$

$$= \cos(x)$$

Thus we have established the following results, which we first introduced in Section 6.4.

Sine Is Cosine Shifted

$$\sin(x) = \cos\left(x - \frac{\pi}{2}\right)$$

Cosine Is Sine Shifted

$$\cos(x) = \sin\left(x + \frac{\pi}{2}\right)$$

We now present all the addition formulas.

Addition Formulas

10. $\cos(\alpha - \beta) = \cos\alpha\cos\beta + \sin\alpha\sin\beta$

11. $\cos(\alpha + \beta) = \cos\alpha\cos\beta - \sin\alpha\sin\beta$

12. $\sin(\alpha - \beta) = \sin\alpha\cos\beta - \cos\alpha\sin\beta$

13. $\sin(\alpha + \beta) = \sin\alpha\cos\beta + \cos\alpha\sin\beta$

14. $\tan(\alpha - \beta) = \dfrac{\tan\alpha - \tan\beta}{1 + \tan\alpha\tan\beta}$

15. $\tan(\alpha + \beta) = \dfrac{\tan\alpha + \tan\beta}{1 - \tan\alpha\tan\beta}$

PROOF OF 11: $\cos(\alpha + \beta) \overset{?}{=} \cos\alpha\cos\beta - \sin\alpha\sin\beta$

$\cos(\alpha - (-\beta))$

10 implies: $\cos\alpha\cos(-\beta) + \sin\alpha\sin(-\beta)$

8 implies: $\cos\alpha\cos\beta - \sin\alpha\sin\beta$

Note that we use only previously established identities in the proof.

PROOF OF 12: In this proof we shall use identities 10 and 11 and the fact that sine is cosine shifted, $\sin\theta = \cos(\theta - \pi/2)$, and cosine is sine shifted, $\cos\theta = \sin(\theta + \pi/2)$. All these results have been established above.

$\sin(\alpha - \beta) \overset{?}{=} \sin\alpha\cos\beta - \cos\alpha\sin\beta$

Sine is cosine shifted: $\cos\left((\alpha - \beta) - \dfrac{\pi}{2}\right)$

$\cos\left(\alpha - \left(\beta + \dfrac{\pi}{2}\right)\right)$

10 implies: $\cos\alpha\cos\left(\beta + \dfrac{\pi}{2}\right)$
$+ \sin\alpha\sin\left(\beta + \dfrac{\pi}{2}\right)$

Cosine is sine shifted: $\cos\alpha\cos\left(\beta + \dfrac{\pi}{2}\right) + \sin\alpha\cos\beta$

11 implies: $\cos\alpha\left(\cos\beta\cos\dfrac{\pi}{2} - \sin\beta\sin\dfrac{\pi}{2}\right)$
$+ \sin\alpha\cos\beta$

$\cos\alpha(-\sin\beta) + \sin\alpha\cos\beta$

$-\cos\alpha\sin\beta + \sin\alpha\cos\beta$

$\sin\alpha\cos\beta - \cos\alpha\sin\beta$

We leave the proofs of identities 13 to 15 for the exercises. ∎

Example 2

Find the exact value of (a) $\sin(105°)$, (b) $\tan\left(-\dfrac{\pi}{12}\right)$.

SOLUTION   (a) We notice that $105° = 60° + 45°$, a sum of standard angles. Using identity 13, we have

$$\sin(105°) = \sin(60° + 45°)$$

$$= \sin(60°)\cos(45°) + \cos(60°)\sin(45°)$$

$$= \frac{\sqrt{3}}{2}\frac{1}{\sqrt{2}} + \frac{1}{2}\frac{1}{\sqrt{2}}$$

$$= \frac{\sqrt{3} + 1}{2\sqrt{2}} \quad \text{or} \quad \frac{\sqrt{6} + \sqrt{2}}{4}$$

(b) We note that $-\dfrac{\pi}{12}$ is equivalent to $-15°$, which can be written as $30° - 45°$, a difference of standard angles. Using identity 14, we have

$$\tan\left(-\frac{\pi}{12}\right) = \tan(30° - 45°)$$

$$= \frac{\tan(30°) - \tan(45°)}{1 + \tan(30°)\tan(45°)}$$

$$= \frac{\dfrac{1}{\sqrt{3}} - 1}{1 + \dfrac{1}{\sqrt{3}}}$$

$$= \frac{1 - \sqrt{3}}{\sqrt{3} + 1} \quad \blacksquare$$

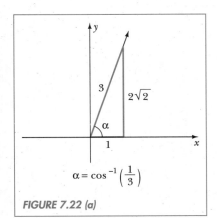

$$\alpha = \cos^{-1}\left(\frac{1}{3}\right)$$

FIGURE 7.22 (a)

$$\beta = \tan^{-1}(-2)$$

FIGURE 7.22 (b)

Example 3

Find the exact value of $\sin\left(\cos^{-1}\left(\dfrac{1}{3}\right) + \tan^{-1}(-2)\right)$.

SOLUTION   Let $\alpha = \cos^{-1}\left(\dfrac{1}{3}\right)$ and $\beta = \tan^{-1}(-2)$. We draw the reference triangles for α and β illustrated in Figures 7.22(a) and 7.22(b). Now,

$$\sin\left(\cos^{-1}\left(\frac{1}{3}\right) + \tan^{-1}(-2)\right) = \sin(\alpha + \beta)$$

$$= \sin\alpha\cos\beta + \cos\alpha\sin\beta$$

$$= \left(\frac{2\sqrt{2}}{3}\right)\left(\frac{1}{\sqrt{5}}\right) + \left(\frac{1}{3}\right)\left(\frac{-2}{\sqrt{5}}\right)$$

$$= \frac{2\sqrt{2} - 2}{3\sqrt{5}} \quad \blacksquare$$

Example 4

Given $F(x) = \cos x$ and $h \neq 0$. Find $\dfrac{F(x+h) - F(x)}{h}$ when $x = \pi/3$.

SOLUTION We have

$$\frac{F(x+h) - F(x)}{h} = \frac{F\left(\dfrac{\pi}{3} + h\right) - F\left(\dfrac{\pi}{3}\right)}{h}$$

$$= \frac{\cos\left(\dfrac{\pi}{3} + h\right) - \cos\left(\dfrac{\pi}{3}\right)}{h}$$

Applying identity 13 to $\cos\left(\dfrac{\pi}{3} + h\right)$, we have

$$= \frac{\cos\left(\dfrac{\pi}{3}\right)\cos h - \sin\left(\dfrac{\pi}{3}\right)\sin h - \cos\left(\dfrac{\pi}{3}\right)}{h}$$

Substituting $\cos(\pi/3) = 1/2$ and $\sin(\pi/3) = \sqrt{3}/2$, we obtain

$$= \frac{\cos h - \sqrt{3}\sin h - 1}{2h} \quad \blacksquare$$

EXERCISES 7.4

1. Prove identity 13. [*Hint:* You can write $\sin(\alpha + \beta)$ as $\sin(\alpha - (-\beta))$ and then use identity 12.]

2. (a) Prove identity 14. Use only identities 1 to 13.
   (b) Prove identity 15. You may use identity 14.

In Exercises 3 to 12, use identities 10 to 15 to find the exact value of the given expression.

3. $\cos\left(\dfrac{3\pi}{4} - \dfrac{\pi}{3}\right)$

4. $\cos\left(\dfrac{\pi}{4} + \dfrac{\pi}{3}\right)$

5. $\sin(15°)$

6. $\sin(75°)$

7. $\tan\left(\dfrac{\pi}{12}\right)$

8. $\tan\left(\dfrac{11\pi}{12}\right)$

9. $\sin 70° \cos 80° + \cos 70° \sin 80°$

10. $\cos\left(\dfrac{3\pi}{8}\right)\cos\left(\dfrac{\pi}{8}\right) + \sin\left(\dfrac{3\pi}{8}\right)\sin\left(\dfrac{\pi}{8}\right)$

11. (a) $\sin(285°)$ (b) $\cos(285°)$

12. (a) $\sin(13\pi/12)$ (b) $\cos(13\pi/12)$

In Exercises 13 to 21, prove that the given equation is an identity.

13. (a) $\sin\left(\dfrac{\pi}{2} - \theta\right) = \cos\theta$, (b) $\cos\left(\dfrac{\pi}{2} - \theta\right) = \sin\theta$

14. $\sin(\alpha + \beta)\sin(\alpha - \beta) = \sin^2\alpha - \sin^2\beta$

15. $\cos(\alpha + \beta)\cos(\alpha - \beta) = \cos^2\beta - \sin^2\alpha$

16. $\cos(\alpha - \beta)\sin(\alpha + \beta) = \sin\alpha\cos\alpha + \sin\beta\cos\beta$

17. $\cot(\alpha + \beta) = \dfrac{\cot\alpha\cot\beta - 1}{\cot\alpha + \cot\beta}$

18. $\cot(\alpha - \beta) = \dfrac{1 + \cot\alpha\cot\beta}{\cot\beta - \cot\alpha}$

19. $\sin(2\theta) = 2\sin\theta\cos\theta$

20. $\cos(2\theta) = \cos^2\theta - \sin^2\theta$

21. $\sin\theta + \cos\theta = \sqrt{2}\sin\left(\theta + \dfrac{\pi}{4}\right)$

In Exercises 22 to 27, find the exact value of the given expression.

22. $\sin\left(\cos^{-1}\left(\dfrac{1}{3}\right) + \sin^{-1}\left(\dfrac{1}{3}\right)\right)$

23. $\cos(\tan^{-1}(2) - \tan^{-1}(-2))$

24. $\cos\left(\sin^{-1}\left(\dfrac{4}{5}\right) - \sin^{-1}\left(-\dfrac{4}{5}\right)\right)$

25. $\sin\left(\cos^{-1}\left(-\dfrac{1}{3}\right) + \sin^{-1}\left(-\dfrac{1}{3}\right)\right)$

26. $\tan\left(\cos^{-1}\left(\dfrac{2}{\sqrt{5}}\right) - \cos^{-1}\left(\dfrac{-2}{\sqrt{5}}\right)\right)$

27. $\tan\left(\sin^{-1}\left(\dfrac{3}{5}\right) - \sin^{-1}\left(-\dfrac{3}{5}\right)\right)$

28. Use the result from Exercise 21 to graph $F(x) = \sin x + \cos x$. Determine the period and range of F.

29. Consider $F(x) = \dfrac{1}{2}\cos x + \dfrac{\sqrt{3}}{2}\sin x$.

   (a) Use identity 13 to write F as one sine function.
   (b) Use your result from part (a) to graph F.
   (c) Determine the period and range of F.

30. Write $\sin(3x)$ in terms of $\sin x$ and $\cos x$.

31. Write $\cos(3x)$ in terms of $\sin x$ and $\cos x$.

In Exercises 32 and 33, find $\dfrac{F(x + h) - F(x)}{h}$ in terms

of h, $\sin h$, and $\cos h$ for the given function F and value for x. Assume $h \neq 0$.

32. $F(x) = \sin x$, $x = \pi/4$

33. $F(x) = \cos x$, $x = \pi/6$

34. Prove $\dfrac{F(x + h) - F(x)}{h}$

$$= \cos x\,\dfrac{\sin h}{h} + \sin x\left(\dfrac{\cos h - 1}{h}\right)$$

when $F(x) = \sin x$. Assume $h \neq 0$.

In Exercises 35 to 42, prove or disprove with a counter-example.

35. $\cos(\theta + \pi/3) = \cos\theta + 1/2$

36. $\tan(\theta + \pi/4) = \tan\theta + 1$

37. $\tan(\theta + \pi/2)\tan(\theta - \pi/2) = \tan^2\theta$

38. $\cos(\theta + \pi)\cos(\theta - \pi) = \cos^2\theta$

39. $\sin\theta - 1/2 = \sin(\theta - \pi/6)$

40. $\sin(\theta + \pi)\cos(\theta - \pi) = \sin^2\theta$

41. $\dfrac{\sin(\alpha^2 - \beta^2)}{\sin(\alpha - \beta)} = \sin(\alpha + \beta)$

42. $\sin(\theta + \theta) = 2\sin\theta$

43. Let $\alpha = \tan^{-1}\left(\dfrac{1}{3}\right)$ and $\beta = \tan^{-1}\left(\dfrac{1}{2}\right)$.

   (a) Explain why $0 < \alpha + \beta < \pi$.
   (b) Evaluate $\sin(\alpha + \beta)$ and $\cos(\alpha + \beta)$.
   (c) Use (a) and (b) to determine the value of $\alpha + \beta$.

44. Let l_1 and l_2 be two perpendicular lines that are not vertical or horizontal. Suppose the slopes of l_1 and l_2 are m_1 and m_2, respectively. This exercise outlines the proof that $m_2 = -1/m_1$, as promised in Section 1.7.

   (a) Consider the angles α and β in Figure 7.23. Show that $\cos(\alpha - \beta) = 0$. [*Hint:* why is $|\alpha - \beta| = \pi/2$?]
   (b) From part (a), deduce $\cos\alpha\cos\beta + \sin\alpha\sin\beta = 0$.
   (c) From part (b), deduce $\tan\alpha\tan\beta = -1$.
   (d) From part (c), deduce that $m_2 = -1/m_1$.

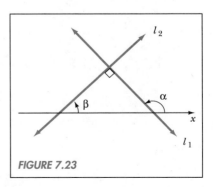

FIGURE 7.23

7.5 DOUBLE- AND HALF-ANGLE FORMULAS

In this section we discuss special cases of the addition formulas. We begin with identities involving functions of twice an angle; we call these the **double-angle formulas**.

Double-Angle Formulas

16. $\sin(2\theta) = 2 \sin \theta \cos \theta$
17. $\cos(2\theta) = \cos^2 \theta - \sin^2 \theta$

PROOF OF 16: Since 2θ can be rewritten as $\theta + \theta$, apply identity 13 to $\sin(\theta + \theta)$. The details are left as an exercise.

PROOF OF 17: We apply identity 11 to the left side of the equation.

$$\cos(2\theta) \overset{?}{=} \cos^2 \theta - \sin^2 \theta$$

$$\cos(\theta + \theta)$$

$$\cos \theta \cos \theta - \sin \theta \sin \theta$$

$$\cos^2 \theta - \sin^2 \theta$$

This completes the proof. ∎

Example 1

Suppose θ is in quadrant II and $\sin \theta = \dfrac{3}{4}$. Find $\sin(2\theta)$ and $\cos(2\theta)$. In what quadrant is the terminal side of 2θ?

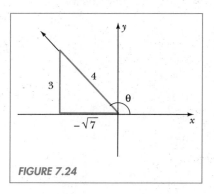

FIGURE 7.24

SOLUTION We begin the solution by drawing a reference triangle for θ. See Figure 7.24. The adjacent leg is computed using the Pythagorean Theorem.

$$\text{Adjacent leg} = -\sqrt{4^2 - 3^2} = -\sqrt{7}$$

From the reference triangle, $\sin \theta = 3/4$ and $\cos \theta = -\sqrt{7}/4$. Now we are ready to apply the double-angle formulas.

From identity 16,

$$\sin(2\theta) = 2 \sin \theta \cos \theta$$

$$= 2\left(\frac{3}{4}\right)\left(-\frac{\sqrt{7}}{4}\right)$$

$$= \frac{-3\sqrt{7}}{8}$$

From identity 17,

$$\cos(2\theta) = \cos^2 \theta - \sin^2 \theta$$

$$= \left(-\frac{\sqrt{7}}{4}\right)^2 - \left(\frac{3}{4}\right)^2$$

$$= \frac{7}{16} - \frac{9}{16}$$

$$= -\frac{1}{8}$$

Note that 2θ must be a quadrant III angle since both $\sin(2\theta)$ and $\cos(2\theta)$ are negative. ■

The following identities are just alternate forms of identity 17. However, they are important enough to state separately.

18.(a) $\sin^2 \theta = \dfrac{1 - \cos(2\theta)}{2}$,   (b) $\cos^2 \theta = \dfrac{1 + \cos(2\theta)}{2}$

PROOF OF 18(a):

$$\sin^2 \theta \overset{?}{=} \frac{1 - \cos(2\theta)}{2}$$

$$\frac{1 - (\cos^2 \theta - \sin^2 \theta)}{2} \qquad \text{by identity 17}$$

$$\frac{1 - \cos^2 \theta + \sin^2 \theta}{2}$$

$$\frac{\sin^2 \theta + \sin^2 \theta}{2} \qquad \text{by identity 5(b)}$$

$$\frac{2 \sin^2 \theta}{2}$$

$$\sin^2 \theta$$

PROOF OF 18(b):   The proof is similar to the one given for identity 18(a). We leave this for an exercise. ■

Example 2

Use identity 18(a) to find the graph of $y = \sin^2 x$.

SOLUTION We wish to graph the equation

$$y = \sin^2 x$$

Using identity 18(a), we can rewrite this as

$$y = \frac{1 - \cos(2x)}{2}$$

This is equivalent to

$$y = \frac{1}{2} - \frac{1}{2}\cos(2x)$$

or

$$y - \frac{1}{2} = -\frac{1}{2}\cos(2x)$$

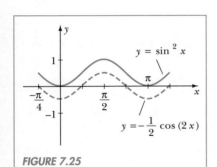

FIGURE 7.25

We recognize this last equation as the curve $y = -\frac{1}{2}\cos(2x)$ shifted $+\frac{1}{2}$ unit in the y-direction. Figure 7.25 shows the graph of $y = -\frac{1}{2}\cos(2x)$, which has period π and amplitude 1/2. By shifting this curve up 1/2 unit in the y-direction, we obtain the graph of $y = \sin^2 x$. ∎

By substituting $\theta = x/2$ into identities 18(a) and 18(b), we obtain the **half-angle formulas**.

Half-Angle Formulas

19. $\sin^2\left(\dfrac{x}{2}\right) = \dfrac{1 - \cos x}{2}$

20. $\cos^2\left(\dfrac{x}{2}\right) = \dfrac{1 + \cos x}{2}$

Example 3

Find the exact value of $\sin(22.5°)$.

SOLUTION Since $22.5°$ is half the standard angle $45°$, identity 19 applies.

We have

$$\sin^2(22.5°) = \sin^2\left(\frac{45°}{2}\right)$$

$$= \frac{1 - \cos(45°)}{2}$$

$$= \frac{1 - \sqrt{2}/2}{2}$$

Simplifying the compound fraction, we get

$$\sin^2(22.5°) = \frac{2 - \sqrt{2}}{4}$$

Now, apply the square root operation to both sides.

$$|\sin(22.5°)| = \frac{\sqrt{2 - \sqrt{2}}}{2}$$

Finally, since $\sin(22.5°)$ must be positive, we conclude

$$\sin(22.5°) = \frac{\sqrt{2 - \sqrt{2}}}{2} \quad \blacksquare$$

Example 4

Let $\theta = \tan^{-1}(x)$. Find $\sin(2\theta)$ in terms of x.

SOLUTION Since θ is the angle (between $-\pi/2$ and $\pi/2$) whose tangent is x, we can label a reference triangle for θ as follows:

$$\text{Opposite leg} = x$$

$$\text{Adjacent leg} = 1$$

Then the hypotenuse is given by $\sqrt{x^2 + 1}$ from the Pythagorean Theorem. The situation is illustrated in Figure 7.26. Now,

$$\sin(2\theta) = 2 \sin\theta\cos\theta \qquad \text{by identity 16}$$

$$= 2\left(\frac{x}{\sqrt{x^2 + 1}}\right)\left(\frac{1}{\sqrt{x^2 + 1}}\right) \qquad \text{using the reference triangle}$$

$$= \frac{2x}{x^2 + 1} \quad \blacksquare$$

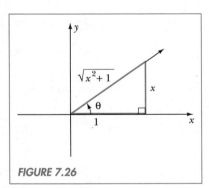

FIGURE 7.26

EXERCISES 7.5

1. Prove identity 16.

2. Prove identity 18(b).

In Exercises 3 to 6, find $\sin(2\theta)$, $\cos(2\theta)$, and $\tan(2\theta)$ from the given information.

3. θ in quadrant I, $\sin\theta = 1/3$

4. θ in quadrant IV, $\cos\theta = 3/5$

5. $\dfrac{\pi}{2} < \theta < \pi$, $\sin \theta = 4/5$

6. $\cos \theta < 0$, $\sin \theta = -5/6$

7. Use identity 18(b) to find the graph of $y = \cos^2 x$.

8. Use identity 18(a) to find the graph of $y = -2 \sin^2 x$.

In Exercises 9 to 12, use identity 19 or 20 to find the exact value of the given expression.

9. (a) $\sin\left(\dfrac{\pi}{12}\right)$ (b) $\sin\left(\dfrac{\pi}{24}\right)$

10. (a) $\cos(15°)$ (b) $\cos(7.5°)$

11. (a) $\cos(22.5°)$ (b) $\cos(11.25°)$

12. (a) $\sin\left(\dfrac{\pi}{8}\right)$ (b) $\sin\left(\dfrac{\pi}{16}\right)$

In Exercises 13 to 24, prove that the given equation is an identity.

13. $\cos(2\theta) = 1 - 2 \sin^2 \theta$ 14. $\cos(2\theta) = 2 \cos^2 \theta - 1$

15. $\tan(2\theta) = \dfrac{2 \tan \theta}{1 - \tan^2 \theta}$ 16. $\cot(2\theta) = \dfrac{\cot^2 \theta - 1}{2 \cot \theta}$

17. $\dfrac{\sin 2\theta}{2 \sin \theta} = \cos \theta$ 18. $2 \csc(2\theta) = \csc \theta \sec \theta$

19. $\sec(2\theta) = \dfrac{\sec^2 \theta}{2 - \sec^2 \theta}$ 20. $\sec(2\theta) = \dfrac{\csc^2 \theta}{\csc^2 \theta - 2}$

21. $\cos^4 \theta - \sin^4 \theta = \cos(2\theta)$

22. $\sin^4 \theta + \cos^4 \theta = \dfrac{1 + \cos^2 (2\theta)}{2}$

23. $\tan \theta + \cot \theta = 2 \csc(2\theta)$

24. $\dfrac{\csc^2 \theta \sec^2 \theta}{\cot^2 \theta - \tan^2 \theta} = \sec(2\theta)$

25. (a) Prove that $(\sin x + \cos x)^2 = 1 + \sin(2x)$ is an identity.
    (b) Use the result in part (a) to find the graph of $F(x) = (\sin x + \cos x)^2$. What are the period and range of this function?

26. (a) Prove that $\sin x \cos x = \dfrac{1}{2} \sin(2x)$ is an identity.
    (b) Use the result in part (a) to find the graph of

$F(x) = \sin x \cos x$. What are the period and range of this function?

27. Graph $F(x) = \sin^2 \pi x - \cos^2 \pi x$. Find the period and range. (*Hint:* use identity 17.)

28. Rewrite $\sin(4x)$ in terms of $\sin x$ and $\cos x$.

29. Rewrite $\cos(4x)$ in terms of $\sin x$ and $\cos x$.

30. Assume $0 < \theta < \pi$. Prove that $\sqrt{\dfrac{1 - \cos \theta}{1 + \cos \theta}} = \tan\left(\dfrac{\theta}{2}\right)$.

31. Rewrite $\sin^4(x)$ in the form $a + b \cos(2x) + c \cos(4x)$.

32. Rewrite $\cos^4(x)$ in the form $a + b \cos(2x) + c \cos(4x)$.

In Exercises 33 and 34, prove that the given equation is an identity.

33. $\cos(2x) + \cos(4x) = 8 \cos^4 x - 6 \cos^2 x$

34. $\dfrac{\sin(2x)}{2} + \dfrac{\sin(4x)}{4} = 2 \sin x \cos^3 x$

35. Let $\theta = \sin^{-1}(x)$. Find $\sin(2\theta)$ and $\cos(2\theta)$ in terms of x.

36. Let $\theta = \tan^{-1}(x)$. Find $\cos(2\theta)$ in terms of x.

37. Suppose $\theta = \cos^{-1}(y)$. Find $\sin(2\theta)$ and $\cos(2\theta)$ in terms of y.

38. Suppose α and β are the acute angles in a right triangle. Show that $\sin(2\alpha) = \sin(2\beta)$.

In Exercises 39 to 42, prove or disprove with a counter-example.

39. $\dfrac{\sin(x/2)}{x} \cdot \dfrac{2}{2} = \dfrac{\sin(x)}{2x}$

40. $\dfrac{\cos(\theta)}{\cos(2)} = \cos(\theta/2)$

41. $\dfrac{\tan(2\theta)}{2} = \tan \theta$

42. $\dfrac{\sin(2\theta)}{\cos(2\theta)} = \tan(1)$

PRODUCT FORMULAS AND REVIEW

Our final set of identities expresses an equivalence between products and sums of the trigonometric functions. We refer to these identities as the **product formulas**.

Product Formulas

21. $\sin \alpha \cos \beta = \dfrac{1}{2}[\sin(\alpha - \beta) + \sin(\alpha + \beta)]$

22. $\sin \alpha \sin \beta = \dfrac{1}{2}[\cos(\alpha - \beta) - \cos(\alpha + \beta)]$

23. $\cos \alpha \cos \beta = \dfrac{1}{2}[\cos(\alpha - \beta) + \cos(\alpha + \beta)]$

PROOF OF 21: Apply the addition formulas 12 and 13 to the right side of the statement.

$$\sin \alpha \cos \beta \stackrel{?}{=} \frac{1}{2}[\sin(\alpha - \beta) + \sin(\alpha + \beta)]$$

$$\frac{1}{2}[\sin \alpha \cos \beta - \cos \alpha \sin \beta + \sin \alpha \cos \beta + \cos \alpha \sin \beta]$$

$$\frac{1}{2}[2 \sin \alpha \cos \beta]$$

$$\sin \alpha \cos \beta$$

This completes the proof of identity 21. ■

Identities 22 and 23 can be proved by applying the addition formulas 10 and 11. The details are left for the exercises.

Example 1

Write $\sin(3x)\cos(2x)$ as a sum of functions.

SOLUTION Using identity 21 with $\alpha = 3x$ and $\beta = 2x$, we have

$$\sin(3x)\cos(2x) = \frac{1}{2}[\sin(3x - 2x) + \sin(3x + 2x)]$$

$$= \frac{1}{2}\sin(x) + \frac{1}{2}\sin(5x) \quad\blacksquare$$

Example 1 illustrates how it is possible to *split* a simple product of two functions, in this case sine and cosine, into a sum of functions. We now ask if it is possible to split *any* product of sines and cosines into a sum. Of course, the answer is yes. If we can split a product of two, then we can split a product of any number of sines and cosines by doing two at a time. Example 2 demonstrates the method.

Example 2

Write $4\sin(x)\sin(2x)\sin(3x)$ as a sum of functions without products.

SOLUTION Begin by applying identity 22 to the product $\sin(x)\sin(2x)$.

$$4\sin(x)\sin(2x)\sin(3x)$$

$$4\left(\frac{1}{2}\right)[\cos(x - 2x) - \cos(x + 2x)]\sin(3x)$$

$$2[\cos(-x) - \cos(3x)]\sin(3x)$$

We note that $\cos(-x) = \cos(x)$ by identity 8(b).

$$2[\cos(x) - \cos(3x)]\sin(3x)$$

Distribute 2 and $\sin(3x)$.

$$2\sin(3x)\cos(x) - 2\sin(3x)\cos(3x)$$

Now apply identity 21 to $\sin(3x)\cos(x)$ and $\sin(3x)\cos(3x)$.

$$2\left(\frac{1}{2}\right)[\sin(3x - x) + \sin(3x + x)] - 2\left(\frac{1}{2}\right)[\sin(3x - 3x) + \sin(3x + 3x)]$$

$$[\sin(2x) + \sin(4x)] - [\sin(0) + \sin(6x)]$$

$$\sin(2x) + \sin(4x) - \sin(6x) \quad \blacksquare$$

Example 3

Prove that $\cos(x) + \cos(y) = 2\cos\left(\dfrac{x + y}{2}\right)\cos\left(\dfrac{x - y}{2}\right)$.

SOLUTION We use identity 23 on the right side of this equation, with $\alpha = (x + y)/2$ and $\beta = (x - y)/2$.

$$\cos x + \cos y \overset{?}{=} 2\cos\left(\frac{x + y}{2}\right)\cos\left(\frac{x - y}{2}\right)$$

$$2\left(\frac{1}{2}\right)\left[\cos\left(\frac{x + y}{2} - \frac{x - y}{2}\right) + \cos\left(\frac{x + y}{2} + \frac{x - y}{2}\right)\right]$$

$$\cos(y) + \cos(x) \quad \blacksquare$$

This completes our discussion of the trigonometric identities. For reference, here is a list of all the identities.

Fundamental Identities

1. $\csc\theta = \dfrac{1}{\sin\theta}$

2. $\sec\theta = \dfrac{1}{\cos\theta}$

3. (a) $\tan\theta = \dfrac{\sin\theta}{\cos\theta}$

   (b) $\tan\theta = \dfrac{1}{\cot\theta}$

4. (a) $\cot\theta = \dfrac{\cos\theta}{\sin\theta}$

   (b) $\cot\theta = \dfrac{1}{\tan\theta}$

5. (a) $\sin^2\theta + \cos^2\theta = 1$

   (b) $\sin^2\theta = 1 - \cos^2\theta$

   (c) $\cos^2\theta = 1 - \sin^2\theta$

6. $1 + \tan^2\theta = \sec^2\theta$

7. $1 + \cot^2\theta = \csc^2\theta$

8. (a) $\sin(-\theta) = -\sin\theta$

   (b) $\cos(-\theta) = \cos\theta$

   (c) $\tan(-\theta) = -\tan\theta$

9. For all $k \in \mathbf{Z}$,

   (a) $\sin(\theta + 2\pi k) = \sin(\theta)$

   (b) $\cos(\theta + 2\pi k) = \cos(\theta)$

   (c) $\tan(\theta + \pi k) = \tan(\theta)$

Addition Formulas

10. $\cos(\alpha - \beta) = \cos\alpha\cos\beta + \sin\alpha\sin\beta$

11. $\cos(\alpha + \beta) = \cos\alpha\cos\beta - \sin\alpha\sin\beta$

12. $\sin(\alpha - \beta) = \sin\alpha\cos\beta - \cos\alpha\sin\beta$

13. $\sin(\alpha + \beta) = \sin\alpha\cos\beta + \cos\alpha\sin\beta$

14. $\tan(\alpha - \beta) = \dfrac{\tan\alpha - \tan\beta}{1 + \tan\alpha\tan\beta}$

15. $\tan(\alpha + \beta) = \dfrac{\tan\alpha + \tan\beta}{1 - \tan\alpha\tan\beta}$

Double-Angle Formulas

16. $\sin(2\theta) = 2\sin\theta\cos\theta$

17. $\cos(2\theta) = \cos^2\theta - \sin^2\theta$

18. (a) $\sin^2\theta = \dfrac{1 - \cos(2\theta)}{2}$

   (b) $\cos^2\theta = \dfrac{1 + \cos(2\theta)}{2}$

Half-Angle Formulas

19. $\sin^2\left(\dfrac{x}{2}\right) = \dfrac{1 - \cos x}{2}$

20. $\cos^2\left(\dfrac{x}{2}\right) = \dfrac{1 + \cos x}{2}$

Product Formulas

21. $\sin\alpha\cos\beta = \dfrac{1}{2}[\sin(\alpha - \beta) + \sin(\alpha + \beta)]$

22. $\sin\alpha\sin\beta = \dfrac{1}{2}[\cos(\alpha - \beta) - \cos(\alpha + \beta)]$

23. $\cos\alpha\cos\beta = \dfrac{1}{2}[\cos(\alpha - \beta) + \cos(\alpha + \beta)]$

EXERCISES 7.6

1. Prove identity 22.

2. Prove identity 23.

In Exercises 3 to 12, write the given product as a sum of functions without products.

3. $\sin(5x)\cos(3x)$

4. $\sin(2x)\cos(3x)$

5. $\sin x \sin(2x)$

6. $\sin x \sin(3x)$

7. $\cos 4x \cos 2x$

8. $\cos x \cos 2x$

9. $4 \cos x \cos\left(\dfrac{x}{2}\right)\sin\left(\dfrac{3x}{2}\right)$

10. $4 \cos x \cos(2x) \cos(3x)$

11. $\sin 2x \sin 4x \sin 6x$

12. $\sin x \cos\left(\dfrac{x}{2}\right)\sin\left(\dfrac{3x}{2}\right)$

In Exercises 13 to 18, use identities 21 to 23 to help prove the identities.

13. $\sin x + \sin y = 2 \sin\left(\dfrac{x + y}{2}\right)\cos\left(\dfrac{x - y}{2}\right)$

14. $\sin x - \sin y = 2 \sin\left(\dfrac{x - y}{2}\right)\cos\left(\dfrac{x + y}{2}\right)$

15. $\cos x - \cos y = -2 \sin\left(\dfrac{x + y}{2}\right)\sin\left(\dfrac{x - y}{2}\right)$

16. $\cos(x + y)\cos(x - y) = \cos^2 x - \sin^2 y$

17. $2 \sin \alpha \cos(\alpha + \beta) = \sin(2\alpha + \beta) - \sin \beta$

18. $\sin(2\alpha)\sin(2\beta) = \cos^2(\alpha - \beta) - \cos^2(\alpha + \beta)$

In Exercises 19 to 24, prove that the given equation is an identity.

19. $1 + \cos\left(\theta + \dfrac{\pi}{2}\right)\cos\left(\theta - \dfrac{\pi}{2}\right) = \cos^2 \theta$

20. $1 + \sin\left(\theta + \dfrac{\pi}{2}\right)\sin\left(\theta - \dfrac{\pi}{2}\right) = \sin^2 \theta$

21. $\tan\left(\dfrac{\pi}{2} - \theta\right) = \cot \theta$

22. $\dfrac{\cos 2\theta}{\cos \theta - \sin \theta} = \cos \theta + \sin \theta$

23. $\dfrac{\cos \theta}{1 - \tan \theta} + \dfrac{\sin \theta}{1 - \cot \theta} = \cos \theta + \sin \theta$

24. $\dfrac{2 \cos \theta + 2 \sin \theta}{\sin 2\theta} = \sec \theta + \csc \theta$

In Exercises 25 to 32, simplify the given expression.

25. $\sqrt{9 - 9 \sin^2 \theta}$

26. $\sqrt{4 + 4 \tan^2 \theta}$

27. $\csc^2 \theta - \cot^2 \theta$

28. $\cos \theta \cot \theta - \csc \theta$

29. $1 - \dfrac{\tan \theta}{\tan \theta + \cot \theta}$

30. $\dfrac{\csc \theta}{\csc \theta + \sec \theta} + \dfrac{\sec \theta}{\csc \theta - \sec \theta}$

31. (a) $\sin(10\pi - \theta)$
    (b) $\cos(\theta + 49\pi)$
    (c) $\tan(5\pi - \theta)$

32. (a) $\sin(\theta - 100\pi)$
    (b) $\cos(99\pi - \theta)$
    (c) $\tan(\theta + 15\pi)$

In Exercises 33 to 38, answer true or false.

33. $\dfrac{\sin 2x}{2} = \sin x$

34. $(\sin x + \cos x)^2 = 1$

35. $\sin(x + y) + \sin(x - y) = \sin(2xy)$

36. $\tan^{-1}x = \dfrac{\sin^{-1} x}{\cos^{-1} x}$

37. $\cot \theta + \tan \theta = \dfrac{1}{2}\csc(2\theta)$

38. $\sin^4 \theta - \cos^4 \theta = -\cos(2\theta)$

In Exercises 39 and 40, evaluate the given expression.

39. $\sin(2 \tan^{-1}(2))$

40. $\cos\left(2 \sin^{-1}\left(-\dfrac{4}{5}\right)\right)$

41. Evaluate in terms of x: $\sin(2 \cos^{-1} x)$.

The main idea for solving a general trigonometric equation is to convert the problem into one or more of the following basic equations:

$$\sin(bx) = r \qquad \cos(bx) = r \qquad \tan(bx) = r$$

Using the methods studied in Section 7.2, we can then find the desired solutions.

How do we go about converting a general trigonometric equation into one of the basic forms? There are three methods we will discuss: (1) quadratic formula, (2) factoring and the Zero Product Rule, and (3) using identities.

Our first example involves only the tangent function appearing to the first and second power. This reminds us of a quadratic equation.

Example 1

Solve the equation $\tan^2 x - 4\tan x + 2 = 0$ for x. Represent solutions using degrees accurate to two decimal places.

SOLUTION Instead of trying to find x, we first concentrate on solving for the quantity $\tan x$. Letting $u = \tan x$, we have $\tan^2 x - 4\tan x + 2 = 0$ is equivalent to

$$u^2 - 4u + 2 = 0$$

This is an ordinary quadratic equation. Using the quadratic formula,

$$u = \frac{4 \pm \sqrt{(-4)^2 - 4(1)(2)}}{2(1)}$$

$$= \frac{4 \pm 2\sqrt{2}}{2}$$

$$= 2 \pm \sqrt{2}$$

Therefore we must solve the basic equations $\tan x = 2 \pm \sqrt{2}$. Recall the steps given in Section 7.2.

$\tan x = 2 + \sqrt{2}$	*or*	$\tan x = 2 - \sqrt{2}$

Step 1: $x_0 = \tan^{-1}(2 + \sqrt{2})$

$\approx 73.68°$

Step 2: $x = 73.68° + 180° = 253.68°$ is the only other solution between $0°$ and $360°$ not coterminal with x_0.

Step 3:

$$x = \begin{cases} 73.68° + 360k° \\ 253.68° + 360k° \end{cases} \quad k \in \mathbf{Z}$$

Step 1: $x_0 = \tan^{-1}(2 - \sqrt{2})$

$\approx 30.36°$

Step 2: $x = 30.36° + 180° = 210.36°$ is the only other solution between $0°$ and $360°$ not coterminal with x_0.

Step 3:

$$x = \begin{cases} 30.36° + 360k° \\ 210.36° + 360k° \end{cases} \quad k \in \mathbf{Z}$$

All these answers give the required solutions to the original equation. If we are interested in only those solutions between 0° and 360°, then $x = 30.36°$, 73.68°, 210.36°, and 253.68°. ∎

Example 2 illustrates the method of factoring and the Zero Product Rule.

Example 2

Find all x satisfying $\sin x = \sin x \cos x$.

SOLUTION   It is tempting to divide both sides of this equation by $\sin x$, but doing so would cause us to lose some of the solutions (when $\sin x = 0$). As a general rule, it is not advisable to divide both sides of an equation by an unknown quantity unless you are certain the quantity cannot be zero. Instead, we shall get everything onto one side of the equation, factor, and use the Zero Product Rule.

$$\sin x = \sin x \cos x$$

$$\sin x - \sin x \cos x = 0$$

$$\sin x (1 - \cos x) = 0$$

| $\sin x = 0$ | or | $1 - \cos x = 0$ |

$$x_0 = \sin^{-1}(0)$$

$$= 0$$

$$-\cos x = -1$$

$$\cos x = 1$$

$$x_0 = \cos^{-1}(1)$$

$$= 0$$

The only other solution between 0 and 2π is $x = \pi$.

Thus

$$x = \begin{cases} 0 + 2\pi k \\ \pi + 2\pi k \end{cases} \quad k \in \mathbf{Z}$$

There are no other solutions between 0 and 2π. Thus

$$x = 0 + 2\pi k \quad k \in \mathbf{Z}$$

Hence all solutions are given by

$$x = \begin{cases} 0 + 2\pi k \\ \pi + 2\pi k \end{cases} \quad k \in \mathbf{Z}$$

Note that this can be simplified to read $x = \pi k, k \in \mathbf{Z}$. ∎

Examples 3 to 5 illustrate how trigonometric identities may be used to simplify a general trigonometric equation into one or more basic equations. We use degrees approximated to two decimal places when exact answers are not possible.

Example 3

Find all x satisfying $\sin x - \cos^2 x = 0$.

SOLUTION Use the fundamental identity $\cos^2 x = 1 - \sin^2 x$ to change the equation so that it involves only $\sin x$.

$$\sin x - \cos^2 x = 0$$
$$\sin x - (1 - \sin^2 x) = 0$$
$$\sin x - 1 + \sin^2 x = 0$$

Write this in descending powers of sine.

$$\sin^2 x + \sin x - 1 = 0$$

Let $u = \sin x$.

$$u^2 + u - 1 = 0$$

Using the quadratic formula,

$$u = \frac{-1 \pm \sqrt{5}}{2}$$

Therefore we must solve the basic equations

$$\sin x = \frac{-1 + \sqrt{5}}{2} \quad or \quad \sin x = \frac{-1 - \sqrt{5}}{2}$$

$$x_0 = \sin^{-1}\left(\frac{-1 + \sqrt{5}}{2}\right) \qquad x_0 = \sin^{-1}\left(\frac{-1 - \sqrt{5}}{2}\right)$$

$$\approx 38.17° \qquad \text{No solution, since}$$

This implies

$$\frac{-1 - \sqrt{5}}{2} < -1$$

$$x = \begin{cases} 38.17° + 360k° \\ 141.83° + 360k° \end{cases}$$

The answers on the left are all the values for x satisfying the original equation. ■

Example 4

Find all x satisfying $\sin x \cos x = .25$.

SOLUTION Recall identity 16, $2\sin \theta \cos \theta = \sin(2\theta)$. We have

$$\sin x \cos x = .25$$

Multiply by 2,

$$2 \sin x \cos x = .5$$

Apply identity 16:

$$\sin(2x) = .5$$

Let $\theta = 2x$.

$$\sin(\theta) = .5 \quad \text{implies} \quad \theta_o = \sin^{-1}(.5) = \frac{\pi}{6}$$

It follows that

$$\theta = \begin{cases} \dfrac{\pi}{6} + 2\pi k \\[2ex] \dfrac{5\pi}{6} + 2\pi k \end{cases} \quad k \in \mathbf{Z}$$

Dividing these results by 2, we obtain

$$x = \frac{\theta}{2} = \begin{cases} \dfrac{\pi}{12} + \pi k \\[2ex] \dfrac{5\pi}{12} + \pi k \end{cases} \quad k \in \mathbf{Z} \quad \blacksquare$$

Example 5

Find all x satisfying $2 \tan x = \sec x$.

SOLUTION
$$2 \tan x = \sec x$$

$$2\frac{\sin x}{\cos x} = \frac{1}{\cos x}$$

This is a fractional equation. We may remove the fractions by multiplying by $\cos x$, provided we remember to check our answers in the original equation.

$$\cos x \left(2\frac{\sin x}{\cos x} \right) = \cos x \left(\frac{1}{\cos x} \right)$$

$$2 \sin x = 1$$

$$\sin x = \frac{1}{2}$$

Thus

$$x_0 = \sin^{-1}\left(\frac{1}{2} \right) = \frac{\pi}{6}$$

This implies

$$x = \begin{cases} \dfrac{\pi}{6} + 2\pi k \\[2ex] \dfrac{5\pi}{6} + 2\pi k \end{cases} \quad k \in \mathbf{Z}$$

You can verify that these answers check in the original equation. $\blacksquare$

Example 6 illustrates how the ability to graph simple trigonometric functions can help us solve a trigonometric equation.

Example 6

Find all $x \in [0, 2\pi)$ satisfying $\sin x + \cos x = 1$.

SOLUTION We shall use the following idea:

> To solve $F(x) = G(x)$ graphically, look for the intersection points of the two curves $y = F(x)$ and $y = G(x)$. The x-coordinates of these points are the solutions to the equation $F(x) = G(x)$.

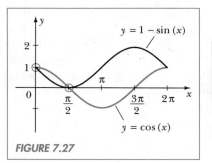

FIGURE 7.27

Rearranging the equation,

$$\sin x + \cos x = 1$$
$$\cos x = 1 - \sin x$$

This requires that we find the intersection points of $y = \cos x$ and $y = 1 - \sin x$. Figure 7.27 shows the graphs of these functions. From the graphs we find two solutions in the interval $[0, 2\pi)$: $x = 0$ and $\pi/2$. ∎

Remark

An alternate method for solving this equation without graphing is as follows:

$$\sin x + \cos x = 1$$
$$\sin x = 1 - \cos x$$
$$(\sin x)^2 = (1 - \cos x)^2$$
$$\sin^2 x = 1 - 2\cos x + \cos^2 x$$
$$0 = 1 - \sin^2 x - 2\cos x + \cos^2 x$$
$$0 = \cos^2 x - 2\cos x + \cos^2 x$$
$$0 = 2\cos^2 x - 2\cos x$$
$$0 = 2\cos x(\cos x - 1)$$

$$\cos x = 0 \qquad or \qquad \cos x - 1 = 0$$
$$\cos x = 1$$

implies $\quad x = \dfrac{\pi}{2} \; or \; \dfrac{3\pi}{2} \qquad$ implies $\quad x = 0$

Since we used the squaring operation, we must check our answers. Doing so, we find that $3\pi/2$ does not work. Thus $x = \pi/2$ and 0 are the only solutions in $[0, 2\pi)$. This agrees with our previous result.

When the functions in the equation are relatively simple, as in this case, we believe the graphical approach to solving the equation is more fun. Of course, graphing only works when the intersection points are easily observed.

In Exercises 1 to 30, find all x satisfying the given equation. Also, determine which solutions are in the interval $[0, 2\pi)$. Approximate answers may be given in degrees accurate to two decimal places. Give exact answers in radians.

1. $\sin^2 x + 3 \sin x - 3 = 0$ 2. $\tan^2 x = 3(1 - \tan x)$

3. $\tan^2 x = 3 \tan x$ 4. $\tan^2 x - 5 \tan x + 6 = 0$

5. $\sqrt{2} \cos^2 x = \cos x$ 6. $\sin^2 x = 2 \sin x$

7. $1 + \cos x = \sin x + \sin x \cos x$

8. $2 \sin x \cos x = \sin x$

9. $\sin^2 x + \cos x = 0$ 10. $\cos^2 x + \sin x = 1$

11. $\sec^2 x + \tan^2 x = 3$ 12. $\sqrt{2} \cos^2 x = \sin x$

13. $\cos(2x) + \sin x + 2 = 0$ 14. $\sin(2x) + \sin x = 0$

15. $\cos x \tan^2 x = \sin x$ 16. $\cos(2x) + \cos x + 1 = 0$

17. $\tan x = \cos x$ 18. $\tan x = \sin x$

19. $\sin x \cos x = .5$ 20. $\sin x \cos x = .225$

21. $\cos^2 x = \sin^2 x + .9511$ 22. $\cos^2 x = \sin^2 x$

23. $3 \cos x - \sin x = 0$ 24. $2 \sin x + \cos x = 0$

25. $\sin x \cos x = 0$ 26. $\sin x \cos x = 1$

27. $\sin x + 1 = \cos x$ 28. $\sin x - \cos x = 1$

29. $\sec x \csc x = \tan x$ 30. $\sec x + \csc x = 1$

In Exercises 31 to 34, find all x in the interval $[0, 2\pi]$ satisfying the inequality. Graph the two functions involved on the same axes to aid in finding the solution.

31. $\sin x < \sin(2x)$

32. $\sin x < \cos(2x)$

33. $\cos(2x) < \cos x$

34. $\sin(2x) > \cos x$

35. Graph $F(x) = \tan x$ and $G(x) = \sec x$ over the interval $\left(-\dfrac{\pi}{2}, \dfrac{\pi}{2} \right)$ on the same axes. Do these curves intersect? Explain.

36. Graph $F(x) = \csc x$ and $G(x) = \cot x$ over the interval $(0, \pi)$ on the same axes. Do these curves intersect? Explain.

37. Point P moves along the vertical line $x = 2$. Let θ be the measure of the angle in standard position whose terminal side passes through P.
    (a) Find the distance d between P and the point $(0, 1)$ as a function of θ, $d = F(\theta)$. See Figure 7.28. (*Hint:* find the coordinates of P in terms of θ and use the distance formula.)
    (b) Find θ when $d = 3$.

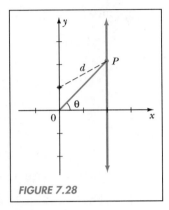

FIGURE 7.28

38. Point P moves along the horizontal line $y = 2$. Let θ be the measure of the angle in standard position whose terminal side passes through P.
    (a) Find the distance d between P and the point $(2, 1)$ as a function of θ, $d = F(\theta)$. See Figure 7.29.
    (b) Find θ when $d = 2$.

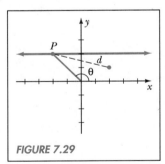

FIGURE 7.29

In Exercises 1 to 6, evaluate the expression without using a calculator. Some expressions may be undefined.

1. $\sin^{-1}(-1) + \cos^{-1}(1) + \tan^{-1}(1)$

2. $\tan^{-1}(-\sqrt{3})$

3. $\cos^{-1}(-1/2)$ 4. $\sin^{-1}(\sqrt{2})$

5. $\sin\left(\tan^{-1}\left(\dfrac{5}{12}\right)\right)$ 6. $\tan\left(\cos^{-1}\left(\dfrac{-1}{5}\right)\right)$

7. Evaluate in terms of x: $\sec(\sin^{-1} x)$

In Exercises 8 and 9, solve for all x. Determine which solutions are in the interval $[0, 2\pi)$.

8. $\sin(3x) = \dfrac{1}{2}$ 9. $\tan(-x) = 2$

In Exercises 10 to 17, simplify the expression given in the left-hand column. Find the correct answer among the choices in the right-hand column.

10. $\sqrt{16 + 16 \tan^2 \theta}$

11. $(\sin \theta + \cos \theta)^2$

12. $\sqrt{16 - 16 \sin^2 \theta}$

13. $(\sin \theta - \cos \theta)(\csc \theta + \sec \theta)$

14. $\sec^2 \theta - \tan^2 \theta$

15. $\dfrac{\sin \theta}{1 - \cos \theta}$

16. $2 \sin^2 \theta$

17. $\dfrac{\cot^2 \theta - \tan^2 \theta}{\csc^2 \theta \sec^2 \theta}$

(A) 0
(B) 1
(C) $\csc \theta + \cot \theta$
(D) $\tan \theta + \cot \theta$
(E) $\tan \theta - \cot \theta$
(F) $1 + \sin(2\theta)$
(G) $1 + \cos(2\theta)$
(H) $1 - \cos(2\theta)$
(I) $\cos(2\theta)$
(J) $4|\cos \theta|$
(K) $4|\sec \theta|$
(L) $4|\csc \theta|$

In Exercises 18 to 22, prove each identity.

18. $\dfrac{\cot \theta - \tan \theta}{\cot \theta + \tan \theta} = \cos(2\theta)$

19. $\sec^2 \theta + \csc^2 \theta = \sec^2 \theta \csc^2 \theta$

20. $\tan \alpha + \tan \beta = \dfrac{\sin(\alpha + \beta)}{\cos \alpha \cos \beta}$

21. $\sin^2 x - \sin^2 y = \sin(x + y)\sin(x - y)$

22. $\csc^2 \theta - \sec^2 \theta = 4 \cot(2\theta)\csc(2\theta)$

In Exercises 23 to 25, find the exact value of the given expression.

23. $\cos(165°)$

24. $\sin\left(\tan^{-1}(3) + \cos^{-1}\left(\dfrac{3}{5}\right)\right)$

25. $\tan\left(\dfrac{\pi}{8}\right)$

26. Suppose $\sin \theta = .6$ and $\cos \theta < 0$. Find $\tan(2\theta)$.

27. Write $\cos^3 x$ in the form $a \cos x + b \cos(3x)$. (*Hint:* use identity 18(b) and then identity 23.

28. Write $\sin^3 x$ in the form $a \sin x + b \sin(3x)$. (*Hint:* use identity 18(a) and then identity 21.

29. Write $\tan(3x)$ in terms of $\tan(x)$.

In Exercises 30 to 33, find all x satisfying the given equation. Determine which solutions are in the interval $[0, 2\pi)$. Give exact answers in radians, approximate answers to two decimal places in degrees.

30. $7 \sin x + 5 = 2 \cos^2 x$ 31. $\cos^2 x = 3 \cos x - 1$

32. $\sin x = 1 + \cos x$ 33. $\tan x = \sec x$

34. Express each function in terms of the sine function.
   (a) $\csc x$ (b) $\cos x$
   (c) $\sec x$ (d) $\tan x$
   (e) $\cot x$

35. Answer true or false.
   (a) $\sin(2xy) = 2 \sin x \cos y$
   (b) $\sin^{-1}(.3) = \dfrac{1}{\sin(.3)}$
   (c) $\sec^2 \theta = 1 - \csc^2 \theta$
   (d) The equation $\sin x + \cos x = 2$ has no solutions.
   (e) $\sin(-\theta)\cos(-\theta)\tan(-\theta) = -\sin^2 \theta$
   (f) $\cos^2 \theta = \dfrac{1}{2} + \dfrac{1}{2}\cos(2\theta)$

CHAPTER 8 Applications of Trigonometry

8.1 RIGHT TRIANGLE APPLICATIONS

In any right triangle, if we know one of the sides and one of the acute angles, then we can easily determine any of the other sides by using the trigonometric functions. In general, given a right triangle with side s and acute angle θ, we can find an unknown side x as follows:

- Step 1. Form the ratio $\dfrac{x}{s}$ and determine which trigonometric function F of θ this represents.

- Step 2. Solve $\dfrac{x}{s} = F(\theta)$ for x, $x = sF(\theta)$.

Example 1

Find the sides x and y for the right triangle in Figure 8.1. Approximate to two decimal places.

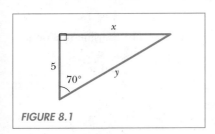

FIGURE 8.1

SOLUTION The ratio $\dfrac{x}{5}$ is $\dfrac{\text{opposite leg}}{\text{adjacent leg}}$ for the angle 70°. This is the tangent function. Therefore

$$\frac{x}{5} = \tan(70°)$$

Solving for x,

$$x = 5\tan(70°) \approx 13.74$$

Now there are two options available for finding y. We may use trigonometry again or the Pythagorean Theorem.

Method 1. The ratio $\dfrac{y}{5}$ is $\dfrac{\text{hypotenuse}}{\text{adjacent leg}}$ for the angle 70°. This is the secant function. Therefore

$$\frac{y}{5} = \sec(70°)$$

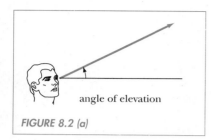

FIGURE 8.2 (a)

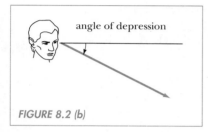

FIGURE 8.2 (b)

Solving for y,

$$y = 5 \sec(70°) = \frac{5}{\cos(70°)} \approx 14.62$$

Method 2.   By the Pythagorean Theorem,

$$y = \sqrt{5^2 + x^2} \approx \sqrt{25 + (13.74)^2} = \sqrt{213.7876} \approx 14.62 \ \blacksquare$$

The exercises contain the phrases *angle of elevation* and *angle of depression*. The angle of elevation is the angle between the horizontal and the line of sight from the observer's eye to an object *above* the horizontal. This situation is illustrated in Figure 8.2(a). Figure 8.2(b) shows an angle of depression, which is measured between the horizontal and the line of sight from the observer to an object *below* the horizontal.

Example   2

From the top of a 232-foot-high observation tower, an object appears at a 13° angle of depression on the ground below. Assuming the ground is level, find the distance from the base of the tower to the object. Approximate the answer to the nearest foot.

SOLUTION   We begin by defining the quantity we wish to find. Thus we let x represent the distance from the base of the tower to the object. Next, draw a picture of the situation and label (Figure 8.3).

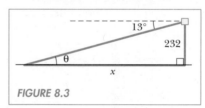

FIGURE 8.3

Note that the horizontal from the observation point at the top of the tower is parallel to the ground below. Thus $\theta = 13°$ since parallel lines cut by a transversal form equal alternate interior angles. Now, the ratio $\frac{x}{232}$ is $\frac{\text{adjacent leg}}{\text{opposite leg}}$ for the angle θ. Therefore

$$\frac{x}{232} = \cot(\theta)$$

Substituting 13° for θ and solving for x, we have

$$x = 232 \cot(13°)$$

$$= 232\left(\frac{1}{\tan(13°)}\right)$$

$$\approx 1005 \text{ feet} \ \blacksquare$$

Example 3

Find the area of triangle ABC if $A = 65°$, $AB = 15$, and $AC = 8$.

SOLUTION   Recall that the area of a triangle is given by

$$\text{Area} = \frac{1}{2}(\text{base})(\text{height})$$

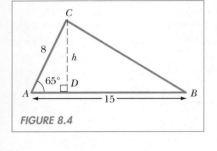

FIGURE 8.4

If we choose to use side AB for the base, then the height h is given by the length of the perpendicular CD drawn from the vertex C to the base. See Figure 8.4. To find h, we note in right triangle ADC that

$$\frac{h}{8} = \frac{\text{opposite leg}}{\text{hypotenuse}} \quad \text{for angle } A$$

Thus

$$\frac{h}{8} = \sin(65°)$$

$$h = 8\sin(65°)$$

It follows that

$$\text{Area} = \frac{1}{2}(15)8\sin(65°) \approx 54.38 \ \blacksquare$$

Example 3 can be generalized to the following result.

> Given two sides a and b and the included angle θ in a triangle, then
> $$\text{Area} = \frac{1}{2}ab \sin \theta$$

Example 4

An unidentified flying object (UFO) is sighted in line with two observation points A and B on the ground. From A, the angle of elevation of the UFO is $25°$, and from B, the angle of elevation is $27°$. If 900 feet separate A and B, find the altitude of the UFO.

SOLUTION   Let y be the altitude of the UFO. Figure 8.5 is a diagram of the situation. To find y, we introduce the letter x to represent the distance BC in the diagram. Now we can write two equations for the unknowns y and x. In the right triangle BCU, we have

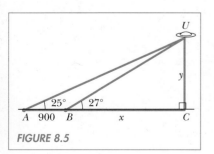

FIGURE 8.5

$$\frac{y}{x} = \tan(27°) \tag{1}$$

In the right triangle *ACU*, we have

$$\frac{y}{x + 900} = \tan(25°) \tag{2}$$

From Equation (1) we find that $x = y/\tan(27°)$. Substituting this into Equation (2),

$$\frac{y}{\dfrac{y}{\tan(27°)} + 900°} = \tan(25°)$$

Simplify the left side.

$$\frac{y \tan(27°)}{y + 900 \tan(27°)} = \tan(25°)$$

Now solve for *y*.

$$y \tan(27°) = y \tan(25°) + 900 \tan(27°) \tan(25°)$$

$$y \tan(27°) - y \tan(25°) = 900 \tan(27°) \tan(25°)$$

$$y\big(\tan(27°) - \tan(25°)\big) = 900 \tan(27°) \tan(25°)$$

$$y = \frac{900 \tan(27°) \tan(25°)}{\tan(27°) - \tan(25°)}$$

Using our calculator, we obtain to the nearest foot,

$$y \approx 4948 \text{ feet } \blacksquare$$

EXERCISES 8.1

Approximate all answers to two decimal places.

In Exercises 1 to 6, find *x* and *y* in the right triangles shown. Write the equation first and then compute.

1.

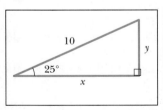

2.

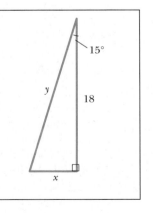

3.

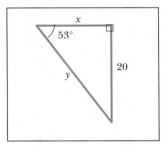

4.

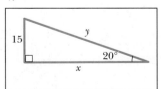

5.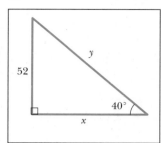

6.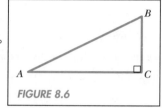

In Exercises 7 to 10, find the missing sides and angles in triangle ABC in Figure 8.6.

7. $AC = 25$, $A = 28.28°$

8. $AB = 144$, $B = 70.2°$

9. $BC = 7.91$, $B = 81.35°$

10. $BC = 40$, $A = 32.25°$

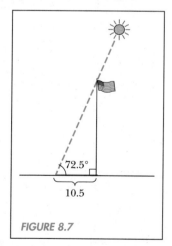

FIGURE 8.6

11. The angle of elevation of the top of a building is found to be 62.65° as measured from a point on the ground 30 feet from its base. How tall is this building?

12. In Figure 8.7, the flagpole casts a shadow 10.5 feet in length on level ground when the angle of elevation of the sun is 72.5°. Find the height of the flagpole.

FIGURE 8.7

13. A tree casts a shadow 23 feet long on level ground when the angle of elevation of the sun is 40.29°. How tall is the tree?

14. A 20-foot ladder is leaning against a building. If the bottom of the ladder makes a 77.2° angle with respect to the ground, how high up the building is the top of the ladder?

15. Suppose that a plane flies 1000 feet at a 5° angle with respect to the horizontal. How many feet does the plane gain in altitude?

16. A car travels 1/4 mile along a highway that has a 3° grade (it makes an angle of 3° with respect to the horizontal). How many feet has the car gained in altitude?

17. An isosceles triangle has base angles 1 radian and sides 14.32 centimeters. Find the base and altitude.

18. An isosceles triangle has base angles 50° and base 26 centimeters. Find the sides.

19. An isosceles triangle has altitude 100 centimeters and vertex angle 2°. Find the base.

20. The angle of depression of a point A, as measured from the top of a 212-foot-tall building, is 0.3135 radians. How far is A from the base of the building?

21. From the top of a 150-foot lighthouse, a boat is sighted with angle of depression 10°. How far is the boat from the base of the lighthouse?

22. A rocket is being tracked at all times by a camera 6000 feet from the launch pad. At time $t = 0$, the rocket blasts off. At $t = 5$ seconds, the camera angle is 20°. At $t = 10$ seconds, the camera angle is 40°. Assume that the rocket travels straight up from the ground.
    (a) What is the altitude of the rocket at $t = 5$?
    (b) What is the altitude of the rocket at $t = 10$?
    (c) What is the average speed of the rocket for the time interval $0 \leq t \leq 5$?
    (d) What is the average speed of the rocket for the time interval $5 \leq t \leq 10$?

23. A weather balloon is being tracked at all times from a point situated 100 feet from the launching site. Four seconds after the balloon has been launched, the balloon's angle of elevation is 40°. After 10 seconds, the angle of elevation is 70°. Assuming the balloon floats exactly straight up, find its average speed during the time period from 4 to 10 seconds.

In Exercises 24 and 25, find the area of triangle ABC.

24. $A = 50°$, $AB = 30$, $AC = 20$

25. $C = 66°$, $AC = 18.4$, $BC = 31$

26. Find the area of triangle PQR in the x, y-plane if P and Q have coordinates $P(-2, 1)$, $Q(3, 2)$, length $QR = 6$, and angle $PQR = 76°$.

In Exercises 27 to 29, find the area of the trapezoid $ABCD$ in Figure 8.8, with $AB = 16$, $AD = 10$, and the given information.

27. $A = B = 105°$

28. $A = 110°$, $B = 125°$

29. $A = 1.1868$,
    $B = 2.2689$

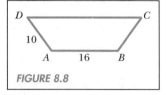

FIGURE 8.8

30. In Example 4, suppose the UFO is sighted a second time in line with A and B so that the angles of elevation are 29° from A and 31° from B. Find the altitude of the UFO in this case.

31. Repeat Exercise 30 for angles of elevation 33° from A and 34° from B.

In Exercises 32 to 35, in Figure 8.9 find the x- and y-coordinates of P, given α, β, and c.

32. $\alpha = 32°$, $\beta = 40°$,
    $c = 5$

33. $\alpha = 20°$, $\beta = 35°$,
    $c = 5$

34. $\alpha = .9273$,
    $\beta = 1.3258$, $c = 10$

35. $\alpha = 50°$, $\beta = 100°$,
    $c = 10$

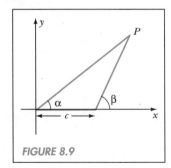

FIGURE 8.9

36. Find the x- and y-coordinates of point P in Figure 8.9 in terms of α, β, and c. Assume α and β are between 0° and 180°.

37. You have just finished a long climb to the top of a mountain. As you look down below, you see in the same line of sight two buildings, A and B. The angle of depression for A is 50° and for B, 30°. If A and B are at sea level (elevation 0), and the distance between them is 2000 feet, find the elevation of the mountain top.

38. The great pyramid in Egypt has a square base with sides 130.56 meters long. (See Figure 8.10.) The slant angle of each side (how much it leans from the vertical) is 38.17194°. Find the height of the pyramid.

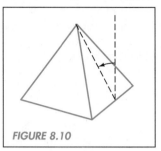

FIGURE 8.10

39. Two objects A and B move in the x, y-plane as follows. A leaves the origin at an angle of 25° at 3 units per second. Two seconds later, B leaves the origin at an angle of 35° at 4 units per second. Find the distance between A and B 10 seconds after A leaves the origin. (*Hint:* find the coordinates of A and B and then use the distance formula.)

40. An airplane is flying with constant velocity and fixed altitude of 1000 feet. An observer on the ground in line with the flight of the plane measures the plane's angle of elevation as 25°; 6 seconds later the angle of elevation is found to be 40°. Find the velocity of the plane.

41. Consider triangle AOB shown in the circle $x^2 + y^2 = r^2$ in Figure 8.11. Show that the area of triangle AOB is given by $\frac{1}{2}r^2 \sin \theta$ provided $0 \le \theta \le \pi$.

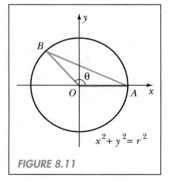

FIGURE 8.11

42. Show that the area of the shaded region inside the circle $x^2 + y^2 = r^2$ in Figure 8.11 is given by $\frac{1}{2}\theta r^2 - \frac{1}{2}r^2 \sin\theta$, where θ is expressed in radians, $0 < \theta < \pi$.

43. A 100-foot-long crane is raised at a 50° angle (Figure 8.12). From the top of the crane two lines are dropped to the ground below, each line 80 feet in length. How far apart can we place the ends of these lines so that they remain on the ground?

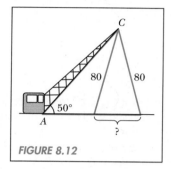

FIGURE 8.12

44. In Figure 8.13, find $d_1 + d_2$ as a function of θ. Assume a, b, and c are constants, $0 < \theta < \pi/2$.

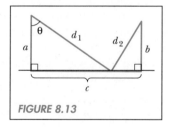

FIGURE 8.13

45. Find the area of triangle ACD in Figure 8.14 as a function of θ. Angle ACB is a right angle, $AC = 10$, $BC = 10$.

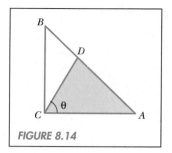

FIGURE 8.14

46. Repeat Exercise 45 assuming $AC = 20$ and $BC = 10$.

47. Five points A, B, C, D, and E (Figure 8.15) are equally spaced on the circle of radius a. Find the area of the star in terms of a. *Hint:* In a circle any inscribed angle is half of its intercepted arc.

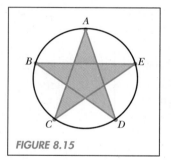

FIGURE 8.15

8.2 SOLVING TRIANGLES

A triangle consists of six parts: three angles and three sides. We say that a triangle is *solved* when all six parts are completely determined. The congruence theorems from plane geometry tell us that a minimum of three parts of a triangle must be known for the remaining parts to be determined.

For our purposes we shall interpret these theorems in terms of a single triangle. For example, if we know two angles and the included side of a triangle, then it follows from the ASA congruence theorem that the remaining parts of the triangle must be uniquely determined. We call such a triangle an **ASA triangle**. We also define an **SAS triangle** when two sides and an included angle are given, and an **SSS triangle** when all three sides are given. Once again, the SAS and SSS congruence theorems tell us that the remaining parts in these triangles are uniquely determined. We shall also consider **SSA triangles** when two sides and a nonincluded angle are given. In this case, since there is no SSA congruence theorem, we may or may not be able to determine the remaining parts.

   Our approach to solving a triangle involves two steps. First, we classify the triangle according to the information given, ASA, SAS, SSS, or SSA. Second, we use the method that corresponds to our given triangle as indicated in the following table.

Triangle	Method for Solution
ASA	Law of Sines
SAS	Law of Cosines
SSS	Law of Cosines
SSA	Law of Sines

   As shown in the table, there are two methods for solving triangles: the Law of Sines and the Law of Cosines. We shall explain these methods shortly, but first we must agree on some notation. For the remainder of this section we label both the vertices and the angles of a triangle with the capital letters A, B, and C. The lengths of the sides opposite angles A, B, and C will be denoted by a, b, and c, respectively. See Figure 8.16. Our discussion consists of three cases: ASA triangles, SAS and SSS triangles, and, finally, SSA triangles.

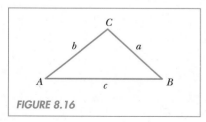

FIGURE 8.16

ASA Triangles. These triangles are solved using the Law of Sines. We begin with the statement and proof of this law.

> **Law of Sines**
>
> In any triangle, the ratio formed by the sine of one of its angles divided by the length of the opposite side is the same for all three angles.
>
> 1. $\dfrac{\sin A}{a} = \dfrac{\sin B}{b}$
>
> 2. $\dfrac{\sin A}{a} = \dfrac{\sin C}{c}$
>
> 3. $\dfrac{\sin B}{b} = \dfrac{\sin C}{c}$

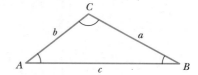

PROOF: Draw an altitude h from vertex C to side AB. Figure 8.17(a) to (c) shows the various possible situations.

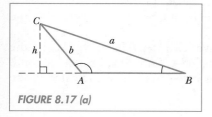

FIGURE 8.17 (a)

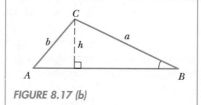

FIGURE 8.17 (b)

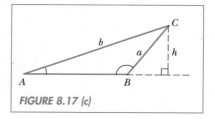

FIGURE 8.17 (c)

In any case, using the reference triangle definition of sine,

$$\sin A = \frac{h}{b} \quad \text{and} \quad \sin B = \frac{h}{a}$$

This implies

$$b \sin A = h \quad \text{and} \quad a \sin B = h$$

It follows that

$$b \sin A = a \sin B$$

Now, dividing by ab,

$$\frac{\sin A}{a} = \frac{\sin B}{b}$$

This establishes Equation (1). Equations (2) and (3) are proven similarly. This completes the proof of the Law of Sines. ■

Example 1

Solve triangle ABC given $A = 34°$, $B = 67°$, and $a = 100$.

SOLUTION Recall that the sum of the angles in any triangle is 180°. Thus

$$A + B + C = 180° \quad \text{implies} \quad 34° + 67° + C = 180°$$
$$101° + C = 180°$$
$$C = 79°$$

Therefore we know two angles and an included side, as shown in Figure 8.18.

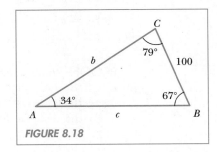

FIGURE 8.18

Since we have an ASA triangle, we use the Law of Sines.

$$\frac{\sin(34°)}{100} = \frac{\sin(67°)}{b}$$

This implies

$$b = \frac{100\sin(67°)}{\sin(34°)} \approx 164.61$$

Next, we solve for c.

$$\frac{\sin(34°)}{100} = \frac{\sin(79°)}{c}$$

This implies

$$c = \frac{100\sin(79°)}{\sin(34°)} \approx 175.54$$

We summarize all the results as follows:

$$A = 34° \qquad a = 100$$
$$B = 67° \qquad b = 164.61$$
$$C = 79° \qquad c = 175.54 \quad \blacksquare$$

SAS and SSS Triangles. We solve SAS and SSS triangles using the Law of Cosines.

> ## Law of Cosines
>
> In a triangle, the square of any side is equal to the sum of the squares of the two remaining sides minus twice their product with the cosine of the included angle.
>
> 1. $c^2 = a^2 + b^2 - 2ab \cos C$
> 2. $a^2 = b^2 + c^2 - 2bc \cos A$
> 3. $b^2 = a^2 + c^2 - 2ac \cos B$

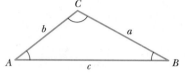

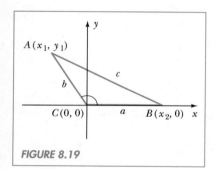

FIGURE 8.19

PROOF: For triangle ABC, place angle C in standard position with initial side CB. Let vertices A and B have coordinates (x_1, y_1) and $(x_2, 0)$, respectively. See Figure 8.19.

By the distance formula,

$$c^2 = (AB)^2$$
$$= (x_2 - x_1)^2 + (0 - y_1)^2$$
$$= x_2^2 - 2x_1x_2 + x_1^2 + y_1^2$$

Note that $x_2 = a$ and $x_1^2 + y_1^2 = (AC)^2 = b^2$. Also, by the definition of cosine,

$$\cos C = \frac{x_1}{b}$$

It follows that $x_1 = b \cos C$. Substituting these values into the equation for c^2, we have

$$c^2 = x_2^2 - 2x_1x_2 + x_1^2 + y_1^2$$
$$= a^2 - 2(b \cos C)a + b^2$$
$$= a^2 + b^2 - 2ab \cos C$$

This establishes Equation (1). Similar arguments may be used to derive Equations (2) and (3). This completes the proof of the Law of Cosines. ∎

Example 2

Solve triangle ABC in Figure 8.20.

SOLUTION   Since two sides and the included angle are given, we have an SAS triangle, which requires the Law of Cosines. Let us begin by finding a.

FIGURE 8.20

$$a^2 = b^2 + c^2 - 2bc \cos A$$
$$= 14^2 + 20^2 - 2(14)(20)\cos(25°)$$
$$= 596 - 560 \cos(25°)$$

Using the calculator, we compute a accurate to three decimal places.

$$a = \sqrt{596 - 560 \cos 25°} \approx 9.406$$

Next, we use the Law of Cosines once again to find angle C. From the equation

$$c^2 = a^2 + b^2 - 2ab \cos C$$

we have

$$\cos C = \frac{c^2 - a^2 - b^2}{-2ab}$$

Therefore

$$C = \cos^{-1}\left(\frac{c^2 - a^2 - b^2}{-2ab}\right)$$
$$\approx \cos^{-1}\left(\frac{20^2 - 9.406^2 - 14^2}{-2(9.406)(14)}\right)$$

Note that this number is approximate since we are using an approximate value for a in the calculation. With the help of our calculator, we find that

$$C \approx 116.02°$$

Finally, we obtain angle B using the fact that the sum of the angles in any triangle must be 180°. Thus

$$A + B + C = 180°$$
$$B = 180° - A - C$$
$$\approx 180° - 25° - 116.02°$$
$$= 38.98°$$

We summarize our results as follows:

$$A = 25° \qquad a = 9.406$$
$$B = 38.98° \qquad b = 14$$
$$C = 116.02° \qquad c = 20 \quad \blacksquare$$

Remark

After finding side a in Example 2, why didn't we use the Law of Sines to determine angle C instead of the Law of Cosines? Indeed, it would seem that the calculations might be easier. Let us see what happens.

Applying the Law of Sines to find angle C in Figure 8.21, we have

$$\frac{\sin C}{20} = \frac{\sin(25°)}{9.406}$$

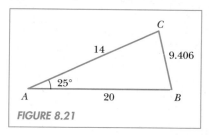

FIGURE 8.21

Solving for C,

$$\sin C = \frac{20 \sin(25°)}{9.406}$$

$$C = \sin^{-1}\left(\frac{20 \sin(25°)}{9.406}\right) \approx 63.98°$$

The computation was made using the inverse sine function on our calculator. Unfortunately, this answer cannot be correct because it implies that angle B measures $91.02°$, which would make angle B the largest angle in the triangle. However, angle C, being opposite the largest side in the triangle, must be the largest angle. Therefore we have a contradiction.

Why does the Law of Sines lead us to an incorrect answer? Actually, there is nothing wrong with the Law of Sines; we just neglected to find *all* possible solutions for angle C. Since C could be between $0°$ and $180°$, there are two solutions to the equation

$$\sin C = \left(\frac{20 \sin(25°)}{9.406}\right)$$

We use the inverse sine function to obtain the first solution of $63.98°$. The other solution is found in the second quadrant with the same reference angle of $63.98°$; this is $180° - 63.98° = 116.02°$. As indicated above, the first quadrant angle cannot be correct. We conclude that $C \approx 116.02°$.

Compare this solution with the one using the Law of Cosines given in Example 2. Using the inverse cosine function in the example, we were able to find the correct value of C without having to check any other possibilities. From the definitions of $\cos^{-1}(x)$ and $\sin^{-1}(x)$ we know

- $\cos^{-1}(x)$ detects angles between $0°$ and $180°$.
- $\sin^{-1}(x)$ detects angles between $-90°$ and $90°$.

Thus, if we are looking for an angle that happens to be larger than $90°$, the inverse cosine will find it directly, whereas the inverse sine will not, which is why we recommend using the Law of Cosines (and therefore the inverse cosine function) when attempting to find an unknown angle in an SAS or SSS triangle.

Example 3

Solve the triangle determined by the vertices $A(-2, 4)$, $B(6, -2)$, and $C(5, 1)$.

SOLUTION   We use the distance formula to find the lengths of the sides.

$$a = BC = \sqrt{(5-6)^2 + (1+2)^2} = \sqrt{10}$$

$$b = AC = \sqrt{(5+2)^2 + (1-4)^2} = \sqrt{58}$$

$$c = AB = \sqrt{(6+2)^2 + (-2-4)^2} = \sqrt{100} = 10$$

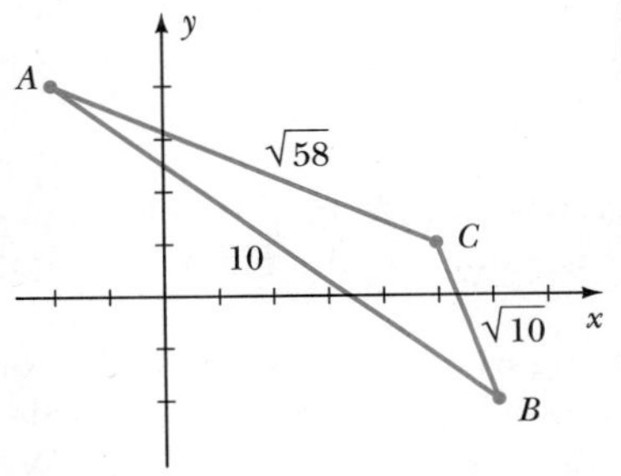

Thus we have the SSS triangle shown in Figure 8.22. Using the Law of Cosines, we find angle A as follows:

$$\cos A = \frac{a^2 - b^2 - c^2}{-2bc}$$

FIGURE 8.22

This implies

$$A = \cos^{-1}\left(\frac{10 - 58 - 100}{-2(\sqrt{58})(10)}\right) \approx 13.67°$$

Next, we find angle B. Again, from the Law of Cosines,

$$\cos B = \frac{b^2 - a^2 - c^2}{-2ac}$$

This implies

$$B = \cos^{-1}\left(\frac{58 - 10 - 100}{-2(\sqrt{10})(10)}\right) \approx 34.70°$$

Finally,

$$C = 180 - A - B \approx 180° - 13.67° - 34.70° = 131.63°$$

In summary, we have

$$A = 13.67° \qquad a = \sqrt{10}$$
$$B = 34.70° \qquad b = \sqrt{58}$$
$$C = 131.63° \qquad c = 10 \quad \blacksquare$$

SSA Triangles.   When two sides and a *nonincluded* angle are given for a triangle, a solution may or may not exist. For example, suppose we know angle A and sides a and b. Figure 8.23 on the next page shows the situation when $A \geq 90°$. We have two possibilities: either $a \leq b$ and no triangle exists, or $a > b$ and one triangle exists.

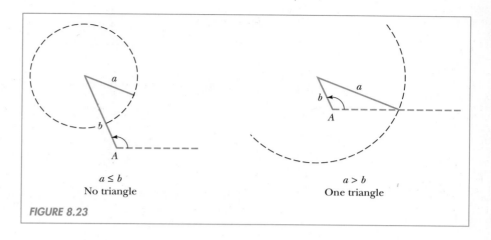

FIGURE 8.23

When $A < 90°$, we have three possible outcomes consisting of two triangles (the ambiguous case), exactly one triangle, or none at all. Figure 8.24(a) to (c) illustrates these possibilities.

To determine the parts for a given SSA triangle, we shall use the Law of Sines, which means we will be using the inverse sine function to find an angle. As mentioned, this procedure requires checking two possible solutions. We must keep in mind that it is possible to find as many as two different yet valid solutions or none at all.

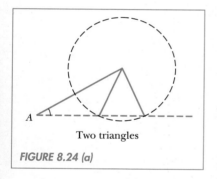

FIGURE 8.24 (a)

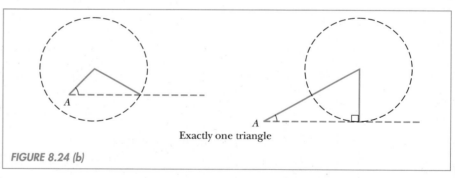

FIGURE 8.24 (b)

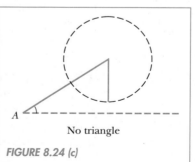

FIGURE 8.24 (c)

Example 4

Solve triangle ABC if $A = 35°$, $b = 20$, and (a) $a = 15$, (b) $a = 30$, (c) $a = 10$.

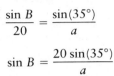

FIGURE 8.25

SOLUTION   Figure 8.25 shows that for each given value of a, triangle ABC will be an SSA triangle. Using the Law of Sines to find angle B, we have

$$\frac{\sin B}{20} = \frac{\sin(35°)}{a}$$

$$\sin B = \frac{20\sin(35°)}{a}$$

We shall use this equation to begin the solution of triangle ABC for each of the given values of a.

(a) $a = 15$.   Using the inverse sine function on our calculator, the equation

$$\sin B = \frac{20\sin(35°)}{15}$$

implies $B \approx 49.89°$. We also know that the second quadrant angle, given by $180° - 49.89° = 130.11°$, also satisfies the equation. We now check that each one of these answers for B corresponds to a valid solution for triangle ABC.

Using $B = 49.89°$, we find

$$C = 180° - A - B \approx 180° - 35° - 49.89° = 95.11°$$

Now, for side c we use the Law of Sines once again.

$$\frac{\sin(35°)}{15} = \frac{\sin(95.11°)}{c}$$

This implies

$$c = \frac{15\sin(95.11°)}{\sin(35°)} \approx 26.05$$

In this case, the solution for triangle ABC is given by

$$A = 35° \qquad a = 15$$
$$B = 49.89° \qquad b = 20$$
$$C = 95.11° \qquad c = 26.05$$

Next, using $B = 130.11°$, we find

$$C = 180° - A - B \approx 180° - 35° - 130.11° = 14.89°$$

You can verify that the Law of Sines now implies

$$c = \frac{15\sin(14.89°)}{\sin(35°)} \approx 6.72$$

In this case, the solution for triangle ABC is given by

$$A = 35° \qquad a = 15$$

$$B = 130.11° \qquad b = 20$$

$$C = 14.89° \qquad c = 6.72$$

Thus we have two solutions for triangle ABC when $a = 15$.

(b) $a = 30$. Using the inverse sine function on our calculator, the equation

$$\sin B = \frac{20 \sin(35°)}{30}$$

implies $B \approx 22.48°$. The other possible solution is given by $180° - 22.48° = 157.52°$. However, if $B = 157.52°$, then $A + B = 35 + 157.52 = 192.52°$. This contradicts the fact that the sum of the angles in a triangle must equal $180°$. Therefore we have only one solution for triangle ABC corresponding to $B = 22.48°$.

To determine angle C, we have

$$C = 180 - A - B \approx 180° - 35° - 22.48° = 122.52°$$

Now, it follows from the Law of Sines that

$$c = \frac{30 \sin(122.52°)}{\sin(35°)} \approx 44.10$$

The solution for triangle ABC is therefore

$$A = 35° \qquad a = 30$$

$$B = 22.48° \qquad b = 20$$

$$C = 122.52° \qquad c = 44.10$$

(c) $a = 10$. We note that the equation for B,

$$\sin B = \frac{20 \sin(35°)}{10}$$

yields

$$\sin B \approx 1.15$$

which is not possible. (Recall that $-1 \le \sin\theta \le 1$ for any value of θ.) In this case we must conclude that there is no solution for triangle ABC. ∎

We close this section with the application in Example 5.

Example 5

The angle of elevation of a window on a building is 65° as measured from a point on the ground below. From the same point we find that the angle of elevation for the top of the building is 70°. If the distance from the window to the top of the building is 10 feet, how tall is the building? Assume the ground is level.

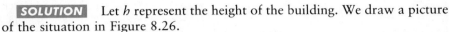 Let h represent the height of the building. We draw a picture of the situation in Figure 8.26.

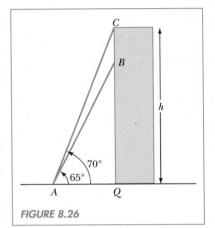

FIGURE 8.26

We can find h if we know the length AC as follows. Using the right triangle AQC,

$$\frac{h}{AC} = \sin 70° \quad \text{or} \quad h = AC \sin 70° \qquad (*)$$

Therefore we shall first try to find the length AC. To do this, we consider triangle ABC. Can it be solved? We have $A = 70° - 65° = 5°$ and $BC = 10$. What else is known? Since triangle AQC is a right triangle with $\angle CAQ = 70°$ and $\angle CQA = 90°$, we must have

$$C = 180° - \angle CAQ - \angle CQA = 180° - 70° - 90° = 20°$$

It follows that angle B in triangle ABC is given by

$$B = 180° - A - C = 180° - 5° - 20° = 155°$$

Thus, triangle ABC is an ASA triangle, as shown in Figure 8.27.

Using the Law of Sines in triangle ABC, we have

$$\frac{\sin 5°}{10} = \frac{\sin 155°}{AC}$$

Solving for AC, we get

$$AC = \frac{10 \sin 155°}{\sin 5°}$$

Substituting this expression for AC into Equation $(*)$, we find

$$h = (AC)\sin 70° = \frac{10 \sin 155°}{\sin 5°} \sin 70° \approx 45.57 \text{ feet} \quad \blacksquare$$

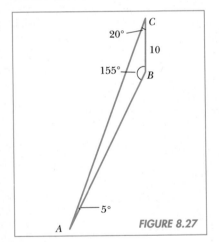

FIGURE 8.27

Remark

It is possible to solve this problem without the Law of Sines by using only right triangles. The solution is similar to what was done in Example 4, Section 8.1. However, we believe the method shown here, using the Law of Sines, requires less work.

EXERCISES 8.2

In Exercises 1 to 12, from the given information, classify triangle ABC as ASA, SAS, SSS, or SSA. Then solve triangle ABC.

1. $A = 50°$
   $B = 95°$
   $c = 100$

2. $A = 50°$
   $C = 60°$
   $c = 20$

3. $C = 2.5$ radians
   $a = 123.4$
   $b = 95$

4. $C = 33°$
   $a = 15$
   $b = 10$

5. $a = 8$
   $b = 5$
   $c = 10$

6. $a = 30$
   $b = 40$
   $c = 60$

7. $C = 19°$
   $a = 30$
   $c = 20$

8. $A = 28°$
   $a = 25$
   $c = 45$

9. $A = 15°$
   $a = 40$
   $c = 10$

10. $A = 50°$
    $a = 15$
    $c = 45$

11. A has coordinates $A(-3, 4)$, B has coordinates $B(5, 0)$, $A = 65°$, $B = 54°$.

12. The coordinates of A, B, and C are $A(-3, 0)$, $B(1, 2)$, and $C(5, -1)$.

In Exercises 13 and 14, find the coordinates of $P(x, y)$ from the given information.

13. $A(5, 0)$, $B(0, 0)$, angle $ABP = 20°$, angle $PAB = 145°$.

14. $A(5, 0)$, $B(-5, 0)$, angle $ABP = 50°$, angle $PAB = 100°$.

15. The angle of elevation of the top of a building is $70°$ as measured from ground level. If we move 19.2 feet closer to the building, the angle of elevation becomes $75°$. How tall is the building?

16. From your observation point P in the forest, you see two trees A and B being cut down. Three seconds after you see tree A fall, you hear it crash; 2 seconds after you see tree B fall, you hear it crash. If the angle APB is $35°$, find the distance between A and B. Assume the speed of sound is 1100 feet per second.

17. Triangle ABC has vertices $A(-6, 3)$, $B(2, 18)$, and $C(26, 11)$. Find the largest angle.

18. The sides of a triangle measure 180, 200, and 215. Find the area of the triangle.

19. A triangular-shaped garden has sides measuring 58 feet, 120 feet, and 112 feet. Find the area of the garden.

20. The quadrilateral $ABCD$ in Figure 8.28 has angles $A = 74°$, $B = 80°$, and $C = 102°$ and sides $AB = 105$, $BC = 95$. Find the area of the quadrilateral.

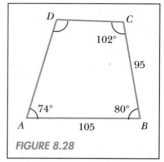

FIGURE 8.28

21. Point P in Figure 8.29 moves along the circle $x^2 + y^2 = 16$. Let θ be the measure of angle POA with initial side OA, where O is the origin and A has coordinates $(4, 0)$.

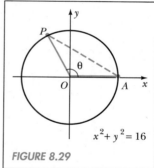

FIGURE 8.29

(a) Find the length d of the segment PA as a function of θ, $d = F(\theta)$.
(b) Find θ when $d = 4$.
(c) Find θ when $d = 6$.
(d) Suppose P moves with constant speed, completing 60 revolutions per second. If P starts at point A when $t = 0$, find d as a function of time t. Assume t is measured in seconds. [*Hint:* find θ in terms of t and then use part (a).]

22. Figure 8.30 shows a rope of fixed length attached at one end to a swivel at P. The rope passes through a guide ring at A and is hooked to a weight at point B. As P moves around the circle of radius a, the rope remains taut and the weight moves up and down. Let θ be the measure of angle PCA with initial side CA.

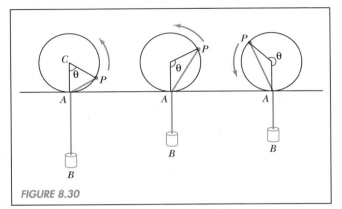

FIGURE 8.30

(a) Let y = height of point B. Assuming $y = 0$ when $\theta = 0$, find y as a function of θ. (*Hint:* how are y and the length of PA related?)
(b) Find θ when $y = a$.
(c) Find θ when $y = a/2$.
(d) Suppose P moves at constant speed, completing one revolution around the circle every 3 seconds. Find y as a function of time t (measured in seconds). Assume $y = 0$ when $t = 0$.

23. Point A moves along the circle $x^2 + y^2 = 9$ in Figure 8.31. A rod of length 12 is attached to point A on one end and to a piston at B on the other end. As A moves around the circle, the rod moves and drives the piston back and forth.

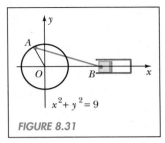

FIGURE 8.31

(a) If θ is the measure of angle AOB with initial side OB, find x, the x-coordinate of B, as a function of θ, $x = F(\theta)$. (*Hint:* use the Law of Cosines to write an equation for x and then use the quadratic formula to solve for x.)
(b) Find θ when $x = 12$.
(c) Suppose A moves with constant speed, completing four revolutions per second. If A starts at the point $(3, 0)$ when $t = 0$, find x as a function of time t (measured in seconds).

DEMOIVRE'S THEOREM

Recall from Section 4.2 that a complex number z has the form $a + bi$, where a and b are real numbers called the real and imaginary parts of z, respectively, and i is an imaginary number satisfying $i^2 = -1$. We know that the real and imaginary parts of a complex number are unique. In other words, two complex numbers are the same if and only if their real and imaginary parts are equal.

> If $z_1 = a_1 + b_1 i$ and $z_2 = a_2 + b_2 i$, then $z_1 = z_2$ if and only if $a_1 = a_2$ and $b_1 = b_2$.

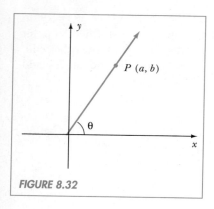

FIGURE 8.32

For this reason, every complex number $z = a + bi$ corresponds to a unique point P in the plane whose coordinates are (a, b). We shall use this point to rewrite z in a special way.

Suppose $z = a + bi$ is any nonzero complex number. Let θ be the angle in standard position whose terminal side passes through the point $P(a, b)$. See Figure 8.32. Letting $r = \sqrt{a^2 + b^2}$, the distance from the origin to P, we have, by the definition of the cosine function,

$$\frac{a}{r} = \cos \theta \qquad \text{or} \qquad a = r \cos \theta$$

Similarly, the definition of the sine function yields $b = r \sin \theta$. Thus we may write

$$z = a + bi = r \cos \theta + ir \sin \theta$$

or $z = r(\cos \theta + i \sin \theta)$. We call this a **polar representation** for z.

Polar Representation of a Complex Number
Let $z = a + bi$ be nonzero. Then

$$z = r(\cos \theta + i \sin \theta)$$

where $r = \sqrt{a^2 + b^2}$ and θ is an angle in standard position whose terminal side passes through $P(a, b)$.

Remark

A polar representation of a given complex number is not unique since $\theta + 2\pi k$ can be used in place of θ for any integer k.

Example 1

Find a polar representation for (a) $z = -1 + \sqrt{3}i$, (b) $z = 5$.

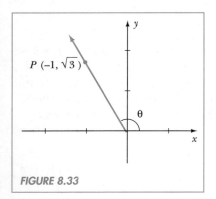

FIGURE 8.33

SOLUTION (a) We have $z = a + bi = -1 + \sqrt{3}i$, so $a = -1$ and $b = \sqrt{3}$. Thus

$$r = \sqrt{a^2 + b^2} = \sqrt{(-1)^2 + (\sqrt{3})^2} = \sqrt{1 + 3} = 2$$

Next, Figure 8.33 shows an angle θ in standard position with terminal side passing through $P(-1, \sqrt{3})$. We recognize θ as the standard angle $2\pi/3$. Therefore

$$z = r(\cos \theta + i \sin \theta) = 2\left(\cos \frac{2\pi}{3} + i \sin \frac{2\pi}{3} \right)$$

This gives us a polar representation for z.

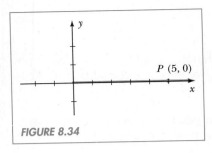

FIGURE 8.34

(b) We have $z = a + bi = 5$, so $a = 5$ and $b = 0$. Thus

$$r = \sqrt{a^2 + b^2} = \sqrt{5^2} = 5$$

Next, Figure 8.34 shows that the angle θ in standard position with terminal side passing through $P(5, 0)$ has measure zero. Therefore

$$z = r(\cos \theta + i \sin \theta) = 5(\cos 0 + i \sin 0)$$

This gives us a polar representation for z. ∎

Why do we want to rewrite a complex number using a polar representation? After all, a polar representation seems more complicated than the original expression for the number. Although this is true, the following theorem shows how polar representations can reduce what would otherwise be a difficult calculation into something relatively easy.

DeMoivre's Theorem

If n is a positive integer, then

$$[r(\cos \theta + i \sin \theta)]^n = r^n(\cos n\theta + i \sin n\theta)$$

You will prove this theorem in Chapter 11 using a proof technique called mathematical induction. (See Exercise 21, Section 11.4.)

Example 2

Use DeMoivre's Theorem to find $(-1 + \sqrt{3}i)^6$.

SOLUTION   Our first step is to find a polar representation for $-1 + \sqrt{3}i$. This was done in Example 1(a). We have

$$-1 + \sqrt{3}i = 2\left(\cos \frac{2\pi}{3} + i \sin \frac{2\pi}{3}\right)$$

Next, we apply DeMoivre's Theorem.

$$(-1 + \sqrt{3}i)^6 = \left[2\left(\cos \frac{2\pi}{3} + i \sin \frac{2\pi}{3}\right)\right]^6$$

$$= 2^6\left(\cos 6\left(\frac{2\pi}{3}\right) + i \sin 6\left(\frac{2\pi}{3}\right)\right)$$

$$= 64(\cos 4\pi + i \sin 4\pi)$$

$$= 64 \quad ∎$$

Example 3

Find $(1 + i)^5$.

SOLUTION We find a polar representation for $1 + i$ and then apply DeMoivre's Theorem. For $a + bi = 1 + i$, we have $a = 1$ and $b = 1$. Thus

$$r = \sqrt{a^2 + b^2} = \sqrt{1^2 + 1^2} = \sqrt{2}$$

Figure 8.35 shows an angle θ in standard position with its terminal side passing through the point $P(1, 1)$. Evidently, $\theta = \pi/4$. Therefore

$$1 + i = r(\cos \theta + i \sin \theta) = \sqrt{2}\left(\cos \frac{\pi}{4} + i \sin \frac{\pi}{4}\right)$$

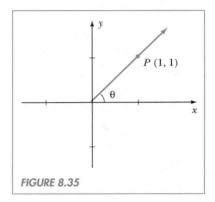

FIGURE 8.35

Now we apply DeMoivre's Theorem.

$$(1 + i)^5 = \left[\sqrt{2}\left(\cos \frac{\pi}{4} + i \sin \frac{\pi}{4}\right)\right]^5$$

$$= (\sqrt{2})^5\left(\cos \frac{5\pi}{4} + i \sin \frac{5\pi}{4}\right)$$

$$= 4\sqrt{2}\left(\frac{-1}{\sqrt{2}} + i\frac{-1}{\sqrt{2}}\right)$$

$$= -4 - 4i \quad \blacksquare$$

We saw that DeMoivre's Theorem helps us find powers of complex numbers. We can also use it to find roots.

Definition of nth Root

Suppose c is any complex number. We say that $z \in \mathbf{C}$ is an **nth root** of c if $z^n = c$.

As a simple example, we note that $-i$ is a third root of i since

$$(-i)^3 = -(i^2)i = -(-1)i = i$$

Also, Example 2 shows that $-1 + \sqrt{3}i$ is a sixth root of 64.

Given a fixed number c, how may we find the nth roots of c? Indeed, do nth roots always exist? If so, how many are there? Let us look at a specific example.

Example 4

Find the third roots of i.

SOLUTION We wish to find all $z \in \mathbf{C}$ satisfying

$$z^3 = i$$

We start by rewriting each side of this equation using polar representations. Since z is unknown, we write

$$z = r(\cos \theta + i \sin \theta)$$

For i we have

$$i = 1\left(\cos \frac{\pi}{2} + i \sin \frac{\pi}{2}\right)$$

Therefore the equation $z^3 = i$ becomes

$$[r(\cos \theta + i \sin \theta)]^3 = 1\left(\cos \frac{\pi}{2} + i \sin \frac{\pi}{2}\right)$$

Applying DeMoivre's Theorem to the left side, we have

$$r^3(\cos 3\theta + i \sin 3\theta) = 1\left(\cos \frac{\pi}{2} + i \sin \frac{\pi}{2}\right)$$

From this we conclude that $r^3 = 1$, $\cos 3\theta = \cos \pi/2$, and $\sin 3\theta = \sin \pi/2$. Now,

$$r^3 = 1 \quad \text{implies} \quad r = 1$$

Also, $\cos 3\theta = \cos \pi/2$ and $\sin 3\theta = \sin \pi/2$ imply 3θ must be coterminal with $\pi/2$. Hence

$$3\theta = \frac{\pi}{2} + 2\pi k \qquad k \in \mathbf{Z}$$

or

$$\theta = \frac{\pi}{6} + \frac{2\pi}{3}k \qquad k \in \mathbf{Z}$$

Therefore, for the third roots of i we have

$$z = r(\cos \theta + i \sin \theta)$$

$$= 1\left(\cos\left(\frac{\pi}{6} + \frac{2\pi}{3}k\right) + i \sin\left(\frac{\pi}{6} + \frac{2\pi}{3}k\right)\right) \quad k \in \mathbb{Z}$$

Evaluating this expression when $k = 0$, 1, and 2, we find

k	r	$\theta = \dfrac{\pi}{6} + \dfrac{2\pi}{3}k$	$z = r(\cos \theta + i \sin \theta)$
0	1	$\dfrac{\pi}{6}$	$\dfrac{\sqrt{3}}{2} + \dfrac{1}{2}i$
1	1	$\dfrac{5\pi}{6}$	$-\dfrac{\sqrt{3}}{2} + \dfrac{1}{2}i$
2	1	$\dfrac{3\pi}{2}$	$-i$

If we evaluate the general expression for z at any other integers k, we will repeat the values appearing on this list. We conclude that the third roots of i are $\dfrac{\sqrt{3}}{2} + \dfrac{1}{2}i$, $-\dfrac{\sqrt{3}}{2} + \dfrac{1}{2}i$, and i. ■

Consider the following proposition.

> The complex number z is an nth root of c if and only if z is a zero of $F(x) = x^n - c$.

PROOF: By definition, z is an nth root of c if and only if

$$z^n = c$$

But this is equivalent to $z^n - c = 0$. Now, this is true if and only if $F(z) = z^n - c = 0$. Finally, this last statement is true if and only if z is a zero of F. This completes the proof. ■

Thus finding the nth roots of c is equivalent to finding the zeros of a degree n polynomial function. Recall that the Fundamental Theorem of Algebra (Section 4.3) implies that a degree n polynomial function has n zeros among the complex numbers (counting multiplicities). Therefore it is not surprising to find that any nonzero complex number will have precisely n distinct nth roots.

Example 5

Find all the zeros of $F(x) = x^6 + 64$.

SOLUTION We know that z is a zero of F if and only if $F(z) = z^6 + 64 = 0$. Thus we want to find a $z \in \mathbf{C}$ satisfying $z^6 + 64 = 0$, or, equivalently

$$z^6 = -64$$

As in Example 4, we start the solution by writing polar representations for both sides of the equation. We have

$$[r(\cos \theta + i \sin \theta)]^6 = 64(\cos \pi + i \sin \pi)$$

where $z = r(\cos \theta + i \sin \theta)$. Applying DeMoivre's Theorem to the left side, we write

$$r^6(\cos 6\theta + i \sin 6\theta) = 64(\cos \pi + i \sin \pi)$$

Thus we have $r^6 = 64$, $\cos 6\theta = \cos \pi$, and $\sin 6\theta = \sin \pi$. Now,

$$r^6 = 64 \quad \text{implies} \quad r = 2$$

Also, $\cos 6\theta = \cos \pi$ and $\sin 6\theta = \sin \pi$ imply 6θ is coterminal with π. Hence

$$6\theta = \pi + 2\pi k \qquad k \in \mathbf{Z}$$

or

$$\theta = \frac{\pi}{6} + \frac{\pi}{3}k \qquad k \in \mathbf{Z}$$

Choosing $k = 0$, 1, 2, 3, 4, and 5, we find six corresponding values for z:

k	r	$\theta = \dfrac{\pi}{6} + \dfrac{\pi}{3}k$	$z = r(\cos \theta + i \sin \theta)$
0	2	$\dfrac{\pi}{6}$	$2\left(\dfrac{\sqrt{3}}{2} + \dfrac{1}{2}i\right) = \sqrt{3} + i$
1	2	$\dfrac{\pi}{2}$	$2(0 + i) = 2i$
2	2	$\dfrac{5\pi}{6}$	$2\left(-\dfrac{\sqrt{3}}{2} + \dfrac{1}{2}i\right) = -\sqrt{3} + i$
3	2	$\dfrac{7\pi}{6}$	$2\left(-\dfrac{\sqrt{3}}{2} - \dfrac{1}{2}i\right) = -\sqrt{3} - i$
4	2	$\dfrac{3\pi}{2}$	$2(0 - i) = -2i$
5	2	$\dfrac{11\pi}{6}$	$2\left(\dfrac{\sqrt{3}}{2} - \dfrac{1}{2}i\right) = \sqrt{3} - i$

We know F cannot have more than six zeros, so our list exhausts all the possibilities. We conclude that the zeros of F are given by $\sqrt{3} \pm i$, $-\sqrt{3} \pm i$, and $\pm 2i$. ∎

EXERCISES 8.3

In Exercises 1 to 6, find a polar representation for z.

1. (a) $z = 1 + \sqrt{3}i$
   (b) $z = 1 - i$

2. (a) $\sqrt{3} + i$
   (b) $-1 + i$

3. (a) $z = 6$
   (b) $z = -6$

4. (a) $z = 9$
   (b) $z = -9$

5. $-4i$

6. $2i$

In Exercises 7 to 14, use DeMoivre's Theorem to evaluate the given expression.

7. $\left(\cos \dfrac{\pi}{12} + i \sin \dfrac{\pi}{12} \right)^6$

8. $\left(\cos \dfrac{\pi}{10} + i \sin \dfrac{\pi}{10} \right)^{10}$

9. $(\sqrt{3} + i)^6$

10. $(1 - i)^4$

11. $(-\sqrt{2} - \sqrt{2}i)^9$

12. $(-\sqrt{3} + i)^8$

13. $-\left(\dfrac{1}{2} - \dfrac{1}{2}i \right)^5$

14. $\left(\cos \dfrac{\pi}{8} - i \sin \dfrac{\pi}{8} \right)^{12}$

15. Find the third roots of $27i$.

16. Find the third roots of $-8i$.

17. Find the sixth roots of 1.

18. Find the fourth roots of 1.

In Exercises 19 to 24, find all the complex zeros for the given function.

19. $F(x) = x^4 + 16$

20. $F(x) = x^6 - 27$

21. $F(x) = x^3 - 5$

22. $F(x) = x^6 + 2$

23. $F(x) = x^7 - 729x$

24. $F(x) = x^5 + 81x$

25. Prove $(\overline{z^n}) = (\bar{z})^n$ for any $z \in \mathbf{C}$. Recall that if $z = a + bi$, then $\bar{z} = a - bi$. [*Hint:* Use DeMoivre's Theorem and the facts $\cos(-\theta) = \cos \theta$ and $\sin(-\theta) = -\sin \theta$.]

26. Let z_1 and z_2 be complex numbers with polar representations $z_1 = a(\cos \alpha + i \sin \alpha)$ and $z_2 = b(\cos \beta + i \sin \beta)$.
   (a) Show that a polar representation for $z_1 \cdot z_2$ is $ab(\cos(\alpha + \beta) + i \sin(\alpha + \beta))$.
   (b) Show that if z_1 and z_2 are nonzero, then $z_1 \cdot z_2 \neq 0$.

8.4 VECTORS

Many physical quantities are characterized by two pieces of data: an amount or *magnitude*, and *direction*. For example, the position of an object relative to the origin is determined by its distance and direction from the origin. If the object is moving, then it has a velocity that consists of both the speed and direction of motion. Finally, there may be some force acting on the object, which is characterized by the amount of the force and the direction in which it acts.

To analyze physical quantities such as position, velocity, and force, the concept of a **vector** has been devised. A vector is a single mathematical entity that carries with it two pieces of data: magnitude and direction. We represent vectors by using boldface letters, such as **u**, **v**, and **w**. Let us formally define a vector.

> **Definition of Vector**
>
> A vector **v** consists of the following two things:
> 1. A nonnegative real number called the magnitude of **v**, denoted by $\|\mathbf{v}\|$.
> 2. A direction in the plane.

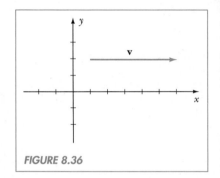

FIGURE 8.36

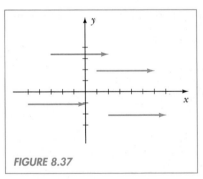

FIGURE 8.37

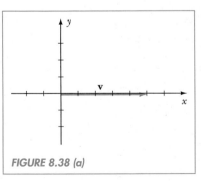

FIGURE 8.38 (a)

Notice that we required the direction of a vector to be restricted to the plane. Generally, vectors may have any direction in three dimensions. However, in this chapter we limit our discussion to vectors in the plane (two dimensions), although all the results we obtain for two-dimensional vectors can be extended to three dimensions if desired.

For our first example of a vector, let us describe the velocity of a hypothetical wind blowing across the x, y-plane. Suppose the wind blows at 5 mph in the positive x-direction. Let **v** represent the velocity of the wind. Then

Magnitude of **v**:   5

Direction of **v**:   positive x-direction

We may picture **v** by drawing an arrow of length 5 pointing in the positive x-direction, as in Figure 8.36. We refer to the pointed end of the arrow as the **head** of **v** and the opposite end as the **tail**.

In general, we may give any vector with nonzero magnitude a geometric representation by drawing a directed line segment, or arrow, in the plane. The length of the arrow corresponds to the magnitude of the vector, and the direction of the arrow is the same as the direction defining the vector. In case a vector has magnitude zero, we represent it with a single point in the plane.

Any vector of magnitude zero is called a **zero vector**, denoted by 0. The direction of a zero vector is undefined since it makes no difference in what direction we point something of magnitude zero. With this fact in mind, we agree that all zero vectors are the same.

We say that two nonzero vectors are the same, or *equal*, if and only if their magnitudes and directions are the same. Geometrically, this means that two arrows of the same length in the plane and pointing in the same parallel direction represent the same vector. Thus an arrow used to represent a vector is not unique; it can be moved around in the plane as long as its length and direction remain constant. This concept is illustrated in Figure 8.37, where each of the four arrows represents the vector **v** in Figure 8.36.

If we draw an arrow representing a vector so that its tail is at the origin, then we say the vector is in **standard position**. The angle formed by a vector in a standard position as measured counterclockwise from the positive x-axis is called the **direction angle** of the vector. We write θ_v to designate the direction angle for the vector **v**.

Figure 8.38(a) shows the vector **v** with magnitude 5 and direction angle 0° in standard position. This vector is the same vector illustrated in Figures 8.36

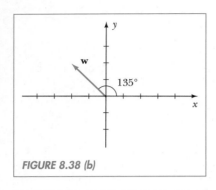

FIGURE 8.38 (b)

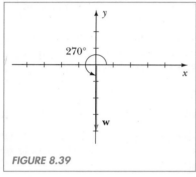

FIGURE 8.39

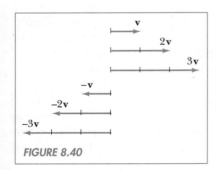

FIGURE 8.40

and 8.37. Also, in Figure 8.38(b), the vector **w** is drawn with magnitude 3 and direction angle 135° in standard position.

We shall restrict the direction angle of any vector to less than 360°. Thus, the direction angle for any vector **v** must satisfy $0° \leq \theta_v < 360°$. It follows that θ_v is unique whenever **v** is nonzero.

Example 1

Let **w** be the velocity of a 4 mph wind blowing in the negative y-direction. Draw an arrow representing **w** in standard position and find its direction angle.

SOLUTION The desired arrow is shown in Figure 8.39. We have $\|\mathbf{w}\| = 4$ and $\theta_w = 270°$. ∎

Our next objective is to describe how to multiply a vector by a real number. Some scientists, especially physicists, like to call real numbers **scalars** in order to distinguish them from vectors. Thus, "scalar" is just another word for real number, meaning a pure number without direction. We refer to the multiplication of a vector **v** by a real number r as **scalar multiplication**. The result will be a new vector, denoted $r\mathbf{v}$, described as follows.

Definition of Scalar Multiplication

If **v** is a vector and r is a real number, then $r\mathbf{v}$ is the vector satisfying

(1) Magnitude of $r\mathbf{v}$: $\|r\mathbf{v}\| = |r| \|\mathbf{v}\|$

(2) Direction of $r\mathbf{v}$: $\begin{cases} \text{same as } \mathbf{v} \text{ if } r > 0 \\ \text{opposite } \mathbf{v} \text{ if } r < 0 \end{cases}$

In the special case when $r = 0$, we have $0\mathbf{v} = \mathbf{0}$. Also, we shall write $-\mathbf{v}$ to mean $-1\mathbf{v}$.

Figure 8.40 shows the geometric interpretation of scalar multiplication. If the vector **v** is multiplied by a positive number r, we simply change the length of the arrow representing **v** by the factor r. If r is negative, the length is again changed by the factor $|r|$, but the arrow now points in the opposite direction of **v**.

Example 2

Given the vector **v** with $\|\mathbf{v}\| = 2$ and $\theta_v = 225°$. Draw in standard position the vectors $2\mathbf{v}$, $-\mathbf{v}$, $-\frac{1}{2}\mathbf{v}$, $0\mathbf{v}$. Also, determine the magnitude and direction angle for each vector.

The solution is shown in Figure 8.41. ■

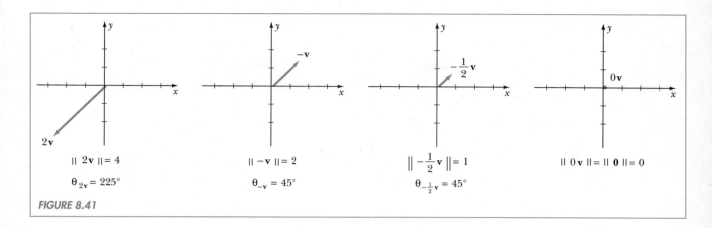

$$\| 2\mathbf{v} \| = 4$$

$$\theta_{2\mathbf{v}} = 225°$$

$$\| -\mathbf{v} \| = 2$$

$$\theta_{-\mathbf{v}} = 45°$$

$$\left\| -\tfrac{1}{2}\mathbf{v} \right\| = 1$$

$$\theta_{-\frac{1}{2}\mathbf{v}} = 45°$$

$$\| 0\mathbf{v} \| = \| \mathbf{0} \| = 0$$

FIGURE 8.41

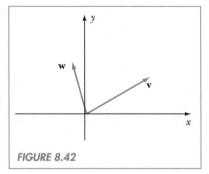

FIGURE 8.42

We shall now discuss how to add two vectors, say, **v** and **w**, forming a new vector **v** + **w**. Suppose we are given **v** and **w**, as shown in Figure 8.42. To add **w** to **v**, we keep the position of **v** fixed and move **w** so that its tail is located on the head of **v**. The vector **u** obtained by drawing an arrow from the tail of **v** to the head of **w** in this arrangement is called the **resultant** of adding **w** to **v**. See Figure 8.43. We write **v** + **w** = **u**.

We note that vector addition is commutative; that is, adding **w** to **v** is the same as adding **v** to **w**,

$$\mathbf{v} + \mathbf{w} = \mathbf{w} + \mathbf{v}$$

Figure 8.44 illustrates this property. By taking **v** and **w** in standard position and adding **w** to **v** (placing **w** at the head of **v**) and also adding **v** to **w** (placing **v** at the head of **w**), we form a parallelogram. The resultant in either case is the diagonal vector **u** in the parallelogram.

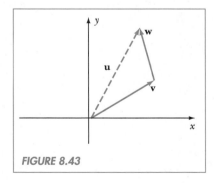

FIGURE 8.43

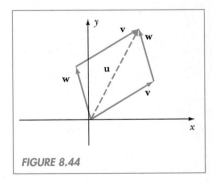

FIGURE 8.44

Example 3

Construct the vector **v** + **w** and find its magnitude and direction angle given $\|\mathbf{v}\| = 10$, $\theta_v = 45°$, and $\|\mathbf{w}\| = 10$, $\theta_w = 135°$.

SOLUTION The vectors **v** and **w** are shown in standard position in Figure 8.45(a). In Figure 8.45(b) we moved **w** to the head of **v** to construct the resultant **v** + **w**. The triangle formed by **v**, **w**, and **v** + **w** in this figure is a 45–45–90 triangle. It follows that **v** + **w** has magnitude $10\sqrt{2}$ and direction angle 90°. ∎

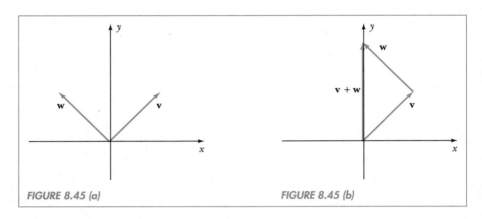

FIGURE 8.45 (a)　　　　　　　　　　　　FIGURE 8.45 (b)

Now we discuss how to subtract a vector from another one. This procedure is defined in terms of addition as follows.

$$\mathbf{v} - \mathbf{w} = \mathbf{v} + (-\mathbf{w})$$

Figures 8.46(a) to (c) illustrate the geometric construction of **v** − **w**. We begin in Figure 8.46(a) with the given vectors **v** and **w**. In Figure 8.46(b) we draw −**w**, which has the same length as **w** but points in the opposite direction, and then move it so that the tail of −**w** is on the head of **v**. The resultant vector **v** + (−**w**) is then drawn from the tail of **v** to the head of −**w**, as shown in Figure 8.46(c).

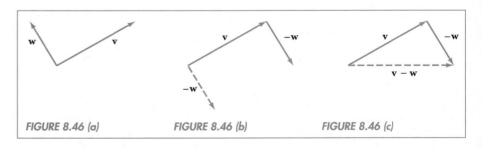

FIGURE 8.46 (a)　　　　　FIGURE 8.46 (b)　　　　　FIGURE 8.46 (c)

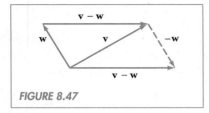

FIGURE 8.47

Now consider Figure 8.47, which shows that the vector $\mathbf{v} - \mathbf{w}$ can be obtained by drawing an arrow *from* the head of $\mathbf{w}$ *to* the head of $\mathbf{v}$ when $\mathbf{v}$ and $\mathbf{w}$ are left in standard position. For this reason we often say that *$\mathbf{v} - \mathbf{w}$ is the vector from $\mathbf{w}$ to $\mathbf{v}$*. Finding $\mathbf{v} - \mathbf{w}$ in this way is geometrically easier since we do not have to move any vectors around to obtain the result. Note also from Figure 8.47 that

$$\mathbf{w} + (\mathbf{v} - \mathbf{w}) = \mathbf{v}$$

as one would expect!

Example 4

Find $\mathbf{v} - \mathbf{w}$ geometrically and determine its magnitude and direction angle given vectors $\mathbf{v}$ and $\mathbf{w}$ with $\|\mathbf{v}\| = 3$, $\theta_v = 90°$, and $\|\mathbf{w}\| = 4$, $\theta_w = 0°$.

SOLUTION   We begin by drawing $\mathbf{v}$ and $\mathbf{w}$ in standard position, as shown in Figure 8.48(a). Next, simply draw the arrow from the head of $\mathbf{w}$ to the head of $\mathbf{v}$ to obtain $\mathbf{v} - \mathbf{w}$. We note that the length of $\mathbf{v} - \mathbf{w}$ is the hypotenuse of a right triangle with legs 3 and 4. Thus

$$\|\mathbf{v} - \mathbf{w}\| = \sqrt{3^2 + 4^2} = \sqrt{25} = 5$$

Finally, Figure 8.48(b) shows $\mathbf{v} - \mathbf{w}$ in standard position. If θ is the direction angle for $\mathbf{v} - \mathbf{w}$, we have

$$\tan \theta = -\frac{3}{4} \quad \text{and} \quad 90° < \theta < 180°$$

Using the calculator, we find that $\theta = \tan^{-1}(-3/4) + 180° \approx -36.87° + 180° = 143.13°$. ∎

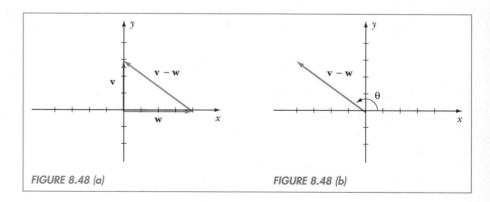

FIGURE 8.48 (a)   **FIGURE 8.48 (b)**

So far our examples illustrating addition and subtraction of vectors have been geometrically simple. In the next section we develop a more powerful analytic technique that will handle difficult geometric situations with ease.

EXERCISES 8.4 _____

In Exercises 1 and 2, determine how many different vectors are represented. Which vectors are the same?

1.
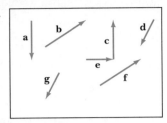

2.
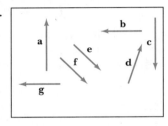

In Exercises 3 to 6, draw the vector described in standard position.

3. $\|\mathbf{v}\| = 3\sqrt{2}$, $\theta_v = 45°$ 4. $\|\mathbf{v}\| = 5$, $\theta_v = 0°$

5. $\|\mathbf{v}\| = 2$, $\theta_v = 180°$ 6. $\|\mathbf{v}\| = 1$, $\theta_v = 315°$

In Exercises 7 to 10, draw all possible vectors described in standard position (there are two). Find the direction angle for each vector.

7. The direction of $\mathbf{v}$ is parallel to the line $y = x + 1$ and $\|\mathbf{v}\| = 2$.

8. The direction of $\mathbf{n}$ is perpendicular to the line $y = x + 1$ and $\|\mathbf{n}\| = 1$.

9. The direction of $\mathbf{u}$ is perpendicular to the line $x + \sqrt{3}y = 3$ and $\|\mathbf{u}\| = 1$.

10. The direction of $\mathbf{v}$ is parallel to the line $x + \sqrt{3}y = 3$ and $\|\mathbf{v}\| = 1$.

11. Suppose $\mathbf{v}$ has magnitude 4 and direction angle 135°. Draw the given vector in standard position and find its magnitude and direction angle: (a) $2\mathbf{v}$, (b) $-\mathbf{v}$, (c) $-\frac{1}{2}\mathbf{v}$, (d) $0\mathbf{v}$, (e) $\frac{3}{2}\mathbf{v}$.

12. Repeat Exercise 11 for the vector $\mathbf{v}$ with magnitude 5 and direction angle 60°.

In Exercises 13 to 18, use the given vectors $\mathbf{v}$ and $\mathbf{w}$ to construct $\mathbf{v} + \mathbf{w}$ and $\mathbf{v} - \mathbf{w}$. In each case, find the magnitude and direction angle of the resultant vector.

13.
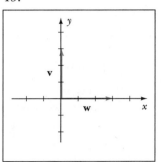

14.
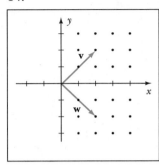

15.
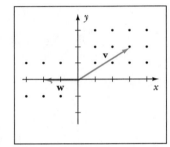

16.
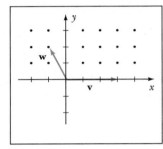

17.
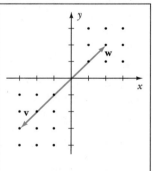

18.
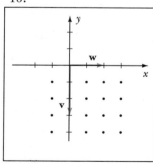

In Exercises 19 and 20, find the magnitude and direction of $\mathbf{v} + \mathbf{w}$ from the given information. You need use only elementary geometry.

19. $\|\mathbf{v}\| = 5$, $\theta_v = 10°$, and $\|\mathbf{w}\| = 5$, $\theta_w = 70°$

20. $\|\mathbf{v}\| = 3$, $\theta_v = 140°$, and $\|\mathbf{w}\| = 6$, $\theta_w = 20°$

In Exercises 21 to 24, suppose $\|\mathbf{v}\| = 10$, $\theta_v = 0°$, and $\|\mathbf{w}\| = 6$, $\theta_w = 90°$. Find the magnitude and direction angle for the given vector.

21. $\dfrac{1}{2}\mathbf{v} + 2\mathbf{w}$

22. $2\mathbf{v} - 3\mathbf{w}$

23. $2\mathbf{v} - \mathbf{w}$

24. $\dfrac{1}{2}\mathbf{w} - \mathbf{v}$

25. Suppose $\mathbf{v} + \mathbf{w} = 0$. What is the relationship between the magnitude and direction of $\mathbf{v}$ and $\mathbf{w}$?

In Exercises 26 to 30, answer true or false.

26. $\mathbf{v} + \mathbf{v} = 2\mathbf{v}$

27. $\|\mathbf{v} + \mathbf{w}\| = \|\mathbf{v}\| + \|\mathbf{w}\|$

28. $\mathbf{v} - \mathbf{v} = 0$

29. $\|3\mathbf{v} + 3\mathbf{w}\| = 3\|\mathbf{w} + \mathbf{v}\|$

30. If $\mathbf{v} \neq 0$, then $\dfrac{1}{\|\mathbf{v}\|}\mathbf{v}$ is a vector of magnitude 1.

8.5 COMPONENT REPRESENTATION OF VECTORS

In Section 8.4 we learned that a vector $\mathbf{v}$ can be represented by an arrow whose length and direction is the same as the magnitude and direction of $\mathbf{v}$. Now we shall use an ordered pair of numbers to represent a vector; this way we can perform all the analysis of vectors algebraically without resorting to the geometry of arrows.

Definition of Component Representation
Let $\mathbf{v}$ be a vector and $P(a, b)$ the point at the head of the arrow representing $\mathbf{v}$ in standard position. We write

$$\langle a, b \rangle$$

to represent the vector $\mathbf{v}$. We call this notation the **component representation** of $\mathbf{v}$, where a is the **x-component** and b is the **y-component**.

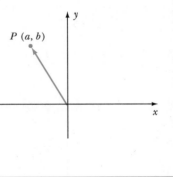

Remarks

1. The component representation of a vector is unique; that is, two vectors are the same if and only if their components are equal.

$$\langle a, b \rangle = \langle c, d \rangle \qquad \text{if and only if} \qquad a = c \text{ and } b = d$$

2. The zero vector has component representation $0 = \langle 0, 0 \rangle$.

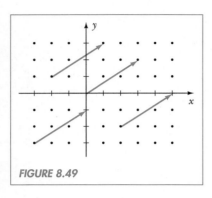

FIGURE 8.49

We emphasize once again that an arrow representing the vector $\langle a, b \rangle$ can be positioned anywhere in the plane as long as the length and direction remain constant. For example, Figure 8.49 illustrates the vector $\langle 3, 2 \rangle$ in standard position and also at three other locations. The same symbol, $\langle 3, 2 \rangle$, corresponds to each arrow.

The following theorem gives the relationship between the components of a vector and its magnitude and direction.

Component Theorem

Suppose vector $\mathbf{v}$ has direction angle θ_v and magnitude $\|\mathbf{v}\|$. Then $\mathbf{v} = \langle a, b \rangle$, where

$$a = \|\mathbf{v}\| \cos \theta_v \qquad b = \|\mathbf{v}\| \sin \theta_v$$

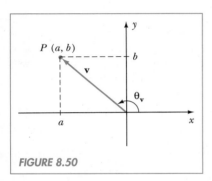

FIGURE 8.50

PROOF: First suppose that $\mathbf{v} \neq 0$. Placing $\mathbf{v}$ in standard position, let $P(a, b)$ be the point at the head of $\mathbf{v}$. See Figure 8.50. By definition, the component representation of $\mathbf{v}$ is $\langle a, b \rangle$. Note that $\sqrt{a^2 + b^2} = OP = \|\mathbf{v}\|$. Therefore, using $P(a, b)$ as a point on the terminal side of θ_v, the definitions of $\cos \theta_v$ and $\sin \theta_v$ say that

$$\cos \theta_v = \frac{a}{\|\mathbf{v}\|} \qquad \sin \theta_v = \frac{b}{\|\mathbf{v}\|}$$

Multiplying by $\|\mathbf{v}\|$, we find that $a = \|\mathbf{v}\| \cos \theta_v$ and $b = \|\mathbf{v}\| \sin \theta_v$ as desired.

If $\mathbf{v} = 0$, then the component representation of $\mathbf{v}$ is $\langle 0, 0 \rangle$. Therefore we must show that $a = \|\mathbf{v}\| \cos \theta_v$ and $b = \|\mathbf{v}\| \sin \theta_v$ are both zero. Since $\|\mathbf{v}\| = \|0\| = 0$, this follows immediately. This completes the proof. ∎

Example 1

Find the component representation for the vector $\mathbf{v}$ with magnitude 6 and direction angle 110°. Approximate to two decimal places.

> **SOLUTION** We have $\|\mathbf{v}\| = 6$ and $\theta_v = 110°$. By the Component Theorem, $\mathbf{v} = \langle a, b \rangle$, where

$$a = \|\mathbf{v}\| \cos \theta_v \qquad b = \|\mathbf{v}\| \sin \theta_v$$
$$= 6 \cos 110° \qquad = 6 \sin 110°$$
$$\approx -2.05 \qquad \approx 5.64$$

Thus, accurate to two decimal places, $\mathbf{v} = \langle -2.05, 5.64 \rangle$. A picture of $\mathbf{v}$ is in Figure 8.51. ∎

FIGURE 8.51

The following properties allow us to analyze and manipulate vectors using only their component representations.

Properties of Components

1. If $\mathbf{v} = \langle a, b \rangle$ and $a \neq 0$, then $\tan \theta_v = \dfrac{b}{a}$.
2. $\|\langle a, b \rangle\| = \sqrt{a^2 + b^2}$
3. $r\langle a, b \rangle = \langle ra, rb \rangle$
4. $\langle a, b \rangle + \langle c, d \rangle = \langle a + c, b + d \rangle$
5. $\langle a, b \rangle - \langle c, d \rangle = \langle a - c, b - d \rangle$

PROOF OF 1: If $\mathbf{v} = \langle a, b \rangle$ and $a \neq 0$, then $P(a, b)$ is on the terminal side of θ_v. The definition of tangent says $\tan \theta_v = b/a$ as desired.

PROOF OF 2: Let $\mathbf{v} = \langle a, b \rangle$. Then $P(a, b)$ is the point at the head of $\mathbf{v}$ in standard position. Therefore $\|\langle a, b \rangle\| = \|\mathbf{v}\| = OP = \sqrt{a^2 + b^2}$ by the distance formula.

PROOF OF 3: We leave this for an exercise.

PROOF OF 4: Let $\mathbf{v} = \langle a, b \rangle$, $\mathbf{w} = \langle c, d \rangle$, and $\mathbf{v} + \mathbf{w} = \langle x, y \rangle$. We wish to show that $x = a + c$ and $y = b + d$. Figure 8.52 illustrates the geometric construction of $\mathbf{v} + \mathbf{w}$ in the case when a, b, c, and d are all positive. It is clear from the figure that $x = a + c$ and $y = b + d$ as desired. This result remains valid in case any of the components a, b, c, or d are negative. We omit the details.

PROOF OF 5: We have

$$\langle a, b \rangle - \langle c, d \rangle = \langle a, b \rangle + (-1\langle c, d \rangle) \qquad \text{by definition}$$
$$= \langle a, b \rangle + \langle -c, -d \rangle \qquad \text{by property 3}$$
$$= \langle a - c, b - d \rangle \qquad \text{by property 4} \ \blacksquare$$

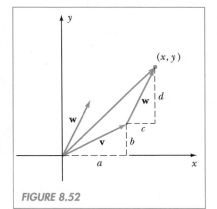

FIGURE 8.52

Example 2

Find the magnitude and direction angle for $\mathbf{v} = \langle 4, -3 \rangle$.

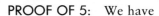 By property 2,

$$\|\mathbf{v}\| = \|\langle 4, -3 \rangle\| = \sqrt{4^2 + (-3)^2} = \sqrt{25} = 5$$

Figure 8.53 shows $\mathbf{v}$ in standard position. We see that the direction angle for $\mathbf{v}$ must satisfy $270° < \theta_v < 360°$. By property 1,

$$\tan \theta_v = -\frac{3}{4}$$

Solving this equation for θ_v between $270°$ and $360°$, we find that $\theta_v = \tan^{-1}(-3/4) + 360° \approx -36.87° + 360° = 323.13°$. $\blacksquare$

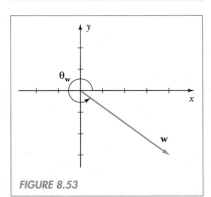

FIGURE 8.53

Example 3

Suppose $\mathbf{v} = \langle 7, -2 \rangle$ and $\mathbf{w} = \langle 1, 5 \rangle$. Find the component representation, direction angle, and magnitude for $-\mathbf{v} + 2\mathbf{w}$.

SOLUTION Note that $-\mathbf{v}$ means $-1\mathbf{v}$. Using properties 3 and 4, we have

$$
\begin{aligned}
-\mathbf{v} + 2\mathbf{w} &= -1\mathbf{v} + 2\mathbf{w} \\
&= -1\langle 7, -2 \rangle + 2\langle 1, 5 \rangle \\
&= \langle -7, 2 \rangle + \langle 2, 10 \rangle \\
&= \langle -7 + 2, 2 + 10 \rangle \\
&= \langle -5, 12 \rangle
\end{aligned}
$$

This gives us $\langle -5, 12 \rangle$ for the component representation of $-\mathbf{v} + 2\mathbf{w}$. Figure 8.54 shows $\langle -5, 12 \rangle$ in standard position. We see that the direction angle θ satisfies

$$
\tan \theta = \frac{12}{-5} \quad \text{and} \quad 90° < \theta < 180°
$$

It follows that $\theta = \tan^{-1}(-12/5) + 180° \approx -67.38° + 180° = 112.62°$. Finally, the magnitude of $-\mathbf{v} + 2\mathbf{w}$ is given by

$$
\|-\mathbf{v} + 2\mathbf{w}\| = \|\langle -5, 12 \rangle\| = \sqrt{(-5)^2 + (12)^2} = \sqrt{169} = 13 \quad \blacksquare
$$

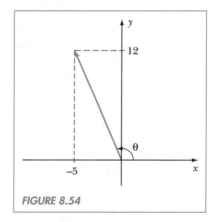

FIGURE 8.54

Example 4

Consider vectors $\mathbf{u}$, $\mathbf{v}$, and $\mathbf{w}$, where $\|\mathbf{u}\| = 12$, $\theta_\mathbf{u} = 0°$, $\|\mathbf{v}\| = 2$, $\theta_\mathbf{v} = 120°$, and $\|\mathbf{w}\| = 8$, $\theta_\mathbf{w} = 225°$. Find the component representation, magnitude, and direction angle for $\mathbf{u} + \mathbf{v} + \mathbf{w}$.

SOLUTION Figure 8.55(a) shows $\mathbf{u}$, $\mathbf{v}$, and $\mathbf{w}$ in standard position. Figure 8.55(b) shows the geometric construction for the sum $\mathbf{u} + \mathbf{v} + \mathbf{w}$.

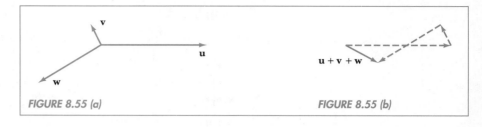

FIGURE 8.55 (a)   FIGURE 8.55 (b)

To find the magnitude and direction of $\mathbf{u} + \mathbf{v} + \mathbf{w}$, we need to find its x- and y-components. This involves summing the components of the individual vectors $\mathbf{u}$, $\mathbf{v}$, and $\mathbf{w}$. By the Component Theorem,

$$\mathbf{u} = \langle \|\mathbf{u}\| \cos \theta_u, \|\mathbf{u}\| \sin \theta_u \rangle$$

$$= \langle 12 \cos 0°, 12 \sin 0° \rangle$$

$$= \langle 12, 0 \rangle$$

Similarly, $\mathbf{v} = \langle 2 \cos 120°, 2 \sin 120° \rangle = \langle -1, \sqrt{3} \rangle$ and $\mathbf{w} = \langle 8 \cos 225°,$ $8 \sin 225° \rangle = \langle -4\sqrt{2}, -4\sqrt{2} \rangle$. To add these vectors, we simply add their components. Thus

$$\mathbf{u} + \mathbf{v} + \mathbf{w} = \langle 12, 0 \rangle + \langle -1, \sqrt{3} \rangle + \langle -4\sqrt{2}, -4\sqrt{2} \rangle$$

$$= \langle 12 - 1 - 4\sqrt{2}, 0 + \sqrt{3} - 4\sqrt{2} \rangle$$

$$= \langle 11 - 4\sqrt{2}, \sqrt{3} - 4\sqrt{2} \rangle$$

Note that the x-component is positive and the y-component is negative, so the arrow in standard position representing $\mathbf{u} + \mathbf{v} + \mathbf{w}$ must be in the fourth quadrant. This agrees with the picture in Figure 8.55(b).

For the magnitude of $\mathbf{u} + \mathbf{v} + \mathbf{w}$, we have, with the help of our calculator,

$$\|\mathbf{u} + \mathbf{v} + \mathbf{w}\| = \sqrt{(11 - 4\sqrt{2})^2 + (\sqrt{3} - 4\sqrt{2})^2}$$

$$\approx 6.63$$

Finally, the direction angle θ for $\mathbf{u} + \mathbf{v} + \mathbf{w}$ must satisfy

$$\tan \theta = \frac{\sqrt{3} - 4\sqrt{2}}{11 - 4\sqrt{2}} \quad \text{and} \quad 270° < \theta < 360°$$

It follows that $\theta \approx 323.7°$. ∎

A vector whose magnitude happens to be 1 unit is called a **unit vector**. Given any nonzero vector $\mathbf{v}$, we can use scalar multiplication to find a unit vector whose direction is the same as $\mathbf{v}$.

Example 5

Suppose $\mathbf{v} = \langle 6, -2 \rangle$. Find a unit vector with the same direction as $\mathbf{v}$.

SOLUTION We note that $\mathbf{v}$ is not a unit vector since

$$\|\mathbf{v}\| = \|\langle 6, -2 \rangle\| = \sqrt{6^2 + (-2)^2} = \sqrt{40} = 2\sqrt{10}$$

Recall that scalar multiplication of $\mathbf{v}$ by a positive constant changes its length but not its direction. Therefore, consider

$$\mathbf{u} = \frac{1}{\|\mathbf{v}\|}\mathbf{v} = \frac{1}{2\sqrt{10}}\langle 6, -2 \rangle = \left\langle \frac{3}{\sqrt{10}}, \frac{-1}{\sqrt{10}} \right\rangle$$

Since $\mathbf{u}$ is a positive scalar multiple of $\mathbf{v}$, it points in the same direction as $\mathbf{v}$. Furthermore,

$$\|\mathbf{u}\| = \left\| \left\langle \frac{3}{\sqrt{10}}, \frac{-1}{\sqrt{10}} \right\rangle \right\| = \sqrt{\left(\frac{3}{\sqrt{10}}\right)^2 + \left(\frac{-1}{\sqrt{10}}\right)^2} = \sqrt{\frac{9}{10} + \frac{1}{10}} = \sqrt{1} = 1$$

Thus $\mathbf{u} = \left\langle \dfrac{3}{\sqrt{10}}, \dfrac{-1}{\sqrt{10}} \right\rangle$ is a unit vector that meets our requirements. ∎

Example 5 illustrates the following general fact.

> If $\mathbf{v}$ is a nonzero vector, then $\mathbf{u} = \dfrac{1}{\|\mathbf{v}\|}\mathbf{v}$ is a unit vector with the same direction as $\mathbf{v}$.

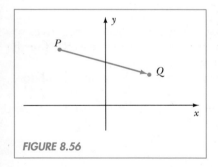

FIGURE 8.56

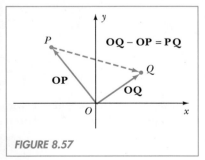

FIGURE 8.57

Vectors are useful for keeping track of the position of an object as measured from a particular reference point. For example, suppose we wish to describe the distance and direction from the reference point P to another point Q in the plane. To do this, we use the vector defined geometrically as the arrow from P to Q. See Figure 8.56. We shall use the notation $\mathbf{PQ}$ to designate *the vector from the point P to the point Q.*

Let us find the component representation for $\mathbf{PQ}$. Suppose P and Q have coordinates $P(x_1, y_1)$ and $Q(x_2, y_2)$. If O is the origin, it follows that $\mathbf{OP}$ and $\mathbf{OQ}$ have component representations $\langle x_1, y_1 \rangle$ and $\langle x_2, y_2 \rangle$, respectively. Now consider Figure 8.57, which shows that $\mathbf{PQ}$ is obtained by drawing an arrow from the head of $\mathbf{OP}$ to the head of $\mathbf{OQ}$. We recall from Section 8.4 that this vector is found by subtracting $\mathbf{OP}$ from $\mathbf{OQ}$. Thus

$$\mathbf{PQ} = \mathbf{OQ} - \mathbf{OP}$$
$$= \langle x_2, y_2 \rangle - \langle x_1, y_1 \rangle$$
$$= \langle x_2 - x_1, y_2 - y_1 \rangle$$

We summarize our results as follows.

> The vector from the point $P(x_1, y_1)$ to the point $Q(x_2, y_2)$ is given by
>
> $$PQ = \langle x_2 - x_1, y_2 - y_1 \rangle$$

Example 6

Suppose P and Q have coordinates $P(-4, 9)$ and $Q(6, -1)$. Find the component representation for the vector **PQ**. Draw **PQ** in standard position and find its magnitude and direction angle.

SOLUTION The vector **PQ** is established in Figure 8.58(a) by drawing an arrow from $P(-4, 9)$ to $Q(6, -1)$. We have

$$\mathbf{PQ} = \langle 6 - (-4), -1 - 9 \rangle$$
$$= \langle 10, -10 \rangle$$

Thus we get **PQ** in standard position by drawing an arrow from the origin to the point $(10, -10)$. See Figure 8.58(b). Using the component representation of **PQ**, we find the magnitude and direction angle θ as follows:

$$\|\mathbf{PQ}\| = \sqrt{10^2 + (-10)^2} = \sqrt{200} = 10\sqrt{2}$$

$$\tan \theta = \frac{-10}{10} = -1$$

Figure 8.58(b) shows that $270° < \theta < 360°$. Therefore $\tan \theta = -1$ implies $\theta = 315°$. ■

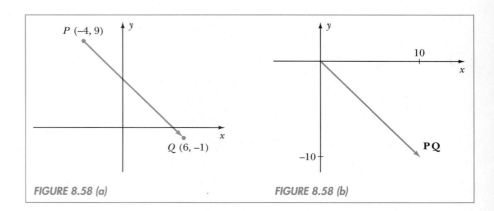

FIGURE 8.58 (a)   FIGURE 8.58 (b)

Example 7

From a control tower situated at point P, we locate a commercial airliner at Q and a small private plane at T. For simplicity, assume P, Q, and T are all in the x, y-plane. We wish to inform the pilot of the airliner about the position of the small plane. We do this by giving her the vector $\mathbf{QT}$, which represents the position of the plane relative to the airliner. If $\mathbf{PQ} = \langle 15, 8 \rangle$ and $\mathbf{PT} = \langle 14, 6 \rangle$, find the magnitude and direction of $\mathbf{QT}$. Assume the components are given in miles.

SOLUTION Figure 8.59(a) illustrates the relative positions of P, Q, and T. We find the component representation of $\mathbf{QT}$ as follows:

$$\mathbf{QT} = \mathbf{PT} - \mathbf{PQ}$$
$$= \langle 14, 6 \rangle - \langle 15, 8 \rangle$$
$$= \langle -1, -2 \rangle$$

Figure 8.59(b) shows $\mathbf{QT}$ in standard position. We have

$$\|\mathbf{QT}\| = \sqrt{(-1)^2 + (-2)^2} = \sqrt{10} \approx 3.16$$

The direction angle θ must satisfy

$$\tan \theta = \frac{-2}{-1} = 2 \quad \text{and} \quad 180° < \theta < 270°$$

This implies that $\theta = \tan^{-1}(2) + 180° \approx 63.43° + 180° = 243.43°$. Thus $\mathbf{QT}$ has magnitude 3.16 miles and direction angle 243.43°, accurate to two decimal places. ∎

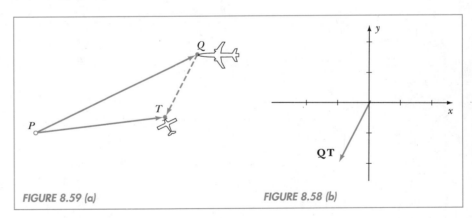

FIGURE 8.59 (a) FIGURE 8.58 (b)

Example 8

Consider points $A(-2, 3)$ and $B(9, 7)$. Find the coordinates of the point P that is two-thirds of the distance from A to B along the line segment joining A and B.

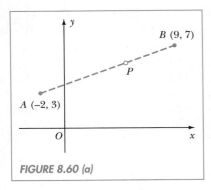

FIGURE 8.60 (a)

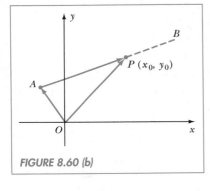

FIGURE 8.60 (b)

SOLUTION Let P have coordinates (x_0, y_0). Figure 8.60(a) shows the approximate position of P. We can use vectors to express the fact that P is two-thirds the distance from A to B along the line segment joining A and B. This is done as follows:

$$\mathbf{AP} = \frac{2}{3}\mathbf{AB}$$

We know $A(-2, 3)$ and $B(9, 7)$. Therefore $\mathbf{AB} = \langle 9 - (-2), 7 - 3 \rangle = \langle 11, 4 \rangle$. Thus

$$\mathbf{AP} = \frac{2}{3}\mathbf{AB} = \frac{2}{3}\langle 11, 4 \rangle = \left\langle \frac{22}{3}, \frac{8}{3} \right\rangle$$

Now consider Figure 8.60(b). We see that

$$\mathbf{OP} = \mathbf{OA} + \mathbf{AP}$$

We know $\mathbf{OP} = \langle x_0, y_0 \rangle$, $\mathbf{OA} = \langle -2, 3 \rangle$, and $\mathbf{AP} = \langle 22/3, 8/3 \rangle$. Substituting these into our last equation,

$$\langle x_0, y_0 \rangle = \langle 2, 3 \rangle + \left\langle \frac{22}{3}, \frac{8}{3} \right\rangle$$

$$= \left\langle -2 + \frac{22}{3}, 3 + \frac{8}{3} \right\rangle$$

$$= \left\langle \frac{16}{3}, \frac{17}{3} \right\rangle$$

Thus the coordinates of P are $(16/3, 17/3)$. ∎

EXERCISES 8.5

In Exercises 1 to 6, find the component representation for the vector $\mathbf{v}$ with the given magnitude and direction angle. Approximate to two decimal places where necessary.

1. $\|\mathbf{v}\| = 100$, $\theta_v = 34°$
2. $\|\mathbf{v}\| = 18$, $\theta_v = 280°$
3. $\|\mathbf{v}\| = 14$, $\theta_v = 240°$
4. $\|\mathbf{v}\| = 6$, $\theta_v = 135°$
5. $\mathbf{v}$ has magnitude 10 in the positive y-direction.
6. $\mathbf{v}$ has magnitude 4 in the negative x-direction.

In Exercises 7 to 10, find the magnitude and direction angle for the given vector. Approximate to two decimal places where necessary.

7. (a) $\langle 4, 3 \rangle$
   (b) $\langle -.7, -2.4 \rangle$

8. (a) $\langle 12, 5 \rangle$
   (b) $\langle -60, 80 \rangle$

9. (a) $\langle 5, -5 \rangle$
   (b) $\langle 0, -8 \rangle$

10. (a) $\langle -1, -1 \rangle$
    (b) $\langle -13, 0 \rangle$

In Exercises 11 to 14, suppose $\mathbf{v}$ has magnitude 50 and direction angle 300°. Find the component representation for the given vector. Do not approximate.

11. (a) $\frac{1}{5}\mathbf{v}$
    (b) $-2\mathbf{v}$

12. (a) $3\mathbf{v}$
    (b) $-\frac{1}{2}\mathbf{v}$

13. A unit vector with the same direction as $\mathbf{v}$

14. A unit vector with the opposite direction of $\mathbf{v}$

In Exercises 15 to 18, suppose $\mathbf{v} = \langle 6, 2 \rangle$ and $\mathbf{w} = \langle -3, 4 \rangle$. Find the component representation, magnitude, and direction angle for the given vector.

15. $\mathbf{v} + \mathbf{w}$

16. $-\mathbf{v} + 2\mathbf{w}$

17. $\frac{1}{2}\mathbf{v} - \mathbf{w}$

18. $\frac{3}{2}\mathbf{v} - \mathbf{w}$

In Exercises 19 to 22, find the component representation for $\mathbf{u} + \mathbf{v} + \mathbf{w}$ and approximate its magnitude and direction angle to two decimal places.

19. $\|\mathbf{u}\| = 10$, $\theta_\mathbf{u} = 30°$; $\|\mathbf{v}\| = 20$, $\theta_\mathbf{v} = 120°$; $\|\mathbf{w}\| = 15$, $\theta_\mathbf{w} = 225°$.

20. $\|\mathbf{u}\| = 10$, $\theta_\mathbf{u} = 0°$; $\|\mathbf{v}\| = 8$, $\theta_\mathbf{v} = 120°$; $\|\mathbf{w}\| = 6$, $\theta_\mathbf{w} = 240°$.

21. $\|\mathbf{u}\| = 8$, $\theta_\mathbf{u} = 60°$; $\|\mathbf{v}\| = 10$, $\theta_\mathbf{v} = 180°$; $\|\mathbf{w}\| = 8$, $\theta_\mathbf{w} = 300°$.

22. $\|\mathbf{u}\| = 12$, $\theta_\mathbf{u} = 45°$; $\|\mathbf{v}\| = 20$, $\theta_\mathbf{v} = 180°$; $\|\mathbf{w}\| = 10$, $\theta_\mathbf{w} = 270°$.

In Exercises 23 to 28, find the component representation for a unit vector with (a) the same direction as $\mathbf{v}$; (b) the opposite direction of $\mathbf{v}$.

23. $\mathbf{v} = \langle -3, -4 \rangle$

24. $\mathbf{v} = \langle 12, -5 \rangle$

25. $\mathbf{v} = \left\langle \frac{1}{2}, \frac{1}{3} \right\rangle$

26. $\mathbf{v} = \left\langle -\frac{1}{4}, -\frac{1}{2} \right\rangle$

27. $\mathbf{v} = \langle 0, -6 \rangle$

28. $\mathbf{v} = \langle 3, 0 \rangle$

In Exercises 29 to 32, for the given points P and Q, draw the vector $\mathbf{PQ}$ in standard position and find its magnitude and direction angle.

29. $P(2, -3)$, $Q(3, -2)$

30. $P(-2, 1)$, $Q(-5, -3)$

31. $P(-1, 8)$, $Q(1, 8)$

32. $P(4, -5)$, $Q(-5, 4)$

In Exercises 33 and 34, find the component representation of $\mathbf{QT}$ from the given information.

33. $\mathbf{PQ} = \langle 6, 10 \rangle$, $\mathbf{PT} = \langle 4, 5 \rangle$

34. $\mathbf{PQ} = \langle -8, 3 \rangle$, $\mathbf{PT} = \langle 1, -2 \rangle$

35. Suppose that in Example 7 $\mathbf{PQ} = \langle 12, 7 \rangle$ and $\mathbf{PT} = \langle 14, 8 \rangle$. Find the magnitude and direction angle for $\mathbf{QT}$.

36. From an observation point P, two ships are located at sea at points Q and T. If $\mathbf{PQ} = \langle 2, 4 \rangle$ and $\mathbf{PT} = \langle 3, -1 \rangle$, find the magnitude and direction angle for $\mathbf{QT}$. Assume all points are in the plane and the components are given in miles.

In Exercises 37 and 38, for the given points A and B, find the coordinates of the point P, which is two-thirds the distance from A to B along the line segment joining A and B.

37. $A(-3, 2)$ $B(9, 9)$

38. $A(-4, 6)$ $B(10, 2)$

39. Repeat Exercise 37 for the point P, which is one-fourth the distance from A to B.

40. Repeat Exercise 38 for the point P, which is three-fourths the distance from A to B.

41. Given points $A(a, b)$ and $B(c, d)$,
    (a) Use vectors to find the coordinates of the point P, which is two-thirds the distance from A to B along the line segment joining A and B.
    (b) Use vectors to find the midpoint of the line segment joining A and B.

42. Let P and Q have coordinates $P(x_1, y_1)$ and $Q(x_2, y_2)$. Find the component representation for $\mathbf{PQ}$ and $\mathbf{QP}$. What is the relationship between $\mathbf{PQ}$ and $\mathbf{QP}$?

43. Suppose $\mathbf{v} \neq 0$ and $\mathbf{v} = \langle a, b \rangle$. Use the properties of components to prove that $\mathbf{u} = \dfrac{1}{\|\mathbf{v}\|} \mathbf{v}$ is a unit vector.

44. Prove property 3 of the Properties of Components as follows.
    (a) Show that vectors $r\langle a, b \rangle$ and $\langle ra, rb \rangle$ have the same magnitude. Use property 2 and the definition of scalar multiplication given in Section 8.4.
    (b) Show that vectors $r\langle a, b \rangle$ and $\langle ra, rb \rangle$ have the same direction angle.
    (c) Conclude from parts (a) and (b) that $r\langle a, b \rangle = \langle ra, rb \rangle$ as desired.

When appropriate, approximate answers to two decimal places.

1. From the top of a building a point is spotted on the ground below at a 16° angle of depression. If the point is 500 feet from the base of the building, how tall is the building? Assume the ground is level.

2. Find the area of triangle ABC in the x, y-plane if A and B have coordinates $A(1, 2)$, $B(5, -1)$, $BC = 10$, and angle $ABC = 54°$.

3. One end of a 90-foot cable is attached to the top of a vertical tower and the other end, after being pulled taut, is attached to a point A on the ground. If the cable makes an angle of 48.7° with respect to the ground, how far is A from the base of the tower? Assume the ground is level.

4. Find x in Figure 8.61.

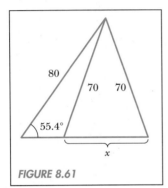

FIGURE 8.61

5. In Figure 8.62, find the x- and y-coordinates of P in terms of α and β.

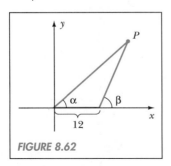

FIGURE 8.62

6. In triangle ABC, $A = 18°$, $B = 105°$, and $AB = 10$. Find AC.

7. The vertices of triangle ABC are $A(-4, 1)$, $B(8, 2)$, and $C(-1, 5)$. Find angles A, B, and C.

8. In triangle ABC, $AB = 21$, $AC = 32.4$, and $A = 1$ radian. Find angle B.

9. A particle starts from point $A(50, 0)$ and moves in a straight line path in the x, y-plane at the constant speed of 12 units per second. After 3 seconds the particle stops at point B. Find the distance from the origin O to B if angle $AOB = 40°$. There are two possible answers!

10. Find a polar representation for the given complex number z.
    (a) $-2 + 2i$ (b) $-3i$

11. Use DeMoivre's Theorem to evaluate the given expression.
    (a) $(2 - 2i)^6$ (b) $\left(\cos\left(\frac{7\pi}{12}\right) + i \sin\left(\frac{7\pi}{12}\right)\right)^8$

12. Find the third roots of 1.

13. Find all the complex zeros of $F(x) = x^6 + 1$.

14. Suppose vector $\mathbf{v}$ has magnitude 6 and direction angle 45°. Draw the given vector in standard position and find its magnitude and direction angle.
    (a) $2\mathbf{v}$ (b) $-\mathbf{v}$ (c) $-\frac{1}{2}\mathbf{v}$

15. Use the vectors $\mathbf{v}$ and $\mathbf{w}$ in Figure 8.63 to construct $\mathbf{v} + \mathbf{w}$ and $\mathbf{v} - \mathbf{w}$. In each case, find the magnitude and direction angle of the resultant vector.

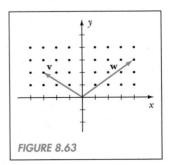

FIGURE 8.63

16. Find the component representation of vector **v** if $\|\mathbf{v}\| = 20$, $\theta_v = 150°$.

17. Find the magnitude and direction angle for vector **v** = $\langle 15, -8 \rangle$.

18. Suppose **v** = $\langle -2, 1 \rangle$ and **w** = $\langle 5, 3 \rangle$. Find the component representation, magnitude, and direction angle for the vector described.

    (a) $3\mathbf{v} - \mathbf{w}$

    (b) A unit vector in the direction of **v**

    (c) A unit vector in the opposite direction of **w**

19. Consider vectors **u**, **v**, and **w** with $\|\mathbf{u}\| = 10$, $\theta_u = 45°$, $\|\mathbf{v}\| = 20$, $\theta_v = 150°$, $\|\mathbf{w}\| = 10$, $\theta_w = 300°$. Find the component representation, magnitude, and direction angle for $\mathbf{u} + \mathbf{v} + \mathbf{w}$.

20. Consider points $P(-3, -5)$, $Q(6, -2)$, and $O(0, 0)$. Find the component representation for (a) **PQ**, (b) **PO**, (c) **OP** + **PQ**.

21. Suppose **PQ** = $\langle -2, 7 \rangle$ and **PT** = $\langle 5, 3 \rangle$. Find **QT**.

CHAPTER 9 Systems of Equations

9.1 ELEMENTARY SYSTEMS

A **system of equations** consists of two or more equations containing two or more unknowns. A **solution** for a given system of equations is a set of real numbers, one for each unknown involved, which satisfies all the equations in the system simultaneously. For example, the following system consists of two equations in two unknowns:

$$x^2 + y = 3$$
$$-x + y^2 = 5$$

The numbers $x = -1$ and $y = 2$ satisfy both equations simultaneously and therefore qualify as a solution of the system. We may write this solution as an ordered pair: $(-1, 2)$. Note that an ordered pair such as $(3, -6)$, which satisfies one equation (the first) but not the other, does not qualify as a solution for the system.

In this section we concentrate entirely on systems consisting of two equations in two unknowns. Solutions for these systems correspond to points of intersection on the graphs of the given equations.

Example 1

Show that $(3, 2)$ is a solution for the system $8y = (x + 1)^2$ and $x = y^2 - 1$. Graph both equations on the same axes. How many solutions does this system have?

SOLUTION We must show that $(3, 2)$ satisfies both equations. This is established by substituting $x = 3$ and $y = 2$ into each equation as follows:

$$8y \overset{?}{=} (x + 1)^2 \qquad\qquad x \overset{?}{=} y^2 - 1$$

$8(2)$	$(3 + 1)^2$		3	$(2)^2 - 1$
16	$(4)^2$			$4 - 1$
	16			3
Yes			Yes	

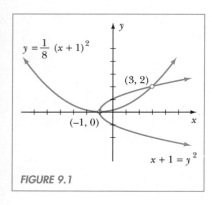

$y = \frac{1}{8}(x + 1)^2$

$(3, 2)$

$(-1, 0)$

$x + 1 = y^2$

FIGURE 9.1

Notice that $8y = (x + 1)^2$ is equivalent to $y = \frac{1}{8}(x + 1)^2$, which is $y = \frac{1}{8}x^2$ shifted -1 unit in the x-direction. Also, $x = y^2 - 1$ is equivalent to $x + 1 = y^2$, which is $x = y^2$ shifted -1 unit in the x-direction. Thus we obtain the graphs shown in Figure 9.1.

From the graph, we see that there are two points of intersection, $(3, 2)$ and $(-1, 0)$. Therefore the system must have two solutions corresponding to these points. We have verified that $(3, 2)$ is a solution. You can check that $(-1, 0)$ also satisfies both equations. ■

Graphing a system of two equations in two unknowns gives us a nice geometric interpretation of the solutions for a system. However, graphing does not provide us with a practical general approach to finding solutions. Indeed, points of intersection may not have integer coordinates and, in some cases, may be impossible to determine from a graph. Therefore we shall concentrate on solving systems algebraically; following is an outline of our procedure.

Solving Two Equations in Two Unknowns
Use one of the following two methods.
1. Addition: Add a multiple of one equation to the other to eliminate one of the unknowns.
2. Substitution: Use one equation to find an expression for one unknown in terms of the other. Substitute this expression into the other equation.
With either method we obtain one equation in one unknown. Solve this equation and then return to a previous equation to find the corresponding value of the second unknown. Check answers in both of the original equations.

Example 2

Solve the system $2x - 3y = 8$ and $3x + y = 1$.

SOLUTION   Arrange the equations over each other, with x's in the first column, y's in the second column, and constants on the right side of the equal sign in the third column. We use *addition*, multiplying the second equation by 3 and adding it to the first to eliminate y.

$$
\begin{array}{ccc}
2x - 3y = 8 & \longrightarrow & 2x - 3y = 8 \\
3x + y = 1 & \xrightarrow{\times (3)} & 9x + 3y = 3 \\
\hline
& & 11x = 11 \\
& & x = 1
\end{array}
$$

Thus we obtain $x = 1$. To find the corresponding value of y, we substitute $x = 1$ into a previous equation, such as $3x + y = 1$.

$$3(1) + y = 1$$

$$3 + y = 1$$

$$y = -2$$

Therefore the solution is $x = 1$ and $y = -2$ or, written as an ordered pair, $(1, -2)$. You should check that this answer works in both the original equations. ∎

Example 3

Solve the system $\dfrac{y^2}{4} + \dfrac{x^2}{8} = 1$ and $y^2 - x = 0$.

SOLUTION We use *substitution*. From the second equation, we have

$$y^2 - x = 0 \quad \text{implies} \quad y^2 = x$$

Substitute this into the first equation, replacing y^2 with x.

$$\left.\begin{array}{c} \dfrac{y^2}{4} + \dfrac{x^2}{8} = 1 \\[2mm] y^2 = x \end{array}\right\} \quad \text{implies} \quad \dfrac{x}{4} + \dfrac{x^2}{8} = 1$$

Next, we solve this equation for x. Begin by multiplying both sides by 8 to eliminate fractions.

$$8\left(\frac{x}{4} + \frac{x^2}{8}\right) = 8(1)$$

$$2x + x^2 = 8$$

$$x^2 + 2x - 8 = 0$$

$$(x + 4)(x - 2) = 0$$

$$x + 4 = 0 \quad or \quad x - 2 = 0$$

$$x = -4 \quad \bigg| \quad x = 2$$

Finally, we return to a previous equation, such as $y^2 = x$, to find the corresponding values of y.

$x = -4$	$x = 2$
$y^2 = x$	$y^2 = x$
$y^2 = -4$	$y^2 = 2$
No solution	$y = \sqrt{2} \quad or \quad y = -\sqrt{2}$

Thus we have two solutions; $(2, \sqrt{2})$ and $(2, -\sqrt{2})$. You may verify that each answer works in both the original equations. ∎

Remark

When we substituted $y^2 = x$ into the first equation, we could have chosen to replace x with y^2, which would have given us the equation

$$\frac{y^2}{4} + \frac{y^4}{8} = 1$$

This is equivalent to $y^4 + 2y^2 - 8 = 0$, which factors: $(y^2 + 4) \times (y^2 - 2) = 0$. You may verify that this leads to the same solutions found above.

Any equation of the form $ax + by = c$ (where not both a and b are zero) is called a **linear equation** in two unknowns. We say that a system of equations in two unknowns is a **linear system** if it consists entirely of linear equations. The system in Example 2 qualifies as a linear system. On the other hand, if a system in two unknowns contains an equation that is not linear, then we call it a **nonlinear system**. The systems in Examples 1 and 3 are types of nonlinear systems.

Recall from Section 1.7 that the graph of a linear equation is just a straight line. We know that two lines may intersect in no points (if the lines are parallel and distinct), one point (if the lines are not parallel), or an infinite number of points (if both lines are the same). See Figure 9.2. Corresponding to these three geometric situations are the following conclusions about the solution set for a linear system of two equations in two unknowns.

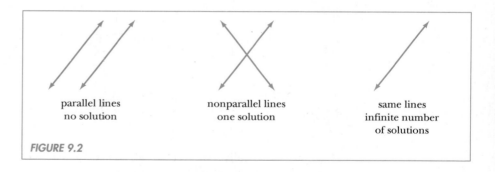

parallel lines
no solution

nonparallel lines
one solution

same lines
infinite number
of solutions

FIGURE 9.2

The solution set for a linear system consists of exactly one of the following:
1. No solutions
2. Exactly one solution
3. An infinite number of solutions

Unfortunately, we cannot make a similar statement about the solution set for a nonlinear system. Since the graphs for these equations are, in general, curves and not straight lines, the intersections could consist of any number of points. It follows that a nonlinear system may have any number of solutions.

Examples 4 to 7 illustrate the different possibilities for solution sets.

Example 4

Solve the system $2x + y = 1$
$2y = 1 - 4x$.

SOLUTION Before starting to work, we note that this is a linear system of equations, so we may end up with none, one, or an infinite number of solutions. To solve linear systems we tend to favor the method of *addition*. Thus we line up the x's and y's on the left side, multiply the first equation by -2, and then add.

$$
\begin{array}{llll}
2x + y = 1 & \longrightarrow\ 2x + y = 1 & \xrightarrow{\times(-2)} & -4x - 2y = -2 \\
2y = 1 - 4x & \longrightarrow\ 4x + 2y = 1 & \longrightarrow & \underline{4x + 2y = 1} \\
& & & 0 = -1
\end{array}
$$

The statement $0 = -1$ is nonsense; it means that we must conclude that there is *no solution* to this system.

As a check on our algebra, let us consider the given system of equations from a geometric point of view. Rewriting the equations in slope-intercept form, we have

$$
\begin{array}{ll}
2x + y = 1 & \longrightarrow\quad y = -2x + 1 \\
2y = 1 - 4x & \longrightarrow\quad y = -2x + \dfrac{1}{2}
\end{array}
$$

We recognize these equations as representing two different parallel lines (each has slope -2). Since the lines do not intersect, the system cannot have a solution. Thus the geometry agrees with our algebra above. ■

Example 5

Solve the system $\quad 4y - 12x = 8$
$27x - 9y = -18$.

SOLUTION Once again we have a linear system, so our solution set will consist of none, one, or an infinite number of solutions. To solve this system, we choose the method of *addition*. After lining up x's and y's on the left side, we attempt to eliminate the y's by multiplying the first equation by 9 and the second equation by 4.

$$
\begin{array}{llll}
4y - 12x = 8 & \longrightarrow\ -12x + 4y = 8 & \xrightarrow{\times(9)} & -108x + 36y = 72 \\
27x - 9y = -18 & \longrightarrow\ 27x - 9y = -18 & \xrightarrow{\times(4)} & \underline{108x - 36y = -72} \\
& & & 0 = 0
\end{array}
$$

Both x's and y's are eliminated, but this time we are left with a true statement, $0 = 0$. When this occurs in a linear system, it means that both the original equations are equivalent. In geometric terms, the equations represent the same line, so we have an *infinite number* of solutions. We can verify this by rewriting each equation in slope-intercept form.

$$4y - 12x = 8 \longrightarrow 4y = 12x + 8 \longrightarrow y = 3x + 2$$

$$27x - 9y = -18 \longrightarrow -9y = -27x - 18 \longrightarrow y = 3x + 2$$

Thus the solution set for the system can be written as $\{(x, y): y = 3x + 2\}$. ∎

Example 6

Solve the system
$$x^2 - y = -1$$
$$2x - 2y = 3.$$

SOLUTION   This is a nonlinear system of equations, so we may have any number of solutions. Note that the y's may be eliminated using either *addition* or *substitution*. Let us use *addition*.

$$
\begin{array}{ll}
x^2 - y = -1 & \xrightarrow{\times (2)} \quad 2x^2 - 2y = -2 \\
2x - 2y = 3 & \xrightarrow{\times (-1)} \quad \underline{-2x + 2y = -3} \\
 & \qquad\qquad\; 2x^2 - 2x = -5
\end{array}
$$

This gives us a quadratic equation in x to solve. We have

$$2x^2 - 2x = -5$$

$$2x^2 - 2x + 5 = 0$$

Using the quadratic formula, we find

$$x = \frac{2 \pm \sqrt{(-2)^2 - 4(2)(5)}}{2(2)}$$

$$= \frac{2 \pm \sqrt{-36}}{4} \qquad \text{no real solution}$$

We conclude that there is *no solution* to the original system. ∎

Remark

The answer to Example 6 of no solution makes sense geometrically. If we graph $x^2 - y = -1$ (the shifted parabola $y - 1 = x^2$) and $2x - 2y = 3$ (a straight line) on the same axes, we find that they do not intersect. See Figure 9.3.

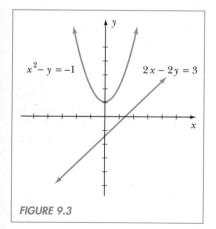

FIGURE 9.3

Example 7

Solve the system $x^2 - y^2 = 7$

$$\frac{4}{x^2} + \frac{27}{4y^2} = 1.$$

SOLUTION   We have a nonlinear system of equations, so any number of solutions is possible. To solve this system we shall use *substitution*. From the first equation, we find

$$x^2 - y^2 = 7 \quad \text{implies} \quad x^2 = y^2 + 7$$

Substituting for x^2 in the second equation, we have

$$\left.\begin{array}{c} x^2 = y^2 + 7 \\[2mm] \dfrac{4}{x^2} + \dfrac{27}{4y^2} = 1 \end{array}\right\} \quad \text{implies} \quad \frac{4}{y^2 + 7} + \frac{27}{4y^2} = 1$$

Now we have one equation in y to solve. Multiply both sides by $4y^2(y^2 + 7)$ to eliminate the fractions.

$$4y^2(y^2 + 7)\left(\frac{4}{y^2 + 7} + \frac{27}{4y^2}\right) = 4y^2(y^2 + 7)(1)$$

$$4y^2(4) + (y^2 + 7)27 = 4y^4 + 28y^2$$

$$16y^2 + 27y^2 + 189 = 4y^4 + 28y^2$$

$$0 = 4y^4 - 15y^2 - 189$$

$$0 = (4y^2 + 21)(y^2 - 9)$$

$$4y^2 + 21 = 0 \qquad or \qquad y^2 - 9 = 0$$

$$y^2 = \frac{-21}{4} \qquad\qquad\qquad y^2 = 9$$

No real solution | $y = 3 \quad or \quad y-3$

Now return to a previous equation, such as $x^2 = y^2 + 7$, to find the corresponding values of x.

$\underline{y = 3}$	$\underline{y = -3}$
$x^2 = y^2 + 7$	$x^2 = y^2 + 7$
$x^2 = 3^2 + 7$	$x^2 = (-3)^2 + 7$
$x^2 = 16$	$x^2 = 16$
$x = 4 \quad or \quad x = -4$	$x = 4 \quad or \quad x = -4$

Thus we obtain four solutions: $(4, 3)$, $(-4, 3)$, $(4, -3)$, and $(-4, -3)$. You should check that all four answers satisfy both the original equations in the system. ∎

Remark

Notice that both the original equations in the system are symmetric with respect to the y-axis, x-axis, and origin. Hence, if the graphs intersect at (a, b), then they must also intersect at the symmetric images of (a, b) with respect to the y-axis, x-axis, and origin, namely, $(-a, b)$, $(a, -b)$, and $(-a, -b)$. This agrees with our results above.

We close this section with an application in Example 8.

Example 8

Find the coordinates of the point P that is equidistant from the points $A(0, 0)$, $B(4, 1)$, and $C(1, 3)$.

SOLUTION Figure 9.4 illustrates the situation. For P to be equidistant from A, B, and C, we must have the distances PA, PB, and PC all equal. Let P have coordinates (x, y). By the distance formula,

$$PA = PB \quad \text{means} \quad \sqrt{x^2 + y^2} = \sqrt{(x - 4)^2 + (y - 1)^2}$$

$$PA = PC \quad \text{means} \quad \sqrt{x^2 + y^2} = \sqrt{(x - 1)^2 + (y - 3)^2}$$

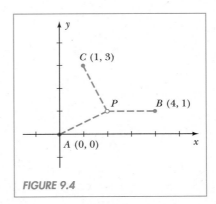

FIGURE 9.4

This gives us two equations in the two unknowns x and y. Before we solve this system, we can simplify the equations quite a bit. Consider the first equation. Squaring both sides yields

$$x^2 + y^2 = (x - 4)^2 + (y - 1)^2$$

Next, expand the right side.

$$x^2 + y^2 = x^2 - 8x + 16 + y^2 - 2y + 1$$

After putting the unknowns on the left side and simplifying, we obtain

$$8x + 2y = 17$$

You may verify in a similar fashion that the second equation

$$\sqrt{x^2 + y^2} = \sqrt{(x - 1)^2 + (y - 3)^2}$$

simplifies to

$$x + 3y = 5$$

Now we have a much simpler system of equations to solve. We use *addition* as follows:

$$
\begin{array}{ll}
8x + 2y = 17 \quad\longrightarrow & 8x + 2y = 17 \\
x + 3y = 5 \quad\xrightarrow{\times(-8)} & \underline{-8x - 24y = -40} \\
& -22y = -23 \\
& y = \dfrac{23}{22}
\end{array}
$$

To find the corresponding value of x, we return to the equation $x + 3y = 5$ and substitute $y = 23/22$.

$$x + 3y = 5$$

$$x + 3\left(\frac{23}{22}\right) = 5$$

$$x + \frac{69}{22} = 5$$

$$x = 5 - \frac{69}{22}$$

$$= \frac{41}{22}$$

Therefore the desired point P has coordinates $\left(\dfrac{41}{22}, \dfrac{23}{22}\right)$. ■

Remarks

1. To find two unknowns, we usually need two equations to solve the problem. In Example 8 we used the two equations $PA = PB$, yielding $8x + 2y = 17$, and $PA = PC$, giving us $x + 3y = 5$. We could have written a third equation, namely, $PB = PC$, but there is no new information in this equation since it can be obtained from the first two:

    $$PA = PB \text{ and } PA = PC \quad \text{implies} \quad PB = PC$$

2. Geometrically speaking, the point P is the intersection point of the perpendicular bisectors of AB, AC, and BC.

EXERCISES 9.1

In Exercises 1 to 6, use *addition* to solve the given system. Graph.

1. $4x - y = 13$
   $x + 2y = 1$

2. $4x + y = 1$
   $4x - y = 2$

3. $3x = 5y + 9$ *and* $2y + 7x + 4 = 0$

4. $15x = 3y + 4$ *and* $y - 5x = 0$

5. (a) $x - y = 2$
   $y - x = 3$
   (b) $6(y + 1) = -4(x + 2)$
   $2x + 3y + 7 = 0$

6. (a) $2x = 4y$
   $2\left(y - \dfrac{1}{2}\right) = x + 1$
   (b) $3x - 4y = 8$
   $2y - \dfrac{3}{2}x = -4$

In Exercises 7 to 12, use *substitution* to solve the given system. Graph.

7. $y = x^2$
   $y = 3x$

8. $x = y^2$
   $x + y = 2$

9. $x^2 + y^2 = 1$
   $x + y = 0$

10. $(x - 1)^2 + y^2 = 1$
    $x - y = 0$

11. $xy = 2$ *and* $y = \dfrac{|x|}{x}$

12. $2y = x^3 - x$ *and* $3x - 2y = 0$

In Exercises 13 to 28, use either *addition* or *substitution* to solve the given system.

13. $\dfrac{x^2}{16} + \dfrac{y^2}{4} = 1$
    $2x - y = 2$

14. $\dfrac{x^2}{4} - \dfrac{y^2}{12} = 1$
    $2y - x = 2$

15. $x^2 + y^2 = 1$
    $x^2 - y^2 = 1$

16. $x^2 + y^2 = 4$
    $y^2 - x^2 = 2$

17. $\dfrac{x^2}{25} + \dfrac{y^2}{9} = 1$
    $3x^2 - 25y = 75$

18. $\dfrac{x^2}{9} - \dfrac{y^2}{3} = 1$
    $x^2 - 9y = 9$

19. $4x^2 + y^2 = 8$
    $5x^2 + 2y^2 = 13$

20. $4x^2 + 4y^2 = 17$
    $xy = 1$

21. $y = e^x - e^{-x}$
    $y = e^{-x}$

22. $e^x + e^{2y} = 7$
    $e^x - e^y = 1$

23. $x - 2y = 4$
    $y + \sqrt{x + 4} = 0$

24. $xy = 4$
    $\sqrt{x} - y = 0$

25. $\ln(x^2) + \ln(y^2) = 2$
    $\ln x - \ln y = 0$

26. $\ln(xy) = 4$
    $\ln\left(\dfrac{x}{y}\right) = -1$

27. $9y^2 = x^2 - 2x + 6$
    $2x - 3y = 3$

28. $4x^2 + y^2 = 2 - 3x$
    $3y - 2x = -4$

In Exercises 29 to 32, solve for a and b.

29. $\dfrac{4}{a^2} - \dfrac{48}{b^2} = 1$
    $a^2 + b^2 = 17$

30. $\dfrac{4}{a^2} + \dfrac{20}{9b^2} = 1$
    $a^2 - b^2 = 5$

31. $\dfrac{3}{a} + \dfrac{2}{b} = 1$
    $\dfrac{1}{a} + \dfrac{3}{b} = \dfrac{1}{6}$

32. $\dfrac{4}{a} - \dfrac{5}{b} = 2$
    $\dfrac{3}{a} + \dfrac{1}{b} = 1$

In Exercises 33 and 34, find the coordinates of the point P that is equidistant from A, B, and C.

33. $A(0, 0)$ $B(3, 2)$ $C(-1, 4)$ 34. $A(0, 0)$ $B(5, 1)$ $C(-2, 6)$

35. Find all points on the line $y = 2x + 1$ that are 3 units from $A(2, 1)$.

36. Find all points on the curve $y + 1 = x^2$ that are 4 units from $A(0, 1)$.

In Exercises 37 and 38, find the point on the given line that is equidistant from A and B.

37. $5x + 4y = 20$, $A(0, 3)$, $B(9, 0)$

38. $3x + 2y = 4$, $A(-2, 3)$, $B(4, 5)$

39. Find the points that are equidistant from $A(1, 0)$, $B(3, 2)$, and the y-axis.

40. Find the equation of the line parallel to the x-axis passing through the point of intersection of the lines $x + 2y = 8$ and $2x - y = 1$.

41. Tangent lines are drawn to the circle $(x - 3)^2 + (y - 2)^2 = 29$ at the points $(-2, 0)$ and $(5, 7)$. Find where these tangent lines intersect.

42. A point $P(x, y)$ in the plane has the property that the sum of its coordinates is 3.3 and the product of the coordinates is 2. Find the point.

43. The coordinates of a point in the plane have the property that their sum is 11 and the sum of their squares is 65. Find the point. (Two points are possible.)

44. A wire 8 feet long is cut into two pieces. One piece is bent to form a square, and the other piece is bent to form the legs of an isosceles right triangle. If the sum of the areas of the square and the triangle is 19 square feet, find the length of each piece of wire.

45. Solve the following system for x and y in terms of a, b, c, α, and β.

$$\frac{y}{x} = a \tan \alpha \qquad \frac{y}{x + c} = b \tan \beta$$

46. Consider two distinct lines ℓ_1 and ℓ_2 with equations $y = m_1 x + b_1$ and $y = m_2 x + b_2$, respectively.
   (a) Show that the point of intersection of ℓ_1 and ℓ_2 is given by $\left(\dfrac{b_2 - b_1}{m_1 - m_2}, \dfrac{m_1 b_2 - m_2 b_1}{m_1 - m_2} \right)$, provided $m_1 \neq m_2$.
   (b) Use the result in part (a) to explain why two distinct nonvertical lines are parallel if and only if their slopes are equal. (Recall that two lines in the plane are parallel if and only if they do not intersect.)

9.2 GRAPHING REGIONS

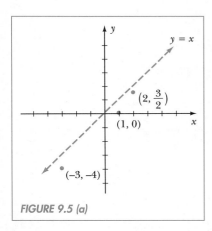

FIGURE 9.5 (a)

FIGURE 9.5 (b)

In this section we graph regions in the plane that are described by inequalities in two unknowns. We begin by graphing single inequalities, and then we examine systems involving two or more inequalities.

To graph a single inequality in two unknowns, we must graph the set of all ordered pairs (x, y) that are solutions of the inequality. Let us start with a simple example.

Example 1

Graph the solution set for $y - x < 0$.

SOLUTION Since $y - x < 0$ is equivalent to $y < x$, we wish to graph the set

$$\{(x, y): y < x\}$$

A point $P(a, b)$ is in this set if and only if its y-coordinate is less than its x-coordinate, $b < a$. Therefore points such as $(1, 0)$, $(2, 3/2)$, and $(-3, -4)$ are solutions; these are graphed in Figure 9.5(a). Of course, there are many other solution points. How do we go about finding *all* the solutions?

We begin by changing the inequality $y < x$ into the equation $y = x$. We then graph this line dashed, as shown in Figure 9.5(a). Notice that any point below this line will have its y-coordinate less than its corresponding x-coordinate. Therefore, any point below the line $y = x$ will be in the solution set for $y < x$. On the other hand, any point above the line $y = x$ is not in the solution set since its y-coordinate will be greater than its corresponding x-coordinate. We conclude that the solution set for $y < x$ is exactly the region below the line $y = x$. This is graphed in Figure 9.5(b) by shading the appropriate region. ∎

Remark

We call the line $y = x$ the **boundary** of the solution set. When the inequality symbol is strict ($<$ or $>$), as in this example, the points in the boundary are not part of the solution set. Thus we graph the boundary dashed. However, if the inequality symbol is not strict ($\leq$ or $\geq$), then the boundary points are included in the solution set. In this case, we draw the boundary solid.

The arguments in Example 1 generalize into the following procedure for graphing inequalities.

Graphing Inequalities in Two Variables

Step 1. Graph the boundary equation obtained by changing the inequality symbol to $=$. Graph the boundary dashed if the inequality symbol is $<$ or $>$, solid if $\leq$ or $\geq$.

Step 2. If the boundary divides the plane into two regions, test a point $P(a, b)$ not on the boundary in the original inequality. If P is a solution, shade in the region on the same side of the boundary that P is on; if P is not a solution, shade in the opposite side.

Note: if one of the expressions in the inequality is a function whose domain is restricted, then we must also take this restriction into account.

The purpose of the note at the bottom of the procedure will be made clear in Example 3.

Example 2

Graph the solution set for $x - y^2 \geq -1$.

SOLUTION   Step 1.   The boundary curve is $x - y^2 = -1$. This is equivalent to $x + 1 = y^2$, which is the basic curve $x = y^2$ shifted to the left 1 unit. We draw this with a solid curve since the inequality symbol includes equals, $\geq$. See Figure 9.6(a).

Step 2.   We need to test a point in the inequality that is not on the boundary. We use the point $P(0, 0)$.

$$x - y^2 \overset{?}{\geq} -1$$
$$0 - (0)^2$$
$$0 \quad\Big|$$
$$\text{Yes}$$

Therefore we shade the region on the same side that contains $P(0, 0)$. Figure 9.6(b) shows the solution set. ∎

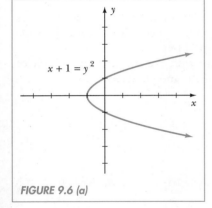

FIGURE 9.6 (a)

$x + 1 = y^2$

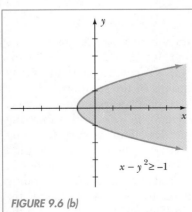

FIGURE 9.6 (b)

$x - y^2 \geq -1$

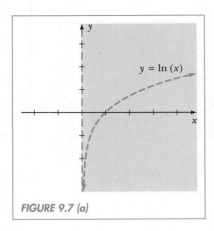

FIGURE 9.7 (a)

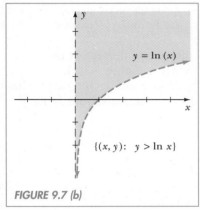

{(x, y): y > ln x}

FIGURE 9.7 (b)

Example 3

Graph the solution set for $y > \ln x$.

SOLUTION This inequality involves the function $F(x) = \ln x$, whose domain is restricted to $\{x: x > 0\}$. According to the note in our graphing procedure, we must take this into account. This means that we can consider only solution points for the inequality that lie in the set $\{(x, y): x > 0\}$. This is the half-plane illustrated by the shaded region in Figure 9.7(a). Thus we apply our two-step procedure for graphing $y > \ln x$ restricted to this half-plane.

Step 1. Graph the boundary curve $y = \ln x$. Notice that this is contained in the half-plane shown in Figure 9.7(a).

Step 2. Test a point not on the boundary curve, say, $(e, 0)$. Note that we must also choose this point in the half-plane $x > 0$.

$$y \overset{?}{>} \ln x$$
$$0 \mid \ln e$$
$$\mid 1$$
$$\text{No}$$

Since $(e, 0)$ is below the boundary curve, we shade in the opposite side above $y = \ln x$ for the final solution set, as shown in Figure 9.7(b). ∎

We now turn our attention to graphing systems of inequalities, which will involve finding ordered pairs (x, y) that satisfy two or more inequalities simultaneously. To do this, we first graph the solution set for each individual inequality in the system and then find the intersection of these sets. This gives us the final solution set for the entire system.

Example 4

Graph the solution set for the system

$$\begin{cases} x - y^2 \geq -1 \\ x^2 + y^2 < 1 - 2x \end{cases}$$

SOLUTION The solution set for this system is the intersection of the solutions to $x - y^2 > -1$ and $x^2 + y^2 < 1 - 2x$.

$$\{(x, y): x - y^2 \geq -1\} \cap \{(x, y): x^2 + y^2 < 1 - 2x\}$$

We shall graph each set individually and then take their intersection. The first set, $\{(x, y): x - y^2 \geq -1\}$, was graphed in Example 2. See Figure 9.6(b). For the second set, $\{(x, y): x^2 + y^2 < 1 - 2x\}$, we follow our two-step procedure.

Step 1. The equation for the boundary curve is $x^2 + y^2 = 1 - 2x$. Rearranging terms and completing the square, we have

$$x^2 + 2x + y^2 = 1$$
$$x^2 + 2x + 1 + y^2 = 1 + 1$$
$$(x + 1)^2 + y^2 = 2$$

We recognize this as a circle with center at $(-1, 0)$ and radius $\sqrt{2}$.

Step 2. Testing the point $P(0, 0)$ on the inside of the circle, we find

$$x^2 + y^2 \overset{?}{<} 1 - 2x$$

$$
\begin{array}{c|c}
0^2 + 0^2 & 1 - 2(0) \\
0 & 1
\end{array}
$$
$$\text{Yes}$$

Therefore the solution set for $x^2 + y^2 < 1 - 2x$ consists of the region inside the circle shown in Figure 9.8(a). Note that the boundary is drawn dashed since it is not included in the solution set.

Now, the solution set for the system is obtained by intersecting the regions drawn in Figures 9.6(b) and 9.8(a); this is shown in Figure 9.8(b).

Notice that we labeled where the two boundary curves intersect. These points were found by solving the two boundary equations simultaneously.

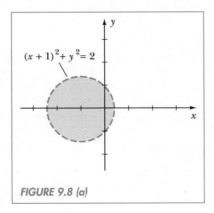

FIGURE 9.8 (a)

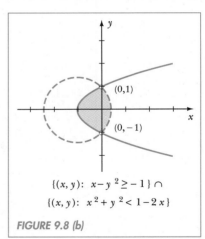

$\{(x, y): \ x - y^2 \geq -1\} \cap$
$\{(x, y): \ x^2 + y^2 < 1 - 2x\}$

FIGURE 9.8 (b)

Remark

The points on the boundary curve $x^2 + y^2 = 1 - 2x$ are not included in the solution set. Therefore we drew open circles at the points $(0, 1)$ and $(0, -1)$ to indicate that these points are not in the solution.

If an inequality in two variables contains absolute values, then we may have two or more boundary curves dividing the plane into three or more regions. This requires that we check points from all the regions in the original inequality.

Example 5

Graph $\{(x, y): |x + y| \geq 1\}$.

SOLUTION The boundary curve has the equation $|x + y| = 1$. But this is equivalent to

$$x + y = 1 \quad or \quad x + y = -1$$

Thus the boundary for the solution set consists of two parallel lines dividing the plane into three regions, as shown in Figure 9.9(a). Now, we test a point from each region in the original inequality.

$$
\begin{array}{ccc}
P_1(-2, 0) & P_2(0, 0) & P_3(2, 0) \\[4pt]
|x + y| \overset{?}{\geq} 1 & |x + y| \overset{?}{\geq} 1 & |x + y| \overset{?}{\geq} 1 \\[4pt]
|-2 + 0| & |0 + 0| & |2 + 0| \\[4pt]
2 & 0 & 2 \\[2pt]
\text{Yes} & \text{No} & \text{Yes}
\end{array}
$$

Therefore the solution set consists of the regions containing P_1 and P_3. We shade in these regions as shown in Figure 9.9(b).

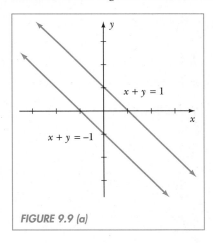

FIGURE 9.9 (a)

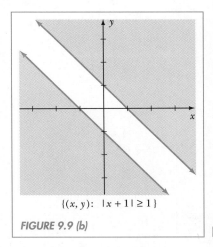

$\{(x, y): |x + 1| \geq 1\}$

FIGURE 9.9 (b)

We close this section with a simple application in Example 6.

Example 6

A rectangle is to be constructed so that its interior will contain a circle of radius 10. The perimeter of the rectangle cannot exceed 100. Write a system of inequalities that describes all the possible choices for the length and width of the rectangle. Graph the system.

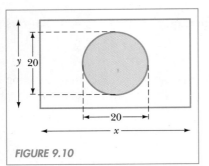

FIGURE 9.10

SOLUTION We wish to describe the possible values for the length and width of the rectangle. Therefore we define two variables x and y as follows:

$$x = \text{length of rectangle}$$

$$y = \text{width of rectangle}$$

To fit a circle of radius 10 inside the rectangle, both x and y must be at least 20 (the diameter of the circle). See Figure 9.10. Therefore we must have $x \geq 20$ and $y \geq 20$. Furthermore, the perimeter of the rectangle cannot exceed 100. It follows that $2x + 2y \leq 100$ or, more simply, $x + y \leq 50$. In conclusion, we have the following system:

$$\begin{cases} x \geq 20 \\ y \geq 20 \\ x + y \leq 50 \end{cases}$$

To graph this system, we first draw the boundary line and shade the appropriate side for each inequality. The intersection of all these sets gives us the final solution set for the system. We have

Set	Description
$\{(x, y): x \geq 20\}$	Points on and to the right of the vertical line $x = 20$
$\{(x, y): y \geq 20\}$	Points on and above the horizontal line $y = 20$
$\{(x, y): x + y \leq 50\}$	Points on and below the line $x + y = 50$

We obtain the intersection of these sets in three steps, as shown in Figure 9.11(a) to (c).

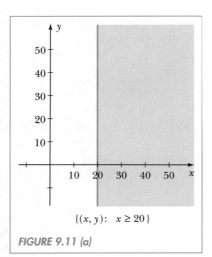

$\{(x, y): x \geq 20\}$

FIGURE 9.11 (a)

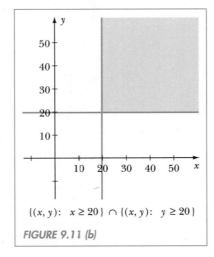

$\{(x, y): x \geq 20\} \cap \{(x, y): y \geq 20\}$

FIGURE 9.11 (b)

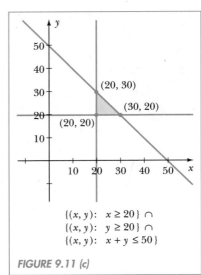

$\{(x, y): x \geq 20\} \cap$
$\{(x, y): y \geq 20\} \cap$
$\{(x, y): x + y \leq 50\}$

FIGURE 9.11 (c)

EXERCISES 9.2

In Exercises 1 to 20, graph the solution set for the given inequality.

1. (a) $x - y > 1$
   (b) $x > -3$

2. (a) $2x + y < 4$
   (b) $y \geq -2$

3. (a) $3y + 2x \leq -6$
   (b) $y \leq 0$

4. (a) $x \geq 2y$
   (b) $x < 0$

5. $y < x^3 + 1$

6. $x > y^3 - 1$

7. $x \leq (y + 1)^2$

8. $2y \geq x^2 + 2$

9. $(x - 1)^2 + (y + 1)^2 \leq 2$

10. $x^2 + (y - 1)^2 \geq 1$

11. $x^2 + y^2 > y$

12. $x^2 + y^2 < 2(x + y)$

13. $y \geq x^2 - 4x + 3$

14. $x^2 + 4x + y \leq 0$

15. (a) $y > |x - 2|$
    (b) $|x - y| > 1$
    (c) $|2x + y| \leq 2$

16. (a) $x \leq |y + 1|$
    (b) $|3x + 2y| < 6$
    (c) $|x - 2y| > 4$

17. (a) $y > \sqrt{x}$
    (b) $x > \ln y$

18. (a) $y < \sqrt{x + 1}$
    (b) $y > \log_2(x + 2)$

19. $xe^x - ye^x > 0$

20. $xe^y + e^y < 0$

In Exercises 21 to 40, graph the solution set for the given system. Label intersection points of the boundary curves.

21. $x < -y^2, \; x \geq -4$

22. $y + 1 > x^2, \; y - 1 < -x^2$

23. $y > x^2, \; x > y^2$

24. $y \geq x^3, \; y \leq x^2$

25. $\begin{cases} 4x - 3y > 9 \\ 3x + 4y < -12 \end{cases}$

26. $\begin{cases} y > (x - 1)^2 \\ x - y > -1 \end{cases}$

27. $\begin{cases} x^2 + 2x + y^2 \leq 0 \\ x + 1 \geq y^2 \end{cases}$

28. $\begin{cases} y > |x - 1| \\ y - 2 < -(x - 1)^2 \end{cases}$

29. $\begin{cases} |x| > 1 \\ |y| \leq 2 \end{cases}$

30. $\begin{cases} x^2 + y^2 \geq 1 \\ x^2 + y^2 \leq 4 \end{cases}$

31. $\begin{cases} y > 2^x \\ y < e^x \end{cases}$

32. $\begin{cases} y < e^x \\ y > e^{-x} \end{cases}$

33. $\begin{cases} y > \sin x \\ y < \cos x \end{cases}$

34. $\begin{cases} y < \cos \pi x \\ y > -\sqrt{\dfrac{1}{4} - x^2} \end{cases}$

35. $\begin{cases} y \leq x \\ x + y \leq 2 \\ x - 3y \leq 2 \end{cases}$

36. $\begin{cases} x^2 + y^2 < 25 \\ 4y + 3x > 0 \\ y + 7x < 25 \end{cases}$

37. $\begin{cases} xy \geq 1 \\ 2x + 2y \leq 5 \\ y \geq 0 \end{cases}$

38. $\begin{cases} xy \geq 4 \\ y \geq |x| \\ y \leq 4 \end{cases}$

39. $\begin{cases} y \leq \log_2 x \\ y \geq \log_{10} x \\ x \leq 10 \end{cases}$

40. $\begin{cases} y < \ln x \\ y > \dfrac{1 - x}{x} \\ x \geq 1 \end{cases}$

41. Graph (a) $\{(x, y): xy > 0\}$, (b) $\{(x, y): xy < 0\}$

42. Graph (a) $\{(x, y): \ln(xy) \geq 0\}$, (b) $\{(x, y): \ln(xy) \leq 0\}$

In Exercises 43 to 48, graph the solution set for the inequalities.

43. $|x| > y^2$

44. $|y| > |x|$

45. $|xy| \leq 1$

46. $y \geq |x^3|$

47. $|x| + |y| \leq 1$

48. $|x| - |y| \geq 1$

In Exercises 49 to 53, graph the set described.

49. $\{(x, y): y \leq |x| \text{ and } x \leq |y|\}$

50. $\{(x, y): |x| = 2 \text{ and } |y| \leq 1\} \cup \{(x, y): |y| = 1 \text{ and } |x| \leq 2\}$

51. $\{(x, y): |x - 1| \leq 1 \text{ or } |y| \leq 1\}$

52. $\{(x, y): x < |x| \text{ or } y < |y|\}$

53. $\{(x, y): x^2y < 1 \text{ and } y \geq 0\}$

In Exercises 54 to 60, you are asked for the possible values of two quantities. Label the two quantities x and y. Translate the conditions described into a system of inequalities and then graph the system.

54. A rancher wishes to use some fencing to enclose a rectangular area. If the maximum length of fencing available is 400 feet, describe all possible choices for the length and width of the rectangle.

55. A farmer wishes to use 240 feet or less of fencing to enclose a rectangular area. However, the diagonal of the rectangle cannot exceed 100 feet. Describe all possible choices for the length and width of the rectangle.

56. The area of a rectangle must be at least 100 square inches, but the perimeter cannot exceed 50 inches. Describe all possible choices for the length and width.

57. Three sides of a rectangular area are to be established with chain link fencing. The area of the rectangle must be at least 450 square feet. The length of the fencing cannot exceed 100 feet. Describe all possible choices for the length and width.

58. A right triangle is to be constructed so that its area is at least 80 square inches. The hypotenuse cannot exceed 20 inches. Describe all possible choices for the lengths of the legs of the triangle.

59. An artist needs to purchase at least 30 square yards of canvas. The store has two kinds: regular weave and fine weave. The regular weave costs $2 per square yard, and the fine weave costs $3 per square yard. The artist cannot spend more than $80. Describe all possible choices for the purchase.

60. A student wishes to retain a job while going to school. She must spend at least 20 hours each week for study, and her work schedule cannot exceed 20 hours each week. Describe all possible choices for the number of hours spent each week for study and for work, assuming the total is less than 60 hours.

9.3 LINEAR SYSTEMS: THREE UNKNOWNS

An equation in two or more unknowns is called **linear** if each term in the equation contains at most one of the unknowns to the first power. Following are typical examples of linear equations:

$$\text{Two unknowns:} \quad 2x - 3y = 4$$

$$\text{Three unknowns:} \quad x + 2y - 6z = 18$$

$$\text{Four unknowns:} \quad x_1 - \frac{1}{2}x_2 + \frac{2}{3}x_3 - x_4 = 0$$

On the other hand, equations such as $x + xy = 1$ or $x^2 + y = 3$ do not qualify as linear; these contain the nonlinear terms xy and x^2.

A linear *system* of equations consists of two or more linear equations in two or more unknowns. In this section we concentrate on solving linear systems with exactly three equations in three unknowns. This means that we want to find three real numbers corresponding to the three unknowns involved that satisfy all three equations in the system simultaneously. Systems involving other numbers of equations and unknowns are discussed in Section 9.4.

Example 1

Solve the system of equations $x + y + z = 0$, $2y - z = 6$, and $2z = -8$.

SOLUTION   We write the equations over one another as follows:

$$x + y + z = 0 \qquad (1)$$

$$2y - z = 6 \qquad (2)$$

$$2z = -8 \qquad (3)$$

Starting with Equation (3), we find z.

$$2z = -8 \longrightarrow z = -4$$

Moving up to Equation (2), we substitute $z = -4$ to get y.

$$2y - z = 6 \longrightarrow 2y - (-4) = 6 \longrightarrow y = 1$$

Finally, we move up to Equation (1) and substitute $y = 1$ and $z = -4$ to get x.

$$x + y + z = 0 \longrightarrow x + 1 + (-4) = 0 \longrightarrow x = 3$$

For convenience, we may write the solution as an ordered triple (x, y, z): $(3, 1, -4)$. ∎

The system in Example 1 was particularly easy to solve since all three unknowns did not appear in each equation. Equation (2) did not contain the first unknown, x, and Equation (3) did not contain the first or second unknowns, x or y. When this occurs, we say the system is in **echelon form**.

Figure 9.12 illustrates what echelon form means for a linear system of three equations in three unknowns x_1, x_2, and x_3. First, we note that the equations are written in three rows, so that x_1 is in the first column, x_2 is in the second column, and x_3 is in the third column. Furthermore, the second row does not contain x_1, and the third row does not contain x_1 or x_2. We assume that the first coefficient at the top, a, is nonzero; otherwise we would not have three unknowns. Finally, it may happen that coefficient b is zero. In this special case, either coefficient c is also zero or there is no third equation.

Let us consider examples of systems that are not in echelon form. As we show, each system is easily rearranged into the correct form.

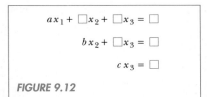

$$ax_1 + \square x_2 + \square x_3 = \square$$
$$bx_2 + \square x_3 = \square$$
$$cx_3 = \square$$

FIGURE 9.12

Not in Echelon Form	Reason	Rewritten in Echelon Form
(1) $\ x_2 + 2x_1 + 3x_3 = 1$	Equation (1) is not written in correct order.	$2x_1 + x_2 + 3x_3 = 1$
(2) $\qquad\qquad x_2 + x_3 = 2$		$x_2 + x_3 = 2$
(3) $\qquad\qquad\qquad x_3 = 3$		$x_3 = 3$
(1) $\qquad s + t = 1$	Equation (2) contains the first variable. Interchange Equations (1) and (2).	$r + s + t = 2$
(2) $\ r + s + t = 2$		$s + t = 1$
(3) $\qquad\qquad t = 3$		$t = 3$
(1) $\ x + y + z = 1$	Equation (3) must not contain the first or second variable. Interchange Equations (2) and (3).	$x + y + z = 1$
(2) $\qquad\qquad z = 2$		$y - z = 3$
(3) $\qquad y - z = 3$		$z = 2$

Of course, every linear system that we wish to solve will not be given to us in echelon form. Our task will be to transform a given system into an equivalent system that is in echelon form. If we can do this, then the solution will be relatively easy to find, as demonstrated in Example 1.

Example 2

Solve the following system of equations: $y + 2z = 4$, $x - y + 2z = -2$, and $3y + 4z = 8$.

SOLUTION We begin by writing the equations over one another, with x's in the first column, y's in the second column, and z's in the third column.

$$y + 2z = 4$$
$$x - y + 2z = -2$$
$$3y + 4z = 8$$

We wish to transform this system into an equivalent system that is in echelon form. By simply interchanging the equations appearing in the first two rows, we have

$$x - y + 2z = -2$$
$$y + 2z = 4$$
$$3y + 4z = 8$$

This certainly gives us an equivalent system since it does not matter in what order we write the equations. Furthermore, the system is now closer to echelon form. All that remains for us to do is eliminate the y term from the last row. We accomplish this by taking -3 times the equation in the second row and adding this to the equation in the last row.

$$y + 2z = 4 \xrightarrow{\times(-3)} -3y - 6z = -12$$
$$3y + 4z = 8 \longrightarrow \underline{\quad 3y + 4z = 8 \quad}$$
$$-2z = -4$$

Thus we may replace the last row with the new equation $-2z = -4$.

$$x - y + 2z = -2 \tag{1}$$
$$y + 2z = 4 \tag{2}$$
$$-2z = -4 \tag{3}$$

Now we have a system of equations, equivalent to the original system, in echelon form. Equation (3) tells us that $z = 2$. Substituting this into Equation (2) gives us $y = 0$. Finally, substituting $y = 0$ and $z = 2$ into Equation (1) yields $x = -6$. Therefore the solution is $(-6, 0, 2)$. ∎

Remark

When we choose x to be in the first column, y in the second column, and z in the third column, we are saying that x is the *first variable* of the system, y is the *second*, and z is the *third*. We could have written the equations with some other order, such as z first, y second, and x third—this is a matter of taste. However, whatever order is chosen, it must remain consistent throughout the solution.

To transform the system in Example 2 into an equivalent one in echelon form, we used two *operations*: interchanging two equations, and adding a multiple of one equation to another. There is a third operation that we can perform: simplifying an equation. For example, suppose the following equation appears in a system:

$$2x + 4y - 2z = 6$$

We can multiply this equation by 1/2 to obtain the equivalent equation,

$$x + 2y - z = 3$$

Thus we have three operations that can be used to transform a given linear system of equations into an equivalent one in echelon form.

Operations for Transforming a Linear System into Echelon Form
1. Interchange two equations.
2. Add a multiple of one equation to another.
3. Multiply an equation by a nonzero constant.

Example 3

Solve the system of equations $2r + 6s + t = -10$, $r + 2s + 2t = -1$, and $5r - s + 6t = -2$.

SOLUTION   We begin by writing the equations over one another so that r's appear in the first column, s's in the second, and t's in the third. We number each row of the system so that we may refer to it more easily.

$$(1) \quad 2r + 6s + t = -10$$

$$(2) \quad r + 2s + 2t = -1$$

$$(3) \quad 5r - s + 6t = -2$$

We will use the equation in row 1 to eliminate the r terms from the equations in rows 2 and 3. For this reason it is advisable for the equation in row 1 to have its r coefficient equal to 1 (if possible). Therefore we interchange equations in rows 1 and 2.

$$(1) \quad r + 2s + 2t = -1$$
$$(2) \quad 2r + 6s + t = -10$$
$$(3) \quad 5r - s + 6t = -2.$$

We now describe each operation and its effect in transforming the system into echelon form. The procedure is to eliminate the first unknown from the equations in rows 2 and 3 and then the second unknown from the equation in row 3.

Add $(-2) \times$ row 1 to row 2:

$$
\begin{aligned}
-2r - 4s - 4t &= 2 \\
\underline{2r + 6s + t = -10} \\
2s - 3t &= -8
\end{aligned}
\qquad
\begin{cases}
(1) & r + 2s + 2t = -1 \\
(2) & 2s - 3t = -8 \\
(3) & 5r - s + 6t = -2
\end{cases}
$$

Add $(-5) \times$ row 1 to row 3:

$$
\begin{aligned}
-5r - 10s - 10t &= 5 \\
\underline{5r - s + 6t = -2} \\
-11s - 4t &= 3
\end{aligned}
\qquad
\begin{cases}
(1) & r + 2s + 2t = -1 \\
(2) & 2s - 3t = -8 \\
(3) & -11s - 4t = 3
\end{cases}
$$

Add $(11/2) \times$ row 2 to row 3:

$$
\begin{aligned}
11s - \frac{33}{2}t &= -44 \\
\underline{-11s - 4t = 3} \\
-\frac{41}{2}t &= -41
\end{aligned}
\qquad
\begin{cases}
(1) & r + 2s + 2t = -1 \\
(2) & 2s - 3t = -8 \\
(3) & -\frac{41}{2}t = -41
\end{cases}
$$

Multiply row 3 by $(-2/41)$:

$$
\begin{cases}
(1) & r + 2s + 2t = -1 \\
(2) & 2s - 3t = -8 \\
(3) & t = 2
\end{cases}
$$

Thus $t = 2$. Substituting this into row 2, we find that $s = -1$. Finally, substituting $s = -1$ and $t = 2$ into row 1, we get $r = -3$. Therefore the solution is $(-3, -1, 2)$. ∎

The individual steps used in solving a linear system of equations are quite simple. However, much work is involved in writing out the equations after each operation is performed. To cut down on all this writing, we may use a certain shorthand notation to represent the linear system. The notation consists of an array of numbers corresponding to the coefficients of the unknowns and the constants appearing in the equations. We enclose this array by a large set of square brackets and call it the **matrix** *for the system*. Following is the matrix for the system in Example 3:

$$2r + 6s + t = -10$$
$$r + 2s + 2t = -1$$
$$5r - s + 6t = -2$$

$$\begin{bmatrix} 2 & 6 & 1 & -10 \\ 1 & 2 & 2 & -1 \\ 5 & -1 & 6 & -2 \end{bmatrix}$$

Example 4

Write the matrix for the system $4x - 2y + z = 1$, $y + 3z = 2$, and $-5x + y = 3$.

SOLUTION Write the equations over one another, being careful to line up the x terms in the first column, y's in the second column, and z's in the third column.

$$4x - 2y + z = 1$$
$$y + 3z = 2$$
$$-5x + y = 3$$

Now we can write the matrix for this system as follows:

$$\begin{bmatrix} 4 & -2 & 1 & 1 \\ 0 & 1 & 3 & 2 \\ -5 & 1 & 0 & 3 \end{bmatrix}$$

Note the use of zeros corresponding to the x term missing in the second equation and the z term missing in the third equation. ■

Example 5

Write the system corresponding to the matrix

$$\begin{bmatrix} 1 & 2 & 3 & 4 \\ 0 & 5 & 6 & 7 \\ 0 & 0 & 3 & -2 \end{bmatrix}$$

SOLUTION This matrix represents the linear system

$$x + 2y + 3z = 4$$
$$5y + 6z = 7$$
$$3z = -2$$

Note that this system of equations is in echelon form. Thus we say that the matrix for this system is also in *echelon form*. ■

Now we are ready to solve a linear system of equations using matrices. Since each row in the matrix for a system represents an equation, our three operations for transforming a system into echelon form correspond to **elementary row operations** in the matrix.

> ### Elementary Row Operations
> 1. Interchange two rows.
> 2. Add a multiple of one row to another.
> 3. Multiply a row by a nonzero constant.

Example 6

Consider the linear system of equations $x + y + 2z = 7$, $x = 1 + 2y + z$, and $y + z - x = 0$. Write the matrix for this system. Solve the system by transforming the matrix into echelon form using elementary row operations.

SOLUTION We begin by writing the equations over one another. We must rearrange terms, however, to get the x's in the first column, y's in the second column, and z's in the third column.

$$x + y + 2z = 7 \longrightarrow x + y + 2z = 7$$
$$x = 1 + 2y + z \longrightarrow x - 2y - z = 1$$
$$y + z - x = 0 \longrightarrow -x + y + z = 0$$

Now, we can write the matrix for this system and proceed to transform it into echelon form.

$$\begin{bmatrix} 1 & 1 & 2 & 7 \\ 1 & -2 & -1 & 1 \\ -1 & 1 & 1 & 0 \end{bmatrix}$$

The first objective is to get zeros below the first row in column 1. We use row 1 to accomplish this. Start by adding $(-1) \times$ row 1 to row 2.

$$\longrightarrow \begin{bmatrix} 1 & 1 & 2 & 7 \\ 0 & -3 & -3 & -6 \\ -1 & 1 & 1 & 0 \end{bmatrix}$$

Now add row 1 to row 3.

$$\longrightarrow \begin{bmatrix} 1 & 1 & 2 & 7 \\ 0 & -3 & -3 & -6 \\ 0 & 2 & 3 & 7 \end{bmatrix}$$

Our next objective is to get a zero in the second column of row 3. We use row 2 to accomplish this. First, we simplify row 2 by multiplying it by $-1/3$.

$$\longrightarrow \begin{bmatrix} 1 & 1 & 2 & 7 \\ 0 & 1 & 1 & 2 \\ 0 & 2 & 3 & 7 \end{bmatrix}$$

Now add $(-2) \times$ row 2 to row 3.

$$\longrightarrow \begin{bmatrix} 1 & 1 & 2 & 7 \\ 0 & 1 & 1 & 2 \\ 0 & 0 & 1 & 3 \end{bmatrix}$$

The matrix is now in echelon form. Writing the equations corresponding to this matrix, we have

$$x + y + 2z = 7 \tag{1}$$

$$y + z = 2 \tag{2}$$

$$z = 3 \tag{3}$$

Starting with Equation (3) and working upward, we find that $z = 3$, then $y = -1$, and, finally, $x = 2$. Therefore the solution is $(2, -1, 3)$. ∎

The matrix method introduced here has several advantages. First, we can follow each step in the solution rather clearly from beginning to end, which makes errors easier to find and correct. Second, the method is quite general, being applicable to a linear system containing any number of equations and unknowns (we pursue this aspect in the next section). Third, and most important, the method is programmable; we can tell a machine how to carry out the entire procedure, which is most helpful in practical applications where very large systems of equations need to be solved.

For reference, following is a summary of the steps involved in the matrix method.

Matrix Method for Solving Linear Systems
Step 1. Write a matrix for the given system.
Step 2. Use elementary row operations to transform the matrix into echelon form.
Step 3. Translate the echelon form matrix back into its corresponding equations and solve.

Step 2 contains the heart of the method: using elementary row operations, transform the matrix for the system into echelon form. To do this for a system of three equations in three unknowns, we suggest the following:

- Step 2a. By adding multiples of row 1 to the other rows, get zeros down the first column below row 1. It helps to try to arrange things so that the first number in row 1 is a 1.
- Step 2b. Use row 2 to get a zero below row 2 in the second column.

Example 7

Use the matrix method to solve the system $3x + y - 5z = 4$, $7x - 6y - 20z = 1$, and $2x + 3y - 4z = 0$.

SOLUTION   We begin by writing the matrix for the system.

$$
\begin{array}{c}
3x + y - 5z = 4 \\
7x - 6y - 20z = 1 \\
2x + 3y - 4z = 0
\end{array}
\longrightarrow
\begin{bmatrix}
3 & 1 & -5 & 4 \\
7 & -6 & -20 & 1 \\
2 & 3 & -4 & 0
\end{bmatrix}
$$

Now we use elementary row operations to transform the matrix into echelon form. Notice that the first number in row 1 is not a 1 as we would like; however, we shall manipulate the rows to force a 1 to appear there. For brevity, we write $-2R_1 + R_2$ to mean add -2 times row 1 to row 2 and $R_1 \longleftrightarrow R_2$ to mean interchange rows 1 and 2.

$$
\begin{bmatrix}
3 & 1 & -5 & 4 \\
7 & -6 & -20 & 1 \\
2 & 3 & -4 & 0
\end{bmatrix}
\xrightarrow{-2R_1 + R_2}
\begin{bmatrix}
3 & 1 & -5 & 4 \\
1 & -8 & -10 & -7 \\
2 & 3 & -4 & 0
\end{bmatrix}
$$

$$
\xrightarrow{R_1 \longleftrightarrow R_2}
\begin{bmatrix}
1 & -8 & -10 & -7 \\
3 & 1 & -5 & 4 \\
2 & 3 & -4 & 0
\end{bmatrix}
\xrightarrow[-2R_1 + R_3]{-3R_1 + R_2}
\begin{bmatrix}
1 & -8 & -10 & -7 \\
0 & 25 & 25 & 25 \\
0 & 19 & 16 & 14
\end{bmatrix}
$$

$$
\xrightarrow{\frac{1}{25} \times R_2}
\begin{bmatrix}
1 & -8 & -10 & -7 \\
0 & 1 & 1 & 1 \\
0 & 19 & 16 & 14
\end{bmatrix}
\xrightarrow{-19R_2 + R_3}
\begin{bmatrix}
1 & -8 & -10 & -7 \\
0 & 1 & 1 & 1 \\
0 & 0 & -3 & -5
\end{bmatrix}
$$

The matrix is now in echelon form. Translating back to the corresponding equations, we have

$$x - 8y - 10z = 7 \tag{1}$$

$$y + z = 1 \tag{2}$$

$$-3z = -5 \tag{3}$$

Equation (3) implies $z = 5/3$. Substituting this into Equation (2), we find that $y = -2/3$. Finally, substituting $y = -2/3$ and $z = 5/3$ into Equation (1) yields $x = 13/3$. Thus the solution is $(13/3, -2/3, 5/3)$. ∎

Remark

How one uses the elementary row operations to transform a matrix into echelon form is not unique. Consider the following alternate approach for the matrix in Example 7:

$$\begin{bmatrix} 3 & 1 & -5 & 4 \\ 7 & -6 & -20 & 1 \\ 2 & 3 & -4 & 0 \end{bmatrix} \xrightarrow{-R_3 + R_1} \begin{bmatrix} 1 & -2 & -1 & 4 \\ 7 & -6 & -20 & 1 \\ 2 & 3 & -4 & 0 \end{bmatrix}$$

$$\xrightarrow[-2R_1 + R_3]{-7R_1 + R_2} \begin{bmatrix} 1 & -2 & -1 & 4 \\ 0 & 8 & -13 & -27 \\ 0 & 7 & -2 & -8 \end{bmatrix} \xrightarrow{-R_3 + R_2} \begin{bmatrix} 1 & -2 & -1 & 4 \\ 0 & 1 & -11 & -19 \\ 0 & 7 & -2 & -8 \end{bmatrix}$$

$$\xrightarrow{-7R_2 + R_3} \begin{bmatrix} 1 & -2 & -1 & 4 \\ 0 & 1 & -11 & -19 \\ 0 & 0 & 75 & 125 \end{bmatrix} \longrightarrow \begin{cases} x - 2y - z = 4 \\ y - 11z = -19 \\ 75z = 125 \end{cases}$$

You may check that this result yields the same solution, $(13/3, -2/3, 5/3)$.

So far all the systems that we solved in this section had exactly one solution. This is not always going to happen. For example, in Section 9.1 we found that some linear systems in two unknowns may have no solution or an infinite number of solutions. Similar results are true for linear systems containing three or more unknowns. We look at these general cases in the next section, but for now let us consider an application.

Example 8

Find a quadratic function $Q(x)$ whose graph passes through the points $(2, -1/2)$, $(1, 1)$, $(-1, 7)$.

SOLUTION We know that any quadratic function has the rule

$$Q(x) = ax^2 + bx + c$$

We need to find the coefficients a, b, and c. Since $(2, -1/2)$ is on the graph of Q, we have $Q(2) = -1/2$. Thus $\left(2, -\dfrac{1}{2}\right)$ on the graph implies

$$Q(2) = -\frac{1}{2} \quad \text{or} \quad a(2)^2 + b(2) + c = -\frac{1}{2} \quad \text{or} \quad 4a + 2b + c = -\frac{1}{2} \quad (1)$$

Similarly, $(1, 1)$ on the graph implies

$$Q(1) = 1 \quad \text{or} \quad a(1)^2 + b(1) + c = 1 \quad \text{or} \quad a + b + c = 1 \qquad (2)$$

Finally, $(-1, 7)$ on the graph implies

$$Q(-1) = 7 \quad \text{or} \quad a(-1)^2 + b(-1) + c = 7 \quad \text{or} \quad a - b + c = 7 \qquad (3)$$

Thus we have the following linear system of equations to solve:

$$4a + 2b + c = -\frac{1}{2} \qquad (1)$$

$$a + b + c = 1 \qquad (2)$$

$$a - b + c = 7 \qquad (3)$$

We write the matrix for this system and proceed to use elementary row operations to transform the matrix into echelon form.

$$\begin{bmatrix} 4 & 2 & 1 & -\dfrac{1}{2} \\ 1 & 1 & 1 & 1 \\ 1 & -1 & 1 & 7 \end{bmatrix} \xrightarrow{R_1 \longleftrightarrow R_2} \begin{bmatrix} 1 & 1 & 1 & 1 \\ 4 & 2 & 1 & -\dfrac{1}{2} \\ 1 & -1 & 1 & 7 \end{bmatrix}$$

$$\xrightarrow[{-R_1 + R_3}]{-4R_1 + R_2} \begin{bmatrix} 1 & 1 & 1 & 1 \\ 0 & -2 & -3 & -\dfrac{9}{2} \\ 0 & -2 & 0 & 6 \end{bmatrix} \xrightarrow{-R_2 + R_3} \begin{bmatrix} 1 & 1 & 1 & 1 \\ 0 & -2 & -3 & -\dfrac{9}{2} \\ 0 & 0 & 3 & \dfrac{21}{2} \end{bmatrix}$$

The matrix is now in echelon form. Translating back to the corresponding equations, we have

$$a + b + c = 1 \qquad (1)$$

$$-2b - 3c = -\frac{9}{2} \qquad (2)$$

$$3c = \frac{21}{2} \qquad (3)$$

Starting with Equation (3) and working upward, we find that $c = 7/2$, then $b = -3$, and, finally, $a = 1/2$. Therefore the quadratic function we seek is $Q(x) = \dfrac{1}{2}x^2 - 3x + \dfrac{7}{2}$. ∎

EXERCISES 9.3

In Exercises 1 and 2, solve the given system of equations.

1. $3x + 2y + z = 3$
$y - 2z = 8$
$3z = 12$

2. $2x - y - z = 0$
$4y - 3z = -6$
$\frac{2}{5}z = 4$

In Exercises 3 to 6, (a) write the system corresponding to the given matrix; (b) use elementary row operations to transform the matrix into echelon form. (Answers may vary.)

3. $\begin{bmatrix} 1 & -1 & 2 & 3 \\ 1 & -2 & 1 & 4 \\ 0 & 3 & 2 & -6 \end{bmatrix}$

4. $\begin{bmatrix} 1 & 3 & 2 & 1 \\ 1 & 2 & 3 & 4 \\ 0 & 1 & 1 & -2 \end{bmatrix}$

5. $\begin{bmatrix} 0 & 2 & 1 & 2 \\ 1 & 0 & -1 & 1 \\ 2 & -3 & 1 & 4 \end{bmatrix}$

6. $\begin{bmatrix} 0 & -1 & -1 & 1 \\ 1 & 2 & 1 & 0 \\ 3 & -1 & 1 & 2 \end{bmatrix}$

In Exercises 7 to 22, use the matrix method to solve the given system.

7. $x + y = 2, x + 2y + z = 1, y + 4z = 2$

8. $x + y + z = 1, y + z = 2, y + 2z = 3$

9. $4x - y + z = 10, x + y + z = -2,$
$-2x - 2y - 3z = -7$

10. $3x + 2y + z = 0, x - y + z = -2,$
$-2x + y - z = 4$

11. $2r - 4s + 3t = 1, r + 2s - 3t = 0,$
$r + 6s + 6t = 4$

12. $2r - 2s - 3t = -11, r + 2s + 3t = 5,$
$r - 2s + 6t = 3$

13. $a + b + c = -9, 2a - b - c = -3,$
$3a + b + 3c = -1$

14. $a + b + c = -7, 2a + 3b - c = 22,$
$3a + 4b + 2c = -9$

15. $r + t = 2$
$s + t = 1$
$r + s = -3$

16. $a + 3c = 1$
$2b + c = 1$
$3a + b = 1$

17. $5x + y + 3z = -2$
$-3x + 2y - 2z = 0$
$7x - 3y + 5z = 4$

18. $7x + 4y + 9z = 14$
$3x + y + 3z = 5$
$4x + 3y - 10z = -10$

19. $5a + 4b = 14$
$3a + 2b + 6c = 4$
$3a - 3b - 3c = 9$

20. $3r + 6s + 2t = 16$
$2r - s - t = 1$
$2r - 4s = 6$

21. $12x + 12y + 12z = 1$
$3x + 4y - 2z = 4$
$6x - 4y - 18z = 0$

22. $16x + 4y + z = 2$
$4x + 2y + z = 1$
$-12x + y - z = -6$

In Exercises 23 and 24, solve the given system.

23. $\dfrac{1}{x + y} = 2$
$\dfrac{1}{y + z} = 4$
$\dfrac{1}{x + y + z} = 1$

24. $\dfrac{y}{x + 1} = 3$
$\dfrac{z}{y + 1} = 4$
$\dfrac{x}{z - 1} = 1$

In Exercises 25 and 26, find the quadratic function $Q(x)$, whose graph passes through the given points.

25. $(-2, 7), (2, 3), \left(3, \dfrac{9}{2}\right)$

26. $\left(1, \dfrac{1}{2}\right), (2, 5), (-2, -1)$

In Exercises 27 and 28, find the quadratic function $F(x)$ with the given properties.

27. $F(1) = -1, F(2) = 3, F(3) = 1$

28. $F(1) = 2, F(2) = 3, F(3) = -4$

29. A three-digit number has the following properties. The sum of the first two digits (100s and 10s digits) is 5 more than the third digit. The sum of all the digits is 15. Finally, if we reverse the order of the digits, we get a number that is 198 more than the original. Find the original number.

30. Two points P_1 and P_2 are located on the x-axis and the line $y - x = 2$, respectively. The slope of the line through P_1 and P_2 is 3. The sum of the coordinates of P_1 and P_2 is 12. Find the coordinates of P_1 and P_2.

LINEAR SYSTEMS: GENERAL CASES

In Section 9.3 we solved linear systems containing three equations in three unknowns. In each case we found exactly one solution for the system. As we will see in this section, it is also possible for a linear system to have an infinite number of solutions or no solutions at all. Examples 1 and 2 here demonstrate the infinite solutions and no solutions cases when the linear system consists of three equations in three unknowns.

Example 1

Solve the system $x + 2y - z = 2$, $-x + 2y - 2z = -1$, and $x + 6y - 4z = 5$.

SOLUTION   We write the matrix for this system and then transform the matrix into echelon form using elementary row operations.

$$
\begin{aligned}
x + 2y - z &= 2 \\
-x + 2y - 2z &= -1 \\
x + 6y - 4z &= 5
\end{aligned}
\longrightarrow
\begin{bmatrix}
1 & 2 & -1 & 2 \\
-1 & 2 & -2 & -1 \\
1 & 6 & -4 & 5
\end{bmatrix}
$$

$$
\xrightarrow[\; -R_1 + R_3 \;]{R_1 + R_2}
\begin{bmatrix}
1 & 2 & -1 & 2 \\
0 & 4 & -3 & 1 \\
0 & 4 & -3 & 3
\end{bmatrix}
\xrightarrow{-R_2 + R_3}
\begin{bmatrix}
1 & 2 & -1 & 2 \\
0 & 4 & -3 & 1 \\
0 & 0 & 0 & 2
\end{bmatrix}
$$

Translating back to the corresponding equations, we have

$$x + 2y - z = 2 \tag{1}$$
$$4y - 3z = 1 \tag{2}$$
$$0z = 2 \tag{3}$$

Equation (3) says that $0z = 2$, which is nonsense! No matter what value z is, $0z$ will be 0, not 2. Therefore we conclude that the system has *no solution*. ∎

Example 2

Solve the system $x + y - z = 1$, $x + 2y + z = 2$, and $-2x + y + 8z = 1$.

SOLUTION   We proceed with the matrix method.

$$
\begin{aligned}
x + y - z &= 1 \\
x + 2y + z &= 2 \\
-2x + y + 8z &= 1
\end{aligned}
\longrightarrow
\begin{bmatrix}
1 & 1 & -1 & 1 \\
1 & 2 & 1 & 2 \\
-2 & 1 & 8 & 1
\end{bmatrix}
$$

$$
\xrightarrow[\; 2R_1 + R_3 \;]{-R_1 + R_2}
\begin{bmatrix}
1 & 1 & -1 & 1 \\
0 & 1 & 2 & 1 \\
0 & 3 & 6 & 3
\end{bmatrix}
\xrightarrow{-3R_2 + R_3}
\begin{bmatrix}
1 & 1 & -1 & 1 \\
0 & 1 & 2 & 1 \\
0 & 0 & 0 & 0
\end{bmatrix}
$$

Translating back into the corresponding equations,

$$x + y - z = 1 \tag{1}$$

$$y + 2z = 1 \tag{2}$$

$$0z = 0 \tag{3}$$

Equation (3) says that $0z = 0$, which is *true* for any value of z. Therefore we may *not* conclude that the system has no solutions. Moving up to Equation (2), we write

$$y + 2z = 1 \longrightarrow y = 1 - 2z$$

Using this result in Equation (1), we have

$$x + y - z = 1 \longrightarrow x + (1 - 2z) - z = 1$$

$$\longrightarrow x = 3z$$

Thus we are able to write both x and y in terms of z, but what is z? Since there is no restriction on z, we may let it be any real number, say, $z = t$ for $t \in \mathbf{R}$. It follows that $y = 1 - 2t$ and $x = 3t$. Therefore we write the solution for the system as follows:

$$\left. \begin{array}{l} x = 3t \\ y = 1 - 2t \\ z = t \end{array} \right\} t \in \mathbf{R}$$

We call this a **parametric representation** for the solution set with **parameter** t. Note that it represents an infinite number of solutions. For example, when $t = 0$, we get $(0, 1, 0)$; when $t = 1$, we get $(3, -1, 1)$; when $t = -1$, we get $(-3, 3, -1)$; and so on. You may verify that each answer satisfies all three equations in the original system. ■

In Section 9.1 we found that a linear system consisting of two equations in two unknowns must have none, one, or an infinite number of solutions. The same is true for a linear system containing any number of equations and unknowns. We will not prove this fact here since this is usually done in an introductory linear algebra course. However, let us formally state the result for reference.

> Any linear system of equations satisfies exactly one of the following:
> 1. The system has no solutions.
> 2. The system has exactly one solution.
> 3. The system has an infinite number of solutions.

Occasionally we may be required to find a solution for a system that contains more unknowns than equations. Example 3 illustrates this case.

Example 3

Solve the system $3x - y + 2z = 8$ and $x - y - 3z = 5$. Give two specific solutions as well as the general result.

SOLUTION In all our previous examples of linear systems with three unknowns we had three equations. This time we have only two equations. We begin the solution as usual by writing the matrix for the system. Of course, the matrix contains just two rows.

$$\begin{array}{c} 3x - y + 2z = 8 \\ x - y - 3z = 5 \end{array} \longrightarrow \begin{bmatrix} 3 & -1 & 2 & 8 \\ 1 & -1 & -3 & 5 \end{bmatrix}$$

We proceed to transform this matrix into echelon form, just as we did before. The fact that there is no third row does not matter. (We can imagine the third row as consisting entirely of zeros.)

$$\begin{bmatrix} 3 & -1 & 2 & 8 \\ 1 & -1 & -3 & 5 \end{bmatrix} \xrightarrow{R_1 \longleftrightarrow R_2} \begin{bmatrix} 1 & -1 & -3 & 5 \\ 3 & -1 & 2 & 8 \end{bmatrix}$$

$$\xrightarrow{-3R_1 + R_2} \begin{bmatrix} 1 & -1 & -3 & 5 \\ 0 & 2 & 11 & -7 \end{bmatrix}$$

Translating back to the corresponding equations, we have

$$x - y - 3z = 5 \tag{1}$$

$$2y + 11z = -7 \tag{2}$$

From Equation (2), we find that

$$y = \frac{-7 - 11z}{2}$$

Using this result in Equation (1), we have

$$x - y - 3z = 5 \longrightarrow x - \left(\frac{-7 - 11z}{2} \right) - 3z = 5$$

$$\longrightarrow x = 5 + \frac{-7 - 11z}{2} + 3z$$

$$\longrightarrow x = \frac{3 - 5z}{2}$$

Since there is no restriction on z, we let $z = t$ for $t \in \mathbf{R}$. Therefore we write the solution set as follows:

$$\left. \begin{array}{l} x = \dfrac{3 - 5t}{2} \\[2mm] y = \dfrac{-7 - 11t}{2} \\[2mm] z = t \end{array} \right\} \; t \in \mathbf{R}$$

Two specific solutions can be found by substituting two values for t into the parametric representation. Thus, corresponding to $t = 0$ and $t = 1$, we get the solutions $(3/2, -7/2, 0)$ and $(-1, -9, 1)$, respectively. ▪

So far we have been looking at linear systems involving only three unknowns. Our matrix method is applicable to other linear systems as well. Example 4 illustrates the situation when four equations are given in four unknowns.

Example 4

Solve the system $x_1 + x_2 - x_4 = 2$, $x_1 + 2x_2 + x_3 = 0$, $-x_2 + x_3 + x_4 = 4$, and $x_1 + x_3 + x_4 = 5$.

SOLUTION We begin by translating the system into a matrix.

$$
\begin{aligned}
x_1 + x_2 \quad - x_4 &= 2 \\
x_1 + 2x_2 + x_3 \quad &= 0 \\
-x_2 + x_3 + x_4 &= 4 \\
x_1 \quad + x_3 + x_4 &= 5
\end{aligned}
\longrightarrow
\begin{bmatrix}
1 & 1 & 0 & -1 & 2 \\
1 & 2 & 1 & 0 & 0 \\
0 & -1 & 1 & 1 & 4 \\
1 & 0 & 1 & 1 & 5
\end{bmatrix}
$$

Next, we transform this matrix into echelon form. As before, the first objective is to get zeros below the first entry in column 1. Of course, we may use only elementary row operations in our work. Thus

$$
\begin{array}{c}
-R_1 + R_2 \\
\xrightarrow{} \\
-R_1 + R_4
\end{array}
\begin{bmatrix}
1 & 1 & 0 & -1 & 2 \\
0 & 1 & 1 & 1 & -2 \\
0 & -1 & 1 & 1 & 4 \\
0 & -1 & 1 & 2 & 3
\end{bmatrix}
$$

Our next objective is to get zeros below the second entry in column 2. We use row 2 to accomplish this.

$$
\begin{array}{c}
R_2 + R_3 \\
\xrightarrow{} \\
R_2 + R_4
\end{array}
\begin{bmatrix}
1 & 1 & 0 & -1 & 2 \\
0 & 1 & 1 & 1 & -2 \\
0 & 0 & 2 & 2 & 2 \\
0 & 0 & 2 & 3 & 1
\end{bmatrix}
$$

Finally, we get a zero below the third entry in column 3.

$$
\begin{array}{c}
-R_3 + R_4 \\
\xrightarrow{}
\end{array}
\begin{bmatrix}
1 & 1 & 0 & -1 & 2 \\
0 & 1 & 1 & 1 & -2 \\
0 & 0 & 2 & 2 & 2 \\
0 & 0 & 0 & 1 & -1
\end{bmatrix}
$$

The matrix is now in echelon form. However, we can perform one further simplification by multiplying row 3 by 1/2. This gives us the following matrix, which we translate back to the corresponding equations.

$$\xrightarrow{\frac{1}{2} \times R_3} \begin{bmatrix} 1 & 1 & 0 & -1 & 2 \\ 0 & 1 & 1 & 1 & -2 \\ 0 & 0 & 1 & 1 & 1 \\ 0 & 0 & 0 & 1 & -1 \end{bmatrix} \longrightarrow \begin{array}{r} x_1 + x_2 \quad\quad - x_4 = 2 \\ x_2 + x_3 + x_4 = -2 \\ x_3 + x_4 = 1 \\ x_4 = -1 \end{array}$$

Working from the bottom equation upward, we find that $x_4 = -1$, $x_3 = 2$, $x_2 = -3$, and, finally, $x_1 = 4$. We may represent the solution as an ordered quadruple: $(4, -3, 2, -1)$. ∎

As an application of solving linear systems, we shall look at a procedure that enables us to rewrite a rational function as a sum of simpler rational functions. The process is called the decomposition of a rational function into its partial fractions, or just *partial fractions* for short. Before presenting the method, we recall an important fact derived in Chapter 4.

> Two polynomial functions are equal if and only if their corresponding coefficients are equal.

Example 5

Find constants A and B so that
$$\frac{x}{(x + 1)(x + 2)} = \frac{A}{x + 1} + \frac{B}{x + 2}$$

SOLUTION We begin by combining the right side of the equation.
$$\frac{A}{x + 1} + \frac{B}{x + 2} = \frac{A(x + 2) + B(x + 1)}{(x + 1)(x + 2)} = \frac{(A + B)x + (2A + B)}{(x + 1)(x + 2)}$$

Notice that we collected like terms together in the numerator. Therefore the original equation can be rewritten as
$$\frac{x}{(x + 1)(x + 2)} = \frac{(A + B)x + (2A + B)}{(x + 1)(x + 2)}$$

If this is true, then the two numerators must be equal. Thus
$$x = (A + B)x + (2A + B)$$

Each side of this equation consists of a polynomial function in x. Hence, the numerators are equal if and only if their corresponding coefficients are equal.

Noting that x can be expressed as $1x + 0$, we have

$$\text{Equating coefficients of } x\!: \quad A + B = 1$$

$$\text{Equating constant coefficients:} \quad 2A + B = 0$$

This gives us a linear system of two equations for the two unknowns A and B. Although this system could be solved by inspection, we shall use matrices, to be consistent with our previous work.

$$
\begin{aligned}
A + B &= 1 \\
2A + B &= 0
\end{aligned}
\qquad \longrightarrow \qquad
\begin{bmatrix} 1 & 1 & 1 \\ 2 & 1 & 0 \end{bmatrix}
$$

$$
\xrightarrow{\;-2R_1 + R_2\;}
\begin{bmatrix} 1 & 1 & 1 \\ 0 & -1 & -2 \end{bmatrix}
\xrightarrow{\;-1 \times R_2\;}
\begin{bmatrix} 1 & 1 & 1 \\ 0 & 1 & 2 \end{bmatrix}
$$

Translating back to the corresponding equations,

$$
\begin{aligned}
A + B &= 1 \\
B &= 2
\end{aligned}
$$

It follows that $B = 2$ and $A = -1$. We conclude

$$\frac{x}{(x + 1)(x + 2)} = \frac{-1}{x + 1} + \frac{2}{x + 2}$$

You may check that combining the fractions on the right side yields the rational function in the left. ■

Example 6

Find constants A, B, C, and D so that

$$\frac{x^2 + x - 1}{x^4 - 1} = \frac{A}{x + 1} + \frac{B}{x - 1} + \frac{Cx + D}{x^2 + 1}$$

SOLUTION   We begin by combining the right side of the equation. You should verify the following result:

$$\frac{A}{x + 1} + \frac{B}{x - 1} + \frac{Cx + D}{x^2 + 1}$$

$$= \frac{(A + B + C)x^3 + (-A + B + D)x^2 + (A + B - C)x + (-A + B - D)}{x^4 - 1}$$

We want this last fraction to equal $\dfrac{x^2 + x - 1}{x^4 - 1}$. This implies that the numerators must be equal. Therefore

$$x^2 + x - 1$$

$$= (A + B + C)x^3 + (-A + B + D)x^2 + (A + B - C)x + (-A + B - D)$$

For these polynomials to be equal, their corresponding coefficients must be the same. Hence

$$\text{Equating} \quad x^3 \quad \text{coefficients:} \quad A + B + C \qquad = 0$$

$$\text{Equating} \quad x^2 \quad \text{coefficients:} \quad -A + B \qquad + D = 1$$

$$\text{Equating} \quad x \quad \text{coefficients:} \quad A + B - C \qquad = 1$$

$$\text{Equating constant coefficients:} \quad -A + B \qquad - D = -1$$

We write the matrix for this system and transform it into echelon form.

$$\begin{bmatrix} 1 & 1 & 1 & 0 & 0 \\ -1 & 1 & 0 & 1 & 1 \\ 1 & 1 & -1 & 0 & 1 \\ -1 & 1 & 0 & -1 & -1 \end{bmatrix} \xrightarrow[\begin{array}{c} R_1 + R_2 \\ -R_1 + R_3 \\ R_1 + R_4 \end{array}]{} \begin{bmatrix} 1 & 1 & 1 & 0 & 0 \\ 0 & 2 & 1 & 1 & 1 \\ 0 & 0 & -2 & 0 & 1 \\ 0 & 2 & 1 & -1 & -1 \end{bmatrix}$$

$$\xrightarrow[-R_2 + R_4]{} \begin{bmatrix} 1 & 1 & 1 & 0 & 0 \\ 0 & 2 & 1 & 1 & 1 \\ 0 & 0 & -2 & 0 & 1 \\ 0 & 0 & 0 & -2 & -2 \end{bmatrix}$$

This gives us the following equations.

$$A + B + C \qquad = 0 \qquad (1)$$

$$2B + C + D = 1 \qquad (2)$$

$$-2C \qquad = 1 \qquad (3)$$

$$-2D = -2 \qquad (4)$$

Equation (4) implies $D = 1$. Equation (3) yields $C = -1/2$. Substituting C and D into Equation (2), we find that $B = 1/4$. Substituting B and C into Equation (1), we get $A = 1/4$. Therefore

$$\frac{x^2 + x - 1}{x^4 - 1} = \frac{1/4}{x + 1} + \frac{1/4}{x - 1} + \frac{-\frac{1}{2}x + 1}{x^2 + 1} \quad \blacksquare$$

EXERCISES 9.4

In Exercises 1 to 10, use the matrix method to solve each system. If there are infinitely many solutions, give a parametric representation.

1. $x + y - z = 0$
    $x - y + z = 2$
    $ y - z = 1$

2. $x + y + z = 1$
    $x - y + z = 0$
    $-x + y - z = 2$

3. $2x + y - z = 3$
    $x + 2y + 3z = -1$
    $x - y - 4z = 4$

4. $x + 2y + 3z = 4$
    $-x + 2y = -2$
    $x + 6y + 6z = 6$

5. $a - 2b + c = 2$
    $2a + b + 2c = -1$
    $a + 3b + c = 7$

6. $a + 2b + 3c = 1$
    $2a + 3b + 4c = 2$
    $3a + 2b + c = -1$

7. $\begin{aligned} x - y + z &= 4 \\ x + y - z &= 2 \\ 2x - y + z &= 7 \end{aligned}$

8. $\begin{aligned} x + 2y + z &= 6 \\ x + y + z &= 4 \\ 3x - y + 3z &= 4 \end{aligned}$

9. $\begin{aligned} a + 2b + 3c &= 1 \\ 3a + b + 2c &= 1 \\ 2a + 3b + c &= 1 \end{aligned}$

10. $\begin{aligned} r + s + t &= 1 \\ 3r + s - t &= 1 \\ r - s + 3t &= 1 \end{aligned}$

In Exercises 11 to 14, solve the system. Find two specific solutions as well as the general result.

11. $\begin{aligned} x + y + 2z &= 2 \\ 2x - y - 5z &= 10 \end{aligned}$

12. $\begin{aligned} 3x - y - 2z &= 8 \\ x + y - 2z &= 0 \end{aligned}$

13. $\begin{aligned} x - 2y + 3z &= 8 \\ 3x + 2y - z &= -4 \end{aligned}$

14. $\begin{aligned} 2x - 3y + z &= 2 \\ x + 2y + z &= 6 \end{aligned}$

In Exercises 15 to 18, use the matrix method to solve the given system.

15. $\begin{aligned} x_1 + x_2 + x_3 + x_4 &= 14 \\ x_1 - x_2 + x_3 - x_4 &= 2 \\ x_1 - 2x_2 + 2x_3 + x_4 &= 5 \\ 2x_1 - x_2 + x_3 &= 9 \end{aligned}$

16. $\begin{aligned} x_1 + x_2 + x_3 + 2x_4 &= -6 \\ x_1 - x_3 - x_4 &= -2 \\ x_1 - 2x_2 + x_3 + 3x_4 &= 0 \\ 2x_1 + x_2 - 4x_3 + 4x_4 &= -4 \end{aligned}$

17. $\begin{aligned} -a + b + 2c - 2d &= 4 \\ 2a + b - 2c &= -5 \\ a - 4b - 3c + 8d &= -5 \\ -a + 4b + 5c - 2d &= 10 \end{aligned}$

18. $\begin{aligned} a + b + c + d &= 12 \\ 2a + b + 3c - d &= 4 \\ 3a - b + c - 2d &= 4 \\ 4a + 2b + 7c + d &= 50 \end{aligned}$

In Exercises 19 to 22, find constants A and B so that the statement is true.

19. $\dfrac{x}{(x - 2)(x + 3)} = \dfrac{A}{x - 2} + \dfrac{B}{x + 3}$

20. $\dfrac{2x + 1}{(x + 1)(x + 2)} = \dfrac{A}{x + 1} + \dfrac{B}{x + 2}$

21. $\dfrac{4}{x^2 - 4} = \dfrac{A}{x + 2} + \dfrac{B}{x - 2}$

22. $\dfrac{5x - 2}{(2x - 1)(x + 3)} = \dfrac{A}{2x - 1} + \dfrac{B}{x + 3}$

In Exercises 23 and 24, find constants A, B, and C so that the statement is true.

23. $\dfrac{1 - x + x^2}{x(x^2 - 1)} = \dfrac{A}{x + 1} + \dfrac{B}{x - 1} + \dfrac{C}{x}$

24. $\dfrac{2 + x - x^2}{x(x^2 - 9)} = \dfrac{A}{x + 3} + \dfrac{B}{x - 3} + \dfrac{C}{x}$

In Exercises 25 to 28, find constants A, B, C, and D so that the statement is true.

25. $\dfrac{x^3 - x + 1}{x^4 - 1} = \dfrac{A}{x + 1} + \dfrac{B}{x - 1} + \dfrac{Cx + D}{x^2 + 1}$

26. $\dfrac{1 + x + x^2}{x^4 - 16} = \dfrac{A}{x + 2} + \dfrac{B}{x - 2} + \dfrac{Cx + D}{x^2 + 4}$

27. $\dfrac{x^2 + 3x + 3}{(x + 1)^2(x + 2)^2} = \dfrac{A}{x + 1} + \dfrac{B}{(x + 1)^2} + \dfrac{C}{x + 2} + \dfrac{D}{(x + 2)^2}$

28. $\dfrac{x^3 - x^2 + 1}{x^2(2x + 1)^2} = \dfrac{A}{x} + \dfrac{B}{x^2} + \dfrac{C}{2x + 1} + \dfrac{D}{(2x + 1)^2}$

29. Two points P_1 and P_2 in the plane have the following properties. The sum of the coordinates of P_1 is 15 and the sum of the coordinates of P_2 is 9. The symmetric image of P_2 across the diagonal $y = x$ is also the symmetric image of P_1 across the y-axis. Find the coordinates of P_1 and P_2.

30. Two points P_1 and P_2 are on the lines $x + 2y = 2$ and $2x - y = -1$, respectively. The sum of the x-coordinates of P_1 and P_2 is -10, and the sum of their y-coordinates is 12. Find the coordinates of P_1 and P_2.

In previous sections we used matrices to help solve systems of linear equations. It turns out that matrices are quite useful mathematical objects possessing numerous applications beyond the solution of linear systems. We will not pursue these applications here (many can be found in a field of mathematics called linear algebra). However, we shall introduce basic matrix operations that will supply a framework for later study.

We begin with the definition of a real matrix.

> **Definition of Matrix**
> A **real matrix** is a rectangular array of real numbers enclosed by square brackets.

As an example, consider the following matrix consisting of three rows and four columns:

$$\begin{bmatrix} 1 & 2 & 3 & 0 \\ -2 & 1 & 1 & 6 \\ 4 & -5 & 7 & -1 \end{bmatrix}$$

We call this a three by four, or 3×4 matrix. Each number inside the matrix is called an **entry**. We identify an entry by its location according to the row and column in which it appears. For example, -5 is located in row 3 and column 2 of the matrix, so we call -5 the 3, 2 entry.

In general, we refer to a matrix with m rows and n columns as an **m × n matrix**, and we say that its **size** is $m \times n$. Matrices can be practically any size, as the following examples illustrate:

$$\begin{array}{ccccc} 1 \times 3 & 1 \times 1 & 2 \times 2 & 4 \times 1 & 4 \times 4 \end{array}$$

$$\begin{bmatrix} 3 & -2 & 4 \end{bmatrix} \quad \begin{bmatrix} 5 \end{bmatrix} \quad \begin{bmatrix} 1 & 2 \\ 3 & 4 \end{bmatrix} \quad \begin{bmatrix} 6 \\ 0 \\ 6 \\ -1 \end{bmatrix} \quad \begin{bmatrix} a & 0 & 0 & 0 \\ 0 & b & 0 & 0 \\ 0 & 0 & c & 0 \\ 0 & 0 & 0 & d \end{bmatrix}$$

If a matrix has the same number of rows and columns, it is called a **square matrix**. The 1×1, 2×2, and 4×4 matrices above are examples of square matrices.

We shall use capital letters such as A, B, and C to designate matrices. If A is a matrix, then we write a_{ij} to represent its i, j entry. Thus, for the matrix

$$A = \begin{bmatrix} 1 & 2 \\ 3 & 4 \end{bmatrix}$$

we have $a_{11} = 1$, $a_{12} = 2$, $a_{21} = 3$, and $a_{22} = 4$.

We say that two matrices A and B are *equal* if and only if they have the same size and all their corresponding entries are equal, $a_{ij} = b_{ij}$, for each i and j.

We are now ready to describe various arithmetic operations for matrices. We begin with the multiplication of a matrix by a real number, which is called *scalar multiplication*.

Scalar Multiplication of Matrices

Let A denote a matrix with entries a_{ij}. If $r \in \mathbf{R}$, then rA is the matrix with entries ra_{ij}.

Thus, multiplying a matrix by a real number r corresponds to multiplying each entry in the matrix by r.

Example 1

If $A = \begin{bmatrix} 3 & 0 & -2 \\ 1 & 6 & 1 \end{bmatrix}$, find $4A$.

SOLUTION We have

$$4A = 4 \begin{bmatrix} 3 & 0 & -2 \\ 1 & 6 & 1 \end{bmatrix} = \begin{bmatrix} 12 & 0 & -8 \\ 4 & 24 & 4 \end{bmatrix} \blacksquare$$

The addition of matrices is accomplished in the obvious way.

Addition of Matrices

Let A and B be matrices of the same size with entries a_{ij} and b_{ij}, respectively. Then $A + B$ is the matrix with entries $a_{ij} + b_{ij}$.

Example 2

If $A = \begin{bmatrix} 1 & 2 \\ 3 & 4 \end{bmatrix}$ and $B = \begin{bmatrix} 1 & 3 \\ 0 & -2 \end{bmatrix}$, find (a) $A + B$, (b) $-A + 2B$.

(a) We have

$$A + B = \begin{bmatrix} 1 & 2 \\ 3 & 4 \end{bmatrix} + \begin{bmatrix} 1 & 3 \\ 0 & -2 \end{bmatrix} = \begin{bmatrix} 1+1 & 2+3 \\ 3+0 & 4-2 \end{bmatrix} = \begin{bmatrix} 2 & 5 \\ 3 & 2 \end{bmatrix}$$

(b) We think of $-A$ as $-1A$. Thus

$$-A + 2B = -1\begin{bmatrix} 1 & 2 \\ 3 & 4 \end{bmatrix} + 2\begin{bmatrix} 1 & 3 \\ 0 & -2 \end{bmatrix} = \begin{bmatrix} -1 & -2 \\ -3 & -4 \end{bmatrix} + \begin{bmatrix} 2 & 6 \\ 0 & -4 \end{bmatrix}$$

$$= \begin{bmatrix} 1 & 4 \\ -3 & -8 \end{bmatrix} \blacksquare$$

Remark

Note that the sum $A + B$ is defined only when both matrices have the same size.

Our next objective is to define matrix multiplication. If A and B are matrices, how should we define their product, AB? Let us consider the 2×2 matrices A and B in Example 2.

$$AB = \begin{bmatrix} 1 & 2 \\ 3 & 4 \end{bmatrix}\begin{bmatrix} 1 & 3 \\ 0 & -2 \end{bmatrix} = \begin{bmatrix} & ? & \end{bmatrix}$$

We want the product to be another matrix. In fact, AB will be a 2×2 matrix whose entries are obtained as follows.

1, 1 Entry of AB.   To get the entry in the first row and first column of AB, we use the first row of A and the first column of B.

$$\begin{bmatrix} 1 & 2 \\ 3 & 4 \end{bmatrix}\begin{bmatrix} 1 & 3 \\ 0 & -2 \end{bmatrix} = \begin{bmatrix} 1(1) + 2(0) & \\ & \end{bmatrix}$$

We call $1(1) + 2(0)$ the **inner product** of row 1 of A with column 1 of B. Thus the 1, 1 entry of AB is $1(1) + 2(0) = 1$.

1, 2 Entry of AB.   In this case we take the inner product of row 1 of A with column 2 of B.

$$\begin{bmatrix} 1 & 2 \\ 3 & 4 \end{bmatrix}\begin{bmatrix} 1 & 3 \\ 0 & -2 \end{bmatrix} = \begin{bmatrix} 1 & 1(3) + 2(-2) \\ & \end{bmatrix}$$

Thus the 1, 2 entry of AB is $1(3) + 2(-2) = -1$.

2, 1 Entry of AB.   Now we take the inner product of row 2 of A with column 1 of B.

$$\begin{bmatrix} 1 & 2 \\ 3 & 4 \end{bmatrix} \begin{bmatrix} 1 & 3 \\ 0 & -2 \end{bmatrix} = \begin{bmatrix} 1 & -1 \\ 3(1) + 4(0) & \end{bmatrix}$$

Thus the 2, 1 entry of AB is $3(1) + 4(0) = 3$.

2, 2 Entry of AB. Finally, we take the inner product of row 2 of A with column 2 of B.

$$\begin{bmatrix} 1 & 2 \\ 3 & 4 \end{bmatrix} \begin{bmatrix} 1 & 3 \\ 0 & -2 \end{bmatrix} = \begin{bmatrix} 1 & -1 \\ 3 & 3(3) + 4(-2) \end{bmatrix}$$

So, the 2, 2 entry of AB is $3(3) + 4(-2) = 1$.

We conclude,

$$AB = \begin{bmatrix} 1 & 2 \\ 3 & 4 \end{bmatrix} \begin{bmatrix} 1 & 3 \\ 0 & -2 \end{bmatrix} = \begin{bmatrix} 1 & -1 \\ 3 & 1 \end{bmatrix}$$

In summary, the i, j entry for AB is obtained by taking the inner product of row i in matrix A with column j in matrix B. This is how we will define matrix multiplication in general, except for one important stipulation: each row in A must have the same length as each column in B; otherwise the inner product does not exist. Note that the length of a row in A equals the length of a column in B if and only if the number of columns in A equals the number of rows in B. This brings us to the following definition.

Matrix Multiplication
Suppose A and B are $m \times n$ and $n \times k$ matrices, respectively, so that the number of columns in A equals the number of rows in B. Then the product AB is an $m \times k$ matrix whose i, j entry equals the inner product of row i in A with column j in B.

Example 3

If $A = \begin{bmatrix} 2 & -1 & 3 \\ 1 & 4 & -2 \end{bmatrix}$ and $B = \begin{bmatrix} -3 & 1 \\ -1 & 2 \\ 2 & 1 \end{bmatrix}$, find (a) AB, (b) BA.

SOLUTION (a) The number of columns in A (three) equals the number of rows in B, so the product AB exists. Furthermore, since A is a 2×3 matrix and B is a 3×2 matrix, AB will have size 2×2. We now demonstrate how each entry in the product is obtained.

1, 1 entry of AB. We take the inner product of row 1 in A with column 1 in B: $2(-3) + (-1)(-1) + 3(2) = 1$.

$$\begin{bmatrix} 2 & -1 & 3 \\ 1 & 4 & -2 \end{bmatrix} \begin{bmatrix} -3 & 1 \\ -1 & 2 \\ 2 & 1 \end{bmatrix} = \begin{bmatrix} 1 & \\ & \end{bmatrix}$$

1, 2 entry of AB. We take the inner product of row 1 in A with column 2 in B: $2(1) + (-1)(2) + 3(1) = 3$.

$$\begin{bmatrix} 2 & -1 & 3 \\ 1 & 4 & -2 \end{bmatrix} \begin{bmatrix} -3 & 1 \\ -1 & 2 \\ 2 & 1 \end{bmatrix} = \begin{bmatrix} 1 & 3 \end{bmatrix}$$

Notice that there is no 1, 3 entry for AB since B does not have a third column. Thus we move on to row 2 entries.

2, 1 entry of AB. We take the inner product of row 2 in A with column 1 in B: $1(-3) + 4(-1) + (-2)(2) = -11$.

$$\begin{bmatrix} 2 & -1 & 3 \\ 1 & 4 & -2 \end{bmatrix} \begin{bmatrix} -3 & 1 \\ -1 & 2 \\ 2 & 1 \end{bmatrix} = \begin{bmatrix} 1 & 3 \\ -11 & \end{bmatrix}$$

2, 2 entry of AB. We take the inner product of row 2 in A with column 2 in B: $1(1) + 4(2) + (-2)(1) = 7$.

$$\begin{bmatrix} 2 & -1 & 3 \\ 1 & 4 & -2 \end{bmatrix} \begin{bmatrix} -3 & 1 \\ -1 & 2 \\ 2 & 1 \end{bmatrix} = \begin{bmatrix} 1 & 3 \\ -11 & 7 \end{bmatrix}$$

We conclude,

$$AB = \begin{bmatrix} 1 & 3 \\ -11 & 7 \end{bmatrix}$$

(b) The number of columns in B (two) equals the number of rows in A, so the product BA exists. We have

$$BA = \begin{bmatrix} -3 & 1 \\ -1 & 2 \\ 2 & 1 \end{bmatrix} \begin{bmatrix} 2 & -1 & 3 \\ 1 & 4 & -2 \end{bmatrix}$$

$$= \begin{bmatrix} -6+1 & 3+4 & -9-2 \\ -2+2 & 1+8 & -3-4 \\ 4+1 & -2+4 & 6-2 \end{bmatrix}$$

$$= \begin{bmatrix} -5 & 7 & -11 \\ 0 & 9 & -7 \\ 5 & 2 & 4 \end{bmatrix} \blacksquare$$

Remark

Example 3 clearly illustrates that if A and B are matrices, then AB and BA are not necessarily equal.

Example 4

If $A = \begin{bmatrix} 1 & -2 & 3 \\ 3 & 4 & 5 \end{bmatrix}$ and $B = \begin{bmatrix} 4 & 3 \\ 2 & 1 \end{bmatrix}$, find (a) AB, (b) BA.

SOLUTION (a) We cannot compute the product AB since A has three columns and B has only two rows. We must say that AB *is undefined*.

(b) B has two columns and A has two rows, so the product BA exists. We have

$$BA = \begin{bmatrix} 4 & 3 \\ 2 & 1 \end{bmatrix} \begin{bmatrix} 1 & -2 & 3 \\ 3 & 4 & 5 \end{bmatrix}$$

$$= \begin{bmatrix} 4+9 & -8+12 & 12+15 \\ 2+3 & -4+4 & 6+5 \end{bmatrix}$$

$$= \begin{bmatrix} 13 & 4 & 27 \\ 5 & 0 & 11 \end{bmatrix} \blacksquare$$

For the remainder of this section we focus our attention on square matrices only. Given a square matrix A, we assign to it a unique real number called the **determinant** of A, denoted det(A). We can think of the determinant as a real function whose domain is the set of all square matrices. Thus

$$A \xrightarrow{\text{det}} r$$

where r is a real number. How does the determinant function assign a real number to a given square matrix? We shall answer this question in stages: first we consider 1×1 matrices, then 2×2 matrices, then 3×3 matrices, and, finally, general $n \times n$ matrices.

For any 1×1 matrix $[a]$, we define det$[a] = a$. The 1×1 case is not terribly exciting, so let us move on to the definition of the determinant of a 2×2 matrix.

Determinant of a 2 × 2 Matrix

$$\det \begin{bmatrix} a & b \\ c & d \end{bmatrix} = ad - bc$$

Example 5

Evaluate the determinant of the given matrix:

(a) $A = \begin{bmatrix} 1 & 2 \\ 3 & 4 \end{bmatrix}$; (b) $B = \begin{bmatrix} 3 & -4 \\ -6 & 8 \end{bmatrix}$.

SOLUTION

(a)
$$\det(A) = \det\begin{bmatrix} 1 & 2 \\ 3 & 4 \end{bmatrix} = 1(4) - 2(3) = -2$$

(b)
$$\det(B) = \det\begin{bmatrix} 3 & -4 \\ -6 & 8 \end{bmatrix} = 3(8) - (-4)(-6) = 0 \quad \blacksquare$$

Next, we define the determinant of a 3×3 matrix.

Determinant of a 3×3 Matrix

$$\det\begin{bmatrix} a_1 & a_2 & a_3 \\ b_1 & b_2 & b_3 \\ c_1 & c_2 & c_3 \end{bmatrix} = a_1(b_2c_3 - b_3c_2) - a_2(b_1c_3 - b_3c_1) + a_3(b_1c_2 - b_2c_1)$$

This formula is not as random as it might seem. Notice that we have three terms corresponding to a_1, a_2, and a_3. These are the entries in the first row of the matrix. Furthermore, the factors appearing with the a's turn out to be determinants of 2×2 matrices. Let us examine the formula more closely.

First Term: $a_1(b_2c_3 - b_3c_2)$. If we delete the row and column in which a_1 appears, we are left with a 2×2 matrix whose determinant is $b_2c_3 - b_3c_2$. Thus

$$\begin{bmatrix} a_1 & a_2 & a_3 \\ b_1 & b_2 & b_3 \\ c_1 & c_2 & c_3 \end{bmatrix} \qquad a_1 \det\begin{bmatrix} b_2 & b_3 \\ c_2 & c_3 \end{bmatrix} = a_1(b_2c_3 - b_3c_2)$$

Second Term: $-a_2(b_1c_3 - b_3c_1)$. If we delete the row and column in which a_2 appears, we are left with a 2×2 matrix whose determinant is $b_1c_3 - b_3c_1$. Thus

$$\begin{bmatrix} a_1 & a_2 & a_3 \\ b_1 & b_2 & b_3 \\ c_1 & c_2 & c_3 \end{bmatrix} \qquad -a_2 \det\begin{bmatrix} b_1 & b_3 \\ c_1 & c_3 \end{bmatrix} = -a_2(b_1c_3 - b_3c_1)$$

Third Term: $a_3(b_1c_2 - b_2c_1)$. If we delete the row and column in which a_3 appears, we are left with a 2×2 matrix whose determinant is $b_1c_2 - b_2c_1$. Thus

$$\begin{bmatrix} a_1 & a_2 & a_3 \\ b_1 & b_2 & b_3 \\ c_1 & c_2 & c_3 \end{bmatrix} \qquad a_3 \det \begin{bmatrix} b_1 & b_2 \\ c_1 & c_2 \end{bmatrix} = -a_3(b_1c_2 - b_2c_1)$$

Putting all this together, we can rewrite our formula for the determinant of a 3×3 matrix as follows:

$$\det \begin{bmatrix} a_1 & a_2 & a_3 \\ b_1 & b_2 & b_3 \\ c_1 & c_2 & c_3 \end{bmatrix} = a_1 \det \begin{bmatrix} b_2 & b_3 \\ c_2 & c_3 \end{bmatrix} - a_2 \det \begin{bmatrix} b_1 & b_3 \\ c_1 & c_3 \end{bmatrix} + a_3 \det \begin{bmatrix} b_1 & b_2 \\ c_1 & c_2 \end{bmatrix}$$

Written in this form, we say that we are evaluating the determinant by *expanding along the first row*.

Remark

The negative sign in front of the a_2 term is not a mistake. It allows for certain desirable properties of the determinant function.

Example 6

Let $A = \begin{bmatrix} 2 & 3 & 1 \\ 1 & 4 & 3 \\ -5 & -2 & 5 \end{bmatrix}$. Evaluate $\det(A)$ by expanding along the first row.

SOLUTION We have

$$\det \begin{bmatrix} 2 & 3 & 1 \\ 1 & 4 & 3 \\ -5 & -2 & 5 \end{bmatrix}$$

$$= 2 \det \begin{bmatrix} 4 & 3 \\ -2 & 5 \end{bmatrix} - 3 \det \begin{bmatrix} 1 & 3 \\ -5 & 5 \end{bmatrix} + 1 \det \begin{bmatrix} 1 & 4 \\ -5 & -2 \end{bmatrix}$$

$$= 2(20 - (-6)) - 3(5 - (-15)) + 1(-2 - (-20))$$

$$= 52 - 60 + 18$$

$$= 10 \quad \blacksquare$$

As you might guess, we can evaluate the determinant of a 3×3 matrix by expanding along *any* row or column. Let us explain how this is done. Suppose A is an arbitrary 3×3 matrix. Let a_{ij} represent the i, j entry of A. Thus

$$A = \begin{bmatrix} a_{11} & a_{12} & a_{13} \\ a_{21} & a_{22} & a_{23} \\ a_{31} & a_{32} & a_{33} \end{bmatrix}$$

We write M_{ij} to represent the matrix obtained from A by deleting its ith row and jth column. For example,

$$\begin{bmatrix} a_{11} & a_{12} & a_{13} \\ a_{21} & a_{22} & a_{23} \\ a_{31} & a_{32} & a_{33} \end{bmatrix} \quad \text{implies} \quad M_{21} = \begin{bmatrix} a_{12} & a_{13} \\ a_{32} & a_{33} \end{bmatrix}$$

Similarly, we have

$$M_{22} = \begin{bmatrix} a_{11} & a_{13} \\ a_{31} & a_{33} \end{bmatrix} \quad \text{and} \quad M_{23} = \begin{bmatrix} a_{11} & a_{12} \\ a_{31} & a_{32} \end{bmatrix}$$

Now, if we wish to evaluate the determinant of matrix A by expanding along the *second* row, we have

$$\det(A) = -a_{21} \det(M_{21}) + a_{22} \det(M_{22}) - a_{23} \det(M_{23})$$

The only difference in the previous pattern is the signs appearing in front of each term. The rule for how we determine these signs is as follows.

Sign Rule for Expanding a Determinant

If the ith row and jth column are deleted for a term in the expansion of a determinant, then the sign for this term is given by

$$\text{Sign of } (-1)^{i+j} = \begin{cases} + & \text{if } i + j \text{ is even} \\ - & \text{if } i + j \text{ is odd} \end{cases}$$

Thus, for a 3×3 matrix we would have the following signs corresponding to each possible term in an expansion of the determinant:

$$\begin{bmatrix} + & - & + \\ - & + & - \\ + & - & + \end{bmatrix}$$

Example 7

Evaluate the determinant of the 3×3 matrix A given in Example 6 by expanding along the (a) second row, (b) third column.

SOLUTION (a) Expanding along the second row, we have

$$\det \begin{bmatrix} 2 & 3 & 1 \\ 1 & 4 & 3 \\ -5 & -2 & 5 \end{bmatrix} = -1 \det(M_{21}) + 4 \det(M_{22}) - 3 \det(M_{23})$$

$$= -1 \det \begin{bmatrix} 3 & 1 \\ -2 & 5 \end{bmatrix} + 4 \det \begin{bmatrix} 2 & 1 \\ -5 & 5 \end{bmatrix} - 3 \det \begin{bmatrix} 2 & 3 \\ -5 & -2 \end{bmatrix}$$

$$= -1(17) + 4(15) - 3(11)$$

$$= 10$$

(b) Expanding along the third column, we have

$$\det \begin{bmatrix} 2 & 3 & 1 \\ 1 & 4 & 3 \\ -5 & -2 & 5 \end{bmatrix} = 1 \det(M_{13}) - 3 \det(M_{23}) + 5 \det(M_{33})$$

$$= 1 \det \begin{bmatrix} 1 & 4 \\ -5 & -2 \end{bmatrix} - 3 \det \begin{bmatrix} 2 & 3 \\ -5 & -2 \end{bmatrix} + 5 \det \begin{bmatrix} 2 & 3 \\ 1 & 4 \end{bmatrix}$$

$$= 1(18) - 3(11) + 5(5)$$

$$= 10 \blacksquare$$

Remark

In Examples 6 and 7 we calculated the determinant of the same matrix in three ways: expanding along row 1, row 2, and column 3. In each case we found the same answer for the determinant. In general, no matter what row or column we use for calculating the determinant, the answer will always be the same.

Example 8

Evaluate $\det \begin{bmatrix} 0 & 3 & 1 \\ 6 & 4 & 3 \\ 0 & -2 & 5 \end{bmatrix}$.

SOLUTION   We choose to expand along the first column since it contains two zeros.

$$\det \begin{bmatrix} 0 & 3 & 1 \\ 6 & 4 & 3 \\ 0 & -2 & 5 \end{bmatrix} = 0 \det(M_{11}) - 6 \det(M_{21}) + 0 \det(M_{31})$$

$$= 0 - 6 \det \begin{bmatrix} 3 & 1 \\ -2 & 5 \end{bmatrix} + 0$$

$$= -6(17)$$

$$= -102 \; \blacksquare$$

Next, we extend the definition of the determinant to any $n \times n$ matrix.

Determinant of an $n \times n$ Matrix, $n \geq 2$

Let A be any $n \times n$ matrix ($n \geq 2$) and let M_{ij} denote the matrix obtained from A by deleting the ith row and jth column. To evaluate $\det(A)$, choose a row (or column) of A and form a sum as follows: for each entry a_{ij} in the chosen row (or column), add $(-1)^{i+j} a_{ij} \det(M_{ij})$.

Note that the factor $(-1)^{i+j}$ gives us the correct sign for each term in the expansion according to the sign rule stated earlier.

Example 9

Evaluate $\det(A)$ if $A = \begin{bmatrix} 2 & 1 & 0 & 0 \\ 0 & 2 & 3 & 1 \\ 6 & 1 & 4 & 3 \\ 0 & -5 & -2 & 5 \end{bmatrix}.$

SOLUTION   We choose to expand along the first row. Thus

$$\det(A) = 2 \det(M_{11}) - 1 \det(M_{12}) + 0 \det(M_{13}) - 0 \det(M_{14})$$

$$= 2 \det \begin{bmatrix} 2 & 3 & 1 \\ 1 & 4 & 3 \\ -5 & -2 & 5 \end{bmatrix} - 1 \det \begin{bmatrix} 0 & 3 & 1 \\ 6 & 4 & 3 \\ 0 & -2 & 5 \end{bmatrix} + 0 + 0$$

The first determinant has value 10, as calculated in Examples 6 and 7; the second determinant is -102, as found in Example 8. Thus

$$\det(A) = 2(10) - 1(-102) = 122 \; \blacksquare$$

In Example 9 we were fortunate that the matrix contained several zeros. In general, however, the computation for the determinant of a large matrix may get rather long and messy. The following properties help alleviate some of the difficulties.

Properties of the Determinant

Let A be a square matrix.
1. If A' is obtained from A by adding a multiple of one row to another, then $\det(A) = \det(A')$.
2. If A' is obtained from A by factoring out a constant c from each entry in one of the rows of A, then $\det(A) = c \det(A')$.

We omit the proofs of these properties, except in the case of 2×2 matrices; these proofs are outlined in the exercises.

Example 10

Evaluate $\det \begin{bmatrix} -12 & -20 & -31 \\ -25 & -30 & 10 \\ 2 & 4 & 9 \end{bmatrix}$.

SOLUTION First, we notice that the entries in row 2 have the common factor 5. Thus, using property 2,

$$\det \begin{bmatrix} -12 & -20 & -31 \\ -25 & -30 & 10 \\ 2 & 4 & 9 \end{bmatrix} = 5 \det \begin{bmatrix} -12 & -20 & -31 \\ -5 & -6 & 2 \\ 2 & 4 & 9 \end{bmatrix}$$

Before computing, we choose to introduce a zero in the matrix by adding $6 \times$ row 3 to row 1. By property 1, this does not change the value of the determinant. Thus

$$= 5 \det \begin{bmatrix} 0 & 4 & 23 \\ -5 & -6 & 2 \\ 2 & 4 & 9 \end{bmatrix}$$

Now we expand along the first row.

$$= 5 \left(0 - 4 \det \begin{bmatrix} -5 & 2 \\ 2 & 9 \end{bmatrix} + 23 \det \begin{bmatrix} -5 & -6 \\ 2 & 4 \end{bmatrix} \right)$$

$$= 5(-4(-49) + 23(-8))$$

$$= 5(196 - 184)$$

$$= 60 \ \blacksquare$$

Remark

There is a third property of determinants corresponding to interchanging two rows in a matrix.

3. If A' is obtained from A by interchanging two rows, then $\det(A) = -\det(A')$.

For our purposes, it is not necessary to use this property.

EXERCISES 9.5

In Exercises 1 and 2, for the given matrix A, find the (a) number of rows in A, (b) number of columns in A, (c) size of A, (d) 1, 2 entry, (e) a_{31}.

1. $\begin{bmatrix} 6 & 5 \\ 4 & 3 \\ -2 & 0 \end{bmatrix}$

2. $\begin{bmatrix} 1 & 2 & 1 \\ 4 & 5 & -1 \\ 7 & 8 & 9 \\ -3 & -2 & 6 \end{bmatrix}$

In Exercises 3 to 8, for the given matrices A and B, find each of the following expressions or indicate that the expression is undefined: (a) $3A$, (b) $A + B$, (c) $-A + 2B$, (d) AB.

3. $A = \begin{bmatrix} 1 & 0 \\ 1 & 1 \end{bmatrix}$

   $B = \begin{bmatrix} 1 & 1 \\ 0 & 1 \end{bmatrix}$

4. $A = \begin{bmatrix} 1 & 0 \\ 0 & 1 \end{bmatrix}$

   $B = \begin{bmatrix} 0 & 1 \\ 1 & 0 \end{bmatrix}$

5. $A = \begin{bmatrix} 1 & -2 & 0 \\ -3 & 1 & 2 \end{bmatrix}$

   $B = \begin{bmatrix} 4 & 5 & -6 \\ 2 & -1 & 1 \end{bmatrix}$

6. $A = \begin{bmatrix} -3 & 1 \\ -2 & 4 \\ 0 & 4 \end{bmatrix}$

   $B = \begin{bmatrix} 1 & 8 \\ 4 & 2 \\ -3 & -7 \end{bmatrix}$

7. $A = \begin{bmatrix} 2 & 3 \\ -1 & 2 \\ 0 & 0 \end{bmatrix}$

   $B = \begin{bmatrix} 1 & 2 \\ 3 & 4 \end{bmatrix}$

8. $A = \begin{bmatrix} 1 & 0 & 1 \\ 1 & 1 & 0 \end{bmatrix}$

   $B = \begin{bmatrix} 1 & 1 \\ 0 & 1 \\ 1 & 0 \end{bmatrix}$

In Exercises 9 to 12, for the given matrices A and B, find AB and BA.

9. $A = \begin{bmatrix} -2 & 1 \\ 1 & 2 \\ 0 & 3 \end{bmatrix}$

   $B = \begin{bmatrix} 4 & 2 & 1 \\ -3 & 0 & 2 \end{bmatrix}$

10. $A = \begin{bmatrix} 1 \\ 2 \\ 3 \end{bmatrix}$

    $B = [4 \quad 5 \quad 6]$

11. $A = \begin{bmatrix} 1 & 0 & 1 \\ 0 & 2 & 0 \\ 1 & 0 & 1 \end{bmatrix}$

    $B = \begin{bmatrix} 1 & 1 & 0 \\ 1 & 0 & 1 \\ 0 & 1 & 1 \end{bmatrix}$

12. $A = \begin{bmatrix} 3 & 3 & 3 \\ 2 & 2 & 2 \\ 1 & 1 & 1 \end{bmatrix}$

    $B = \begin{bmatrix} 0 & 0 & 1 \\ 0 & 1 & 0 \\ 1 & 0 & 0 \end{bmatrix}$

In Exercises 13 to 18, find (a) $A^2 = AA$, (b) $A^3 = AA^2$, (c) A^n, n a positive integer.

13. $A = \begin{bmatrix} 1 & 1 \\ 0 & 1 \end{bmatrix}$

14. $A = \begin{bmatrix} 1 & 0 \\ 2 & 1 \end{bmatrix}$

15. $A = \begin{bmatrix} a & 0 & 0 \\ 0 & b & 0 \\ 0 & 0 & c \end{bmatrix}$

16. $A = \begin{bmatrix} 0 & 0 & x \\ 0 & 1 & 0 \\ x & 0 & 0 \end{bmatrix}$

17.
$$A = \begin{bmatrix} 1 & 0 & 0 & 1 \\ 0 & 1 & 0 & 0 \\ 0 & 0 & 1 & 0 \\ 1 & 0 & 0 & 1 \end{bmatrix}$$

18.
$$A = \begin{bmatrix} -1 & 0 & 0 & 0 \\ 0 & -1 & 0 & 0 \\ 0 & 0 & -1 & 0 \\ 0 & 0 & 0 & -1 \end{bmatrix}$$

In Exercises 19 to 30, evaluate the determinant of the given matrix.

19. (a) $[-8]$ (b) $\begin{bmatrix} 4 & 5 \\ 6 & 7 \end{bmatrix}$

20. (a) $[\cos \pi]$ (b) $\begin{bmatrix} 8 & -5 \\ 3 & 2 \end{bmatrix}$

21. $\begin{bmatrix} 4 & -6 \\ -14 & -21 \end{bmatrix}$ 22. $\begin{bmatrix} a & b \\ b & b \end{bmatrix}$

23. $\begin{bmatrix} -3 & 2 & 0 \\ 0 & -1 & 4 \\ 0 & 4 & 1 \end{bmatrix}$ 24. $\begin{bmatrix} 1 & 2 & 3 \\ 3 & 1 & 2 \\ 2 & 3 & 1 \end{bmatrix}$

25. $\begin{bmatrix} 1 & 1 & 0 \\ 1 & 0 & 1 \\ 0 & 1 & 1 \end{bmatrix}$ 26. $\begin{bmatrix} 0 & 1 & 2 \\ 1 & 0 & 1 \\ 2 & 1 & 0 \end{bmatrix}$

27. $\begin{bmatrix} 1 & 2 & 0 & 2 \\ 0 & 1 & 1 & 0 \\ 2 & 1 & 2 & 1 \\ 1 & 0 & 1 & 1 \end{bmatrix}$

28. $\begin{bmatrix} -2 & 0 & 2 & -2 \\ 1 & 0 & 1 & -1 \\ 0 & 2 & 1 & -1 \\ -1 & 1 & 0 & 1 \end{bmatrix}$

29. $\begin{bmatrix} a & x & y & z \\ 0 & b & w & s \\ 0 & 0 & c & t \\ 0 & 0 & 0 & d \end{bmatrix}$ 30. $\begin{bmatrix} 91 & 92 & 93 & w \\ 94 & 95 & z & 0 \\ 96 & y & 0 & 0 \\ x & 0 & 0 & 0 \end{bmatrix}$

In Exercises 31 to 36, evaluate the determinant of each matrix. Use determinant property 1 or 2 to help simplify the computations.

31. $\begin{bmatrix} 22 & -44 & 66 \\ 4 & 1 & 2 \\ 15 & 12 & 3 \end{bmatrix}$ 32. $\begin{bmatrix} 30 & 12 & 6 \\ 20 & -50 & 20 \\ 15 & 30 & 75 \end{bmatrix}$

33. $\begin{bmatrix} 1 & 1 & 1 & 1 \\ 1 & 2 & 2 & 2 \\ 1 & 2 & 3 & 3 \\ 1 & 2 & 3 & 4 \end{bmatrix}$ 34. $\begin{bmatrix} 1 & 1 & 1 & 1 \\ 1 & -1 & 1 & -1 \\ 2 & 2 & -2 & -2 \\ 3 & -3 & 3 & 3 \end{bmatrix}$

35. $\begin{bmatrix} 1 & 1 & 1 \\ x & y & z \\ x^2 & y^2 & z^2 \end{bmatrix}$ 36. $\begin{bmatrix} x & y & z \\ x^2 & y^2 & z^2 \\ x^3 & y^3 & z^3 \end{bmatrix}$

In Exercises 37 to 39, let A, B, and C be arbitrary 2×2 matrices. Prove the following:

37. $\det(AB) = \det(A)\det(B)$ 38. $A(BC) = (AB)C$

39. $\det(rA) = r^2 \det(A)$ for any $r \in \mathbf{R}$.

40. This exercise establishes properties 1 to 3 of the determinant for 2×2 matrices. Let a, b, c, d, and r be real numbers.

(a) Show that $\det \begin{bmatrix} a & b \\ c & d \end{bmatrix} = \det \begin{bmatrix} a & b \\ c+ra & d+rb \end{bmatrix}$
$$= \det \begin{bmatrix} a+rc & b+rd \\ c & d \end{bmatrix}$$

(b) Show that $r \det \begin{bmatrix} a & b \\ c & d \end{bmatrix} = \det \begin{bmatrix} ra & rb \\ c & d \end{bmatrix}$
$$= \det \begin{bmatrix} a & b \\ rc & rd \end{bmatrix}$$

(c) Show that $\det \begin{bmatrix} a & b \\ c & d \end{bmatrix} = -\det \begin{bmatrix} c & d \\ a & b \end{bmatrix}$

In Exercises 41 to 44, refer to the matrices D, E, and I:

$$D = \begin{bmatrix} \sin\theta & \cos\theta \\ -\sin\theta & \cos\theta \end{bmatrix}$$

$$E = \begin{bmatrix} \cos\theta & \sin\theta \\ \sin\theta & \cos\theta \end{bmatrix} \qquad I = \begin{bmatrix} 1 & 0 \\ 0 & 1 \end{bmatrix}$$

41. Find DEI, DIE, and IDE. Compute $\det(DEI)$. [*Note:* By Exercise 38, the product of three 2×2 matrices can be found by one of two ways: $A(BC)$ or $(AB)C$.]

42. Find DE and ED. Compute $\det(DE)$ and $\det(ED)$.

43. Let $r = 1/\det(D)$. Find rD and $\det(rD)$.

44. Let $r = 1/\det(E)$. Find rE and $\det(rE)$.

CHAPTER 9 / REVIEW

In Exercises 1 and 2, solve the given system using *addition* or *substitution*.

1. $\dfrac{x}{4} - \dfrac{y}{5} = 5$

$\dfrac{x}{3} + \dfrac{y}{2} = -1$

2. $\dfrac{x^2}{4} - \dfrac{y^2}{9} = 1$

$\dfrac{x^2}{64} + \dfrac{y^2}{36} = 1$

3. Find the coordinates of the point on the line $x - 3y = 3$ that is equidistant from $A(1, 2)$ and $B(5, -1)$.

In Exercises 4 and 5, graph the solution set for the given inequality.

4. $x^2 + y^2 \geq 2(x - y)$

5. $|x + 2y| < 3$

In Exercises 6 to 8, graph the solution set for the given system of inequalities. Label the intersection points of the boundary curves.

6. $\begin{cases} y - 2x < 1 \\ y + 2 > x^2 \end{cases}$

7. $\begin{cases} y \leq \dfrac{e}{x} \\ y \geq \ln x \end{cases}$

8. $\begin{cases} y \leq \sqrt{x + 4} \\ x + 2y \geq 4 \\ 3x - y \leq 12 \end{cases}$

9. A rectangle must have area at least 12 square units but its diagonal cannot exceed 5 units. Describe all possible values for the length and width of the rectangle with a system of inequalities and a graph.

In Exercises 10 to 13, use the matrix method to solve the given system.

10. $\begin{aligned} 3x - y + 2z &= 2 \\ 5x + 2y - 7z &= 1 \\ 7x - 6y - z &= -1 \end{aligned}$

11. $\begin{aligned} 3x - 2y + z &= -10 \\ -x + 3y - 5z &= 1 \\ 9x - 8y + 7z &= -12 \end{aligned}$

12. $\begin{aligned} x + 4y + 7z &= 4 \\ 3x + 14y - z &= 2 \\ 2x + 10y - 8z &= -2 \end{aligned}$

13. $\begin{aligned} 2x + 3y - 4z + 5w &= -10 \\ x + 3y + 2z + 10w &= -8 \\ 4x - 6y + 12z - 38w &= 19 \\ 10x - 9y - 8z - 79w &= 18 \end{aligned}$

14. Find the quadratic function $F(x) = ax^2 + bx + c$ whose graph passes through the points $(-1, -3)$, $(2, -8)$, and $(-4, -34)$.

15. Find constants A, B, and C so that the following statement is true:

$$\frac{4x^2 + 1}{(2x - 1)(x + 1)^2} = \frac{A}{x + 1} + \frac{B}{(x + 1)^2} + \frac{C}{2x - 1}$$

In Exercises 16 to 19, given

$$A = \begin{bmatrix} -3 & 1 \\ 2 & 4 \end{bmatrix} \quad B = \begin{bmatrix} 1 & 2 \\ 0 & -6 \end{bmatrix} \quad C = \begin{bmatrix} -5 & 4 & 3 \\ 1 & 2 & -1 \end{bmatrix}$$

find each of the following expressions or indicate that the expression is undefined:

16. (a) $-3C$

(b) $2A - B$

17. (a) AB

(b) BC

18. (a) ABC
    (b) CA

19. (a) $\det(A)$
    (b) $\det(C)$

In Exercises 20 to 22, given

$$D = \begin{bmatrix} 3 & 2 & 1 \\ 0 & 3 & 2 \\ 0 & 0 & 3 \end{bmatrix} \quad E = \begin{bmatrix} 1 & 0 & 0 \\ 2 & 2 & 0 \\ 3 & 3 & 3 \end{bmatrix} \quad I = \begin{bmatrix} 1 & 0 & 0 \\ 0 & 1 & 0 \\ 0 & 0 & 1 \end{bmatrix}$$

find each of the following:

20. (a) $D(EI)$
    (b) $(DE)I$

21. (a) $\det(D)$
    (b) $\det(E)$
    (c) $\det(DE)$

22. (a) $I^2 = II$
    (b) $I^3 = III$
    (c) $\det(I)$

23. Evaluate $\det \begin{bmatrix} -1 & 3 & -5 \\ 7 & 9 & 8 \\ 6 & -4 & 2 \end{bmatrix}$

24. Evaluate $\det \begin{bmatrix} 1 & 2 & 1 & 2 \\ 2 & 3 & 2 & 3 \\ 3 & 2 & 3 & -2 \\ 4 & 5 & 5 & 4 \end{bmatrix}$

25. Answer true or false: if all the entries in a row or column of a square matrix A are zero, then $\det(A) = 0$.

10 Elementary Curves in the Plane

10.1 PARABOLAS

We consider two categories of elementary curves in this chapter: the conic sections and the basic polar curves. In Sections 10.1 to 10.5 we concentrate on the conic sections, which derive their name from the fact that intersecting a double cone with a plane at various angles will produce these curves. Figure 10.1 illustrates three of the possible intersections: a parabola, an ellipse, and a hyperbola. As we shall see, each curve has an equation of the form

$$Ax^2 + Bxy + Cy^2 + Dx + Ey + F = 0$$

We call this a *general second-degree equation* in x and y.

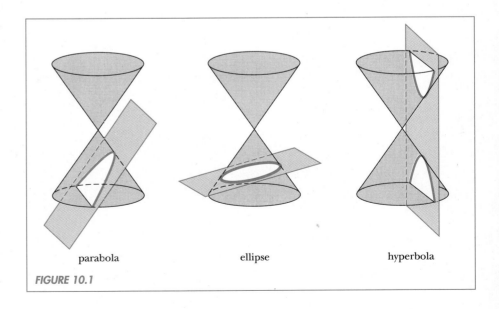

parabola   ellipse   hyperbola

FIGURE 10.1

We begin our study of the conic sections with the parabola. We present a geometric definition of the parabola and then see how the geometry leads us to a second-degree equation for the curve.

> **Geometric Definition of Parabola**
> A **parabola** consists of the set of all points $P(x, y)$ in the plane that are equidistant from a given point F, called the **focus**, and a given line l, called the **directrix**.

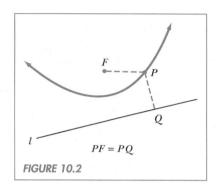

$$PF = PQ$$

FIGURE 10.2

Figure 10.2 shows a typical parabola with focus F and directrix l. By definition, every point P on the parabola has the property that its distance to the focus, PF, is equal to its distance to the directrix, PQ.

$$PF = PQ$$

This geometric requirement determines an equation that the points $P(x, y)$ on the parabola must satisfy. Our objective is to find this equation.

For simplicity, we consider a parabola whose focus F is located on the x-axis at $(p, 0)$, p a nonzero constant, and the directrix l is the vertical line $x = -p$. Figure 10.3 shows the situation when $p > 0$. A typical point $P(x, y)$ on the parabola is also drawn in Figure 10.3. Note that the distance from P to l is given by the length of the perpendicular line segment PQ, where Q has coordinates $(-p, y)$. Therefore, by the definition of the parabola, we have

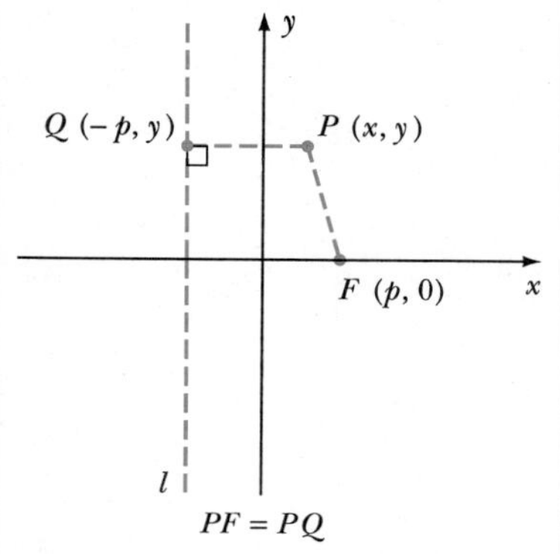

$$PF = PQ$$

$$\sqrt{(x - p)^2 + (y - 0)^2} = \sqrt{(x + p)^2 + (y - y)^2} \qquad \text{by the distance formula}$$

$$(x - p)^2 + y^2 = (x + p)^2 \qquad \text{squaring both sides}$$

$$x^2 - 2px + p^2 + y^2 = x^2 + 2px + p^2$$

$$y^2 = 4px$$

FIGURE 10.3

This last equation is called the **standard form** for the equation of the parabola. Note that replacing y with $-y$ in the equation results in no change.

$$(-y)^2 = 4px \quad \text{is equivalent to} \quad y^2 = 4px$$

Thus the parabola is symmetric with respect to the x-axis. We call this the axis of symmetry, or just the **axis** of the parabola. The axis will always pass through the focus and be perpendicular to the directrix. The point of intersection of the parabola with its axis is called the **vertex**. Figure 10.4 shows what the graph of $y^2 = 4px$ looks like when $p > 0$. Following is a summary of the important parts of this parabola:

Axis: $y = 0$ (the x-axis)

Vertex: $(0, 0)$

Focus: $(p, 0)$

Directrix: $x = -p$

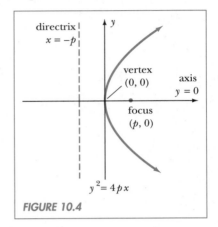

FIGURE 10.4

The number p is the key to finding all parts of the parabola. Note that because p can be either positive or negative, we say that p is the *directed distance* from the vertex to the focus of the parabola. When $p < 0$, the parts are still as given above; however, the graph will curve toward the left.

Example 1

Find the axis, vertex, focus, directrix, and graph of $4y^2 + x = 0$.

SOLUTION First, we put the equation into standard form.

$$4y^2 + x = 0 \quad \text{is equivalent to} \quad y^2 = -\frac{1}{4}x$$

We recognize this equation as the parabola $y^2 = 4px$. Therefore

$$4p = -\frac{1}{4} \quad \text{implies} \quad p = -\frac{1}{16}$$

It follows,

Axis: $y = 0$ (the x-axis)

Vertex: $(0, 0)$

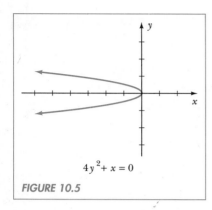

$4y^2 + x = 0$

FIGURE 10.5

Focus: $\left(-\dfrac{1}{16}, 0\right)$

Directrix: $x = \dfrac{1}{16}$

Since p is negative, we know that this parabola will curve toward the left. The following short table of values helps us determine the graph shown in Figure 10.5:

x	0	-4	-16
y	0	± 1	± 2

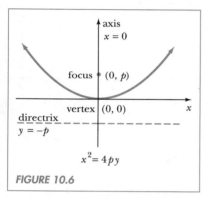

$x^2 = 4py$

FIGURE 10.6

Now, suppose the focus of our parabola is located along the y-axis instead of the x-axis and the directrix is a horizontal line instead of vertical. Let us say the focus is at $(0, p)$ and the directrix is $y = -p$. Figure 10.6 shows the situation when $p > 0$. The geometric definition of the parabola in this case leads to the following equation:

$$x^2 = 4py$$

We leave the derivation of this equation for an exercise. Whether p is positive or negative, we have the following summary of parts:

Axis: $x = 0$ (the y-axis)

Vertex: $(0, 0)$

Focus: $(0, p)$

Directrix: $y = -p$

Example 2

Find the axis, vertex, focus, directrix, and graph of $x^2 = 2y$.

SOLUTION The equation $x^2 = 2y$ represents a parabola of the form $x^2 = 4py$. We have

$$4p = 2 \quad \text{implies} \quad p = \dfrac{1}{2}$$

It follows,

Axis: $x = 0$ (the y-axis)

Vertex: $(0, 0)$

Focus: $\left(0, \dfrac{1}{2}\right)$

Directrix: $y = -\dfrac{1}{2}$

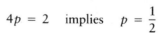

$x^2 = 2y$

FIGURE 10.7

The following table of values helps us determine the graph which was shown in Figure 10.7:

x	0	± 1	± 2	± 3
y	0	$\dfrac{1}{2}$	2	$\dfrac{9}{2}$

Let us summarize our results.

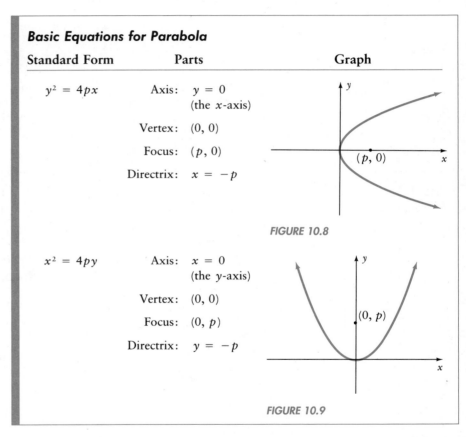

Basic Equations for Parabola

Standard Form	Parts		Graph
$y^2 = 4px$	Axis:	$y = 0$ (the x-axis)	
	Vertex:	$(0, 0)$	
	Focus:	$(p, 0)$	
	Directrix:	$x = -p$	

FIGURE 10.8

$x^2 = 4py$	Axis:	$x = 0$ (the y-axis)	
	Vertex:	$(0, 0)$	
	Focus:	$(0, p)$	
	Directrix:	$y = -p$	

FIGURE 10.9

So far we have been given an equation for a parabola and asked to find the parts. Now we wish to solve the converse problem: given information about the parts of a parabola, find its equation.

Example 3

Find an equation for the parabola with vertex at the origin and focus $(-5, 0)$.

SOLUTION Figure 10.10 shows the given parts. Since the axis of a parabola always passes through the focus and vertex, we know that the axis in this case is $y = 0$. The standard form for the equation of a parabola with

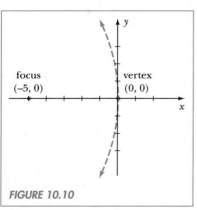

focus
$(-5, 0)$

vertex
$(0, 0)$

FIGURE 10.10

vertex at the origin and axis $y = 0$ is given by

$$y^2 = 4px$$

Therefore we need only find p. As Figure 10.10 indicates, the directed distance from the vertex to the focus is -5. So $p = -5$, and the particular equation we seek is

$$y^2 = 4(-5)x \qquad \text{or} \qquad y^2 = -20x \quad \blacksquare$$

Our next example is an application involving a parabola. In this type of problem we usually are asked to find some quantity that is related to a certain parabola; the following steps often lead to a solution:

1. Begin by drawing the parabola involved in an x, y-coordinate system so that the vertex of the parabola is at the origin and its axis is either the x- or y-axis.
2. Write the equation of the parabola: $x^2 = 4py$ or $y^2 = 4px$.
3. Define and label the unknown quantity.
4. Label coordinates of any known points on the parabola or points that are related to the unknown.
5. Use the fact that the coordinates of any point on the parabola must satisfy the equation of the parabola. In this way, each point on the parabola produces a corresponding equation.

These five steps should produce one or more equations that help find the unknown. If more equations are needed, look for other relationships between the given and unknown quantities.

Now we turn our attention to a specific application. Parabolas have reflective properties that are important in the design of certain mirrors and antenna. For example, consider the cross section of a parabolic mirror shown in Figure 10.11. A beam of light emitted from the focus will be reflected by the parabolic surface along a line parallel to the axis of the parabola. Conversely, any light rays approaching the mirror parallel to the axis will be reflected into the focus.

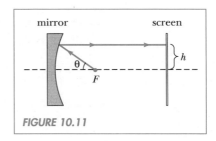

FIGURE 10.11

Example 4

A beam of light is emitted from the focus of the parabolic mirror shown in Figure 10.11. The beam is reflected off the mirror and hits a screen at the right at height h above the axis. If the focus is 2 centimeters from the vertex and $\theta = 60°$, find h.

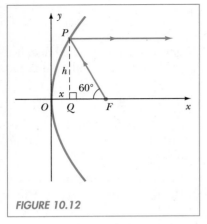

FIGURE 10.12

SOLUTION   We draw the parabola with its vertex at the origin in the x, y-coordinate system shown in Figure 10.12. Since the major axis for this parabola is along the x-axis, its equation is given by

$$y^2 = 4px$$

We know that the focus is 2 centimeters from the vertex; therefore $p = 2$. Hence the equation of the parabola becomes

$$y^2 = 8x$$

Let P be the point where the beam of light hits the parabola. Comparing Figures 10.12 and 10.11, it is clear that the y-coordinate of P is the unknown height h. Let x represent the x-coordinate of P. Since P is a point on the parabola, its coordinates must satisfy the equation of the parabola. Thus

$$P(x, h) \text{ on parabola} \quad \text{implies} \quad h^2 = 8x \tag{1}$$

Next, draw the perpendicular PQ from P to the x-axis. From Figure 10.12 we see that

$$QF = OF - OQ$$

$$= 2 - x$$

Thus, by considering the right triangle PQF, we find that

$$\tan 60° = \frac{PQ}{QF} = \frac{h}{2-x}$$

$$\sqrt{3} = \frac{h}{2-x}$$

$$2\sqrt{3} - \sqrt{3}x = h \tag{2}$$

At this point we have the following system of equations:

$$h^2 = 8x \tag{1}$$

$$2\sqrt{3} - \sqrt{3}x = h \tag{2}$$

Using *substitution*, we solve the system for h. From Equation (1) we find that $x = h^2/8$. Substituting this into Equation (2), we have

$$2\sqrt{3} - \sqrt{3}\frac{h^2}{8} = h$$

Multiply by 8:

$$16\sqrt{3} - \sqrt{3}h^2 = 8h$$

$$0 = \sqrt{3}h^2 + 8h - 16\sqrt{3}$$

Now apply the quadratic formula.

$$h = \frac{-8 \pm \sqrt{64 + 64(3)}}{2\sqrt{3}} = \frac{-4 \pm 8}{\sqrt{3}}$$

Thus

$$h = \frac{4}{\sqrt{3}} \quad or \quad h = \frac{-12}{\sqrt{3}}$$

$$= \frac{4\sqrt{3}}{3} \qquad\qquad = -4\sqrt{3}$$

The required answer must be positive. Hence $h = 4\sqrt{3}/3$. ∎

EXERCISES 10.1

In Exercises 1 to 6, find the axis of symmetry, vertex, focus, directrix, and graph.

1. $x^2 + y = 0$
2. $y^2 + x = 0$
3. $-2y^2 + 8x = 0$
4. $6x^2 - 4y = 0$
5. $y^2 + 10x = 0$
6. $x^2 + 8y = 0$

In Exercises 7 to 14, find an equation for the parabola with the given properties.

7. Vertex $(0, 0)$, focus $(-3, 0)$

8. Vertex $(0, 0)$, focus $(0, 1)$

9. Vertex $(0, 0)$, directrix $y = -2$

10. Vertex $(0, 0)$, directrix $x = 1/2$

11. Vertex $(0, 0)$, axis of symmetry $y = 0$, graph passing through $(3, 4)$

12. Vertex $(0, 0)$, graph passing through $(-2, -2)$ and $(2, -2)$

13. Vertex $(0, 0)$, graph passing through $(-3, -12)$ and $(3, -12)$

14. Focus $(2, 0)$, directrix $x = -2$.

15. Find all points on the parabola $y^2 = 4x$ that are at a distance of 4 units from $A(4, 0)$.

16. Find all the points on the parabola $y^2 = -8x$ that are at a distance of 5 units from the focus.

17. Find all points on the parabola $x^2 = 4y$ that are at a distance of 3 units from the focus.

18. Find an equation for the line that passes through the foci of $x^2 + 5y = 0$ and $3y^2 - x = 0$.

19. Let P be a point on the parabola $x^2 - 2y = 0$ and F the focus of the parabola. If P has x-coordinate -3, find an equation for the line that passes through P and F.

20. In Figure 10.13, a line segment AB is drawn perpendicular to the y-axis with its endpoints on the parabola $x^2 = 4y$. Let B have coordinates (x, y).

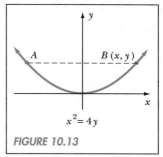

FIGURE 10.13

(a) Find the length of AB in terms of x.
(b) Find the length of AB in terms of y.
(c) Find the length of AB if AB passes through the focus.

21. Let $P(x, y)$ be on the parabola $x = y^2$ and PB be perpendicular to the x-axis (Figure 10.14). Find a formula for the area of triangle OPB in terms of (a) x; (b) y.

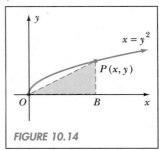

FIGURE 10.14

22. Figure 10.15 shows the cross section of a parabolic mirror. If the diameter of the mirror is 12 inches and its depth is 1 inch, find the focus.

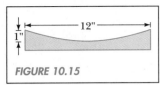

FIGURE 10.15

23. Repeat Exercise 22 if the diameter of the mirror is 16 inches and the depth is 1/2 inch.

24. A parabolic mirror has diameter l and depth h. Find the distance from the vertex to its focus in terms of l and h.

25. A parabolic arch is to be constructed so that its height will be 20 feet. Also, we want the focus 6 feet above ground level. See Figure 10.16. How wide is the base of the arch?

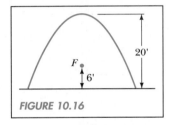

FIGURE 10.16

26. An object thrown horizontally off the top of a 200-foot-high building falls to the ground along a parabolic path, the vertex being the point of departure at the top of the building. See Figure 10.17. After the object has fallen 50 feet vertically, its distance from the building is 12 feet. How far from the building will the object be when it hits the ground?

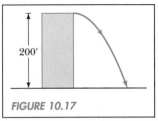

FIGURE 10.17

27. Refer to Figure 10.11, which shows a beam of light emitted from the focus of a parabolic mirror at the angle θ. The beam is reflected off the mirror and hits a screen at the height h above the axis. If $\theta = 45°$ and the focus is 4 centimeters from the vertex, find h.

28. Repeat Exercise 27 if the angle for the beam of light is θ and the focus is 2 centimeters from the vertex. In other words, find h as a function of θ, $h(\theta)$.

29. Let $P(x, y)$ be a point on the parabola $x^2 = 4py$. Let PF be the distance between P and the focus F.
    (a) Find a formula for the distance PF in terms of x and p only.
    (b) At what point P on the parabola is the distance PF a minimum? Prove your assertion.

30. Suppose $P(x, y)$ is equidistant from the point $(0, p)$ and line $y = -p$. Show that (x, y) must satisfy $x^2 = 4py$.

10.2 ELLIPSES

Geometric Definition of Ellipse
An **ellipse** consists of the set of all points $P(x, y)$ in the plane, the sum of whose distances from two fixed points, called the foci, is a constant.

Figure 10.18 shows a typical ellipse with foci F_1 and F_2. By definition, every point P on the ellipse has the property that the sum of its distances to F_1 and

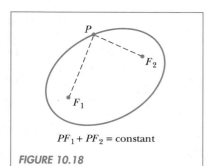

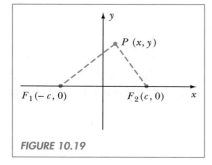

FIGURE 10.18

FIGURE 10.19

F_2 is a constant:

$$PF_1 + PF_2 = \text{constant}$$

This geometric requirement determines an equation for the ellipse that we would like to find. We begin by considering a relatively simple ellipse, leaving the more general situations for later discussion.

In Figure 10.19, the two foci for the ellipse are situated along the x-axis at $F_1(-c, 0)$ and $F_2(c, 0)$, where c is a positive constant. If $P(x, y)$ is any point on the ellipse, then by definition $PF_1 + PF_2 = \text{constant}$. Without loss of generality, we let this constant sum equal $2a$. (We use $2a$ instead of a so that the algebra in our derivation works out easier.) Thus

$$PF_1 + PF_2 = 2a$$

By the distance formula, this becomes

$$\sqrt{(x + c)^2 + y^2} + \sqrt{(x - c)^2 + y^2} = 2a$$

Next, we isolate the first square root and then square both sides.

$$\sqrt{(x + c)^2 + y^2} = 2a - \sqrt{(x - c)^2 + y^2}$$

$$(x + c)^2 + y^2 = 4a^2 - 4a\sqrt{(x - c)^2 + y^2} + (x - c)^2 + y^2$$

After some work, this equation simplifies to

$$a^2 - cx = a\sqrt{(x - c)^2 + y^2}$$

Now, square both sides once again.

$$a^4 - 2a^2cx + c^2x^2 = a^2(x - c)^2 + a^2y^2$$

After more simplification, we find that

$$a^2(a^2 - c^2) = (a^2 - c^2)x^2 + a^2y^2$$

Now, divide by $a^2(a^2 - c^2)$.

$$1 = \frac{x^2}{a^2} + \frac{x^2}{a^2 - c^2}$$

In the exercises you are asked to show that $a^2 - c^2 > 0$. Assuming this fact, we may let $b^2 = a^2 - c^2$. Thus our equation becomes

$$\frac{x^2}{a^2} + \frac{y^2}{b^2} = 1$$

This is the *standard form* for the equation of an ellipse with foci at $(\pm c, 0)$, where $c^2 = a^2 - b^2$. Note that replacing x with $-x$ or y with $-y$ will not change the equation; therefore the graph of this equation is symmetric with respect to both the x- and y-axes. The axis of symmetry containing the foci (in this case, the x-axis) is called the **major axis** of the ellipse. The other axis of symmetry is called the **minor axis**. The point where the two axes intersect is the **center** of the ellipse.

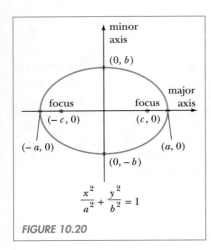

minor axis

$(0, b)$

major axis

focus | focus
$(-c, 0)$ | $(c, 0)$

$(-a, 0)$ | $(a, 0)$

$(0, -b)$

$$\frac{x^2}{a^2} + \frac{y^2}{b^2} = 1$$

FIGURE 10.20

What do the numbers a and b in the equation $x^2/a^2 + y^2/b^2 = 1$ correspond to geometrically? If we substitute $y = 0$ into the equation and solve for x, we find that $x = a$ or $x = -a$. Thus $(a, 0)$ and $(-a, 0)$ are the points where the graph of the ellipse intersects the major axis. Similarly, if we substitute $x = 0$ into the equation and solve for y, we find that $(0, b)$ and $(0, -b)$ are the points where the graph intersects the minor axis. These four points are called the **vertices** of the ellipse. In summary, we have (see Figure 10.20)

Equation: $\quad \dfrac{x^2}{a^2} + \dfrac{y^2}{b^2} = 1$

Major axis: $\quad y = 0$ (the x-axis)

Minor axis: $\quad x = 0$ (the y-axis)

Vertices: $\quad (\pm a, 0) \quad (0, \pm b)$

Foci: $\quad (\pm c, 0), \quad c^2 = a^2 - b^2$

One final remark concerning a and b: by definition, $b^2 = a^2 - c^2$. It follows that a^2 is larger than b^2 unless $c = 0$. If $c = 0$, then both foci are at the center of the ellipse and $a^2 = b^2$. The equation becomes

$$\frac{x^2}{a^2} + \frac{y^2}{a^2} = 1$$

This is equivalent to $x^2 + y^2 = a^2$, which is a circle of radius a. Thus a circle is a special case of an ellipse.

Example 1

Graph the ellipse $\dfrac{x^2}{16} + \dfrac{y^2}{4} = 1$. Find the major and minor axes, vertices, and foci.

SOLUTION The equation $x^2/16 + y^2/4 = 1$ is in the standard form $x^2/a^2 + y^2/b^2 = 1$. It follows that

$$a^2 = 16 \qquad \text{implies} \quad a = 4$$
$$b^2 = 4 \qquad \text{implies} \quad b = 2$$
$$c^2 = a^2 - b^2 \quad \text{implies} \quad c^2 = 16 - 4 = 12, \text{ so } c = 2\sqrt{3}$$

Therefore we have

Major axis: $\quad y = 0$ (the x-axis)

Minor axis: $\quad x = 0$ (the y-axis)

Vertices: $\quad (\pm 4, 0) \quad (0, \pm 2)$

Foci: $\quad (\pm 2\sqrt{3}, 0)$

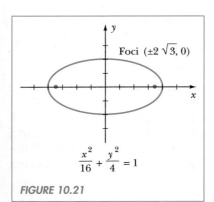

y

Foci $(\pm 2\sqrt{3}, 0)$

x

$$\frac{x^2}{16} + \frac{y^2}{4} = 1$$

FIGURE 10.21

The graph is obtained by drawing a smooth curve through the four vertices, as shown in Figure 10.21. ∎

Instead of beginning with the foci of the ellipse on the x-axis, we could have put them on the y-axis at $(0, \pm c)$, in which case, the y-axis would become the major axis. The derivation of the equation for this ellipse would be the same as before, except that the roles of x and y would be interchanged. Rather than go through this derivation, we leave it for an exercise to show that the result is given by

$$\frac{y^2}{a^2} + \frac{x^2}{b^2} = 1$$

This is the standard form for the equation of an ellipse centered at the origin with its major axis the y-axis and foci at $(0, \pm c)$. The constants a, b, and c obey the same properties as before, namely, $a^2 \geq b^2$ and $c^2 = a^2 - b^2$. Furthermore, the vertices occur at $(0, \pm a)$ and $(\pm b, 0)$.

We now have two basic equations for ellipses:

$$\text{Major axis: } x\text{-axis} \quad \frac{x^2}{a^2} + \frac{y^2}{b^2} = 1$$

$$\text{Major axis: } y\text{-axis} \quad \frac{y^2}{a^2} + \frac{x^2}{b^2} = 1$$

The distinguishing characteristic for each equation is where the larger of the two numbers a^2 and b^2 occurs. When the major axis is the x-axis, the larger number a^2 occurs under x^2. When the major axis is the y-axis, the larger number a^2 occurs under y^2.

Example 2

Find the major and minor axes, the vertices, foci, and graph $25x^2 + 9y^2 = 225$.

SOLUTION   We begin by putting the equation into standard form.

$$25x^2 + 9y^2 = 225 \quad \text{implies} \quad \frac{x^2}{9} + \frac{y^2}{25} = 1$$

Observe that a^2 is the larger number 25. We have

$$a^2 = 25 \qquad \text{implies} \quad a = 5$$
$$b^2 = 9 \qquad \text{implies} \quad b = 3$$
$$c^2 = a^2 - b^2 \quad \text{implies} \quad c^2 = 25 - 9 = 16, \text{ so } c = 4$$

It follows,

$$\text{Major axis: } \quad x = 0 \text{ (the } y\text{-axis)}$$
$$\text{Minor axis: } \quad y = 0 \text{ (the } x\text{-axis)}$$
$$\text{Vertices: } \quad (0, \pm 5) \quad (\pm 3, 0)$$
$$\text{Foci: } \quad (0, \pm 4)$$

The graph is shown in Figure 10.22. ■

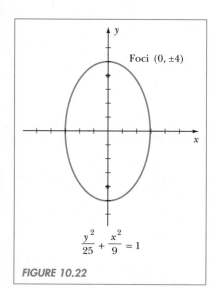

Foci $(0, \pm 4)$

$$\frac{y^2}{25} + \frac{x^2}{9} = 1$$

FIGURE 10.22

Let us summarize all our results for the ellipse.

Basic Equations for Ellipse

Standard Form	Parts	Graphs

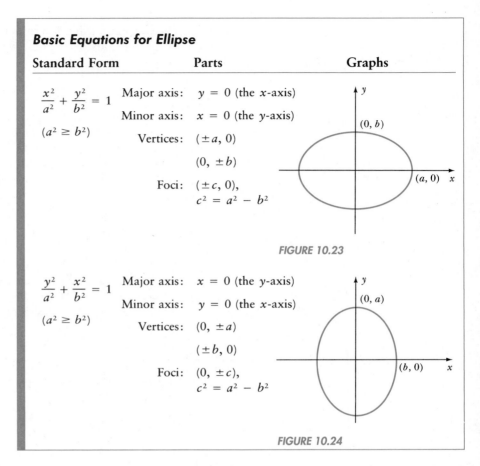

$$\frac{x^2}{a^2} + \frac{y^2}{b^2} = 1$$

$(a^2 \geq b^2)$

Major axis: $y = 0$ (the x-axis)

Minor axis: $x = 0$ (the y-axis)

Vertices: $(\pm a, 0)$

$(0, \pm b)$

Foci: $(\pm c, 0)$,

$c^2 = a^2 - b^2$

$(0, b)$

$(a, 0)$

FIGURE 10.23

$$\frac{y^2}{a^2} + \frac{x^2}{b^2} = 1$$

$(a^2 \geq b^2)$

Major axis: $x = 0$ (the y-axis)

Minor axis: $y = 0$ (the x-axis)

Vertices: $(0, \pm a)$

$(\pm b, 0)$

Foci: $(0, \pm c)$,

$c^2 = a^2 - b^2$

$(0, a)$

$(b, 0)$

FIGURE 10.24

Examples 3 and 4 illustrate how to find the equation of an ellipse given some of its parts.

Example 3

The foci of a certain ellipse are located at $(\pm 3, 0)$, and two of its vertices occur at $(0, \pm 1)$. Find the equation.

SOLUTION The fact that the foci are located at $(\pm 3, 0)$ tells us that this ellipse is centered at the origin with the x-axis as its major axis. (The foci are always located along the major axis.) Hence the equation must look like

$$\frac{x^2}{a^2} + \frac{y^2}{b^2} = 1$$

Since the vertices $(0, \pm 1)$ are on the minor axis, we know that $b = 1$. This also follows by substituting the solution point $(0, 1)$ into the equation.

Now we need to find a^2. We use the relationship $c^2 = a^2 - b^2$. Since the foci are at $(\pm 3, 0)$, we know that $c = 3$. Therefore

$$c^2 = a^2 - b^2 \quad \text{implies} \quad 3^2 = a^2 - 1^2, \text{ or } a^2 = 10$$

Thus the equation for the ellipse is

$$\frac{x^2}{10} + \frac{y^2}{1} = 1 \quad \blacksquare$$

Example 4

An ellipse has two vertices at $(\pm 8, 0)$, and its graph passes through the point $P(4, 6)$. Find the equation.

SOLUTION Because two of the vertices are at $(\pm 8, 0)$, we know that this ellipse is centered at the origin and its major axis is either the x- or y-axis. Therefore we have two choices for the equation:

$$(1) \ \frac{x^2}{a^2} + \frac{y^2}{b^2} = 1 \quad \text{or} \quad (2) \ \frac{x^2}{b^2} + \frac{y^2}{a^2} = 1$$

Let us assume for the moment that Equation (1) is the right choice. We need to find a^2 and b^2. First, use the fact that the vertex $(8, 0)$ must satisfy the equation.

$$\frac{8^2}{a^2} + \frac{0^2}{b^2} = 1 \quad \text{implies} \quad a^2 = 64$$

Hence the equation is $\frac{x^2}{64} + \frac{y^2}{b^2} = 1$. Now use the fact that $P(4, 6)$ must satisfy this equation.

$$\frac{4^2}{64} + \frac{6^2}{b^2} = 1 \quad \text{implies} \quad b^2 = 48$$

Therefore the equation for the ellipse is

$$\frac{x^2}{64} + \frac{y^2}{48} = 1$$

If we had assumed that Equation (2) was the correct choice instead of Equation (1), the analysis still would have yielded the same equation for the ellipse. You may prove this by carrying out the required computations. $\blacksquare$

The planets revolve around the sun in roughly elliptical orbits. (Johannes Kepler (1571–1630) first discovered this fact; Sir Isaac Newton (1642–1727) later established it as a consequence of his universal law of gravitation.) Similarly, satellites launched into earth orbit rotate along an elliptical path, with the center of the earth at one focus. We call the point nearest the earth in the orbit of the satellite the *perigee*, and the point farthest away from the earth the *apogee*. Although we shall not prove it here, these points occur at the vertices along the major axis of the ellipse.

Example 5

The distance from the center of the earth to the perigee of a satellite's orbit is $2R_E$ and the distance to the apogee is $7R_E$, where R_E represents the radius of the earth (approximately 4000 miles). Write an equation for the orbit of this satellite. How wide is the orbit along the minor axis?

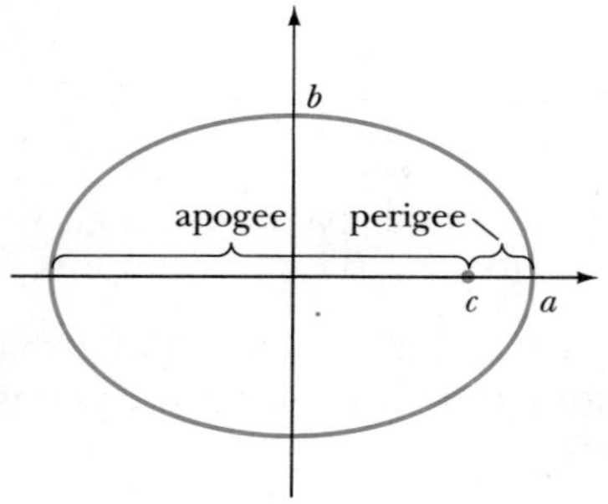

SOLUTION   As mentioned, the orbit of the satellite follows an elliptical path, so we introduce an x, y-coordinate system with the center of this ellipse at the origin. See Figure 10.25. Therefore an equation for the satellite's orbit is given by

$$\frac{x^2}{a^2} + \frac{y^2}{b^2} = 1$$

FIGURE 10.25

Next, we need to find a^2 and b^2. As shown in Figure 10.25, the sum of the distances to the perigee and apogee equals the length of the ellipse along the major axis. Since the distance from the center of the ellipse to either vertex along the major axis is a, we have

$$\left(\begin{array}{c}\text{Distance} \\ \text{to perigee}\end{array}\right) + \left(\begin{array}{c}\text{Distance} \\ \text{to apogee}\end{array}\right) = 2a$$

$$2R_E + 7R_E = 2a$$

$$9R_E = 2a$$

$$\frac{9R_E}{2} = a$$

Thus $a^2 = 81R_E^2/4$. Now we must find b^2.

Recall that the center of the earth is located at one of the foci for the ellipse. Hence we may denote the distance from the origin to the center of the earth by c. From Figure 10.25, it follows that

$$\left(\begin{array}{c}\text{Distance from} \\ \text{origin to center} \\ \text{of earth}\end{array}\right) + \left(\begin{array}{c}\text{Distance} \\ \text{to perigee}\end{array}\right) = a$$

$$c + 2R_E = \frac{9R_E}{2}$$

Thus $c = 9R_E/2 - 2R_E = 5R_E/2$. Finally,

$$c^2 = a^2 - b^2 \quad \text{implies} \quad b^2 = a^2 - c^2$$

$$= \left(\frac{9R_E}{2}\right)^2 - \left(\frac{5R_E}{2}\right)^2$$

$$= \frac{81R_E^2}{4} - \frac{25R_E^2}{4}$$

$$= 14R_E^2$$

Therefore the equation for the satellite's orbit is given by

$$\frac{x^2}{81R_E^2/4} + \frac{y^2}{14R_E^2} = 1$$

We note that the width of the orbit along the minor axis equals $2b$. Since $b^2 = 14R_E^2$, we have $b = \sqrt{14}R_E$. Thus

$$2b = 2(\sqrt{14}R_E) \approx 29933 \text{ miles} \quad \blacksquare$$

EXERCISES 10.2

In Exercises 1 to 6, find the major axis, minor axis, vertices, foci, and graph.

1. $\dfrac{x^2}{9} + \dfrac{y^2}{4} = 1$

2. $\dfrac{x^2}{25} + \dfrac{y^2}{16} = 1$

3. $12x^2 + 3y^2 - 3 = 0$

4. $4x^2 + 2y^2 - 1 = 0$

5. $x^2 + 9y^2 - 144 = 0$

6. $4 - 4x^2 - y^2 = 0$

In Exercises 7 and 8, graph the following special ellipses. Where are the foci?

7. $\dfrac{x^2}{4} + \dfrac{y^2}{4} = 1$ 8. $2x^2 + 2y^2 - 18 = 0$

In Exercises 9 and 10, the following equations are called *degenerate* cases of the ellipse. Describe the graph of each equation.

9. (a) $9x^2 + 4y^2 = 0$
   (b) $4x^2 + 9y^2 + 36 = 0$

10. (a) $16x^2 + 25y^2 + 400 = 0$
    (b) $25x^2 + 16y^2 = 0$

In Exercises 11 and 12, graph the given equation.

11. (a) $x = 2\sqrt{1 - y^2}$ (b) $y = -\dfrac{1}{2}\sqrt{4 - x^2}$

12. (a) $x = -\dfrac{1}{2}\sqrt{9 - y^2}$ (b) $y = \sqrt{9 - 4x^2}$

In Exercises 13 to 16, graph the set.

13. $\{(x, y): x^2 + 3y^2 < 12\}$

14. $\{(x, y): 4x^2 + y^2 > 100\}$

15. $\{(x, y): x^2 + 4y^2 \le 4 \text{ and } 4x^2 + y^2 \ge 4\}$

16. $\{(x, y): x^2 + 4y^2 \ge 16 \text{ and } 9x^2 + 4y^2 \le 144\}$

In Exercises 17 to 23, find an equation for the ellipse with the given properties.

17. Vertices at $(\pm 5, 0)$ and $(0, \pm 4)$

18. Foci at $(0, \pm 6)$, two vertices at $(0, \pm 8)$

19. Two vertices at $(0, \pm 10)$, graph passes through the point $P(2, 8)$

20. Two vertices at $(\pm 2, 0)$, graph passes through the point $P(1, 1)$

21. Foci at $(\pm 3, 0)$, two vertices at $(\pm 4, 0)$

22. Foci at $(\pm 1, 0)$, graph passes through the point $P(1, 1/\sqrt{2})$

23. Foci at $(0, \pm\sqrt{3})$, graph passes through the point $P(1, 2)$

24. Consider the collection of all points $P(x, y)$ in the plane that satisfy the following condition: the distance from P to the line $x = 4$ is twice the distance from P to $(1, 0)$. Find an equation for this set of points. Is this an ellipse? If it is, find the foci.

25. The ellipse $\dfrac{x^2}{9} + \dfrac{y^2}{4} = 1$ shown in Figure 10.26 has an inscribed rectangle whose sides are parallel with the x- and y-axes. Let $P(x, y)$ be the vertex of the rectangle in the first quadrant.
    (a) Find the area of the rectangle in terms of x.
    (b) Find the area of the rectangle in terms of y.
    (c) Find the perimeter of the rectangle in terms of x.
    (d) Find the perimeter of the rectangle in terms of y.

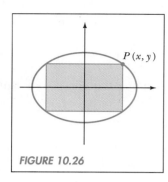

FIGURE 10.26

26. Repeat Exercise 25 for the ellipse $\dfrac{x^2}{a^2} + \dfrac{y^2}{b^2} = 1$.
(Assume a and b are positive constants.)

27. Figure 10.27 shows a bridge with an elliptical base. How high is the arch in the middle?

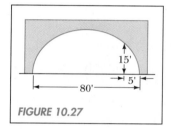

FIGURE 10.27

28. The distance from the center of the earth to the perigee of a satellite's orbit is $3R_E$ ($R_E \approx 4000$ miles), and the distance to the apogee is $4R_E$. Write an equation for the orbit of this satellite. Find the width of the orbit along the minor axis.

29. Repeat Exercise 28 if the distance to the perigee is $3R_E/2$ and the distance to the apogee is $2R_E$.

30. The width of a satellite's orbit along its major axis is $8R_E$ and the width along its minor axis is $5R_E$. Find an equation for the orbit. Where is the earth located?

31. Repeat Exercise 30 if the widths along the major and minor axes are $5R_E$ and $3R_E$, respectively.

32. A line perpendicular to the x-axis is drawn through the focus of the ellipse $x^2/25 + y^2/16 = 1$ with the positive x-coordinate. If this line intersects the ellipse at points A and B, find the distance AB.

33. Repeat Exercise 32 for the ellipse $x^2/a^2 + y^2/b^2 = 1$. (The answer should be in terms of a and b.)

34. A line is drawn from $(0, 1)$ through the focus of the ellipse $x^2/5 + y^2 = 1$ with the positive x-coordinate. Find where the line intersects the ellipse.

35. Repeat Exercise 34 for the ellipse $x^2/10 + y^2 = 1$.

36. A line is drawn from $(0, b)$ through the focus $(c, 0)$ of the ellipse $x^2/a^2 + y^2/b^2 = 1$. Show that this line also intersects the ellipse at (x_0, y_0), where $x_0 = 2a^2\sqrt{a^2 - b^2}/(2a^2 - b^2)$ and $y_0 = -b^3/(2a^2 - b^2)$.

37. Refer to Figure 10.19, which shows $P(x, y)$, $F_1(-c, 0)$, and $F_2(c, 0)$. Assume that $PF_1 + PF_2 = 2a$, and a and c are positive constants.
(a) By considering triangle PF_1F_2, show that $2a > 2c$. (Use the fact that the sum of the lengths of two sides of a triangle is greater than the length of the third side.)
(b) Use the result in part (a) to show that $a^2 - c^2 > 0$.

38. Consider the set of points $\left\{(x, y): \dfrac{x^2}{a^2} + \dfrac{y^2}{b^2} = 1\right\}$, where a and b are positive constants. Prove (a) $|x| \le a$, (b) $|y| \le b$.

10.3 HYPERBOLAS

Geometric Definition of Hyperbola
A **hyperbola** consists of the set of all points $P(x, y)$ in the plane, the difference of whose distances from two fixed points, called the foci, is a constant.

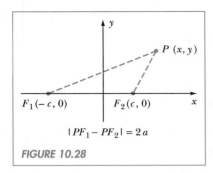

$P(x, y)$

$F_1(-c, 0)$ $F_2(c, 0)$ x

$|PF_1 - PF_2| = 2a$

FIGURE 10.28

Let us derive an equation for the hyperbola from the geometric definition. As usual, we begin by considering the relatively simple case where the foci are on the x-axis at $F_1(-c, 0)$ and $F_2(c, 0)$, c a positive constant. See Figure 10.28.

If $P(x, y)$ is any point on the hyperbola, then the definition says that the difference between the two distances PF_1 and PF_2 must be a constant. We shall let $2a$ represent this constant, where a is positive. Since we do not know which distance is larger, PF_1 or PF_2, we measure their difference with the absolute value,

$$\text{Difference of distances } PF_1 \text{ and } PF_2 = |PF_1 - PF_2|$$

Therefore, by the definition of a hyperbola, we have

$$|PF_1 - PF_2| = 2a$$

Using the distance formula, this says that

$$\left|\sqrt{(x + c)^2 + y^2} - \sqrt{(x - c)^2 + y^2}\right| = 2a$$

After squaring this equation once, then simplifying, then squaring again and simplifying, we obtain

$$\frac{x^2}{a^2} - \frac{y^2}{c^2 - a^2} = 1$$

The details of this derivation are outlined in the exercises. We also leave for an exercise the proof that $c^2 - a^2$ is positive. Assuming this fact, we may let $b^2 = c^2 - a^2$; then our equation becomes

$$\frac{x^2}{a^2} - \frac{y^2}{b^2} = 1$$

This is the *standard form* for the equation of a hyperbola centered at the origin with foci at $(\pm c, 0)$, where $c^2 = a^2 + b^2$. It is important to note that a^2 may be smaller than, equal to, or larger than b^2.

Observe that the equation remains unchanged if x is replaced with $-x$ or y is replaced with $-y$. Therefore the graph of this hyperbola is symmetric with respect to the x- and y-axes. The axis containing the foci (in this case, the x-axis) is called the **major axis**. The other axis of symmetry is called the **minor axis**. The **center** of the hyperbola occurs at the intersection of the two axes. By substituting $y = 0$ into the equation, we find that the curve intersects the major axis at $(\pm a, 0)$. We call these points the **vertices** of the hyperbola. Figure 10.29 shows the position of the axes, center, vertices, and foci. Notice that the foci are farther from the center than the vertices.

Let us determine what the graph of the hyperbola looks like. First, there is *no* graph between the vertical lines $x = -a$ and $x = a$. We can see this by solving the equation for x.

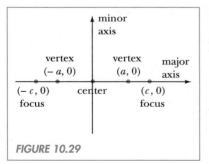

minor axis

vertex $(-a, 0)$ vertex $(a, 0)$ major axis

$(-c, 0)$ focus   center $(c, 0)$ focus

FIGURE 10.29

$$\frac{x^2}{a^2} - \frac{y^2}{b^2} = 1 \quad \text{implies} \quad x^2 = a^2\left(1 + \frac{y^2}{b^2}\right)$$

Notice that $1 + \dfrac{y^2}{b^2} \geq 1$ for all y. Hence

$$x^2 = a^2\left(1 + \frac{y^2}{b^2}\right) \quad \text{implies} \quad x^2 \geq a^2$$

$$\text{implies} \quad |x| \geq a$$

$$\text{implies} \quad x \leq -a \quad \text{or} \quad a \leq x$$

Thus the graph of the hyperbola splits into two branches, one to the left of $x = -a$ and the other to the right of $x = a$.

Now, let us determine what the graph of the hyperbola will look like far to the right or left of the origin when $|x|$ is large. First, we solve the equation for y.

$$\frac{x^2}{a^2} - \frac{y^2}{b^2} = 1 \quad \text{implies} \quad y^2 = \frac{b^2}{a^2}x^2 - b^2$$

$$\text{implies} \quad y = \pm\sqrt{\frac{b^2}{a^2}x^2 - b^2}$$

If $|x|$ is large, then $\dfrac{b^2}{a^2}x^2 - b^2 \approx \dfrac{b^2}{a^2}x^2$. (When $|x|$ is large, a polynomial in x behaves like its highest-degree term.) Thus

$$|x|\,\text{large} \quad \text{implies} \quad y \approx \pm\sqrt{\frac{b^2}{a^2}x^2} \quad \text{or} \quad y \approx \pm\frac{b}{a}x$$

We conclude that the hyperbola will look like the lines $y = \pm bx/a$ when $|x|$ is large. In fact, the distance between the graph of the hyperbola and the lines $y = \pm bx/a$ will approach zero as we move farther and farther away from the origin. We outline a proof of this fact in the exercises. Therefore we say that $y = \pm bx/a$ are **asymptotes** for the hyperbola.

As a result of the above arguments, the graph of the hyperbola $x^2/a^2 - y^2/b^2 = 1$ may be found by locating its vertices, $(\pm a, 0)$, and asymptotes, $y = \pm bx/a$, and then drawing a smooth curve from each vertex branching out toward an asymptote, getting closer to the asymptote as we move farther away from the origin. This gives us a graph such as the one in Figure 10.30.

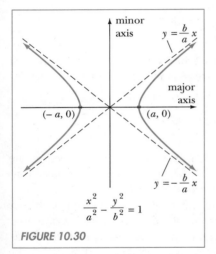

FIGURE 10.30

Example 1

Find the major and minor axes, vertices, foci, asymptotes, and graph of $x^2 - 4y^2 - 4 = 0$.

SOLUTION First, we put the equation into standard form.

$$\frac{x^2}{4} - \frac{y^2}{1} = 1$$

We recognize this as a hyperbola of the form $x^2/a^2 - y^2/b^2 = 1$. It follows that the major axis is the x-axis and the minor axis is the y-axis. We also note that

$$a^2 = 4 \qquad \text{implies} \quad a = 2$$
$$b^2 = 1 \qquad \text{implies} \quad b = 1$$
$$c^2 = a^2 + b^2 \quad \text{implies} \quad c^2 = 4 + 1 = 5, \text{ so } c = \sqrt{5}$$

Therefore we have

$$\text{Major axis:} \quad y = 0 \text{ (the } x\text{-axis)}$$
$$\text{Minor axis:} \quad x = 0 \text{ (the } y\text{-axis)}$$
$$\text{Vertices:} \quad (\pm 2, 0)$$
$$\text{Foci:} \quad (\pm \sqrt{5}, 0)$$

Now, we must find the asymptotes. Instead of using a formula, we just solve the equation for y and then approximate for $|x|$ large.

$$\frac{x^2}{4} - \frac{y^2}{1} = 1 \quad \text{implies} \quad y^2 = \frac{x^2}{4} - 1$$

$$\text{implies} \quad y = \pm \sqrt{\frac{x^2}{4} - 1}$$

$$\text{implies} \quad y \approx \pm \frac{x}{2} \quad \text{when } |x| \text{ is large}$$

Therefore the asymptotes are $y = \pm \dfrac{1}{2} x$.

Figure 10.31 shows the parts of the hyperbola; the graph is drawn in Figure 10.32. Notice how the asymptotes serve as guidelines for drawing the curve.

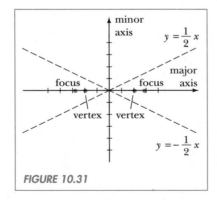

FIGURE 10.31

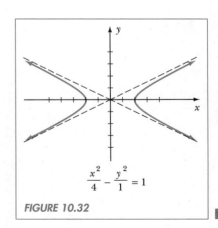

$$\frac{x^2}{4} - \frac{y^2}{1} = 1$$

FIGURE 10.32

Suppose we choose the foci of a hyperbola to be on the y-axis at $(0, \pm c)$ instead of the x-axis. This just interchanges the roles of x and y in all our previous work. It follows that

$$\frac{y^2}{a^2} - \frac{x^2}{b^2} = 1$$

represents the equation in standard form for the hyperbola centered at the origin with major axis the y-axis.

Following is a summary of all our results.

Basic Equations for Hyperbola

Standard Form	Parts	Graph
$\dfrac{x^2}{a^2} - \dfrac{y^2}{b^2} = 1$	Major axis: $y = 0$ (the x-axis) Minor axis: $x = 0$ (the y-axis) Vertices: $(\pm a, 0)$ Foci: $(\pm c, 0)$, $c^2 = a^2 + b^2$ Asymptotes: $y = \pm \dfrac{b}{a}x$	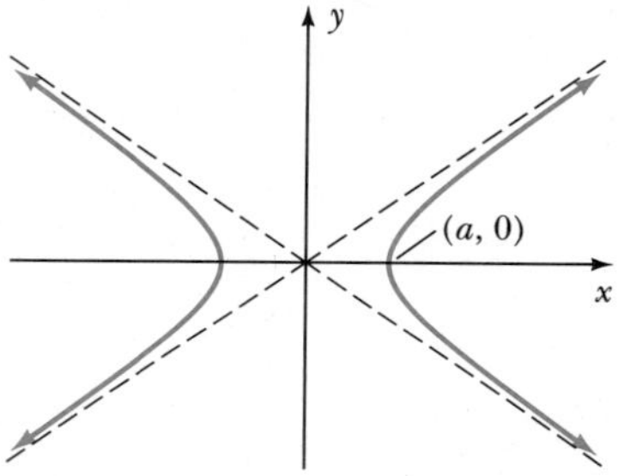 FIGURE 10.33
$\dfrac{y^2}{a^2} - \dfrac{x^2}{b^2} = 1$	Major axis: $x = 0$ (the y-axis) Minor axis: $y = 0$ (the x-axis) Vertices: $(0, \pm a)$ Foci: $(0, \pm c)$, $c^2 = a^2 + b^2$ Asymptotes: $y = \pm \dfrac{a}{b}x$	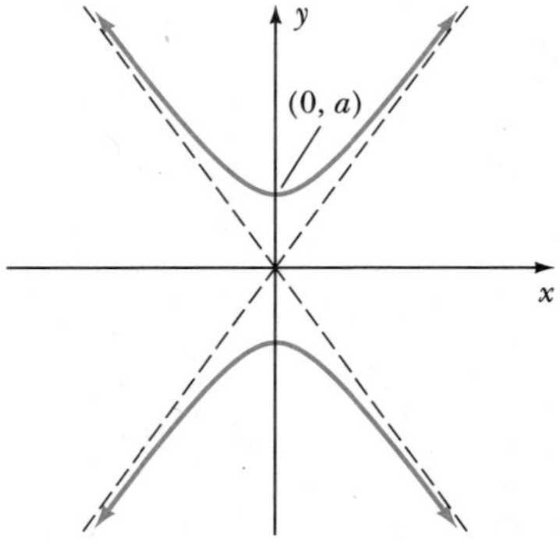 FIGURE 10.34

Remarks

1. The relationship between a, b, and c says that $a^2 + b^2 = c^2$. This is different from the ellipse.

2. The value of a^2 may be less than, equal to, or greater than b^2.

3. When the equation is in standard form, the positive term determines the major axis:

   x^2 term positive implies x-axis is major.
   y^2 term positive implies y-axis is major.

The asymptotes do not contain points on the hyperbola, so we draw them with dashed lines. A quick geometric procedure for drawing the asymptotes without referring to their equations is described below.

Drawing the Asymptotes for a Hyperbola
Locate two points P_1 and P_2 as follows. From the center of the hyperbola, move out along the major axis a units (to a vertex) and then move parallel to the minor axis $+b$ units for P_1 and $-b$ units for P_2. Draw dashed lines through the center and P_1 and through the center and P_2 for the asymptotes (Figures 10.35 and 10.36).

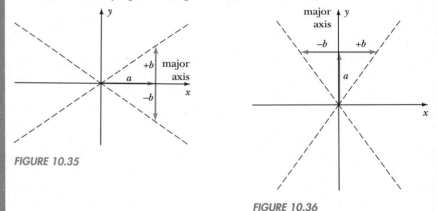

FIGURE 10.35

FIGURE 10.36

Example 2

Find the major and minor axes, vertices, foci, asymptotes, and graph of $25y^2 - 9x^2 = 225$.

SOLUTION First, we divide both sides of the equation by 225 to put it in standard form.

$$\frac{y^2}{9} - \frac{x^2}{25} = 1$$

We recognize this as a hyperbola of the form $y^2/a^2 - x^2/b^2 = 1$. It follows that the major axis is the y-axis and the minor axis the x-axis. We also note that

$$a^2 = 9 \qquad \text{implies} \quad a = 3$$
$$b^2 = 25 \qquad \text{implies} \quad b = 5$$
$$c^2 = a^2 + b^2 \quad \text{implies} \quad c^2 = 9 + 25 = 34, \text{ so } c = \sqrt{34}$$

Therefore we have

$$\text{Major axis:} \quad x = 0 \text{ (the } y\text{-axis)}$$
$$\text{Minor axis:} \quad y = 0 \text{ (the } x\text{-axis)}$$
$$\text{Vertices:} \quad (0, \pm 3)$$
$$\text{Foci:} \quad (0, \pm \sqrt{34})$$

Next, we find the asymptotes by solving the equation for y and approximating for $|x|$ large.

$$\frac{y^2}{9} - \frac{x^2}{25} = 1 \quad \text{implies} \quad y^2 = \frac{9}{25}x^2 + 9$$

$$\text{implies} \quad y = \pm \sqrt{\frac{9}{25}x^2 + 9}$$

$$\text{implies} \quad y \approx \pm \frac{3}{5}x \quad \text{when } |x| \text{ is large}$$

Therefore the asymptotes are $y = \pm \dfrac{3}{5}x$. The graph is shown in Figure 10.37.

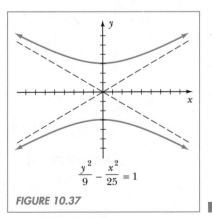

$$\frac{y^2}{9} - \frac{x^2}{25} = 1$$

FIGURE 10.37

Examples 3 and 4 illustrate how we may determine the equation of a hyperbola from some of its parts.

Example 3

A hyperbola has vertices at $(\pm 6, 0)$, and its graph also passes through the point $P(8, 7)$. Find an equation for this hyperbola.

SOLUTION Since the vertices are at $(\pm 6, 0)$, the hyperbola is centered at the origin, with major axis the x-axis. Therefore the equation in standard form is given by

$$\frac{x^2}{a^2} - \frac{y^2}{b^2} = 1$$

Now, a is the distance from the center to either vertex, so $a = 6$. Thus

$$\frac{x^2}{36} - \frac{y^2}{b^2} = 1$$

To find b^2, we use the fact that $P(8, 7)$ is on the curve. This means that $(8, 7)$ must satisfy the equation. Hence

$$\frac{8^2}{36} - \frac{7^2}{b^2} = 1 \quad \text{implies} \quad \frac{64}{36} - 1 = \frac{49}{b^2}$$

$$\frac{28}{36} = \frac{49}{b^2}$$

$$b^2 = \frac{(49)(36)}{28} = 63$$

Thus the desired equation is

$$\frac{x^2}{36} - \frac{y^2}{63} = 1 \quad \blacksquare$$

Example 4

Find an equation for the hyperbola with foci at $(0, \pm 5\sqrt{3})$ and asymptotes $y = \pm \frac{4}{3}x$.

SOLUTION Since the foci are at $(0, \pm 5\sqrt{3})$, the hyperbola is centered at the origin with major axis the y-axis. Therefore the equation is given by

$$\frac{y^2}{a^2} - \frac{x^2}{b^2} = 1$$

We know that c, the distance from the center to the foci, has value $5\sqrt{3}$. Thus

$$a^2 + b^2 = c^2 \text{ and } c = 5\sqrt{3} \quad \text{implies} \quad a^2 + b^2 = 75$$

This gives us one equation for a and b.

Next we use the information concerning the asymptotes. Recall that the equations for the asymptotes of $y^2/a^2 - x^2/b^2 = 1$ are $y = \pm ax/b$. (If you have forgotten, you can always derive the equations by solving the equation of the hyperbola for y and approximating for $|x|$ large.) However, we were given the equations for the asymptotes, namely, $y = \pm 4x/3$. It follows that $a/b = 4/3$.

Now we have a system of two equations for a and b:

$$a^2 + b^2 = 75 \tag{1}$$

$$\frac{a}{b} = \frac{4}{3} \tag{2}$$

From Equation (2), $a = 4b/3$. Substituting this into Equation (1),

$$\left(\frac{4}{3}b\right)^2 + b^2 = 75$$

$$\frac{25}{9}b^2 = 75$$

$$b^2 = 27$$

Returning to Equation (1), we have $a^2 + 27 = 75$, so $a^2 = 48$. Therefore the equation for the hyperbola is

$$\frac{y^2}{48} - \frac{x^2}{27} = 1 \quad \blacksquare$$

Warning

The fact that $\dfrac{a}{b} = \dfrac{4}{3}$ does *not* imply $a = 4$ and $b = 3$!

EXERCISES 10.3

In Exercises 1 to 6, find the major and minor axes, vertices, foci, asymptotes, and graph.

1. $\dfrac{x^2}{36} - \dfrac{y^2}{64} = 1$ 2. $\dfrac{x^2}{8} - \dfrac{y^2}{9} = -1$

3. $6 - 10x^2 + 5y^2 = 16$ 4. $16x^2 - 9y^2 - 1 = 0$

5. $4x^2 = y^2 + 1$ 6. $y^2 = x^2 + 1$

In Exercises 7 to 10, graph the given equation. Determine any asymptotes.

7. (a) $x = \dfrac{3}{2}\sqrt{y^2 + 4}$ (b) $x = -\dfrac{3}{2}\sqrt{y^2 + 4}$

8. (a) $y = 2\sqrt{x^2 + 4}$ (b) $y = -2\sqrt{x^2 + 4}$

9. (a) $y = \sqrt{x^2 - 1}$ (b) $y = -\sqrt{x^2 - 1}$

10. (a) $x = \sqrt{y^2 - 9}$ (b) $x = -\sqrt{y^2 - 9}$

In Exercises 11 to 16, graph the set.

11. $\{(x, y): x^2 - 4y^2 = 0\}$ 12. $\{(x, y): x^2 - y^2 \geq 1\}$

13. $\{(x, y): x^2 - y^2 < 4\}$

14. $\{(x, y): x^2 - 4y^2 + 36 < 0\}$

15. $\{(x, y): y^2 - 4x^2 \geq 1 \text{ and } y^2 - x^2 \leq 4\}$

16. $\{(x, y): x^2 - y^2 \geq 25 \text{ and } x^2 - 4y^2 \leq 1\}$

In Exercises 17 to 26, find an equation for the hyperbola with the given properties.

17. Vertices at $(\pm 1, 0)$, graph passes through the point $P(4, 8)$

18. Vertices at $(\pm 2, 0)$, foci at $(\pm 3, 0)$

19. Vertices at $(0, \pm 4)$, foci at $(0, \pm 5)$

20. Vertices at $(0, \pm 10)$, asymptotes $y = \pm 2x$

21. Vertices at $(\pm 3, 0)$, asymptotes $y = \pm x/2$

22. Foci at $(\pm \sqrt{13}, 0)$, graph passes through the point $P(5, 8/3)$

23. Foci at $(0, \pm 5)$, graph passes through the point $P(3, 4\sqrt{2})$.

24. One focus at $(0, 4\sqrt{5})$, asymptotes $y = \pm x/3$.

25. One focus at $(-2\sqrt{29}, 0)$, asymptotes $y = \pm 2x/5$.

26. Asymptotes $y = \pm 3x/2$, graph passes through the point $P(3, -2)$.

27. Consider the collection of all points $P(x, y)$ in the plane that satisfy the following condition: the distance from P to the line $x = 1$ is one-half the distance from P to $(4, 0)$. Find an equation for this set of points. Is this a hyperbola? If it is, find the foci.

28. Figure 10.38 is the hyperbola $\dfrac{x^2}{4} - \dfrac{y^2}{1} = 1$ with a rectangle whose sides are parallel to the axes. The vertices of the rectangle are on the hyperbola. Let $P(x, y)$ be the coordinates of the vertex in the first quadrant.

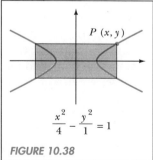

$$\frac{x^2}{4} - \frac{y^2}{1} = 1$$

FIGURE 10.38

(a) Find the area of the rectangle in terms of x.
(b) Find the area of the rectangle in terms of y.
(c) Find the perimeter of the rectangle in terms of x.
(d) Find the perimeter of the rectangle in terms of y.

29. Repeat Exercise 28 for the hyperbola $\dfrac{x^2}{a^2} - \dfrac{y^2}{b^2} = 1$. (Answers will be in terms of a and b.)

30. Consider the equation $\dfrac{y^2}{a^2} - \dfrac{x^2}{b^2} = 1$. Assume that a and b are positive. Show that $|y| \geq a$. What are the possible values for x?

31. This exercise outlines some of the algebra details omitted from the text for deriving a basic equation for the hyperbola. Begin with the equation

$$\left| \sqrt{(x + c)^2 + y^2} - \sqrt{(x - c)^2 + y^2} \right| = 2a$$

(a) Square both sides and simplify, showing that the equation becomes

$$x^2 + y^2 + c^2 - 2a^2$$
$$= \sqrt{(x^2 - c^2)^2 + 2x^2 y^2 + 2y^2 c^2 + y^4}$$

(b) Square both sides of the equation in part (a), then simplify, showing that the equation becomes

$$(c^2 - a^2)x^2 - a^2 y^2 = a^2(c^2 - a^2)$$

(c) Divide the equation in part (b) by $a^2(c^2 - a^2)$, obtaining

$$\frac{x^2}{a^2} - \frac{y^2}{c^2 - a^2} = 1$$

32. Refer to Figure 10.28, where $F_1(-c, 0)$, $F_2(c, 0)$, and $|PF_1 - PF_2| = 2a$, where a and c are positive constants.
(a) Using the fact that the sum of two sides of a triangle is greater than its third side, show that $F_1 F_2 > PF_1 - PF_2$ and $F_1 F_2 > PF_2 - PF_1$.
(b) Use the result in part (a) to show that $F_1 F_2 > |PF_1 - PF_2|$.
(c) Deduce from the result in part (b) that $c > a$.
(d) Use the result of part (c) to show that $c^2 - a^2 > 0$.

33. This exercise establishes the fact that $y = \pm bx/a$ are asymptotes for the hyperbola $x^2/a^2 - y^2/b^2 = 1$. Assume that a and b are positive.
(a) Show that the part of the hyperbola above the x-axis has equation $y = \dfrac{b}{a}\sqrt{x^2 - a^2}$.
(b) Let $P_1(x, y_1)$ be a point on $y = bx/a$ and $P_2(x, y_2)$ be a point on the upper half of the hyperbola. P_1 and P_2 both have the same x-coordinate, x.

Show that

$$y_1 - y_2 = \frac{b}{a}(x - \sqrt{x^2 - a^2})$$

(c) Multiplying the numerator and denominator of the expression for $y_1 - y_2$ given in part (b) by $x + \sqrt{x^2 - a^2}$, show that

$$y_1 - y_2 = \frac{ab}{x + \sqrt{x^2 - a^2}}$$

(d) Use part (c) to explain why $\lim_{x \to +\infty} (y_1 - y_2) = 0$.

(e) Part (d) says that the graph of the hyperbola gets arbitrarily close to $y = bx/a$ as $x \to +\infty$. Use the symmetry of the hyperbola to show that the graph also approaches $y = -bx/a$ in quadrants II and IV and $y = bx/a$ in quadrant III.

10.4 SHIFTED SECOND-DEGREE CURVES

In this section we consider parabolas, ellipses, and hyperbolas that are shifted away from the origin so that the major axis remains parallel to the x- or y-axis. From our work in Chapter 1, we know that replacing x with $(x - h)$ and y with $(y - k)$ in an x, y equation will shift its graph h units in the x-direction and k units in the y-direction. Following are several examples of this fact:

Unshifted Equation		Shifted Equation
$y^2 = 4px$	$\longrightarrow$	$(y - k)^2 = 4p(x - h)$
$\dfrac{x^2}{a^2} + \dfrac{y^2}{b^2} = 1$	$\longrightarrow$	$\dfrac{(x - h)^2}{a^2} + \dfrac{(y - k)^2}{b^2} = 1$
$\dfrac{x^2}{a^2} - \dfrac{y^2}{b^2} = 1$	$\longrightarrow$	$\dfrac{(x - h)^2}{a^2} - \dfrac{(y - k)^2}{b^2} = 1$

Column 1 contains standard equations for a parabola, ellipse, and hyperbola, each one centered at the origin with its major axis the x-axis ($y = 0$). In column 2, the curve is shifted, with its center at (h, k) and major axis $y = k$.

To analyze a shifted second-degree curve, we first determine the parts of its unshifted counterpart and then shift everything. The process of completing the square is the key to finding the unshifted equation.

Example 1

Analyze and graph $y^2 - 2x + 6y + 11 = 0$.

SOLUTION   First, notice that the equation has a y^2 term but no x^2 term, which reminds us of the basic parabola $y^2 = 4px$. Indeed, we show that our equation is a shift of this curve by completing the square on the y-terms.

$$y^2 - 2x + 6y + 11 = 0$$
$$y^2 + 6y = 2x - 11$$
$$y^2 + 6y + 9 = 2x - 11 + 9$$
$$(y + 3)^2 = 2(x - 1)$$

We recognize this last equation as a shift by $+1$ unit in the x-direction and -3 units in the y-direction of the basic parabola $y^2 = 2x$. Our next step is to find the parts (vertex, axis, focus, directrix) of the unshifted curve $y^2 = 2x$ and then shift them to obtain the parts for $(y + 3)^2 = 2(x - 1)$. We have

$$y^2 = 2x \quad \text{implies} \quad 4p = 2, \text{ so } p = \frac{1}{2}$$

This gives us the following results:

Unshifted Equation		Shifted Equation
$y^2 = 2x$	$\longrightarrow$	$(y + 3)^2 = 2(x - 1)$
$(0, 0)$	vertex	$(1, -3)$
$y = 0$	axis	$y = -3$
$\left(\dfrac{1}{2}, 0\right)$	focus	$\left(\dfrac{3}{2}, -3\right)$
$x = \dfrac{-1}{2}$	directrix	$x = \dfrac{1}{2}$

We add to this information the x- and y-intercepts, which are important points for graphing. These are most easily computed from the original equation, $y^2 - 2x + 6y + 11 = 0$.

x-intercepts	y-intercepts
Setting $y = 0$, we have	Setting $x = 0$, we have
$-2x + 11 = 0$	$y^2 + 6y + 11 = 0$
$x = \dfrac{11}{2}$	$y = \dfrac{-6 \pm \sqrt{36 - 44}}{2}$
	No real solutions.

Thus we have an x-intercept at $(11/2, 0)$ and no y-intercept.

Finally, we graph the curve in Figure 10.39. Notice that the point $(11/2, 0)$ has a symmetric partner with respect to the axis of the parabola at $(11/2, -6)$. ∎

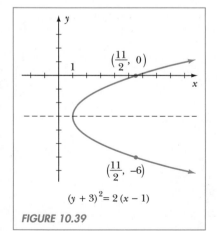

$(y + 3)^2 = 2(x - 1)$

FIGURE 10.39

Example 2

Analyze and graph $3x^2 + y^2 + 12x - 6y + 9 = 0$.

SOLUTION   Notice that because the equation contains an x^2 term and a y^2 term, both positive, we suspect that the equation represents an ellipse. We shall complete the square to show that this indeed is the case.

$$3x^2 + y^2 + 12x - 6y + 9 = 0$$

$$3x^2 + 12x \quad + y^2 - 6y \quad = -9$$

$$3(x^2 + 4x \quad) + y^2 - 6y \quad = -9$$

$$3(x^2 + 4x + 4) + y^2 - 6y + 9 = -9 + 12 + 9$$

$$3(x + 2)^2 + (y - 3)^2 = 12$$

$$\frac{(x + 2)^2}{4} + \frac{(y - 3)^2}{12} = 1$$

$$\frac{(y - 3)^2}{12} + \frac{(x + 2)^2}{4} = 1$$

This is a shift by -2 units in the x-direction and $+3$ units in the y-direction of the basic ellipse $\dfrac{y^2}{a^2} + \dfrac{x^2}{b^2} = 1$, where $a^2 = 12$ and $b^2 = 4$. We have

$$a^2 = 12 \qquad \text{implies} \qquad a = 2\sqrt{3}$$

$$b^2 = 4 \qquad \text{implies} \qquad b = 2$$

$$c^2 = a^2 - b^2 \quad \text{implies} \quad c^2 = 12 - 4 = 8, \text{ so } c = 2\sqrt{2}$$

Now, we find the parts for the unshifted ellipse and then shift.

Unshifted Equation		Shifted Equation
$\dfrac{y^2}{12} + \dfrac{x^2}{4} = 1$	$\longrightarrow$	$\dfrac{(y - 3)^2}{12} + \dfrac{(x + 2)^2}{4} = 1$
$(0, 0)$	center	$(-2, 3)$
$x = 0$	major axis	$x = -2$
$y = 0$	minor axis	$y = 3$
$(0, \pm 2\sqrt{3})$	vertices	$(-2, 3 \pm 2\sqrt{3})$
$(\pm 2, 0)$		$(-2 \pm 2, 3)$
$(0, \pm 2\sqrt{2})$	foci	$(-2, 3 \pm 2\sqrt{2})$

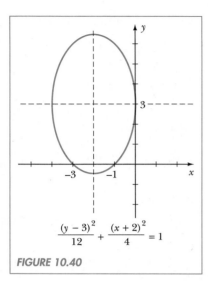

$$\frac{(y-3)^2}{12} + \frac{(x+2)^2}{4} = 1$$

FIGURE 10.40

We add to this information the x- and y-intercepts. Using the original equation $3x^2 + y^2 + 12x - 6y + 9 = 0$, we find

x-intercepts	y-intercepts
$3x^2 + 12x + 9 = 0$	$y^2 - 6y + 9 = 0$
$x^2 + 4x + 3 = 0$	$(y - 3)^2 = 0$
$(x + 3)(x + 1) = 0$	$y = 3$
$x = -3$ or $x = -1$	

Thus we have x-intercepts at $(-3, 0)$ and $(-1, 0)$ and y-intercept at $(0, 3)$. The graph is shown in Figure 10.40. ∎

Example 3

Analyze and graph $x^2 - 4y^2 - 4x + 8y - 16 = 0$.

SOLUTION   As usual, we begin by completing the square so as to write the equation in a familiar form.

$$x^2 - 4y^2 - 4x + 8y - 16 = 0$$

$$x^2 - 4x \quad\quad - 4y^2 + 8y \quad\quad = 16$$

$$x^2 - 4x \quad\quad - 4(y^2 - 2y \quad\ ) = 16$$

$$x^2 - 4x + 4 - 4(y^2 - 2y + 1) = 16 + 4 - 4$$

$$(x - 2)^2 - 4(y - 1)^2 = 16$$

$$\frac{(x - 2)^2}{16} - \frac{(y - 1)^2}{4} = 1$$

We recognize this as a shift by $+2$ units in the x-direction and $+1$ unit in the y-direction of the basic hyperbola $\dfrac{x^2}{a^2} - \dfrac{y^2}{b^2} = 1$, where $a^2 = 16$ and $b^2 = 4$. Thus

$$a^2 = 16 \quad\quad \text{implies} \quad a = 4$$

$$b^2 = 4 \quad\quad \text{implies} \quad b = 2$$

$$c^2 = a^2 + b^2 \quad \text{implies} \quad c^2 = 16 + 4 = 20, \text{ so } c = 2\sqrt{5}$$

Now, we find the parts for the unshifted hyperbola and then shift.

10.4 / Shifted Second-Degree Curves   ▪   **527**

Unshifted Equation		Shifted Equation
$\dfrac{x^2}{16} - \dfrac{y^2}{4} = 1$	$\longrightarrow$	$\dfrac{(x-2)^2}{16} - \dfrac{(y-1)^2}{4} = 1$
$(0, 0)$	center	$(2, 1)$
$y = 0$	major axis	$y = 1$
$x = 0$	minor axis	$x = 2$
$(\pm 4, 0)$	vertices	$(2 \pm 4, 1)$
$(\pm 2\sqrt{5}, 0)$	foci	$(2 \pm 2\sqrt{5}, 1)$
$y = \pm\dfrac{1}{2}x$	asymptotes	$y - 1 = \pm\dfrac{1}{2}(x - 2)$

We add to this information the x- and y-intercepts. Using the original equation $x^2 - 4y^2 - 4x + 8y - 16 = 0$, we find

x-intercepts	y-intercepts
$x^2 - 4x - 16 = 0$	$-4y^2 + 8y - 16 = 0$
$x = \dfrac{4 \pm \sqrt{16 + 4(16)}}{2}$	$y^2 - 2y + 4 = 0$
$x = 2 \pm 2\sqrt{5}$	$y = \dfrac{2 \pm \sqrt{4 - 16}}{2}$
	No real solution

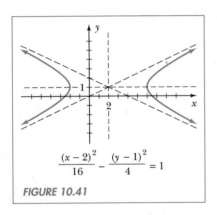

$$\frac{(x-2)^2}{16} - \frac{(y-1)^2}{4} = 1$$

FIGURE 10.41

Thus we have x-intercepts at $(2 + 2\sqrt{5}, 0)$ and $(2 - 2\sqrt{5}, 0)$ and no y-intercepts.

The graph is shown in Figure 10.41. ∎

Examples 1 to 3 illustrate how the process of completing the square enables us to rewrite the equation

$$Ax^2 + Cy^2 + Dx + Ey + F = 0$$

as a shift of one of the basic equations for the parabola, ellipse, or hyperbola. Except for special cases, called *degenerate cases*, the equation $Ax^2 + Cy^2 + Dx + Ey + F = 0$ must represent a parabola, ellipse, or hyperbola. Which curve the equation represents depends on the value of the constants A, C, D, E, and F. A degenerate case occurs when the choices for the constants cause the equation to reduce to a line, a pair of lines, a point, or the empty set. For example,

A line:	$x + y + 1 = 0$	$A = C = 0, D = E = F = 1$
Two lines:	$x^2 - 4 = 0$	$A = 1, F = -4, C = D = E = 0$
A point:	$x^2 + y^2 = 0$	$A = C = 1, D = E = F = 0$
Empty set:	$x^2 + y^2 + 1 = 0$	$A = C = F = 1, D = E = 0$

The degenerate cases will not be of primary importance. Having said this, we are now ready for the following theorem, which tells us how to classify the equation

$$Ax^2 + Cy^2 + Dx + Ey + F = 0$$

Classification Theorem

Except for degenerate cases, the equation $Ax^2 + Cy^2 + Dx + Ey + F = 0$ represents

1. A parabola if A or C is zero (but not both).
2. An ellipse if A and C have the same sign.
3. A hyperbola if A and C have opposite signs.

We shall omit the proof of this theorem, which requires completing the square in each situation.

Example 4

By observation, determine what type of curve the given equation represents, assuming it is not a degenerate case.
(a) $4x - 5y^2 - x^2 + 3 = 0$
(b) $4x - 5y^2 + y - 1 = 0$
(c) $12 - 6x + 5x^2 = 4y + y^2$

SOLUTION We rewrite each equation in the form $Ax^2 + Cy^2 + Dx + Ey + F = 0$ and apply the Classification Theorem.

Equation	Type of Curve	Reason
(a) $-x^2 - 5y^2 + 4x + 3 = 0$	Ellipse	$A = -1$ and $C = -5$ have the same sign.
(b) $-5y^2 + 4x + y - 1 = 0$	Parabola	$A = 0$.
(c) $5x^2 - y^2 - 6x - 4y + 12 = 0$	Hyperbola	$A = 5$ and $C = -1$ have opposite signs.

Example 5 illustrates how we find an equation for a shifted parabola, ellipse, or hyperbola given certain information about its parts.

Example 5

Find an equation for the ellipse with vertices at $(-6, 3)$, $(2, 3)$, $(-2, 5)$, and $(-2, 1)$.

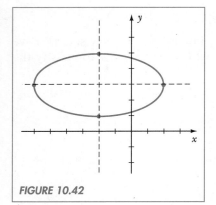

FIGURE 10.42

SOLUTION Figure 10.42 is a sketch of the ellipse. Since the curve is elongated in the x-direction, we know that the major axis is parallel to the x-axis. Therefore the basic unshifted equation must be

$$\frac{x^2}{a^2} + \frac{y^2}{b^2} = 1$$

As we can see from the graph, $a = 4$ and $b = 2$. Furthermore, the center of the ellipse is at $(-2, 3)$, which tells us that we must shift the basic equation -2 units in the x-direction and $+3$ units in the y-direction. Hence

$$\frac{x^2}{16} + \frac{y^2}{4} = 1 \quad\longrightarrow\quad \frac{(x+2)^2}{16} + \frac{(y-3)^2}{4} = 1$$

The shifted equation is the desired result. ∎

EXERCISES 10.4

In Exercises 1 to 6, determine by observation what type of curve the given equation represents, assuming it is not a degenerate case. Do not graph.

1. $x^2 + 2y^2 = y + 2x^2 - 5$
2. $x^2 + 2x + 3 = 3y^2 + 2y + 1$
3. $(x + y)^2 + (x - y)^2 = x + y^2$
4. $3x - 2y^2 + 4 = 1 + x^2$
5. $2 + y = 5x - y^2$ 6. $x^2 + y^2 = (y + 1)^2$

In Exercises 7 to 16, graph the given equation. For a parabola, find the axis, vertex, focus, and directrix. For an ellipse or hyperbola, find the center, major and minor axes, vertices, and foci. Also, include the asymptotes for a hyperbola.

7. $x^2 + 4y^2 - 4x - 8y + 4 = 0$
8. $16x^2 + 9y^2 - 96x - 18y + 9 = 0$
9. $x^2 + 6x - 9y^2 = 0$
10. $3x^2 - 12x - 3y + 9 = 0$
11. $2y^2 - 12y + 3x = 0$
12. $x^2 + 4y^2 + 2x + 16y + 1 = 0$
13. $9y^2 - 4x^2 + 36y + 24x - 36 = 0$
14. $y^2 - 3x - 9 = 0$
15. $x^2 + 8x + 8y + 24 = 0$
16. $16x^2 - y^2 + 32x + 8y - 16 = 0$

In Exercises 17 to 22, find an equation for the curve described.

17. The ellipse with vertices at $(7, -1)$, $(1, -1)$, $(4, 1)$, and $(4, -3)$
18. The ellipse with vertices at $(-1, 1)$, $(-1, 13)$, $(-4, 7)$, and $(2, 7)$
19. The hyperbola with vertices at $(1, 3)$ and $(5, 3)$ and foci at $(3 \pm 2\sqrt{5}, 3)$
20. The hyperbola with vertices at $(0, 0)$ and $(0, 6)$ and foci at $(0, -1)$ and $(0, 7)$
21. The parabola with vertex $(-5, -2)$ and focus $(-5, -6)$
22. The parabola with vertex $(2, 5)$ and focus $(0, 5)$

In Exercises 23 to 30, write an equation for the points $P(x, y)$ described. If possible, put the equation in standard form and name the curve.

23. P is equidistant from the line $x = 3$ and the point $(-1, 6)$.

24. P is equidistant from the line $y = 1$ and the point $(4, 5)$.

25. The distance from P to the point $(-1, 0)$ is three times the distance from P to the line $y = 8$.

26. The distance from P to the point $(0, 0)$ is two times the distance from P to the line $x = 6$.

27. The distance from P to the point $(4, 1)$ is one-half the distance from P to the line $x = -2$.

28. The distance from P to the point $(-3, 2)$ is one-third the distance from P to the line $y = -6$.

29. The sum of the distances from P to $(1, -2)$ and $(1, 2)$ is 8.

30. The difference of the distances from P to $(-2, 4)$ and $(6, 4)$ is 6.

In Exercises 31 to 36, each equation represents a circle or a degenerate case of a second degree curve. Describe the graph. (*Hint:* Begin by rewriting the equation in the usual standard form.)

31. $3x^2 + 3y^2 - 42y + 135 = 0$

32. $2x^2 + 2y^2 + 8x - 12y - 6 = 0$

33. $6x^2 + 5y^2 - 12x - 20y + 26 = 0$

34. $4x^2 + y^2 - 8x + 8 = 0$

35. $x^2 - y^2 - 2x + 4y - 3 = 0$

36. $4x^2 + 9y^2 + 24x - 18y + 45 = 0$

37. (a) Find an equation for the right branch of the hyperbola $\dfrac{x^2}{a^2} - \dfrac{y^2}{b^2} = 1$. (*Hint:* Solve the equation for x.)

(b) Find an equation for a parabola with vertex at $(a, 0)$ and focus at $(a + p, 0)$.

(c) Can one branch of a hyperbola be the same as a parabola? Compare your results from (a) and (b) to answer this question.

10.5 ROTATED SECOND-DEGREE CURVES

When the major axis of a parabola, ellipse, or hyperbola is parallel to either the x- or y-axis, we say that the curve is *unrotated*. For example, Figure 10.43 shows a typical unrotated parabola. On the other hand, if the major axis is not parallel to either the x- or y-axis, then we say that the curve is *rotated*. Figure 10.44 shows a rotated parabola whose major axis makes an angle of 45° with respect to the x-axis.

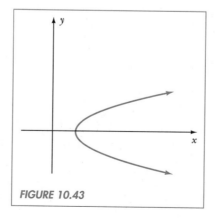

FIGURE 10.43

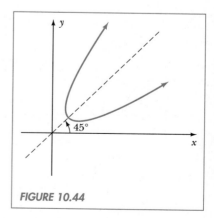

FIGURE 10.44

In Section 10.4 we found that any unrotated parabola, ellipse, or hyperbola has an equation of the form

$$Ax^2 + Cy^2 + Dx + Ey + F = 0$$

Conversely, by completing the square we can show that any equation of this form represents (except for degenerate cases) an unrotated parabola, ellipse, or hyperbola. Now we wish to extend these results to the rotated case. We begin with the following theorem.

Rotated Parabola, Ellipse, or Hyperbola Theorem

Any rotated parabola, ellipse, or hyperbola in the plane has an equation of the form

$$Ax^2 + Bxy + Cy^2 + Dx + Ey + F = 0 \qquad (*)$$

Notice that Equation (∗) is nearly identical with an equation for an unrotated parabola, ellipse, or hyperbola except for the new term Bxy.

Example 1

Consider the parabola $y^2 - x + 1 = 0$, whose graph is shown in Figure 10.45. If we rotate this curve through the angle $\theta = 45°$, we get the dashed curve also shown in Figure 10.45. Find an equation for this rotated curve.

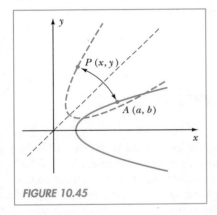

FIGURE 10.45

SOLUTION   Let $A(a, b)$ be a point on the unrotated parabola. When the curve is rotated, point A then moves to the new point $P(x, y)$, as shown in Figure 10.45. Let β be the angle in standard position whose terminal side passes through P, and let $r = OA = OP$. As Figure 10.46 shows, $\beta - 45°$ is the angle whose terminal side passes through point A. Now, by the definition of cosine, we have

$$\frac{a}{r} = \cos(\beta - 45°)$$

Thus

$$a = r\cos(\beta - 45°)$$
$$= r(\cos\beta\cos 45° + \sin\beta\sin 45°)$$
$$= r\cos\beta\cos 45° + r\sin\beta\sin 45°$$
$$= x\cos 45° + y\sin 45°$$

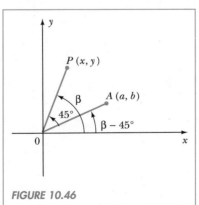

FIGURE 10.46

The last step follows from the definition of cosine and sine of β, $x/r = \cos\beta$ and $y/r = \sin\beta$. Now, substituting $\cos 45° = 1/\sqrt{2}$ and $\sin 45° = 1/\sqrt{2}$, we find that

$$a = \frac{x}{\sqrt{2}} + \frac{y}{\sqrt{2}}$$

Similarly, we have

$$b = r \sin(\beta - 45°)$$
$$= r(\sin\beta\cos 45° - \cos\beta\sin 45°)$$
$$= r\sin\beta\cos 45° - r\cos\beta\sin 45°$$
$$= \frac{y}{\sqrt{2}} - \frac{x}{\sqrt{2}}$$

Since (a, b) is a point on the unrotated parabola, it must satisfy the equation for this parabola.

$$b^2 - a + 1 = 0$$

Substituting $a = x/\sqrt{2} + y/\sqrt{2}$ and $b = y/\sqrt{2} - x/\sqrt{2}$, we find that (x, y) on the rotated parabola must satisfy

$$\left(\frac{y}{\sqrt{2}} - \frac{x}{\sqrt{2}}\right)^2 - \left(\frac{x}{\sqrt{2}} + \frac{y}{\sqrt{2}}\right) + 1 = 0$$

After multiplying and simplifying this equation, we obtain

$$x^2 - 2xy + y^2 - \sqrt{2}x - \sqrt{2}y + 2 = 0$$

This is the desired equation for the rotated parabola. Note that it has the form (∗). ∎

Using the same arguments given in Example 1, we may find the equation of any rotated curve in the plane.

Equation for a Rotated Curve
Suppose we are given an x, y equation for a curve in the plane. If we rotate this curve about the origin through the angle θ, then an x, y equation for the rotated curve can be found from the unrotated equation as follows:
1. Replace x with $x\cos\theta + y\sin\theta$.
2. Replace y with $y\cos\theta - x\sin\theta$.

Example 2

Suppose the curve $x^2 - y^2 = 1$ is rotated about the origin through the angle $\theta = 30°$. Find an equation for this rotated curve.

SOLUTION To find the desired equation, we must replace x with

$$x \cos \theta + y \sin \theta = x \cos 30° + y \sin 30°$$

$$= \frac{\sqrt{3}}{2} x + \frac{1}{2} y$$

and y with

$$y \cos \theta - x \sin \theta = y \cos 30° - x \sin 30°$$

$$= \frac{\sqrt{3}}{2} y - \frac{1}{2} x$$

Substituting these expressions for x and y into the unrotated equation $x^2 - y^2 = 1$, we have

$$\left(\frac{\sqrt{3}}{2} x + \frac{1}{2} y\right)^2 - \left(\frac{\sqrt{3}}{2} y - \frac{1}{2} x\right)^2 = 1$$

After multiplying and simplifying, we obtain

$$x^2 + 2\sqrt{3} xy - y^2 - 2 = 0$$

Again, notice that this equation has the form (*). ∎

Now consider an arbitrary unrotated parabola, ellipse, or hyperbola. We know such a curve has the equation

$$A_1 x^2 + C_1 y^2 + D_1 x + E_1 y + F_1 = 0 \qquad (1)$$

If we rotate this curve, then the new equation for the rotated curve is found from Equation (1) by replacing x and y with expressions of the form $ax + by$, a and b constants. Clearly, this gives us an equation with the same form as (1) except for the possible introduction of some xy terms. Thus we have established our theorem, which says that any rotated parabola, ellipse, or hyperbola in the plane has an equation of the form

$$Ax^2 + Bxy + Cy^2 + Dx + Ey + F = 0$$

Next, we consider the converse of this theorem.

General Second-Degree Equation Theorem

The general second-degree equation

$$Ax^2 + Bxy + Cy^2 + Dx + Ey + F = 0 \qquad (*)$$

represents a parabola, ellipse, or hyperbola (except for degenerate cases).

We already know that this theorem is true if the coefficient B is zero; this was established in Section 10.4. Therefore we will complete the proof for this

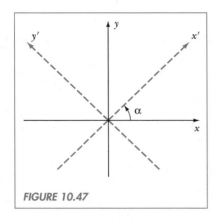

FIGURE 10.47

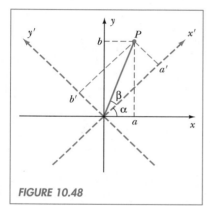

FIGURE 10.48

theorem by showing that Equation (∗) represents a parabola, ellipse, or hyperbola when B is nonzero.

Let us begin the task by demonstrating how one may transform an x, y equation into an equivalent equation written with respect to what is called a *rotated* x', y' *system*. (The symbol x' is read "x prime" and y' is read "y prime.") The x', y' system is a second coordinate system superimposed on the x, y system, with the x'-axis rotated about the origin through the angle α with respect to the positive x-axis. Figure 10.47 is an example.

Let us consider a fixed point P located in the plane, such as the point shown in Figure 10.48. P has two different sets of coordinates, depending on which coordinate system we wish to use. As indicated in Figure 10.48, one pair, (a, b), represents P in the x, y system, and another pair, (a', b'), represents P in the x', y' system. Let us determine the relationship between (a, b) and (a', b'). If r is the distance from the origin to P and β is the angle OP makes with the x'-axis, then

$$\frac{a}{r} = \cos(\alpha + \beta)$$

by the definition of cosine. Hence

$$a = r\cos(\alpha + \beta)$$
$$= r(\cos\alpha\cos\beta - \sin\alpha\sin\beta)$$
$$= r\cos\beta\cos\alpha - r\sin\beta\sin\alpha$$
$$= a'\cos\alpha - b'\sin\alpha$$

The last step follows from the fact that $a'/r = \cos\beta$ and $b'/r = \sin\beta$. In a similar manner, we have

$$b = r\sin(\alpha + \beta)$$
$$= r(\sin\alpha\cos\beta + \cos\alpha\sin\beta)$$
$$= r\cos\beta\sin\alpha + r\sin\beta\cos\alpha$$
$$= a'\sin\alpha + b'\cos\alpha$$

These relationships between x, y-coordinates and x', y'-coordinates give us the following result.

Rotational Transformation Equations

To transform an x, y equation into an equivalent equation in a rotated x', y' system, replace x and y with

$$x = x'\cos\alpha - y'\sin\alpha$$
$$y = x'\sin\alpha + y'\cos\alpha$$

where α is the angle that the x'-axis makes with respect to the x-axis.

Example 3

Consider the x', y' system whose x'-axis makes the angle $\alpha = 45°$ with respect to the x-axis. Transform the x, y equation $x + y = \sqrt{2}$ into its equivalent equation in the x', y' system.

SOLUTION Figure 10.49 shows the graph of $\{(x, y): x + y = \sqrt{2}\}$. Figure 10.50 shows the same set of points in the x', y' system. Our job is to find an x', y' equation representing this same set of points. By the transformation equations, we have

$$x = x' \cos 45° - y' \sin 45° \qquad \text{and} \qquad y = x' \sin 45° + y' \cos 45°$$

$$= \frac{x'}{\sqrt{2}} - \frac{y'}{\sqrt{2}} \qquad\qquad\qquad = \frac{x'}{\sqrt{2}} + \frac{y'}{\sqrt{2}}$$

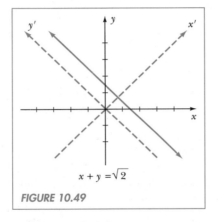

FIGURE 10.49

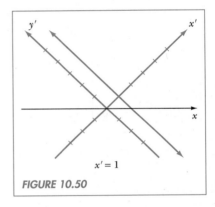

FIGURE 10.50

Therefore $x + y = \sqrt{2}$ is equivalent to the x', y' equation given by

$$\left(\frac{x'}{\sqrt{2}} - \frac{y'}{\sqrt{2}} \right) + \left(\frac{x'}{\sqrt{2}} + \frac{y'}{\sqrt{2}} \right) = \sqrt{2}$$

$$\frac{2x'}{\sqrt{2}} = \sqrt{2}$$

$$x' = 1$$

Thus the desired equation is $x' = 1$. ■

Remark

We emphasize once again that both x, y and x', y' equations represent the *same* set of points,

$$\{(x, y): x + y = \sqrt{2}\} = \{(x', y'): x' = 1\}$$

By using the transformation equations, a given x, y equation can be rewritten in any x', y' system desired. As Example 3 illustrates, a transformation may result in a much simpler equation. In fact, it is possible to eliminate any xy terms in a second-degree equation by choosing the x', y' system carefully. The next theorem tells us exactly how this is done.

Eliminating the xy Term

Consider the second-degree equation

$$Ax^2 + Bxy + Cy^2 + Dx + Ey + F = 0$$

If $B \neq 0$, then the xy term will be eliminated by rewriting the equation in the x', y' system chosen so that α satisfies

$$\cot(2\alpha) = \frac{A - C}{B}$$

PROOF: Let the x'-axis make the angle α with respect to the x-axis. By the transformation equations, the second-degree equation $Ax^2 + Bxy + Cy^2 + Dx + Ey + F = 0$ is equivalent to the x', y' equation

$$A(x' \cos \alpha - y' \sin \alpha)^2 + B(x' \cos \alpha - y' \sin \alpha)(x' \sin \alpha + y' \cos \alpha)$$
$$+ C(x' \sin \alpha + y' \cos \alpha)^2 + D(x' \cos \alpha - y' \sin \alpha)$$
$$+ E(x' \sin \alpha + y' \cos \alpha) + F = 0$$

After some multiplication and then simplification, we may rewrite this equation in the form

$$A'x'^2 + B'x'y' + C'y'^2 + D'x' + E'y' + F' = 0$$

where the coefficient of the $x'y'$ term is given by

$$B' = (C - A)2 \cos \alpha \sin \alpha + B(\cos^2 \alpha - \sin^2 \alpha)$$

If the x', y' equation is to have no $x'y'$ term, then we must choose α so that $B' = 0$. Thus we want

$$(C - A)2 \cos \alpha \sin \alpha + B(\cos^2 \alpha - \sin^2 \alpha) = 0$$

Using the double-angle formulas, we have

$$(C - A)\sin(2\alpha) + B \cos(2\alpha) = 0$$
$$B \cos(2\alpha) = (A - C)\sin(2\alpha)$$
$$\frac{\cos(2\alpha)}{\sin(2\alpha)} = \frac{A - C}{B}$$
$$\cot(2\alpha) = \frac{A - C}{B}$$

Therefore, by choosing α satisfying this last condition, the $x'y'$ term will vanish. This completes the proof. ∎

Example 4

Analyze the equation $5x^2 - 6xy + 5y^2 - 8 = 0$ as follows.
(a) Rewrite the equation in an x', y' system so that it has no $x'y'$ term. Name the curve.
(b) Find the x', y'-coordinates of the vertices and foci.
(c) Graph the curve.

SOLUTION (a) To eliminate the xy term, we choose an angle α satisfying

$$\cot(2\alpha) = \frac{A - C}{B} = \frac{5 - 5}{-6} = 0$$

This implies that

$$\frac{\cos(2\alpha)}{\sin(2\alpha)} = 0$$

Multiplying by $\sin(2\alpha)$, we find that

$$\cos(2\alpha) = 0$$

It follows that we may choose $2\alpha = 90°$, so $\alpha = 45°$.

Now, we rewrite the given equation in the x', y' system whose x'-axis makes the angle $\alpha = 45°$ with respect to the x-axis. By the transformation equations,

$$x = x' \cos \alpha - y' \sin \alpha \qquad \text{and} \qquad y = x' \sin \alpha + y' \cos \alpha$$

$$= \frac{x'}{\sqrt{2}} - \frac{y'}{\sqrt{2}} \qquad\qquad\qquad = \frac{x'}{\sqrt{2}} + \frac{y'}{\sqrt{2}}$$

Therefore the equation $5x^2 - 6xy + 5y^2 - 8 = 0$ is equivalent to

$$5\left(\frac{x'}{\sqrt{2}} - \frac{y'}{\sqrt{2}}\right)^2 - 6\left(\frac{x'}{\sqrt{2}} - \frac{y'}{\sqrt{2}}\right)\left(\frac{x'}{\sqrt{2}} + \frac{y'}{\sqrt{2}}\right) + 5\left(\frac{x'}{\sqrt{2}} + \frac{y'}{\sqrt{2}}\right)^2 - 8 = 0$$

After multiplying terms and simplifying, we obtain

$$\frac{x'^2}{4} + \frac{y'^2}{1} = 1$$

We recognize this as an ellipse centered at the origin with major axis the x'-axis.

(b) From the standard equation $\dfrac{x'^2}{4} + \dfrac{y'^2}{1} = 1$, we find that

$$a^2 = 4 \qquad \text{implies} \qquad a = 2$$

$$b^2 = 1 \qquad \text{implies} \qquad b = 1$$

$$c^2 = a^2 - b^2 \quad \text{implies} \quad c^2 = 4 - 1 = 3, \text{ so } c = \sqrt{3}$$

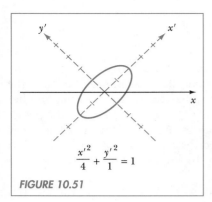

$$\frac{x'^2}{4} + \frac{y'^2}{1} = 1$$

FIGURE 10.51

It follows that

x', y'-coordinates
Vertices: $(\pm 2, 0)$ and $(0, \pm 1)$
Foci: $(\pm \sqrt{3}, 0)$

(c) The graph is shown in Figure 10.51. ■

Eliminating the xy term allows us to transform any second-degree equation into one with no xy term. Thus the set of points in the plane satisfying

$$Ax^2 + Bxy + Cy^2 + Dx + Ey + F = 0$$

will be the same as the set of points satisfying an x', y' equation of the form

$$A'x'^2 + C'y'^2 + D'x' + E'y' + F' = 0$$

We know that this last equation represents a parabola, ellipse, or hyperbola (except for degenerate cases). This establishes the General Second-Degree Equation Theorem.

EXERCISES 10.5

In Exercises 1 to 6, suppose that the given curve is rotated about the origin through the angle θ. Find an equation for the rotated curve.

1. $x^2 - y^2 = 1$, $\theta = 45°$ 2. $x = y^2$, $\theta = 45°$
3. $4x^2 + 16y^2 = 16$, $\theta = 30°$ 4. $xy = 1$, $\theta = 60°$
5. $x^2 + y^2 = 2\sqrt{2}x$, $\theta = 45°$
6. $y = 0$, $\theta = 30°$

In Exercises 7 to 12, transform the x, y equation into its equivalent equation in the x', y' system whose x'-axis makes the angle α with respect to the x-axis.

7. $x - y = 1$, $\alpha = 45°$ 8. $x + y = 1$, $\alpha = 45°$
9. $xy = 1$, $\alpha = 45°$ 10. $x = 2$, $\alpha = 30°$
11. $y = 1$, $\alpha = 60°$ 12. $x^2 + y^2 = 1$, $\alpha = 60°$

In Exercises 13 to 20, analyze the given equation as follows.
(a) Rewrite the equation in an x', y' system so that it has no $x'y'$ term. Name the curve.
(b) Find the x', y'-coordinates of the vertices and foci.
(c) Graph the curve.

13. $3x^2 - 10xy + 3y^2 + 8 = 0$
14. $13x^2 + 10xy + 13y^2 - 72 = 0$
15. $34x^2 + 32xy + 34y^2 - 450 = 0$

16. $2xy - 1 = 0$
17. $x^2 - 2\sqrt{3}xy + 3y^2 - 2\sqrt{3}x - 2y + 8 = 0$
18. $3x^2 + 2\sqrt{3}xy + y^2 + 2x - 2\sqrt{3}y = 0$
19. $xy + x + y - 1 = 0$
20. $7x^2 - 2\sqrt{3}xy + 5y^2 - 8 = 0$

In Exercises 21 and 22, rewrite the given equation in an x', y' system so that it has no $x'y'$ term. Graph. Find the x, y-*coordinates* of the vertices. (Find the x', y'-coordinates first, and then use the transformation equations to determine the x, y-coordinates.)

21. $13x^2 - 6\sqrt{3}xy + 7y^2 - 8x - 8\sqrt{3}y = 0$
22. $y^2 + 2\sqrt{3}xy - x^2 - 2x - 2\sqrt{3}y = 0$

23. Transform $y = x + \dfrac{1}{x}$ into its equivalent equation in the x', y' system whose x'-axis makes the angle $\alpha = -22.5°$ with respect to the x-axis. Name the curve. [*Hint:* use the half-angle formulas to find exact expressions for $\cos(-22.5°)$ and $\sin(-22.5°)$].

24. Transform $x^2 + y^2 = 1$ into its equivalent equation in the x', y' system whose x'-axis makes the angle α with respect to the x-axis.

10.6 POLAR COORDINATES

The Cartesian coordinate system enables us to describe the position of any point in the plane by using an ordered pair of numbers (a, b), where a is the x-coordinate and b is the y-coordinate. In this section we describe another coordinate system that also uses ordered pairs for locating points in the plane: the **polar coordinate** system. A point is represented in polar coordinates by an ordered pair (r, θ), where r is a directed distance from the origin and θ is a direction angle. Before we explain how to locate (r, θ), we introduce a preliminary construction.

Given a fixed x, y-coordinate system, let us draw another coordinate axis coinciding with the x-axis and call it an x'-*axis*. We allow the x'-axis to be rotated about the origin through any angle θ with respect to the positive x-axis. As usual, we rotate in the counterclockwise direction if θ is positive and clockwise if θ is negative. Figure 10.52 shows two rotations of the x'-axis, one through the positive angle $3\pi/4$ and the other through the negative angle $-\pi/2$. Notice that the arrow on one end of the rotated x'-axis designates the positive direction of the x'-axis.

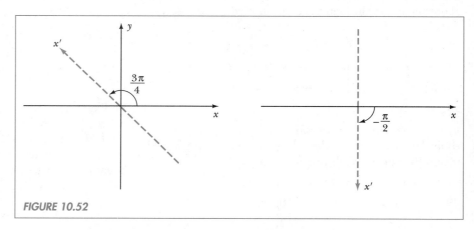

FIGURE 10.52

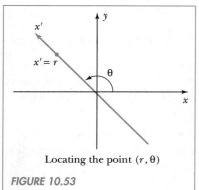

Locating the point (r, θ)

FIGURE 10.53

> **Graphing Polar Coordinates (r, θ)**
> Use the following two-step procedure to graph the point P in the plane with polar coordinates (r, θ):
>
> Step 1   Rotate an x'-axis about the origin through the angle θ.
>
> Step 2   Move out from the origin along the rotated x'-axis to the x'-coordinate r; this is the position of P. See Figure 10.53.

Example 1

Graph the points that are represented by the polar coordinates $A(2, 3\pi/4)$, $B(-2, 3\pi/4)$, $C(3, -\pi/2)$, $D(-1, \pi)$, and $E(0, \pi/4)$.

SOLUTION   The points are graphed in Figure 10.54. In each case, pay close attention to which direction along the rotated x'-axis is positive and which is negative. For instance, A has coordinates $(2, 3\pi/4)$, so we must move out from the origin in the *positive x'-direction* 2 units; this locates A in quadrant II as shown. However, B, with coordinates $(-2, 3\pi/4)$, must be located by moving 2 units out from the origin in the *negative x'-direction*, so B is found in quadrant IV.

Notice that point E with coordinates $(0, \pi/4)$ is simply the origin. Indeed, all points of the form $(0, \theta)$ for any value of θ will be located at the origin. ∎

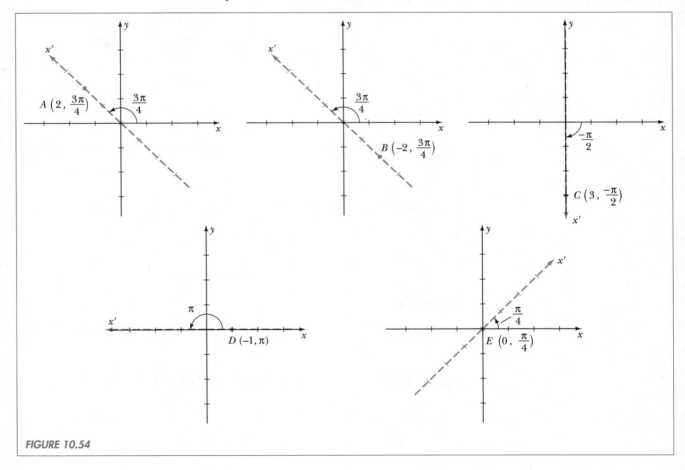

FIGURE 10.54

Every point in the plane has exactly one Cartesian coordinate pair representing it. Thus we say that Cartesian coordinates are *unique*. On the other hand, polar coordinates are *not* unique. In fact, as Example 2 illustrates, a point in the plane has many different polar coordinate representations.

Example 2

Let P be the point with Cartesian coordinates $(1, 1)$. Find three different polar coordinate pairs representing P.

SOLUTION Figure 10.55 shows that the polar coordinates $(\sqrt{2}, \pi/4)$, $(\sqrt{2}, 9\pi/4)$, and $(-\sqrt{2}, 5\pi/4)$ all represent P. ■

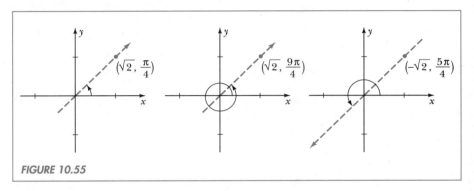

FIGURE 10.55

In general, any point in the plane has infinitely many polar coordinate representations.

> The point P with polar coordinates (r, θ) is also represented by
> $$(r, \theta + 2\pi k) \quad \text{and} \quad (-r, \theta + (2k + 1)\pi)$$
> where $k \in \mathbf{Z}$.

Example 3

Suppose P has polar coordinates $(2, \pi/3)$. Find two other polar coordinates for P, one with positive r and one with negative r. Also, list all possible polar coordinates for P.

SOLUTION Two other possible polar coordinates for P would be

$$\left(2, \frac{\pi}{3} + 2\pi\right) \quad \text{and} \quad \left(-2, \frac{\pi}{3} + \pi\right)$$

which simplify to $(2, 7\pi/3)$ and $(-2, 4\pi/3)$. The list of all possible polar coordinates for P is given by

$$\left(2, \frac{\pi}{3} + 2\pi k\right) \quad \text{and} \quad \left(-2, \frac{\pi}{3} + (2k + 1)\pi\right) \qquad k \in \mathbf{Z} \quad ■$$

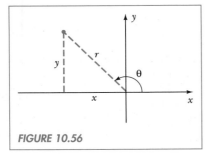

FIGURE 10.56

We now establish some equations that relate polar and Cartesian coordinates. Consider Figure 10.56, which shows a point P with polar coordinates (r, θ) and Cartesian coordinates (x, y). By definition of $\cos \theta$, $\sin \theta$, and $\tan \theta$, we have

$$\cos \theta = \frac{x}{r} \qquad \sin \theta = \frac{y}{r} \qquad \tan \theta = \frac{y}{x}$$

Therefore $x = r \cos \theta$, $y = r \sin \theta$, and $y/x = \tan \theta$. Furthermore,

$$x^2 + y^2 = r^2 \cos^2 \theta + r^2 \sin^2 \theta = r^2(\cos^2 \theta + \sin^2 \theta) = r^2$$

Although Figure 10.56 illustrates the situation when r is positive, the equations we found above remain valid even if r is negative. We call these equations polar-Cartesian transformation equations because they allow us to transform from one coordinate system to the other.

> **Polar-Cartesian Transformation Equations**
> The relationships between polar coordinates (r, θ) and Cartesian coordinates (x, y) are expressed by the following equations:
>
> $$x = r \cos \theta \qquad x^2 + y^2 = r^2$$
> $$y = r \sin \theta \qquad \tan \theta = \frac{y}{x}$$

Example 4

Find the Cartesian coordinates of the point with the given polar coordinates: (a) $A(-2, 3\pi/4)$, (b) $B(3, 2)$.

SOLUTION (a) For the point $A(-2, 3\pi/4)$ we have $r = -2$ and $\theta = 3\pi/4$. Thus

$$
\begin{array}{ll}
x = r \cos \theta & y = r \sin \theta \\
\quad = -2 \cos(3\pi/4) & \quad = -2 \sin(3\pi/4) \\
\quad = -2(-\sqrt{2}/2) & \quad = -2(\sqrt{2}/2) \\
\quad = \sqrt{2} & \quad = -\sqrt{2}
\end{array}
$$

So A has x, y-coordinates $(\sqrt{2}, -\sqrt{2})$.

(b) For the point $B(3, 2)$ we have $r = 3$ and $\theta = 2$ radians. Thus

$$
\begin{array}{ll}
x = r \cos \theta & y = r \sin \theta \\
\quad = 3 \cos(2) & \quad = 3 \sin(2) \\
\quad \approx -1.2484 & \quad \approx 2.7279
\end{array}
$$

So B has x, y-coordinates $(3 \cos(2), 3 \sin(2))$, or approximately $(-1.2484, 2.7279)$. ∎

Example 5

Find polar coordinates for the point with Cartesian coordinates $P(-3, -4)$.

SOLUTION When changing from Cartesian to polar coordinates, we use the equations

$$x^2 + y^2 = r^2 \quad \text{and} \quad \tan \theta = \frac{y}{x}$$

Since $x = -3$ and $y = -4$, we have

$$r^2 = x^2 + y^2 = (-3)^2 + (-4)^2 = 25$$

Thus r could be either 5 or -5. Let us choose $r = 5$. Referring to Figure 10.57, we see that if r is positive, we need θ to be in quadrant III. It follows,

$$180° < \theta < 270° \quad \text{and} \quad \tan \theta = \frac{y}{x} = \frac{-4}{-3} = \frac{4}{3}$$

Using the calculator, we find that $\tan^{-1}(4/3) \approx 53.13°$. Adding $180°$ to this gives us $\theta \approx 233.13°$. We conclude that a set of polar coordinates for P is exactly $[5, \tan^{-1}(4/3) + \pi]$, or approximately $(5, 233.13°)$. ∎

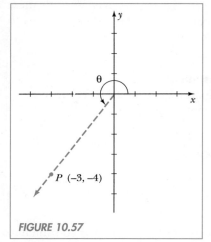

FIGURE 10.57

Example 6

Transform the Cartesian equation $x^2 + y^2 - 2x + 2y = 0$ into an equivalent polar equation. If possible, write the polar equation in the form $r = F(\theta)$, where $F(\theta)$ is a function of θ.

SOLUTION By the transformation equations, we may replace $x^2 + y^2$ with r^2, x with $r \cos \theta$, and y with $r \sin \theta$. Thus

$$x^2 + y^2 - 2x + 2y = 0$$
$$r^2 - 2r \cos \theta + 2r \sin \theta = 0$$
$$r^2 = 2r \cos \theta - 2r \sin \theta$$
$$r^2 = 2r(\cos \theta - \sin \theta)$$

We can get r as a function of θ if we divide both sides of the last equation by r. Thus

$$r_{\bullet} = 2(\cos \theta - \sin \theta)$$

We must be careful about dividing any equation by a variable expression since this expression might be zero. In this case, the expression r could be zero because this corresponds to the origin, which is a solution point in the original equation $x^2 + y^2 - 2x + 2y = 0$. However, $r = 0$ is also a solution point for the new equation $r = 2(\cos \theta - \sin \theta)$ (take $\theta = \pi/4$, for example). Therefore we have not lost any solutions by dividing by r, so our final equation is still equivalent to the original. ∎

Example 7

Transform the polar equation $r = \cos \theta$ into an equivalent Cartesian equation.

SOLUTION   We multiply both sides of the equation by r, obtaining

$$r^2 = r \cos \theta$$

Now, substitute $x^2 + y^2$ for r^2 and x for $r \cos \theta$.

$$x^2 + y^2 = x$$

This is the desired Cartesian equation. ∎

Remark

It is interesting to note that the equation in Example 7 represents a circle. In fact, we may write it in standard form as follows:

$$x^2 - x + y^2 = 0$$

$$x^2 - x + \frac{1}{4} + y^2 = \frac{1}{4}$$

$$\left(x - \frac{1}{2} \right)^2 + y^2 = \frac{1}{4}$$

Thus we have a circle of radius 1/2 centered at $\left(\frac{1}{2}, 0 \right)$.

Example 8

Transform $r = \dfrac{3}{2 + \cos \theta}$ into an equivalent Cartesian equation.

SOLUTION   Multiply both sides of the equation by $2 + \cos \theta$, obtaining

$$2r + r \cos \theta = 3$$

Now, substitute $r \cos \theta = x$.

$$2r + x = 3$$

Putting x on the other side and squaring, we have

$$2r = 3 - x$$

$$4r^2 = (3 - x)^2$$

Now substitute $x^2 + y^2$ for r^2,

$$4(x^2 + y^2) = (3 - x)^2$$

We leave it for you to verify that this equation can be written in the form

$$\frac{(x + 1)^2}{4} + \frac{y^2}{3} = 1$$

This is the desired Cartesian equation for $r = 3/(2 + \cos \theta)$. We recognize this as an ellipse centered at $(-1, 0)$ with major axis the x-axis. ∎

We end this section by graphing sets of points in the plane that are described in terms of polar coordinates. This will be an introduction to Section 10.7, which covers graphing polar curves.

Example 9

Graph the set of points (r, θ) described.
(a) $\{(r, \theta): r = 2 \text{ and } \pi/2 \le \theta \le \pi\}$
(b) $\{(r, \theta): r = -2 \text{ and } \pi/2 \le \theta \le \pi\}$
(c) $\{(r, \theta): r = 2\}$

SOLUTION (a) As θ varies from $\pi/2$ to π, r remains equal to 2. The graph will be the quarter circle in the second quadrant shown in Figure 10.58.

(b) As θ varies from $\pi/2$ to π, r remains equal to -2. The result is the quarter circle in the fourth quadrant shown in Figure 10.59.

(c) The equation $r = 2$ has no restrictions on θ, so r is 2 for *any* value of θ. The graph is the complete circle shown in Figure 10.60. ∎

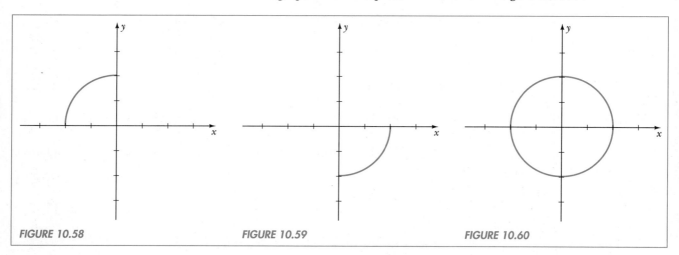

FIGURE 10.58   FIGURE 10.59   FIGURE 10.60

Example 10

Graph the set of points (r, θ) described.
(a) $\{(r, \theta): -1 \le r \le 2 \text{ and } \theta = \pi/4\}$
(b) $\{(r, \theta): \theta = \pi/4\}$
(c) $\{(r, \theta): \tan \theta = 1\}$

SOLUTION (a) In this case, θ remains fixed at π/4 and r varies from −1 to 2. The graph is the line segment shown in Figure 10.61.

(b) The equation θ = π/4 has no restrictions on r. Thus while θ remains fixed at π/4, r varies from −∞ to +∞. The graph is the diagonal line shown in Figure 10.62.

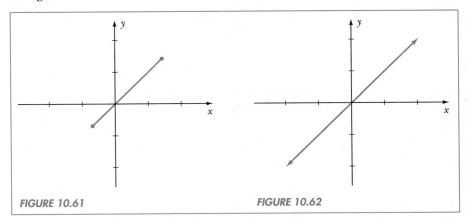

FIGURE 10.61   FIGURE 10.62

(c) Note that the condition tan θ = 1 is equivalent to saying θ = π/4, 5π/4, or any angle coterminal to these. Therefore, since there are no restrictions on r, the graph consists of all points (r, θ) satisfying

$$\theta = \frac{\pi}{4}, \quad -\infty < r < +\infty \qquad \text{or} \qquad \theta = \frac{5\pi}{4}, \quad -\infty < r < +\infty$$

Both conditions have the same graph, which is the diagonal shown in Figure 10.62. Thus graphing tan θ = 1 is equivalent to graphing θ = π/4. ∎

EXERCISES 10.6

In Exercises 1 and 2, graph the given polar coordinates.

1. (a) $A(2, \pi/4)$ (b) $B(-2, \pi/4)$ (c) $C(1, \pi)$
   (d) $D(2, -5\pi/6)$ (e) $E(3, 0)$ (f) $F(0, -3)$

2. (a) $A(1, \pi/3)$ (b) $B(-1, \pi/3)$ (c) $C(-2, -\pi)$
   (d) $D(-3/2, 3\pi/2)$ (e) $E(0, 1)$ (f) $F(1, 0)$

3. Find the Cartesian coordinates for each point given in Exercise 1.

4. Find the Cartesian coordinates for each point given in Exercise 2.

In Exercises 5 and 6, find polar coordinates for the point with the given Cartesian coordinates.

5. (a) $A(-3, 3)$ (b) $B(-2, 0)$ (c) $C(4, -5)$
   (d) $D(\pi/2, \pi)$ (e) $E(-2, -4)$

6. (a) $A(-3, -1)$ (b) $B(-2, 2)$ (c) $C(3, 4)$
   (d) $D(0, 5)$ (e) $E(\pi, -\pi)$

In Exercises 7 and 8, suppose P has the polar coordinates given. Find two other polar coordinates for P, one with positive r and one with negative r. Also, list *all* possible polar coordinates for P.

7. (a) $P\left(5, \frac{2\pi}{3}\right)$ (b) $P(-3, 0)$ (c) $P(\sqrt{2}, -\pi/4)$

8. (a) $P\left(-2, \frac{\pi}{8}\right)$ (b) $P(0, -3)$ (c) $P(4, -\pi/3)$

In Exercises 9 and 10, answer true or false.

9. The polar coordinates $(-6, 2)$ and $(6, 2 - \pi)$ represent the same point.

10. The polar coordinates $(-\pi, \pi)$ and $(\pi, -\pi)$ represent the same point.

In Exercises 11 and 12, use the transformation equations to rewrite the given expression in polar coordinates.

11. (a) $(x + y)^2$
    (b) $(1 - x^2 - y^2)^{3/2}$
    (c) $1 + \dfrac{y^2}{x^2}$

12. (a) $(x - y)^2$
    (b) $\sqrt{x^2 + y^2 + 1}$
    (c) $\dfrac{x}{y} + \dfrac{y}{x}$

In Exercises 13 to 26, transform the given Cartesian equation into an equivalent polar equation. If possible, write the equation in the form $r = F(\theta)$, where $F(\theta)$ is a function of θ.

13. $x = 4$

14. $x = -1$

15. $y = -2$

16. $y = 3$

17. $x + 2y = 3$

18. $2x - 3y = 4$

19. (a) $x^2 + y^2 = 9$
    (b) $(x - 3)^2 + y^2 = 9$

20. (a) $x^2 + y^2 = 4$
    (b) $x^2 + (y - 2)^2 = 4$

21. $(x^2 + y^2)^3 = 4x^2y^2$

22. $(x^2 + y^2)^2 = x^2 - y^2$

23. $x = y^2$

24. $y = x^2$

25. $\ln\sqrt{x^2 + y^2} = \dfrac{y}{x}$

26. $y^2 = 1 - 2x$

In Exercises 27 to 36, transform the given polar equation into an equivalent Cartesian equation.

27. $r = \sin\theta$

28. $r = -6\cos\theta$

29. $r = 1$

30. $3r\cos\theta - 4r\sin\theta = 5$

31. $r = \dfrac{1}{1 - \sin\theta}$

32. $r = \dfrac{2}{3 + \cos\theta}$

33. $r^2 = \sec(2\theta)$

34. $r^2 = \csc(2\theta)$

35. $\cos\theta = 1$

36. $\theta = \pi/2$

In Exercises 37 to 44, graph the set of points (r, θ) described.

37. (a) $\{(r, \theta): r = -3 \text{ and } 0 \le \theta \le \pi/2\}$
    (b) $\{(r, \theta): r = -3\}$

38. (a) $\{(r, \theta): r \le 2 \text{ and } \theta = 3\pi/4\}$
    (b) $\{(r, \theta): \theta = 3\pi/4\}$

39. (a) $\{(r, \theta): |r| \le 2 \text{ and } \theta = 5\pi/4\}$
    (b) $\left\{(r, \theta): r \ge 2 \text{ and } \theta = \dfrac{5\pi}{4}\right\}$

40. (a) $\{(r, \theta): 1 \le r \le 2 \text{ and } 0 \le \theta \le \pi\}$
    (b) $\{(r, \theta): 1 \le r \le 2\}$

41. (a) $\{(r, \theta): -1 \le r \le 2 \text{ and } \pi \le \theta \le 3\pi/2\}$
    (b) $\{(r, \theta): r > 1 \text{ and } \pi \le \theta \le 3\pi/2\}$
    (c) $\{(r, \theta): r < 1 \text{ and } \pi \le \theta \le 3\pi/2\}$

42. (a) $\{(r, \theta): |r| < 1 \text{ and } \pi/2 \le \theta \le \pi\}$
    (b) $\{(r, \theta): |r| < 1\}$

43. $\left\{(r, \theta): \tan\theta = \dfrac{2}{3}\right\}$

44. $\left\{(r, \theta): \sin\theta = \dfrac{1}{2}\right\}$

In Exercises 45 and 46, transform into an equivalent Cartesian equation and then graph.

45. $r = -2\csc\theta$

46. $r = \dfrac{1}{2}\sec\theta$

In Exercises 47 and 48, transform into an equivalent polar equation and then graph.

47. $x^2 + y^2 = 2xy$

48. $xy = x^2 - y^2$

10.7 GRAPHING POLAR CURVES

In this section we concentrate on graphing four basic types of polar curves:

$$\text{Circles:}\quad r = a\cos\theta, \quad r = a\sin\theta$$

$$\text{Limaçons and cardioids:}\quad r = a + b\cos\theta, \quad r = a + b\sin\theta$$

$$\text{Roses:}\quad r = a\cos(n\theta), \quad r = a\sin(n\theta)$$

$$\text{Spirals:}\quad r = a\theta, \quad r = a^\theta$$

Let us begin with the circles.

Example 1

Graph $r = \cos\theta$.

SOLUTION   Assuming that we have no previous knowledge of this curve, we begin by making a detailed table of values of r and θ. Taking θ varying from 0 to π, we compute the corresponding values of r approximated to two decimal places:

θ	0	$\dfrac{\pi}{6}$	$\dfrac{\pi}{4}$	$\dfrac{\pi}{3}$	$\dfrac{\pi}{2}$	$\dfrac{2\pi}{3}$	$\dfrac{3\pi}{4}$	$\dfrac{5\pi}{6}$	π
$r = \cos\theta$	1	.87	.71	.5	0	$-.5$	$-.71$	$-.87$	-1

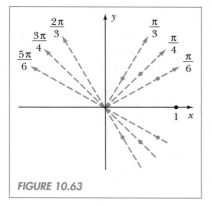

FIGURE 10.63

The points from the table are graphed in Figure 10.63. Although it is not entirely obvious from these few points, it turns out that a smooth curve connecting the points produces the circle shown in Figure 10.64. What happens as θ continues to vary from π to 2π? In this case, r takes on the same values that we found in the table for θ between 0 and π, but in reverse order. This means that the graph of $r = \cos\theta$ for $\pi \le \theta \le 2\pi$ traces out the same circle shown in Figure 10.64. Of course, for all other values of θ, we continue to get points on the same curve. Therefore Figure 10.64 shows the complete graph of $r = \cos\theta$.

How do we know that the graph of $r = \cos\theta$ is exactly a circle? Recall that in the remark following Example 7, Section 10.6, we transformed the polar equation $r = \cos\theta$ into its equivalent Cartesian equation, obtaining

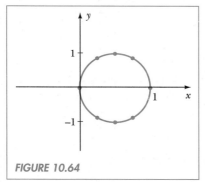

FIGURE 10.64

$$\left(x - \frac{1}{2}\right)^2 + y^2 = \frac{1}{4}$$

We recognize this as the standard form for the equation of a circle. This proves that the graph of $r = \cos\theta$ must be a circle centered at $(1/2, 0)$ with radius $1/2$. ∎

Remark

We could have approached the problem of graphing $r = \cos\theta$ by first translating this equation into its equivalent Cartesian form and then graphing the x, y equation. However, this would not tell us *how* the graph is formed as θ varies. For example, the entire circle is traced out when θ varies from 0 to π. The circle is then repeated a *second* time as θ varies from π to 2π. This type of information will be important for future work in calculus.

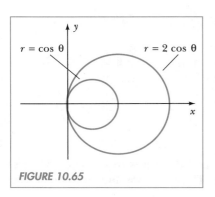

$r = \cos\theta$ $r = 2\cos\theta$

FIGURE 10.65

What happens if we graph $r = 2\cos\theta$ instead of $r = \cos\theta$? Of course, we get the same type of graph, a circle, except that the values of r are doubled. Figure 10.65 shows the graph of $r = 2\cos\theta$ in relation to $r = \cos\theta$. Notice that the graph remains tangent to the y-axis at the origin.

In general, the graph of $r = a \cos \theta$ will be a circle tangent to the y-axis at the origin with its center at $(a/2, 0)$. Similarly, $r = a \sin \theta$ will be a circle centered at $(0, a/2)$ and tangent to the x-axis at the origin. You will verify these results in the exercises. Figures 10.66 to 10.69 summarize the graphs:

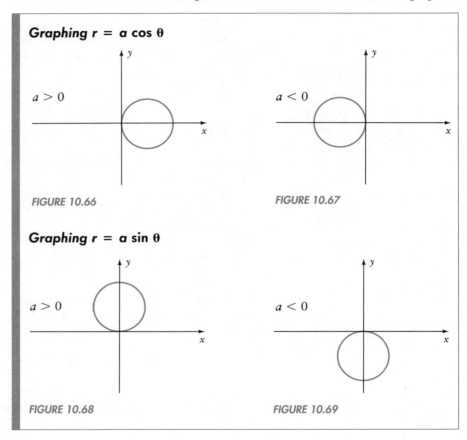

Graphing $r = a \cos \theta$

$a > 0$

FIGURE 10.66

$a < 0$

FIGURE 10.67

Graphing $r = a \sin \theta$

$a > 0$

FIGURE 10.68

$a < 0$

FIGURE 10.69

Next we consider limaçons and cardioids.

Example 2

Graph $r = 1 + 2 \cos \theta$.

SOLUTION We begin with a table of values, letting θ vary from 0 to π. The values for r are approximated to one decimal place.

θ	0	$\dfrac{\pi}{6}$	$\dfrac{\pi}{4}$	$\dfrac{\pi}{3}$	$\dfrac{\pi}{2}$	$\dfrac{2\pi}{3}$	$\dfrac{3\pi}{4}$	$\dfrac{5\pi}{6}$	π
$r = 1 + 2 \cos \theta$	3	2.7	2.4	2	1	0	$-.4$	$-.7$	-1

As θ varies from 0 to $2\pi/3$, r starts at 3 and decreases to 0. When θ varies from $2\pi/3$ to π, r takes on negative values. The corresponding graph is shown

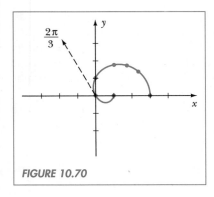

FIGURE 10.70

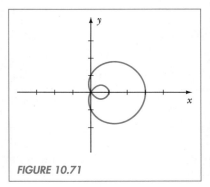

FIGURE 10.71

in Figure 10.70. When θ varies from π to 2π, r takes on the same values that are in the table for θ from 0 to π, but in reverse order. Thus we complete the graph as shown in Figure 10.71. We call this curve a *limaçon*. Note that for values of θ greater than 2π or less than 0, we continue to get points on the same curve. ▪

Example 2 illustrates a typical case of $r = a + b \cos \theta$ with $|a| < |b|$. We have other possibilities corresponding to $|a| = |b|$ and $|a| > |b|$; Figures 10.72 to 10.75 illustrate representative cases:

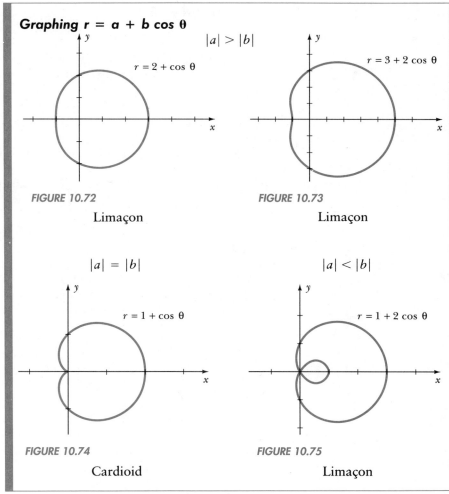

Graphing $r = a + b \cos \theta$

$|a| > |b|$

$r = 2 + \cos \theta$

FIGURE 10.72

Limaçon

$r = 3 + 2 \cos \theta$

FIGURE 10.73

Limaçon

$|a| = |b|$

$r = 1 + \cos \theta$

FIGURE 10.74

Cardioid

$|a| < |b|$

$r = 1 + 2 \cos \theta$

FIGURE 10.75

Limaçon

Figures 10.72 and 10.73 show that when $|a| > |b|$, the graph of $r = a + b \cos \theta$ goes around the origin without touching it. Notice that when $1 < |a/b| < 2$, the graph develops a slight indentation. (It takes a little calculus to show why this is true, so we omit the proof.) When $|a| = |b|$, the indentation is enough to make the curve just touch the origin, producing a heart-shaped figure called a *cardioid*. Figure 10.74 shows a typical example. Finally, when $|a| < |b|$, the curve goes through the origin, forming an inside loop such as the graph appearing in Figure 10.75. Notice that all the graphs for $r = a + b \cos\theta$ are symmetric to the x-axis.

Similar results hold for the graphs of $r = a + b \sin\theta$, except that these curves will be symmetric to the y-axis.

Next we consider graphs of roses. Our first example has the equation $r = \sin(3\theta)$, which turns out to be a *three-leaved rose*.

Example 3

Graph $r = \sin(3\theta)$.

SOLUTION We proceed to make a table and plot points. The values of r are approximated to two decimal places.

θ	0	10°	15°	20°	30°	40°	45°	50°	60°
$r = \sin(3\theta)$	0	.5	.71	.87	1	.87	.71	.5	0

As θ varies from 0° to 60°, the first *leaf* of the rose is traced out, as shown in Figure 10.76.

When θ varies over the next 60° interval, from 60° to 120°, r takes on negative values, from 0 to -1 then back to 0.

θ	60°	75°	90°	115°	120°
$r = \sin(3\theta)$	0	$-.71$	-1	$-.71$	0

Thus we get the second leaf traced out in Figure 10.77.

Finally, the rose is completed when the third leaf is traced out as θ varies from 120° to 180°, as shown in Figure 10.78.

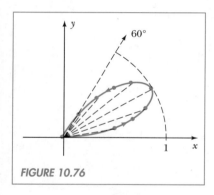

FIGURE 10.76

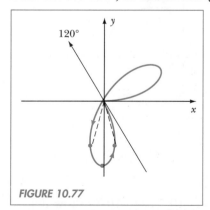

FIGURE 10.77

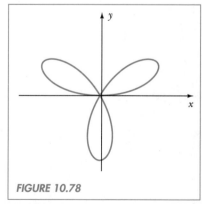

FIGURE 10.78

Notice that when θ varies from 180° to 360°, the same curve is again traced out. Also, the graph is symmetric with respect to the y-axis. ■

Since the graph of $r = \sin(3\theta)$ has three leaves, we might expect $r = \sin(2\theta)$ to have just two leaves. As it turns out, however, $r = \sin(2\theta)$ has *four* leaves. Figure 10.79 shows the construction of the graph for $r = \sin(2\theta)$.

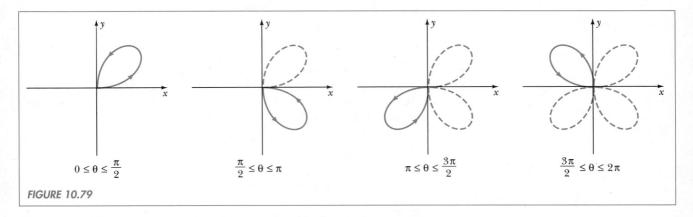

$$0 \leq \theta \leq \frac{\pi}{2} \qquad \frac{\pi}{2} \leq \theta \leq \pi \qquad \pi \leq \theta \leq \frac{3\pi}{2} \qquad \frac{3\pi}{2} \leq \theta \leq 2\pi$$

FIGURE 10.79

We have the following general rules for roses.

> **Graphing Roses**
> If n is a positive integer, $n \geq 2$, then the graph of $r = a\cos(n\theta)$ or $r = a\sin(n\theta)$ will be a rose with
>
> $$n \text{ leaves} \quad \text{if } n \text{ is odd}$$
> $$2n \text{ leaves} \quad \text{if } n \text{ is even}$$

Examples 4 and 5 cover spirals.

Example 4

Graph $r = e^{\theta/\pi}$.

SOLUTION We start by making a table of values for r as θ varies from 0 to 2π. We use our calculator and approximate r to two decimal places.

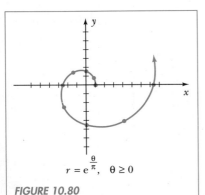

$$r = e^{\frac{\theta}{\pi}}, \quad \theta \geq 0$$

FIGURE 10.80

θ	0	$\dfrac{\pi}{4}$	$\dfrac{\pi}{2}$	$\dfrac{3\pi}{4}$	π	$\dfrac{5\pi}{4}$	$\dfrac{3\pi}{2}$	$\dfrac{7\pi}{4}$	2π
$r = e^{\theta/\pi}$	1	1.28	1.65	2.12	2.72	3.49	4.48	5.75	7.39

Connecting these points with a smooth curve, we obtain the graph shown in Figure 10.80. Note that as θ continues to increase, the curve will continue to spiral outward.

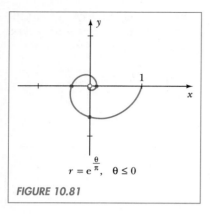

$r = e^{\frac{\theta}{\pi}}, \quad \theta \le 0$

FIGURE 10.81

There are no restrictions on θ, so we must also consider the behavior of the graph as θ takes on negative values. Consider the following table:

θ	0	$-\dfrac{\pi}{2}$	$-\pi$	$-\dfrac{3\pi}{2}$	-2π
$r = e^{\theta/\pi}$	1	.61	.37	.22	.14

As θ takes on negative values of larger and larger magnitude, $r = e^{\theta/\pi}$ will become a smaller and smaller positive number. Thus the curve spirals in toward the origin, as shown in Figure 10.81. Note, however, that the curve will never reach the origin. ∎

Example 5

Graph $r = \dfrac{\theta}{\pi}$ for $-\pi \le \theta \le \pi$.

> **SOLUTION**   First consider the portion of the graph corresponding to values of θ between $-\pi$ and 0. We have the following table:

θ	$-\pi$	$-\dfrac{3\pi}{4}$	$-\dfrac{\pi}{2}$	$-\dfrac{\pi}{4}$	0
$r = \dfrac{\theta}{\pi}$	-1	$-\dfrac{3}{4}$	$-\dfrac{1}{2}$	$-\dfrac{1}{4}$	0

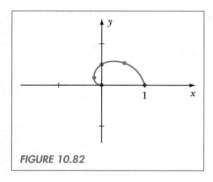

FIGURE 10.82

Notice that the values of r are negative, which produces the curve in quadrants I and II shown in Figure 10.82.

Now consider the points corresponding to $0 \le \theta \le \pi$. We have

θ	0	$\dfrac{\pi}{4}$	$\dfrac{\pi}{2}$	$\dfrac{3\pi}{4}$	π
$r = \dfrac{\theta}{\pi}$	0	$\dfrac{1}{4}$	$\dfrac{1}{2}$	$\dfrac{3}{4}$	1

Using these points, we complete the graph as shown in Figure 10.83. ∎

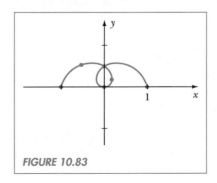

FIGURE 10.83

1. Let a be a nonzero constant. Find an equivalent Cartesian equation for $r = a \sin \theta$. Write this equation in standard form.

2. Repeat Exercise 1 for $r = a \cos \theta$.

In Exercises 3 to 18, graph the given polar equation.

3. (a) $r = 6 \cos \theta$
   (b) $r = -2 \sin \theta$

4. (a) $r = 4 \sin \theta$
   (b) $r = -2 \cos \theta$

5. $r = 1 - \sin \theta$

6. $r = 1 - \cos \theta$

7. $r = 3 - 2 \cos \theta$

8. $r = 3 + 2 \sin \theta$

9. $r = 2 + 4 \sin \theta$

10. $r = 2 - 4 \cos \theta$

11. $r = \cos(3\theta)$

12. $r = \cos(2\theta)$

13. $r = -4 \sin(2\theta)$

14. $r = -2 \sin(3\theta)$

15. $r = \dfrac{\theta}{\pi}$

16. $r = \dfrac{\pi}{\theta}$

17. $r = 2^{\theta/\pi}$

18. $r = e^{-\theta}$

In Exercises 19 to 30, graph all points (r, θ) satisfying the given conditions.

19. $r = 2 \sin \theta,\ \pi \le \theta \le 3\pi/2$

20. $r = 2 \cos \theta,\ \pi/2 \le \theta \le \pi$

21. $r = 2 + 4 \cos \theta,\ 0 \le \theta \le \pi$

22. $r = 2 - 4 \sin \theta,\ 0 \le \theta \le \pi$

23. $r = -\sin(3\theta),\ 0 \le \theta \le \pi/2$

24. $r = \cos(3\theta),\ 0 \le \theta \le \pi/2$

25. $r = 2 \cos(2\theta),\ \pi/2 \le \theta \le \pi$

26. $r = -\sin(2\theta),\ \pi/2 \le \theta \le \pi$

27. $r = 2 \cos(4\theta),\ 0 \le \theta \le \pi/2$

28. $r = 2 \sin(4\theta),\ 0 \le \theta \le \pi/2$

29. $r = \dfrac{2\theta}{\pi},\ -\pi \le \theta \le \pi$

30. $r = 2^{\theta/\pi},\ -2\pi \le \theta \le 0$

In Exercises 31 to 38, graph the set described.

31. $\{(r, \theta): r \le \cos \theta,\ 0 \le \theta \le \pi\}$

32. $\{(r, \theta): r \le \sin \theta,\ 0 \le \theta \le \pi\}$

33. $\{(r, \theta): r \le \sin 2\theta,\ 0 \le \theta \le \pi\}$

34. $\{(r, \theta): r \le \cos 2\theta,\ 0 \le \theta \le \pi\}$

35. $\{(r, \theta): |r| < \sin \theta\}$

36. $\{(r, \theta): |r| < \cos \theta\}$

37. $\{(r, \theta): r = |\cos \theta|\}$

38. $\{(r, \theta): r = |\sin \theta|\}$

In Exercises 39 and 40, graph the given equation. (*Hint:* First transform to an equivalent polar equation.)

39. $(x^2 + y^2)^{3/2} = y^2$

40. $(x^2 + y^2)^{3/2} = x^2$

10.8 PARAMETRIC EQUATIONS

Suppose a certain bee flies along a circular path in the x, y-plane, perhaps a circle of radius 1 centered at the origin (see Figure 10.84). If we wish to describe this situation mathematically, we might say that the bee follows the path $x^2 + y^2 = 1$. However, this does not tell us where the bee is at any particular time, nor does it tell us the direction of its flight.

Time is the key ingredient here; the x- and y-coordinates of the bee will vary with time, which means that both x and y are *functions of time*, say,

$$x = X(t) \quad \text{and} \quad y = Y(t)$$

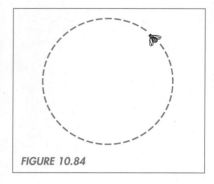

FIGURE 10.84

We call the equations $x = X(t)$ and $y = Y(t)$ **parametric equations** for x and y; we call t the **parameter**. As t varies, the corresponding values for x and y will vary, thereby tracing out a path in the x, y-plane.

Example 1

Suppose a bee moves in the x, y-plane according to the parametric equations

$$x = \cos(\pi t) \quad \text{and} \quad y = \sin(\pi t)$$

where the parameter t is measured in seconds. Determine the motion of the bee as t varies from 0 to 2 seconds.

SOLUTION For each value of t there are corresponding values for x and y that determine the bee's position in the plane. We make a table to keep track of the different positions as t varies from 0 to 2 seconds. Our table consists of three columns, one for time, and the other two for the corresponding values of x and y:

t	$x = \cos(\pi t)$	$y = \sin(\pi t)$
0	1	0
1/4	$1/\sqrt{2}$	$1/\sqrt{2}$
1/2	0	1
3/4	$-1/\sqrt{2}$	$1/\sqrt{2}$
1	-1	0
5/4	$-1/\sqrt{2}$	$-1/\sqrt{2}$
3/2	0	-1
7/4	$1/\sqrt{2}$	$-1/\sqrt{2}$
2	1	0

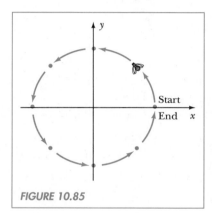

FIGURE 10.85

Figure 10.85 shows the x, y-position of the bee (computed in our table) after each quarter second interval of time. The arrows indicate the direction of the motion.

The bee is flying counterclockwise along what appears to be a circular path. We can prove that the path is indeed circular by *eliminating the parameter t* from the parametric equations. This means that we shall reduce the two equations involving x, y, and t to one equation in x and y alone. First, recall the fundamental identity $\sin^2 \theta + \cos^2 \theta = 1$. Beginning with the original parametric equations,

$$x = \cos(\pi t) \quad \text{and} \quad y = \sin(\pi t)$$

we square them, obtaining

$$x^2 = \cos^2(\pi t) \quad \text{and} \quad y^2 = \sin^2(\pi t)$$

Now, adding these equations, we find that

$$x^2 + y^2 = \cos^2(\pi t) + \sin^2(\pi t)$$

The right side is just the constant 1, so we have

$$\begin{cases} x = \cos(\pi t) \\ y = \sin(\pi t) \end{cases} \quad \text{implies} \quad x^2 + y^2 = 1$$

This means that for any value of t, the corresponding x- and y-coordinates for the bee must satisfy the equation $x^2 + y^2 = 1$. In other words, the motion of the bee is restricted to the unit circle. ■

In general, we will be given a set of parametric equations

$$x = X(t) \quad \text{and} \quad y = Y(t)$$

with certain conditions on the parameter t, say, $t \in [a, b]$. To graph the motion described by these equations, we follow a two-step procedure.

Graphing Parametric Equations
Given $x = X(t)$, $y = Y(t)$, and $t \in [a, b]$.
Step 1   If possible, eliminate the parameter t from the parametric equations to obtain an x, y equation. Graph this equation with a dashed line, which we call the *guide curve* for the motion. The motion described by the parametric equations must be restricted to the guide curve at all times.
Step 2   Make a table consisting of three columns: t, x, and y. For values of t varying from a to b, compute the corresponding values of $x = X(t)$ and $y = Y(t)$. Graph the x, y-coordinates from the table (these should fall on the guide curve). Indicate the starting point (when $t = a$), the direction of motion, and the end point (when $t = b$).

Example 2

An object moves according to the parametric equations $x = -4\cos(\pi t)$, $y = 2\sin(\pi t)$, and $t \in [0, 1]$. Graph the motion.

SOLUTION   Step 1.   We eliminate the parameter t as follows:

$$x = -4\cos(\pi t) \quad \text{is equivalent to} \quad -\frac{x}{4} = \cos(\pi t)$$

$$y = 2\sin(\pi t) \quad \text{is equivalent to} \quad \frac{y}{2} = \sin(\pi t)$$

Squaring these equations, we have

$$\frac{x^2}{16} = \cos^2(\pi t) \qquad \text{and} \qquad \frac{y^2}{4} = \sin^2(\pi t)$$

Adding and replacing $\cos^2(\pi t) + \sin^2(\pi t)$ with 1, we obtain

$$\frac{x^2}{16} + \frac{y^2}{4} = 1$$

Thus our guide curve is an ellipse centered at the origin with major axis the x-axis. The motion of the object must be restricted to this ellipse.

Step 2.   We construct the following table of values:

t		$x = -4\cos(\pi t)$	$y = 2\sin(\pi t)$
Start	0	-4	0
	$\frac{1}{4}$	$-2\sqrt{2}$	$\sqrt{2}$
	$\frac{1}{2}$	0	2
	$\frac{3}{4}$	$2\sqrt{2}$	$\sqrt{2}$
End	1	4	0

Thus we find that the object moves clockwise along the upper half of the guide curve found in step 1, as shown in Figure 10.86.

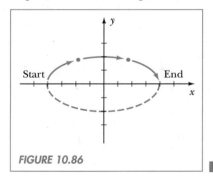

FIGURE 10.86

Remark

As Example 2 illustrates, the motion described by a set of parametric equations may traverse only a portion of the guide curve, not its entire length.

Next, we consider the physical situation of a ball rolling up an inclined plane. The ball slows down and stops, from the force of gravity, and rolls back downhill, regaining most of its original speed. Parametric equations may be used to describe this type of motion.

Example 3

An object moves according to the parametric equations $x = 2t - t^2$, $y = 2t - t^2$, and $t \in [0, 2]$. Graph the motion.

SOLUTION Step 1. Both expressions for x and y are identical. Thus

$$\begin{cases} x = 2t - t^2 \\ y = 2t - t^2 \end{cases} \quad \text{implies} \quad y = x$$

Therefore our guide curve is the diagonal $y = x$.

Let us make a further observation that will be helpful. Note that

$$2t - t^2 \le 1 \qquad \text{for all } t$$

[You can show this by graphing the quadratic function $F(t) = 2t - t^2$.] This means that for any t, the maximum value possible for x or y is 1. It follows that $(1, 1)$ is the point farthest up the diagonal that the object may go.

Step 2. We construct a table and graph the motion as shown in Figure 10.87.

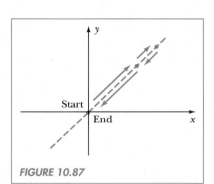

FIGURE 10.87

t		$x = 2t - t^2$	$y = 2t - t^2$
Start	0	0	0
	1/2	3/4	3/4
	1	1	1
	3/2	3/4	3/4
End	2	0	0

EXERCISES 10.8

In Exercises 1 to 20, graph the motion of an object that moves according to the given parametric equations. Use the two-step procedure described in the text.

1. $x = -\cos(\pi t)$, $y = \sin(\pi t)$, $t \in [0, 2]$

2. $x = 3 \cos\left(\dfrac{\pi t}{2}\right)$, $y = -3 \sin\left(\dfrac{\pi t}{2}\right)$, $t \in [0, 2]$

3. $x = \sin(\pi t)$, $y = 3 \cos(\pi t)$, $t \in [0, 2]$

4. $x = 2 \cos(\pi t)$, $y = \sin(\pi t)$, $t \in [0, 1]$

5. $x = 4 \cos(t)$, $y = |2 \sin(t)|$, $t \in [0, 2\pi]$

6. $x = \cos(t)$, $y = -|\sin(t)|$, $t \in [0, 2\pi]$

7. $x = t^2$, $y = 4 - t^4$, $t \in [-2, 2]$

8. $x = 4 - t^2$, $y = t^2 + 3$, $t \in [-2, 2]$

9. $x = \sqrt{1 + t^2}$, $y = t$, $t \in (-\infty, +\infty)$

10. $y = \sqrt{4 + t^2}$, $x = 2t$, $t \in (-\infty, +\infty)$

11. $x = t^2 - 2t$, $y = 2t - t^2$, $t \in [0, +\infty)$

12. $x = 8 - t$, $y = \sqrt{t + 1}$, $t \in [0, +\infty)$

13. $x = 2^t$, $y = 2^{-t}$, $t \in [0, +\infty)$

14. $x = \sin\left(\dfrac{\pi t}{4}\right)$, $y = \csc\left(\dfrac{\pi t}{4}\right)$, $t \in (0, 4)$

15. $x = \cos(2t)$, $y = \sin(t)$, $t \in [0, 2\pi]$

16. $x = 2\cos(2t)$, $y = \sin(t)$, $t \in [-\pi/2, \pi/2]$

17. $x = 2t$, $y = 2t - t^2$, $t \in [0, +\infty)$

18. $x = 4\cos^2 t$, $y = 4\cos(t)\sin(t)$, $t \in [0, \pi]$

19. $x = 2t$, $y = \sin(\pi t)$, $t \in [0, 4]$

20. $x = \cos(t)$, $y = t$, $t \in [0, 3\pi]$

21. Each of the following parametric equations describes the motion of a bee in the plane.
Bee 1: $x = 2(1 - t)$, $y = 2\sqrt{2t - t^2}$, $t \in [0, 2]$
Bee 2: $x = 2\cos(\pi t/2)$, $y = 2\sin(\pi t/2)$, $t \in [0, 2]$
Bee 3: $x = -2\cos(\pi t/2)$, $y = 2\sin(\pi t/2)$, $t \in [0, 2]$
(a) Eliminate t to determine the guide curve for each bee.
(b) Even though the bees have the same guide curve, their motions are not the same. For example, each

bee passes through the point $(1, \sqrt{3})$ at different times. Find the time that each bee arrives at $(1, \sqrt{3})$.

22. Three bees move according to the following parametric equations.
Bee 1: $x = \sqrt{t}$, $y = t^{-1}$, $t \in (0, +\infty)$
Bee 2: $x = t^2$, $y = t^{-4}$, $t \in (0, +\infty)$
Bee 3: $x = e^t$, $y = e^{-2t}$, $t \in (-\infty, +\infty)$
(a) Eliminate t to determine the guide curve for each bee.
(b) Find the time that each bee arrives at $(2, 1/4)$.

In Exercises 23 and 24, the motion of a pair of bees is described by the given parametric equations. Graph the motion for each bee on the same axes. Will the bees collide? If so, where and when?

23. Bee 1: $x = t^{1/3}$, $y = t^{2/3}$, $t \in [0, +\infty)$
Bee 2: $x = t/4$, $y = \sqrt{t + 8}$, $t \in [0, +\infty)$

24. Bee 1: $x = 6 - 2t$, $y = t$, $t \in [0, +\infty)$
Bee 2: $x = 2\sqrt{5}\cos(t)$, $y = \sqrt{5}\sin(t)$, $t \in [0, +\infty)$

CHAPTER 10 / REVIEW

In Exercises 1 to 6, graph the given equation. For a parabola, find the axis, vertex, focus, and directrix. For an ellipse or hyperbola, find the center, major and minor axes, vertices, and foci. Also, include the asymptotes for a hyperbola.

1. $2x^2 + y = 0$

2. $4x^2 + y^2 - 1 = 0$

3. $4x^2 - 9y^2 + 36 = 0$

4. $y^2 - 4x - 8y - 8 = 0$

5. $4x^2 + 9y^2 + 24x - 18y + 9 = 0$

6. $4x^2 - y^2 - 8x + 2y - 1 = 0$

7. A parabolic mirror of diameter 8 inches is ground to a depth of 1/4 inch. How far is the focus from the vertex of the mirror?

8. Find an equation for the ellipse with foci at $(\pm 3\sqrt{3}, 0)$ and whose graph passes through the point $(2\sqrt{5}, -2)$.

9. Find an equation for the hyperbola with foci at $(0, \pm 2\sqrt{15})$ and asymptotes $y = \pm 2x$.

10. Find an equation for the ellipse with vertices at $(-3, 0)$, $(0, 6)$, $(-3, 12)$, and $(-6, 6)$.

In Exercises 11 to 13, find an equation for the set of all points $P(x, y)$ described. Write the equation in standard form and name the curve.

11. The distance from P to the line $x = -1$ equals the distance from P to the point $(3, 0)$.

12. The distance from P to the line $y = -6$ is three times its distance from $(0, 2)$.

13. The distance from P to the line $x = -3$ is one-half its distance from $(0, 1)$.

14. A quadrilateral is determined by the four vertices of the ellipse $y^2/a^2 + x^2/b^2 = 1$. Find the area of this quadrilateral in terms of a and b.

15. Find the equation of the curve obtained by rotating $y = x^2$ about the origin $45°$.

In Exercises 16 and 17, rewrite the given equation in an x', y' system so that it has no $x'y'$ term. Name the curve, find the x', y'-coordinates of the vertices and foci, and graph.

16. $5x^2 + 6xy + 5y^2 - 8 = 0$

17. $x^2 + 2\sqrt{3}xy - y^2 - 2\sqrt{3}x - 2y = 0$

18. Given polar coordinates $A(4, 5\pi/4)$, $B(-3, 2\pi/3)$, $C(2, \pi)$ and $D(0, 1)$.
    (a) Graph each point.
    (b) Find Cartesian coordinates for each point.
    (c) Find two other polar coordinates for A, one with positive r and one with negative r.
    (d) List all possible polar coordinates for B.
    (e) List all possible polar coordinates for D.

19. Given Cartesian coordinates $P(1, 2)$ and $Q(-3\sqrt{3}, 3)$. Find polar coordinates for P and Q.

In Exercises 20 and 21, transform the Cartesian equation into an equivalent polar equation. If possible, write the equation in the form $r = F(\theta)$, where $F(\theta)$ is a function of θ.

20. $(x^2 + y^2)^{3/2} = x^2 - y^2$ 21. $(x - 1)^2 + (y - 2)^2 = 5$

In Exercises 22 and 23, transform the given polar equation into an equivalent Cartesian equation.

22. $r = \dfrac{3}{1 + 2\cos\theta}$ 23. $r = \sec\theta - \csc\theta$

24. Graph the set $\{(r, \theta): -2 \le r \le 1, \pi/4 \le \theta \le \pi/2\}$.

In Exercises 25 to 27, graph the given polar equation.

25. $r = -3\cos\theta$

26. $r = 1 + 2\sin\theta$

27. $r = \left(\dfrac{1}{2}\right)^{\theta/\pi}$

In Exercises 28 and 29, graph the points (r, θ) satisfying the given conditions.

28. $r = 1 - \sin\theta, 0 \le \theta \le \pi$

29. $r = 4\cos(2\theta), 0 \le \theta \le \pi$

In Exercises 30 to 32, graph the motion of an object that moves according to the given parametric equations.

30. $x = 6\cos t, y = 3\sin t, t \in [0, 3\pi/2]$

31. $x = \sin \pi t, y = \sin^2 \pi t, t \in [0, 2]$

32. $x = 2\sqrt{1 + t^2}, y = t, t \in [0, +\infty)$

CHAPTER 11 Sequences and Sums

SEQUENCES

When the word **sequence** is used in a phrase such as "sequence of events" or "sequence of objects," it means a following of one thing after another. In mathematics, the word sequence refers to a set of numbers, one following another by a given order of succession. As an example, consider the following sequence:

$$1, \frac{1}{2}, \frac{1}{4}, \frac{1}{8}, \frac{1}{16}, \cdots$$

Each number is called a *term* of the sequence. Thus 1 is the first term, 1/2 is the second term, 1/4 is the third term, and so on. Commas are used to separate terms, and three dots at the end indicate that the sequence goes on forever. In this case we say that the sequence is *infinite* because it never ends. However, if a sequence stops after a finite number of terms, then we say that the sequence is *finite*.

Given an infinite sequence, we may define a function F corresponding to the sequence as follows:

Domain of F: **N**, the set of positive integers

Rule for F: $F(n) = n$th term of the sequence

Thus for the sequence

$$1, \frac{1}{2}, \frac{1}{4}, \frac{1}{8}, \frac{1}{16}, \cdots$$

we have

$$F(1) = 1, \ F(2) = \frac{1}{2}, \ F(3) = \frac{1}{4}, \ F(4) = \frac{1}{8}, \ F(5) = \frac{1}{16}$$

and so on. Notice that there is an obvious pattern to these terms, namely, 1 divided by a power of 2. It is not difficult to see that $F(n) = 1/2^{n-1}$ is a formula for the nth term of this sequence.

The idea of associating a function with every sequence leads us to the formal definition of an infinite sequence.

Definition of Infinite Sequence
An infinite sequence is a function with domain **N**.

Thus, the most obvious way for us to refer to a sequence is to indicate the function that defines it. We would say

Consider the sequence defined by the function $F(n)$.

The letter n instead of x is normally used for the variable of a sequence because the domain of the function consists of the positive integers **N**.

In our discussion of sequences, we shall use four methods of describing a sequence. We explore these methods in Examples 1 to 10.

Methods for Describing Sequences
1. Write the formula for the nth term, $F(n)$.
2. List the terms.
3. Use a graph.
4. Use a recurrence relation.

Example 1

Consider the sequence defined by the function $F(n) = \dfrac{1}{n}$. Describe this sequence by (a) listing its terms and (b) displaying its graph.

SOLUTION (a) Listing the terms of the sequence $F(n) = \dfrac{1}{n}$, we have

$$1, \frac{1}{2}, \frac{1}{3}, \frac{1}{4}, \frac{1}{5}, \cdots$$

Of course, it is impossible to list all the terms of this sequence. However, since the numbers follow a nice pattern, it is obvious what the rest of the terms should be.

(b) In general, the graph of a sequence defined by the function F consists of the points

$$\{(n, F(n)): n \in \mathbf{N}\}$$

Figure 11.1 shows a portion of the graph for the sequence $F(n) = 1/n$; we graphed the points

$$(1, 1) \quad \left(2, \frac{1}{2}\right) \quad \left(3, \frac{1}{3}\right) \quad \left(4, \frac{1}{4}\right) \quad \left(5, \frac{1}{5}\right) \quad \blacksquare$$

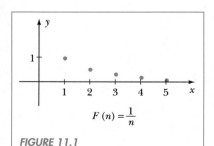

$$F(n) = \frac{1}{n}$$

FIGURE 11.1

Example 2

Consider the sequence defined by the function $F(n) = \dfrac{2n + 1}{n + 1}$. List the first five terms and graph.

SOLUTION We begin by computing the values of the first five terms:

$$F(1) = \frac{2(1) + 1}{1 + 1} = \frac{3}{2}$$

$$F(2) = \frac{2(2) + 1}{2 + 1} = \frac{5}{3}$$

$$F(3) = \frac{2(3) + 1}{3 + 1} = \frac{7}{4}$$

$$F(4) = \frac{2(4) + 1}{4 + 1} = \frac{9}{5}$$

$$F(5) = \frac{2(5) + 1}{5 + 1} = \frac{11}{6}$$

Now we may list the terms as follows:

$$\frac{3}{2}, \frac{5}{3}, \frac{7}{4}, \frac{9}{5}, \frac{11}{6}$$

Figure 11.2 is a graph of these terms. ■

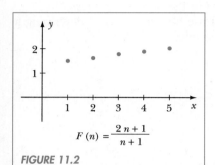

$$F(n) = \frac{2n + 1}{n + 1}$$

FIGURE 11.2

Example 3

Consider the sequence 2, 5, 8, 11, 14, ... Assuming the terms follow the obvious pattern, find a formula for the function defining this sequence.

SOLUTION Notice that 3 is added each time we move from one term to the next. Thus we can get to any term we desire by adding an appropriate multiple of 3 to the first term.

n No. of Term	$F(n)$ Value of Term	
1	2	
2	2 + (1)3	= 5
3	2 + (2)3	= 8
4	2 + (3)3	= 11
⋮	⋮	
n	2 + $(n - 1)$3	

Hence the function for this sequence is $F(n) = 2 + (n - 1)3$, or, more simply, $F(n) = 3n - 1$. ∎

An **arithmetic sequence** has the property that there exists a constant d such that every term after the first may be obtained by adding d to the previous term. Thus if a is the first term, a list of the terms is

$$a, a + d, a + 2d, a + 3d, \ldots$$

It is not hard to see that a formula for the nth term of this sequence is given by

$$F(n) = a + (n - 1)d$$

Note that the sequence in Example 3 is an arithmetic sequence with $a = 2$ and $d = 3$.

Example 4

Consider the sequence $2, \dfrac{2}{3}, \dfrac{2}{9}, \dfrac{2}{27}, \dfrac{2}{81}, \ldots$. Assuming that the terms follow the obvious pattern, find a formula for the function defining this sequence.

SOLUTION   In this sequence, each time we move from one term to the next we multiply by the factor 1/3. Thus we can get to any term desired by multiplying the first term by an appropriate power of 1/3.

n No. of Term	$F(n)$ Value of Term	
1	2	
2	$2\left(\dfrac{1}{3}\right)$	$= \dfrac{2}{3}$
3	$2\left(\dfrac{1}{3}\right)^2$	$= \dfrac{2}{9}$
4	$2\left(\dfrac{1}{3}\right)^3$	$= \dfrac{2}{27}$
5	$2\left(\dfrac{1}{3}\right)^4$	$= \dfrac{2}{81}$
$\vdots$	$\vdots$	
n	$2\left(\dfrac{1}{3}\right)^{n-1}$	

Therefore the function for this sequence is given by the formula $F(n) = 2(1/3)^{n-1}$. ∎

A **geometric sequence** has the property that there exists a constant r, called the **ratio**, such that every term after the first may be obtained by multiplying the previous term by r. If a is the first term, then a list of the terms is

$$a, ar, ar^2, ar^3, \ldots$$

In this case, a formula for the nth term is given by

$$F(n) = ar^{n-1}$$

Note that the sequence in Example 4 is a geometric sequence with $a = 2$ and $r = 1/3$.

It will be important to recognize whether or not a given sequence is geometric and, if it is, to be able to find the ratio r. Usually one can tell by simply listing the terms of the sequence. In addition, note that if a sequence defined by $F(n)$ is geometric, then

$$\frac{F(n+1)}{F(n)} = \frac{ar^n}{ar^{n-1}} = r$$

Thus the ratio of any two successive terms in a geometric sequence must equal the constant ratio r.

Example 5

Determine if the sequence defined by the function F is geometric. If it is, find the ratio r.

(a) $F(n) = (-1)^{n-1}\dfrac{3^n}{4}$, (b) $F(n) = 2^{n-1} + 1$

SOLUTION (a) Listing the terms of $F(n) = (-1)^{n-1}\dfrac{3^n}{4}$, we have

$$\frac{3}{4}, \quad -\frac{3^2}{4}, \quad \frac{3^3}{4}, \quad -\frac{3^4}{4}, \quad \ldots$$

This appears to be geometric with ratio $r = -3$. As a check, consider the ratio of successive terms.

$$\frac{F(n+1)}{F(n)} = \frac{(-1)^n 3^{n+1}/4}{(-1)^{n-1} 3^n/4} = (-1)^1 3^1 = -3$$

This proves the sequence is geometric with ratio -3.

(b) Listing the terms of $F(n) = 2^{n-1} + 1$, we have

$$2, 3, 5, 9, \ldots$$

This is *not* a geometric sequence because $F(2) = \dfrac{3}{2}F(1)$, but $F(3) = \dfrac{5}{3}F(2)$. ∎

Next, we define a special sequence generated by the **factorial** function, $F(n) = n!$ The notation $n!$ is read "n factorial." We define it as follows:

$$1! = 1$$
$$2! = 1 \cdot 2 \qquad\;\;\; = 2$$
$$3! = 1 \cdot 2 \cdot 3 \quad\;\; = 6$$
$$4! = 1 \cdot 2 \cdot 3 \cdot 4 = 24$$
$$\vdots \qquad\quad \vdots \qquad\quad \vdots$$

In general, $n!$ is the product of all the positive integers $\leq n$ for each $n \geq 1$. It is convenient to extend the definition to zero so that $0! = 1$.

Example 6

Evaluate the given expression: (a) $6!$, (b) $\dfrac{10!}{8!}$.

SOLUTION (a) By definition,

$$6! = 1 \cdot 2 \cdot 3 \cdot 4 \cdot 5 \cdot 6 = 720$$

(b) In this case we cancel common factors in the quotient before multiplying.

$$\frac{10!}{8!} = \frac{1 \cdot 2 \cdot 3 \cdot 4 \cdot 5 \cdot 6 \cdot 7 \cdot 8 \cdot 9 \cdot 10}{1 \cdot 2 \cdot 3 \cdot 4 \cdot 5 \cdot 6 \cdot 7 \cdot 8} = 9 \cdot 10 = 90 \quad \blacksquare$$

Example 7

Simplify the expression $\dfrac{n!}{(n+1)!}$.

SOLUTION We assume that n represents a nonnegative integer. Thus

$$\frac{n!}{(n+1)!} = \frac{1 \cdot 2 \cdot 3 \cdots n}{1 \cdot 2 \cdot 3 \cdots n \cdot (n+1)} = \frac{1}{n+1} \quad \blacksquare$$

In Example 8 we describe a sequence by giving the value of its first term and then defining a rule that explains how to go from one term to the next.

Example 8

Consider the sequence defined by

$$F(1) = 1 \quad \text{and} \quad F(n) = \frac{F(n-1)}{F(n-1) + 1} \quad \text{for } n \geq 2$$

List the first five terms of this sequence.

SOLUTION We are given the first term of the sequence, $F(1) = 1$. To find the second term, we use the formula

$$F(n) = \frac{F(n-1)}{F(n-1)+1}$$

with $n = 2$. Thus

$$F(2) = \frac{F(1)}{F(1)+1} = \frac{1}{1+1} = \frac{1}{2}$$

Now that we know $F(2)$, we may use the formula again to find $F(3)$.

$$F(3) = \frac{F(2)}{F(2)+1} = \frac{1/2}{1/2+1} = \frac{1}{3}$$

Each time we find the value of a term, we use it in the formula to find the next term. The process repeats until we reach the desired term.

$$F(4) = \frac{F(3)}{F(3)+1} = \frac{1/3}{1/3+1} = \frac{1}{4}$$

$$F(5) = \frac{F(4)}{F(4)+1} = \frac{1/4}{1/4+1} = \frac{1}{5}$$

Thus the first five terms of the sequence are 1, 1/2, 1/3, 1/4, 1/5. Notice that this is the same sequence discussed in Example 1. ■

When we define a sequence by giving the value of the first term and then writing a formula that tells how to find the nth term, $F(n)$, using the previous term, $F(n-1)$, the process is called a **recursive definition**. (A recursive definition was presented in Example 8.) In Example 9 we show how recursive definitions may be used to describe arithmetic and geometric sequences.

Example 9

Find a recursive definition for the given sequence:
(a) $a, a + d, a + 2d, a + 3d, \ldots$ (arithmetic)
(b) $a, ar, ar^2, ar^3, \ldots$ (geometric)

SOLUTION (a) Since we add d to get from one term to the next, we have $F(n) = F(n-1) + d$ for $n \geq 2$. Therefore we write the recursive definition as follows:

$$F(1) = a$$

$$F(n) = F(n-1) + d \qquad \text{for } n \geq 2$$

(b) Since we multiply by r to get from one term to the next, we write

$$F(1) = a$$

$$F(n) = rF(n-1) \qquad \text{for } n \geq 2 \quad ■$$

Recursive definitions can be more general in the sense that we may write a formula for the nth term using several previous terms.

Example 10

Consider the sequence defined by $F(1) = 1$, $F(2) = 1$, and $F(n) = F(n-1) + F(n-2)$ for $n \geq 3$. List the first six terms.

SOLUTION We have

$$F(1) = 1$$

$$F(2) = 1$$

$$F(3) = F(2) + F(1) = 1 + 1 = 2$$

$$F(4) = F(3) + F(2) = 2 + 1 = 3$$

$$F(5) = F(4) + F(3) = 3 + 2 = 5$$

$$F(6) = F(5) + F(4) = 5 + 3 = 8$$

Therefore the first six terms of the sequence are 1, 1, 2, 3, 5, 8. ∎

Remark

The sequence in Example 10 is called the *Fibonacci sequence*; it has several interesting properties related to the geometry of nature.

EXERCISES 11.1

In Exercises 1 to 10, list the first six terms of the sequence defined by the given function and graph.

1. $F(n) = 2n - 1$

2. $F(n) = \dfrac{1}{2} n(n - 1)$

3. $F(n) = \sin\left(\dfrac{n\pi}{2}\right)$

4. $F(n) = \cos(n\pi)$

5. $F(n) = \dfrac{n-1}{n+1}$

6. $F(n) = 3 - \dfrac{1}{n}$

7. $F(n) = 2^{-n}$

8. $F(n) = n \log_2\left(\dfrac{1}{n}\right)$

9. $F(n) = \begin{cases} 2 - \dfrac{1}{n}, & n \text{ odd} \\ 2 + \dfrac{1}{n}, & n \text{ even} \end{cases}$

10. $F(n) = (-1)^{n+1} \dfrac{n}{n+1}$

In Exercises 11 and 12, list the first twelve terms of the sequence defined by the given function.

11. $F(n) = $ smallest prime number $\geq n$.

12. $F(n) = $ the number of prime numbers $\leq n$.

In Exercises 13 to 24, find a formula for the nth term of the given sequence. Assume that the numbers follow the obvious pattern. State whether the sequence is arithmetic or geometric. In the case it is geometric determine the ratio r.

13. $-1, 1, 3, 5, 7, \ldots$

14. $14, 7, 0, -7, -14, \ldots$

15. $3, 1, 1/3, 1/9, 1/27, \ldots$

16. $10, 20, 40, 80, 160, \ldots$

17. $5, -5, 5, -5, 5, \ldots$

18. $-1, 2, -3, 4, -5, \ldots$

19. $2, 3/2, 4/3, 5/4, 6/5, \ldots$

20. $4, -2, 1, -1/2, 1/4, \ldots$

21. $15, 21, 27, 33, 39, \ldots$

22. $3, 5, 7, 9, 11, \ldots$

23. $4, 5, 6, 7, 8, 9, \ldots$

24. $2, 6, 12, 20, 30, 42, \ldots$

In Exercises 25 and 26, list the first five terms of the sequence indicated by the graph. Assuming that the numbers follow the obvious pattern, find a formula for the nth term.

25.

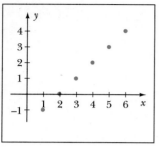

26.

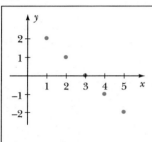

In Exercises 27 to 32, determine if the sequence defined by the given function is geometric. If it is, find the ratio r.

27. $F(n) = 4^{n-1}$

28. $F(n) = \dfrac{5}{2^n}$

29. $F(n) = (-1)^{n-1}n$

30. $F(n) = 3n^2$

31. $F(n) = \dfrac{3^n}{5^{n-1}}$

32. $F(n) = (-1)^{n-1}4(3^n)$

In Exercises 33 to 42, a sequence is defined by the given recursive definition. List the first six terms.

33. $F(1) = 3$, $F(n) = 2F(n-1)$ for $n \geq 2$

34. $F(1) = 3$, $F(n) = 2 + F(n-1)$ for $n \geq 2$

35. $F(1) = 1$, $F(n) = F(n-1) + (-1)^{n-1}2$ for $n \geq 2$

36. $F(1) = 1$, $F(n) = (n-2)F(n-1) + (-1)^{n-1}$ for $n \geq 2$

37. $F(1) = 1$, $F(2) = 2$, $F(n) = F(n-1) - F(n-2)$ for $n \geq 3$

38. $F(1) = 1$, $F(2) = 2$, $F(n) = 2F(n-1) - F(n-2)$ for $n \geq 3$

39. $F(1) = 1$, $F(2) = 1$, $F(n) = (n-1)F(n-1) + (n-2)F(n-2)$ for $n \geq 3$

40. $F(1) = 1$, $F(2) = 2$, $F(n) = F(n-1) + 2F(n-2)$ for $n \geq 3$

41. $F(1) = 1$, $F(2) = 2$, $F(n) = F(n-1)F(n-2)$ for $n \geq 3$

42. $F(1) = 2$, $F(2) = 5$, $F(n) = \dfrac{F(n-1)}{F(n-2)}$ for $n \geq 3$

43. List the first eight terms of the sequence defined by $F(1) = a$, $F(2) = b$, $F(n) = \dfrac{F(n-1)}{F(n-2)}$ for $n \geq 3$. Will the sequence repeat itself?

44. Consider the sequence defined by $F(1) = 1$ and $F(n) = nF(n-1)$ for $n \geq 2$. List the first six terms of this sequence. Verify that $F(n) = n!$ for $n = 1, 2, \ldots, 6$.

In Exercises 45 to 50, list the first six terms of the sequence defined by the given function.

45. $F(n) = (n+1)!$

46. $F(n) = \dfrac{(n+2)!}{n!}$

47. $F(n) = 2(n!)$

48. $F(n) = (2n)!$

49. $F(n) = \dfrac{(2n)!}{n!}$

50. $F(n) = \dfrac{n!(n-1)!}{(n+1)!}$

In Exercises 51 to 54, evaluate the expression $\dfrac{n!}{k!(n-k)!}$ for the given values of n and k.

51. $n = 5$, $k = 3$

52. $n = 8$, $k = 2$

53. $n = 9$, $k = 1$

54. $n = 6$, $k = 6$

We begin this section by introducing a shorthand notation for writing the sum of a finite sequence of numbers. The Greek letter sigma, denoted by $\sum$, is used in the notation to indicate the sum.

Summation Notation
If n is a positive integer, then

$$\sum_{k=1}^{n} F(k) = F(1) + F(2) + F(3) + \cdots + F(n)$$

Example 1

Expand and evaluate the given sum (a) $\sum_{k=1}^{5} 2k$, (b) $\sum_{k=1}^{4} (3k - 1)$, (c) $\sum_{k=1}^{6} 3$.

SOLUTION

(a)
$$\sum_{k=1}^{5} 2k = 2(1) + 2(2) + 2(3) + 2(4) + 2(5)$$
$$= 2 + 4 + 6 + 8 + 10$$
$$= 30$$

(b) $\sum_{k=1}^{4} (3k - 1) = (3(1) - 1) + (3(2) - 1) + (3(3) - 1) + (3(4) - 1)$

$$= 2 + 5 + 8 + 11$$
$$= 26$$

(c) In this case we are evaluating the sum $\sum_{k=1}^{6} F(k)$, where $F(k) = 3$ for all k. Thus

$$\sum_{k=1}^{6} 3 = 3 + 3 + 3 + 3 + 3 + 3$$
$$= 6(3)$$
$$= 18 \quad\blacksquare$$

The variable k in the sum $\sum_{k=1}^{n} F(k)$ is called the **index** of the summation. Any letter may be used for the index. Furthermore, the starting value of the index may be different from 1.

Example 2

Expand and evaluate the given sum: (a) $\sum_{i=0}^{3} \dfrac{2^i}{i!}$, (b) $\sum_{j=3}^{6} \dfrac{(-1)^{j-1}}{j}$.

SOLUTION (a) Recall that $0! = 1$. Thus

$$\sum_{i=0}^{3} \frac{2^i}{i!} = \frac{2^0}{0!} + \frac{2^1}{1!} + \frac{2^2}{2!} + \frac{2^3}{3!}$$

$$= \frac{1}{1} + \frac{2}{1} + \frac{4}{2} + \frac{8}{6}$$

$$= 1 + 2 + 2 + \frac{4}{3}$$

$$= \frac{19}{3}$$

(b)
$$\sum_{j=3}^{6} \frac{(-1)^{j-1}}{j} = \frac{(-1)^2}{3} + \frac{(-1)^3}{4} + \frac{(-1)^4}{5} + \frac{(-1)^5}{6}$$

$$= \frac{1}{3} - \frac{1}{4} + \frac{1}{5} - \frac{1}{6}$$

$$= \frac{7}{60} \quad \blacksquare$$

It is possible to write a given sum using summation notation in more than one way. For instance, the sum in Example 2(b) could also be written as follows:

$$\sum_{k=1}^{4} \frac{(-1)^{k-1}}{k+2} \quad \text{or} \quad \sum_{k=0}^{3} \frac{(-1)^{k}}{k+3}$$

Each form represents the same sum, $1/3 - 1/4 + 1/5 - 1/6$.

Now we consider the problem of how to rewrite a given sum of numbers using the shorthand summation notation. The solution depends on whether or not we can find a function that describes the sequence of numbers being added.

Example 3

Use summation notation to rewrite the sum $1 + 4 + 9 + 16 + 25$.

SOLUTION We need to find a function for the sequence 1, 4, 9, 16, 25. We recognize a pattern: the first number is 1^2, the second number is 2^2, the third is 3^2, and so on. The function that describes this sequence is clearly $F(k) = k^2$. Thus

$$1 + 4 + 9 + 16 + 25 = \sum_{k=1}^{5} k^2 \quad \blacksquare$$

Example 4

Use summation notation to rewrite the sum $3 + .3 + .03 + .003 + .0003 + .00003$.

SOLUTION We need to find a function for the sequence

$$3, .3, .03, .003, .0003, .00003$$

We can rewrite these decimals as fractions.

$$\frac{3}{1}, \frac{3}{10}, \frac{3}{10^2}, \frac{3}{10^3}, \frac{3}{10^4}, \frac{3}{10^5}$$

The function that describes this sequence is given by

$$F(k) = \frac{3}{10^{k-1}}$$

Thus

$$3 + .3 + .03 + .003 + .0003 + .00003 = \sum_{k=1}^{6} \frac{3}{10^{k-1}} \quad\blacksquare$$

Remark

If we let the index in the summation notation start at zero, we get an equally valid solution for Example 4 given by

$$\sum_{k=0}^{5} \frac{3}{10^k}$$

Example 5

Use summation notation to rewrite the expression $1 - 2x + 3x^2 - 4x^3 + 5x^4$. Assume that $x \neq 0$.

SOLUTION We need to find a function for the sequence

$$1, -2x, 3x^2, -4x^3, 5x^4$$

Forgetting about the sign of each term, we see that kx^{k-1} will generate the correct numbers starting with $k = 1$ and going to $k = 5$. Now, how do we get the alternating $+$ and $-$ signs into the formula? Notice that powers of (-1) will do this.

k	1	2	3	4	5
$(-1)^k$	-1	1	-1	1	-1
$(-1)^{k-1}$	1	-1	1	-1	1

We have two choices, either $(-1)^k$ or $(-1)^{k-1}$. Since our sequence begins with a positive term when $k = 1$, we use $(-1)^{k-1}$. Thus the function we want is

$$F(k) = (-1)^{k-1}kx^{k-1}$$

Hence

$$1 - 2x + 3x^2 - 4x^3 + 5x^4 = \sum_{k=1}^{5} (-1)^{k-1}kx^{k-1} \quad \blacksquare$$

Now we turn our attention to the problem of evaluating some special types of finite sums. For example, suppose we wish to evaluate

$$\sum_{k=1}^{100} k = 1 + 2 + 3 + 4 + \cdots + 99 + 100$$

It would be a lot of work to grind out all this addition directly, so let us try to find a shortcut. For convenience, we let S represent the desired sum,

$$S = \sum_{k=1}^{100} k$$

Now, write out the sum twice, first "forward" and then "backward."

$$S = 1 + 2 + 3 + 4 + \cdots + 99 + 100$$
$$S = 100 + 99 + 98 + 97 + \cdots + 2 + 1$$

Next, adding these two equations by columns, we find that

$$2S = 101 + 101 + 101 + 101 + \cdots + 101 + 101$$

On the right side we have 100 copies of 101. Thus

$$2S = 100(101)$$
$$S = \frac{100(101)}{2}$$

We conclude,

$$\sum_{k=1}^{100} k = \frac{100(101)}{2} = 5050$$

The same idea can be used to evaluate $\sum_{k=1}^{n} k$, for any positive integer n. This gives us the following formula.

The Sum of the First n Integers

$$\sum_{k=1}^{n} k = \frac{n(n + 1)}{2}$$

Example 6

Evaluate the sum of the even integers from 2 to 100, $2 + 4 + 6 + \cdots + 98 + 100$.

SOLUTION We can rewrite this sum as follows:

$$2 + 4 + 6 + \cdots + 98 + 100 = 2(1 + 2 + 3 + \cdots + 49 + 50)$$

$$= 2 \left(\sum_{k=1}^{50} k \right)$$

Now apply the formula.

$$2 \left(\sum_{k=1}^{50} k \right) = 2 \left(\frac{50(51)}{2} \right) = 2(1275) + 2550 \quad \blacksquare$$

We can use the formula for the sum of the first n integers to help us evaluate any sum of the form $\sum_{k=1}^{n} (ak + b)$, where a and b are constants. The idea is to split the sum as follows:

$$\sum_{k=1}^{n} (ak + b) = (a \cdot 1 + b) + (a \cdot 2 + b) + \cdots + (a \cdot n + b)$$

$$= a \cdot 1 + a \cdot 2 + \cdots + a \cdot n \ + \ b + b + \cdots + b$$

$$= a(1 + 2 + \cdots + n) \ + \ b + b + \cdots + b$$

$$= a \sum_{k=1}^{n} k + \sum_{k=1}^{n} b$$

At this point we know how to evaluate each of these sums. Following is a summary of the results.

Evaluating $\sum_{k=1}^{n} (ak + b)$

$$\sum_{k=1}^{n} (ak + b) = a \sum_{k=1}^{n} k + \sum_{k=1}^{n} b$$

$$= a \, \frac{n(n + 1)}{2} + bn$$

Example 7

Evaluate $\displaystyle\sum_{k=1}^{20} \frac{k-3}{2}$.

SOLUTION   We have

$$\sum_{k=1}^{20} \frac{k-3}{2} = \sum_{k=1}^{20} \left(\frac{1}{2}k - \frac{3}{2} \right)$$

$$= \frac{1}{2} \sum_{k=1}^{20} k + \sum_{k=1}^{20} \left(-\frac{3}{2} \right)$$

$$= \frac{1}{2} \frac{20(21)}{2} + \left(-\frac{3}{2} \right) 20$$

$$= 105 - 30$$

$$= 75 \quad \blacksquare$$

Recall that a geometric sequence has the form $a, ar, ar^2, ar^3, \ldots,$ where a and r are constants. If we add the first n terms of this sequence,

$$a + ar + ar^2 + \cdots + ar^{n-1}$$

we get what is called a *finite geometric sum*. Let us derive a formula for this sum as follows. Letting S represent the sum, we write out S and, under it, rS.

$$S = a + ar + ar^2 + \cdots + ar^{n-2} + ar^{n-1}$$

$$rS = ar + ar^2 + ar^3 + \cdots + ar^{n-1} + ar^n$$

Subtracting rS from S, we find that all but the first term of S and the last term of rS on the right side cancel out. Thus

$$S - rS = a - ar^n$$

If $r \neq 1$, then

$$S(1 - r) = a(1 - r^n)$$

$$S = \frac{a}{1 - r} (1 - r^n)$$

We summarize this result below.

Evaluating a Finite Geometric Sum
If $r \neq 1$, then

$$\sum_{k=1}^{n} ar^{k-1} = a + ar = ar^2 + \cdots + ar^{n-1}$$

$$= \frac{a}{1 - r} (1 - r^n)$$

Example 8

Evaluate the given finite geometric sum: (a) $\sum_{k=1}^{100} \dfrac{2^k}{5^{k-1}}$, (b) $1 - 2 + 4 - 8 + 16 - \cdots + 256$.

SOLUTION (a) We must find a, r, and n in order to use the formula for a finite geometric sum. First, $n = 100$ because there are 100 terms in the sum. To find a and r, we write out the first few terms of the sum.

$$\sum_{k=1}^{100} \frac{2^k}{5^{k-1}} = 2 + \frac{2^2}{5} + \frac{2^3}{5^2} + \cdots + \frac{2^{100}}{5^{99}}$$

Obviously, $a = 2$. Furthermore, we see that $r = 2/5$. Thus

$$\sum_{k=1}^{100} \frac{2^k}{5^{k-1}} = \frac{a}{1-r}(1 - r^n)$$

$$= \frac{2}{1 - \dfrac{2}{5}}\left(1 - \left(\frac{2}{5}\right)^{100}\right)$$

$$= \frac{10}{3}\left(1 - \left(\frac{2}{5}\right)^{100}\right)$$

It is interesting to note that $(2/5)^{100}$ is roughly 1.6×10^{-40}, or a decimal point followed by 39 zeros and then 16. This is an extremely small number, so $\dfrac{10}{3}\left(1 - \left(\dfrac{2}{5}\right)^{100}\right)$ is very close to $\dfrac{10}{3}$.

(b) We note that

$$1 - 2 + 4 - 8 + 16 - \cdots + 256$$

may be rewritten as

$$1 + (-2) + (-2)^2 + (-2)^3 + (-2)^4 + \cdots + (-2)^8$$

It follows that $a = 1$, $r = -2$, and $n = 9$. Thus the sum is given by

$$\frac{a}{1-r}(1 - r^n) = \frac{1}{1 - (-2)}(1 - (-2)^9)$$

$$= \frac{1}{3}(1 + 512)$$

$$= 171 \ \blacksquare$$

EXERCISES 11.2

In Exercises 1 to 14, expand and evaluate the given sum.

1. $\displaystyle\sum_{k=1}^{5} 2k$

2. $\displaystyle\sum_{k=1}^{5} (k + 2)$

3. $\displaystyle\sum_{k=1}^{5} (-1)^{k-1} k$

4. $\displaystyle\sum_{k=1}^{5} \frac{(-1)^{k-1}}{k}$

5. $\displaystyle\sum_{k=0}^{10} \cos(k\pi)$

6. $\displaystyle\sum_{k=0}^{5} \sin\left(\frac{k\pi}{2}\right)$

7. $\displaystyle\sum_{k=2}^{4} \frac{k}{k+1}$

8. $\displaystyle\sum_{k=2}^{4} \frac{k-1}{k}$

9. $\displaystyle\sum_{k=1}^{4} 5$

10. $\displaystyle\sum_{k=0}^{4} 4$

11. $\displaystyle\sum_{i=1}^{4} (i-1)2^{i-1}$

12. $\displaystyle\sum_{j=1}^{6} (-1)^j \sin\left((2j-1)\frac{\pi}{2}\right)$

13. $\displaystyle\sum_{k=0}^{3} 2k!$

14. $\displaystyle\sum_{k=1}^{4} \frac{1}{(k-1)!}$

In Exercises 15 to 20, rewrite the sum with the index k starting at 1.

15. $\displaystyle\sum_{k=10}^{12} k$

16. $\displaystyle\sum_{k=10}^{12} (k+4)$

17. $\displaystyle\sum_{k=5}^{8} (2k-1)$

18. $\displaystyle\sum_{k=5}^{8} (3k-2)$

19. $\displaystyle\sum_{k=0}^{4} 3^k$

20. $\displaystyle\sum_{k=0}^{4} \frac{k}{k+1}$

In Exercises 21 to 28, use summation notation to rewrite the sum.

21. $1 + 1 + 1 + 1 + 1$

22. $2 + 2 + 2 + 2 + 2$

23. $2 + 4 + 6 + 8 + 10$

24. $1 + \frac{1}{2} + \frac{1}{3} + \frac{1}{4} + \frac{1}{5}$

25. $.4 + .04 + .004 + .0004$

26. $.1 + .01 + .001 + .0001 + .00001$

27. $1 - \frac{1}{3} + \frac{1}{9} - \frac{1}{27} + \frac{1}{81}$

28. $5 - \frac{5}{2} + \frac{5}{4} - \frac{5}{8}$

In Exercises 29 to 42, use summation notation to rewrite the following sums. Assume the terms follow the obvious pattern.

29. $1 + \frac{1}{2} + \frac{1}{4} + \frac{1}{8} + \cdots + \frac{1}{1024}$

30. $\frac{1}{2} + \frac{1}{4} + \frac{1}{6} + \frac{1}{8} + \cdots + \frac{1}{100}$

31. $\frac{1}{3} + \frac{1}{6} + \frac{1}{9} + \frac{1}{12} + \cdots + \frac{1}{150}$

32. $1 + \frac{3}{2} + \frac{9}{4} + \frac{27}{8} + \cdots + \frac{2187}{128}$

33. $1 - 2 + 3 - 4 + \cdots - 200$

34. $1 - \frac{2}{3} + \frac{3}{5} - \frac{4}{7} + \frac{5}{9} - \cdots + \frac{101}{201}$

35. $1 - 4 + 9 - 16 + 25 - \cdots - 100$

36. $2 - 4 + 6 - 8 + \cdots - 20$

37. $1 - \frac{x}{2} + \frac{x^2}{3} - \frac{x^3}{4} + \cdots + \frac{x^{10}}{11}$ (Assume $x \neq 0$.)

38. $1 + \frac{x}{2} + \frac{x^2}{6} + \frac{x^3}{24} + \cdots + \frac{x^{19}}{20!}$ (Assume $x \neq 0$.)

39. $2 + 5 + 8 + 11 + 14 + \cdots + 101$

40. $1 + 7 + 13 + 19 + 25 + \cdots + 61$

41. $1 - 4 - 9 - 14 - 19 - \cdots - 99$

42. $5 + 1 - 3 - 7 - 11 - \cdots - 75$

In Exercises 43 to 52, evaluate the given sum.

43. $\displaystyle\sum_{k=1}^{200} k$

44. $\displaystyle\sum_{k=1}^{500} k$

45. $\displaystyle\sum_{k=1}^{100} 5k$

46. $\displaystyle\sum_{k=1}^{200} \frac{k}{5}$

47. $\displaystyle\sum_{k=1}^{50} (6k+3)$

48. $\displaystyle\sum_{k=1}^{100} (3k+2)$

49. $\displaystyle\sum_{k=1}^{100} \frac{7k-2}{4}$

50. $\displaystyle\sum_{k=1}^{50} \frac{1-2k}{3}$

51. $\displaystyle\sum_{k=0}^{60} (7k-10)$

52. $\displaystyle\sum_{k=10}^{60} (10k-6)$

53. Find the sum of all the odd integers from 1 to 99, $1 + 3 + 5 + 7 + \cdots + 97 + 99$.

54. Find the sum of all the even integers from 100 to 200, $100 + 102 + 104 + \cdots + 198 + 200$.

55. Find the sum of all the odd integers between 100 and 200.

56. Find the sum of all the integers divisible by 3 between 1 and 100.

In Exercises 57 to 64, evaluate the given finite geometric sum.

57. $\displaystyle\sum_{k=1}^{10} \frac{2}{3^k}$

58. $\displaystyle\sum_{k=1}^{10} \frac{1}{4^k}$

59. $\displaystyle\sum_{k=1}^{20} \left(\frac{3}{4}\right)^k$

60. $\displaystyle\sum_{k=1}^{20} 5\left(\frac{2}{3}\right)^{k-1}$

61. $\displaystyle\sum_{k=1}^{50} \frac{(-1)^{k-1}}{2^k}$

62. $\displaystyle\sum_{k=1}^{50} (-1)^{k-1}5^{-k}$

63. $6 - 2 + \dfrac{2}{3} - \dfrac{2}{9} + \cdots - \dfrac{2}{729}$

64. $\dfrac{1}{2} - 2 + 8 - 32 + \cdots + 2048$

11.3 INFINITE SUMS

In this section we discuss the problem of finding the sum of an infinite number of real numbers. Two examples of this type of problem are:

1. $\displaystyle\sum_{k=1}^{\infty} k = 1 + 2 + 3 + 4 + 5 + \cdots$

2. $\displaystyle\sum_{k=1}^{\infty} \frac{1}{k^2 + k} = \frac{1}{2} + \frac{1}{6} + \frac{1}{12} + \frac{1}{20} + \frac{1}{30} + \cdots$

Each sum is called an **infinite sum**; the infinity symbol, ∞, appearing in the summation notation indicates that the sum goes on forever. (We also indicate this by using three dots after writing out the first few terms of the sum.) Both sums contain only positive terms, so our initial reaction might be that each sum must be arbitrarily large. This is obviously what happens in sum 1; however, sum 2 is not so obvious. In fact, as we will see in Example 1, sum 2 has a finite value.

When faced with the task of evaluating an infinite sum, we do not perform an infinite number of additions, which is certainly impossible. Instead, we try to see what happens as we *approach* doing an infinite number of additions. Thus the value of an infinite sum will be defined in terms of a limit.

Example 1

Evaluate $\displaystyle\sum_{k=1}^{\infty} \frac{1}{k^2 + k}$.

SOLUTION To find the value of the infinite sum

$$\sum_{k=1}^{\infty} \frac{1}{k^2 + k} = \frac{1}{2} + \frac{1}{6} + \frac{1}{12} + \frac{1}{20} + \frac{1}{30} + \cdots$$

we begin by finding *partial sums*: the sum of the first term, denoted S_1, the sum of the first two terms, denoted S_2, the sum of the first three terms, denoted S_3, and so on. We have

$$S_1 = \frac{1}{2} \qquad\qquad\qquad = \frac{1}{2}$$

$$S_2 = \frac{1}{2} + \frac{1}{6} \qquad\qquad = \frac{2}{3}$$

$$S_3 = \frac{1}{2} + \frac{1}{6} + \frac{1}{12} \qquad = \frac{3}{4}$$

$$S_4 = \frac{1}{2} + \frac{1}{6} + \frac{1}{12} + \frac{1}{20} \qquad = \frac{4}{5}$$

$$S_5 = \frac{1}{2} + \frac{1}{6} + \frac{1}{12} + \frac{1}{20} + \frac{1}{30} = \frac{5}{6}$$

These partial sums form a sequence of numbers that appear to be following a pattern: the numerator of the sum of the first n terms is n, and the denominator is $n + 1$. This leads us to conjecture that

$$S_n = \sum_{k=1}^{n} \frac{1}{k^2 + k} = \frac{n}{n + 1}$$

We call S_n, the sum of the first n terms of the infinite sum, the *nth partial sum*. To find the sum of all the numbers in the infinite sum, we should let n tend toward positive infinity in the expression for S_n. Thus

$$\sum_{k=1}^{\infty} \frac{1}{k^2 + k} = \lim_{n \to +\infty} (S_n)$$

$$= \lim_{n \to +\infty} \left(\frac{n}{n + 1} \right)$$

To evaluate this limit, recall that $n \to +\infty$ implies $\dfrac{n}{n + 1} \approx \dfrac{n}{n} = 1$. Therefore

$$\lim_{n \to +\infty} \left(\frac{n}{n + 1} \right) = 1$$

We conclude,

$$\sum_{k=1}^{\infty} \frac{1}{k^2 + k} = 1 \quad \blacksquare$$

Let us review what was done in Example 1 to find the value of the infinite sum $\sum_{k=1}^{\infty} \dfrac{1}{k^2 + k}$. First, we added a few terms: the first one, the first two, the first three, and so on. In doing this, we found a pattern. Using this pattern, we were able to determine a formula for the sum of the first n terms,

$$S_n = \sum_{k=1}^{n} \frac{1}{k^2 + k} = \frac{n}{n+1}$$

This tells us the sum of the first n terms for any n without having to do any addition. For example, the sum of the first 10 terms is given by

$$S_{10} = \frac{10}{11}$$

or, the sum of the first 100 terms is

$$S_{100} = \frac{100}{101}$$

It follows that to find the sum of all the terms, we should evaluate the limit

$$\lim_{n \to +\infty} (S_n)$$

We formalize this procedure in the following definition.

Definition of Infinite Sum
The value of the infinite sum

$$\sum_{k=1}^{\infty} F(k) = F(1) + F(2) + F(3) + \cdots$$

is given by

$$\lim_{n \to +\infty} (S_n)$$

where S_n is the nth partial sum, $S_n = \sum_{k=1}^{n} F(k)$.

Thus there are two problems to overcome when attempting to evaluate an infinite sum: (1) finding a formula for S_n, and (2) finding the limit of S_n as $n \to +\infty$. In the case of an infinite geometric sum, both problems have been completely solved, as we now show.

Let us consider an arbitrary infinite geometric sum,

$$\sum_{k=1}^{\infty} ar^{k-1} = a + ar + ar^2 + ar^3 + \cdots$$

The problem of finding a formula for the nth partial sum was solved in Section 11.2. We have

$$S_n = \sum_{k=1}^{n} ar^{k-1} = \frac{a}{1-r}(1 - r^n)$$

Now we must examine the limit of this expression as $n \to +\infty$. Note that if r is a fraction, $-1 < r < 1$, then r^n will tend to get closer to zero as n tends toward positive infinity. For example, if $r = .9$, we have

$$(.9)^{10} \approx .349$$

$$(.9)^{100} \approx 2.66 \times 10^{-5}$$

$$(.9)^{1000} \approx 1.75 \times 10^{-46}$$

Thus $(.9)^n \to 0$ as $n \to +\infty$. In general, whenever $-1 < r < 1$, $r^n \to 0$ as $n \to +\infty$, so that

$$\lim_{n \to +\infty} (S_n) = \frac{a}{1-r}(1-0) = \frac{a}{1-r}$$

It can be shown that if $|r| \geq 1$, then the limit for S_n either does not exist or is infinite. We leave the proofs for calculus. This gives us the following result.

Evaluating an Infinite Geometric Sum

If $-1 < r < 1$, then

$$\sum_{k=1}^{\infty} ar^{k-1} = a + ar + ar^2 + ar^3 + \cdots$$

$$= \frac{a}{1-r}$$

For all other values of r, the sum either does not exist or is infinite.

Example 2

Determine a and r for the given infinite geometric sum. Evaluate the sum if possible.

(a) $\displaystyle\sum_{k=1}^{\infty} \frac{1}{2^{k-1}}$ (b) $\displaystyle\sum_{k=1}^{\infty} \frac{1}{5}\left(\frac{3}{2}\right)^{k-1}$ (c) $1 - \frac{1}{3} + \frac{1}{9} - \frac{1}{27} + \frac{1}{81} - \cdots$

SOLUTION (a) Writing out the first few terms,

$$\sum_{k=1}^{\infty} \frac{1}{2^{k-1}} = 1 + \frac{1}{2} + \frac{1}{4} + \frac{1}{8} + \cdots$$

we see that $a = 1$ and $r = 1/2$. Since $-1 < r < 1$, we may apply the formula for the infinite sum.

$$\sum_{k=1}^{\infty} \frac{1}{2^{k-1}} = \frac{a}{1-r} = \frac{1}{1-1/2} = 2$$

(b) Writing out the first few terms,

$$\sum_{k=1}^{\infty} \frac{1}{5}\left(\frac{3}{2}\right)^{k-1} = \frac{1}{5} + \frac{1}{5}\left(\frac{3}{2}\right) + \frac{1}{5}\left(\frac{3}{2}\right)^2 + \cdots$$

we see that $a = 1/5$ and $r = 3/2$. Since r is not between -1 and 1, the value of this sum is not finite. It is important to realize that the formula $a/(1 - r)$ does not apply in this case.

(c) By examining the terms of the sum

$$1 - \frac{1}{3} + \frac{1}{9} - \frac{1}{27} + \frac{1}{81} - \cdots$$

we see that $a = 1$ and $r = -1/3$. Since $-1 < r < 1$, the sum has the value

$$\frac{a}{1 - r} = \frac{1}{1 - \left(-\dfrac{1}{3}\right)} = \frac{3}{4} \blacksquare$$

We may use infinite geometric sums to show that a repeating decimal represents a rational number. Example 3 illustrates how this is done.

Example 3

Find a rational number that represents $.\overline{45}$.

SOLUTION First, we rewrite the repeating decimal as an infinite sum.

$$.\overline{45} = .45454545 \cdots$$

$$= .45 + .45(.01) + .45(.01)^2 + .45(.01)^3 + \cdots$$

We recognize this as an infinite geometric sum with $a = .45$ and $r = .01$. Since $-1 < r < 1$, the value of this sum is given by

$$\frac{a}{1 - r} = \frac{.45}{1 - .01} = \frac{.45}{.99} = \frac{5}{11}$$

Thus $.\overline{45} = 5/11$. $\blacksquare$

EXERCISES 11.3

In Exercises 1 to 4, for the given infinite sum, (a) evaluate the partial sums S_1, S_2, S_3, S_4, S_5. (b) From the results in part (a), guess the formula for the nth partial sum, S_n. (c) Evaluate the infinite sum by determining $\lim\limits_{n \to +\infty} (S_n)$.

1. $\displaystyle\sum_{k=1}^{\infty} \frac{1}{4k^2 - 1}$

2. $\displaystyle\sum_{k=1}^{\infty} \frac{1}{(3k - 2)(3k + 1)}$

3. $\displaystyle\sum_{k=1}^{\infty} \frac{2}{(k + 1)(k + 2)}$

4. $\displaystyle\sum_{k=1}^{\infty} \frac{3}{4k^2 + 8k + 3}$

In Exercises 5 to 12, determine a and r for the given infinite geometric sum. If possible, evaluate the sum.

5. $\displaystyle\sum_{k=1}^{\infty} \frac{5}{2^k}$

6. $\displaystyle\sum_{k=1}^{\infty} \frac{1}{3^k}$

7. $\displaystyle\sum_{k=1}^{\infty} \frac{3^k}{5^{k-1}}$

8. $\displaystyle\sum_{k=1}^{\infty} \frac{2^{k+1}}{3^{k-1}}$

9. $9 - 3 + 1 - \dfrac{1}{3} + \dfrac{1}{9} - \cdots$

10. $16 - 4 + 1 - \dfrac{1}{4} + \dfrac{1}{16} - \cdots$

11. $\displaystyle\sum_{k=1}^{\infty} \left(\frac{4}{3}\right)^k$

12. $\displaystyle\sum_{k=1}^{\infty} (-1)^{k-1}5^k$

In Exercises 13 to 16, for the given repeating decimal, (a) rewrite the decimal as an infinite geometric sum, and (b) find a rational number representing the decimal by evaluating the sum in part (a).

13. $.\overline{36}$

14. $.\overline{21}$

15. $.\overline{9}$

16. $.\overline{6}$

In Exercises 17 and 18, find a rational number representing the given repeating decimal.

17. $1.23\overline{81}$

18. $1.6\overline{621}$

19. If $p > 1$ is a constant, evaluate $\displaystyle\sum_{k=1}^{\infty} \frac{1}{p^k}$.

20. Find a formula for $\displaystyle\sum_{k=1}^{n} \ln\left(\frac{k+1}{k}\right)$. (*Hint:* Recall $\ln(a/b) = \ln a - \ln b$.) Is $\displaystyle\sum_{k=1}^{\infty} \ln\left(\frac{k+1}{k}\right)$ finite?

21. If $-1 < x < 1$ and $x \neq 0$, evaluate $\displaystyle\sum_{k=0}^{\infty} x^k$ in terms of x.

22. If $-1 < x < 1$ and $x \neq 0$, evaluate $\displaystyle\sum_{k=0}^{\infty} x^{2k}$ in terms of x.

23. An infinite collection of rectangles is constructed under the curve $y = 2^{-x}$ as shown in Figure 11.3. Let $F(k)$ = area of kth rectangle (counting from the left).
   (a) Verify that $F(1) = 1/2$ and $F(2) = 1/4$.
   (b) Find a formula in terms of k for $F(k)$.
   (c) Find a formula in terms of n for $\displaystyle\sum_{k=1}^{n} F(k)$.
   (d) Evaluate the sum of the areas of all the rectangles, $\displaystyle\sum_{k=1}^{+\infty} F(k)$.

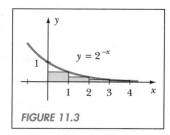

FIGURE 11.3

24. Repeat Exercise 23, parts (b), (c), and (d), for the curve $y = 3^{-x}$.

11.4 THE PRINCIPLE OF MATHEMATICAL INDUCTION

In this section we introduce the principle of mathematical induction, a powerful and elegant proof technique fundamental to all fields of mathematics. The principle is related to sequences, but not the usual numerical kind that we saw previously. Instead, we will be concerned with sequences of mathematical statements and the problem of how to prove that each statement is true.

Let us begin with the problem of trying to prove the validity of each statement in the following sequence:

$$4^1 - 1 \quad \text{is a multiple of} \quad 3$$
$$4^2 - 1 \quad \text{is a multiple of} \quad 3$$
$$4^3 - 1 \quad \text{is a multiple of} \quad 3$$
$$4^4 - 1 \quad \text{is a multiple of} \quad 3$$
$$4^5 - 1 \quad \text{is a multiple of} \quad 3$$
$$\vdots$$

Let $P(n)$ represent the statement $4^n - 1$ *is a multiple of 3*. Our problem is to prove that $P(n)$ is true for every positive integer n. Of course, the first few of these statements are easily shown to be true by direct computation. For example,

$$4^1 - 1 = 3 = 3(1)$$
$$4^2 - 1 = 16 - 1 = 15 = 3(5)$$
$$4^3 - 1 = 64 - 1 = 63 = 3(21)$$

So $P(1)$, $P(2)$, and $P(3)$ are true. However, it is impossible to show the computation for *every* positive integer n since there are an infinite number of them. Instead, we give a complete proof that requires only two steps.

> **PROOF:** Step 1. Show that $P(1)$ is true. As indicated above, $4^1 - 1 = 3 = 3(1)$.
>
> Step 2. Let n be an arbitrary positive integer. Show that if $P(n)$ is true, then this implies that $P(n + 1)$ will be true. The idea here is to *assume* (without proof) that $P(n)$ is true; this means that $4^n - 1$ is a multiple of 3. Under this assumption, our job is to prove that $P(n + 1)$ must also be true; that is, $4^{n+1} - 1$ is a multiple of 3. We shall do this as follows:
>
> $$4^{n+1} - 1 = 4^{n+1} - 4 + 3$$
> $$= 4 \cdot 4^n - 4 + 3$$
> $$= 4(4^n - 1) + 3$$
>
> At this point we use the assumption that $P(n)$ is true: $4^n - 1$ is a multiple of 3. This says that there is an integer q such that $4^n - 1 = 3q$. Therefore
>
> $$4^{n+1} - 1 = 4(4^n - 1) + 3$$
> $$= 4(3q) + 3$$
> $$= 3(4q + 1)$$
>
> Hence $4^{n+1} - 1$ is a multiple of 3. We have thus shown that if $P(n)$ is true, then $P(n + 1)$ is true. ■

Why does this argument, consisting of only two steps, prove that $P(n)$ is true for every positive integer n? Step 1 tells us that $P(1)$ is true. Next, step 2, which says that *if $P(n)$ is true then $P(n + 1)$ is true*, can be applied when $n = 1$,

$$P(1) \text{ is true and this implies } P(2) \text{ is true}$$

But now we know that $P(2)$ is true, so we can apply step 2 once again, only this time with $n = 2$. This would say

$$P(2) \text{ is true implies } P(3) \text{ is true}$$

Now that $P(3)$ is true, we can apply step 2 with $n = 3$ to show that $P(4)$ is true, and so on. Obviously, we may repeat this argument over and over to establish the validity of $P(n)$ for every positive integer n.

Notice that this method of proof is not direct. For example, we did not actually calculate $4^{100} - 1$ and then find that this number was a multiple of 3. Instead, the method of proof was indirect; $P(1)$ is true eventually implies the validity of $P(100)$. This indirect technique is where the power of mathematical induction lies. We were able to prove an infinite number of results in just two steps.

Step 1 of the proof is called the **beginning step**; it establishes the validity of $P(1)$ directly, giving us a starting point to work from. Step 2 is called the **induction step**; it gets us from the truth of one statement, $P(n)$, to the truth of the next statement, $P(n + 1)$. The assumption that $P(n)$ is true in this part of the proof is called the **induction hypothesis**. Establishing the beginning step and the induction step constitutes a proof by mathematical induction that $P(n)$ is true for every positive integer n. Let us formally state this principle, but without proof because it is very closely related to the axioms of arithmetic.

The Principle of Mathematical Induction
Let $P(n)$ be a statement concerning the positive integer n. Then $P(n)$ is true for every positive integer n if the following two steps can be done:
Beginning step: Show that $P(1)$ is true.
Induction step: Show that if $P(n)$ is true, then this implies that $P(n + 1)$ is true.

Example 1

Use the principle of mathematical induction to prove the formula for the sum of the first n integers:

$$1 + 2 + 3 + \cdots + n = \frac{n(n + 1)}{2}$$

SOLUTION   Let $P(n)$ be the statement $1 + 2 + 3 + \cdots + n = n(n + 1)/2$.
Beginning step: We wish to show that $P(1)$ is true.

$$1 = \frac{1(1 + 1)}{2}$$

This is clearly true.
Induction step: Let n be an arbitrary positive integer and assume that $P(n)$ is true (this is the induction hypothesis). We must show that this implies $P(n + 1)$ is true.

$$1 + 2 + 3 + \cdots + n + (n + 1) \stackrel{?}{=} \frac{(n + 1)(n + 2)}{2}$$

The question mark indicates that we need to prove this statement. Working on the left side of the statement,

$$1 + 2 + 3 + \cdots + n + (n + 1) = (1 + 2 + 3 + \cdots + n) + (n + 1)$$

Using the induction hypothesis,

$$= \frac{n(n + 1)}{2} + (n + 1)$$

$$= \frac{n(n + 1) + 2(n + 1)}{2}$$

$$= \frac{(n + 1)(n + 2)}{2}$$

Thus $P(n + 1)$ is true. Notice that the crucial step in the proof was made possible by the induction hypothesis that $P(n)$ is true: $1 + 2 + 3 + \cdots + n = n(n + 1)/2$.

This completes the proof that $P(n)$ is true for all positive integers n by the principle of mathematical induction. ■

In Section 11.3 we evaluated certain infinite sums by guessing the value of their nth partial sums. We now have the ability to prove that those guesses were correct.

Example 2

Prove $\displaystyle\sum_{k=1}^{n} \frac{1}{k^2 + k} = \frac{n}{n + 1}$ for all positive integers n.

SOLUTION Let $P(n)$ be the statement $\displaystyle\sum_{k=1}^{n} \frac{1}{k^2 + k} = \frac{n}{n + 1}$.
Beginning step: We wish to show that $P(1)$ is true.

$$\sum_{k=1}^{1} \frac{1}{k^2 + k} = \frac{1}{1 + 1}$$

The left side is identical to the right side, so $P(1)$ is true.
Induction step: Let n be an arbitrary positive integer and assume $P(n)$ is true. We must show that this implies $P(n + 1)$ is true,

$$\sum_{k=1}^{n+1} \frac{1}{k^2 + k} \overset{?}{=} \frac{n + 1}{n + 2}$$

Working on the left side, we have

$$\sum_{k=1}^{n+1} \frac{1}{k^2 + k} = \sum_{k=1}^{n} \frac{1}{k^2 + k} + \frac{1}{(n+1)^2 + (n+1)}$$

Using the induction hypothesis,

$$= \frac{n}{n+1} + \frac{1}{(n+1)^2 + (n+1)}$$

$$= \frac{n}{n+1} + \frac{1}{(n+1)(n+1+1)}$$

$$= \frac{n(n+2)}{(n+1)(n+2)} + \frac{1}{(n+1)(n+2)}$$

$$= \frac{n^2 + 2n + 1}{(n+1)(n+2)}$$

$$= \frac{(n+1)(n+1)}{(n+1)(n+2)}$$

$$= \frac{n+1}{n+2}$$

Thus $P(n+1)$ is true.

This completes the proof that $P(n)$ is true for all positive integers n by the principle of mathematical induction. ∎

Our statement of the principle of mathematical induction can be generalized by saying that the beginning step in the proof need not start with the integer 1. Suppose, for example, we wish to prove a statement $P(n)$ for all integers n greater than or equal to k, where k is some fixed integer (not necessarily 1). In this case the beginning step would be to prove that $P(k)$ is true [instead of $P(1)$]. The induction step would be the same, except we assume $n \geq k$ (instead of $n \geq 1$).

Example 3

Use the principle of mathematical induction to prove $n! > 2^n$ for all integers $n \geq 4$.

SOLUTION   Let $P(n)$ be the statement $n! > 2^n$.
Beginning step: We must show that $P(4)$ is true,

$$4! > 2^4$$

We have $4! = 1 \cdot 2 \cdot 3 \cdot 4 = 24$ and $2^4 = 16$. Thus $4! > 2^4$ is true.

Induction step: Let n be an arbitrary positive integer, $n \geq 4$, and assume that $P(n)$ is true. We wish to show that this implies $P(n+1)$ is true,

$$(n+1)! \overset{?}{>} 2^{n+1}$$

Working on the left side, we have

$$(n + 1)! = n!(n + 1)$$
$$> 2^n(n + 1) \qquad \text{induction hypothesis}$$
$$> 2^n(2) \qquad \text{since } n \geq 4, \ n + 1 \geq 5 > 2$$
$$= 2^{n+1}$$

Thus $P(n + 1)$ is true.

This completes the proof that $P(n)$ is true for all integers $n \geq 4$ by the principle of mathematical induction. ∎

EXERCISES 11.4

In Exercises 1 to 10, use the principle of mathematical induction to prove that the given statement is true for every positive integer n.

1. $5^n - 1$ is a multiple of 4.

2. $6^n - 1$ is a multiple of 5.

3. $n^3 + 2n$ is a multiple of 3.

4. $3^{n+1} - 2n - 3$ is a multiple of 4.

5. $1 + 3 + 5 + 7 + \cdots + (2n - 1) = n^2$

6. $2 + 4 + 6 + 8 + \cdots + 2n = n^2 + n$

7. $1^2 + 2^2 + 3^2 + 4^2 + \cdots + n^2 = \dfrac{1}{6}n(n + 1)(2n + 1)$

8. $1^3 + 2^3 + 3^3 + 4^3 + \cdots + n^3 = \dfrac{1}{4}n^2(n + 1)^2$

9. $2^0 + 2^1 + 2^2 + 2^3 + \cdots + 2^{n-1} = 2^n - 1$

10. $e^0 + e^1 + e^2 + e^3 + \cdots + e^{n-1} = e^n - 1/(e - 1)$

In Exercises 11 to 14, use the principle of mathematical induction to establish the validity of the given formula for all positive integers n.

11. $\displaystyle\sum_{k=1}^{n} \dfrac{1}{4k^2 - 1} = \dfrac{n}{2n + 1}$

12. $\displaystyle\sum_{k=1}^{n} \dfrac{1}{(3k - 2)(3k + 1)} = \dfrac{n}{3n + 1}$

13. $\displaystyle\sum_{k=1}^{n} \dfrac{2}{(k + 1)(k + 2)} = \dfrac{n}{n + 2}$

14. $\displaystyle\sum_{k=1}^{n} \dfrac{3}{4k^2 + 8k + 3} = \dfrac{n}{2n + 3}$

15. Prove $e^n > 2^n$ for all positive integers n.

16. Prove $n! > n^2$ for all integers $n \geq 4$.

17. Prove $n! > 3^n$ for all integers $n \geq 7$.

18. Prove $2^n > n^2$ for all integers $n \geq 5$.

19. Prove $\left(1 - \dfrac{1}{2}\right)\left(1 - \dfrac{1}{3}\right)\left(1 - \dfrac{1}{4}\right) \cdots \left(1 - \dfrac{1}{n}\right) = \dfrac{1}{n}$ for all integers $n \geq 2$.

20. Prove $\left(1 - \dfrac{1}{4}\right)\left(1 - \dfrac{1}{9}\right)\left(1 - \dfrac{1}{16}\right) \cdots \left(1 - \dfrac{1}{n^2}\right) = \dfrac{n + 1}{2n}$ for all integers $n \geq 2$.

21. Prove DeMoivre's Theorem: $(\cos \theta + i \sin \theta)^n = \cos(n\theta) + i \sin(n\theta)$ for every positive integer n.

22. Prove $1(1!) + 2(2!) + 3(3!) + \cdots + n(n!) = (n + 1)! - 1$ for all integers $n \geq 1$.

23. Prove that if z is any complex number, then $\overline{(z^n)} = (\bar{z})^n$ for all integers $n \geq 2$.

24. (a) Prove that
$$a(\cos \alpha + i \sin \alpha) \cdot b(\cos \beta + i \sin \beta)$$
$$= ab[\cos(\alpha + \beta) + i \sin(\alpha + \beta)].$$
(b) Prove that if z and w are nonzero complex numbers, then $z \cdot w \neq 0$. [*Hint:* Use polar representations for z and w, say, $z = a(\cos \alpha + i \sin \alpha)$ and $w = b(\cos \beta + i \sin \beta)$, and then apply the result of part (a).]
(c) Use the principle of mathematical induction to prove that if $z_1, z_2, \ldots, z_n$ are nonzero complex numbers, $n \geq 2$, then the product $z_1 \cdot z_2 \cdots z_n$ is nonzero. [*Hint:* Use the result from part (b).]

THE BINOMIAL THEOREM

In this section we find a formula for the expansion of

$$(A + B)^n$$

for any nonnegative integer n. We begin by expanding $(A + B)^n$ directly for $n = 0, 1, 2, 3, 4$.

$$(A + B)^0 = 1$$
$$(A + B)^1 = A + B$$
$$(A + B)^2 = A^2 + 2AB + B^2$$
$$(A + B)^3 = A^3 + 3A^2B + 3AB^2 + B^3$$
$$(A + B)^4 = A^4 + 4A^3B + 6A^2B^2 + 4AB^3 + B^4$$

Notice that the exponents on A and B in each term of the expansion of $(A + B)^n$ add up to n. For example, in the expansion of $(A + B)^4$, we have terms corresponding to

$$A^4 = A^4B^0, \ A^3B^1, \ A^2B^2, \ A^1B^3, \text{ and } A^0B^4 = B^4$$

In general, the expansion of $(A + B)^n$ will have $n + 1$ terms corresponding to

$$A^n = A^nB^0, \ A^{n-1}B^1, \ A^{n-2}B^2, \ldots, \ A^1B^{n-1}, \text{ and } A^0B^n = B^n$$

Therefore the expansion of $(A + B)^n$ must look like the following:

$$(A + B)^n = \underline{\ ?\ }\ A^n + \underline{\ ?\ }\ A^{n-1}B + \underline{\ ?\ }\ A^{n-2}B^2 + \cdots + \underline{\ ?\ }\ AB^{n-1} + \underline{\ ?\ }\ B^n$$

The question marks indicate the missing numerical coefficients. Let us introduce some notation for these numbers.

Definition of Binomial Coefficient
The **binomial coefficient** denoted by

$$C(n, k)$$

is the numerical coefficient of the term corresponding to $A^{n-k}B^k$ in the expansion of $(A + B)^n$.

Using this notation, we may write the expansion of $(A + B)^n$ as

$$(A + B)^n = C(n, 0)A^n + C(n, 1)A^{n-1}B + C(n, 2)A^{n-2}B^2 + \cdots + C(n, n)B^n$$

or, using summation notation,

$$(A + B)^n = \sum_{k=0}^{n} C(n, k)A^{n-k}B^k$$

What are the numbers $C(n, k)$? There is an interesting procedure that allows us to determine these binomial coefficients recursively. The idea is to use the coefficients for the expansion of $(A + B)^n$ to find the coefficients for $(A + B)^{n+1}$. The procedure may be illustrated by what we call **Pascal's triangle**, a portion of which is shown below.

$$
\begin{array}{c}
n = 0 \rightarrow \qquad\qquad\qquad\qquad 1 \\
n = 1 \rightarrow \qquad\qquad\qquad 1 \qquad 1 \\
n = 2 \rightarrow \qquad\qquad 1 \qquad 2 \qquad 1 \\
n = 3 \rightarrow \qquad 1 \qquad 3 \qquad 3 \qquad 1 \\
n = 4 \rightarrow \quad 1 \qquad 4 \qquad 6 \qquad 4 \qquad 1
\end{array}
$$

On the left, each row of the triangle is labeled beginning with $n = 0$. We see that each number (other than the 1s) is obtained by adding the two numbers diagonally above it. Thus the fifth row of the triangle can be obtained from the fourth row as follows:

$$
\begin{array}{c}
n = 4 \rightarrow \quad 1 \quad\searrow\; 4 \;\nearrow\;\searrow\; 6 \;\nearrow\;\searrow\; 4 \;\nearrow\;\searrow\; 1 \;\nearrow \\
n = 5 \rightarrow \quad 1 \qquad 5 \qquad 10 \qquad 10 \qquad 5 \qquad 1
\end{array}
$$

It turns out that the numbers in the nth row of Pascal's triangle are exactly the binomial coefficients $C(n, k)$ that we seek. More specifically, to find $C(n, k)$, we go to the nth row of the triangle and take the number in the kth column over from the left (starting with $k = 0$).

Example 1

Use Pascal's triangle to expand $(A + B)^5$.

SOLUTION The fifth row of Pascal's triangle will tell us the coefficients $C(5, k)$ in the expansion of $(A + B)^5$. The fifth row is shown below, together with the binomial coefficient that it represents:

$$
\begin{array}{cccccc}
1 & 5 & 10 & 10 & 5 & 1 \\
C(5, 0) & C(5, 1) & C(5, 2) & C(5, 3) & C(5, 4) & C(5, 5)
\end{array}
$$

Thus

$$(A + B)^5 = \sum_{k=0}^{5} C(5, k) A^{5-k} B^k$$

$$= A^5 + 5A^4B + 10A^3B^2 + 10A^2B^3 + 5AB^4 + B^5 \quad\blacksquare$$

Example 2

Use Pascal's triangle to expand $(A + B)^6$.

SOLUTION We need to use the sixth row of Pascal's triangle. As shown below, we obtain the sixth row from the fifth:

$$n = 5 \rightarrow \qquad 1 \quad 5 \quad 10 \quad 10 \quad 5 \quad 1$$

$$n = 6 \rightarrow \quad 1 \quad 6 \quad 15 \quad 20 \quad 15 \quad 6 \quad 1$$

$$\quad C(6, 0) \quad C(6, 1) \quad C(6, 2) \quad C(6, 3) \quad C(6, 4) \quad C(6, 5) \quad C(6, 6)$$

Thus

$$(A + B)^6 = \sum_{k=0}^{6} C(6, k) A^{6-k} B^k$$

$$= A^6 + 6A^5B + 15A^4B^2 + 20A^3B^3 + 15A^2B^4 + 6AB^5 + B^6 \quad \blacksquare$$

If we want to find the expansion of $(A + B)^{20}$ or $(A + B)^{100}$, Pascal's triangle is a bit cumbersome to deal with in such cases. We need a formula for $C(n, k)$, which is precisely what the Binomial Theorem gives us.

Binomial Theorem

$$(A + B)^n = \sum_{k=0}^{n} C(n, k) A^{n-k} B^k$$

where $C(n, k) = \dfrac{n!}{k!(n-k)!}$

We present a proof of the Binomial Theorem at the end of this section. Note that $0!$ appears in some of the coefficients; recall that $0! = 1$, by definition.

Example 3

Find the coefficient of $A^{12}B^8$ in the expansion of $(A + B)^{20}$.

SOLUTION Each term in the expansion of $(A + B)^{20}$ looks like $C(20, k) A^{20-k} B^k$. Therefore the $A^{12}B^8$ term corresponds to $k = 8$. By the Binomial Theorem,

$$C(20, 8) = \frac{20!}{8!(20-8)!} = \frac{20!}{8!12!} = 125970$$

So the desired coefficient is 125970. $\quad \blacksquare$

Example 4

Use the Binomial Theorem to expand $(x - 2y)^5$.

SOLUTION First, we use the formula for $(A + B)^5$:

$$(A + B)^5 = \sum_{k=0}^{5} C(5, k) A^{5-k} B^k$$

$$= A^5 + 5A^4B + 10A^3B^2 + 10A^2B^3 + 5AB^4 + B^5$$

Now substitute $A = x$ and $B = -2y$, obtaining

$$(x - 2y)^5 = x^5 + 5x^4(-2y) + 10x^3(-2y)^2 + 10x^2(-2y)^3 + 5x(-2y)^4 + (-2y)^5$$

$$= x^5 - 10x^4y + 40x^3y^2 - 80x^2y^3 + 80xy^4 - 32y^5 \quad \blacksquare$$

Example 5

Find the coefficient of x^2y^{13} in the expansion of $(3x - y)^{15}$.

SOLUTION $(3x - y)^{15}$ can be obtained from the formula

$$(A + B)^{15} = \sum_{k=0}^{15} C(15, k) A^{15-k} B^k$$

with $A = 3x$ and $B = -y$. The term containing x^2y^{13} corresponds to $k = 13$. Thus

$$C(15, 13) A^2 B^{13} = \frac{15!}{13!(15 - 13)!} (3x)^2 (-y)^{13}$$

$$= -\frac{15!(9)}{13!(2)!} x^2 y^{13}$$

$$= -945 x^2 y^{13}$$

Therefore the desired coefficient is -945. $\blacksquare$

We close this section with a proof of the Binomial Theorem.

PROOF OF THE BINOMIAL THEOREM: We use the principle of mathematical induction. Let $P(n)$ be the statement

$$(A + B)^n = \sum_{k=0}^{n} C(n, k) A^{n-k} B^k \qquad \text{where } C(n, k) = \frac{n!}{k!(n - k)!}$$

Beginning step: We wish to show that $P(1)$ is true,

$$(A + B)^1 = C(1, 0) A + C(1, 1) B$$

where $C(1, 0) = \dfrac{1!}{0!1!} = 1$ and $C(1, 1) = \dfrac{1!}{1!0!} = 1$. Thus $P(1)$ says that $(A + B)^1 = A + B$, which is true.

Induction step: Let n be an arbitrary positive integer and assume $P(n)$ is true. We want to show that this implies $P(n + 1)$ is true,

$$(A + B)^{n+1} \stackrel{?}{=} \sum_{k=0}^{n+1} C(n + 1, k) A^{n+1-k} B^k \quad \text{where } C(n + 1, k) = \frac{(n + 1)!}{k!(n + 1 - k)!}$$

Let us start to work on the left side. We write

$$(A + B)^{n+1} = (A + B)(A + B)^n$$

$$= (A + B) \sum_{k=0}^{n} C(n, k) A^{n-k} B^k$$

Note that we have replaced $(A + B)^n$ with $\sum_{k=0}^{n} C(n, k) A^{n-k} B^k$, where $C(n, k) = \dfrac{n!}{k!(n - k)!}$. This is permissible by the induction hypothesis. Continuing, we multiply the sum by the factor $(A + B)$ to get

$$= A \sum_{k=0}^{n} C(n, k) A^{n-k} B^k + B \sum_{k=0}^{n} C(n, k) A^{n-k} B^k$$

$$= \sum_{k=0}^{n} C(n, k) A^{n+1-k} B^k + \sum_{k=0}^{n} C(n, k) A^{n-k} B^{k+1}$$

$$= C(n, 0) A^{n+1} + C(n, 1) A^n B + \cdots + C(n, n) AB^n$$
$$C(n, 0) A^n B + \cdots + C(n, n - 1) AB^n + C(n, n) B^{n+1}$$

The first row corresponds to the first sum, and the second row corresponds to the second sum. Collecting like terms, we have

$$(A + B)^{n+1} = C(n, 0) A^{n+1} + [C(n, 1) + C(n, 0)] A^n B + \cdots$$

$$+ [C(n, n) + C(n, n - 1)] AB^n + C(n, n) B^{n+1}$$

Now, computing each coefficient, we find

Term	Coefficient	Value of Coefficient
A^{n+1}	$C(n, 0)$	$\dfrac{n!}{0!n!} = 1$
$A^n B$	$C(n, 1) + C(n, 0)$	$\dfrac{n!}{1!(n - 1)!} + \dfrac{n!}{0!n!} = n + 1$
$\vdots$	$\vdots$	$\vdots$
$A^{n+1-k} B^k$	$C(n, k) + C(n, k - 1)$	$\dfrac{n!}{k!(n - k)!} + \dfrac{n!}{(k - 1)!(n - k + 1)!} = \dfrac{(n + 1)!}{k!(n + 1 - k)!}$
$\vdots$	$\vdots$	$\vdots$
AB^n	$C(n, n) + C(n, n - 1)$	$\dfrac{n!}{n!0!} + \dfrac{n!}{(n - 1)!1!} = n + 1$
B^{n+1}	$C(n, n)$	$\dfrac{n!}{n!0!} = 1$

We omitted the details of how $C(n, k) + C(n, k - 1)$ works out to be $(n + 1)!/(k!(n + 1 - k)!)$ for $1 \le k \le n$; this is left for an exercise. Now, if we compare the coefficients we just calculated for the expansion of $(A + B)^{n+1}$ with those given by the statement $P(n + 1)$, we see that they are identical. Thus $P(n + 1)$ is true.

This completes the proof of the Binomial Theorem by the principle of mathematical induction. ∎

EXERCISES 11.5

In Exercises 1 to 6, use Pascal's triangle to expand the given expression.

1. $(1 + x)^6$

2. $(x - 2)^5$

3. $(2x + y)^5$

4. $(x + 2y)^6$

5. $(x^2 - y)^5$

6. $(x^3 - 2y)^4$

In Exercises 7 to 10, use the Binomial Theorem to expand the given expression.

7. $\left(1 - \dfrac{1}{x}\right)^7$

8. $\left(x - \dfrac{1}{x}\right)^7$

9. $(x^2 + 1)^6$

10. $(x^2 + y^2)^5$

In Exercises 11 to 20, find the coefficient for the given term.

11. x^3 in $(x - 2)^{10}$

12. x^2 in $(1 + x)^{10}$

13. $x^{30}y^{10}$ in $(x + y)^{40}$

14. $x^5 y^{35}$ in $(x - y)^{40}$

15. x^3 in $(1 - 2x)^{13}$

16. $x^9 y^{11}$ in $\left(2x + \dfrac{y}{2}\right)^{20}$

17. y^{-1} in $(x + y^{-1})^{11}$

18. x^{-1} in $(x^2 - x^{-1})^{10}$

19. x^4 in $(1 + \sqrt{x})^{20}$

20. x^2 in $(1 - \sqrt{x})^{12}$

In Exercises 21 to 24, use the formula for $C(n, k)$ to prove each statement. Assume n is a positive integer.

21. (a) $C(n, 0) = 1$
    (b) $C(n, 1) = n$

22. (a) $C(n, n) = 1$
    (b) $C(n, n - 1) = n$

23. $C(n, k) = C(n, n - k)$, where k is a nonnegative integer, $0 \le k \le n$.

24. $C(n, k) + C(n, k - 1) = \dfrac{(n + 1)!}{k!(n + 1 - k)!}$ for $1 \le k \le n$.

25. Show that $\displaystyle\sum_{k=0}^{n} C(n, k) = 2^n$. [*Hint:* let $x = y = 1$ in the expansion of $(x + y)^n$.]

26. Show that $\displaystyle\sum_{k=0}^{n} (-1)^k C(n, k) = 0$. [*Hint:* let $x = 1$ and $y = -1$ in the expansion of $(x + y)^n$.]

In Exercises 27 and 28, by choosing appropriate values for x and y in the expansion of $(x + y)^n$, evaluate the given sum.

27. $\displaystyle\sum_{k=0}^{n} C(n, k)2^k$

28. $\displaystyle\sum_{k=0}^{n} (-1)^{n-k} C(n, k)2^k$

In Exercises 29 to 31, for the given function F, let $P(x) = \dfrac{F(x + h) - F(x)}{h}$.

(a) Expand and simplify $P(x)$.
(b) What is the degree of the polynomial $P(x)$?
(c) How many terms does $P(x)$ have?
(d) How many terms in $P(x)$ contain h?

29. $F(x) = x^6$

30. $F(x) = x^{100}$

31. $F(x) = x^n$, n a positive integer.

In Exercises 1 and 2, list the first six terms of the sequence defined by the given function and graph.

1. $F(n) = 3 - n$

2. $F(n) = \cos\left(n\dfrac{\pi}{4}\right)$

In Exercises 3 and 4, find a formula for the nth term of the given sequence, assuming that the obvious pattern is followed.

3. $17, 12, 7, 2, -3, \ldots$

4. $4, -6, 9, -\dfrac{27}{2}, \dfrac{81}{4}, \ldots$

In Exercises 5 to 7, a sequence is defined by the given recursive definition. List the first six terms.

5. $F(1) = 2$, $F(n) = \dfrac{F(n-1)}{1 + F(n-1)}$ for $n \geq 2$

6. $F(1) = 1$, $F(n) = 3F(n-1)$ for $n \geq 2$

7. $F(1) = 0$, $F(2) = 1$,
   $F(n) = F(n-1) + 2F(n-2)$ for $n \geq 3$

8. Evaluate $(2n)!$ for $n = 0, 1, 2, 3$.

In Exercises 9 and 10, expand and evaluate the given sum.

9. $\displaystyle\sum_{k=1}^{5} \dfrac{(-1)^{k-1}}{k^2 + k}$

10. $\displaystyle\sum_{k=0}^{5} 7$

In Exercises 11 to 13, use summation notation to rewrite the given sum.

11. $3 - 4 + 5 - 6 + 7 - 8$

12. $\dfrac{1}{2} + \dfrac{2}{3} + \dfrac{3}{4} + \cdots + \dfrac{51}{52}$

13. $1 + x^2 + x^4 + x^6 + \cdots + x^{80}$ (Assume $x \neq 0$.)

In Exercises 14 and 15, evaluate the given sum.

14. $\displaystyle\sum_{k=1}^{90} \dfrac{k+1}{2}$

15. $\displaystyle\sum_{k=1}^{30} (-1)^{k-1} \dfrac{2}{3^k}$

16. Consider the infinite sum $\displaystyle\sum_{k=1}^{\infty} \dfrac{2}{4k^2 - 1}$.
    (a) Evaluate the partial sums S_1, S_2, S_3, S_4, S_5.
    (b) From the results in part (a), guess the formula for the nth partial sum, S_n.
    (c) Evaluate the infinite sum by determining $\lim_{n \to +\infty} (S_n)$.

17. Evaluate $\displaystyle\sum_{k=1}^{\infty} 3\left(\dfrac{2}{5}\right)^k$.

18. Rewrite the repeating decimal $.\overline{84}$ as an infinite geometric sum. Find a rational number representing $.\overline{84}$.

19. True or false: $\displaystyle\sum_{k=1}^{\infty} \dfrac{(1.01)^{k-1}}{2} = \dfrac{1/2}{1 - 1.01} = -50$. Explain your answer.

In Exercises 20 and 21, use the principle of mathematical induction to prove that the given statement is true for all positive integers n.

20. $4 + 8 + 12 + 16 + \cdots + 4n = 2n(n+1)$

21. $7^n - 1$ is a multiple of 6.

22. Use the principle of mathematical induction to prove that $n! > 4^n$ for all integers $n \geq 9$.

23. Use Pascal's triangle to expand $(2x + 1)^5$.

24. Use the Binomial Theorem to expand $\left(\dfrac{x}{y} - \dfrac{y}{x}\right)^6$.

25. Find the coefficient of $x^{14}y^5$ in $\left(x - \dfrac{1}{2}y\right)^{19}$.

26. Evaluate $C(13, 6)$.

27. Simplify $\dfrac{C(n, k)}{C(n-1, k)}$.

Appendix

FACTORING

When we **factor** a given algebraic expression, we rewrite it as a product of terms. Typical examples of factoring are

$$4h^2 - 2xh = 2h(2h - x)$$
$$x^2 - y^2 = (x + y)(x - y)$$
$$3x^2 - x - 2 = (3x + 2)(x - 1)$$

Following is a summary of the basic methods of factoring.

Methods of Factoring

Step 1   Factor out any common factors.

Step 2   Categorize the expression to be factored, and then try the corresponding method.

Type of Expression	Method
Two terms	$a^2 - b^2 = (a + b)(a - b)$ $a^3 - b^3 = (a - b)(a^2 + ab + b^2)$ $a^3 + b^3 = (a + b)(a^2 - ab + b^2)$
Three terms	Guess two binomial factors: $(\underline{\quad}\ \underline{\quad})(\underline{\quad}\ \underline{\quad})$
Four terms	Use grouping.

Example 1

Factor $36 - 16x^2$.

SOLUTION   The first step is to check for common factors. Thus

$$36 - 16x^2 = 4(9 - 4x^2)$$

Next, we attempt to factor $9 - 4x^2$. This is a two-term expression consisting of a difference of squares, $3^2 - (2x)^2$. Therefore we use the formula

$$a^2 - b^2 = (a + b)(a - b)$$

$$3^2 - (2x)^2 = (3 + 2x)(3 - 2x)$$

Putting these results together, we have

$$36 - 16x^2 = 4(9 - 4x^2)$$

$$= 4(3 + 2x)(3 - 2x) \blacksquare$$

Example 2

Factor $8x^3 - 27y^3$.

SOLUTION First, we note that $8x^3 - 27y^3$ does not contain any common factors. Next, we categorize $8x^3 - 27y^3$ as a two-term expression consisting of a difference of cubes, $(2x)^3 - (3y)^3$. Therefore we use the formula

$$a^3 - b^3 = (a - b)(a^2 + ab + b^2)$$

$$(2x)^3 - (3y)^3 = (2x - 3y)[(2x)^2 + (2x)(3y) + (3y)^2]$$

We conclude,

$$8x^3 - 27y^3 = (2x - 3y)(4x^2 + 6xy + 9y^2) \blacksquare$$

Example 3

Factor $1 + x^3$.

SOLUTION First, we notice that there are no common factors. Next, we categorize $1 + x^3$ as a two-term expression consisting of a sum of cubes, $1^3 + x^3$. Therefore we use the formula

$$a^3 + b^3 = (a + b)(a^2 - ab + b^2)$$

$$1^3 + x^3 = (1 + x)(1^2 - 1x + x^2)$$

We conclude,

$$1 + x^3 = (1 + x)(1 - x + x^2) \blacksquare$$

Example 4

Factor $2x^3 + 5x^2 - 12x$.

SOLUTION First, we check for common factors. Thus

$$2x^3 + 5x^2 - 12x = x(2x^2 + 5x - 12)$$

Now we attempt to factor $2x^2 + 5x - 12$. We categorize this as a three-term expression, so we try to guess two binomial factors.

$$2x^2 + 5x - 12 = (\underline{\ ?\ }\ \underline{\ ?\ })(\underline{\ ?\ }\ \underline{\ ?\ })$$

The product of the first terms must be $2x^2$, so we use $2x$ and x.

$$2x^2 + 5x - 12 = (2x\ \underline{\ ?\ })(x\ \underline{\ ?\ })$$

The product of the last terms must be -12. There are many possibilities using the numbers 1 and 12, 2 and 6, or 3 and 4. Without worrying about signs, we try 3 and 4.

$$2x^2 + 5x - 12 = (2x\ ?\ 3)(x\ ?\ 4)$$

Now we attempt to choose the appropriate signs so that two things happen:

1. The product of the outside terms plus the product of the inside terms equals $+5x$.
2. The product of the last terms equals -12.

Both properties will be true if we choose the following signs:

$$
\begin{array}{c}
+8x \\
\overbrace{} \\
2x^2 + 5x - 12 = (2x - 3)(x + 4) \\
\underbrace{} \\
-3x
\end{array}
$$

To conclude, we have

$$2x^3 + 5x^2 - 12x = x(2x^2 + 5x - 12)$$
$$= x(2x - 3)(x + 4)\ \blacksquare$$

Remark

If the numbers 3 and 4 had *not* worked as just shown, we could have changed the order in which we used them. Thus, instead of $(2x\ ?\ 3)(x\ ?\ 4)$, we could have tried $(2x\ ?\ 4)(x\ ?\ 3)$. If that had not worked, we still would have had the numbers 2 and 6 or 1 and 12 to try.

Example 5

Factor $x^3 + x - x^2 - 1$.

SOLUTION First, we check for common factors and notice that there are none. Next, we categorize $x^3 + x - x^2 - 1$ as a four-term expression, so we use the method of **grouping**. We describe this method in the following four steps:

Step 1. Split the expression into two groups consisting of the first two terms and the last two terms.

Step 2. Factor out any common factors from the first group.

Step 3. Factor out any common factors from the second group.

Step 4. If a common factor appears in the entire expression, factor it out.

Thus

$$\text{Step 1} \quad x^3 + x - x^2 - 1$$

$$\text{Step 2} \quad x(x^2 + 1) - x^2 - 1$$

$$\text{Step 3} \quad x(x^2 + 1) - 1(x^2 + 1)$$

$$\text{Step 4} \quad (x^2 + 1)(x - 1)$$

We conclude that $x^3 + x - x^2 - 1 = (x^2 + 1)(x - 1)$. ∎

Warning

Notice that $x^2 + 1$ does *not* factor. Indeed, any expression of the form $a^2 + b^2$, a *sum* of squares, does not factor (when restricted to using real numbers).

Example 6

Factor $a^2b - x^2 + a^2 - bx^2$.

SOLUTION Checking for common factors, we find none. The expression involves four terms, so we try grouping.

$$a^2b - x^2 + a^2 - bx^2$$

Unfortunately, neither the first two terms nor the last two terms factor. However, we may try rearranging the order of the terms. For example, interchanging the second and third terms, we have

$$a^2b + a^2 - x^2 - bx^2$$

Now we try grouping once more.

$$a^2b + a^2 - x^2 - bx^2$$

$$a^2(b + 1) - x^2(1 + b)$$

$$(b + 1)(a^2 - x^2)$$

$$(b + 1)(a + x)(a - x)$$

Thus $a^2b - x^2 + a^2 - bx^2 = (b + 1)(a + x)(a - x)$. ∎

EXERCISES A.1

In Exercises 1 to 44, factor the given expression if possible.

1. $4x^2 - y^2$

2. $9 - x^2$

3. $16 - x^4$

4. $x^4 - 5x^2 + 4$

5. $x - a^3x$

6. $y^6 - 64$

7. $1 - \dfrac{1}{x^2}$

8. $\dfrac{1}{4} - x^4$

9. $x^6 - y^6$

10. $8x^3 + 1$

11. $4y^2 + 9$

12. $a^2 + 16$

13. $xy^3 + 27x$

14. $a^3 + b^6$

15. $3x^2 - \dfrac{1}{3}$

16. $\dfrac{4}{5}x^2 - \dfrac{1}{5}y^2$

17. $2x^2 - 7x - 15$

18. $2x^2 + 3x - 20$

19. $2y^2 - 11y + 6$

20. $b^2 - 3b + 15$

21. $4x^2y - 11xy + 6y$

22. $6x^2y - 25xy + 21y$

23. $r^2 + s^2 - 2rs$

24. $c^2 + 4d^2 - 4cd$

25. $xy + y + x + 1$

26. $2ax + bx - 4a - 2b$

27. $60h^2 - 94hk - 120k^2$

28. $60r^2 - 116r - 21$

29. $ab + c^2 - ac - bc$

30. $ax - by + ay - bx$

31. $x^2y - 3 + 3x^2 - y$

32. $xy^3 - y^3 + x - 1$

33. $x^4 + 2x^2y^2 + y^4$

34. $x^4 - y^4$

35. $(a - b)^2 - 1$

36. $25 - (b - 4)^2$

37. $x^3 - x^2 - xy^2 + y^2$

38. $m^3 - n^3 + nm^2 - mn^2$

39. $a^2 + 2ab + b^2 - c^2$

40. $x^4 + 2x^3 + x^2 - 1$

41. $x^3 + y^3 + 3x^2y + 3xy^2$

42. $b^3 + 1 + 3b + 3b^2$

43. $x^2 + y^2 + x + y + 2xy$

44. $4a^2 + b^2 + 2a + b + 4ab$

A.2 FRACTIONS

Recall the definition of a fraction.

Definition of $\dfrac{a}{b}$

If $b \neq 0$, then $\quad \dfrac{a}{b} = c \quad$ if and only if $\quad a = b \cdot c$

As a simple example, we have $\dfrac{21}{7} = 3$ because $21 = 7 \cdot 3$. Note that the definition of a/b purposely avoids the case when $b = 0$. The reason for this may be explained as follows. Suppose we wish to define $a/0$ to be equal to a real number c. This would mean that

$$\frac{a}{0} = c \quad \text{if and only if} \quad a = 0 \cdot c$$

But $0 \cdot c = 0$ for any real number c. Thus if $a \neq 0$, then there is no c for which $a = 0 \cdot c$. On the other hand, if $a = 0$, then there is no *unique* c for which $a = 0 \cdot c$; that is, $0/0$ could be anything. Thus expressions such as $2/0$, $x/(3 - 3)$, and $0/0$ are said to be **undefined**, or meaningless. We conclude,

Division by zero is undefined.

Example 1

For what values of x is the expression $\dfrac{x-1}{x+1}$ defined?

SOLUTION Since division by zero is undefined, we cannot allow the denominator of the fraction to be zero. Thus $(x-1)/(x+1)$ is defined whenever $x + 1 \neq 0$ or, more simply, for all $x \neq -1$. ∎

When dealing with fractional expressions, we assume that the variables involved do not take on values that result in division by zero.

Now we turn our attention to the simplification of fractional expressions. Our work is based on the following principle.

> **The Fundamental Principle of Fractions**
> If b and c are nonzero, then
> $$\frac{c(a)}{c(b)} = \frac{a}{b}$$

This says that we may reduce a fraction by canceling a common factor from the numerator and denominator. For instance,

$$\frac{2x}{2y} = \frac{x}{y} \qquad \frac{3xy^2}{5y} = \frac{3xy}{5}$$

Example 2

Simplify $\dfrac{4x}{4 + 2x}$.

SOLUTION We reduce this fraction by canceling the common factor 2.

$$\frac{4x}{4 + 2x} + \frac{2(2x)}{2(2 + x)} = \frac{2x}{2 + x}$$

No further simplification is possible. ∎

Warning

Be careful not to cancel anything unless it factors completely out of the entire numerator and denominator. Since 2 does not factor out of the denominator $2 + x$, we may not cancel it with the 2 in the numerator of the fraction $2x/(2 + x)$.

The fundamental principle of fractions also says that we can *multiply* the numerator and denominator of a fraction by the same nonzero number without changing its value. For example,

$$\frac{2}{3} = \frac{5(2)}{5(3)} = \frac{10}{15} \qquad \frac{4n}{3} = \frac{3x(4n)}{3x(3)} = \frac{12xn}{9x}$$

The operation of multiplying the numerator and denominator of a fraction by the same quantity is helpful in simplifying certain **compound fractions**. Any fraction that contains one or more fractions in either its numerator or denominator is a compound fraction. Typical examples of compound fractions include:

$$\frac{\dfrac{a}{b}}{\dfrac{c}{d}} \qquad\qquad \frac{\dfrac{1}{x+1} - \dfrac{1}{x}}{2 + \dfrac{1}{x+1}}$$

We say that a compound fraction is simplified if it is rewritten in an equivalent form with no fractions appearing in its numerator or denominator; in other words, the fraction is no longer compound. As an example, suppose we want to simplify $\dfrac{a/b}{c/d}$. Multiply both the numerator and denominator by the reciprocal of the denominator, d/c. Thus

$$\frac{\dfrac{a}{b}}{\dfrac{c}{d}} = \frac{\dfrac{a}{b} \cdot \dfrac{d}{c}}{\dfrac{c}{d} \cdot \dfrac{d}{c}} = \frac{\dfrac{ad}{bc}}{1} = \frac{ad}{bc}$$

The end result is the product of the numerator with the reciprocal of the denominator. We summarize this as follows.

If b, c, and d are nonzero, then

$$\frac{\dfrac{a}{b}}{\dfrac{c}{d}} = \frac{a}{b} \cdot \frac{d}{c}$$

Example 3

Simplify (a) $\dfrac{\dfrac{2}{3}}{\dfrac{1}{2}}$; (b) $\dfrac{\dfrac{2}{x}}{b}$; (c) $\dfrac{\dfrac{3}{x+b}}{\dfrac{6}{x}}$.

(a) $\dfrac{\dfrac{2}{3}}{\dfrac{1}{2}} = \dfrac{2}{3} \cdot \dfrac{2}{1} = \dfrac{4}{3}$

(b) $\dfrac{\dfrac{2}{x}}{h} = \dfrac{2}{x} \cdot \dfrac{1}{h} = \dfrac{2}{xh}$

(c) $\dfrac{\dfrac{3}{x+h}}{\dfrac{6}{x}} = \dfrac{3}{x+h} \cdot \dfrac{x}{6} = \dfrac{3x}{6(x+h)} = \dfrac{x}{2(x+h)}$ ∎

If a compound fraction is more complicated, perhaps containing several fractions in either the numerator or denominator, then we multiply the numerator and denominator by the least common denominator of all these fractions, thus reducing the compound fraction to just one fraction.

Example 4

Simplify (a) $\dfrac{\dfrac{1}{x} - \dfrac{1}{a}}{x - a}$, (b) $\dfrac{\dfrac{1}{x+1} - \dfrac{1}{x}}{2 + \dfrac{1}{x+1}}$.

SOLUTION   (a) This compound fraction contains two fractions, $1/x$ and $1/a$, both in the numerator. To eliminate these fractions, we multiply by their least common denominator, ax. Of course, by the fundamental principle of fractions, we must multiply *both* the numerator and denominator of the compound fraction by ax.

$$\frac{\dfrac{1}{x} - \dfrac{1}{a}}{x - a} = \frac{ax\left(\dfrac{1}{x} - \dfrac{1}{a}\right)}{ax(x-a)} = \frac{\dfrac{ax}{x} - \dfrac{ax}{a}}{ax(x-a)} = \frac{a - x}{ax(x-a)}$$

At this point we have an interesting situation, with $a - x$ in the numerator and $x - a$ in the denominator. These are not identical factors, however they are opposites. We have

$$\frac{a - x}{ax(x-a)} = \frac{-1(x-a)}{ax(x-a)} = \frac{-1}{ax}$$

Thus factors appearing in the numerator and denominator that are opposites reduce to -1.

(b) This compound fraction contains three fractions, $1/(x + 1)$ and $1/x$ in the numerator and $1/(x + 1)$ in the denominator. The least common denominator for all these fractions is $x(x + 1)$. Thus

$$\frac{\dfrac{1}{x+1} - \dfrac{1}{x}}{2 + \dfrac{1}{x+1}} = \frac{x(x+1)\left(\dfrac{1}{x+1} - \dfrac{1}{x}\right)}{x(x+1)\left(2 + \dfrac{1}{x+1}\right)}$$

$$= \frac{x - (x+1)}{2x(x+1) + x}$$

$$= \frac{-1}{2x^2 + 3x}$$

$$= \frac{-1}{x(2x+3)} \quad \blacksquare$$

Example 5

Simplify $\dfrac{\sqrt{x} - \dfrac{x}{2\sqrt{x}}}{x^2}$.

SOLUTION   This compound fraction contains only one fraction, $x/2\sqrt{x}$. We eliminate this fraction by multiplying the numerator and denominator by $2\sqrt{x}$.

$$\frac{\sqrt{x} - \dfrac{x}{2\sqrt{x}}}{x^2} = \frac{2\sqrt{x}\left(\sqrt{x} - \dfrac{x}{2\sqrt{x}}\right)}{2\sqrt{x}\, x^2}$$

$$= \frac{2x - x}{2x^2\sqrt{x}}$$

$$= \frac{x}{2x^2\sqrt{x}}$$

$$= \frac{1}{2x\sqrt{x}} \quad \blacksquare$$

Unless required otherwise, we allow square roots to remain in the denominator of an answer.

We close this section by reviewing how to add fractions. If two or more fractions have the same denominator, then we find their sum using the following basic rule.

$$\frac{a}{b} + \frac{c}{b} = \frac{a+c}{b}$$

Of course, not all fractions that we wish to add will have the same denominator. In this case, we use the following procedure:

Step 1. Find a least common denominator for the fractions.
Step 2. Rewrite each fraction over the common denominator.
Step 3. Add.

Example 6

Add

(a) $\dfrac{2}{x + h} - \dfrac{2}{x}$, (b) $\dfrac{-\sqrt{x + 1}}{x^2} + \dfrac{1}{x\sqrt{x + 1}}$.

SOLUTION (a) The common denominator in this case is $x(x + h)$. Thus

$$\frac{2}{x + h} - \frac{2}{x} = \frac{2x}{x(x + h)} - \frac{2(x + h)}{x(x + h)}$$

$$= \frac{2x - 2(x + h)}{x(x + h)}$$

$$= \frac{2x - 2x - 2h}{x(x + h)}$$

$$= \frac{-2h}{x(x + h)}$$

We stop here because this fraction cannot be reduced further.

(b) The common denominator in this case is $x^2\sqrt{x + 1}$. We have

$$\frac{-\sqrt{x + 1}}{x^2} + \frac{1}{x\sqrt{x + 1}} = \frac{-\sqrt{x + 1}\sqrt{x + 1}}{x^2\sqrt{x + 1}} + \frac{x}{x^2\sqrt{x + 1}}$$

$$= \frac{-(x + 1) + x}{x^2\sqrt{x + 1}}$$

$$= \frac{-1}{x^2\sqrt{x + 1}} \quad ∎$$

EXERCISES A.2

In Exercises 1 and 2, for what values of x is the given expression defined?

1. (a) $\dfrac{1}{x}$

   (b) $\dfrac{2x}{x + 3}$

   (c) $\dfrac{2 - \dfrac{1}{x}}{x - 2}$

2. (a) $\dfrac{3}{2 - x}$

   (b) $\dfrac{\dfrac{1}{x} - 2}{x + 4}$

   (c) $\dfrac{x}{x^2 + 1}$

In Exercises 3 and 4, if possible, simplify the given fractions by canceling common factors.

3. (a) $\dfrac{6n}{4m}$

   (b) $\dfrac{6 + n}{4m}$

   (c) $\dfrac{9a - 12b}{6a - 2b}$

   (d) $\dfrac{x - 5}{5 - x}$

4. (a) $\dfrac{3x}{12xy}$

   (b) $\dfrac{3 + x}{12xy}$

   (c) $\dfrac{6x - 10y}{10x - 6y}$

   (d) $\dfrac{2a - 6}{3 - a}$

In Exercises 5 to 12, simplify each fraction by canceling common factors.

5. $\dfrac{6 + 3\sqrt{2}}{6}$

6. $\dfrac{10 - 2x}{8}$

7. $\dfrac{4 - \sqrt{8}}{8}$

8. $\dfrac{4 + \sqrt{12}}{4}$

9. $\dfrac{(x + h)^2 - x^2}{h}$

10. $\dfrac{a^2 - x^2}{x - a}$

11. $\dfrac{3x^2 + 2x - 1 - (3a^2 + 2a - 1)}{x - a}$

12. $\dfrac{(x + h)^2 - 2(x + h) + 3 - (x^2 - 2x + 3)}{h}$

In Exercises 13 to 42, simplify each compound fraction.

13. (a) $\dfrac{\dfrac{3}{4}}{\dfrac{1}{2}}$

(b) $\dfrac{\dfrac{1}{6}}{3}$

(c) $\dfrac{\dfrac{1}{x}}{2x}$

14. (a) $\dfrac{\dfrac{4}{3}}{\dfrac{3}{2}}$

(b) $\dfrac{\dfrac{2}{3}}{6}$

(c) $\dfrac{\dfrac{2x}{y}}{\dfrac{2y}{x}}$

15. $\dfrac{\dfrac{x}{x + 1}}{\dfrac{x}{x - 1}}$

16. $\dfrac{7 + x}{\dfrac{x}{2}}$

17. $\dfrac{\dfrac{1}{x} + 1}{\dfrac{1}{x} - 1}$

18. $\dfrac{\dfrac{1}{x} + \dfrac{2}{x^2}}{\dfrac{4}{x^2} - 1}$

19. $\dfrac{1 + \dfrac{3}{x} + \dfrac{2}{x^2}}{1 + \dfrac{5}{x} + \dfrac{6}{x^2}}$

20. $\dfrac{2 + \dfrac{1}{x} - \dfrac{3}{x^2}}{2 + \dfrac{5}{x} + \dfrac{3}{x^2}}$

21. $\dfrac{\dfrac{1}{a} - \dfrac{1}{x}}{\dfrac{1}{x} - \dfrac{1}{a}}$

22. $\dfrac{\dfrac{1}{ax} - \dfrac{1}{bx}}{\dfrac{1}{by} - \dfrac{1}{ay}}$

23. $\dfrac{\dfrac{1}{2}(x + h) - \dfrac{1}{2}x}{h}$

24. $\dfrac{x + \dfrac{1}{y}}{\dfrac{2}{10}}$

25. $\dfrac{\dfrac{x - \dfrac{1}{y}}{2}}{\dfrac{x + \dfrac{1}{y}}{4}}$

26. $\dfrac{\dfrac{4 - x}{x - \dfrac{1}{4}}}{\dfrac{1}{x} - \dfrac{1}{4}}$

27. $\dfrac{\dfrac{1}{R}}{\dfrac{1}{R_1} + \dfrac{1}{R_2}}$

28. $\dfrac{\dfrac{1}{R}}{\dfrac{1}{R_1} + \dfrac{1}{R_2} + \dfrac{1}{R_3}}$

29. $\dfrac{\dfrac{m_1}{k_1} + \dfrac{m_2}{k_2}}{\dfrac{m_1 + m_2}{k_1 k_2}}$

30. $\dfrac{\dfrac{1}{a} + \dfrac{1}{b} + \dfrac{1}{c}}{3}$

31. $\dfrac{\dfrac{1}{x + h} - \dfrac{1}{x}}{h}$

32. $\dfrac{\dfrac{2}{x} - \dfrac{2}{a}}{x - a}$

33. $\dfrac{x + \dfrac{1}{x} - \left(a + \dfrac{1}{a}\right)}{x - a}$

34. $\dfrac{2 - \dfrac{1}{x + h} - \left(2 - \dfrac{1}{x}\right)}{h}$

35. $\dfrac{x + h - \dfrac{1}{x + h} - \left(x - \dfrac{1}{x}\right)}{h}$

36. $\dfrac{x + h + \dfrac{3}{x + h} - \left(x + \dfrac{3}{x}\right)}{h}$

37. $\dfrac{\dfrac{1}{x^2} - \dfrac{1}{a^2}}{x - a}$

38. $\dfrac{\dfrac{1}{(x + h)^2} - \dfrac{1}{x^2}}{h}$

39. $\dfrac{\dfrac{x}{2\sqrt{x}} - \sqrt{x}}{x^2}$

40. $\dfrac{\sqrt{x + 1} - x\left(\frac{1}{2}\right)\dfrac{1}{\sqrt{x + 1}}}{x + 1}$

41. $\dfrac{\sqrt{x + 1}\left(\frac{1}{2}\right)\dfrac{1}{\sqrt{x}} - \sqrt{x}\left(\frac{1}{2}\right)\dfrac{1}{\sqrt{x + 1}}}{x + 1}$

42. $\dfrac{\sqrt{x^2 + 1} - x\left(\frac{1}{2}\right)\dfrac{2x}{\sqrt{x^2 + 1}}}{x^2 + 1}$

In Exercises 43 to 56, add.

43. $\dfrac{a}{b} + \dfrac{b}{a}$

44. $\dfrac{x}{y} - \dfrac{y}{x}$

45. $\dfrac{1}{C_1} + \dfrac{1}{C_2}$

46. $\dfrac{1}{a} + \dfrac{1}{b} + \dfrac{1}{c}$

47. $\dfrac{1}{x + y} - \dfrac{2}{x}$

48. $\dfrac{3}{x + 1} - \dfrac{2}{x - 1}$

49. $\dfrac{1}{x + 1} - 2$

50. $\dfrac{x}{x - 2} - 1$

51. $\dfrac{x}{x^2 - 4} + \dfrac{2}{x + 2}$

52. $\dfrac{12x}{y^2} - \dfrac{4}{y^2 - 3y}$

53. $\dfrac{2}{x - h} + \dfrac{2}{x + h}$

54. $\dfrac{3}{x - h} + \dfrac{1}{h - x}$

55. $\dfrac{a}{x} + \dfrac{b}{x - 2}$

56. $\dfrac{a}{x} + \dfrac{b}{x^2} + \dfrac{c}{x + 1}$

In Exercises 57 to 64, add. You may leave roots in the denominator.

57. $\dfrac{1}{\sqrt{x}} + \dfrac{1}{3\sqrt{x}}$

58. $\dfrac{\sqrt{x}}{x} + \dfrac{1}{\sqrt{x}}$

59. $\sqrt{x} + \dfrac{1}{\sqrt{x}}$

60. $\sqrt{x^2 + 1} + \dfrac{x^2}{\sqrt{x^2 + 1}}$

61. $\dfrac{x}{2\sqrt{x + 1}} + \sqrt{x + 1}$

62. $\dfrac{x^2}{\sqrt{2x - 3}} + 2x\sqrt{2x - 3}$

63. $\sqrt{\dfrac{x + 1}{x}} - \sqrt{\dfrac{x}{x + 1}}$

64. $\sqrt{\dfrac{x}{x + 1}} + \sqrt{\dfrac{y}{x + 1}}$

In Exercises 65 to 70, answer true or false. Assume that all denominators are nonzero.

65. $\dfrac{2x + 1}{4} = \dfrac{x + 1}{2}$

66. $\dfrac{ax + by}{a + b} = x + y$

67. $\dfrac{\frac{8}{3}}{\frac{1}{2}} = \dfrac{4}{3}$

68. $\dfrac{x}{\frac{a}{b} + 1} = \dfrac{bx}{a + 1}$

69. $\dfrac{1}{\frac{1}{x} + \frac{1}{y}} = x + y$

70. $\dfrac{x - y}{\frac{1}{x} - \frac{1}{y}} = xy$

A.3 COMPLETING THE SQUARE

Consider the equation

$$y = x^2 + 6x + 10 \qquad (*)$$

Suppose we wish to rewrite this equation in the form

$$y - k = (x - h)^2$$

The letters h and k simply represent constant numbers (not necessarily positive). To achieve our goal, we need to manipulate Equation (*) so that its right side will be a perfect square. We call this process **completing the square** on the x terms. First, move the constant term of Equation (*) to the left side.

$$y - 10 = x^2 + 6x$$

Next, add a number to both sides of the equation so that the right side becomes a perfect square. To obtain this number, take half the coefficient of x, which is $6/2 = 3$, and square it, which gives us 9. By adding 9 to both sides,

$$y - 10 + 9 = x^2 + 6x + 9$$

we obtain a perfect square on the right side.

$$y - 1 = (x + 3)^2$$

This accomplishes our goal since the equation now has the form $y - k = (x - h)^2$, where $k = 1$ and $h = -3$.

In general, suppose that we start with an equation of the form

$$y = ax^2 + bx + c$$

and we wish to rewrite it in the form $y - k = a(x - h)^2$. We start by putting the constant term on the left side.

$$y - c = ax^2 + bx$$

Next, factor out a from the right side.

$$y - c = a\left(x^2 + \frac{b}{a}x\right)$$

Now we take half the coefficient of x, which is $b/2a$, and square it, which gives us $b^2/4a^2$. Add this quantity to the expression inside the parentheses on the right side. By doing this, we are really adding $a(b^2/4a^2) = b^2/4a$ to the right side of the equation because of the factor of a in front of the parentheses. Therefore we must also add $b^2/4a$ to the left side so that we still have an equivalent equation.

$$y - c + \frac{b^2}{4a} = a\left(x^2 + \frac{b}{a}x + \frac{b^2}{4a^2}\right)$$

Now we can factor the right side. We have

$$y - \left(c - \frac{b^2}{4a}\right) = a\left(x + \frac{b}{2a}\right)^2$$

Notice that this equation has the desired form, $y - k = a(x - h)^2$.

Example 1

Rewrite $y = -2x^2 + 20x - 54$ in the equivalent form $y - k = a(x - h)^2$.

SOLUTION We begin by moving the constant term, -54, over to the left side.

$$y = -2x^2 + 20x - 54$$

$$y + 54 = -2x^2 + 20x$$

Next, factor out -2 from the right side.

$$y + 54 = -2(x^2 - 10x \qquad)$$

Now we take half the coefficient of x, which is $-10/2 = -5$, and square it, which gives us 25. Add this inside the parentheses on the right side. In doing this, we are really adding $-2(25) = -50$ to the right side (because of the factor of -2 in front of the parentheses). Thus we must add -50 to the left side to keep things balanced.

$$y + 54 - 50 = -2(x^2 - 10x + 25)$$

$$y + 4 = -2(x - 5)^2$$

This gives us the desired form for the equation. ■

Example 2

Use the technique of completing the square to rewrite the equation $x^2 + y^2 - 2x - 4y - 2 = 0$ in the equivalent form $(x - h)^2 + (y - k)^2 = c$.

SOLUTION We can rearrange terms on the left side of the equation so that x's and y's are grouped together.

$$x^2 - 2x \quad + y^2 - 4y \quad - 2 = 0$$

Now move the constant term, -2, over to the right side and complete the square on both the x and y terms.

$$x^2 - 2x \qquad + y^2 - 4y \qquad = 2$$

$$x^2 - 2x \boxed{+1} + y^2 - 4y \boxed{+4} = 2 \boxed{+1} \boxed{+4}$$

$$(x - 1)^2 + (y - 2)^2 \qquad = 7$$

This is the desired form of the equation. ■

Example 3

Rewrite the expression $\sqrt{x^2 - x + 1}$ in the equivalent form $\sqrt{(x - h)^2 + k}$.

SOLUTION In this case it is important to keep in mind that we are dealing with an algebraic expression, $\sqrt{x^2 - x + 1}$, not an equation. In making this expression look like $\sqrt{(x - h)^2 + k}$, we are not allowed to change its value. Let us see how to accomplish this.

$$\sqrt{x^2 - x + 1}$$

$$\overline{\sqrt{x^2 - x \underline{\quad} + 1 \underline{\quad}}}$$

We must add an appropriate number in the first space to complete the square on the x terms. This is half the coefficient of x squared: $(-1/2)^2 = 1/4$. At the same time we add the opposite of 1/4 in the second space so that the value of the entire expression remains unchanged.

$$\sqrt{x^2 - x + \underline{1/4} + 1 \ \underline{-1/4}}$$

$$\sqrt{\left(x - \frac{1}{2}\right)^2 + \frac{3}{4}}$$

This is the desired form. ■

We close this section with a derivation of the quadratic formula.

Example 4

Solve the equation $ax^2 + bx + c = 0$.

SOLUTION   Put the constant term on the right side of the equation and then complete the square.

$$ax^2 + bx + c = 0$$

$$ax^2 + bx = -c$$

$$a\left(x^2 + \frac{b}{a}x \qquad\right) = -c$$

$$a\left(x^2 + \frac{b}{a}x + \frac{b}{4a^2}\right) = \frac{b^2}{4a} - c$$

$$a\left(x + \frac{b}{2a}\right)^2 = \frac{b^2 - 4ac}{4a}$$

Now divide by a and then apply the square root operation.

$$\sqrt{\left(x + \frac{b}{2a}\right)^2} = \sqrt{\frac{b^2 - 4ac}{4a^2}}$$

$$\left|x + \frac{b}{2a}\right| = \frac{\sqrt{b^2 - 4ac}}{2a}$$

$$x + \frac{b}{2a} = \pm\frac{\sqrt{b^2 - 4ac}}{2a}$$

$$x = \frac{-b \pm \sqrt{b^2 - 4ac}}{2a} \quad ■$$

EXERCISES A.3

In Exercises 1 to 20, rewrite the given equation in the equivalent form $y - k = a(x - h)^2$.

1. $y = x^2 + 8x + 1$

2. $y = x^2 + 2x + 2$

3. $y = x^2 - 6x + 3$

4. $y = x^2 - 12x - 1$

5. $y = x^2 - 10 + 3x$

6. $y = x^2 + 7 - 5x$

7. $y = x^2 + x$

8. $y = x^2 - x$

9. $y - 4x = 2x^2 - 3$

10. $y - 9x = 3x^2 + 4$

11. $y = 3x^2 - 4x + 1$

12. $y = 2x^2 - x - 5$

13. $y = 4x - 5x^2$

14. $y = -4x^2 - 3x$

15. $y = 3 + 8x - x^2$

16. $y = -x^2 + 3x - 1$

17. $y = -4x^2 - 4x - 2$

18. $y = 5 + 2x - 3x^2$

19. $y = \dfrac{1}{3}x^2 + 3x + 1$

20. $y = -\dfrac{3}{4}x^2 + x$

In Exercises 21 to 26, rewrite the given equation in the equivalent form $(x - h)^2 + (y - k)^2 = c$.

21. $x^2 + y^2 - 4x + 6y + 9 = 0$

22. $x^2 + y^2 + 18x - 2y + 72 = 0$

23. $4x^2 + 4y^2 + 4x - 32y + 33 = 0$

24. $9x^2 + 9y^2 - 24x + 90y + 187 = 0$

25. $9x^2 + 9y^2 - 108x + 323 = 0$

26. $16x^2 + 16y^2 - 8x - 4y = 0$

In Exercises 27 to 30, rewrite the given equation in the equivalent form $a(x - h)^2 + b(y - k)^2 = c$.

27. $x^2 + 2y^2 + 14x - 12y + 5 = 0$

28. $3x^2 + y^2 - 6x + y - 1 = 0$

29. $5x^2 - 4y^2 - 5x + 8y - 3 = 0$

30. $2x^2 - 4y^2 + 4x + 12y - 15 = 0$

In Exercises 31 to 38, rewrite the given expression in the equivalent form $\sqrt{a(x - h)^2 + k}$.

31. $\sqrt{x^2 - x}$

32. $\sqrt{x^2 + x}$

33. $\sqrt{x^2 + 4x + 3}$

34. $\sqrt{x^2 - 6x + 5}$

35. $\sqrt{2x^2 - 8x}$

36. $\sqrt{3x^2 + 12x}$

37. $\sqrt{3x^2 + 3x + 1}$

38. $\sqrt{2x^2 - 2x + 1}$

In Exercises 39 to 42, solve the given equation by completing the square (see Example 4).

39. $x^2 + 2x - 1 = 0$

40. $x^2 - 6x + 1 = 0$

41. $2x^2 - 20x + 49 = 0$

42. $3x^2 + 6x - 1 = 0$

43. By completing the square, show that the value of $x^2 - x + \dfrac{1}{2}$ is always positive for every real number x.

44. By completing the square, show that the value of $2x - (1 + x^2)$ is always negative for every real number $x \neq 1$.

A.4 EXPONENTS

Recall that an expression of the form a^r is called an **exponential expression** with **base** a and **exponent** r. In Section 5.1 we discuss the meaning of these expressions in detail. Here we review some of the basic algebra involved in manipulating exponential expressions when the exponent consists of an integer or rational number. Throughout the discussion we assume that all symbols appearing in the base of an exponential expression represent positive real numbers.

Rules for Exponents

If $a > 0$ and n is a positive integer, then

$$a^n = \overbrace{a \cdot a \cdot a \cdots a}^{n \text{ factors}}$$

$$a^0 = 1$$

If $a > 0$, $b > 0$, r and s rational numbers, then

$$a^{-r} = \frac{1}{a^r} \qquad (ab)^r = a^r b^r$$

$$a^r a^s = a^{r+s} \qquad \left(\frac{a}{b}\right)^r = \frac{a^r}{b^r}$$

$$(a^r)^s = a^{rs}$$

$$\frac{a^r}{a^s} = a^{r-s} \qquad \left(\frac{a}{b}\right)^{-r} = \left(\frac{b}{a}\right)^r$$

Example 1

Simplify each of the following expressions: (a) $4xxxyy$, (b) $x^2x^3x^{-5}$ (c) $x^{\frac{3}{4}}(x^{\frac{1}{3}} + x^{-\frac{1}{3}})$, (d) $(x^2y)^3$.

SOLUTION

(a) $4xxxyy = 4x^3y^2$

(b) $x^2x^3x^{-5} = x^5x^{-5} = x^0 = 1$

(c) $x^{\frac{3}{4}}(x^{\frac{1}{3}} + x^{-\frac{1}{3}}) = x^{\frac{3}{4}}x^{\frac{1}{3}} + x^{\frac{3}{4}}x^{-\frac{1}{3}}$

$$= x^{\frac{3}{4}+\frac{1}{3}} + x^{\frac{3}{4}-\frac{1}{3}}$$

$$= x^{\frac{9}{12}+\frac{4}{12}} + x^{\frac{9}{12}-\frac{4}{12}}$$

$$= x^{\frac{13}{12}} + x^{\frac{5}{12}}$$

(d) $(x^2y)^3 = (x^2)^3 y^3 = x^6 y^3$ ∎

Example 2

Simplify $\dfrac{10^{12}10^{-3}}{10^{15}}$ to a power of 10 in the numerator.

SOLUTION

$$\frac{10^{12}10^{-3}}{10^{15}} = \frac{10^9}{10^{15}} = 10^{9-15} = 10^{-6} \ \blacksquare$$

Example 3

Simplify to positive exponents: (a) $\dfrac{x^8}{x^{15}}$, (b) $\left(\dfrac{x}{y}\right)^{-4}$.

SOLUTION

(a) $\dfrac{x^8}{x^{15}} = x^{8-15} = x^{-7} = \dfrac{1}{x^7}$

(b) $\left(\dfrac{x}{y}\right)^{-4} = \left(\dfrac{y}{x}\right)^4 = \dfrac{y^4}{x^4}$ ■

When simplifying algebraic fractions containing exponential expressions, the following results may be helpful.

$$\frac{ax^{-n}}{b} = \frac{a}{bx^n} \qquad\qquad \frac{a}{bx^{-n}} = \frac{ax^n}{b}$$

To establish the first result, we have

$$\frac{ax^{-n}}{b} = \frac{a\left(\dfrac{1}{x^n}\right)}{b} = \frac{a\left(\dfrac{1}{x^n}\right)x^n}{bx^n} = \frac{a}{bx^n}$$

The proof of the second result is similar.

Example 4

Simplify $\left(\dfrac{2x^{-3}y}{3y^{-1}}\right)^4$ to positive exponents only.

SOLUTION There are two approaches to this problem.
Method 1: Distribute the power exponent first.

$$\left(\frac{2x^{-3}y}{3y^{-1}}\right)^4 = \frac{2^4x^{-12}y^4}{3^4y^{-4}} = \frac{16x^{-12}y^4}{81y^{-4}} = \frac{16y^4y^4}{81x^{12}}$$

$$= \frac{16y^8}{81x^{12}}$$

Method 2: Simplify inside the parentheses first.

$$\left(\frac{2x^{-3}y}{3y^{-1}}\right)^4 = \left(\frac{2yy}{3x^3}\right)^4 = \left(\frac{2y^2}{3x^3}\right)^4 = \frac{2^4y^8}{3^4x^{12}}$$

$$= \frac{16y^8}{81x^{12}}$$ ■

Example 5

Simplify $\left(\dfrac{x}{x+4}\right)^{-\frac{3}{2}}(x+4)^{-2}$ to positive exponents only.

SOLUTION

$$\left(\frac{x}{x+4}\right)^{-\frac{3}{2}}(x+4)^{-2} = \left(\frac{x+4}{x}\right)^{\frac{3}{2}}\frac{1}{(x+4)^2} = \frac{(x+4)^{\frac{3}{2}}}{x^{\frac{3}{2}}} \cdot \frac{1}{(x+4)^2}$$

$$= \frac{(x+4)^{\frac{3}{2}}}{x^{\frac{3}{2}}(x+4)^2} = \frac{1}{x^{\frac{3}{2}}(x+4)^2(x+4)^{-\frac{3}{2}}}$$

$$= \frac{1}{x^{\frac{3}{2}}(x+4)^{\frac{1}{2}}} \ \blacksquare$$

Warning

We may *not* distribute power exponents over addition. In other words,

$$(a+b)^r \qquad \text{is } not \text{ equal to} \qquad a^r + b^r$$

Thus the expression $(x+4)^{1/2}$ does not simplify further.

Example 6

Simplify $\dfrac{1+x^{-1}}{x+x^{-2}}$ to positive exponents.

SOLUTION

$$\frac{1+x^{-1}}{x+x^{-2}} = \frac{1+\dfrac{1}{x}}{x+\dfrac{1}{x^2}} = \frac{x^2\left(1+\dfrac{1}{x}\right)}{x^2\left(x+\dfrac{1}{x^2}\right)}$$

$$= \frac{x^2+x}{x^3+1} = \frac{x(x+1)}{(x+1)(x^2-x+1)}$$

$$= \frac{x}{x^2-x+1} \ \blacksquare$$

Radical symbols, $\sqrt[n]{}$, are used as another notation for exponents.

Definition of $\sqrt[n]{}$
If n is a positive integer, then

$$\sqrt[n]{a} = a^{\frac{1}{n}}$$

In the radical symbol $\sqrt[n]{a}$, we call n the **index** of the radical. If the index is omitted, then it is understood to be 2. Thus $\sqrt{a}$ means $\sqrt[2]{a}$.

Example 7

Simplify $\sqrt{7}\ \sqrt[3]{7}\ \sqrt[6]{7}$.

SOLUTION

$$\sqrt{7}\ \sqrt[3]{7}\ \sqrt[6]{7} = 7^{\frac{1}{2}}7^{\frac{1}{3}}7^{\frac{1}{6}} = 7^{\frac{1}{2}+\frac{1}{3}+\frac{1}{6}} = 7^1 = 7 \quad \blacksquare$$

Example 8

Simplify $\dfrac{\sqrt{x}}{\sqrt[4]{x^2 y}} \cdot \dfrac{\sqrt[4]{y^3}}{x}$.

SOLUTION

$$\frac{\sqrt{x}}{\sqrt[4]{x^2 y}}\ \frac{\sqrt[4]{y^3}}{x} = \frac{x^{\frac{1}{2}}}{(x^2 y)^{\frac{1}{4}}}\ \frac{(y^3)^{\frac{1}{4}}}{x} = \frac{x^{\frac{1}{2}}y^{\frac{3}{4}}}{x^{\frac{1}{2}}y^{\frac{1}{4}}x}$$

$$= \frac{y^{\frac{3}{4}}y^{\frac{-1}{4}}}{x^{\frac{-1}{2}}x^{\frac{1}{2}}x} = \frac{y^{\frac{1}{2}}}{x} \quad \blacksquare$$

EXERCISES A.4

In Exercises 1 to 12, fill in the box so that the statement is true.

1. $xx^k = x^\square$

2. $\dfrac{x^{k2}}{x^k} = x^\square$

3. $(e^x)^2 = e^\square$

4. $(x^2)^k = x^\square$

5. $\dfrac{2^3}{2^x} = 2^\square$

6. $\left(\dfrac{2^x}{2^y}\right)^{-2} = 2^\square$

7. $\dfrac{(2^x)^2}{2} = 2^\square$

8. $\dfrac{(2 \cdot 2^x)^2}{2} = 2^\square$

9. $\dfrac{1}{e^x} = e^\square$

10. $(e^x)^{-1} = e^\square$

11. $\dfrac{10^{23}10^{-5}}{10^{28}} = 10^\square$

12. $\dfrac{10^6 10^8}{10^9} = 10^\square$

In Exercises 13 to 46, simplify. Write answers so that all exponents are positive. Do not leave compound fractions unsimplified. Assume a, b, x, and y represent positive real numbers.

13. (a) xx^4
    (b) xx^{-4}
    (c) $\dfrac{x^3}{x^7}$

14. (a) y^3y
    (b) y^3y^{-1}
    (c) $\dfrac{y}{y^9}$

15. (a) $\left(\dfrac{x}{y^3}\right)^2$
    (b) $\left(\dfrac{x^2}{y}\right)^{-2}$

16. (a) $\left(\dfrac{a}{b}\right)^3$
    (b) $\left(\dfrac{1}{b^3}\right)^{-3}$

17. $x^{5/4}x^{4/5}x$

18. $x^{1/2}x^{-3}x^{1/4}x^{5/3}$

19. $x^{1/2}(x^{3/2} + x^{-1/2})$

20. $x^{3/4}(x^0y^{1/2} - x^{1/3})$

21. $(x^4y^8)^{3/4}$

22. $(a^3b^6)^{2/3}$

23. $\left(\dfrac{x}{2}\right)^{-2/3}\left(\dfrac{1}{x}\right)^{1/3}$

24. $4x^{-3/2}(4x)^{-1}$

25. $\left(\dfrac{2x^{-1}}{y}\right)^{-2}\left(\dfrac{1}{x^{-2}}\right)^{-1}$

26. $\left(\dfrac{2xy^2}{b^{-3}}\right)^{-2}\left(\dfrac{3x^{-2}}{a^{-2}b^{-2}}\right)^{-1}$

27. $(3x^{-3}yb)^{-3}\dfrac{2y^{-2}}{x^{-1}b^2}$

28. $\dfrac{5y^{-1}}{x^{-1}y}\left(\dfrac{a^3x}{5xy^{-3}}\right)^2$

29. $\left(\dfrac{x^2}{y^{-1/2}}\right)^{\frac{1}{2}}(6^0x^{2/3}y)^{-1}$

30. $\left(\dfrac{x^{-1}y^2}{a^{1/3}x^{1/2}}\right)^{\frac{-1}{2}}(a^{-1/2}xy^{-1})^{-2}$

31. $\left(\dfrac{x}{x+1}\right)^{\frac{-1}{2}}(x+1)^{-2}$

32. $\dfrac{1}{2}\left(\dfrac{x+1}{x}\right)^{\frac{-1}{2}}\left(\dfrac{x^2}{2x+2}\right)^{-1}$

33. $(x^2 - y^2)^{\frac{-1}{2}}(x+y)$

34. $(x^2 + y^2)^{\frac{1}{2}}(x^2 + y^2)^{-1}$

35. $\dfrac{x^{-1} + x}{x^{-2}}$

36. $\dfrac{x^{-1}}{x^{-2} - x^{-1}}$

37. $\dfrac{x^{1/3} - 1}{x^{2/3} + x^{-1/3}}$

38. $\dfrac{y^{-1/4} + y}{1 - y^{3/4}}$

39. $(1 - x^{-1})^{-2}$

40. $(x^2 - x^{-2})^{-2}$

41. $\left(\dfrac{1 - x^{-2}}{1 - x^{-1}}\right)^{-1}$

42. $\left(\dfrac{x^{-2} + 1}{x^{-2}}\right)^{-1}$

43. $\left(\dfrac{x^{-1} + x^{-2}}{y^{-1} + y^{-2}}\right)^{-\frac{1}{2}}$

44. $\left(\dfrac{x - x^{-2}}{1 - x^{-1}}\right)^{-\frac{1}{2}}$

45. $\dfrac{x^{1/2} - x^{-1/2}}{1 - x^{-2}}$

46. $\dfrac{y^{2/3} - y^{-1/3}}{1 - y^{-2}}$

In Exercises 47 to 52, write an equivalent expression using exponents. Simplify when possible.

47. (a) $\sqrt[3]{3}$
    (b) $\sqrt[4]{xy}$
    (c) $\sqrt{\sqrt{x}}$
    (d) $\sqrt{1 + \sqrt{x}}$

48. (a) $\sqrt[3]{a + b}$
    (b) $\sqrt{\dfrac{a}{b^2}}$
    (c) $\sqrt[4]{4}$
    (d) $\sqrt{\dfrac{x^3}{\sqrt{y}}}$

49. (a) $\dfrac{x}{\sqrt{x}}$
    (b) $\sqrt{x^6}\sqrt{x^3}$
    (c) $\sqrt[3]{x^2}\left(\sqrt{x^3} + \dfrac{1}{\sqrt[3]{x}}\right)$

50. (a) $\dfrac{x^2}{2\sqrt{x}}$
    (b) $\sqrt{x}\sqrt[3]{x}$
    (c) $\sqrt[4]{x}\left(x\sqrt[4]{x^3} - \dfrac{1}{\sqrt[4]{x}}\right)$

51. $(x + y)\sqrt{x + y}$

52. $\sqrt{(x + y)^3}\,(x + y)^3$

In Exercises 53 to 58, use exponents to simplify the given expression.

53. (a) $\sqrt[3]{2}\sqrt[4]{2}$
    (b) $\sqrt{12}\sqrt[4]{9}$

54. (a) $\sqrt{2}\sqrt[4]{2}$
    (b) $\sqrt{8}\sqrt[3]{4}$

55. (a) $\sqrt{27}\sqrt[3]{3}\sqrt[6]{9}$
    (b) $\sqrt{3}\sqrt[3]{3}\sqrt[3]{24}$

56. (a) $\sqrt{2}\sqrt[3]{16}\sqrt[6]{4}$
    (b) $\sqrt{18}\sqrt[4]{8}\sqrt[6]{32}$

57. $\sqrt{\sqrt[3]{2}}\sqrt[3]{\sqrt{2}}$

58. $\sqrt[3]{\sqrt[3]{9}}\sqrt[3]{9}\sqrt{9}$

In Exercises 59 to 70, simplify to one fraction with no negative exponents.

59. $x - x^{-2}$

60. $1 + y^{-3}$

61. $\dfrac{1}{\sqrt{x}} + \dfrac{1}{\sqrt{y}}$

62. $\sqrt{x} - \dfrac{1}{x}$

63. $\sqrt{\dfrac{x}{2}} + \sqrt{\dfrac{1}{x}}$

64. $\sqrt{3a} - \sqrt{\dfrac{a}{3}}$

65. $(x + 1)^{1/2} - (x + 1)^{-1/2}$

66. $(x^2 - 1)^{1/2} + (x^2 - 1)^{-1/2}$

67. $\sqrt{1 - x^2} + \dfrac{1}{2}x(1 - x^2)^{-1/2}(-2x)$

68. $\dfrac{1}{2}x(x^2 + 1)^{-1/2}(2x) + \sqrt{x^2 + 1}$

69. $\sqrt[3]{x + 1} + \dfrac{1}{3}x(x + 1)^{-2/3}$

70. $3(x + 1)^2 \sqrt[3]{x} + (x + 1)^3 \dfrac{1}{3}x^{-2/3}$

71. Prove that $x^n\left(1 + \dfrac{y}{x}\right)^n = (x + y)^n$.

A.5 GEOMETRY

This section outlines several basic results from plane geometry that play an important role in our work. To begin, we take the concepts of point, line, and plane as intuitive and so do not attempt to define them. Furthermore, we assume that a line forms an angle of 180° (a **straight angle**) at any point on the line.

Recall the definition of parallel lines.

> **Parallel Lines**
> Two distinct lines in a plane are parallel if and only if they do not intersect.

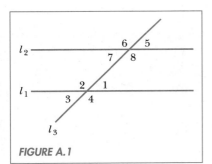

FIGURE A.1

Figure A.1 shows two parallel lines ℓ_1 and ℓ_2, with a third line, ℓ_3, cutting across them. We say that ℓ_1 and ℓ_2 are cut by the **transversal** ℓ_3. Note there are eight angles formed in this situation. For brevity, we write $\angle 1$ when referring to angle 1, $\angle 2$ for angle 2, and so on. The angles formed by the transversal separate into *corresponding* pairs: $\angle 1$ corresponds to $\angle 5$, $\angle 2$ to $\angle 6$, $\angle 3$ to $\angle 7$, and $\angle 4$ to $\angle 8$.

Our first result is a fundamental **axiom** of plane geometry. Recall that an axiom is a statement whose truth we accept without proof.

> **Parallel Axiom**
> If two parallel lines are cut by a transversal, then the corresponding angles are equal.

Thus, for the parallel lines ℓ_1 and ℓ_2 in Figure A.1, we have

$$\angle 1 = \angle 5 \qquad \angle 2 = \angle 6 \qquad \angle 3 = \angle 7 \qquad \angle 4 = \angle 8$$

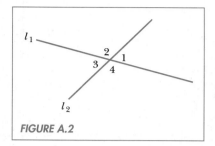

FIGURE A.2

If two distinct lines in a plane are not parallel, then they will intersect in exactly one point. Figure A.2 shows lines ℓ_1 and ℓ_2 intersecting, forming four angles at the point of intersection. We call angles 1 and 3 **vertical angles**. Angles 2 and 4 are also vertical angles.

Vertical Angles Theorem

When two lines intersect, the vertical angles are equal.

PROOF: Consider two intersecting lines ℓ_1 and ℓ_2 as shown in Figure A.2. We wish to prove that $\angle 1 = \angle 3$ and $\angle 2 = \angle 4$. Since ℓ_1 forms a straight angle at the point of intersection, we have

$$\angle 1 + \angle 2 = 180°$$

Similarly, ℓ_2 forms a straight angle at the point of intersection, so

$$\angle 2 + \angle 3 = 180°$$

It follows that

$$\angle 1 + \angle 2 = \angle 2 + \angle 3$$

Subtracting $\angle 2$ from both sides of this equation yields

$$\angle 1 = \angle 3$$

The same type of argument shows that $\angle 2 = \angle 4$. This completes the proof. ■

Example 1

Consider Figure A.1, which shows parallel lines ℓ_1 and ℓ_2 cut by the transversal ℓ_3. Prove that $\angle 1 = \angle 7$.

SOLUTION   We know that $\angle 1 = \angle 5$ by the parallel axiom. Furthermore, $\angle 5 = \angle 7$ since vertical angles are equal. Therefore $\angle 1 = \angle 5 = \angle 7$ as desired. ■

In Figure A.1, we say that $\angle 1$ and $\angle 7$ are **alternate interior angles**. Also, we say that $\angle 2$ and $\angle 8$ are alternate interior angles. An argument similar to the one in Example 1 shows that $\angle 2 = \angle 8$. Thus we have established the following result.

Alternate Interior Angles Theorem

If two parallel lines are cut by a transversal, then the alternate interior angles are equal.

Now we move on to a familiar property of triangles.

Sum of the Angles in a Triangle Theorem

The sum of the angles in a triangle is 180°.

The proof of this theorem may be found in any standard high school geometry textbook.

Example 2

Suppose a triangle has angles measuring 65° and 75°. Find the measure of its third angle.

> SOLUTION Let x be the measure of its third angle in the triangle. Since the sum of the angles in a triangle must be 180°, we have

$$x + 65° + 75° = 180°$$

This implies that

$$x = 180° - 65° - 75° = 40° \blacksquare$$

One of the most important theorems from geometry is the famous Pythagorean Theorem, which gives the relationship between the lengths of the sides in a **right triangle**. By definition, a right triangle is a triangle containing a right angle (90°). We call the side opposite the right angle the **hypotenuse** and the other two sides the **legs**.

Pythagorean Theorem

In a right triangle, the sum of the squares of the legs is equal to the square of the hypotenuse (Figure A.3).

$$a^2 + b^2 = c^2$$

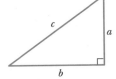

FIGURE A.3

A simple proof of this theorem is outlined in the exercises using the formulas for the area of a rectangle and triangle.

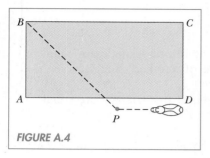

FIGURE A.4

Example 3

Figure A.4 shows a rectangular pool of water $ABCD$, with a duck standing at point D. The duck wants to get to the opposite corner of the pool, point B. If the duck walks x feet along side AD and then jumps into the water and swims in a straight line to point B, find the total distance the duck traveled in terms of x. Assume $AD = 30$ feet, $AB = 12$ feet, and $0 < x < 30$.

SOLUTION Let P be the point where the duck jumps into the water. We wish to find the total distance $BP + PD$. We are given that $PD = x$, so we must find BP. Since triangle ABP is a right triangle, we have

$$(AB)^2 + (AP)^2 = (BP)^2$$

by the Pythagorean Theorem. Now $AB = 12$ and $AP = AD - PD = 30 - x$, so

$$(12)^2 + (30 - x)^2 = (BP)^2$$

It follows that

$$BP = \sqrt{(12)^2 + (30 - x)^2} = \sqrt{x^2 - 60x + 1044}$$

Therefore, the total distance the duck traveled is given by

$$BP + PD = \sqrt{x^2 - 60x + 1044} + x \quad \blacksquare$$

Two triangles are said to be **congruent** (abbreviated $\cong$) if all the parts of one triangle (three angles and three sides) are equal to the corresponding parts of the other triangle. This means that we could pick up one of the triangles and move it so that it would fit exactly on top of the other triangle. Using the symbol $\triangle$ for the word triangle, we can write out the definition of congruent triangles more formally as follows.

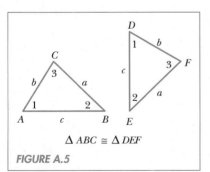

$\triangle ABC \cong \triangle DEF$

FIGURE A.5

Congruent Triangles

$\triangle ABC \cong \triangle DEF$ if and only if $\begin{cases} \angle A = \angle D & AB = DE \\ \angle B = \angle E & AC = DF \\ \angle C = \angle F & BC = EF \end{cases}$

Figure A.5 illustrates two congruent triangles.

Three axioms tell us when two triangles are congruent: ASA (angle-side-angle), SAS (side-angle-side), and SSS (side-side-side).

Example 4

Prove that the diagonal of a rectangle divides the rectangle into two congruent triangles.

SOLUTION   Consider the rectangle $ABCD$ with diagonal AC shown in Figure A.6. We wish to show that $\triangle ABC \cong \triangle CDA$. Since $ABCD$ is a rectangle, we know that $AB = CD$ and $BC = DA$. Furthermore, $AC = CA$. Therefore $\triangle ABC \cong \triangle CDA$ by the SSS axiom. ■

Remark

Alternate proofs exist for this example using either the SAS or the ASA axiom.

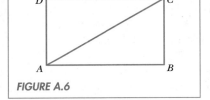

FIGURE A.6

Now we review the formulas for the area of a rectangle and the area of a triangle.

Areas

Rectangle with sides a and b.

$$A = ab$$

Triangle with base b and height h.

$$A = \frac{1}{2}bh$$

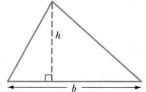

Example 5

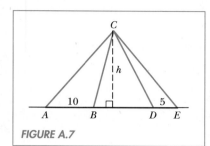

FIGURE A.7

In Figure A.7, the area of $\triangle ABC$ is 40. Find the area of $\triangle DEC$.

SOLUTION Designate DE as the base of $\triangle DEC$. Then

$$\text{Area of } \triangle DEC = \frac{1}{2}(DE) \cdot h = \frac{1}{2}(5)h = \frac{5}{2}h$$

Now we must find the value of h. We can do this using the fact that the area of $\triangle ABC$ is 40. If we designate AB as the base of $\triangle ABC$, then

$$40 = (\text{area of } \triangle ABC) = \frac{1}{2}(AB) \cdot h = \frac{1}{2}(10)h = 5h$$

We have $40 = 5h$, so $h = 8$. Therefore

$$\text{Area of } \triangle DEC = \frac{5}{2}h = \frac{5}{2}(8) = 20 \quad \blacksquare$$

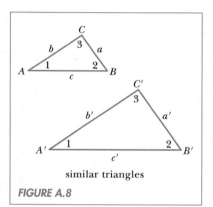

similar triangles

FIGURE A.8

Consider triangles ABC and $A'B'C'$ in Figure A.8. If the angles in $\triangle ABC$ are equal to the angles in $\triangle A'B'C'$, that is,

$$\angle A = \angle A' \qquad \angle B = \angle B' \qquad \angle C = \angle C'$$

then we say that triangle ABC is **similar** to triangle $A'B'C'$. In this situation we have three pairs of corresponding sides:

- Side AB corresponds to side $A'B'$
- Side AC corresponds to side $A'C'$
- Side BC corresponds to side $B'C'$

Similar triangles have the same shape but may differ in size.

Similar Triangles Theorem

In similar triangles, the ratios of the lengths of the corresponding sides are equal. Thus in Figure A.8 we have

$$\frac{a}{a'} = \frac{b}{b'} = \frac{c}{c'}$$

Example 6

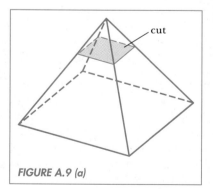

cut

FIGURE A.9 (a)

Consider the pyramid in Figure A.9(a) whose top is 12 inches above the center of its square base. The length of the side of the base is 10 inches. Suppose a cut is made 2 inches from the top, parallel to the base. Find the area of the square surface exposed by the cut.

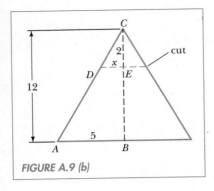

FIGURE A.9 (b)

SOLUTION   Figure A.9(b) shows a side view of the pyramid with altitude $CB = 12$. We labeled half the length of the cut, DE, with x. Comparing $\triangle ABC$ and $\triangle DEC$, we have

$$\angle A = \angle D \qquad \text{by the parallel axiom}$$

$$\angle B = \angle E \qquad \text{both right angles}$$

$$\angle C = \angle C \qquad \text{obvious}$$

Therefore $\triangle ABC$ is similar to $\triangle DEC$. Hence the ratios of the lengths of the corresponding sides are equal. Thus

$$\frac{x}{5} = \frac{2}{12}$$

This implies that $x = 5/6$. Now, the square surface exposed by the cut has side of length $2x = 2(5/6) = 5/3$. Therefore the area of this square is $(5/3)^2 = 25/9$ square inches. ∎

EXERCISES A.5

In Exercises 1 and 2 refer to Figure A.1, where ℓ_1 and ℓ_2 are parallel lines cut by the transversal ℓ_3. From the given angle, find the measure of each of the remaining seven angles.

1. $\angle 1 = 58°$ 2. $\angle 6 = 108°$

3. Two angles in a triangle measure $45°$ and $33°$. Find the third angle.

4. A right triangle contains an angle of measure $29°$. Find the other two angles.

In Exercises 5 and 6, refer to Figure A.10. The boxes indicate right angles at C and D. From the given angle, find the measure of the other four labeled angles.

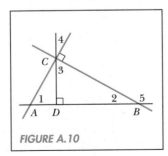

FIGURE A.10

5. $\angle 4 = 15°$ 6. $\angle 5 = 170°$

In Exercises 7 and 8, Figure A.11 shows a box with rectangular base of width 10 inches and length 14 inches. Assume that the height of the box is x inches. An ant starting at point A crawls directly across the bottom of the box to point B and then straight up the edge to point C.

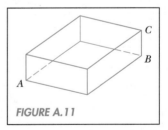

FIGURE A.11

7. Find the total distance that the ant has crawled in terms of x.

8. Find the distance from A to C in terms of x.

In Exercises 9 and 10, Figure A.12 shows a boat at point A anchored 50 feet away from a straight shoreline. A duck sitting in the boat at point A wishes to get to point B on the shoreline. The duck swims in a straight line to the shore at point P and then walks along the shoreline to point B. Let $CP = x$. Use the given information to find the total distance the duck has traveled in terms of x.

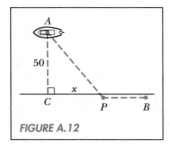

9. $CB = 100$ feet.

10. The distance from A to B is 130 feet.

11. Given ℓ_1 parallel to ℓ_2 and ℓ_3 parallel to ℓ_4 in Figure A.13. Prove that $\triangle ABD \cong \triangle CDB$.

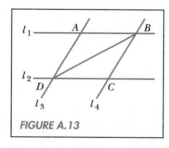

FIGURE A.13

12. Given the right angles shown at points A, C, and E in Figure A.14. Suppose $CB = DC$. Prove that $\triangle ACB \cong \triangle EDC$.

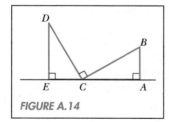

FIGURE A.14

13. Give a proof for Example 4 using the SAS axiom.

14. Give a proof for Example 4 using the ASA axiom.

In Exercises 15 and 16, triangle ABE is cut from rectangle $ABCD$ as shown in Figure A.15. Find the area of the triangle from the given information.

15. $DC = 20$, $BC = 6$

16. $AD = 10$, $AE = 15$, $BE = 12$

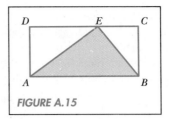

FIGURE A.15

In Exercises 17 and 18, consider the **trapezoid** $ABCD$ shown in Figure A.16. (A trapezoid has four sides, with one pair of opposite sides parallel. In this case AB is parallel to DC.) Find the area of the trapezoid from the given information.

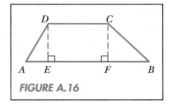

FIGURE A.16

17. $AD = 5$, $AE = 3$, $FB = 2$, $AB = 11$

18. $AB = 15$, $DC = 7$, $CF = 5$

19. Consider $\triangle ABC$ shown in Figure A.17. Line ℓ is drawn parallel to AB, intersecting the sides of $\triangle ABC$ at D and E.

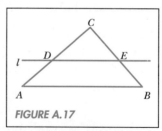

FIGURE A.17

(a) Prove that $\triangle ABC$ is similar to $\triangle DEC$.
(b) Suppose $AB = 10$, $AC = 8$, and $AD = 6$. Find DE.
(c) Suppose $AB = 16$, $DE = 12$, and the area of $\triangle ABC$ is 72. Find the area of $\triangle DEC$.

20. Refer to Figure A.10.
(a) How many similar triangles appear in the figure?
(b) Suppose $AD = 6$ and $DC = 8$. Find BC and DB.
(c) Suppose $BC = 2\sqrt{3}$ and the area of $\triangle CDB$ is 3. Find the area of $\triangle ABC$.

21. Consider the solid wedge shown in Figure A.18. The top, bottom, and left sides are rectangles. Suppose a vertical cut is made parallel to the left side and at a distance x units from the left side. Find the area of the rectangular surface exposed by the cut in terms of x.

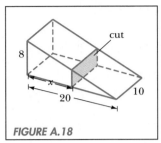

FIGURE A.18

22. Suppose the cut in Example 6 is made y inches from the top instead of 2 inches. Find the area of the square surface exposed by the cut in terms of y.

23. Refer to Example 6. Assume all the conditions are the same except that the base of the pyramid is now a rectangle whose sides measure 10 inches and 15 inches. Find the area of the rectangular surface exposed by the cut.

24. Repeat Exercise 23 if the cut is made y inches from the top. The answer should be in terms of y.

25. This exercise outlines a proof of the Pythagorean Theorem. Suppose a right triangle has legs a and b and hypotenuse c.
    (a) Construct a square with sides $a + b$ as shown in Figure A.19. Note by drawing line segments PQ, QR, RS, and SP we produce four congruent right triangles with legs a and b and hypotenuse c. Prove that $PQRS$ is a square.
    (b) Find the area of the large square in terms of a and b.
    (c) Find the area of the large square by adding the areas of the triangles and the inside square. The answer will be in terms of a, b, and c.
    (d) By equating the answers in (b) and (c), deduce that $a^2 + b^2 = c^2$.

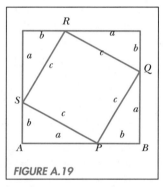

FIGURE A.19

Answers to Odd-Numbered Exercises

Exercises 1.1

1.

(c) (b) (a) (d)
$-\pi$ -1.75 $\frac{3}{4}$ $\sqrt{2}$
-4 -2 0 2 4

3. (a) $\{-6\}$ (b) $\mathbf{R}$ **5.** (a) $\varnothing$ (b) $\{0\}$

7. (a) $\{-2, -1, 0, 1\}$ (b) $\{1\}$
(c) $\{2, 3, 5, 7, 11, 13, 17, 19, 23, 29\}$

9. (a) $\{x: x = 2n + 1, n \in \mathbf{Z}\}$ (b) $\{x: x = 2n, n \in \mathbf{N}\}$

11. $\{x: x = 2\pi n, n \in \mathbf{Z}\}$

13. (a) true (b) true (c) false (d) false

15. (a) $x > 0$ (b) $\{x: x > 0\}$

17. (a) $0 < y < 2$ (b) $\{x: 0 < x < 2\}$

19. (a) rational (b) irrational (c) rational (d) irrational

21. (a) irrational (b) rational (c) irrational

23. Answers may vary: (a) $\frac{3}{2}$ (b) $1 + \pi/10$.

25. Answers may vary: (a) $1.145 = \frac{229}{200}$ (b) $1.14 + \pi/1000$.

27. (a) 4 (b) -5 (c) x^2 (d) $|a|$ (e) $|x + 1| - 1$

29. (a) $2a^2$ (b) $a^2 + 2ab + b^2$

31. (a) $\pi - 3$ (b) $\sqrt{2} - 1$

33. 1 **35.** (a) 2 (b) -2 **37.** (a) $b - a$ (b) 0 (c) -1

39. (a) $-ab$ (b) $1 - a$ (c) $b - a$ **41.** true

43. (a) $x = 4$ or $x = -4$ (b) $x = \frac{5}{2}$ or $x = -\frac{5}{2}$
(c) no solution

45. (a) $x = 4$ or $x = -6$ (b) $x = -\frac{5}{2}$ or $x = \frac{11}{2}$
(c) $x = \frac{2}{5}$

47. (a) $x = -1$ or $x = -\frac{1}{3}$ (b) $x = 0$

49. (a) 8 (b) 6 (c) $|x - 3|$ (d) not real

51. (a) $3\sqrt{10}$ (b) $\sqrt{-x}/2$ (c) 0

53. (a) 12 (b) $|y|/2$ (c) $\sqrt{x^2 + y^2}$

55. (a) $x + 1$ (b) -4 (c) $2\sqrt[3]{5}$

57. (a) $2 < a < 3$ (b) $4 < a < 5$ **59.** (a) $.71$ (b) 1.37

61. Let x be an odd integer, then $x = 2n + 1$ for some
$n \in \mathbf{Z}$. We have $x^2 = (2n + 1)^2 = 4n^2 + 4n + 1 =$
$2(2n^2 + 2n) + 1$. Thus $x^2 = 2m + 1$, where $m =$
$2n^2 + 2n$ is an integer. Because $2m + 1$ is an odd
integer, this proves that x^2 is odd.

63. (a) Suppose $a > 0$ and $b > 0$. Then $ab > 0$ because the
product of two positive numbers is positive. Therefore
$|ab| = ab$ by definition. Next, $|a| = a$ and $|b| = b$ by
definition. Thus $|a||b| = ab$. We conclude that $|ab| =$
$ab = |a||b|$ as desired. (b) Suppose $a < 0$ and $b < 0$.
Then $ab > 0$ because the product of two negative
numbers is positive. Therefore $|ab| = ab$ by definition.
Next, $|a| = -a$ and $|b| = -b$ by definition. Thus
$|a||b| = (-a)(-b) = ab$. We conclude that $|ab| =$
$ab = |a||b|$ as desired. (c) Suppose $a > 0$ and $b < 0$.
Then $ab < 0$ because the product of a positive and
negative number is negative. Therefore $|ab| = -ab$ by
definition. Next, $|a| = a$ and $|b| = -b$ by definition.
Thus $|a||b| = a(-b) = -ab$. We conclude that
$|ab| = -ab = |a||b|$. (d) Suppose $a = 0$ or $b = 0$.
Then $ab = 0$. Therefore $|ab| = 0$. Next, $|a| = 0$ or $|b|$
$= 0$. Thus $|a||b| = 0$. We conclude that $|ab| = 0 =$
$|a||b|$.

Exercises 1.2

1. (a) $\frac{13}{5}$ (b) $\frac{3}{2}$ (c) $-4, -3$ (d) no solution

3. (a) $-\frac{13}{2}, -\frac{1}{2}$ (b) no real solution

5. (a) $(2 \pm \sqrt{2})/2$ (b) $(1 \pm \sqrt{5})/2$ (c) no real solution

7. $1, 3$ **9.** $-\frac{1}{2}$ **11.** $\frac{1}{2}$ **13.** $0, -\frac{3}{2}$

15. no real solution   **17.** no solution   **19.** 1

21. $0, (1 \pm \sqrt{13})/2$ **23.** $2, -2$ **25.** 0

27. $-2, 1, 0, -1$ **29.** $1, -\frac{1}{3}$

31. (a) -2 (b) no solution   **33.** $-\sqrt{2}$ **35.** $\pm 1/\sqrt{2}$

37. (a) 0 (b) $\{x: x \geq 0\}$ **39.** $-\frac{15}{4}$ **41.** $\frac{10}{9}, 2$

43. $\sqrt{5}$ **45.** 1

47. (a) $(d - b)/(a - c)$ (b) $(A - \pi r^2)/(2r)$

49. (a) $\pm\sqrt{2K/m}$ (b) $(-h \pm \sqrt{h^2 + 10\pi})/\pi$

51. x **53.** (a) $AB/(B - A)$ (b) $(3x + 2)/(x - 1)$

55. $y = \pm|b/a| \sqrt{x^2 - a^2}$ **57.** $abc/(a - b)$

59. $b - a, -b - a$ **61.** $k \pm \sqrt{x/a}$

63. divided by variable x losing solution $x = 0$; correct solution: $x^2 - x = x$, $x^2 - 2x = 0$, $x(x - 2) = 0$, $x = 0$ or $x = 2$.

65. $\sqrt{x^2 + 4}$ is not $|x| + 2$. Correct solution: $x^2 + 4 = 9$, $x^2 = 5$, $|x| = \sqrt{5}$, $x = \sqrt{5}$ or $x = -\sqrt{5}$.

67. $\sqrt{(x + 1)^2} = \sqrt{x^2}$ simplifies to $|x + 1| = |x|$, not $x + 1 = x$.

Exercises 1.3

1. (a) $-\frac{9}{20}$ (b) $-1\frac{1}{2}$

3. (a) 225 (b) $\frac{8}{3}$ or $\frac{4}{3}$ (c) If x is input, the output is $9(x - 2)^2$. This is nonnegative because $(x - 2)^2 \geq 0$ for all $x \in \mathbf{R}$.

5. (a) $-\frac{127}{64}$ (b) $2 \pm \sqrt{3}/6$

7. (a) $a + 1/a$ (b) $(a \pm \sqrt{a^2 - 4})/2$

9. 100   11. $12(\sqrt{2} + 1)$ 13. $8 - 4\sqrt{2}$

15. 12 inches from one end (and 16 inches from the other end)   17. 4×100 or 50×8

19. $x = 15\sqrt{3}$, $y = 5\sqrt{3}$ 21. $x = 10\sqrt{2}$, $y = 5\sqrt{2}$

23. 5×8 25. (a) $t = 1$ or 4 (b) $t = 5$ 27. $\frac{5}{2}$

29. $\frac{30}{13}$ 31. $C_1C_2C_3/(C_2C_3 + C_1C_3 + C_1C_2)$

33. (a) $10/\sqrt{\pi}$ (b) $a/\sqrt{\pi}$ 35. $3\sqrt[3]{4}$ feet

Exercises 1.4

1. (a) $(-2, 2)$

(b) $(0, 1]$

(c) $[-\sqrt{3}, +\infty)$

(d) $(-\infty, \pi/2)$

3. (a) $[-1, 2)$ (b) $(-\infty, -1]$

5. (a) $(-4, -2)$

(b) $(-\infty, \sqrt{2}) \cup (\sqrt{3}, +\infty)$

(c) $(0, \frac{1}{2}]$

7. (a) $(-\infty, +\infty)$

(b) $(1, 2)$

9. (a) $[0, +\infty)$

(b) $(-\infty, +\infty)$

11. (a) < (b) < (c) > (d) >   13. true   15. true

17. false   19. $(-10, +\infty)$ 21. $(-15, +\infty)$

23. $[-7, 11]$ 25. $(0, +\infty)$

27. $(-\infty, -\frac{2}{5}) \cup (0, +\infty)$ 29. $[-2, 4)$

31. $(-1, +\infty)$ 33. $(-\infty, -1)$

35. $(-\infty, -1) \cup (0, +\infty)$ 37. $(-\infty, -\frac{1}{2}] \cup [0, 2]$

39. $(-\infty, 4) \cup (4, +\infty)$ 41. $(-\frac{1}{2}, 0) \cup (0, +\infty)$

43. $(-\infty, -3) \cup (-3, 1]$ 45. $(\frac{1}{2}, +\infty)$

47. $[-4, \frac{2}{3}]$ 49. $(-\infty, 0) \cup (1, +\infty)$

51. $(-\infty, -1) \cup (1, +\infty)$ 53. $[-1, 0) \cup [1, +\infty)$

55. $(\frac{1}{5}, 0) \cup (0, \frac{1}{5})$ 57. $(-\infty, (k + 4)/3)$

59. $(-2 - 2k, -2 + 2k)$ 61. $(-\infty, 0) \cup (1/k, +\infty)$

63. (a) $(25, +\infty)$ (b) $[0, \frac{25}{2})$ (c) $[0, +\infty)$ 65. $[-1, +\infty)$

67. $(0, +\infty)$ 69. $(-1, 0) \cup (1, +\infty)$

71. $(-\infty, 0) \cup (0, 1)$

73. (a) -2 is to the left of 1. (b) Needs a parenthesis around $+\infty$, not a bracket. (c) There are no x's satisfying these conditions, $\{x: -3 < x \text{ and } x < -4\} = \emptyset$. (d) $-\infty$ is to the left of zero.

Exercises 1.5

1. (a) $\{x: |x| < 1\}$ (b) $\{x: |x| > 1\}$

3. $(-\infty, -4] \cup [4, +\infty)$ 5. $(-4, -3) \cup (3, 4)$

7. $(1, 7)$ 9. $(-\infty, \frac{3}{2}) \cup (\frac{3}{2}, +\infty)$

11. $[\frac{7}{5}, \frac{8}{5}]$ 13. $[\frac{1}{3}, 1]$ 15. $\emptyset$

17. $[-\frac{1}{2}, 0) \cup (0, \frac{1}{2}]$ 19. $(\frac{19}{20}, \frac{21}{20})$

21. $(-\infty, -1) \cup (-1, -\frac{1}{2})$ 23. $(-3, -1) \cup (1, 3)$

25. $\{x: |x - 5| < 2\}$ 27. $\{x: 0 < |x - 5| < 2\}$

29. $[-\sqrt{3}, \sqrt{3}]$

31. $(-\infty, -1 - \sqrt{3}) \cup (-1 + \sqrt{3}, +\infty)$

33. $(-\sqrt{9.01}, -\sqrt{8.99}) \cup (\sqrt{8.99}, \sqrt{9.01})$

35. (a) $(-\sqrt{3}, \sqrt{3})$ (b) $(-\infty, -\sqrt{3}) \cup (\sqrt{3}, +\infty)$

37. $((-4 - k)/3, (-4 + k)/3)$

39. $(-4 - k, -4) \cup (-4, -4 + k)$ 41. 2   43. .02

Exercises 1.6

1. A in quadrant II, B in quadrant IV
$M(1, -2)$
$AB = 10$

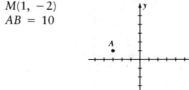

3. *A* in quadrant II, *B* in quadrant III
$M(-2, 1)$
$AB = 4$

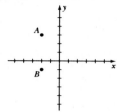

5. $(-3, 3)$, $(5, 0)$, 24
7. $(-7, -2)$, $(4, 6)$, 88
9. (a) 5 (b) 3 (c) $\sqrt{34}$
11. (a) 12 (b) 0 (c) 12
13. $AB = 5 = AC$
15. $(x - 6)^2 + (y + 3)^2 = 4$
17. $(x + 4)^2 + (y + 5)^2 = 41$
19. $(x + \frac{3}{2})^2 + (y - 3)^2 = {}^{137}/_4$
21. (a) $C(0, 0)$, $r = 4$, $(\pm 4, 0)$, $(0, \pm 4)$; four points
    (answers may vary): $(2, \pm 2\sqrt{3})$, $(-2, \pm 2\sqrt{3})$
    (b) $C(-2, 3)$, $r = \sqrt{5}$, no *x*-intercepts, $(0, 4)$, $(0, 2)$;
    four points (answers may vary): $(-1, 5)$, $(-1, 1)$,
    $(-4, 2)$, $(-4, 4)$

23. (a)

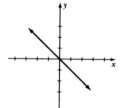

(b)

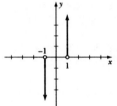

(c)

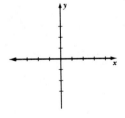

(d)

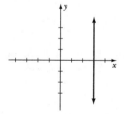

(e)

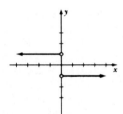

25.

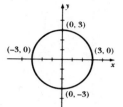

27.

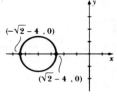

29.

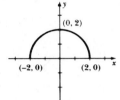

31.

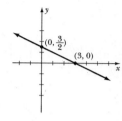

33. Intersection points: $(0,0)$ and $(1,1)$

x	-2	-1	$-\frac{3}{4}$	$-\frac{1}{2}$	$-\frac{1}{4}$	0	$\frac{1}{4}$	$\frac{1}{2}$	$\frac{3}{4}$	1	2
$y = x^2$	4	1	$\frac{9}{16}$	$\frac{1}{4}$	$\frac{1}{16}$	0	$\frac{1}{16}$	$\frac{1}{4}$	$\frac{9}{16}$	1	4
$y = x^3$	-8	-1	$-\frac{27}{64}$	$-\frac{1}{8}$	$-\frac{1}{64}$	0	$\frac{1}{64}$	$\frac{1}{8}$	$\frac{27}{64}$	1	8

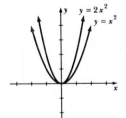

35.

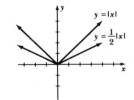

37.

39. (a) 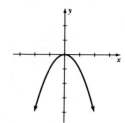 (b)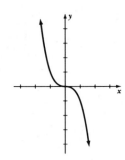

41. (a)  (b)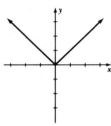

43. (a) 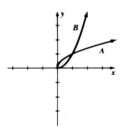 (b)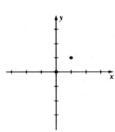

45. (a) 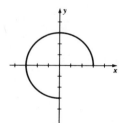 (b)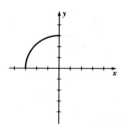

47. (a) 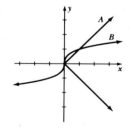 (b)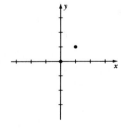

49.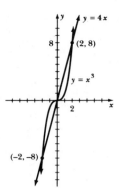

51.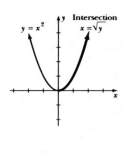

53.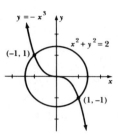

Note: Answers to part b of Exercises 55 to 60 may vary.

55. (a) $(x - 1)^2 + (y - 2)^2 = 9$ (b) $(1, 5)$, $(1, -1)$, $(4, 2)$, $(-2, 2)$

57. (a) $|y| = 2|x|$ (b) $(1, 2)$, $(1, -2)$, $(-1, 2)$, $(-1, -2)$

59. (a) $y^2 - 2x + 1 = 0$ (b) $(1, 1)$, $(1, -1)$, $(2, \sqrt{3})$, $(2, -\sqrt{3})$ **61.** $(6, 0)$

63. (a) $x = 0$ or $x = 2$ (b) $x \in (0, 2)$ (c) $x \in (-\infty, 0) \cup (2, +\infty)$

65. $|a + b|/\sqrt{2}$

Exercises 1.7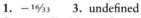

1. $-{}^{16}/_{33}$ **3.** undefined

5. m **7.** 14   **9.** $-{}^{3}/_{2}$

11. $m = 1$, rising   **13.** $m = -{}^{2}/_{3}$, falling

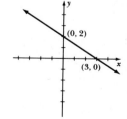

15. $m = \frac{4}{5}$, rising

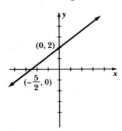
(0, 2)
$(-\frac{5}{2}, 0)$

17. $m = -1$, falling

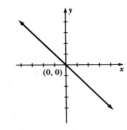
(0, 0)

19. $m = 0$, no x-intercept.

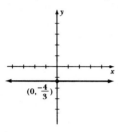
$(0, -\frac{4}{3})$

21. $2x - 9y = 12$

23. $y = -2$ **25.** (a) $3x - 2y = -5$ (b) $3x - 2y = 6$

27. (a) $x = 2$ (b) $x - 2y = 0$ **29.** $3x - 2y = -6$

31. $x = -3$ **33.** (a) $x = -2$ (b) $12x + 17y = 27$

35. $x + y = 0$

37. $3x + 2y = 13$

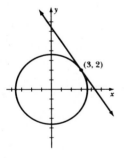
(3, 2)

39. $5x - 2y = 40, 5x + 2y = -10$

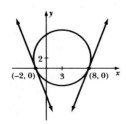
2
$(-2, 0)$ 3 $(8, 0)$

41. $y = x$ **43.** $x + y = 5$ **45.** no   **47.** yes

49. $m_{AB} = -\frac{1}{2}$ and $m_{BC} = 2$ implies $AB \perp BC$. Thus $\triangle ABC$ is a right triangle. Hypotenuse is AC.

51. $x + y = 0$ **53.** (a) $25 + 5\sqrt{13}$ (b) 75

Chapter 1 Review _____

1.
-2.75 $-\sqrt{3}$ $\sqrt{2}$ π
0

2. $\{-1, 1\}$ **3.** $\{0\}$

4. $\varnothing$ **5. R**   **6.** $\{1, 2, 3\}$

7. $\{x: x = 3n, n \in \mathbf{N}\}$

8. $\{x: x = 2n\sqrt{3}, n \in \mathbf{Z}\}$

9. (a) $x > 0$ (b) $\{x: x > 0\}$

10. Answers may vary: (a) $1.45 = \frac{29}{20}$ (b) $1.4 + \pi/100$.

11. (a) 1 (b) -1 **12.** $|x|/3$

13. (a) true (b) true (c) true (d) false

14. (a) $|x + 1|$ (b) $3\sqrt{a^2 + 4}$

15. (a) $x + 1$ (b) $-2\sqrt[3]{x}$

16. $-48, 64$ **17.** no solution

18. 6   **19.** $22, -28$

20. $-5, 1$ **21.** $(3 \pm \sqrt{41})/4$

22. no real solution   **23.** $\pm 1/\sqrt{2}$

24. -3 **25.** $\{x: x \le 0\}$

26. $(2 + 2\sqrt{2})/3$ **27.** $\pm 2/\sqrt{3}$ **28.** $0, \frac{1}{2}, -\frac{3}{2}$

29. 169   **30.** $ab/(a - b)$ **31.** $c/(a - bc)$

32. $(x \pm |x|\sqrt{5})/2$ **33.** $k/100$ **34.** $a \pm b$

35. $\pm\sqrt{r^2 - x^2}$

36. (a) $\frac{1}{24}$ (b) $1/(2a - 2)$ (c) No, because inputting zero causes division by zero. (d) No; input of x produces output $1/(2x) + 1$. There is no solution to $1/(2x) + 1 = 1$.

37. $4(\sqrt{5} + \sqrt{10})$ **38.** 9

39. (a) $[-2, 3)$
-2 0   3

(b) $(-\infty, -\frac{3}{2}) \cup (0, +\infty)$
$-\frac{3}{2}$ 0

40. $(3, +\infty)$ **41.** $(-1, +\infty)$

42. $(-\infty, -1) \cup (\frac{1}{3}, +\infty)$ **43.** $[0, \frac{9}{2})$ **44.** $(-3, 3)$

45. $(-\infty, 2) \cup (3, +\infty)$

46. (a) $\{x: |x| < 2.5\}$ (b) $\{x: |x| \ge \sqrt{3}\}$ (c) $\{x: |x - 6| < 1\}$

47. $(-2, -1) \cup (1, 2)$
-2 -1 0 1 2

48. $[2\frac{1}{2}, 27\frac{1}{2}]$ **49.** $(\frac{59}{3}, \frac{61}{3})$

50. $(-1, -\frac{1}{2}) \cup (-\frac{1}{2}, 0)$ **51.** $(-\infty, -\sqrt{3}) \cup (\sqrt{3}, +\infty)$

52. (a) $(1, 2)$ (b) $\sqrt{34}$ (c) $(x - 1)^2 + (y - 2)^2 = 34$ (d) $(1 \pm \sqrt{30}, 0), (0, 2 \pm \sqrt{33})$

53. true

54.

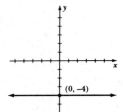

no *x* – or *y* – intercepts

55.

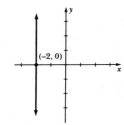

(−2, 0)

56.

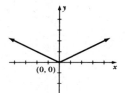

(0, −4)

57.

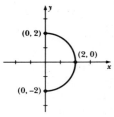

(0, 2)

(2, 0)

(0, −2)

58.

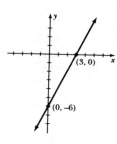

(0, 0)

59.

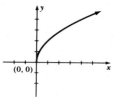

(0, 0)

60.

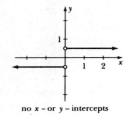

(3, 0)

(0, −6)

61.

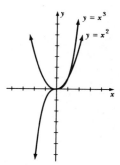

$y = x^3$

$y = x^2$

62. $x^2 - 8y + 16 = 0$
63. (a) $\frac{1}{7}$ (b) $7x + y = -2$ (c) $3x + 4y = 17$
64. $x = -3$

7. No; the rule assigns two numbers to 0.
9. Yes; the rule assigns one number to each element in **R**; range = $\{-2\}$.
11. No; the rule assigns more than one number to elements of **N** (for example, $4 \rightarrow 1, 2, 4$).
13. $F(x) = x^2$ **15.** $F(x) = |x|/x$
17. $F(x) = |x| + 1$ **19.** (a) $2x$ (b) **R**
21. (a) $2(x + 3)$ (b) **R**   **23.** (a) $x^2 + 1$ (b) $[1, +\infty)$
25. (a) -6 (b) $-3x$ (c) $-3a$ (d) $6x$ (e) $6x$ (f) $3x + 3$
(g) $3x + 1$
27. (a) -7 (b) $-2x - 3$ (c) $-2a + 3$ (d) $4x - 3$
(e) $4x - 6$ (f) $2x - 1$ (g) $2x - 2$
29. (a) 3 (b) $x^2 - x + 1$ (c) $-a^2 - a - 1$
(d) $4x^2 + 2x + 1$ (e) $2x^2 + 2x + 2$
(f) $x^2 + 3x + 3$ (g) $x^2 + x + 2$
31. (a) 2 (b) $|x|$ (c) $-|a|$ (d) $2|x|$ (e) $2|x|$ (f) $|x + 1|$
(g) $|x| + 1$
33. (a) -8 (b) $-x^3$ (c) $-a^3$ (d) $8x^3$ (e) $2x^3$
(f) $x^3 + 3x^2 + 3x + 1$ (g) $x^3 + 1$
35. (a) 2 (b) $x/(x - 1)$ (c) $-a/(a + 1)$ (d) $2x/(2x + 1)$
(e) $2x/(x + 1)$ (f) $(x + 1)/(x + 2)$
(g) $(2x + 1)/(x + 1)$
37. (a) $2x^2 - 4x + 3$ (b) $2x^2$ (c) 0 (d) -3 (e) $2(x + a)$
(f) $2(2x + h)$
39. (a) 3 (b) x (c) $-1/(x^2 + x)$ (d) $-1/(x^2 + xh)$
(e) $-1/(ax)$
41. (a) 3 (b) $|x|$ (c) 1 (d) $-\sqrt{1 - x}$
43. (a) $2x$ (b) $(1 - k)/(2 - k)$ (c) -1
45. (a) $-4x$ (b) $(2k + 3)/(2k + 1)$ (c) 2
47. (a) 0 (b) $(k^2 + 2k)/(k^2 - 1)$ (c) $-2x - h$
49. (a) 0 (b) 1 (c) 0
51. (a) 0 (b) $k^2/(k + 1)^2$ (c) $(-2x - h)/(x^2(x + h)^2)$
53. (a) $[5, +\infty)$ (b) $\{x: x \neq -3\}$ (c) **R**
55. $(-\infty, 0) \cup (0, 4]$ **57.** $(-\infty, -1] \cup (0, +\infty)$
59. **R**   **61.** $\{x: x \neq (-1 \pm \sqrt{13})/2\}$
63. (a) $\{x: x \neq 0\}$ (b) **R**   **65.** $[1, +\infty)$
67. $\mathcal{D} =$ **R**; range = **R**.   **69.** $\mathcal{D} =$ **R**; range = $(-\infty, 0]$.
71. $\mathcal{D} =$ **R**; range = $[1, +\infty)$.   **73.** equal
75. Not equal; $\mathcal{D}_F = \{x: x \neq -2, 2\} \neq \mathcal{D}_G = \{x: x \neq 2\}$.
77. Not equal; $F(x) \neq G(x)$ for $x \neq 0$.
79. 4   **81.** 1

Exercises 2.1

1. Yes; the rule assigns one number to each element in the domain; range = $\{4, 6\}$.
3. No; the rule assigns two numbers to 3.
5. Yes; the rule assigns one number to each element in **R**; range = **R**.

Exercises 2.2

1. (a) If F is the function, then the graph of F consists of the set of points $\{(x, y): x \in \mathcal{D}_F \text{ and } y = F(x)\}$. (b) A graph that satisfies the vertical line test determines a function F, where $\mathcal{D}_F = \{x\text{-coordinates on graph}\}$ and $F(x) = y$ if and only if (x, y) is on the graph.

3. (a)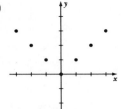

Range = **N** ∪ {0}

(b)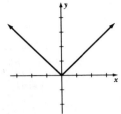

Range = [0, +∞)

5. (a)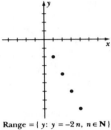

Range = { y: y = −2n, n ∈ **N** }

(b)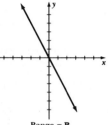

Range = **R**

7. (a)

Range = {−2}

(b)

Range = {−2}

9. (a)

Range = { y: y = n², n ∈ **Z** }

(b)

Range = [0, +∞)

(c)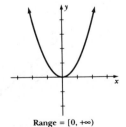

Range = [0, +∞)

11. (a)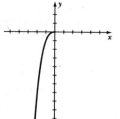

Range = (−∞, 0]

(b)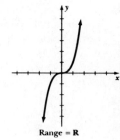

Range = **R**

13. (a)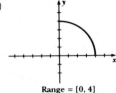

Range = [0, 4]

(b)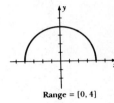

Range = [0, 4]

Function	Domain	Range	Zeros	y-intercept	Increasing	Decreasing
15. F	**R**	**R**	0	(0, 0)	(−∞, +∞)	∅
17. H	[−3, 3]	[0, 3]	−3, 3	(0, 3)	[−3, 0]	[0, 3]
19. G	{x: x ≠ 0}	{−1, 1}	none	none	∅	∅
21. R	(−∞, 0]	[0, +∞)	0	(0, 0)	∅	(−∞, 0]

15.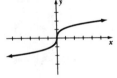

17.

19.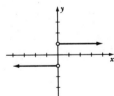

21.

23. (a) 1, 2; (0, 2) (b) −⅞; (0, 7)
25. (a) (−1 ± √13)/2; (0, −3) (b) −3, 3; (0, −3)
27. (a) none; (0, −½) (b) none; (0, −⅓)
29. no
31. yes; increasing on (−∞, −1] ∪ [1, +∞).
33. yes; not increasing or decreasing

35. $\mathcal{D}_F = \{x\text{-coordinates on graph of } F\}$; range of $F = \{y\text{-coordinates on graph of } F\}$; $F(a)$ is the y-coordinate of the point on the graph of F whose x-coordinate is a.

	$\mathcal{D}_F$	Range	$F(0)$	$F(2)$	$F(-\tfrac{3}{2})$
37.	R	$(-\infty, 2]$	2	-1	0
39.	$[-2,3)$	$\{-2, -1, 0, 1, 2\}$	0	2	-2
41.	$\{-2, -\tfrac{3}{2}, -1, 0, 1, 2, 3\}$	$\{-2, -1, 0, 1\}$	1	-1	0
43.	R	$[1, +\infty)$	$\tfrac{3}{2}$	2	2
45.	$[-4, 4]$	$[-1, 2)$	1	1	$-\tfrac{1}{2}$
47.	R	$(-\infty, 1]$	0	0	-1

49.

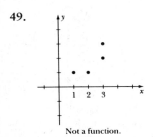

Not a function.

51.

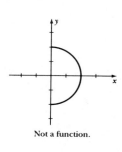

Function; $\mathcal{D} = [0, +\infty)$
Range $= [0, +\infty)$

53.

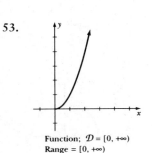

Function; $\mathcal{D} = [0, +\infty)$
Range $= [0, +\infty)$

55.

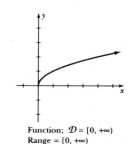

Not a function.

57. $F(x) = \tfrac{3}{2}$ **59.** $F(x) = x$
61. $F(x) = 2x$
63. $F(x) = 1 - x$ **65.** (a) 3 (b) $2 + h$
67. (a) $\sqrt{5} - 2$ (b) $(\sqrt{4 + h} - 2)/h$
69. (a) 11 (b) 1

Exercises 2.3

1. (a) $(5, 2)$ (b) $(-5, -2)$ (c) $(-2, 5)$ (d) $(-5, 2)$
3. (a) $(-2, -3), (3, 2)$ (b) $(2, 3), (-3, -2)$ (c) none
    (d) $(2, -3), (-3, 2)$
5. (a) $(-1, 2), (1, 2)$ (b) none (c) $(-2, -1), (-2, 1)$
    (d) $(1, 2), (-1, 2)$

7.

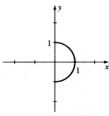

Symmetry: x-axis
Function of x: No

9.

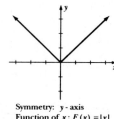

Symmetry: y-axis
Function of x: $F(x) = |x|$

11.

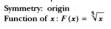
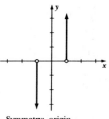

Symmetry: origin
Function of x: $F(x) = \sqrt[3]{x}$

13.

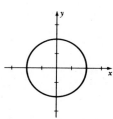

Symmetry: x-axis, y-axis, diagonal, origin
Function of x: No

15.

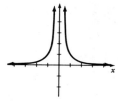

Symmetry: origin
Function of x: No

17.

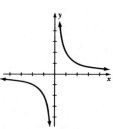

Symmetry: diagonal, origin
Function of x: $F(x) = \dfrac{2}{x}$

19.

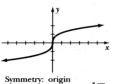

Symmetry: y-axis
Function of x: $F(x) = \dfrac{1}{x^2}$

21.

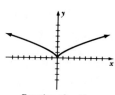

Function of x: Yes

23.

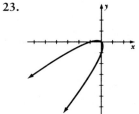

Function of x: No

25.

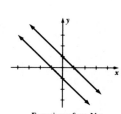

Function of x: No

27.

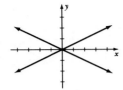

Function of x: No

29. (a) origin (b) y-axis
31. (a) neither (b) y-axis
33. $(4, -5)$
35. (a) (a, b) on curve implies $(a, -b)$ on curve by symmetry with respect to x-axis; $(a, -b)$ on curve implies $(-a, -b)$ on curve by symmetry with respect to y-axis.
(b) Converse: If a curve is symmetric with respect to the origin, then the curve is symmetric with respect to the x- and y-axes. False. (Consider $y = x^3$.)
37. Let $a \in \mathcal{D}_F$ such that $F(a) = b \neq 0$. Then (a, b) is on the graph of F. Now $(a, -b)$ cannot be on the graph of F because this would imply F assigns two numbers to a: b and $-b$. Thus (a, b) is on the graph of F and $(a, -b)$ is not, so the graph of F is not symmetric to the x-axis.

Exercises 2.4

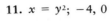

1. $y = \sqrt{x + 1}$ is $y = \sqrt{x}$ shifted -1 unit in the x-direction. $y = \sqrt{x} + 1$ is $y = \sqrt{x}$ shifted $+1$ unit in the y-direction.

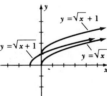

3. $y + 3 = x^2$

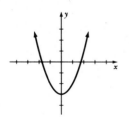

5. $y - 2 = -(x + 1)^3$

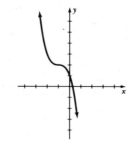

7. $y + 3 = \sqrt{9 - (x - 3)^2}$

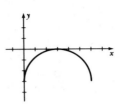

9. $y = \frac{1}{4}x^2$; $+1, +2$

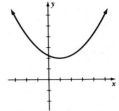

11. $x = y^2$; $-4, 0$

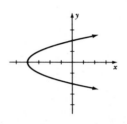

13. shift in x-direction: 0; y-direction: $+2$

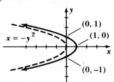

15. shift in x-direction: $+1$; y-direction: 0

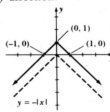

17. shift in x-direction: -2; y-direction: -1

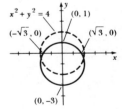

19. shift in x-direction: 0; y-direction: $+1$

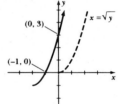

21. shift in x-direction: 0; y-direction: -1

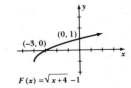

23.

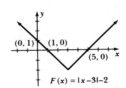

25.

27. (a) 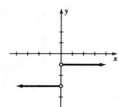 (b) 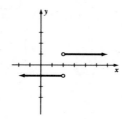 (e)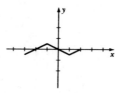

33. $F(x) = |x + 1| - 2$
35. $F(x) = -(x + 1)^2 + 1$

29. (a) 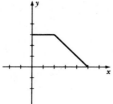 (b)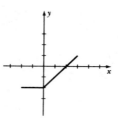

37. (a) (b) 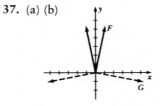 **39.**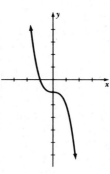

(c) 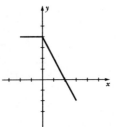 (d)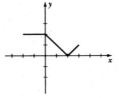

41. (a) 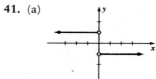 (b)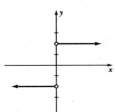

(e)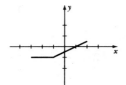

31. (a) 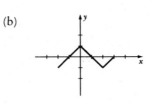 (b)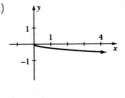

43. (a) (b)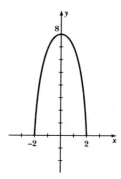

(c) 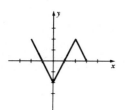 (d)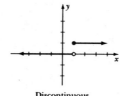

45.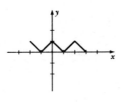

Discontinuous

47.

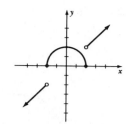

Discontinuous

49.

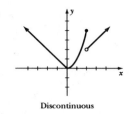

Discontinuous

5.

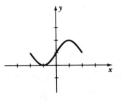

7.

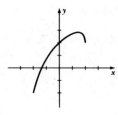

51.

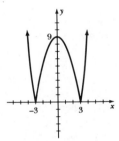

Continuous

53. $F(x) = \begin{cases} |x|, & |x| > 1 \\ \sqrt{1 - x^2}, & |x| \le 1 \end{cases}$

9.

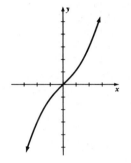

11.

13.

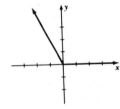

55.

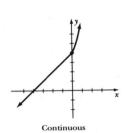

57.

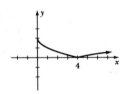

59.

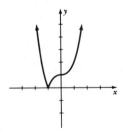

61. $y = ||x| - 2|$

15. (a) $I - K$ (b) K/I (c) $I \cdot I + I$ (d) $K \cdot I \cdot I \cdot I$
17. (a) $6x$ (b) $6x$ (c) $4x$ (d) $9x$
19. (a) $6x + 3$ (b) $6x - 1$ (c) $4x - 3$ (d) $9x + 8$
21. (a) x (b) x (c) $(2x + 9)/8$ (d) $(8x - 9)/2$
23. (a) 33 (b) 13 (c) $4x + 21$ (d) 13
25. (a) $25x^2 - 20x + 4$ (b) $5x^2 - 2$ (c) x^4 (d) $25x - 12$
27. (a) $x^4 + 2x^2$ (b) $x^4 - 2x^2 + 2$ (c) $x^4 - 2x^2$
  (d) $x^4 + 2x^2 + 2$
29. (a) $|x - 1| + 1$ (b) $|x|$ (c) $|x| + 2$ (d) $x - 2$
31. (a) $(3x - 6)/(x + 1)$ (b) $(3 + x)/(3 - 2x)$ (c) x
  (d) $(2x - 1)/(5 - x)$
33. (a) $x/(2x - 1)$ (b) $-x$ (c) $x/(2x + 1)$ (d) x
35. (a) $1/(\sqrt{x} + 1)$; $[0, +\infty)$ (b) $1/\sqrt{x + 1}$; $(-1, +\infty)$
  (c) $(x + 1)/(x + 2)$; $\{x: x \ne -1, -2\}$ (d) $\sqrt{\sqrt{x}}$;
  $[0, +\infty)$
37. $(x + 1)/(x + 3)$; $\{x: x \ne -1, -3\}$
39. $|x + 1|$; R **41.** $\sqrt[3]{x}$, R
43. (a) 0 (b) -2 **45.** -1 **47.** 0
49. $E \circ D$ **51.** $C \circ B$
53. $A \circ E$ **55.** $C \circ C$
57. $A \circ A$ **59.** $E \circ B$

Exercises 2.5 _____

1. (a) 2; R (b) $2x$; R (c) $1 - x^2$; R
  (d) $(1 + x)/(1 - x)$; $\{x: x \ne 1\}$
3. (a) $x^3 + 4$; R (b) $4 - x^3$; R (c) $4x^3$; R
  (d) $4/x^3$; $\{x: x \ne 0\}$

1. If $x \in \mathcal{D}_F = \mathbf{R}$, then $(G \circ F)(x) = G(3x) = 3x/3 = x$. If $x \in \mathcal{D}_G = \mathbf{R}$, then $(F \circ G)(x) = F(x/3) = 3(x/3) = x$.

3. If $x \in \mathcal{D}_F = \{x: x \neq 0\}$, then $(G \circ F)(x) = G(1/x) = 1/(1/x) = x$. If $x \in \mathcal{D}_G = \{x: x \neq 0\}$, then $(F \circ G)(x) = F(1/x) = 1/(1/x) = x$.

5. If $x \in \mathcal{D}_F = [0, +\infty)$, then $(G \circ F)(x) = G(\sqrt{x} + 1) = (\sqrt{x} + 1 - 1)^2 = (\sqrt{x})^2 = x$. If $x \in \mathcal{D}_G = [1, +\infty)$, then $(F \circ G)(x) = F((x - 1)^2) = \sqrt{(x - 1)^2} + 1 = |x - 1| + 1 = x - 1 + 1 = x$. Note $|x - 1| = x - 1$ because $x \geq 1$.

7. $\mathcal{D}_{F^{-1}} = \mathbf{R}$
   $F^{-1}(x) = (x - 3)/2$

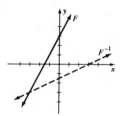

9. $\mathcal{D}_{L^{-1}} = [2, 6]$
   $L^{-1}(x) = 6 - 3x/2$

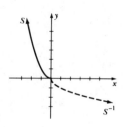

11. $\mathcal{D}_{S^{-1}} = [0, +\infty)$
    $S^{-1}(x) = -\sqrt{x}$

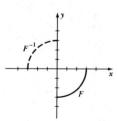

13. no inverse function

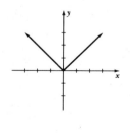

15. $\mathcal{D}_{F^{-1}} = [-3, 0]$
    $F^{-1}(x) = \sqrt{9 - x^2}$

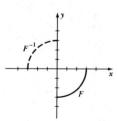

17. no inverse function

19. $\mathcal{D}_{M^{-1}} = \mathbf{R}$
    $M^{-1}(x) = -x$

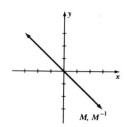

21. $\mathcal{D}_{R^{-1}} = \mathbf{R}$
    $R^{-1}(x) = x^3$

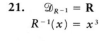
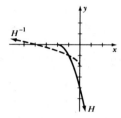

23. $\mathcal{D}_{H^{-1}} = (-\infty, 0]$
    $H^{-1}(x) = \sqrt{-x} - 2$

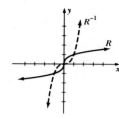

25. Let $x_1, x_2 \in \mathcal{D}_F = \mathbf{R}$. Then $F(x_1) = F(x_2)$ if and only if $4x_1 + 3 = 4x_2 + 3$, $4x_1 = 4x_2$, $x_1 = x_2$. $F^{-1}(x) = (x - 3)/4$.

27. Let $x_1, x_2 \in \mathcal{D}_R = \{x: x \neq 0\}$. Then $R(x_1) = R(x_2)$ if and only if $1/(2x_1) = 1/(2x_2)$, $2x_1 x_2(1/(2x_1)) = 2x_1 x_2(1/(2x_2))$, $x_2 = x_1$. $R^{-1}(x) = 1/(2x)$

29. Let $x_1, x_2 \in \mathcal{D}_F = \{x: x \neq -1\}$. Then $F(x_1) = F(x_2)$ if and only if $x_1/(x_1 + 1) = x_2/(x_2 + 1)$, $x_1(x_2 + 1) = x_2(x_1 + 1)$, $x_1 x_2 + x_1 = x_1 x_2 + x_2$, $x_1 = x_2$. $F^{-1}(x) = x/(1 - x)$

31. (a) F satisfies the horizontal line test. (b) $5, 3, 2, 0, -4$

(c)

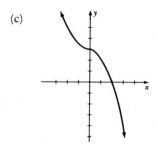

33. true

35. Let $x \in \mathcal{D}_F = [0, 2]$, then $(F \circ F)(x) = F(\sqrt{4 - x^2}) = \sqrt{4 - (\sqrt{4 - x^2})^2} = \sqrt{4 - 4 + x^2} = \sqrt{x^2} = |x| = x$ (since $x \in [0, 2]$).

37. Let $x \in \mathcal{D}_F = \{x: x \neq \frac{2}{3}\}$, then $(F \circ F)(x) =$

$$F((2x + 1)/(3x - 2)) = \frac{2(2x + 1)/(3x - 2) + 1}{3(2x + 1)/(3x - 2) - 2} =$$

$$\frac{2(2x + 1) + (3x - 2)}{3(2x + 1) - 2(3x - 2)} = \frac{7x}{7} = x.$$

39. Let $x \in \mathcal{D}_F = \{x: x \neq 1/b\}$, then $(F \circ F)(x) =$

$$F((x + a)/(bx - 1)) = \frac{(x + a)/(bx - 1) + a}{b(x + a)/(bx - 1) - 1} =$$

$$\frac{(x + a) + a(bx - 1)}{b(x + a) - (bx - 1)} = \frac{x(1 + ab)}{ba + 1} = x.$$

41. $\{y: y \neq 3\}$

43. $\mathcal{D}_{F^{-1}} = (-\infty, 1]$, $F^{-1}(x) = x - 1$

Chapter 2 Review

1. A real function consists of a set of numbers called the domain and a rule that assigns exactly one number to each element in the domain.

2. (a) No; the rule $F(x) = \sqrt{x}$ does not define a function with domain **R**. (b) yes

3. Answers may vary: (a) $F(x) = (-1)^x$ (b) $F(x) = |x| + 1$.

4. $-5, 1, 3, 1 + 2x, 2x - 1, -1 - 2x, -2x, -2,$ $(2k + 1)/(2k - 1)$

5. $-6, 0, -2, -x - x^2, x^2 - x, -x^2 - x, x - x^2,$ $1 - 2x - h, (k + 1)/(k - 1)$

6. $\frac{1}{5}, \frac{1}{2}, 1, 1/(2 - x), -1/(x + 2), 1/(x + 3),$ $(x + 5)/(3x + 6), -1/((x + 2)(x + h + 2)),$ $(k + 2)/(k + 3)$

7. $7, 7, 7, 7, -7, 7, 14, 0, 1$

8. $[0, 1) \cup (1, +\infty)$

9. $(-\infty, 0) \cup [1, +\infty)$

10. $\{x: x \neq -2\}$

11. **R**

12. (a) $F \neq G$ because $\mathcal{D}_F = \mathbf{R} \neq \mathcal{D}_G = \{x: x \neq 1\}$
(b) $F = G$ because $\mathcal{D}_F = \mathbf{R} = \mathcal{D}_G$ and $F(x) = G(x)$ for all $x \in \mathbf{R}$

13. 2

14. true.

15. (a) 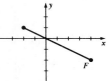 (b)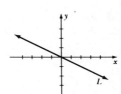

16. (a) 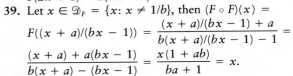 (b)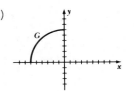

17. (a) 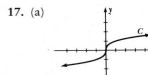 (b)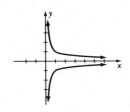

18. (a) It satisfies the vertical line test. (b) $\mathcal{D}_F = [-4, 4]$, range $= [-2, 3]$ (c) -2 (d) 2

19. (a) 5 (b) $4 + h$

20. Symmetry: x-axis
Function of x: no

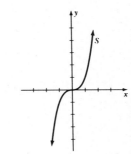

	Function	Domain	Range	Zeros	y-intercept	Increasing	Decreasing
15.	(a) F	$[-2, 4]$	$[-2, 1]$	0	$(0, 0)$	$\varnothing$	$[-2, 4]$
	(b) L	$(-\infty, +\infty)$	$(-\infty, +\infty)$	0	$(0, 0)$	$\varnothing$	$(-\infty, +\infty)$
16.	(a) G	$[-6, 0]$	$[0, 6]$	-6	$(0, 6)$	$[-6, 0]$	$\varnothing$
	(b) H	$[-6, 6]$	$[0, 6]$	$-6, 6$	$(0, 6)$	$[-6, 0]$	$[0, 6]$
17.	(a) C	$(-\infty, +\infty)$	$(-\infty, +\infty)$	0	$(0, 0)$	$(-\infty, +\infty)$	$\varnothing$
	(b) S	$(-\infty, +\infty)$	$(-\infty, +\infty)$	0	$(0, 0)$	$(-\infty, +\infty)$	$\varnothing$

21. Symmetry: diagonal, origin
Function of x: $F(x) = 1/(2x)$

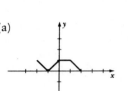

32.

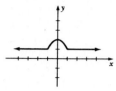

33.

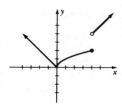

22. Symmetry: origin
Function of x: $F(x) = \sqrt[3]{-x}$

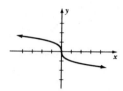

34. (a)

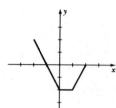

(b)

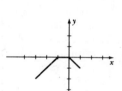

23. origin   **24.** y-axis   **25.** neither

26. $y = |x - 2|/(x - 2)$

27. $y + 4 = \frac{1}{2}(x + 2)^2$

(c)

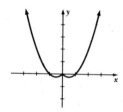

(d)

28. $y = \sqrt{1 - x^2}$; 0; -1

29. $x^2 + y^2 = 4$; 0; $+2$

35.

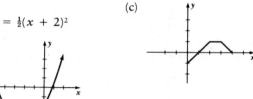

36. (a) $K - I$ (b) K/I (c) $I \cdot I + K \cdot I$
37. (a) $(3x - 1)/(3x - 2)$; $\{x: x \neq \frac{2}{3}\}$ (b) $(3 + x)/x$;
$\{x: x \neq 0\}$ (c) $(2x + 1)/(x + 1)$; $\{x: x \neq 0, -1\}$
(d) $9x - 8$; $\mathbf{R}$

30. $\mathcal{D}_F = [-1, +\infty)$
Range $= (-\infty, 2]$
Zeros: 3
y-intercept: $(0, 1)$

31. $\mathcal{D}_F = (-\infty, +\infty)$
Range $= [0, +\infty)$
Zeros: 1
y-intercept: $(0, 1)$

38. $F^{-1}(x) = (4 - x)/2$
$\mathcal{D}_{F^{-1}} = (-\infty, +\infty)$

39. $F^{-1}(x) = x^2 - 4$
$\mathcal{D}_{F^{-1}} = [0, +\infty)$

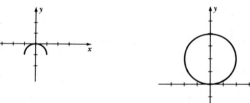

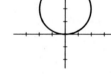

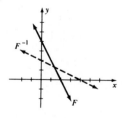

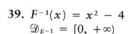

40. no inverse function

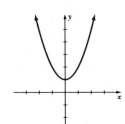

41. $F^{-1}(x) = \sqrt{x} - 1$
$\mathcal{D}_{F^{-1}} = [0, +\infty)$

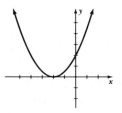

42. *F* cannot have an inverse because it is not $1-1$.

Exercises 3.1 _____

1. zero: $\tfrac{7}{4}$
y-intercept: $(0, -7)$
increasing

3. zero: 0
y-intercept: $(0, 0)$
decreasing

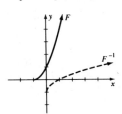

5. $F(x) = 5x - 2$
7. (a) $F(x) = -x/2$ (b) $F(x) = 7x - 6$
   (c) $F(x) = 3x + 4$
9. $L(x) = -x/3 - 1$
11. $F(x) = 8x/3 - 38; x = \tfrac{57}{4}$
13. $L(x) = 2x - 7.8; x = 3.9$
15. (a) $F(t) = -6.75t + 96.8$ (b) $F(t) = -12.5t + 99.1$
   (c) $t = .728$ **17.** (a) $(-4, 3)$ (b) $(3, 2)$

19.
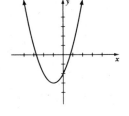

21.
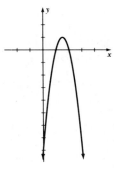

23.
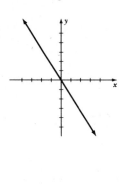

25.
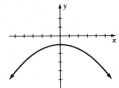

27.
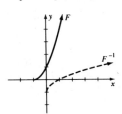

29. $Q(x) = -4(x - \tfrac{1}{2})^2/27 + 4$
31. $Q(x) = (x - 2)^2/25 + 1$
33. $Q(x) = 3(x - 1)^2/2 - 6$
35. $Q(x) = (x - 4)^2/9$
37. $Q(x) = -3(x + 2)^2/16 + 3$
39. $(-\infty, -3) \cup (1, +\infty)$ **41.** $(2 - \sqrt{5}, 2 + \sqrt{5})$
43. $(-\infty, +\infty)$ **45.** $\varnothing$
47. $(-\infty, -\sqrt{7}] \cup [\sqrt{7}, +\infty)$
49. $\left(-4, \dfrac{1 - \sqrt{5}}{2}\right) \cup \left(\dfrac{1 + \sqrt{5}}{2}, +\infty\right)$
51. $(-\infty, -2] \cup (-\sqrt{2}, \sqrt{2})$

	Vertex	Zeros	y-intercept	Line of symmetry	Range	Increasing	Decreasing
19.	$(-2, 0)$	-2	$(0, 2)$	$x = -2$	$[0, +\infty)$	$[-2, +\infty)$	$(-\infty, -2]$
21.	$(\tfrac{3}{2}, 1)$	$1, 2$	$(0, -8)$	$x = \tfrac{3}{2}$	$(-\infty, 1]$	$(-\infty, \tfrac{3}{2}]$	$[\tfrac{3}{2}, +\infty)$
23.	$(-1, -3)$	$-1 \pm \sqrt{3}$	$(0, -2)$	$x = -1$	$[-3, +\infty)$	$[-1, +\infty)$	$(-\infty, -1]$
25.	$(0, -1)$	none	$(0, -1)$	$x = 0$	$(-\infty, -1]$	$(-\infty, 0]$	$[0, +\infty)$
27.	$(4, 4)$	$2, 6$	$(0, -12)$	$x = 4$	$(-\infty, 4]$	$(-\infty, 4]$	$[4, +\infty)$

Exercises 3.2

1. (a) 33 (b) $(5 + \sqrt{33})/4$ **3.** (a) 144 (b) 6
5. $^{225}/_4$; product is $Q(x) = 15x - x^2$, where x is one of
the numbers. Q has maximum when $x = ^{15}/_2$.
7. 200; sum is $Q(x) = 2x^2 - 40x + 400$, where x is
one of the numbers. Q has minimum when $x = 10$.
9. $^5/_2$; $P(^5/_2, 1)$ **11.** 20; $P(^5/_2, 8)$ **13.** $ab/4$; $P(a/2, b/2)$
15. 12; $P(6, 4)$ **17.** $32\frac{1}{2}$ **19.** 22; $P_1(1, 9)$, $P_2(-1, 9)$
21. $\sqrt{37}/2$; $P(9, 3)$ **23.** $\sqrt{11}/2$; $P(\pm \sqrt{5/2}, \frac{1}{2})$
25. 5000 square feet; 50×100
27. $[-1, 2]$; 3

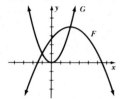

29. $^{32}/_{17}$ feet from one end (used for square) or $^{36}/_{17}$ feet
from other end (used for rectangle)
31. (a) Sum is $Q(x) = 3x^2 - 22x + 91$, which has
minimum value at $x = ^{11}/_3$. (b) Sum is $Q(x) = 3x^2 -
2(x_1 + x_2 + x_3)x + x_1^2 + x_2^2 + x_3^2 + a^2 + b^2 + c^2$,
which has minimum value at $x = (x_1 + x_2 + x_3)/3$.
(c) $(x_1 + x_2 + \cdots + x_n)/n$

Exercises 3.3

1. (a) yes; degree 1, leading coefficient $\sqrt{3}$, constant term
1 (b) no (c) no (d) yes; degree 0, leading coefficient and
constant term, π (e) yes; degree 3, leading coefficient 4,
constant term 2
3. (a) no; it is not a constant function, so the y-coordi-
nates should tend toward $+\infty$ or $-\infty$ as $x \to +\infty$. This
does not happen. (b) yes; leading coefficient positive,
degree even (c) no; not smooth (d) no; not continuous
(e) yes; leading coefficient negative, degree odd

5.

$F(x)$	$x \to +\infty$ implies $F(x) \to$	$x \to -\infty$ implies $F(x) \to$
x	$+\infty$	$-\infty$
x^2	$+\infty$	$+\infty$
x^3	$+\infty$	$-\infty$
x^4	$+\infty$	$+\infty$
x^5	$+\infty$	$-\infty$
x^6	$+\infty$	$+\infty$

From left to right:
$y = x, x^2, x^3, x^4, x^5, x^6$

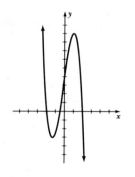

7.

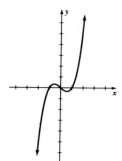

9.

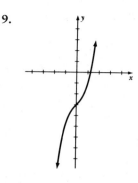

11.

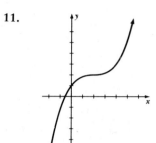

13.

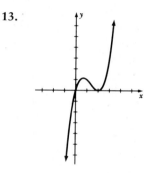

15.

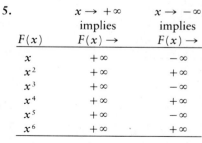

17.

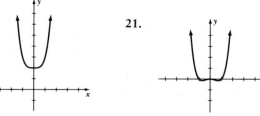

19.

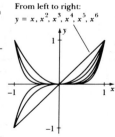

21.

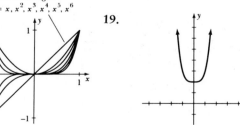

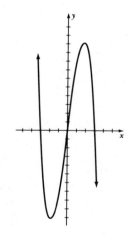

23.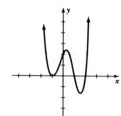

25.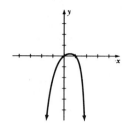

27.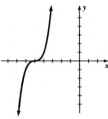

29.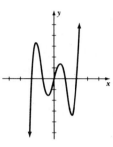

31.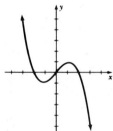

35.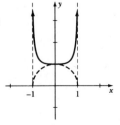

33. true

Exercises 3.4 ——————————————

1. (a) (b)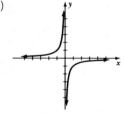

3. symmetry: $x = 0$
 asymptotes: vertical:
 $x = \pm 1$

5. symmetry: no
 asymptotes: horizontal: $y = 0$,
 vertical: $x = 0$

7. symmetry: $x = 0$
 asymptotes: horizontal: $y = 0$,
 vertical: $x = 0$

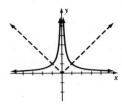

9. symmetry: origin
 asymptotes: horizontal: $y = 0$,
 vertical: $x = 0$

11. symmetry: $x = 0$
 asymptotes:
 horizontal: $y = 0$

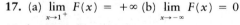

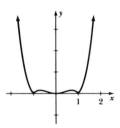

13. $0; 0$ **15.** $-\infty; +\infty$

17. (a) $\lim\limits_{x\to1^+} F(x) = +\infty$ (b) $\lim\limits_{x\to-\infty} F(x) = 0$

19. horizontal asymptote: $y = 0$
 $\lim\limits_{x\to+\infty} F(x) = 0$
 $\lim\limits_{x\to-\infty} F(x) = 0$
 vertical asymptote: $x = 0$
 $\lim\limits_{x\to0^+} F(x) = +\infty$
 $\lim\limits_{x\to0^-} F(x) = -\infty$

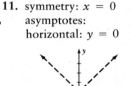

21. horizontal asymptote: $y = 0$
 $\lim\limits_{x\to+\infty} F(x) = 0$
 $\lim\limits_{x\to-\infty} F(x) = 0$
 vertical asymptote: $x = 3$
 $\lim\limits_{x\to3^+} F(x) = +\infty$
 $\lim\limits_{x\to3^-} F(x) = -\infty$

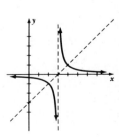
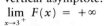

23. horizontal asymptote: $y = 0$
$\lim\limits_{x \to +\infty} F(x) = 0$
$\lim\limits_{x \to -\infty} F(x) = 0$
vertical asymptote: $x = \frac{1}{2}$
$\lim\limits_{x \to \frac{1}{2}^+} F(x) = -\infty$
$\lim\limits_{x \to \frac{1}{2}^-} F(x) = +\infty$

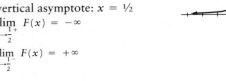

29. horizontal asymptote: $y = 0$
vertical asymptote: $x = 2$, $x = 4$

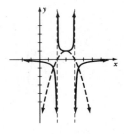

25. (a) horizontal asymptote: $y = 0$
vertical asymptote: none

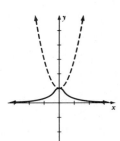

31.

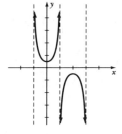

33.

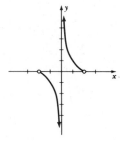

(b) horizontal asymptote: $y = 0$
vertical asymptote: $x = 0$

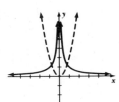
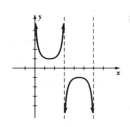

35.

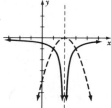

(c) horizontal asymptote: $y = 0$
vertical asymptote: $x = \pm 1$

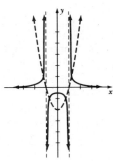

37. $m(x) = -1/(2x)$; $\mathcal{D}_m = \{x : x \neq 0, 2\}$; $\lim\limits_{x \to +\infty} m(x) = 0$, $\lim\limits_{x \to 0^+} m(x) = -\infty$; as $x \to +\infty$, the line through P_1 and P_2 approaches a horizontal; as $x \to 0^+$, the line through P_1 and P_2 approaches a vertical.

Exercises 3.5 _____

1. (a) $F(x) = x$
$\mathcal{D}_F = \{x : x \neq 1\}$

(b) $F(x) = 1/x$
$\mathcal{D}_F = \{x : x \neq 0, 1\}$

27. horizontal asymptote: $y = 0$
vertical asymptote: $x = 2$

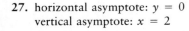

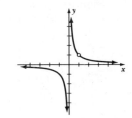

3. $F(x) = 3 - x$
  $\mathcal{D}_F = \{x : x \neq -3\}$

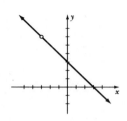

5. $F(x) = 1/(x - 5)$
  $\mathcal{D}_F = \{x : x \neq -1, 5\}$

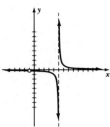

21. (a) vertical asymptotes: $x = -1$
    $\lim\limits_{x \to -1^+} F(x) = +\infty$
    $\lim\limits_{x \to -1^-} F(x) = -\infty$
    (b) slant asymptote: $y = x - 1$

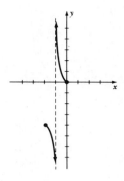

7. $\lim\limits_{x \to 3^+} F(x) = +\infty$
  $\lim\limits_{x \to 3^-} F(x) = -\infty$

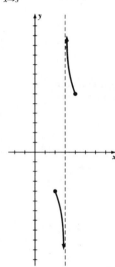

9. $\lim\limits_{x \to 1^+} F(x) = +\infty$
  $\lim\limits_{x \to 1^-} F(x) = -\infty$
  $\lim\limits_{x \to -2^+} F(x) = +\infty$
  $\lim\limits_{x \to -2^-} F(x) = -\infty$

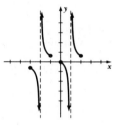

23. (a) vertical asymptotes: $x = 2$
    $\lim\limits_{x \to 2^+} F(x) = +\infty$
    $\lim\limits_{x \to 2^-} F(x) = +\infty$
    (b) horizontal asymptote: $y = 2$

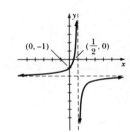

25. (a) $A(x) = 1; 1$ (b) $B(x) = 2x + 2/x; +\infty$

Exercises 3.6

1.

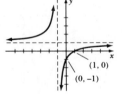

3.

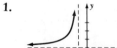

11. $y = {}^3\!/_2$ **13.** $y = 0$ **15.** $y = 2x - 1$
17. (a) vertical asymptotes:
      $x = 0$

  $\lim\limits_{x \to 0^+} F(x) = -\infty$
  $\lim\limits_{x \to 0^-} F(x) = -\infty$
      $x = 2$

  $\lim\limits_{x \to 2^+} F(x) = +\infty$
  $\lim\limits_{x \to 2^-} F(x) = -\infty$
    (b) horizontal asymptote: $y = 0$
19. (a) vertical asymptotes: none
    (b) horizontal asymptote: $y = -1$

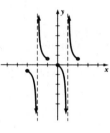

5.

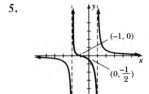

7.

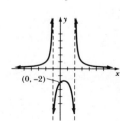

9.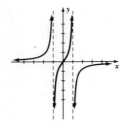

11.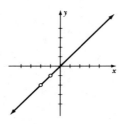

29.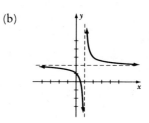

31. (a) $+1$ in x, $+2$ in y (b)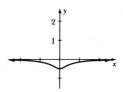

13.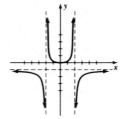

15.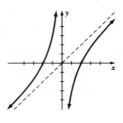

(c) horizontal asymptote: $y = 2$, vertical asymptote: $x = 1$

33. (a) horizontal asymptote: $y = 0$ (b) horizontal asymptote: $y = 0$

17.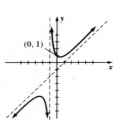

19.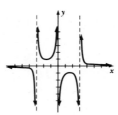

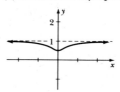

(c) horizontal asymptote: $y = 1$

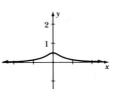

21.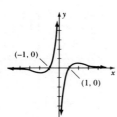

23.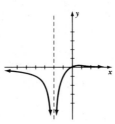

35. slant asymptote: $y = x$, vertical asymptote: $x = 0$

37. slant asymptote: $y = -x$, vertical asymptote: $x = 0$

25.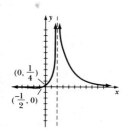

27.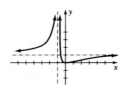

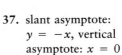

39.

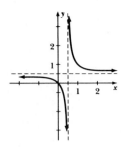

41. (a) $m(x) = (1 - x)/x^2$, $\mathcal{D}_m = \{x : x \neq 0, -1\}$
  (b) $\lim\limits_{x \to +\infty} m(x) = 0$, $\lim\limits_{x \to 0^+} m(x) = +\infty$
  (c)

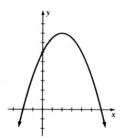

Chapter 3 Review

1. zero: $\frac{7}{2}$ y-intercept: $(0, 7)$ decreasing

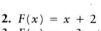

2. $F(x) = x + 2$
3. $F(x) = -3x + 27$; 9 inches
4. $T(t) = t/2 + 17.4$; $t = 5.2$

5.

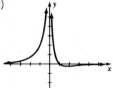

6.

7. $Q(x) = (x - 3)^2/2 + 1$ **8.** $L(x) = -2x/3 + 1$
9. $Q(x) = -2(x + 2)^2/9 + 2$
10. $\left(-\infty, \dfrac{1 - \sqrt{5}}{2}\right) \cup \left(\dfrac{1 + \sqrt{5}}{2}, +\infty\right)$

11. $\left(\dfrac{1 - \sqrt{21}}{2}, \dfrac{1 + \sqrt{21}}{2}\right)$

12. first number 3, second number 1
13. $A(x) = -x^2/3 + 2x$; 3
14. $\sqrt{37}/2$; $P(5, 3)$ **15.** 20; $(\pm 2, 6)$
16. positive, even   **17.** negative, odd   **18.** Polynomial functions have smooth graphs. The graph does not satisfy this condition.
19. For a non-constant polynomial function, as $x \to +\infty$, the y-coordinates on the graph tend toward $+\infty$ or $-\infty$. The graph does not satisfy this condition.

20.

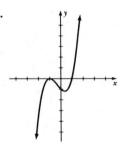

21.

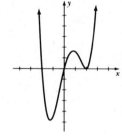

22.

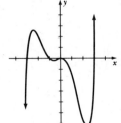

23.

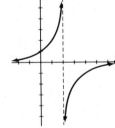

24.

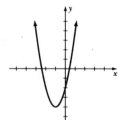

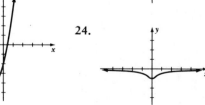

	Vertex	Zeros	y-intercept	Line of symmetry	Range	Increasing	Decreasing
5.	$(2, 8)$	$-2, 6$	$(0, 6)$	$x = 2$	$(-\infty, 8]$	$(-\infty, 2]$	$[2, +\infty)$
6.	$(-1, -4)$	$-1 \pm \sqrt{2}$	$(0, -2)$	$x = -1$	$[-4, +\infty)$	$[-1, +\infty)$	$(-\infty, -1]$

25. symmetry: $x = 0$
horizontal asymptote: $y = 0$
vertical asymptote: $x = 0$

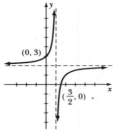

26. horizontal asymptote: $y = 0$
vertical asymptote: $x = -1$

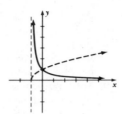

27. symmetry: $x = 2$
horizontal asymptote: $y = 0$
vertical asymptote: $x = 1$, $x = 3$

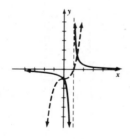

28. 0, 0, $+\infty$, $-\infty$

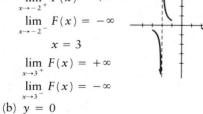

29. **(a)** $x = -2$

$$\lim_{x \to -2^+} F(x) = +\infty$$
$$\lim_{x \to -2^-} F(x) = -\infty$$
$$x = 3$$
$$\lim_{x \to 3^+} F(x) = +\infty$$
$$\lim_{x \to 3^-} F(x) = -\infty$$

(b) $y = 0$

30.

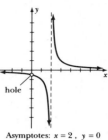

$(0, 3)$ $\left(\frac{3}{2}, 0\right)$

Asymptotes: $x = 1$, $y = 2$

31.

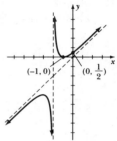

hole

Asymptotes: $x = 2$, $y = 0$

32.

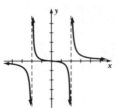

Asymptotes: $x = \pm 2$, $y = 0$

33.

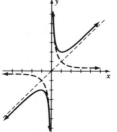

$(-1, 0)$ $\left(0, \frac{1}{2}\right)$

Asymptotes: $x = -2$, $y = x$

34.

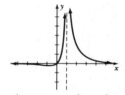

$(0, 3)$ $(3, 0)$

Asymptotes: none

35.

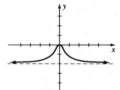

Asymptotes: $x = 1$, $y = 0$

36. (a)

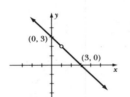

(b)

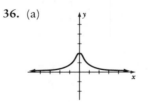

Horizontal asymptote: $y = -2$

37.

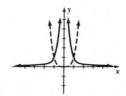

Asymptotes: $x = 0$, $y = x$

38.

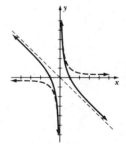

Asymptotes: $x = 0$, $y = -x$

Exercises 4.1

1. $F(x) = P(x)(4x^2 + 4x - 2) - 5x + 12$
3. $F(x) = P(x)(x - 5) + 4x - 13$
5. $F(x) = P(x)(x^4 + x^2 + 1) + 1$
7. $F(x) = P(x)(3x^5 - 2x^3 + x - 2)$
9. $F(x) = P(x)(-\tfrac{2}{3}) + \tfrac{8}{3}$
11. $F(x) = P(x)(x^2 + 1)$
13. $F(x) = P(x)(3x^2 - 2x - 1)$
15. $F(x) = P(x)(3x^2 + 6x + 20) + 19$
17. $F(x) = P(x)(x^2 + 6x + 20) + 60$
19. $F(x) = P(x)(4x^3 + 6x^2 + 6x + 2)$
21. $F(x) = P(x)(4x^2 - 4x + 12) + 2$
23. $F(x) = P(x)(x^5 + x^4 + x^3 + x^2 + x + 1) + 1$
25. $F(x) = P(x)(x^{99} - x^{98} + x^{97} - x^{96} + \cdots + x - 1) + 1$
27. (a) -13 (b) 0, $F(x) = (x - \tfrac{1}{4})(4x + 16)$
29. (a) 3 (b) 0, $F(x) = (x - \tfrac{1}{2})(6x^2 - 2x + 2)$
31. (a) 0, $F(x) = x(4x^2 - 1)$
   (b) 0, $F(x) = (x - \tfrac{1}{2})(4x^2 + 2x)$
33. (a) 0, $F(x) = (x + \sqrt{2}) \times$
   $(x^3 + (1 - \sqrt{2})x^2 + (1 - \sqrt{2})x - \sqrt{2})$ (b) 6
35. (a) 750 (b) 250
37. (a) 3
   (b) 0, $F(x) = x(x^{39} + 2x^{24} + 3x^9 + 4x^4 + 5x)$
39. $F(x) = x^5 + x^3 + 3x - 1$
41. $P(x) = 2x^2 - x + 1$ 43. $P(x) = x - 2$
45. Dividing $F(x) = x^2 + 1$ by $x - c$, the remainder is $c^2 + 1$, which is positive for all $c \in \mathbf{R}$.
47. Let $F(x) = x^{2n-1} + 1$, where $n \in \mathbf{N}$. Then $F(-1) = (-1)^{2n-1} + 1 = -1 + 1 = 0$ because $2n - 1$ is odd. Because -1 is a zero of F, $x - (-1) = x + 1$ must be a factor of $F(x)$, $F(x) = (x + 1)Q(x)$. Synthetic division shows that all the coefficients of $Q(x)$ are integers (1 or -1). It follows that $Q(n)$ is an integer. Thus, letting $x = n$, we have $F(x) = (x + 1)Q(x)$, which implies that $F(n) = (n + 1)Q(n)$, so that $n + 1$ is an integer factor of $F(n) = n^{2n-1} + 1$.

Exercises 4.2

1. (a) $1 - 3i$ (b) $1 + 7i$ (c) $1 - i$ (d) $8 + 6i$
3. (a) $-1 + 3\sqrt{2}i$ (b) 3 (c) $-\tfrac{1}{3} + 2\sqrt{2}i/3$
   (d) $-1 + 2\sqrt{2}i$
5. (a) $-8 + 2i$ (b) $8i$ (c) $i/2$ (d) -4
7. (a) $-5 + 6i$ (b) 29 (c) $\tfrac{21}{29} + (\tfrac{20}{29})i$ (d) $21 + 20i$
9. $3 - 6i$; 45
11. $-4i$; 16 13. 0; 0 15. yes   17. yes
19. $2 \pm i$ 21. $(-1 \pm \sqrt{2}i)/3$ 23. $1 \pm \sqrt{5}i$

25. $x^2 + 3$ 27. $x^2 - 4x + 5$ 29. $x^2 + 6x + 13$
31. $2 - i, 1$ 33. $-i, -\tfrac{1}{2}$ 35. $\sqrt{3}i, \tfrac{1}{2}, -1$
37. $3 - 4i, (-1 \pm \sqrt{5})/2$
39. Let $z = b + ci$ for some $b, c \in \mathbf{R}$. Then

$$
\begin{array}{c|c}
\overline{(az)} & \overset{?}{=} \quad \bar{a}\bar{z} \\
\hline
\overline{(a(b + ci))} & a\overline{(b + ci)} \\
\overline{(ab + aci)} & a(b - ci) \\
ab - aci & ab - aci
\end{array}
$$

$$\text{yes}$$

41. true

Exercises 4.3

1. $2, -5$ (multiplicity two)   3. $0, \sqrt{2}, -\sqrt{2}$
5. $i, -i, 3, -3$
7. $G(x) = ax(x - \tfrac{1}{2})(x + 3)$, $a = $ constant
9. $G(x) = a(x - 3)^3(x^2 - 2x + 2)(x^2 - 4x + 5)$, $a = $ constant
11. $P(x) = -2x^3 + 8x^2 - 2x - 12$
13. $P(x) = 3x^2 + 9$
15. $P(x) = (x^4 - 2x^3 - 22x^2 + 30x + 153)/40$
17. $F(x) = (x - 1)(x - i)(x + i)$; $1, i, -i$
19. $F(x) = (x + 1)(x - (1 + \sqrt{3}i)/2) \times$
   $(x - (1 - \sqrt{3}i)/2)$; $-1, (1 + \sqrt{3}i)/2, (1 - \sqrt{3}i)/2$
21. $F(x) = 3x(x + 1)(x - \tfrac{1}{3})$; $0, -1, \tfrac{1}{3}$
23. $F(x) = (x + 2)(x - (1 + 2i))(x - (1 - 2i))$; $-2, 1 + 2i, 1 - 2i$
25. $F(x) = (x - 3)(x + 3)(x - 3i)(x + 3i)$; $3, -3, 3i, -3i$
27. $F(x) = x^2(x + 4)^2$; $0, -4$ (both multiplicity two)
29. $F(x) = 12x(x - 2\sqrt{3}/3)(x + 2\sqrt{3}/3)(x - i/2) \times (x + i/2)$; $0, 2\sqrt{3}/3, -2\sqrt{3}/3, i/2, -i/2$
31. $F(x) = (x + 1)(x - 2)(x - (-1 + \sqrt{3}i)) \times$
   $(x - (-1 - \sqrt{3}i))(x - (1 + \sqrt{3}i)/2) \times$
   $(x - (1 - \sqrt{3}i)/2)$; $-1, 2, -1 \pm \sqrt{3}i, (1 \pm \sqrt{3}i)/2$
33. $A = 2, B = 0, C = -1$ 35. $A = 0, B = 5, C = 4$
37. $A = 1, B = -1$ 39. $F(x) = (2x^3 - 8x)/15$
41. $F(x) = -5x^3/6 + 5x^2/3 + 25x/6 - 1$
43. $F(x) = x^4/9 - x^3/3 - 7x^2/9 + 3x - 2$
45. (a) 2 (b) 3 (c) n 47. (a) 5 (b) 6 (c) 3 49. true
51. false.

Exercises 4.4

1. $\pm 1, \pm \tfrac{1}{3}, \pm 2, \pm \tfrac{2}{3}, \pm 3, \pm 6$
3. $1, (-9 \pm 3\sqrt{5})/2$; $F(x) = (x - 1) \times$
   $(x - (-9 + 3\sqrt{5})/2)(x - (-9 - 3\sqrt{5})/2)$

Answers to Odd-Numbered Exercises ▪ **649**

5. $0, -1, 1 \pm \sqrt{5}$; $F(x) = x(x + 1)(x - (1 + \sqrt{5})) \times (x - (1 - \sqrt{5}))$

7. $-1, \frac{1}{2}, -\frac{1}{3}, \frac{3}{2}$; $F(x) = (x + 1)(x - \frac{1}{2})(x + \frac{1}{3}) \times (x - \frac{3}{2})$

9. By the Rational Root Theorem, the only possible rational zeros for G are 1 and -1. But $G(1) = 1$ and $G(-1) = -3$. Therefore, all the real zeros for G must be irrational.

11. By the Rational Root Theorem, the only possible rational zeros for G are ± 1, $\pm \frac{1}{2}$, ± 3, and $\pm \frac{3}{2}$. By direct substitution, we find that none of these values is a zero of G. Therefore, all the real zeros of G must be irrational.

13. $(\frac{5}{4}, {}^{171}/_{64})$ **15.** $-1, \frac{1}{2}, \frac{2}{5}$ **17.** $-\frac{1}{2}, \frac{3}{2}, -\frac{4}{3}$

19. 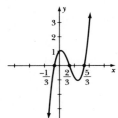   **21.**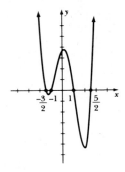

23. $-2, -1, 0$

25. $2 \times 2 \times 5$ or $(5 + \sqrt{105})/4 \times (5 + \sqrt{105})/4 \times (13 - \sqrt{105})/2$

27. Consider $F(x) = x^2 - 2$. By the Rational Root Theorem, the only rational zeros for F are ± 1 and ± 2. By direct substitution, none of these are zeros. Therefore, the only real zeros possible for F must be irrational. Now, $x = \sqrt{2}$ is a real zero of F, so it is irrational.

29. Same proof as in Exercise 27, except replace 2 with p, where p represents a prime number.

31. Let $x =$ first integer. Then we want $x^3 + (x + 1)^3 = (x + 2)^3$. This is equivalent to $x^3 - 3x^2 - 9x - 7 = 0$. By the Rational Root Theorem, the only rational zeros possible for the expression on the left side of the equation are ± 1 and ± 7. By direct substitution, none of these are zeros. Therefore, there are no rational numbers (and hence no integer values) for x satisfying our equation.

Exercises 4.5

1. upper bound: 2, lower bound: -3

3. upper bound: 4, lower bound: -3

5.
$$-1 \,|\; \begin{array}{rrrr} 9 & -18 & 0 & 10 \\ & -9 & 27 & -27 \\ \hline 9 & -27 & 27 & -17 \end{array}$$
Alternating signs implies that $a = -1$ is a lower bound.
$$2 \,|\; \begin{array}{rrrr} 9 & -18 & 0 & 10 \\ & 18 & 0 & 0 \\ \hline 9 & 0 & 0 & 10 \end{array}$$
Same signs implies that $b = 2$ is an upper bound.

7.
$$-2 \,|\; \begin{array}{rrrrr} 5 & -3 & -14 & 9 & 6 \\ & -10 & 26 & -24 & 30 \\ \hline 5 & -13 & 12 & -15 & 36 \end{array}$$
Alternating signs implies that $a = -2$ is a lower bound.
$$2 \,|\; \begin{array}{rrrrr} 5 & -3 & -14 & 9 & 6 \\ & 10 & 14 & 0 & 18 \\ \hline 5 & 7 & 0 & 9 & 24 \end{array}$$
Same signs implies that $b = 2$ is an upper bound.

9. (a)
$$2 \,|\; \begin{array}{rrrr} 1 & -1 & 0 & -1 \\ & 2 & 2 & 4 \\ \hline 1 & 1 & 2 & 3 \end{array}$$
Same signs implies that $b = 2$ is an upper bound.

(b) $F(x) = (x - 1)x^2 - 1$. If $x < 1$, then $x - 1 < 0$ and $x^2 > 0$. Hence $(x - 1)x^2 < 0$ because the product of a negative number and a positive number is negative. Adding -1, the quantity remains negative. Thus $x < 1$ implies that $F(x) = (x - 1)x^2 - 1 < 0$, and we conclude that F has no zeros for $x < 1$.

11. (a)
$$1 \,|\; \begin{array}{rrrr} -9 & 12 & -7 & 2 \\ & -9 & 3 & -4 \\ \hline -9 & 3 & -4 & -2 \end{array}$$
Signs are not all the same, so no conclusion

(b) $F(x) = (x - 1)(-9^2 + 3x - 4) - 2$. $Q(x) = -9x^2 + 3x - 4$ is a downward curving parabola with vertex at $(\frac{1}{6}, -\frac{15}{4})$, hence $Q(x) \le -\frac{15}{4} < 0$ for all x. If $x > 1$, then $x - 1 > 0$ and $Q(x) < 0$. Hence $(x - 1)Q(x) < 0$ because the product of a positive number and a negative number is negative. Adding -2, the quantity remains negative. Thus $x > 1$ implies that $F(x) = (x - 1)Q(x) - 2 < 0$, and we conclude that F has no zeros for $x > 1$.

13. lower bound: -2
$$-2 \,|\; \begin{array}{rrrr} 1 & 1 & -1 & 0 \\ & -2 & 2 & -2 \\ \hline 1 & -1 & 1 & -2 \end{array}$$
alternating signs
upper bound: 1
$$1 \,|\; \begin{array}{rrrr} 1 & 1 & -1 & 0 \\ & 1 & 2 & 1 \\ \hline 1 & 2 & 1 & 1 \end{array}$$
same signs

15. lower bound: -2

$$-2 \,|\, \begin{array}{rrrrr} 4 & 0 & -9 & 0 & 2 \\ & -8 & 16 & -14 & 28 \\ \hline 4 & -8 & 7 & -14 & 30 \end{array}$$

alternating signs

upper bound: 2

$$2 \,|\, \begin{array}{rrrrr} 4 & 0 & -9 & 0 & 2 \\ & 8 & 16 & 14 & 28 \\ \hline 4 & 8 & 7 & 14 & 30 \end{array}$$

same signs

17. Lower bound: 2; we can write $F(x) = (x - 2) \times (x^2 - x + 1)$. $Q(x) = x^2 - x + 1$ is an upward curving parabola with vertex at $(\frac{1}{2}, \frac{3}{4})$; hence $Q(x) > 0$ for all x. If $x < 2$, then $x - 2 < 0$ and $Q(x) > 0$. Hence $(x - 2)Q(x) < 0$ because the product of a negative number with a positive number is negative. Thus $x < 2$ implies that $F(x) = (x - 2)Q(x) < 0$, and we conclude that F has no zeros for $x < 2$.

upper bound: 3

$$3 \,|\, \begin{array}{rrrr} 1 & -3 & 3 & -2 \\ & 3 & 0 & 9 \\ \hline 1 & 0 & 3 & 7 \end{array}$$

same signs

19. $\frac{5}{2}, \pm 6i$ **21.** $\frac{1}{2}, \frac{3}{2}, \pm i$

Exercises 4.6

[a, b]	Midpoint m	Sign of F(a)	F(m)	Sign of F(b)
1. [0, 1]	.5	−	−.75	+
[.5, 1]	.75	−	−.1563	+
[.75, 1]	.875	−	+.3398	+
3. [1, 2]	1.5	−	−.625	+
[1.5, 2]	1.75	−	+1.3594	+
[1.5, 1.75]	1.625	−	+.2910	+

5. The only possible rational zeros are ± 1 and $\pm \frac{1}{2}$. None of these work so that any real zero must be irrational. Zero $\approx .59$.

7. The only possible rational zeros are ± 1 and ± 3. None of these work so that any real zero must be irrational. Zero ≈ 1.21.

9. The only possible rational zeros are ± 1 and ± 2. None of these work so that any real zero must be irrational. Zero ≈ 1.15.

11. $x \approx .68$

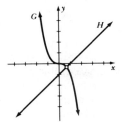

13. $x = -1$ and $x \approx 1.35$

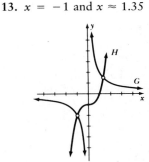

15. $x \approx .57$

17. (a) Slope of line through $(a, F(a))$ and $(b, F(b))$ is $(F(b) - F(a))/(b - a)$, therefore its equation is

$$y - F(a) = \frac{F(b) - F(a)}{b - a}(x - a)$$

Setting $y = 0$ and solving for x we obtain the point where the line intersects the x-axis.

$$-F(a) = \frac{F(b) - F(a)}{b - a}(x - a)$$

$$\frac{-F(a)(b - a)}{F(b) - F(a)} + a = x$$

It follows $x = (aF(b) - bF(a))/(F(b) - F(a))$.

(b)

[a, b]	m	Sign of F(a)	F(m)	Sign of F(b)
[0, 1]	$\frac{1}{3}$	−	$-\frac{8}{27}$	+
[$\frac{1}{3}$, 1]	$\frac{13}{31} \approx .4194$	−	−.0875	+

Chapter 4 Review

1. $F(x) = (x^2 + 1)(2x^3 - 5x + 4) + 3x - 1$

2. $F(x) = (x - \frac{1}{3})(3x^4 - 6x^3 + 3x^2 + 9x - 3) + 1$

3. (a) 0; $F(x) = (x - 2)(x^3 + 3x + 2)$ (b) 426

4. (a) 370 (b) 0; $F(x) = (x + \frac{3}{2})(2x^3 - 2x^2 - 2x + 4)$

5. $F(x) = 2x^5 - x^4 + 3x^3 - 10x^2 + 6x - 13$

6. (a) $7 - 3i$ (b) $5 + i$ (c) $\frac{1}{13} + 5i/13$ (d) $2i$ (e) $1 - i$ (f) 13

7. $1 \pm \sqrt{2}i$

8. $-1 \pm \sqrt{3}i, -1, -3$; $F(x) = (x - (-1 + \sqrt{3}i)) \times (x - (-1 - \sqrt{3}i))(x + 1)(x + 3)$

9. $F(x) = -3(x - 1)^3(x^2 - 4x + 5)/5$

10. (a) $F(x) = x(x + 1)(x - 1)(2x + 3)$; $0, -1, 1, -\frac{3}{2}$
(b) $F(x) = (x + 1)(x - 1)(x + 2)(x - 2)(x + i) \times (x - i)$; $\pm 1, \pm 2, \pm i$

11. (a) $A = 3, B = 1$ (b) $A = 6, B = 0, C = -1$

12. (a) 4 (b) $G(x) = (x^4 - x^3 - 4x^2 + 4x)/3$

13. Let $F(x)$ be a polynomial of degree n, n odd and $n \geq 3$. Any vertical line $x = x_0$ intersects F at $(x_0, F(x_0))$. Now consider the line $y = mx + b$. This line will intersect F at x if and only if $F(x) = mx + b$, or $G(x) = F(x) - mx - b = 0$. Now $G(x)$ is a polynomial function with real coefficients of degree n and has n zeros among the complex numbers. But n is odd and G only can have an even number of complex zeros with nonzero imaginary part. Therefore G must have at least one real zero, say, $x = x_0$. It follows that $y = mx + b$ intersects F at $x = x_0$.

14. (a) ± 1, $\pm\frac{1}{2}$, $\pm\frac{1}{3}$, $\pm\frac{1}{4}$, $\pm\frac{1}{6}$, $\pm\frac{1}{12}$, ± 2, $\pm\frac{2}{3}$, ± 3, $\pm\frac{3}{2}$, $\pm\frac{3}{4}$, ± 6 (b) lower bound: -1, upper bound: 2 (c) $-\frac{1}{2}$, $\frac{2}{3}$, $\frac{3}{2}$ (d) $F(x) = 12(x + \frac{1}{2})(x - \frac{2}{3}) \times (x - \frac{3}{2})$ (e)

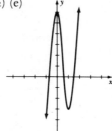

15. (a) ± 1, $\pm\frac{1}{5}$, $\pm\frac{1}{25}$, ± 2, $\pm\frac{2}{5}$, $\pm\frac{2}{25}$, ± 4, $\pm\frac{4}{5}$, $\pm\frac{4}{25}$ (b) lower bound: -2, upper bound: 1 (c) -1, $\frac{2}{5}$ (both multiplicity two) (d) $F(x) = 25(x + 1)^2(x - \frac{2}{5})^2$ (e)

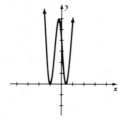

16. Let $F(x) = x^3 - 7$. The only possible rational zeros of F are ± 1 and ± 7. None of these work so that any real zero of F must be irrational. Because $F(\sqrt[3]{7}) = 0$, $\sqrt[3]{7}$ is irrational.

17. We have $F(x) = (x - 2)(2x^2 - 6x + 6) + 1$. $Q(x) = 2x^2 - 6x + 6$ is a parabola curving upward with vertex $(\frac{3}{2}, \frac{3}{2})$. Therefore $Q(x) > 0$ for all x. If $x > 2$, then $x - 2 > 0$ and $Q(x) > 0$, so $(x - 2) \times Q(x) > 0$ because the product of two positive numbers is positive. Adding 1, the quantity remains positive. Thus $x > 2$ implies that $F(x) = (x - 2) \times Q(x) + 1 > 0$.

18. The only possible rational zeros of F are ± 1 and ± 2. None of these work so that any real zero of F must be irrational. Zero $\approx .77$.

Exercises 5.1

1. (a) $\frac{1}{4}$ (b) -3 (c) undefined
3. (a) $-\frac{5}{4}$ (b) 256 (c) undefined
5. (a) $\frac{1}{64}$ (b) undefined (c) 1
7. (a) 1.49535 (b) 1.86607
9. (a) .50297 (b) -1.44225
11. (a) 2^{x+1} (b) b^{21}/a^3 **13.** (a) e^{-2x} (b) e^{x^2+1}
15. (a) x^r (b) x^{1-r} **17.** (a) $(e^{2x} + 2 + e^{-2x})/4$ (b) 1
19. (a) x (b) $2^{(1+x^2)/x}$ **21.** (a) $8x^{-12}$ (b) $a^6b^{8/3}$
23. $(4x^{5/3} - 1)/(3x^{4/3})$ **25.** $(x + 3)/(3(1 + x)^{5/3})$
27. $(4 - 5x)/(5(1 + x)^{2/5}(2 - x)^{3/5})$
29. (a) zero ≈ 2.15443 (b) zero ≈ -2.15443

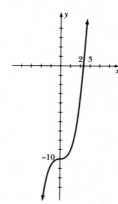

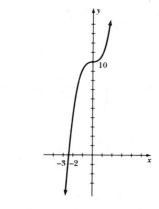

31. zero ≈ 2.11474

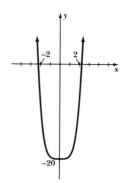

33.

x	3.1	3.14	3.141	3.1415	3.14159	π
2^x	8.5742	8.8152	8.8214	8.8244	8.8250	8.8250

35. (a) 25.95374 (b) 25.95455
37. (a) 36.46216 (b) −1.24687   **39.** true
41. true   **43.** true   **45.** true   **47.** true   **49.** false
51. Let $x = -a^{1/n}$. We wish to prove that $(-a)^{1/n} = x$, but this is equivalent to saying that $x^n = -a$. Now,

$$x^n = (-a^{1/n})^n$$
$$= (-1a^{1/n})^n = (-1)^n(a^{1/n})^n$$
$$= -1(a^{1/n})^n \quad \text{because } n \text{ is odd}$$
$$= -1a \qquad \text{by definition of } a^{1/n}$$
$$= -a \qquad \text{as desired}$$

Exercises 5.2

1. (a) decreasing (b) $\mathcal{D}_F = \mathbf{R}$, range $= (0, +\infty)$ (c) 0
(d) $+\infty$ (e) $y = 0$

3. (a) increasing (b) $\mathcal{D}_F = \mathbf{R}$, range $= (0, +\infty)$ (c) $+\infty$
(d) 0 (e) $y = 0$

5. (a) increasing (b) $\mathcal{D}_F = \mathbf{R}$, range $= (0, +\infty)$ (c) $+\infty$
(d) 0 (e) $y = 0$

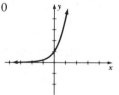

7. (a) decreasing (b) $\mathcal{D}_F = \mathbf{R}$, range $= (0, +\infty)$ (c) 0
(d) $+\infty$ (e) $y = 0$

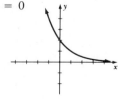

9. (a) decreasing (b) $\mathcal{D}_F = \mathbf{R}$, range $= (-\infty, 0)$ (c) $-\infty$
(d) 0 (e) $y = 0$

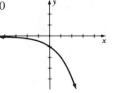

11.

13. 0   **15.** $+\infty$
17. Shift $y = 2^x$ by −1 unit in the y-direction. (a) increasing (b) $(-1, +\infty)$ (c) $+\infty$ (d) $(0, +\infty)$

19. Shift $y = 2^x$ by −1 unit in the x-direction, +1 unit in the y-direction. (a) increasing (b) $(1, +\infty)$ (c) $+\infty$
(d) $(-\infty, +\infty)$

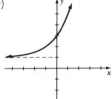

21. Shift $y = -(\frac{1}{3})^x$ by +3 units in the y-direction.
(a) increasing (b) $(-\infty, 3)$
(c) 3 (d) $(-1, +\infty)$

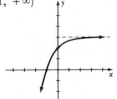

23. (a) 1.3 (b) 2.3   **25.** ⅔   **27.** $(k + 1)/(5k)$
29. (a) $F(n) = 2^n/100$ (b) 27 days
31. (a) $[2, +\infty)$ (b) $+\infty$ **33.** (a) $(-\infty, +\infty)$ (b) $-\infty$

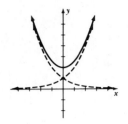   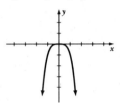

35. symmetry: y-axis
   range = $(0, 1]$
37. symmetry: y-axis
   range = $(-\infty, 0]$

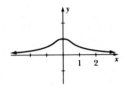

39. (a) symmetric with respect to the y-axis

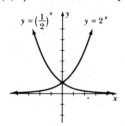

$y = \left(\frac{1}{2}\right)^x$ $y = 2^x$

(b) symmetric with respect to the y-axis

Exercises 5.3

1. (a) 110 (b) 259.37
3. (a) 110.47 (b) 270.70
5. (a) 110.52 (b) 271.83
7. 10 percent compounded daily; 14,774.11
9. 7½ percent compounded annually; 8495.70
11. $Pe^{et/20}$
13. (a) 2.61 (b) 2.71 (c) 2.72 (d) 2.72
15. (a) $e^{1.4}$ (b) $e^{-.01}$

17. $\mathcal{D} = (-\infty, +\infty)$, range = $(0, +\infty)$
   zero: none, y-intercept: $(0, e)$

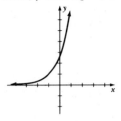

19. $\mathcal{D} = (-\infty, +\infty)$, range = $(-\infty, e)$
   zero: 1, y-intercept: $(0, e - 1)$

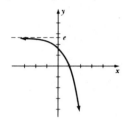

21. (a) 0 (b) $+\infty$ **23.** range = $[1, +\infty)$

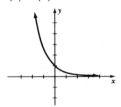   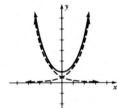

25. true
27. false
29. After t years, bank A yields $P(1 + r/2)^{2t}$ and bank B
   yields $P(1 + R)^t$. Setting these equal implies that
   $(1 + R)^t = (1 + r/2)^{2t}$ or $1 + R = (1 + r/2)^2$. Solving
   the last equation for R yields $R = r + r^2/4$.

Exercises 5.4

1. 3   **3.** 1   **5.** -2 **7.** -1
9. -2 **11.** -3 **13.** 2   **15.** 0
17. 1 and 2   **19.** 2 and 3
21. -3 and -2
23. 0 and 1

25.

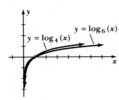

27. decreasing, $\mathcal{D} = (0, +\infty)$, range $= (-\infty, +\infty)$, zeros: 1, $\lim\limits_{x\to 0^+} F(x) = +\infty$, $\lim\limits_{x\to +\infty} F(x) = -\infty$

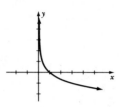

29. increasing, $\mathcal{D} = (0, +\infty)$, range $= (-\infty, +\infty)$, zeros: 1, $\lim\limits_{x\to 0^+} F(x) = -\infty$, $\lim\limits_{x\to +\infty} F(x) = +\infty$

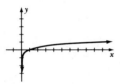

31. increasing, $\mathcal{D} = (0, +\infty)$, range $= (-\infty, +\infty)$, zeros: 1, $\lim\limits_{x\to 0^+} F(x) = -\infty$, $\lim\limits_{x\to +\infty} F(x) = +\infty$

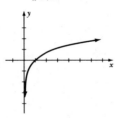

33. decreasing for $x \in (0, 1]$, increasing for $x \in [1, +\infty)$, $\mathcal{D} = (0, +\infty)$, range $= [0, +\infty)$, zeros: 1, $\lim\limits_{x\to 0^+} F(x) = +\infty$, $\lim\limits_{x\to +\infty} F(x) = +\infty$

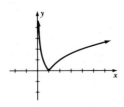

35. decreasing, $\mathcal{D} = (0, 1) \cup (1, +\infty)$, range $= (-\infty, 0) \cup (0, +\infty)$, zeros: none, $\lim\limits_{x\to 0^+} F(x) = 0$, $\lim\limits_{x\to +\infty} F(x) = 0$

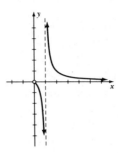

37. decreasing for $x \in (-\infty, 0)$, increasing for $x \in (0, +\infty)$, $\mathcal{D} = (-\infty, 0) \cup (0, +\infty)$, range $= (-\infty, +\infty)$, zeros: $-1, 1$, $\lim\limits_{x\to 0^+} F(x) = -\infty$, $\lim\limits_{x\to +\infty} F(x) = +\infty$

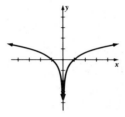

39. decreasing, $\mathcal{D} = (0, +\infty)$, range $= (-\infty, +\infty)$, zeros: 1, $\lim\limits_{x\to 0^+} F(x) = +\infty$, $\lim\limits_{x\to +\infty} F(x) = -\infty$

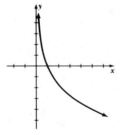

41. $F^{-1}(x) = 4^x$

43. $F^{-1}(x) = \log_{1/3}(x)$

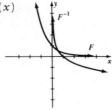

45. (a) $F(x) = \log_2(x)$; shift $+2$ units in the x-direction.
(b)

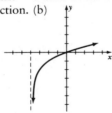

(c) $(2, +\infty)$ (d) $x = 2$ (e) $x \in (3, +\infty)$
47. (a) $F(x) = -\log_2(x)$; shift $+1$ unit in the y-direction.
(b)

(c) $(0, +\infty)$ (d) $x = 0$ (e) $x \in (0, 2)$
49. (a) $F(x) = \log_2(x)$; shift -4 units in the x-direction, -2 units in the y-direction. (b)

(c) $(-4, +\infty)$ (d) $x = -4$ (e) $x \in (0, +\infty)$
51. (a) $+\infty$ (b) $-\infty$ **53.** (a) $+\infty$ (b) $-\infty$
55. (a) $F = G$ (b) equal

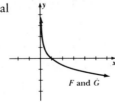

57. false   **59.** (a) 3 (b) x (c) 7 (d) x

1. (a) $4^3 = 64$ (b) $10^4 = x$ (c) $e^y = 3$
3. $10^{-3/2} = x$ **5.** $x + 1 = e^4$
7. (a) $\log_4(2) = \frac{1}{2}$ (b) $\ln(2.5) = x$ (c) $\log a = b$
9. $\ln(x + y) = 3$
11. (a) $F^{-1}(x) = 2^x - 1$ (b) $F^{-1}(x) = 2^{x-1}$
13. (a) $F^{-1}(x) = e^{1-x}$ (b) $F^{-1}(x) = 10^{x-10} + 5$
15. $F^{-1}(x) = e^{-x/2}$ **17.** (a) π (b) 10 (c) 2
19. (a) x^2 (b) $\frac{1}{4}$ (c) 5   **21.** (a) x (b) $1/2^x$
23. (a) $A + B$ (b) $3A$ (c) $C + (A + B)/2$ (d) $-C$
(e) $2C + B - 3A$
25. (a) $1 + A$ (b) $C - 2$
27. (a) 1.2 (b) -3 (c) 9.2 (d) 3.4
29. (a) $4 - 2B$ (b) $B - 1 - A$ (c) $A + B - 3$
31. (a) $\log(x/y^2)$ (b) $\ln(100 y/\sqrt[3]{x})$
33. (a) $\ln(ex)$ (b) $\ln(e^2/x)$
35. (a) 2.3026 (b) 3.3219 (c) .4971 (d) -1.6309
37. .3028   **39.** false   **41.** true   **43.** false
45. false   **47.** false   **49.** false   **51.** false
53.   **55.**

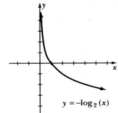
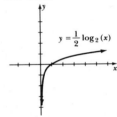

57. Letting $n = \log_a(x)$ and $m = \log_a(y)$, we have $a^n = x$ and $a^m = y$. Now, $x/y = a^n/a^m = a^{n-m}$. This is equivalent to $\log_a(x/y) = n - m$ and $n - m = \log_a(x) - \log_a(y)$ as desired.
59. $a = \sqrt{2}$
61. (a) equal (b) equal (c) We have $\log x(\log y) = \log y(\log x)$. By property 3 of logarithms, we may write $\log(y^{\log x}) = \log(x^{\log y})$. Now, by the $1-1$ property of logarithms, $y^{\log x} = x^{\log y}$.

1. $\log_2(100) \approx 6.6439$ **3.** $2 \log_3(6) \approx 3.2619$
5. $\ln 10/(1 + \ln 2) \approx 1.3599$ **7.** $\ln(\frac{2}{3})/2 \approx -.2027$
9. $\ln(\frac{1}{2})/\ln(\frac{3}{4}) \approx 2.4094$ **11.** $\ln 2/\ln 3 \approx .6309$
13. $\log_2(\frac{5}{4}) \approx .3219$ **15.** $(\ln 3)/2 \approx .5493$ **17.** 0
19. $e^6 \approx 403.4288$ **21.** 6   **23.** 5   **25.** 1 or 3   **27.** 4
29. 1   **31.** $e^{\sqrt{\ln 2 \ln 3}} \approx 2.3932$ or $e^{-\sqrt{\ln 2 \ln 3}} \approx .4178$
33. $10^{1/\sqrt{2}} \approx 5.0946$ **35.** 8 or -8 **37.** $2^{10/9} \approx 2.1601$

39. 0 or -2 **41.** $9^{-5/2} = \frac{1}{243}$ or $-9^{-5/2} = -\frac{1}{243}$
43. $(\frac{1}{4})^{3/5} \approx .4353$ **45.** $-\frac{3}{2}$ **47.** 0, 1, -1
49. -1
51. 0 **53.** no solution

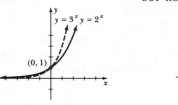

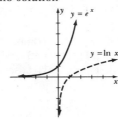

55. 2, 4 **57.** 1

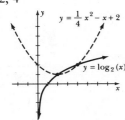

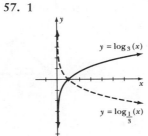

59. $e^x = 10^{x/\ln 10}$, the latter expression uses only 10^x, $\ln x$, and $\div$

Exercises 5.7

1. (a) $Q(t) = 8e^{\ln(5/2)t/2}$ (b) 125
3. approximately .000058 grams
5. approximately 13.97 days
7. (a) $Q(t) = 100e^{\ln(.98)t/4}$ (b) 95.075°C
9. 12.6027 years **11.** 3704.10 **13.** 6.0774%
15. $Q_0 = 1$, $k = \ln(8)/3$ **17.** $Q_0 = 13.5$, $k = \ln(\frac{2}{3})/2$
19. 525.432°C **21.** 5:54 A.M.
23. (a) Given $Q_0 e^{kt_1} = a$ and $Q_0 e^{kt_2} = b$, dividing we have $Q_0 e^{kt_2}/Q_0 e^{kt_1} = b/a$, or $e^{k(t_2 - t_1)} = b/a$, or $k(t_2 - t_1) = \ln(b/a)$, so $k = \ln(b/a)/(t_2 - t_1)$. (b) Substituting $k = \ln(b/a)/(t_2 - t_1)$ into $Q_0 e^{kt_1} = a$, we have $Q_0 e^{\ln(b/a)t_1/(t_2 - t_1)} = a$, or $Q_0(b/a)^{t_1/(t_2 - t_1)} = a$, or $Q_0 = a(a/b)^{t_1/(t_2 - t_1)}$.

Chapter 5 Review

1. (a) $\frac{1}{16}$ (b) not real (c) 16
2. (a) 4.19296 (b) 2.97069
3. (a) $e^{-2x} + 2 + e^{2x}$ (b) $8x^{-3/2}y^{-7}$ **4.** $1/e^2$
5. $(6x^3 + 2x)/(3(x^2 + 1)^{2/3}(x^2 - 1)^{1/3})$

6. decreasing, $\mathcal{D} = (-\infty, +\infty)$, range $= (0, +\infty)$, asymptote: $y = 0$, $\lim_{x \to +\infty} F(x) = 0$, $\lim_{x \to -\infty} F(x) = +\infty$

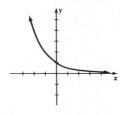

7. decreasing, $\mathcal{D} = (-\infty, +\infty)$, range $= (-\infty, 1)$, asymptote: $y = 1$, $\lim_{x \to +\infty} F(x) = -\infty$, $\lim_{x \to -\infty} F(x) = 1$

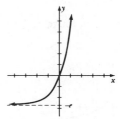

8. increasing, $\mathcal{D} = (-\infty, +\infty)$, range $= (-e, +\infty)$, asymptote: $y = -e$, $\lim_{x \to +\infty} F(x) = +\infty$, $\lim_{x \to -\infty} F(x) = -e$

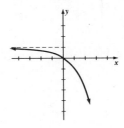

9. (a) 0 (b) $+\infty$
10. (a) range $= [2, +\infty)$ (b) range $(-\infty, +\infty)$

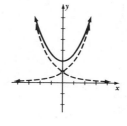

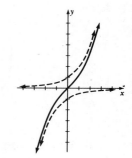

11. (a) 14557.73 (b) 14769.20 (c) 14769.81
12. Bank A yields $P(1 + r/3)^{3t}$ and bank B yields
$P(1 + R)^t$. Setting these equal, we have $P(1 + R)^t = P(1 + r/3)^{3t}$, or $(1 + R)^t = (1 + r/3)^{3t}$, or $1 + R = (1 + r/3)^3$, or $R = r + r^2/3 + r^3/27$.
13. (a) -2 (b) ½ (c) 0   14. 1 and 2
15. increasing, $\mathcal{D} = (-2, +\infty)$, range $= (-\infty, +\infty)$, zero: -1, asymptote: $x = -2$

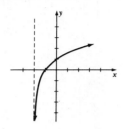

16. increasing, $\mathcal{D} = (0, +\infty)$, range $= (-\infty, +\infty)$, zero: $1/e$, asymptote: $x = 0$

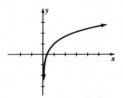

17. decreasing, $\mathcal{D} = (0, +\infty)$, range $= (-\infty, +\infty)$, zero: 1, asymptote: $x = 0$

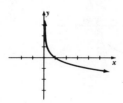

18. decreasing, $\mathcal{D} = (0, +\infty)$, range $= (-\infty, +\infty)$, zero: 1, asymptote: $x = 0$

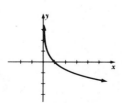

19. (a) $+\infty$ (b) $-\infty$ 20. $x + 1 = e^4$ 21. $P = \log(r - 1)$

22. $F^{-1}(x) = \log_3(x)$ 23. $F^{-1}(x) = e^{x-1} - 1$ 24. 13/4
25. (a) $1 - 2A - B$ (b) $A + 2B + ½$
26. $\ln(4x^3/\sqrt{y})$ 27. 2.0115
28. $\ln 4/\ln(4/5) \approx -6.2126$ 29. $\ln 2/(2 \ln 3) \approx .3155$
30. $5 + 3\sqrt{5} \approx 11.7082$ 31. 7/2, $-9/2$
32. $2^{1/\sqrt{2}} \approx 1.6325$ 33. $-½$ 34. 1, -1
35. 20.6815 million; 2009   36. 11.99 percent
37. (a) $Q(t) = 135\,e^{\ln(2/3)t/10}$ (b) 40°C   38. false   39. true
40. false   41. false   42. true   43. true

Exercises 6.1

1. (a) 5 (b) 10π (c) 25
3. (a)   (b)

(c)   (d)

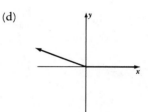

5. (a)   (b)

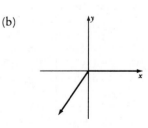

(c)

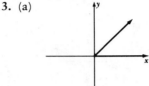

7. (a) 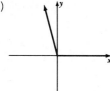   (b) 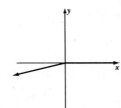   (c) $11\pi/6$

(c)    (d)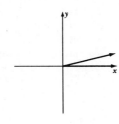

9. (a) ⁵⁄₄ (b) 4   **11.** (a) $\pi/2$; $-3\pi/2$ (b) 2; $2 - 2\pi$
13. (a) 2 (b) 8

15. (a) 60°   (b) 120°

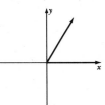   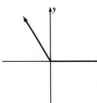

(c) $-120°$

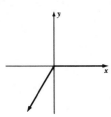

17. (a) $\pi/6$ (b) $-7\pi/6$

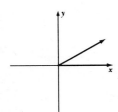   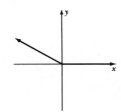

19. (a)

radians:	$\pi/4$	$\pi/2$	$3\pi/4$	π	$5\pi/4$	$3\pi/2$	$7\pi/4$	2π
degrees:	45	90	135	180	225	270	315	360

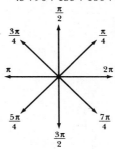

(b)

radians:	$\pi/6$	$\pi/3$	$2\pi/3$	$5\pi/6$	$7\pi/6$	$4\pi/3$	$5\pi/3$	$11\pi/6$
degrees:	30	60	120	150	210	240	300	330

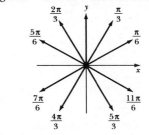

21. (a) $-\pi/15$ (b) 10π (c) $13\pi/12$ (d) $-\pi/8$
23. (a) 18° (b) $-9°$ (c) $-216°$ (d) $(25/2\pi)°$
25. (a) 10π (b) $4\pi/3$
27. $k \in \mathbf{Z}$: (a) $\pi/4 + 2\pi k$ (b) $-2.5 + 2\pi k$
(c) $100° + 360°k$
29. (a) π (b) 90° (c) 305°   **31.** (a) $3\pi/4$ (b) 280° (c) $7\pi/6$
33. (a) $\pi/4$, $5\pi/4$ (b) $\pi/2$, $3\pi/2$ (c) 165°, 345°
35. (a) $\pi/12$, $13\pi/12$, (b) 0, π (c) 150°, 330°
37. (a) $\pi/9$, $7\pi/9$, $13\pi/9$ (b) $7\pi/12$, $15\pi/12$, $23\pi/12$
(c) 30°, 150°, 270°
39. (a)    (b)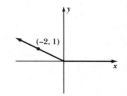

Exercises 6.2

1. $\sin(t) = 2/\sqrt{5}$

Exercise:	3a	3b	3c
$\sin(\theta)$	$3/5$	$-3/5$	$-3/5$
$\cos(\theta)$	$-4/5$	$4/5$	$-4/5$
$\tan(\theta)$	$-3/4$	$-3/4$	$3/4$

Exercise:	5a	5b	5c
$\sin(\theta)$	$-1/\sqrt{10}$	$1/\sqrt{10}$	$1/\sqrt{10}$
$\cos(\theta)$	$3/\sqrt{10}$	$-3/\sqrt{10}$	$3/\sqrt{10}$
$\tan(\theta)$	$-1/3$	$-1/3$	$1/3$

Exercise:	7a	7b	7c
$\sin(\theta)$	1	-1	-1
$\cos(\theta)$	0	0	0
$\tan(\theta)$	und.	und.	und.

9.

t:	3π	-4π	100π
$\sin(t)$	0	0	0
$\cos(t)$	-1	1	1
$\tan(t)$	0	0	0

11.

t:	$5\pi/2$	$-\pi/2$	$99\pi/2$
$\sin(t)$	1	-1	-1
$\cos(t)$	0	0	0
$\tan(t)$	und.	und.	und.

13.

t:	$\pi/4$	$-\pi/4$
$\sin(t)$	$1/\sqrt{2}$	$-1/\sqrt{2}$
$\cos(t)$	$1/\sqrt{2}$	$1/\sqrt{2}$
$\tan(t)$	1	-1

Exercise:	15a	15b
$\sin(t)$	$3/\sqrt{10}$	$-3/\sqrt{10}$
$\cos(t)$	$-1/\sqrt{10}$	$1/\sqrt{10}$
$\tan(t)$	-3	-3

Exercise:	17a	17b
$\sin(t)$	$\pi/\sqrt{1+\pi^2}$	$-\pi/\sqrt{1+\pi^2}$
$\cos(t)$	$1/\sqrt{1+\pi^2}$	$-1/\sqrt{1+\pi^2}$
$\tan(t)$	π	π

Exercise	$\sin(t+\pi/2)$	$\cos(t+\pi/2)$	$\tan(t+\pi/2)$
19.	$2/\sqrt{5}$	$-1/\sqrt{5}$	-2
21.	$-1/\sqrt{2}$	$1/\sqrt{2}$	-1

23. II   **25.** III   **27.** $(12, 9)$
29. $(1/5, -2\sqrt{6}/5)$
31. (a) $x = 8 \cos t$ (b) $y = 8 \sin t$
33. (a) $y = 2 \sin t$ (b) $x = 2 + 2 \cos t$
35. $y = \tan t$ **37.** $-4/3$
39. 0 **41.** $\sin t$
43. For all $k \in \mathbf{Z}$, $t + 2\pi k$ is coterminal with t; therefore we may use the same point to calculate sine, cosine, and tangent at $t + 2\pi k$ and t.

Exercise 6.3

1. (a) 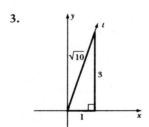   (b)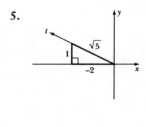

3. 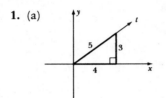   **5.**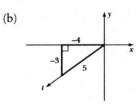

7. 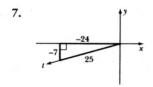   **9.**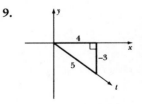

Exercise	$\sin t$	$\cos t$	$\tan t$
3.	$3/\sqrt{10}$	$1/\sqrt{10}$	3
5.	$1/\sqrt{5}$	$-2/\sqrt{5}$	$-1/2$
7.	$-7/25$	$-24/25$	$7/24$
9.	$-3/5$	$4/5$	$-3/4$

11.

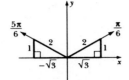

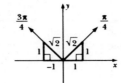

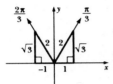

t	$\pi/6$	$\pi/4$	$\pi/3$	$2\pi/3$	$3\pi/4$	$5\pi/6$
$\sin t$	$1/2$	$1/\sqrt{2}$	$\sqrt{3}/2$	$\sqrt{3}/2$	$1/\sqrt{2}$	$1/2$
$\cos t$	$\sqrt{3}/2$	$1/\sqrt{2}$	$1/2$	$-1/2$	$-1/\sqrt{2}$	$-\sqrt{3}/2$
$\tan t$	$1/\sqrt{3}$	1	$\sqrt{3}$	$-\sqrt{3}$	-1	$-1/\sqrt{3}$

13. (a)

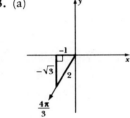

(b)

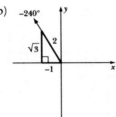

(c)

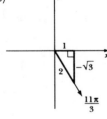

15. (a)

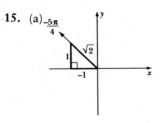

(b)

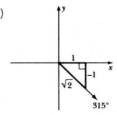

17. (a)

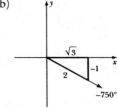

(b)

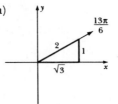

Exercise:	13a	13b	13c	15a	15b	17a	17b
$\sin t$	$-\sqrt{3}/2$	$\sqrt{3}/2$	$-\sqrt{3}/2$	$1/\sqrt{2}$	$-1/\sqrt{2}$	$1/2$	$-1/2$
$\cos t$	$-1/2$	$-1/2$	$1/2$	$-1/\sqrt{2}$	$1/\sqrt{2}$	$\sqrt{3}/2$	$\sqrt{3}/2$
$\tan t$	$\sqrt{3}$	$-\sqrt{3}$	$-\sqrt{3}$	-1	-1	$1/\sqrt{3}$	$-1/\sqrt{3}$

Exercise:	19	21	23	25
$\sin \theta$	$8/17$	$1/3$	$\sqrt{3/7}$	$12/13$
$\cos \theta$	$15/17$	$2\sqrt{2}/3$	$2/\sqrt{7}$	$5/13$
$\tan \theta$	$8/15$	$1/2\sqrt{2}$	$\sqrt{3}/2$	$12/5$

Exercise	Angle θ	Opposite	Adjacent	Hypotenuse	$\sin \theta$	$\cos \theta$	$\tan \theta$
27.	α	3	4	5	$3/5$	$4/5$	$3/4$
	β	4	3	5	$4/5$	$3/5$	$4/3$
29.	α	2	1	$\sqrt{5}$	$2/\sqrt{5}$	$1/\sqrt{5}$	2
	β	1	2	$\sqrt{5}$	$1/\sqrt{5}$	$2/\sqrt{5}$	$1/2$

31. (a) $100/\sin \theta$ (b) $50 \cos \theta$ (c) $12 \tan \theta$
33. (a) $4 \sin \theta$ (b) $4 \cos \theta$ **35.** (a) $6 \tan \theta$ (b) $6/\cos \theta$
37. $x = 10\sqrt{2} \sin \theta$ **39.** $x = 5 \cos \theta, y = 5 \cos \theta$
41. $42 \sin \theta$

Exercises 6.4

1. (a) largest: $\cos(6)$, smallest: $\cos(5)$ (b) largest: $\sin(6)$, smallest: $\sin(5)$
3. (a) $\sin \beta$ (b) $\cos \alpha$
5. (a) $(\sqrt{3}/2, 1/2)$ (b) $(-1/\sqrt{2}, 1/\sqrt{2})$ (c) $(1/2, -\sqrt{3}/2)$

7.

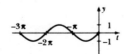

9.

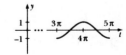

11.

13.

Answers to Odd-Numbered Exercises ▪ **661**

15. 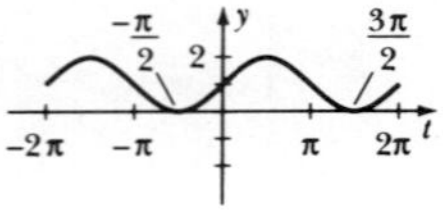

17. 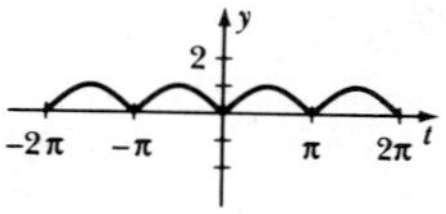

19. 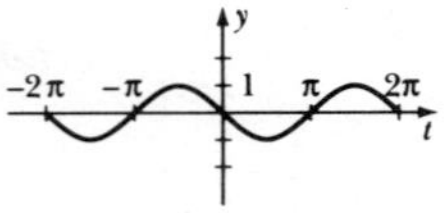

21. $2 \sin \theta$ **23.** $\cos(2)$ **25.** 0 **27.** $2 \cos \theta$
29. 1 **31.** $\cos \theta$ **33.** $2|\sin \theta|$ **35.** 0
37. no; $-1 \le \sin \theta \le 1$ for all θ
39. (a) $-.75680$ (b) .25882 (c) .76604
41. $(-.4161, .9093)$ **43.** $(3.4202, -9.3969)$

45.

t	1	.5	.25	.1	.01	.001
$\sin t/t$	.8415	.9589	.9896	.9983	.99998	.9999998

$$\lim_{t\to 0^+} \frac{\sin t}{t} = 1$$

47. .540, .858, .654, .793, .701; .744

Exercises 6.5

1. 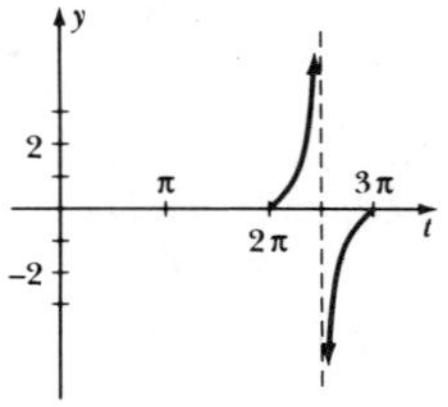

3. 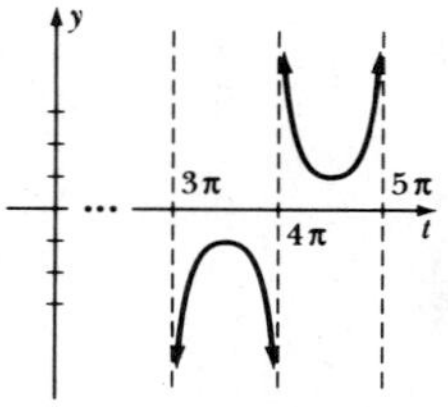

5. 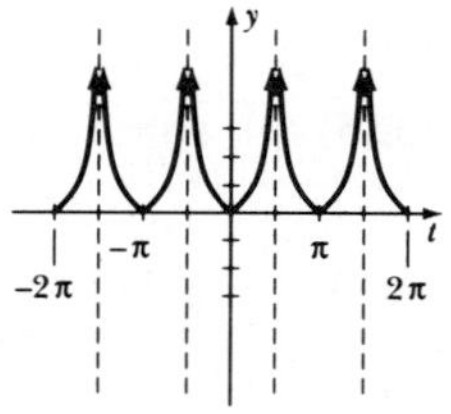

7. 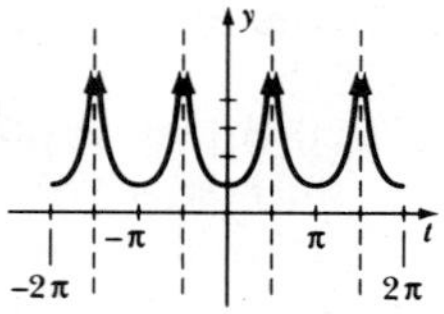

9. 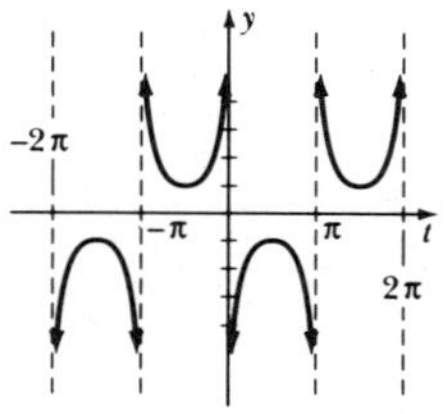

11.

t	0	$\pi/2$	π	$3\pi/2$	2π
$\csc t$	und.	1	und.	-1	und.
$\sec t$	1	und.	-1	und.	1
$\cot t$	und.	0	und.	0	und.

13.

t	$-\pi/6$	$-\pi/4$	$-\pi/3$	$-\pi/2$	$-2\pi/3$	$-3\pi/4$	$-5\pi/6$
$\csc t$	-2	$-\sqrt{2}$	$-2/\sqrt{3}$	-1	$-2/\sqrt{3}$	$-\sqrt{2}$	-2
$\sec t$	$2/\sqrt{3}$	$\sqrt{2}$	2	und.	-2	$-\sqrt{2}$	$-2/\sqrt{3}$
$\cot t$	$-\sqrt{3}$	-1	$-1/\sqrt{3}$	0	$1/\sqrt{3}$	1	$\sqrt{3}$

Exercise:	15	17
$\csc \theta$	41/9	$\sqrt{19/3}$
$\sec \theta$	41/40	$\sqrt{19}/4$
$\cot \theta$	40/9	$4/\sqrt{3}$

Exercise	$\sin t$	$\cos t$	$\tan t$	$\csc t$	$\sec t$	$\cot t$
19.	$2/\sqrt{5}$	$1/\sqrt{5}$	2	$\sqrt{5}/2$	$\sqrt{5}$	½
21.	$-3/\sqrt{13}$	$-2/\sqrt{13}$	3/2	$-\sqrt{13}/3$	$-\sqrt{13}/2$	2/3
23.	$7/5\sqrt{2}$	$-1/5\sqrt{2}$	-7	$5\sqrt{2}/7$	$-5\sqrt{2}$	$-1/7$

25. (a) 3.78485 (b) 3.86370 (c) -75.31966

27.

t	.1	.01	.001	.0001
$\sec t/t$	10.05	100	1000	10,000

$$\lim_{t\to 0^+} \frac{\sec t}{t} = +\infty$$

29. (a) $-3 \csc \theta$ (b) $-3 \cot \theta$
31. (a) $-4 \sec \theta$ (b) $-4 \tan \theta$
33. (a) $81 (\cot \theta)/2$ (b) $8 \tan \theta$ (c) $225 (\sin \theta \cos \theta)/2$
35. 0 **37.** $-\tan \theta$ **39.** 1 **41.** 1 **43.** 2
45. (a) false (b) true (c) false

Exercises 6.6

1. 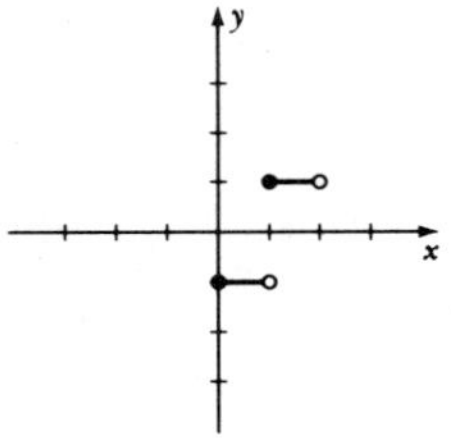

3. not periodic

5. period 2π, amplitude 1

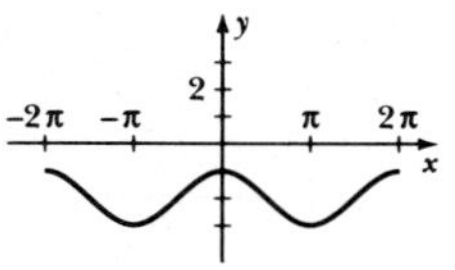

7. period π, amplitude undefined

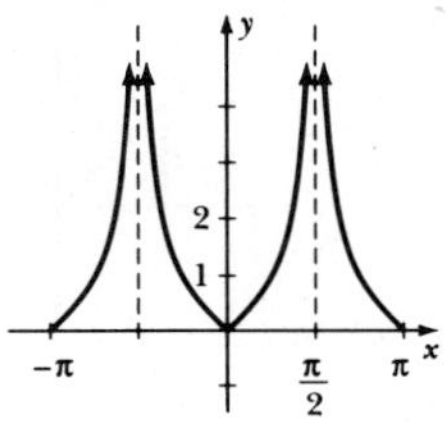

9. period π, amplitude $\frac{1}{2}$

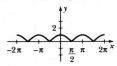

11. (a) period 4, amplitude 2

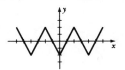

(b) period 4, amplitude $\frac{1}{2}$

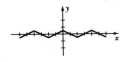

13. period 2π, amplitude 3

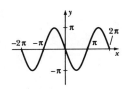

15. period π, amplitude undefined

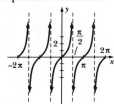

17. period 2π, amplitude π

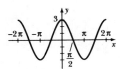

19. maximum points:
$(-\pi/2, 1)$ $(\pi/6, 1)$
minimum points:
$(-\pi/6, -1)$ $(\pi/2, -1)$

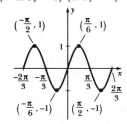

21. maximum points:
$(-4\pi, 3)$ $(0, 3)$ $(4\pi, 3)$
minimum points:
$(-2\pi, -3)$ $(2\pi, -3)$

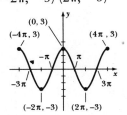

23.

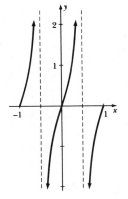

25. maximum points:
$(-\pi/4, 1/2)$ $(3\pi/4, 1/2)$
minimum points:
$(-3\pi/4, -1/2)$ $(\pi/4, -1/2)$

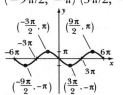

27. maximum points:
$(-2, 1/2)$ $(2, 1/2)$
minimum points:
$(-4, -1/2)$ $(0, -1/2)$
$(4, -1/2)$

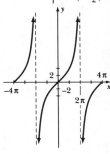

29. maximum points:
$(-3\pi/2, \pi)$ $(9\pi/2, \pi)$
minimum points:
$(-9\pi/2, -\pi)$ $(3\pi/2, -\pi)$

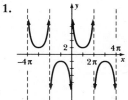

31.

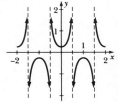

Answers for Exercises 33 to 39 may vary.

33. (a) $F(x) = 5 \sin (2x)$ (b) $F(x) = \sin(4x/3)$
(c) $F(x) = 3 \sin(5x/2)$

35. (a) $F(x) = 10 \sin(\pi x)$ (b) $F(x) = -4 \sin(3x/2)$
(c) $F(x) = -4 \sin (x/3)/\sqrt{3}$

37. (a) $G(x) = 4 \cos(x/2)$ (b) $G(x) = 2 \cos(2x)$
(c) $G(x) = 3 \cos(2x/5)$

39. (a) $G(x) = 5 \cos(4x)$ (b) $G(x) = \cos(\pi x/2)/3$
(c) $G(x) = 4 \cos(\pi x/24)$

Exercises 6.7

1.

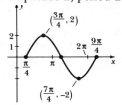

3.

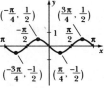

5. maximum point: $(3\pi/4, 2)$
minimum point: $(7\pi/4, -2)$
amplitude 2, period 2π

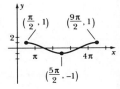

7. maximum points: $(\pi/2, 1)$
$(9\pi/2, 1)$ minimum point:
$(5\pi/2, -1)$ amplitude 1,
period 4π

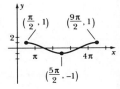

9. maximum point: $(2\pi, 3/2)$
minimum point: $(\pi, 1/2)$
amplitude $\frac{1}{2}$, period 2π

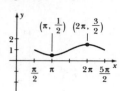

11. maximum points:
$(0, 0)$ $(8, 0)$
minimum point: $(4, -4)$
amplitude 2, period 8

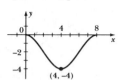

13. amplitude undefined, period $\pi/2$

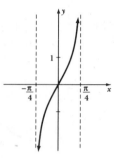

Answers for Exercises 15 and 17 may vary.
15. $F(x) = 2\sin(4x) + 1$
17. $G(x) = -4\cos(\pi x/3) - 1$

19.

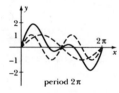

period 2π

21.

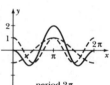

period 2π

23.

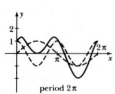

period 2π

25.

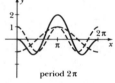

27.

Note: $H(x) = 2\cos x$

29.

Note: $H(x) = 0$

31.

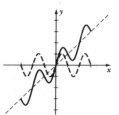

33.

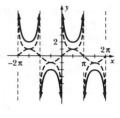

35. (a) $y = 4 + 3\sin\theta$ (b) $x = 6 + 3\cos\theta$

Chapter 6 Review

1.

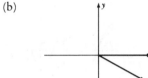

(b)

(c)

2. (a) $7\pi/2$ (b) $\frac{5}{3}$ radians
3. for $k \in \mathbf{Z}$: (a) $17\pi/3 + 2\pi k, 5\pi/3$
(b) $-9 + 2\pi k, 4\pi - 9$ (c) $3000° + 360°k, 120°$
4. (a) $\pi/6, 7\pi/6$ (b) $7\pi/8, 15\pi/8$

5.

θ	$\sin\theta$	$\cos\theta$	$\tan\theta$	$\csc\theta$	$\sec\theta$	$\cot\theta$
(a) t	$-\frac{8}{17}$	$\frac{15}{17}$	$-\frac{8}{15}$	$-\frac{17}{8}$	$\frac{17}{15}$	$-\frac{15}{8}$
(b) $t + \pi$	$\frac{8}{17}$	$-\frac{15}{17}$	$-\frac{8}{15}$	$\frac{17}{8}$	$-\frac{17}{15}$	$-\frac{15}{8}$
(c) $-t$	$\frac{8}{17}$	$\frac{15}{17}$	$\frac{8}{15}$	$\frac{17}{8}$	$\frac{17}{15}$	$\frac{15}{8}$

6. $(-2\sqrt{51}, 14)$

7.

t	$\sin t$	$\cos t$	$\tan t$	$\csc t$	$\sec t$	$\cot t$
0	0	1	0	und.	1	und.
$\pi/2$	1	0	und.	1	und.	0
π	0	-1	0	und.	-1	und.
$3\pi/2$	-1	0	und.	-1	und.	0
2π	0	1	0	und.	1	und.

8. (a)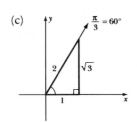

(b)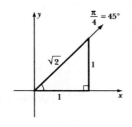

(c)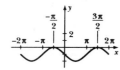

t	$\sin t$	$\cos t$	$\tan t$
$\pi/6$	$1/2$	$\sqrt{3}/2$	$1/\sqrt{3}$
$\pi/4$	$1/\sqrt{2}$	$1/\sqrt{2}$	1
$\pi/3$	$\sqrt{3}/2$	$1/2$	$\sqrt{3}$

9. (a) -1 (b) -1 (c) $-\sqrt{3}$ (d) $\sqrt{2}$ (e) 2 (f) $\sqrt{3}$
(g) undefined (h) 1 (i) 0
10. $AC = 6\cos\theta$, $BC = 6\sin\theta$
11. $BC = 5\tan\theta$, $AB = 5\sec\theta$
12. $AC = 4\cot\theta$, $AB = 4\csc\theta$
13. $50\sin\theta\cos\theta$ **14.** $(-\frac{1}{2}, -\sqrt{3}/2)$
15. (a) $(-7.5175, 2.7362)$ (b) $(6.4721, -4.7023)$ **16.** 6
17. period 2π, amplitude 1 **18.** period π, amplitude $\frac{1}{2}$

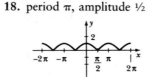

19.

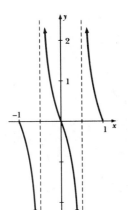

20.

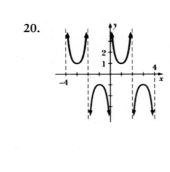

21. (a) no (b) yes (c) no   **22.** $\cos^2\theta$ **23.** 1
24. $2\cos\theta$ **25.** -1 **26.** $\cos\theta$ **27.** $\tan\theta$ **28.** 1
29. (a) $9\cos\theta$ (b) $9\sin\theta$ (c) $\tan\theta$ (d) $-\cot\theta$ (e) $9\sec\theta$
(f) $9\csc\theta$
30. (a) $-10\tan\theta$ (b) $-10\sec\theta$
31. (a) $7\cot\theta$ (b) $7\csc\theta$
32. amplitude 2, period 8π **33.** amplitude $\frac{1}{2}$, period π

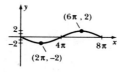

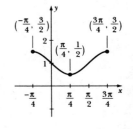

34. amplitude undefined, period π **35.** amplitude undefined, period 4π

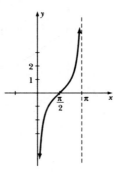

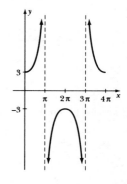

36. period 4

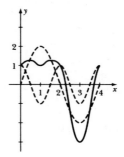

Answers for Exercises 37 and 38 may vary.
37. $F(x) = 4\sin(\pi x/6)$ **38.** $G(x) = 2\cos(2\pi x)$

Exercises 7.1

1.

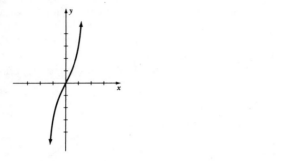

satisfies horizontal line test; $F^{-1}(0) = 0$, $F^{-1}(3) = 1$, $F^{-1}(-12) = -2$

3. 0   **5.** 0   **7.** $2\pi/3$ **9.** $-\pi/4$
11. undefined   **13.** π **15.** $\pi/6$ **17.** undefined
19. 0   **21.** -1 **23.** $\pi/2$ **25.** 1.32582
27. 1.98231   **29.** $-.90334$ **31.** $-.66624$

33. 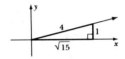   **35.**

Exercise	$\sin\theta$	$\cos\theta$	$\tan\theta$
33.	$\frac{1}{4}$	$\sqrt{15}/4$	$1/\sqrt{15}$
35.	$-\frac{3}{5}$	$\frac{4}{5}$	$-\frac{3}{4}$

37. $-.3$ **39.** $-1/\sqrt{5}$ **41.** $\frac{3}{4}$
43. $\sqrt{1 - x^2}/x$ **45.** $1/x$
47. $\theta = \cos^{-1}(x/\sqrt{16 + x^2})$ **49.** $\pi/2$

Exercises 7.2

In the following answers, $k \in \mathbf{Z}$.
  1. $x = -\pi/2 + 2\pi k$; $3\pi/2$

3. $x = \begin{cases} \pi/3 + 2\pi k \\ 5\pi/3 + 2\pi k \end{cases}$; $\pi/3$, $5\pi/3$

5. $x = \begin{cases} -62.85° + 360°k \\ 117.15° + 360°k \end{cases}$; $117.15°$, $297.15°$

7. $x = \begin{cases} \pi/6 + \pi k \\ -\pi/6 + \pi k \end{cases}$; $\pi/6$, $5\pi/6$, $7\pi/6$, $11\pi/6$

9. $x = \begin{cases} \pi/4 + 2\pi k \\ 5\pi/4 + 2\pi k \end{cases}$; $\pi/4$, $5\pi/4$

11. $x = \begin{cases} \pi/6 + \pi k \\ 5\pi/6 + \pi k \end{cases}$; $\pi/6$, $5\pi/6$, $7\pi/6$, $11\pi/6$

13. $x = \begin{cases} -\pi/8 + \pi k \\ 3\pi/8 + \pi k \end{cases}$; $3\pi/8$, $7\pi/8$, $11\pi/8$, $15\pi/8$

15. $x = \begin{cases} \pi/4 + 3\pi k \\ 11\pi/4 + 3\pi k \end{cases}$; $\pi/4$

17. $x = \begin{cases} 9° + 120°k \\ 51° + 120°k \end{cases}$; $9°$, $51°$, $129°$, $171°$, $249°$, $291°$

19. $x = \pi/4 + \pi k/2$; $\pi/4$, $3\pi/4$, $5\pi/4$, $7\pi/4$
21. no solution

23. $x = \begin{cases} \pi/2 + 4\pi k \\ 7\pi/2 + 4\pi k \end{cases}$; $\pi/2$

25. $\sin x = \frac{3}{5}$; $x = \begin{cases} 36.87° + 360°k \\ 143.13° + 360°k \end{cases}$

27. $\cos x = \frac{3}{10}$; $x = \begin{cases} 72.54° + 360°k \\ 287.46° + 360°k \end{cases}$

29.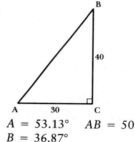

$A = 53.13°$ $AB = 50$
$B = 36.87°$

31.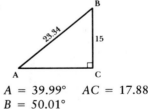

$A = 39.99°$ $AC = 17.88$
$B = 50.01°$

33. $0.52°$
35. $AB = 2\sqrt{2}$, $AC = BC = \sqrt{34}$,
    $A = B \approx 75.96°$, $C \approx 28.08°$
37. (a) If $y = 0$ when $t = 0$, then $0 = a\sin(\alpha)$, which
    implies that $\sin(\alpha) = 0$, so $\alpha = 0 + \pi k$, $k \in \mathbf{Z}$.

    (b) $\alpha = \begin{cases} \pi/6 + 2\pi k \\ 5\pi/6 + 2\pi k \end{cases}$, $k \in \mathbf{Z}$

    (c) $\alpha = \pi/2 + 2\pi k$, $k \in \mathbf{Z}$

1. (a) $(1 + y)(1 - y)$ (b) $(1 + \sin \theta)(1 - \sin \theta)$
3. (a) $(y^2 + x^2)(y + x)(y - x)$
    (b) $(\sin \theta + \cos \theta)(\sin \theta - \cos \theta)$
5. (a) $(y + x)(y^2 - yx + x^2)$
    (b) $(\sin \theta + \cos \theta)(1 - \sin \theta \cos \theta)$
7. (a) $1/(y + 1)$ (b) $1/(\sin \theta + 1)$
9. (a) x/y (b) $\cos \theta/\sin \theta$
11. (a) $1/y$ (b) $1/\sin \theta$
13. (a) $(y^2 + x^2)/xy$ (b) $1/(\cos \theta \sin \theta)$
15. (a) $(1 - y^2)/y^2$ (b) $(1 - \sin^2 \theta)/\sin^2 \theta$
17. (a) $y^2 + 2yx + x^2$ (b) $1 + 2 \sin \theta \cos \theta$
19. (a) $y^2/x + x + y + x^2/y$ (b) $1/\cos \theta + 1/\sin \theta$
21. I **23.** K **25.** H **27.** D **29.** B **31.** A
Note: We present proofs only for certain identity exercises that require unusual maneuvers.

57.

$$\sqrt{\frac{1 + \cos \theta}{1 - \cos \theta}} \stackrel{?}{=} \frac{|\sin \theta|}{1 - \cos \theta}$$

$$\sqrt{\frac{(1 + \cos \theta)(1 - \cos \theta)}{(1 - \cos \theta)(1 - \cos \theta)}}$$

$$\sqrt{\frac{1 - \cos^2 \theta}{(1 - \cos \theta)^2}}$$

$$\sqrt{\frac{\sin^2 \theta}{(1 - \cos \theta)^2}}$$

$$\frac{|\sin \theta|}{1 - \cos \theta}$$

69.

$$\frac{\sin \theta}{1 - \cos \theta} \stackrel{?}{=} \frac{1 + \cos \theta}{\sin \theta}$$

$$\frac{\sin \theta(1 + \cos \theta)}{(1 - \cos \theta)(1 + \cos \theta)}$$

$$\frac{\sin \theta (1 + \cos \theta)}{1 - \cos^2 \theta}$$

$$\frac{\sin \theta(1 + \cos \theta)}{\sin^2 \theta}$$

$$\frac{1 + \cos \theta}{\sin \theta}$$

79. False: $\theta = \pi/3$ implies that the left side is $8/3$ and the right side is $-1/2$.
81. False: $\theta = \pi/2$ implies that the left side is 1 and the right side is -1.
83. False: $\theta = 3\pi/2$ implies that the left side is 1 and the right side is -1.
85. False: $\theta = 1$ implies that the left side is 0 and the right side is $\tan (1)$.
87. true

3. $(\sqrt{3} - 1)/2\sqrt{2}$ **5.** $(\sqrt{3} - 1)/2\sqrt{2}$
7. $(\sqrt{3} - 1)/(\sqrt{3} + 1)$ **9.** $\frac{1}{2}$
11. (a) $(-\sqrt{3} - 1)/2\sqrt{2}$ (b) $(\sqrt{3} - 1)/2\sqrt{2}$
23. $-3/5$ **25.** 1 **27.** $24/7$
29. (a) $F(x) = \sin(x + \pi/6)$
    (b)

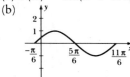

    (c) period 2π, range $[-1, 1]$

31. $\cos^3 x - 3 \cos x \sin^2 x$
33. $\dfrac{\sqrt{3}}{2} \dfrac{\cos h - 1}{h} - \dfrac{1}{2} \dfrac{\sin h}{h}$
35. False: $\theta = 0$ implies that the left side is $\frac{1}{2}$ and the right side is $3/2$.
37. False: $\theta = \pi/6$ implies that the left side is 3 and the right side is $1/3$.
39. False: $\theta = \pi$ implies that the left side is $-\frac{1}{2}$ and the right side is $\frac{1}{2}$.
41. False: $\theta = \pi/2$ implies that the left side is $\sin(\pi^2/4)$ and the right side is 1.
43. (a) By definition, $-\pi/2 < \alpha < \pi/2$ and $-\pi/2 < \beta < \pi/2$. Also, $\tan \alpha = 1/3 > 0$ and $\tan \beta = 1/2 > 0$. It follows that $0 < \alpha < \pi/2$ and $0 < \beta < \pi/2$. Therefore, $0 < \alpha + \beta < \pi/2 + \pi/2 = \pi$. (b) $\sin(\alpha + \beta) = 1/\sqrt{2}$, $\cos(\alpha + \beta) = 1/\sqrt{2}$ (c) $\alpha + \beta = \pi/4$

3. $\sin(2\theta) = 4\sqrt{2}/9$, $\cos(2\theta) = 7/9$, $\tan(2\theta) = 4\sqrt{2}/7$
5. $\sin(2\theta) = -24/25$, $\cos(2\theta) = -7/25$, $\tan(2\theta) = 24/7$
7.

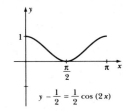

$$y - \frac{1}{2} = \frac{1}{2} \cos (2 x)$$

9. (a) $\sqrt{2 - \sqrt{3}}/2$ (b) $\sqrt{2 - \sqrt{2 + \sqrt{3}}}/2$
11. (a) $\sqrt{2 + \sqrt{2}}/2$ (b) $\sqrt{2 + \sqrt{2 + \sqrt{2}}}/2$

25. (b) period π, range $[0, 2]$ **27.** period 1, range $[-1, 1]$

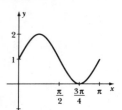

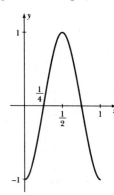

29. $\cos^4 x - 6 \cos^2 x \sin^2 x + \sin^4 x$
31. $\frac{3}{8} - (\frac{1}{2}) \cos(2x) + (\frac{1}{8}) \cos(4x)$
35. $\sin(2\theta) = 2x\sqrt{1 - x^2}$, $\cos(2\theta) = 1 - 2x^2$
37. $\sin(2\theta) = 2y\sqrt{1 - y^2}$, $\cos(2\theta) = 2y^2 - 1$
39. False: $\theta = \pi$ implies that the left side is $1/\pi$ and the right side is 0.
41. False: $\theta = \pi/6$ implies that the left side is $\sqrt{3}/2$ and the right side is $\sqrt{3}/3$.

Exercises 7.6

3. $(\frac{1}{2}) \sin(2x) + (\frac{1}{2}) \sin(8x)$
5. $(\frac{1}{2}) \cos x - (\frac{1}{2}) \cos(3x)$
7. $(\frac{1}{2}) \cos(2x) + (\frac{1}{2}) \cos(6x)$
9. $\sin x + \sin(2x) + \sin(3x)$
11. $(\frac{1}{4}) \sin(4x) + (\frac{1}{4}) \sin(8x) - (\frac{1}{4}) \sin(12x)$
21. change to sines and cosines   **25.** $3|\cos \theta|$ **27.** 1
29. $\cos^2 \theta$ **31.** (a) $-\sin \theta$ (b) $-\cos \theta$ (e) $-\tan \theta$
33. false   **35.** false   **37.** false   **39.** $\frac{4}{5}$
41. $2x\sqrt{1 - x^2}$

Exercises 7.7

In the following answers, $k \in \mathbf{Z}$.
1. $52.31° + 360°k$, $127.69° + 360°k$; $52.31°$, $127.69°$
3. πk, $71.57° + 360°k$, $251.57° + 360°k$; 0, $71.57°$, π, $251.57°$
5. $\pi/2 + \pi k$, $\pi/4 + 2\pi k$, $7\pi/4 + 2\pi k$; $\pi/4$, $\pi/2$, $3\pi/2$, $7\pi/4$
7. $\pi + 2\pi k$, $\pi/2 + 2\pi k$; $\pi/2$, π
9. $128.17° + 360°k$, $231.83° + 360°k$; $128.17°$, $231.83°$
11. $\pi/4 + \pi k$, $3\pi/4 + \pi k$; $\pi/4$, $3\pi/4$, $5\pi/4$, $7\pi/4$

13. $-\pi/2 + 2\pi k$; $3\pi/2$;
15. πk, $\pi/4 + \pi k$; 0, $\pi/4$, π, $5\pi/4$
17. $38.17° + 360°k$, $141.83° + 360°k$; $38.17°$, $141.83°$
19. $\pi/4 + \pi k$; $\pi/4$, $5\pi 4$
21. $9° + 180°k$, $171° + 180°k$; $9°$, $171°$, $189°$, $351°$
23. $71.57° + 360°k$, $251.57° + 360°k$; $71.57°$, $251.57°$
25. πk, $\pi/2 + \pi k$; 0, $\pi/2$, π, 3 $\pi/2$
27. $2\pi k$, $3\pi/2 + 2\pi k$; 0, $3\pi/2$
29. no solutions
31. $x \in (0, \pi/3) \cup (\pi, 5\pi/3)$

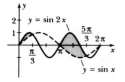

33. $x \in (0, 2\pi/3) \cup (4\pi/3, 2\pi)$

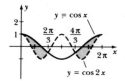

35. The curves do not intersect because the equation $F(x) = G(x)$ has no solutions.

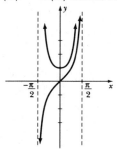

37. (a) $d(\theta) = \sqrt{4 \tan^2 \theta - 4 \tan \theta + 5}$ (b) $58.28°$, $-31.72°$

Chapter 7 Review

1. $3\pi/4$ **2.** $-\pi/3$ **3.** $2\pi/3$ **4.** undefined
5. $\frac{5}{13}$ **6.** $-2\sqrt{6}$ **7.** $1/\sqrt{1 - x^2}$
8. $\pi/18 + 2\pi k/3$, $5\pi/18 + 2\pi k/3$; $\pi/18$, $5\pi/18$, $13\pi/18$, $17\pi/18$, $25\pi/18$, $29\pi/18$
9. $-63.43° + 360°k$, $116.57° + 360°k$; $116.57°$, $296.57°$
10. K **11.** F **12.** J **13.** E **14.** B **15.** C
16. H **17.** I

18. $\dfrac{\cot \theta - \tan \theta}{\cot \theta + \tan \theta} \stackrel{?}{=} \cos (2\theta)$

$\dfrac{\dfrac{\cos \theta}{\sin \theta} - \dfrac{\sin \theta}{\cos \theta}}{\dfrac{\cos \theta}{\sin \theta} + \dfrac{\sin \theta}{\cos \theta}}$ $\quad \cos^2 \theta - \sin^2 \theta$

$\dfrac{\dfrac{\cos^2 \theta - \sin^2 \theta}{\cos^2 \theta + \sin^2 \theta}}{\cos^2 \theta - \sin^2 \theta}$

19. $\sec^2 \theta + \csc^2 \theta \stackrel{?}{=} \sec^2 \theta \csc^2 \theta$

$\dfrac{1}{\cos^2 \theta} + \dfrac{1}{\sin^2 \theta}$ $\qquad \dfrac{1}{\cos^2 \theta} \dfrac{1}{\sin^2 \theta}$

$\dfrac{\sin^2 \theta + \cos^2 \theta}{\cos^2 \theta \sin^2 \theta}$ $\qquad \dfrac{1}{\cos^2 \theta \sin^2 \theta}$

$\dfrac{1}{\cos^2 \theta \sin^2 \theta}$

20. $\tan \alpha + \tan \beta \stackrel{?}{=} \dfrac{\sin (\alpha + \beta)}{\cos \alpha \cos \beta}$

$\dfrac{\sin \alpha}{\cos \alpha} + \dfrac{\sin \beta}{\cos \beta}$ $\qquad \dfrac{\sin \alpha \cos \beta + \cos \alpha \sin \beta}{\cos \alpha \cos \beta}$

$\dfrac{\sin \alpha}{\cos \alpha} + \dfrac{\sin \beta}{\cos \beta}$

21. $\sin^2 x - \sin^2 y \stackrel{?}{=}$

$\sin (x + y) \sin (x - y)$

$(\sin x \cos y + \cos x \sin y) \times$
$(\sin x \cos y - \cos x \sin y)$

$\sin^2 x \cos^2 y - \cos^2 x \sin^2 y$

$\sin^2 x (1 - \sin^2 y) - (1 - \sin^2 x)\sin^2 y$

$\sin^2 x - \sin^2 x \sin^2 y - \sin^2 y +$
$\sin^2 x \sin^2 y$

$\sin^2 x - \sin^2 y$

22. $\csc^2 \theta - \sec^2 \theta \stackrel{?}{=} 4 \cot(2\theta) \csc (2\theta)$

$\dfrac{1}{\sin^2 \theta} - \dfrac{1}{\cos^2 \theta}$ $\qquad 4 \dfrac{\cos(2\theta)}{\sin(2\theta)} \dfrac{1}{\sin(2\theta)}$

$\dfrac{\cos^2 \theta - \sin^2 \theta}{\sin^2 \theta \cos^2 \theta}$ $\qquad \dfrac{4 \cos(2\theta)}{\sin^2(2\theta)}$

$\dfrac{\cos (2\theta)}{\sin^2 \theta \cos^2 \theta}$ $\qquad \dfrac{4 \cos(2\theta)}{(2 \sin \theta \cos \theta)^2}$

$\qquad \dfrac{\cos(2\theta)}{\sin^2\theta \cos^2\theta}$

23. $(-1 - \sqrt{3})/2\sqrt{2}$ **24.** $13/5\sqrt{10}$

25. $\sqrt{(2 - \sqrt{2})/(2 + \sqrt{2})}$ **26.** $-^{24}\!/_7$

27. $(\frac{3}{4})(\cos x + (\frac{1}{4}) \cos(3x)$

28. $(\frac{3}{4}) \sin x - (\frac{1}{4}) \sin(3x)$

29. $(3 \tan x - \tan^3 x)/(1 - 3 \tan^2 x)$

30. $-\pi/6 + 2\pi k, 7\pi/6 + 2\pi k; 7\pi/6, 11\pi/6$

31. $67.54° + 360°k, 292.46° + 360°k; 67.54°, 292.46°$

32. $\pi/2 + 2\pi k, \pi + 2\pi k; \pi/2, \pi$ **33.** no solution

34. (a) $1/\sin x$ (b) $\sin(x + \pi/2)$ (c) $1/\sin(x + \pi/2)$
(d) $\sin x/\sin(x + \pi/2)$ (e) $\sin(x + \pi/2)/\sin x$

35. (a) false (b) false (c) false (d) true (e) false (f) true

Exercises 8.1

1. $x = 10 \cos 25° \approx 9.06, y = 10 \sin 25° \approx 4.23$

3. $x = 20/\tan 53° \approx 15.07, y = 20/\sin 53° \approx 25.04$

5. $x = 20 \sin 33° \approx 10.89, y = 20 \cos 33° \approx 16.77$

7. $B = 61.72°, BC \approx 13.45, AB \approx 28.39$

9. $A = 8.65°, AC \approx 52, AB \approx 52.59$

11. 58 feet **13.** 19.5 feet **15.** 87.16 feet

17. base ≈ 15.47 centimeters, altitude ≈ 12.05 centimeters

19. 3.49 centimeters **21.** 850.69 feet

23. 31.81 feet per second **25.** 260.54 **27.** 179.55

29. 167.05 **31.** 15,705.71 feet

33. $x \approx 10.41, y \approx 3.79$ **35.** $x \approx 8.26, y \approx 9.85$

37. 2239.76 feet **39.** 5.75 **41.** height = y-coordinate of $B = r \sin \theta$; base $= OA = r$; area = $\frac{1}{2} (r \sin \theta)r = \frac{1}{2} r^2 \sin \theta$. **43.** 46.12 feet

45. $50 \sin \theta/(\cos \theta + \sin \theta)$

47. $5a^2 \cos 54° (\sin 54° - \cos 54° \tan 36°) \approx 1.1226 \, a^2$

Exercises 8.2

1. ASA: $C = 35°, a = 133.56, b = 173.68$

3. SAS: $c = 207.45, A = .3640$ radians, $B = .2776$ radians

5. SSS: $A = 52.41°, B = 29.69°, C = 97.9°$

7. SSA, two possible triangles: $A = 29.23°, B = 131.77°, b = 45.82$ or $A = 150.77°, B = 10.23°, b = 10.91$

9. SSA: $C = 3.71°, B = 161.29°, b = 49.58$

11. ASA: $C = 61°, c = 4\sqrt{5}, a = 9.27, b = 8.27$

13. $x = 10.41, y = 3.79$ **15.** 199.96 feet

17. $B = 101.81°$ **19.** 3226.05 square feet

21. (a) $d = 4\sqrt{2 - 2 \cos \theta}$ (b) $\theta = \pi/3 + 2\pi k$ or $5\pi/3 + 2\pi k, k \in \mathbf{Z}$ (c) $\theta = 97.18° + 360°k$ or $262.82° + 360°k, k \in \mathbf{Z}$ (d) $d(t) = 4\sqrt{2 - 2 \cos (120 \pi t)}$

23. (a) $x = 3 \cos \theta + 3\sqrt{\cos^2 \theta + 15}$ (b) $\theta = 82.82° + 360°k$ or $277.18° + 360°k, k \in \mathbf{Z}$ (c) $x = 3 \cos(8\pi t) + 3\sqrt{\cos^2 (8\pi t) + 15}$

Exercises 8.3

1. (a) $z = 2(\cos (\pi/3) + i \sin(\pi/3))$
(b) $z = \sqrt{2}(\cos(7\pi/4) + i \sin(7\pi/4))$

3. (a) $z = 6(\cos 0 + i \sin 0)$ (b) $z = 6(\cos \pi + i \sin \pi)$

4. $z = 4(\cos(3\pi/2) + i \sin(3\pi/2))$

7. i **9.** -64 **11.** $-256\sqrt{2} - 256\sqrt{2}i$

13. $(1 - i)/8$

15. $3\sqrt{3}/2 + 3i/2, -3\sqrt{3}/2 + 3i/2, -3i$

Answers to Odd-Numbered Exercises ▪ **669**

17. $\frac{1}{2} \pm \sqrt{3}i/2$, $-\frac{1}{2} \pm \sqrt{3}i/2$, 1, -1

19. $\sqrt{2} \pm \sqrt{2}i$, $-\sqrt{2} \pm \sqrt{2}i$

21. $\sqrt[3]{5}$, $\sqrt[3]{5}(-\frac{1}{2} \pm \sqrt{3}i/2)$

23. 0, 3, -3, $\frac{3}{2} \pm 3\sqrt{3}i/2$, $-\frac{3}{2} \pm 3\sqrt{3}i/2$

25. Let $z = r(\cos\theta + i\sin\theta)$. We have

$$
\begin{array}{c|c}
\overline{(z^n)} & \overset{?}{=} \quad (\bar{z})^n \\
\hline
\overline{[r(\cos\theta + i\sin\theta)]^n} & (r\cos\theta + ir\sin\theta)^n \\
& (r\cos\theta - ir\sin\theta)^n \\
\overline{r^n(\cos(n\theta) + i\sin(n\theta))} & [r\cos(-\theta) + i\sin(-\theta)]^n \\
\overline{r^n\cos(n\theta) + ir^n\sin(n\theta)} & r^n(\cos(-n\theta) + i\sin(-n\theta)) \\
r^n\cos(n\theta) - ir^n\sin(n\theta) & r^n\cos(n\theta) - ir^n\sin(n\theta)
\end{array}
$$

(c)

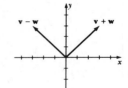

$\theta = 315°$ $\|-\frac{1}{2}\mathbf{v}\| = 2$

(d)

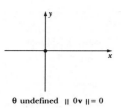

θ undefined $\|0\mathbf{v}\| = 0$

(e)

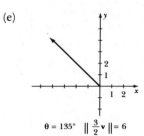

$\theta = 135°$ $\|\frac{3}{2}\mathbf{v}\| = 6$

Exercises 8.4

1. five different vectors; $\mathbf{b} = \mathbf{f}$, $\mathbf{d} = \mathbf{g}$

3.

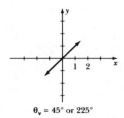

5.

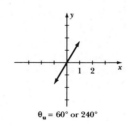

7.

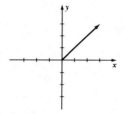

$\theta_\mathbf{v} = 45°$ or $225°$

9.

$\theta_\mathbf{u} = 60°$ or $240°$

11. (a)

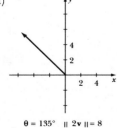

$\theta = 135°$ $\|2\mathbf{v}\| = 8$

(b)

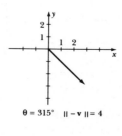

$\theta = 315°$ $\|-\mathbf{v}\| = 4$

13.

	$\mathbf{v} + \mathbf{w}$	$\mathbf{v} - \mathbf{w}$
direction of angle	$45°$	$135°$
magnitude	$3\sqrt{2}$	$3\sqrt{2}$

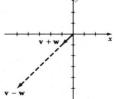

15.

	$\mathbf{v} + \mathbf{w}$	$\mathbf{v} - \mathbf{w}$
direction of angle	$63.4°$	$21.8°$
magnitude	$\sqrt{5}$	$\sqrt{29}$

17.

	$\mathbf{v} + \mathbf{w}$	$\mathbf{v} - \mathbf{w}$
direction of angle	$225°$	$225°$
magnitude	$\sqrt{2}$	$5\sqrt{2}$

19. $5\sqrt{3}$, 40°   **21.** 13, 67.38°   **23.** $2\sqrt{109}$, 343.3°

25. equal magnitudes, opposite directions

27. false   **29.** true

Exercises 8.5

1. $\langle 82.9, 55.92 \rangle$ **3.** $\langle -7, -7\sqrt{3} \rangle$ **5.** $\langle 0, 10 \rangle$

7. (a) 5; 36.87° (b) 2.5; 253.74°

9. (a) $5\sqrt{2}$; 315° (b) 8; 270°

11. (a) $\langle 5, -5\sqrt{3} \rangle$ (b) $\langle -50, 50\sqrt{3} \rangle$

13. $\langle \frac{1}{2}, -\sqrt{3}/2 \rangle$ **15.** $\langle 3, 6 \rangle$; $3\sqrt{5}$; 63.43°

17. $\langle 6, -3 \rangle$; $3\sqrt{5}$; 333.43°

19. $\langle -10 + 5\sqrt{3} - 15\sqrt{2}/2, 5 + 10\sqrt{3} - 15\sqrt{2}/2 \rangle$; 16.73, 135.56°

21. $\langle -2, 0 \rangle$; 2, 180°

23. (a) $\langle -\frac{3}{5}, -\frac{4}{5} \rangle$ (b) $\langle \frac{3}{5}, \frac{4}{5} \rangle$

25. (a) $\langle 3/\sqrt{13}, 2/\sqrt{13} \rangle$ (b) $\langle -3/\sqrt{13}, -2/\sqrt{13} \rangle$

27. (a) $\langle 0, -1 \rangle$ (b) $\langle 0, 1 \rangle$

29. $\langle 1, 1 \rangle$, $\sqrt{2}$, 45°   **31.** $\langle 2, 0 \rangle$, 2, 0°

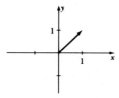
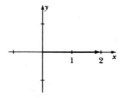

33. $\langle -2, -5 \rangle$ **35.** $\sqrt{5}$; 26.57° ($\mathbf{QT} = \langle 2,1 \rangle$)

37. $P(5, \frac{20}{3})$ **39.** $P(0, \frac{15}{4})$

41. (a) $P((2c + a)/3, (2d + b)/3)$ (b) $((a + c)/2, (b + d)/2)$

43. $\|\mathbf{u}\| = \|\mathbf{v}/\|\mathbf{v}\|\| = \|\mathbf{v}\|/\|\mathbf{v}\| = 1$ as desired

Chapter 8 Review

1. 143.37 feet

2. 20.23

3. 59.4 feet

4. 47.48

5. $x = 12 \tan \beta/(\tan \beta - \tan \alpha)$;
$y = 12 \tan \alpha \tan \beta/(\tan \beta - \tan \alpha)$

6. 11.52

7. $A = 48.37°$, $B = 23.2°$, $C = 108.43°$

8. 82.7°   **9.** 54.52 or 22.08

10. (a) $2\sqrt{2}(\cos(3\pi/4) + i\sin(3\pi/4))$
(b) $3(\cos(3\pi/2) + i\sin(3\pi/2))$

11. (a) $512i$ (b) $-\frac{1}{2} + \sqrt{3}i/2$

12. $1, -\frac{1}{2} \pm \sqrt{3}i/2$

13. $\sqrt{3}/2 \pm i/2, \pm i, -\sqrt{3}/2 \pm i/2$

14. (a) 12, 45°   (b) 6, 225°

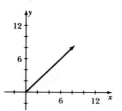
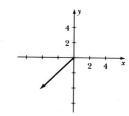

(c) 3,225°

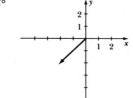

15.

	v + w	v − w
direction of angle	78.69°	188.13°
magnitude	$\sqrt{26}$	$5\sqrt{2}$

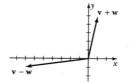

16. $\langle -10\sqrt{3}, 10 \rangle$ **17.** 17; 331.93°

18. (a) $\langle -11, 0 \rangle$; 11; 180° (b) $\langle -2/\sqrt{5}, 1/\sqrt{5} \rangle$; 1; 153.43°
(c) $\langle -5/\sqrt{34}, -3/\sqrt{34} \rangle$; 1; 210.96°

19. $\langle 5 + 5\sqrt{2} - 10\sqrt{3}, 10 + 5\sqrt{2} - 5\sqrt{3} \rangle$; 9.91; 121.97°

20. (a) $\langle 9, 3 \rangle$ (b) $\langle 3, 5 \rangle$ (c) $\langle 6, -2 \rangle$ **21.** $\langle 7, -4 \rangle$

Exercises 9.1

1. $(3, -1)$ **3.** $(-\frac{2}{41}, -\frac{75}{41})$

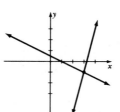
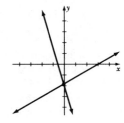

5. (a) no solution

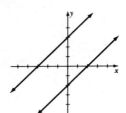

(b) same line, infinite number of solutions

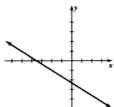

7. (0, 0), (3, 9)

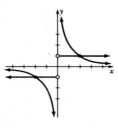

9. $(-1/\sqrt{2}, 1/\sqrt{2})$, $(1/\sqrt{2}, -1/\sqrt{2})$

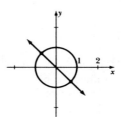

11. $(-2, -1)$, $(2, 1)$

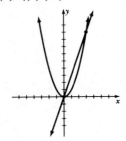

13. $(0, -2)$, $(\,^{32}/_{17}, \,^{30}/_{17})$
15. $(1, 0)$, $(-1, 0)$
17. $(5, 0)$, $(-5, 0)$, $(0, -3)$
19. $(1, \pm 2)$, $(-1, \pm 2)$
21. $(\ln 2/2, 1/\sqrt{2})$
23. $(0, -2)$
25. $(e^{1/2}, e^{1/2})$
27. $(\,^{1}/_{3}, -\,^{7}/_{9})$, $(3, 1)$
29. $(1, \pm 4)$, $(-1, \pm 4)$
31. $(\,^{21}/_{8}, -14)$ **33.** $(\,^{9}/_{14}, \,^{16}/_{7})$
35. $((2 + \sqrt{29})/5, (9 + 2\sqrt{29})/5)$,
       $((2 - \sqrt{29})/5, (9 - 2\sqrt{29})/5)$ **37.** $(4, 0)$
39. $(4 - \sqrt{6}, -1 + \sqrt{6})$, $(4 + \sqrt{6}, -1 - \sqrt{6})$
41. $(-\,^{20}/_{3}, \,^{35}/_{3})$ **43.** $(4, 7)$, $(7, 4)$
45. $x = bc \tan \beta/(a \tan \alpha - b \tan \beta)$
       $y = abc \tan \alpha \tan \beta/(a \tan \alpha - b \tan \beta)$

Exercises 9.2

1. (a)

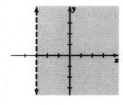

(b)

3. (a)

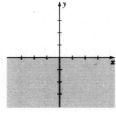

(b)

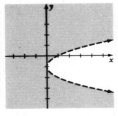

5.

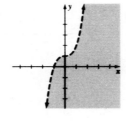

7.

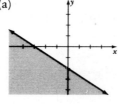

9.

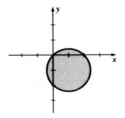

11.

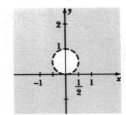

13.

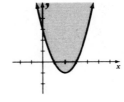

15. (a) 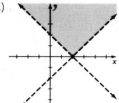   (b)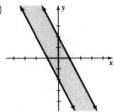

(c)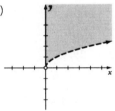

17. (a)   (b)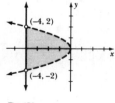

19.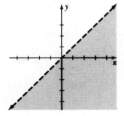

21.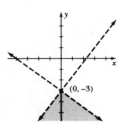

Fig. 459

23.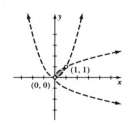

25. ...

27. $\left(\dfrac{-3+\sqrt{5}}{2},\ \sqrt{\dfrac{-1+\sqrt{5}}{2}}\right)$

$\left(\dfrac{-3+\sqrt{5}}{2},\ -\sqrt{\dfrac{-1+\sqrt{5}}{2}}\right)$

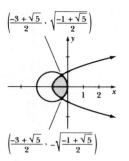

29.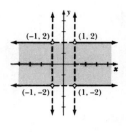

(−1, 2)   (1, 2)

(−1, −2)   (1, −2)

31.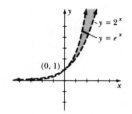

$y = 2^x$

$y = e^x$

(0, 1)

33.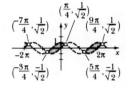

$\left(\dfrac{-7\pi}{4},\ \dfrac{1}{\sqrt{2}}\right)$ $\left(\dfrac{\pi}{4},\ \dfrac{1}{\sqrt{2}}\right)$ $\left(\dfrac{9\pi}{4},\ \dfrac{1}{\sqrt{2}}\right)$

−2π   2π

$\left(\dfrac{-3\pi}{4},\ \dfrac{-1}{\sqrt{2}}\right)$ $\left(\dfrac{5\pi}{4},\ \dfrac{-1}{\sqrt{2}}\right)$

35.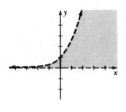

(1, 1)

(−1, −1)   (2, 0)

37.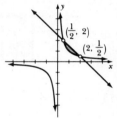

$\left(\dfrac{1}{2},\ 2\right)$

$\left(2,\ \dfrac{1}{2}\right)$

39.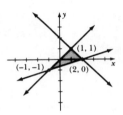

(10, log₂10)

(10, 1)

(1, 0)

41. (a) 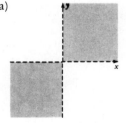   (b)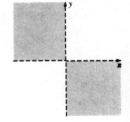

43.

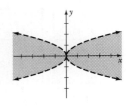

45.

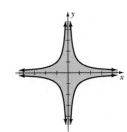

59. x = regular amount
y = fine amount
$x \geq 0, y \geq 0$
$x + y \geq 30$
$2x + 3y \leq 80$

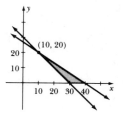

47.

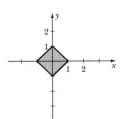

49.

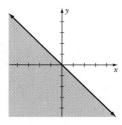

1. $(-11, 16, 4)$

3. (a) $x - y + 2z = 3$
$x - 2y + z = 4$
$3y + 2z = -6$

(b) $\begin{bmatrix} 1 & -1 & 2 & 3 \\ 0 & -1 & -1 & 1 \\ 0 & 0 & -1 & -3 \end{bmatrix}$

5. (a) $2y + z = 2$
$x \quad - z = 1$
$2x - 3y + z = 4$

(b) $\begin{bmatrix} 1 & 0 & -1 & 1 \\ 0 & -1 & 4 & 4 \\ 0 & 0 & 9 & 10 \end{bmatrix}$

7. $(4, -2, 1)$ **9.** $(-\frac{14}{5}, -\frac{51}{5}, 11)$ **11.** $(\frac{1}{2}, \frac{1}{4}, \frac{1}{3})$
13. $(-4, -13, 8)$ **15.** $(-1, -2, 3)$
17. $(-\frac{20}{3}, \frac{1}{3}, 3\frac{1}{3})$ **19.** $(\frac{18}{7}, \frac{2}{7}, -\frac{5}{7})$
21. $(-\frac{2}{3}, \frac{5}{4}, -\frac{1}{2})$ **23.** $(\frac{3}{4}, -\frac{1}{4}, \frac{1}{2})$
25. $Q(x) = x^2/2 - x + 3$
27. $F(x) = -3x^2 + 13x - 11$ **29.** 375

51.

53.

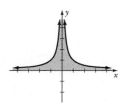

1. no solution
3. $x = (7 + 5t)/3, y = (-5 - 7t)/3, z = t, t \in \mathbf{R}$
5. no solution   **7.** $x = 3, y = t - 1, z = t, t \in \mathbf{R}$
9. $(\frac{1}{6}, \frac{1}{6}, \frac{1}{6})$
11. $x = 4 + t, y = -2 - 3t, z = t, t \in \mathbf{R};$
$(4, -2, 0), (5, -5, 1)$
13. $x = (2 - t)/2, y = (5t - 14)/4, z = t, t \in \mathbf{R};$
$(\frac{1}{2}, -\frac{9}{4}, 1), (0, -1, 2)$
15. $(5, 4, 3, 2)$ **17.** $(-2, 1, 1, \frac{1}{2})$ **19.** $A = \frac{2}{5}, B = \frac{3}{5}$
21. $A = -1, B = 1$ **23.** $A = \frac{3}{2}, B = \frac{1}{2}, C = -1$
25. $A = -\frac{1}{4}, B = \frac{1}{4}, C = 1, D = -\frac{1}{2}$
27. $A = -1, B = 1, C = 1, D = 1$
29. $P_1(3, 12), P_2(12, -3)$

55. x = width, y = length
$x \geq 0, y \geq 0$
$x + y \leq 120$
$x^2 + y^2 \leq 10,000$

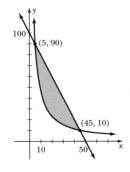

57. x = width, y = length
$x \geq 0, y \geq 0$
$xy \geq 450$
$2x + y \leq 100$

1. (a) 3 (b) 2 (c) 3×2 (d) 5 (e) -2
3. (a) $\begin{bmatrix} 3 & 0 \\ 3 & 3 \end{bmatrix}$ (b) $\begin{bmatrix} 2 & 1 \\ 1 & 2 \end{bmatrix}$ (c) $\begin{bmatrix} 1 & 2 \\ -1 & 1 \end{bmatrix}$ (d) $\begin{bmatrix} 1 & 1 \\ 1 & 2 \end{bmatrix}$

5. (a) $\begin{bmatrix} 3 & -6 & 0 \\ -9 & 3 & 6 \end{bmatrix}$ **(b)** $\begin{bmatrix} 5 & 3 & -6 \\ -1 & 0 & 3 \end{bmatrix}$

(c) $\begin{bmatrix} 7 & 12 & -12 \\ 7 & -3 & 0 \end{bmatrix}$ **(d)** undefined

7. (a) $\begin{bmatrix} 6 & 9 \\ -3 & 6 \\ 0 & 0 \end{bmatrix}$ **(b)** undefined **(c)** undefined

(d) $\begin{bmatrix} 11 & 16 \\ 5 & 6 \\ 0 & 0 \end{bmatrix}$

9. $AB = \begin{bmatrix} -11 & -4 & 0 \\ -2 & 2 & 5 \\ -9 & 0 & 6 \end{bmatrix}$ $BA = \begin{bmatrix} -6 & 11 \\ 6 & 3 \end{bmatrix}$

11. $AB = \begin{bmatrix} 1 & 2 & 1 \\ 2 & 0 & 2 \\ 1 & 2 & 1 \end{bmatrix} = BA$

13. (a) $\begin{bmatrix} 1 & 2 \\ 0 & 1 \end{bmatrix}$ **(b)** $\begin{bmatrix} 1 & 3 \\ 0 & 1 \end{bmatrix}$ **(c)** $\begin{bmatrix} 1 & n \\ 0 & 1 \end{bmatrix}$

15. (a) $\begin{bmatrix} a^2 & 0 & 0 \\ 0 & b^2 & 0 \\ 0 & 0 & c^2 \end{bmatrix}$ **(b)** $\begin{bmatrix} a^3 & 0 & 0 \\ 0 & b^3 & 0 \\ 0 & 0 & c^3 \end{bmatrix}$ **(c)** $\begin{bmatrix} a^n & 0 & 0 \\ 0 & b^n & 0 \\ 0 & 0 & c^n \end{bmatrix}$

17. (a) $\begin{bmatrix} 2 & 0 & 0 & 2 \\ 0 & 1 & 0 & 0 \\ 0 & 0 & 1 & 0 \\ 2 & 0 & 0 & 2 \end{bmatrix}$ **(b)** $\begin{bmatrix} 4 & 0 & 0 & 4 \\ 0 & 1 & 0 & 0 \\ 0 & 0 & 1 & 0 \\ 4 & 0 & 0 & 4 \end{bmatrix}$

(c) $\begin{bmatrix} 2^{n-1} & 0 & 0 & 2^{n-1} \\ 0 & 1 & 0 & 0 \\ 0 & 0 & 1 & 0 \\ 2^{n-1} & 0 & 0 & 2^{n-1} \end{bmatrix}$

19. (a) -8 **(b)** -2 **21.** -168 **23.** 51 **25.** -2
27. 4 **29.** $abcd$ **31.** 924 **33.** 1
35. $(y - x)(z - x)(z - y)$

37. Let $A = \begin{bmatrix} a_1 & a_2 \\ a_3 & a_4 \end{bmatrix}$, $B = \begin{bmatrix} b_1 & b_2 \\ b_3 & b_4 \end{bmatrix}$. Then $\det(AB) =$

$\det \begin{bmatrix} a_1b_1 + a_2b_3 & a_1b_2 + a_2b_4 \\ a_3b_1 + a_4b_3 & a_3b_2 + a_4b_4 \end{bmatrix} = a_1a_3b_1b_2 +$

$a_1a_4b_1b_4 + a_2a_3b_2b_3 + a_2a_4b_3b_4 - a_1a_3b_1b_2 -$
$a_2a_3b_1b_4 - a_1a_4b_2b_3 - a_2a_4b_3b_4 = a_1a_4b_1b_4 +$
$a_2a_3b_2b_3 - a_2a_3b_1b_4 - a_1a_4b_2b_3 =$
$(a_1a_4 - a_2a_3)(b_1b_4 - b_2b_3) = \det(A) \det(B).$

39. Let $A = \begin{bmatrix} a & b \\ c & d \end{bmatrix}$. Then $\det(rA) = \det \begin{bmatrix} ra & rb \\ rc & rd \end{bmatrix} =$
$r^2ad - r^2bc = r^2(ad - bc) = r^2 \det(A).$

41. $DEI = DIE = IDE = \begin{bmatrix} \sin(2\theta) & 1 \\ 0 & \cos(2\theta) \end{bmatrix}$,

$\det(DEI) = \sin(2\theta)\cos(2\theta)$

43.
$rD = \begin{bmatrix} \dfrac{\sin\theta}{\sin 2\theta} & \dfrac{\cos\theta}{\sin 2\theta} \\ -\dfrac{\sin\theta}{\sin 2\theta} & \dfrac{\cos\theta}{\sin 2\theta} \end{bmatrix}$, $\det(rD) = \dfrac{1}{\sin(2\theta)}$

Chapter 9 Review

1. $(12, -10)$ **2.** $(4, \pm 3\sqrt{3}), (-4, \pm 3\sqrt{3})$
3. $(5/2, -1/6)$

4.

5.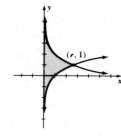

6.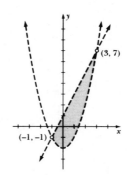

7.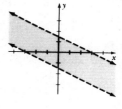

8.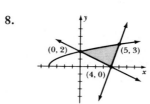

9. $x = $ width, $y = $ length
$x \geq 0, y \geq 0$
$xy \geq 12$
$x^2 + y^2 \leq 25$

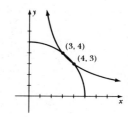

10. $(^{13}/_{22}, {}^{17}/_{22}, ^{1}/_{2})$ **11.** no solution
12. $x = 24 - 51t, y = -5 + 11t, z = t, t \in \mathbf{R}$
13. $(^{17}/_{4}, -^{33}/_{4}, 0, ^{5}/_{4})$ **14.** $F(x) = -2x^2 + x/3 - ^{2}/_{3}$
15. $A = {}^{14}/_{9}, B = -^{5}/_{3}, C = ^{8}/_{9}$
16. (a) $\begin{bmatrix} 15 & -12 & -9 \\ -3 & -6 & 3 \end{bmatrix}$ (b) $\begin{bmatrix} -7 & 0 \\ 4 & 14 \end{bmatrix}$

17. (a) $\begin{bmatrix} -3 & -12 \\ 2 & -20 \end{bmatrix}$ (b) $\begin{bmatrix} -3 & 8 & 1 \\ -6 & -12 & 6 \end{bmatrix}$

18. (a) $\begin{bmatrix} -3 & -36 & 3 \\ -30 & -32 & 26 \end{bmatrix}$ (b) undefined

19. (a) -14 (b) undefined
20. (a) $\begin{bmatrix} 10 & 7 & 3 \\ 12 & 12 & 6 \\ 9 & 9 & 9 \end{bmatrix}$ (b) same as part a

21. (a) 27 (b) 6 (c) 162
22. (a) $\begin{bmatrix} 1 & 0 & 0 \\ 0 & 1 & 0 \\ 0 & 0 & 1 \end{bmatrix}$ (b) same as part a (c) 1

23. 462   **24.** -4 **25.** true

Exercises 10.1

Exercise	Axis	Vertex	Focus	Directrix
1.	$x = 0$	$(0, 0)$	$(0, -^{1}/_{4})$	$y = ^{1}/_{4}$
3.	$y = 0$	$(0, 0)$	$(1, 0)$	$x = -1$
5.	$y = 0$	$(0, 0)$	$(-^{5}/_{2}, 0)$	$x = ^{5}/_{2}$

1. 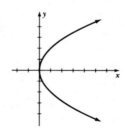   **3.**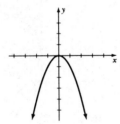

5.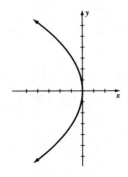

7. $y^2 = -12x$ **9.** $x^2 = 8y$ **11.** $y^2 = 16x/3$
13. $x^2 = -3y/4$ **15.** $(0, 0), (4, \pm 4)$ **17.** $(\pm 2\sqrt{2}, 2)$
19. $8x + 6y = 3$ **21.** (a) $x\sqrt{x}/2$ (b) $y^3/2$
23. 32 inches from vertex   **25.** $8\sqrt{70}$ feet
27. $8(\sqrt{2} - 1)$
29. (a) $PF = |x^2/4p + p|$ (b) PF is minimum when $x = 0$
at point $P(0, 0)$.
Proof: The function $F(x) = |x^2/4p + p|$ is an upward
curving parabola with minimum value occurring at its
vertex when $x = 0$.

Exercises 10.2

Exercise	Major axis	Minor axis	Vertices	Foci
1.	$y = 0$	$x = 0$	$(\pm 3, 0)\ (0, \pm 2)$	$(\pm\sqrt{5}, 0)$
3.	$x = 0$	$y = 0$	$(\pm^{1}/_{2}, 0)\ (0, \pm 1)$	$(0, \pm\sqrt{3}/2)$
5.	$y = 0$	$x = 0$	$(\pm 12, 0)\ (0, \pm 4)$	$(\pm 8\sqrt{2}, 0)$

1. 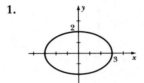   **3.**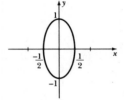

5.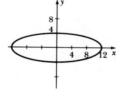

7. foci $(0, 0)$

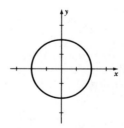

9. (a) point $(0, 0)$ (b) empty set

11. (a) 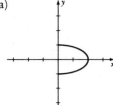 (b)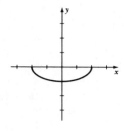

5.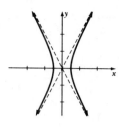

13.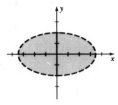

15.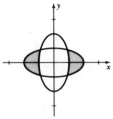

7. (a) asymptotes:
$y = \pm 2x/3$

(b) asymptotes:
$y = \pm 2x/3$

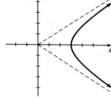

17. $x^2/25 + y^2/16 = 1$ **19.** $9x^2/100 + y^2/100 = 1$
21. $x^2/16 + y^2/7 = 1$ **23.** $y^2/6 + x^2/3 = 1$
25. (a) $8x\sqrt{9 - x^2}/3$ (b) $6y\sqrt{4 - y^2}$
   (c) $4x + 8\sqrt{9 - x^2}/3$ (d) $6\sqrt{4 - y^2} + 4y$
27. $8\sqrt{15}$ feet
29. $16x^2/49R_E^2 + y^2/3R_E^2 = 1$; 13,856 miles
31. $4x^2/25R_E^2 + 4y^2/9R_E^2 = 1$; at focus, $2R_E \approx 8000$ miles
   from the center of orbit along major axis.
33. $2b^2/a$ **35.** $(0, 1), (^{60}/_{19}, -^{1}/_{19})$
37. (a) In triangle PF_1F_2, we know that $PF_1 + PF_2 > F_1F_2$.
   Now, $PF_1 + PF_2 = 2a$ and $F_1F_2 = 2c$. Thus $2a > 2c$.
   (b) $2a > 2c$ implies that $a > c$ implies that $a^2 > c^2$
   (a and c are positive) implies that $a^2 - c^2 > 0$.

9. (a) asymptotes:
$y = \pm x$

(b) asymptotes:
$y = \pm x$

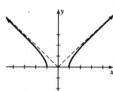

Exercises 10.3

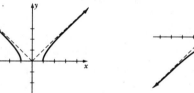

Exercise	Major axis	Minor axis	Vertices	Foci	Asymptotes
1.	$y = 0$	$x = 0$	$(\pm 6, 0)$	$(\pm 10, 0)$	$y = \pm 4x/3$
3.	$x = 0$	$y = 0$	$(0, \pm\sqrt{2})$	$(0, \pm\sqrt{3})$	$y = \pm\sqrt{2}x$
5.	$y = 0$	$x = 0$	$(\pm\frac{1}{2}, 0)$	$(\pm\sqrt{5}/2, 0)$	$y = \pm 2x$

11.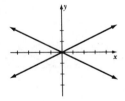

13.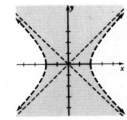

1.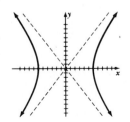

3.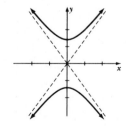

15.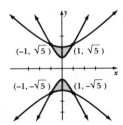

$(-1, \sqrt{5})$ $(1, \sqrt{5})$

$(-1, -\sqrt{5})$ $(1, -\sqrt{5})$

17. $x^2 - 15y^2/64 = 1$ **19.** $y^2/16 - x^2/9 = 1$
21. $x^2/9 - 4y^2/9 = 1$ **23.** $y^2/16 - x^2/9 = 1$
25. $x^2/100 - y^2/16 = 1$
27. $x^2/4 - y^2/12 = 1$; yes, foci $(\pm 4, 0)$
29. (a) $4bx\sqrt{x^2 - a^2}/a$ (b) $4ay\sqrt{y^2 + b^2}/b$
     (c) $4x + 4b\sqrt{x^2 - a^2}/a$ (d) $4a\sqrt{y^2 + b^2}/b + 4y$

---

17. $(x - 4)^2/9 + (y + 1)^2/4 = 1$
19. $(x - 3)^2/4 - (y - 3)^2/16 = 1$
21. $(x + 5)^2 = -16(y + 2)$
23. parabola: $(y - 6)^2 = -8(x - 1)$
25. hyperbola: $(y - 9)^2/9 - (x + 1)^2/72 = 1$
27. ellipse: $(x - 6)^2/16 + (y - 1)^2/12 = 1$
29. ellipse: $y^2/16 + (x - 1)^2/12 = 1$
31. circle: center $(0, 7)$, radius 2
33. point: $(1, 2)$ **35.** two lines: $y - 2 = \pm(x - 1)$
37. (a) $x = a\sqrt{y^2 + b^2}/b$ (b) $y^2 = 4p(x - a)$ (c) no

Exercises 10.4

1. hyperbola   **3.** ellipse   **5.** parabola

7. center: $(2, 1)$
axes: major $y = 1$,
minor $x = 2$
vertices: $(0, 1)(4, 1)(2, 0)(2, 2)$
foci: $(2 \pm \sqrt{3}, 1)$

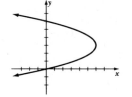

9. center: $(-3, 0)$
axes: major $y = 0$,
minor $x = -3$
vertices: $(-6, 0)$ $(0, 0)$
foci: $(-3 \pm \sqrt{10}, 0)$
asymptotes: $y = \pm(x + 3)/3$

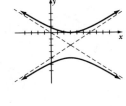

11. axis: $y = 3$
vertex: $(6, 3)$
focus: $(^{45}/_8, 3)$
directrix: $x = {}^{51}/_8$

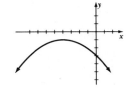

13. center: $(3, -2)$
axes: major $x = 3$,
minor $y = -2$
vertices: $(3, 0)(3, -4)$
foci: $(3, -2 \pm \sqrt{13})$
asymptotes:
$y + 2 = \pm 2(x - 3)/3$

15. axis: $x = -4$
vertex: $(-4, -1)$
focus: $(-4, -3)$
directrix: $y = 1$
"Fig. 511"

Exercises 10.5

1. $2xy = 1$ **3.** $7x^2 - 6\sqrt{3}xy + 13y^2 = 16$
5. $x^2 + y^2 - 2x - 2y = 0$ **7.** $-\sqrt{2}y' = 1$
9. $x'^2 - y'^2 = 2$ **11.** $\sqrt{3}x' + y' = 2$
13. (a) hyperbola: $x'^2/4 - y'^2 = 1$
     (b) vertices: $(\pm 2, 0)$ foci: $(\pm\sqrt{5}, 0)$
     (c)

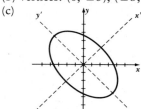

15. (a) ellipse: $y'^2/25 + x'^2/9 = 1$
     (b) vertices: $(0, \pm 5)$, $(\pm 3, 0)$ foci: $(0, \pm 4)$
     (c)

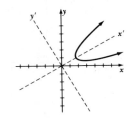

17. (a) parabola: $y'^2 = x' - 2$
     (b) vertex: $(2, 0)$ focus: $(^9/_4, 0)$
     (c)

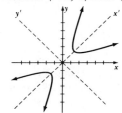

19. (a) hyperbola: $(x' + \sqrt{2})^2/4 - y'^2/4 = 1$
(b) vertices: $(-\sqrt{2} \pm 2, 0)$ foci: $(-\sqrt{2} \pm 2\sqrt{2}, 0)$
(c) .

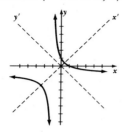

21. using $\alpha = -30°$: $(y' - 2)^2/4 + x'^2 = 1$ vertices
$(x, y\text{-coordinates})$: $(0, 0)$, $(2, 2\sqrt{3})$, $((\sqrt{3} + 2)/2,$
$(2\sqrt{3} - 1)/2)$, $((2 - \sqrt{3})/2, (2\sqrt{3} + 1)/2)$

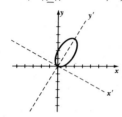

23. hyperbola: $(\sqrt{2} - 1)y'^2/2 - (\sqrt{2} + 1)x'^2/2 = 1$

Exercises 10.6 _____

1.

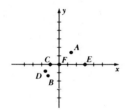

3. $A(\sqrt{2}, \sqrt{2})$, $B(-\sqrt{2}, -\sqrt{2})$, $C(-1, 0)$, $D(-\sqrt{3}, -1)$,
$E(3, 0)$, $F(0, 0)$
5. $A(3\sqrt{2}, 3\pi/4)$, $B(2, \pi)$, $C(\sqrt{41}, \tan^{-1}(-5/4)) \approx$
$(\sqrt{41}, -51.34°)$, $D(\sqrt{5}\pi/2, \tan^{-1}(2)) \approx$
$(\sqrt{5}\pi/2, 63.43°)$, $E(2\sqrt{5}, \tan^{-1}(2) + \pi) \approx$
$(2\sqrt{5}, 243.43°)$
7. $k \in \mathbf{Z}$: (a) $(5, 8\pi/3)$, $(-5, 5\pi/3)$, $(5, 2\pi/3 + 2\pi k)$,
$(-5, 5\pi/3 + 2\pi k)$ (b) $(3, \pi)$, $(-3, 2\pi)$, $(3, \pi + 2\pi k)$,
$(-3, 2\pi k)$ (c) $(\sqrt{2}, 7\pi/4)$, $(-\sqrt{2}, 3\pi/4)$,
$(\sqrt{2}, 7\pi/4 + 2\pi k)$, $(-\sqrt{2}, 3\pi/4 + 2\pi k)$
9. true **11.** (a) $r^2(1 + \sin 2\theta)$ (b) $(1 - r^2)^{3/2}$ (c) $\sec^2 \theta$
13. $r = 4 \sec \theta$ **15.** $r = -2 \csc \theta$

17. $r = 3/(\cos \theta + 2 \sin \theta)$
19. (a) $r = 3$ (b) $r = 6 \cos \theta$ **21.** $r = \sin 2\theta$
23. $r = \csc \theta \cot \theta$ **25.** $|r| = e^{\tan \theta}$
27. $x^2 + (y - 1/2)^2 = 1/4$
29. $x^2 + y^2 = 1$ **31.** $x^2 = 2y + 1$ **33.** $x^2 - y^2 = 1$
35. $y = 0$
37. (a)

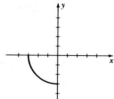

(b)

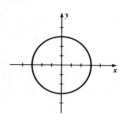

39. (a)

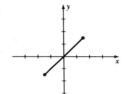

(b)

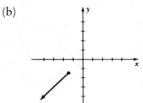

41. (a)

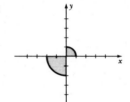

(b)

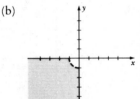

(c)

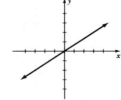

43.

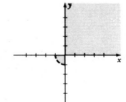

45. $y = -2$

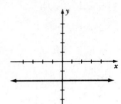

47. $\theta = \pi/4$

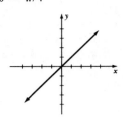

17.

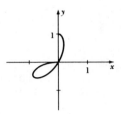

19.

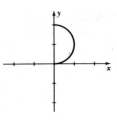

 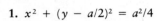
1. $x^2 + (y - a/2)^2 = a^2/4$

3. (a)

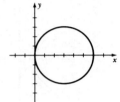

(b)

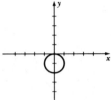

21.

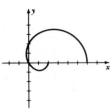

23.

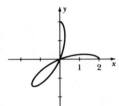

5.

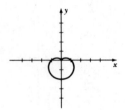

7.

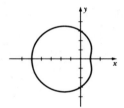

25.

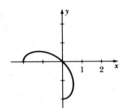

27.

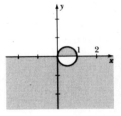

9.

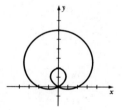

11.

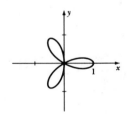

29.

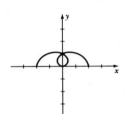

31.

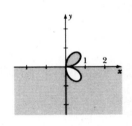

33.

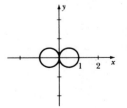

35.

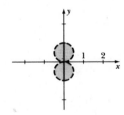

13.

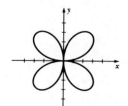

15.

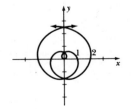

37.

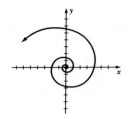

39.

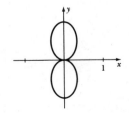

1.

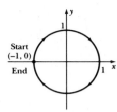
Start (-1, 0)
End

3.

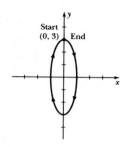
Start (0, 3) End

19.

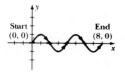
Start (0, 0) End (8, 0)

21. (a) same for each bee: $x^2 + y^2 = 4$ (b) bee 1: $t = \frac{1}{2}$, bee 2: $t = \frac{2}{3}$, bee 3: $t = \frac{4}{3}$

23. collision at (2, 4) when $t = 8$

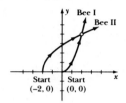
Bee I
Bee II
Start (-2, 0) Start (0, 0)

5.

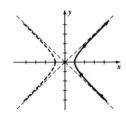
Start (4, 0)
End

7.

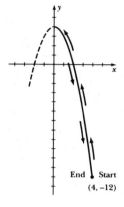
End Start (4, -12)

Chapter 10 Review

1. axis: $x = 0$
vertex: (0, 0)
focus: $(0, -\frac{1}{8})$
directrix: $y = \frac{1}{8}$

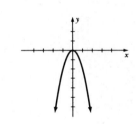

9.

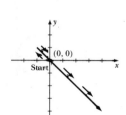

11.

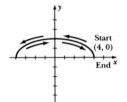
(0, 0)
Start

13.

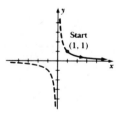
Start (1, 1)

2. center: (0, 0)
axes: major $x = 0$,
minor $y = 0$
vertices: $(0, \pm 1)$, $(\pm \frac{1}{2}, 0)$
foci: $(0, \pm \sqrt{3}/2)$

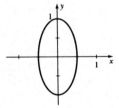

15.

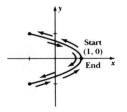
Start (1, 0)
End

17.

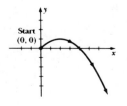
Start (0, 0)

3. center: (0, 0)
axes: major $x = 0$,
minor $y = 0$
vertices: $(0, \pm 2)$
foci: $(0, \pm \sqrt{13})$
asymptotes: $y = \pm 2x/3$

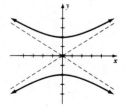

4. axis: $y = 4$
vertex: $(-6, 4)$
focus: $(-5, 4)$
directrix: $x = -7$

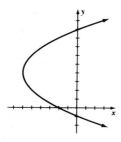

5. center: $(-3, 1)$
axes: major $y = 1$,
minor $x = -3$
vertices: $(-3 \pm 3, 1)$,
$(-3, 1 \pm 2)$
foci: $(-3 \pm \sqrt{5}, 1)$

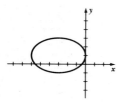

6. center: $(1, 1)$
axes: major $y = 1$,
minor $x = 1$
vertices: $(0, 1)$, $(2, 1)$
foci: $(1 \pm \sqrt{5}, 1)$
asymptotes:
$y - 1 = \pm 2(x - 1)$

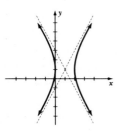

7. 16 inches from vertex   **8.** $x^2/36 + y^2/9 = 1$
9. $y^2/48 - x^2/12 = 1$
10. $(x + 3)^2/9 + (y - 6)^2/36 = 1$
11. parabola: $y^2 = 8(x - 1)$
12. ellipse: $(y - 3)^2/9 + x^2/8 = 1$
13. hyperbola: $(x + 4)^2/4 - (y - 1)^2/12 = 1$
14. $2ab$ **15.** $x^2 + 2xy + y^2 + \sqrt{2}x - \sqrt{2}y = 0$
16. ellipse: $y'^2/4 + x'^2 = 1$
vertices: $(0, \pm 2)$
foci: $(0, \pm \sqrt{5})$

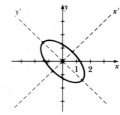

17. hyperbola: $(x' - 1)^2 - y'^2 = 1$
vertices: $(0, 0)$, $(2, 0)$
foci: $(1 \pm \sqrt{2}, 0)$

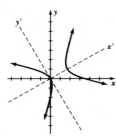

18. (a)

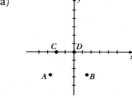

(b) $A(-2\sqrt{2}, -2\sqrt{2})$, $B(\frac{3}{2}, -3\sqrt{3}/2)$, $C(-2, 0)$,
$D(0, 0)$ (c) $(4, 13\pi/4)$, $(-4, \pi/4)$
(d) $(-3, 2\pi/3 + 2\pi k)$, $(3, 5\pi/3 + 2\pi k)$ $k \in \mathbf{Z}$
(e) $(0, \theta)$ $\theta \in \mathbf{R}$

19. $P(\sqrt{5}, \tan^{-1}(2)) \approx (\sqrt{5}, 63.43°)$, $Q(6, 5\pi/6)$
20. $r = \cos(2\theta)$ **21.** $r = 2\cos\theta + 4\sin\theta$
22. $(x - 2)^2 - y^2/3 = 1$
23. $y = x/(1 - x)$

24.

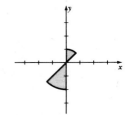

25.

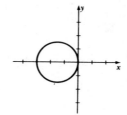

26.

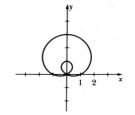

27.

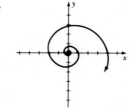

28.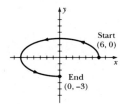

29.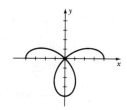

9. $1, \frac{5}{2}, \frac{5}{3}, \frac{9}{4}, \frac{9}{5}, \frac{13}{6}$

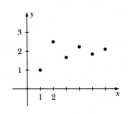

30.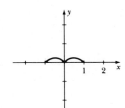

31.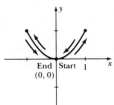

32.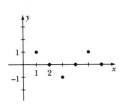

11. 2, 2, 3, 5, 5, 7, 7, 11, 11, 11, 11, 13

13. $F(n) = 2n - 3$; arithmetic

15. $F(n) = 1/3^{n-2}$; geometric, $r = \frac{1}{3}$

17. $F(n) = 5(-1)^{n-1}$; geometric, $r = -1$

19. $F(n) = (n + 1)/n$

21. $F(n) = 6n + 9$; arithmetic

23. $F(n) = n + 3$; arithmetic

25. $-1, 0, 1, 2, 3$; $F(n) = n - 2$

27. geometric; $r = 4$ **29.** not geometric

31. geometric; $r = \frac{3}{5}$ **33.** 3, 6, 12, 24, 48, 96

35. $1, -1, 1, -1, 1, -1$ **37.** $1, 2, 1, -1, -2, -1$

39. 1, 1, 3, 11, 53, 309   **41.** 1, 2, 2, 4, 8, 32

43. $a, b, b/a, 1/a, 1/b, a/b, a, b$; yes

45. 2, 6, 24, 120, 720, 5040

47. 2, 4, 12, 48, 240, 1440

49. 2, 12, 120, 1680, 30240, 665280

51. 10   **53.** 9

Exercises 11.1

1. 1, 3, 5, 7, 9, 11

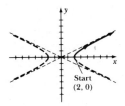

3. $1, 0, -1, 0, 1, 0$

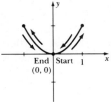

5. $0, \frac{1}{3}, \frac{1}{2}, \frac{3}{5}, \frac{2}{3}, \frac{5}{7}$

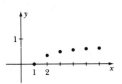

7. $\frac{1}{2}, \frac{1}{4}, \frac{1}{8}, \frac{1}{16}, \frac{1}{32}, \frac{1}{64}$

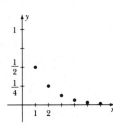

Exercises 11.2

1. $2 + 4 + 6 + 8 + 10 = 30$

3. $1 - 2 + 3 - 4 + 5 = 3$

5. $1 - 1 + 1 - 1 + 1 - 1 + 1 - 1 + 1 - 1 + 1 = 1$

7. $\frac{2}{3} + \frac{3}{4} + \frac{4}{5} = \frac{133}{60}$ **9.** $5 + 5 + 5 + 5 = 20$

11. $0 + 2 + 8 + 24 = 34$ **13.** $2 + 2 + 4 + 12 = 20$

15. $\sum_{k=1}^{3} (k + 9)$ **17.** $\sum_{k=1}^{4} (2k + 7)$ **19.** $\sum_{k=1}^{5} 3^{k-1}$

21. $\sum_{k=1}^{5} 1$ **23.** $\sum_{k=1}^{5} 2k$ **25.** $\sum_{k=1}^{4} 4(\frac{1}{10})^{k}$

27. $\sum_{k=1}^{5} (-\frac{1}{3})^{k-1}$ **29.** $\sum_{k=1}^{11} 1/2^{k-1}$ **31.** $\sum_{k=1}^{50} \frac{1}{3k}$

33. $\sum_{k=1}^{200} (-1)^{k-1}k$ **35.** $\sum_{k=1}^{10} (-1)^{k-1}k^{2}$

37. $\sum_{k=1}^{11} (-1)^{k-1}x^{k-1}/k$ **39.** $\sum_{k=1}^{34} (3k - 1)$

41. $\sum_{k=1}^{21} (6 - 5k)$ **43.** 20,100   **45.** 25,250

47. 7800   **49.** 17,575/2   **51.** 12,200   **53.** 2500

55. 7500   **57.** $1 - 1/3^{10}$ **59.** $3(1 - (\frac{3}{4})^{20})$

61. $(1 - 1/2^{50})/3$ **63.** $9(1 - 1/3^{8})/2$

Exercises 11.3

1. (a) $S_1 = 1/3, S_2 = 2/5, S_3 = 3/7, S_4 = 4/9, S_5 = 5/11$
   (b) $S_n = n/(2n + 1)$ (c) $1/2$
3. (a) $S_1 = 1/3, S_2 = 2/4, S_3 = 3/5, S_4 = 4/6, S_5 = 5/7$
   (b) $S_n = n/(n + 2)$ (c) 1
5. $a = 5/2, r = 1/2$, sum $= 5$
7. $a = 3, r = 3/5$, sum $= 15/2$
9. $a = 9, r = -1/3$, sum $= 27/4$

11. $a = 4/3, r = 4/3$ 13. (a) $\sum_{k=1}^{\infty} .36(.01)^{k-1}$ (b) $4/11$

15. (a) $\sum_{k=1}^{\infty} .9(.1)^{k-1}$ (b) 1 17. $681/550$ 19. $1/(p - 1)$

21. $1/(1 - x)$ 23. (b) $F(k) = 1/2^k$ (c) $1 - 1/2^n$ (d) 1

Exercises 11.4

We give only an outline of the steps for each proof.
1. $P(1)$: $5^1 - 1 = 4 = 4(1)$. $P(n)$ implies $P(n + 1)$: By
   the induction hypothesis, $5^n - 1 = 4q$ for some
   $q \in \mathbf{N}$. Hence $5^{n+1} - 1 = 5(5^n) - 5 + 4 =$
   $5(5^n - 1) + 4 = 5(4q) + 4 = 4(5q + 1)$.
3. $P(1)$: $1^3 + 2(1) = 3 = 3(1)$. $P(n)$ implies
   $P(n + 1)$: By the induction hypothesis, $n^3 + 2n = 3q$
   for some $q \in \mathbf{N}$. Hence $(n + 1)^3 + 2(n + 1) =$
   $n^3 + 3n^2 + 5n + 3 = (n^3 + 2n) + (3n^2 + 3n + 3) =$
   $3q + 3(n^2 + n + 1) = 3(q + n^2 + n + 1)$.
5. $P(1)$: $1 = 1^2$. $P(n)$ implies $P(n + 1)$: $1 + 3 +$
   $5 + \cdots + (2n - 1) + (2(n + 1) - 1) = n^2 +$
   $(2n + 2 - 1) = n^2 + 2n + 1 = (n + 1)^2$
7. $P(1)$: $1^2 = 1(1 + 1)(2(1) + 1)/6 = (2)(3)/6 = 1$. $P(n)$
   implies $P(n + 1)$: $1^2 + 2^2 + 3^2 + \cdots + n^2 +$
   $(n + 1)^2 = n(n + 1)(2n + 1)/6 + (n + 1)^2 =$
   $n(n + 1)(2n + 1)/6 + 6(n + 1)^2/6 = (n + 1)$
   $(n(2n + 1) + 6(n + 1))/6 = (n + 1)(2n^2 + 7n + 6)/6 =$
   $(n + 1)(n + 2)(2(n + 1) + 1)/6$.
9. $P(1)$: $2^0 = 2^1 - 1 = 1$. $P(n)$ implies $P(n + 1)$:
   $2^0 + 2^1 + 2^2 + \cdots + 2^{n-1} + 2^n = (2^n - 1) + 2^n =$
   $2(2^n) - 1 = 2^{n+1} - 1$.

11. $P(1)$: $\sum_{k=1}^{1} \dfrac{1}{4k^2 - 1} = \dfrac{1}{3} = \dfrac{1}{2(1) + 1}$. $P(n)$ implies

$P(n + 1)$: $\sum_{k=1}^{n+1} \dfrac{1}{4k^2 - 1} = \sum_{k=1}^{n} \dfrac{1}{4k^2 - 1} + \dfrac{1}{4(n + 1)^2 - 1}$

$= \dfrac{n}{2n + 1} + \dfrac{1}{(2n + 1)(2n + 3)} = \dfrac{2n^2 + 3n + 1}{(2n + 1)(2n + 3)} =$

$\dfrac{(2n + 1)(n + 1)}{(2n + 1)(2n + 3)} = \dfrac{n + 1}{2(n + 1) + 1}$

13. $P(1)$: $\sum_{k=1}^{1} \dfrac{2}{(k + 1)(k + 2)} = \dfrac{2}{2(3)} = \dfrac{1}{3} = \dfrac{1}{1 + 2}$.
    $P(n)$ implies

$P(n + 1)$: $\sum_{k=1}^{n+1} \dfrac{2}{(k + 1)(k + 2)} = \sum_{k=1}^{n} \dfrac{2}{(k + 1)(k + 2)} +$

$\dfrac{2}{(n + 2)(n + 3)} = \dfrac{n}{n + 2} + \dfrac{2}{(n + 2)(n + 3)} =$

$\dfrac{n^2 + 3n + 2}{(n + 2)(n + 3)} = \dfrac{(n + 2)(n + 1)}{(n + 2)(n + 3)} = \dfrac{n + 1}{(n + 1) + 2}$.

15. $P(1) = e^1 > 2 = 2^1$. $P(n)$ implies $P(n + 1)$:
    $e^{n+1} = e(e^n) > e(2^n) > 2(2^n) = 2^{n+1}$.
17. $P(7)$: $7! = 5040 > 2187 = 3^7$. $P(n)$ implies $P(n + 1)$:
    $(n + 1)! = n!(n + 1) > 3^n(n + 1) > 3^n(3) = 3^{n+1}$
    because $n \geq 7$.
19. $P(2)$: $1 - 1/2 = 1/2$. $P(n)$ implies $P(n + 1)$:
    $(1 - 1/2)(1 - 1/3) \cdots (1 - 1/n)(1 - 1/(n + 1)) =$
    $(1/n)(1 - 1/(n + 1)) = (1/n)(n/(n + 1)) = 1/(n + 1)$.
21. $P(1)$: $(\cos \theta + i \sin \theta)^1 = (\cos (\theta) + i \sin (\theta)$. $P(n)$
    implies $P(n + 1)$: $(\cos \theta + i \sin \theta)^{n+1} =$
    $(\cos \theta + i \sin \theta)^n (\cos \theta + i \sin \theta) = (\cos (n\theta) +$
    $i \sin (n\theta))(\cos \theta + i \sin \theta) = \cos (n\theta) \cos \theta -$
    $\sin (n\theta) \sin \theta + i (\sin (n\theta) \cos \theta + \cos (n\theta) \sin \theta) =$
    $\cos ((n + 1)\theta) + i \sin ((n + 1)\theta)$
23. Let $z = a + bi$. $P(2)$: $\overline{(z^2)} = \overline{(a + bi)^2} =$
    $\overline{(a^2 - b^2) + 2abi} = (a^2 - b^2) - 2abi$. Also, $(\bar{z})^2 =$
    $(a - bi)^2 = (a^2 - b^2) - 2abi$. Hence $\overline{(z^2)} = (\bar{z})^2$. $P(n)$
    implies $P(n + 1)$: First you need to prove that
    $\overline{z_1 z_2} = \bar{z}_1 \bar{z}_2$. (Do this directly, letting $z_1 = a + bi$ and
    $z_2 = c + di$.) Using this fact, it follows that $\overline{(z^n z)} =$
    $\overline{(z^n)} \bar{z}$. Hence $\overline{(z^{n+1})} = \overline{(z^n z)} = \overline{(z^n)} \bar{z} = (\bar{z})^n \bar{z} = (\bar{z})^{n+1}$.

Exercises 11.5

1. $1 + 6x + 15x^2 + 20x^3 + 15x^4 + 6x^5 + x^6$
3. $32x^5 + 80x^4 y + 80x^3 y^2 + 40x^2 y^3 + 10xy^4 + y^5$
5. $x^{10} - 5x^8 y + 10x^6 y^2 - 10x^4 y^3 + 5x^2 y^4 - y^5$
7. $1 - 7/x + 21/x^2 - 35/x^3 + 35/x^4 - 21/x^5 +$
   $7/x^6 - 1/x^7$
9. $x^{12} + 6x^{10} + 15x^8 + 20x^6 + 15x^4 + 6x^2 + 1$
11. $-15,360$ 13. $847,660,528$ 15. -2288
17. $11x^{10}$ 19. $125,970$
21. (a) $C(n, 0) = n!/(0! (n - 0)!) = n!/n! = 1$
    (b) $C(n, 1) = n!/(1!(n - 1)!) = n!/(n - 1)! = n$

23. $C(n, k) = \dfrac{n!}{k!(n - k)!} = \dfrac{n!}{(n - k)! k!} =$

$\dfrac{n!}{(n - k)! (n - (n - k))!} = C(n, n - k)$

25. $2^n = (1 + 1)^n = \sum_{k=0}^{n} C(n, k)1^{n-k}1^k = \sum_{k=0}^{n} C(n, k)$

27. 3^n

29. (a) $P(x) = 6x^5 + 15x^4h + 20x^3h^2 + 15x^2h^3 + 6xh^4 + h^5$ (b) 5 (c) 6 (d) 5

31. (a) $P(x) = C(n, 1)x^{n-1} + C(n, 2)x^{n-2}h + C(n, 3)x^{n-3}h^2 + \cdots + C(n, n-1)xh^{n-2} + C(n, n)h^{n-1}$
(b) $n - 1$ (c) n (d) $n - 1$

Chapter 11 Review

1. $2, 1, 0, -1, -2, -3$

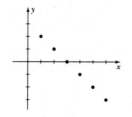

2. $1/\sqrt{2}, 0, -1/\sqrt{2}, -1, -1/\sqrt{2}, 0$

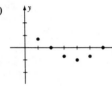

3. $F(n) = 22 - 5n$ **4.** $F(n) = 4(-\frac{3}{2})^{n-1}$

5. $2, \frac{2}{3}, \frac{2}{5}, \frac{2}{7}, \frac{2}{9}, \frac{2}{11}$ **6.** $1, 3, 9, 27, 81, 243$

7. $0, 1, 1, 3, 5, 11$ **8.** $1, 2, 24, 720$

9. $\frac{1}{2} - \frac{1}{6} + \frac{1}{12} - \frac{1}{20} + \frac{1}{30} = \frac{2}{5}$

10. $7 + 7 + 7 + 7 + 7 + 7 = 42$

11. $\sum_{k=3}^{8} (-1)^{k-1}k$ **12.** $\sum_{k=1}^{51} k/(k + 1)$

13. $\sum_{k=0}^{40} x^{2k}$ **14.** $4185/2$ **15.** $(1 - 1/3^{30})/2$

16. (a) $S_1 = \frac{2}{3}, S_2 = \frac{4}{5}, S_3 = \frac{6}{7}, S_4 = \frac{8}{9}, S_5 = \frac{10}{11}$
(b) $S_n = 2n/(2n + 1)$ (c) 1

17. 2   **18.** $\sum_{k=1}^{\infty} .84(.01)^{k-1}; \frac{28}{33}$

19. False; formula for infinite geometric sum does not apply for $r = 1.01 \geq 1$.

20. $P(1): 4(1) = 2(1)(1 + 1)$. $P(n)$ implies $P(n + 1)$:
$4 + 8 + 12 + \cdots + 4n + 4(n + 1) = 2n(n + 1) + 4(n + 1) = (n + 1)(2n + 4) = 2(n + 1)(n + 2)$.

21. $P(1): 7^1 - 1 = 6 = 6(1)$. $P(n)$ implies $P(n + 1)$:
By the induction hypothesis, $7^n - 1 = 6q$ for some $q \in \mathbf{N}$. Hence $7^{n+1} - 1 = 7^n(7) - 7 + 6 = 7(7^n - 1) + 6 = 7(6q) + 6 = 6(7q + 1)$.

22. $P(9): 9! = 362,880 > 262,144 = 4^9$. $P(n)$ implies $P(n + 1): (n + 1)! = n!(n + 1) > 4^n(n + 1) > 4^n(4) = 4^{n+1}$ because $n \geq 9$.

23. $32x^5 + 80x^4 + 80x^3 + 40x^2 + 10x + 1$

24. $x^6/y^6 - 6x^4/y^4 + 15x^2/y^2 - 20 + 15y^2/x^2 - 6y^4/x^4 + y^6/x^6$

25. $-\frac{2907}{8}$ **26.** 1716   **27.** $n!/(n - k)$

Exercises A.1

1. $(2x + y)(2x - y)$ **3.** $(4 + x^2)(2 + x)(2 - x)$

5. $x(1 - a)(1 + a + a^2)$ **7.** $(1 + 1/x)(1 - 1/x)$

9. $(x + y)(x^2 - xy + y^2)(x - y)(x^2 + xy + y^2)$

11. does not factor   **13.** $x(y + 3)(y^2 - 3y + 9)$

15. $3(x + \frac{1}{3})(x - \frac{1}{3})$ **17.** $(2x + 3)(x - 5)$

19. does not factor   **21.** $y(4x - 3)(x - 2)$

23. $(r - s)^2$ **25.** $(x + 1)(y + 1)$

27. $2(5h - 12k)(6h + 5k)$ **29.** $(a - c)(b - c)$

31. $(y + 3)(x + 1)(x - 1)$ **33.** $(x^2 + y^2)^2$

35. $(a - b + 1)(a - b - 1)$

37. $(x - 1)(x + y)(x - y)$

39. $(a + b + c)(a + b - c)$ **41.** $(x + y)^3$

43. $(x + y)(x + y + 1)$

Exercises A.2

1. (a) $x \neq 0$ (b) $x \neq -3$ (c) $x \neq 0, 2$

3. (a) $3n/(2m)$ (b) does not reduce (c) does not reduce (d) -1

5. $(2 + \sqrt{2})/2$ **7.** $(2 - \sqrt{2})/4$ **9.** $2x + h$

11. $3x + 3a + 2$ **13.** (a) $\frac{3}{2}$ (b) $\frac{1}{18}$ (c) $1/(2x^2)$

15. $(x - 1)/(x + 1)$ **17.** $(1 + x)/(1 - x)$

19. $(x + 1)/(x + 3)$ **21.** -1 **23.** $\frac{1}{2}$

25. $2(xy - 1)/(xy + 1)$ **27.** $R_1R_2/(RR_1 + RR_2)$

29. $(m_1k_2 + m_2k_1)/(m_1 + m_2)$ **31.** $-1/(x^2 + xh)$

33. $(ax - 1)/ax$ **35.** $(x^2 + xh + 1)/(x^2 + xh)$

37. $(-a - x)/(a^2x^2)$ **39.** $-1/(2x\sqrt{x})$

41. $1/(2\sqrt{x}\sqrt{x + 1}(x + 1))$ **43.** $(a^2 + b^2)/(ab)$

45. $(C_1 + C_2)/(C_1C_2)$ **47.** $(-x - 2y)/(x^2 + xy)$

49. $(-1 - 2x)/(x + 1)$ **51.** $(3x - 4)/(x^2 - 4)$

53. $4x/(x^2 - h^2)$ **55.** $((a + b)x - 2a)/(x^2 - 2x)$

57. $4/(3\sqrt{x})$ **59.** $(x + 1)/\sqrt{x}$

61. $(3x + 2)/(2\sqrt{x + 1})$ **63.** $1/\sqrt{x^2 + x}$ (assume $x > 0$)

65. false   **67.** false   **69.** false

Exercises A.3

1. $y + 15 = (x + 4)^2$ **3.** $y + 6 = (x - 3)^2$

5. $y + \frac{49}{4} = (x + \frac{3}{2})^2$ **7.** $y + \frac{1}{4} = (x + \frac{1}{2})^2$

9. $y + 5 = 2(x + 1)^2$ **11.** $y + \frac{1}{3} = 3(x - \frac{2}{3})^2$

13. $y - \frac{4}{5} = -5(x - \frac{2}{5})^2$

15. $y - 19 = -(x - 4)^2$

17. $y + 1 = -4(x + \frac{1}{2})^2$ **19.** $y + \frac{23}{4} = (x + \frac{9}{2})^2/3$

21. $(x - 2)^2 + (y + 3)^2 = 4$

23. $(x + \frac{1}{2})^2 + (y - 4)^2 = 8$

25. $(x - 6)^2 + y^2 = \frac{1}{9}$

27. $(x + 7)^2 + 2(y - 3)^2 = 62$

29. $5(x - \frac{1}{2})^2 - 4(y - 1)^2 = \frac{1}{4}$ **31.** $\sqrt{(x - \frac{1}{2})^2 - \frac{1}{4}}$

33. $\sqrt{(x + 2)^2 - 1}$ **35.** $\sqrt{2(x - 2)^2 - 8}$

37. $\sqrt{3(x + \frac{1}{2})^2 + \frac{1}{4}}$

39. $-1 \pm \sqrt{2}$ **41.** $5 \pm 1/\sqrt{2}$

43. $x^2 - x + \frac{1}{2} = (x - \frac{1}{2})^2 + \frac{1}{4}$. We know that $(x - \frac{1}{2})^2 \geq 0$ for all x. It follows that $(x - \frac{1}{2})^2 + \frac{1}{4} > 0$ for all x.

Exercises A.4

1. $k + 1$ **3.** $2x$ **5.** $3 - x$ **7.** $2x - 1$ **9.** $-x$

11. -10 **13.** (a) x^5 (b) $1/x^3$ (c) $1/x^4$

15. (a) x^2/y^6 (b) y^2/x^4 **17.** $x^{61/20}$ **19.** $x^2 + 1$

21. $x^3 y^6$ **23.** $2^{2/3}/x$ **25.** $y^2/4$ **27.** $2x^{10}/(27b^5 y^5)$

29. $x^{1/3}/y^{3/4}$ **31.** $1/(x^{1/2}(x + 1)^{3/2})$

33. $(x + y)^{1/2}/(x - y)^{1/2}$ **35.** $x + x^3$

37. $(x^{2/3} - x^{1/3})/(x + 1)$ **39.** $x^2/(x - 1)^2$

41. $x/(x + 1)$ **43.** $x(y + 1)^{1/2}/(y(x + 1)^{1/2})$

45. $x^{3/2}/(x + 1)$

47. (a) $3^{1/3}$ (b) $(xy)^{1/4}$ (c) $x^{1/4}$ (d) $(1 + x^{1/2})^{1/2}$

49. (a) $x^{1/2}$ (b) $x^{9/2}$ (c) $x^{13/6} + x^{1/3}$ **51.** $(x + y)^{3/2}$

53. (a) $2^{7/12}$ (b) 6 **55.** (a) $9(3^{1/6})$ (b) $2(3^{7/6})$ **57.** $2^{1/3}$

59. $(x^3 - 1)/x^2$ **61.** $(\sqrt{y} + \sqrt{x})/\sqrt{xy}$

63. $(x + \sqrt{2})/\sqrt{2x}$ **65.** $x/(x + 1)^{1/2}$

67. $(1 - 2x^2)/(1 - x^2)^{1/2}$ **69.** $(4x + 3)/(3(x + 1)^{2/3})$

71. $x^n(1 + y/x)^n = x^n((x + y)/x)^n = x^n(x + y)^n/x^n = (x + y)^n$.

Exercises A.5

1. $\angle 1 = \angle 3 = \angle 5 = \angle 7 = 58°$, $\angle 2 = \angle 4 = \angle 6 = \angle 8 = 122°$

3. $102°$ **5.** $\angle 1 = \angle 3 = 75°$, $\angle 2 = 15°$, $\angle 5 = 165°$

7. $x + 2\sqrt{74}$ **9.** $\sqrt{x^2 + 2500} + 100 - x$

11. By the Alternate Interior Angles Theorem for ℓ_1 and ℓ_2, $\angle ABD = \angle CDB$, and for ℓ_3 and ℓ_4, $\angle ADB = \angle CBD$. Also, $BD = DB$. Thus ASA implies that $\triangle ABD \cong \triangle CDB$.

13. By the Alternate Interior Angles Theorem for DC and AB, $\angle DCA = \angle BAC$. Because $ABCD$ is a rectangle, $BA = DC$. Also, $CA = AC$. Thus SAS implies that $\triangle DCA \cong \triangle BAC$.

15. 60 **17.** 34

19. (a) By the parallel axiom, $\angle D = \angle A$ and $\angle E = \angle B$. Of course, $\angle C = \angle C$; thus $\triangle ABC$ is similar to $\triangle DEC$. (b) $\frac{5}{2}$ (c) $8\frac{1}{2}$

21. $80 - 4x$ **23.** $\frac{25}{6}$ square inches

25. (a) Because the four corner triangles are congruent, $PQ = QR = RS = SQ$. In right $\triangle APS$, $\angle APS + \angle ASP = 90°$. Because $\triangle APS \cong \triangle BQP$, $\angle ASP = \angle BPQ$. Therefore $\angle APS + \angle BPQ = 90°$. Now, $\angle P = 180° - (\angle APS + \angle BPQ) = 180° - 90° = 90°$. Similarly, $\angle Q = \angle R = \angle S = 90°$. Thus $PQRS$ is a square. (b) $(a + b)^2$ (c) $2ab + c^2$ (d) $(a + b)^2 = 2ab + c^2$ implies that $a^2 + 2ab + b^2 = 2ab + c^2$ implies that $a^2 + b^2 = c^2$

INDEX

Abscissa, 44
Absolute value
  boundary curves and, 458–459
  definition of, 6
  fundamental properties, 7
  graph of, 50–51, 98
  inequalities and, 40–42
  rules for removing, 41
Addition
  equation solving and, 13
  of fractions, 605–606
  of functions, 102
  of graphs, 102–104
  inequalities and, 33
  of matrices, 483–485
  systems of equations, 446, 449
  of vectors, 429
Addition formulas, trigonometric, 378,
  380, 391
Algebra
  fundamental theorem of, 198
  of logarithms, 260–266
Alternate interior angles theorem, 619
Amplitude, 338–339
Analytic interpretation, of function, 75
And statement, 5, 32
  inequalities and, 41
Angle
  alternate interior angles theorem, 619
  congruence, 407, 622
  coterminal, 291
  definition, 284
  degree measure, 290
  double-angle formula, 384, 391
  half-angle formulas, 386, 391

initial side, 284
  measure of, 284, 289
  radian measure, 285–286, 290
  reference angle, 365
  standard angle, 310
  standard position, 284
  sum of in triangle, 620
  terminal side, 284
  vectors and, 427
  vertical angles theorem, 619
Angle of depression, 401
Angle of elevation, 401
Apogee, 511
Area
  of rectangle, 622
  of triangle, 402, 622
Arithmetic sequence, 565
ASA (angle-side-angle) triangle, 407–
  408
Asymptote
  hyperbola, 516, 519
  rational functions, 166–171, 173–
  179
  reciprocal function, 155
  slant asymptote, 169–170
Axioms, of geometry, 618
Axis
  ellipse, 507, 510
  hyperbola, 515
  parabola, 500
  polar coordinate system, 540

Binomial coefficient, 590
Binomial theorem, 590, 592–593

Boundary equation
  absolute values and, 458–459
  of solution set, 456
Boundary Test, 217

Calculator methods
  exponential functions, 251
  fractional exponents, 231, 233
  inverse trigonometric functions, 359
  log function and, 265
  natural logarithms, 259
  reciprocal functions, 332
  trigonometric functions and, 325–
  326
Capacitor, 27
Carbon 14 decay, 276–277
Cardioids, 548, 550–552
Cartesian coordinate system, 44
  polar-Cartesian transformation, 543–
  547
Center
  ellipse, 507
  hyperbola, 515
Central angle, 285
Changing to base *e* theorem, 265
Circle
  equation for, 47
  graph, 50–51, 548–550
  polar equation, 545
  radian measure and, 285
  tangent line, 58
Classification theorem, for conic sections, 529
Closed interval, 31